STUDY GUIDE

to accompany

LIFE: The Science of Biology

Fourth Edition

Study Guide

to accompany

Purves, Orians, and Heller

LIFE: The Science of Biology

Fourth Edition

Prepared by

Jon C. Glase
Cornell University

Jerry A. Waldvogel
Clemson University

Sinauer Associates, Inc.

W. H. Freeman and Company

THE COVER

Elephants at a water hole in northern Botswana, Africa.
Photograph by Frans Lanting/Minden Pictures.

Study Guide to accompany
LIFE: THE SCIENCE OF BIOLOGY, Fourth Edition

Copyright © 1995 by Sinauer Associates, Inc. All rights reserved. This book may not be reproduced in whole or part without permission.

Address editorial correspondence to
Sinauer Associates, Inc.
23 Plumtree Road
Sunderland, Massachusetts 01375 U.S.A.

Address orders to W. H. Freeman and Co.
Distribution Center, 4419 West 1980 South,
Salt Lake City, Utah 84104 U.S.A.

ISBN 0–7167–2655–6

Printed in U.S.A.

4 3 2

PREFACE

THE PURPOSE OF THIS STUDY GUIDE

A well-written, current textbook is an essential element in the serious study of any complex subject such as biology. Our collective experience in teaching biology during the past 20 years suggests that many students can use their textbooks more effectively if a well-conceived study guide is available to them. Our motivation in writing this guide is to help you learn about biology using the fourth edition of the textbook *Life: The Science of Biology* (1995) by Purves, Orians, and Heller.

To be effective, the study guide that accompanies a biology textbook should facilitate your learning process in two fundamental ways. First, it should help you *clarify* the important concepts and principles in each of the areas of biology. The textbook is designed to explain concepts and principles using a combination of descriptive text, illustrative artwork, and useful examples. Unfortunately, students sometimes lose sight of the major ideas and themes in biology because they fail to distinguish between the essential information and the supportive details. A major goal of our guide is to "unpack" the large quantity of information presented in the textbook, and to restate the most important ideas as a list of *Key Concepts* within each chapter. In writing these concept statements we have attempted to be concise and include only the essence of the idea being considered in the textbook. Sometimes a key concept item will include several closely related concepts which should be learned together. Bear in mind that this study guide is meant to be a *companion* to the textbook; the study guide's key concepts will only be meaningful after you have carefully studied how these ideas are developed in the corresponding chapter of the textbook.

A second major goal of our guide is to provide you with *feedback* about your level of understanding of the key concepts. For each chapter of the textbook, we have developed activities and questions that will test your mastery of the material covered. Some questions test your knowledge of terminology, others examine the relationships between key concepts, and some questions require you to think about biological concepts in a novel way, or perhaps use those concepts to solve a problem. The answers to all questions are presented in the *Answers and Explanations* sections of the guide, with each answer carefully explained and indexed to the key concept(s) related to that question. If you do not understand a question in the study guide, or our proposed answer and its explanation, then you may need to return to the textbook for further review.

HOW TO USE THIS STUDY GUIDE

We suggest that the best approach you can take to the study of biology is a structured approach. Treat the textbook as your primary information source and study it first. Scanning through assigned textbook pages will help to prepare you for lectures, laboratories, or discussion sections covering a specific topic. However, do not expect to completely comprehend all of the ideas you encounter during your first reading of the textbook. Rather, begin by scanning the chapter to understand how it is organized, see what the main topics are, and identify major themes. At the same time, read the corresponding portions of the study guide to further help you identify key concepts as they are presented in the textbook.

After you have attended the lecture, lab, or discussion, you will be in a much better position to clearly identify the material your instructor considers most important. To assist you in this process, we have included a set of *Learning Objectives* at the beginning of each chapter in the study guide. Your instructor may select a subset of objectives from this list, or even add to it. Careful study of the subject on your part should provide you with the necessary understanding to give a complete answer to each of these objectives. Check off each objective as you master it.

After identifying the relevant learning objectives for a specific topic, you should return to the textbook chapter and read the assignment in detail. During this in-depth study session, try to relate information in the textbook to the ideas presented in the study guide, your lecture notes, or laboratory exercises. We have divided each chapter of the study guide into several sections, each of which includes about as much textbook information as you are likely to assimilate during a short, productive study session before taking a short break.

Active involvement will promote your learning, so you should have paper available to write down important terms and to highlight sections of the textbook or study guide. Thoughtfully study the tables and figures included in each chapter. The artwork in your textbook has been carefully designed to help convey information that is difficult to explain using text alone; this includes conceptual relationships as well as the dynamic interactions that take place between the components of biological systems. Sometimes a single diagram can tie together and give meaning to several paragraphs of description. At all times, try to relate new information to what you already know. Use the index of your textbook wisely to locate previously covered information that may need to be reviewed as you take on new material. Remember that the study of biology is a cumulative process, and review is constantly needed as you progress to more complex, abstract material.

After you have actively studied part or all of a chapter in the textbook and the relevant sections of the study guide, you are ready to test your understanding by answering the accompanying study guide questions. We have provided space in the study guide so that you can write your responses to questions directly in the book, circle your answers to multiple choice questions, and make any sketches or calculations required in the various activities. If you encounter difficulty with a question, do not be tempted to short-circuit the learning process by looking at the *Answers and Explanations* section. Instead, you should review that section's key concepts and consult the textbook until you can make a reasonable choice from the possible answers given. Sometimes several choices will seem partially correct. In this case pick the *best*, most specific answer available. When you do make a choice, you should also be able to explain *why* the other choices are incorrect in terms of the key concepts presented in the study guide.

Since your mind works best in short, focused learning sessions, take a break after you have completed one section of the study guide. When you return to your study of biology, it is a good idea to quickly review the most recently studied section before proceeding to new material.

ADDITIONAL AIDS TO STUDYING BIOLOGY

Concept Mapping

It has been estimated that the number of new words encountered in a typical introductory biology course surpasses the vocabulary introduced during the first year of a foreign language course. Rote learning is a very inefficient approach for mastering this terminology, because the act of simply memorizing a new term without connecting it to the things you already know about the topic only promotes short-term learning. To take the analogy further, one does not learn a foreign language simply by memorizing the dictionary. The language can only be learned by understanding the concepts of grammar and syntax that allow you to manipulate the vocabulary in a meaningful fashion. Meaningful, long-term learning therefore occurs when you link new information to the knowledge you already possess. If you attempt to learn a concept meaningfully, then the concept will likely be remembered for a much longer time, and in a more readily accessible form. Educational psychologists have proposed several methods for promoting meaningful learning—one such method is called *concept mapping*.

A concept map is a diagram that shows the relationships between concepts within a topic, making it easier to relate new information to what you already know. As shown in the following generic example, concept maps are usually constructed from top to bottom. Each concept is placed in a box, with the most inclusive concepts near the top, and the more subordinate concepts nearer the bottom. The concept boxes are then connected by labeled lines that indicate the relationship between linked concepts.

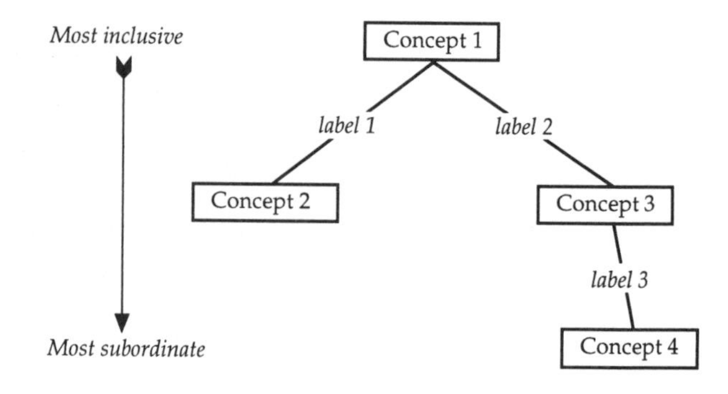

When constructing a concept map, it is best to select a fairly well-defined topic on which to focus. For example, the map at the top of the next page depicts information included under the heading *Characteristics of Living Organisms* found on pages 1–3 of the textbook, and described in key concepts 1–9 on pages 1–2 of the study guide. Notice that the concept map is simply a compact representation of the information found in the two book sources.

The real educational value of concept mapping comes from the actual construction of the map; you have to do it yourself! As you select concepts to include, you must define the relationships between these concepts and look for links to other concepts already present in the map. This makes the information meaningful for you. To make a map, select a group of key concepts that seem to be logically related. Terms in italics in the study guide are good candidates for inclusion as concepts in your map. Connect the concepts together with labels that make the relationships between concepts explicit. Since this is a dynamic learning process, it is best to do your map construction in pencil and to have an eraser handy—a map tends to change and evolve as your knowledge of the topic grows.

Taking Notes

Note taking is best viewed as yet another means of active learning. While this learning tool will most likely be applied to a lecture, laboratory, or discussion setting, it can also help in your efforts to understand the textbook and study guide materials. Like concept mapping, an emphasis should be placed on organizing the material in your notes in a way that has particular meaning for you. Your notes therefore should not simply be a written record of what the lecturer or textbook has to say about a topic. Instead, they should be a running account of your attempts to conceptually link the material you learned previously with the new material currently under study.

Although there is no one "right way" to take effective notes, following a few simple guidelines will increase the utility of your notes as a learning tool.

1. *Identify major themes* in the topic you are studying and use those themes as the section headings within your notes. In the case of lectures, this strategy will be greatly enhanced if you have already previewed the material being covered before coming to class.
2. *Record the key concepts* mentioned by the lecturer, but be sure to spend most of your time concentrating on what

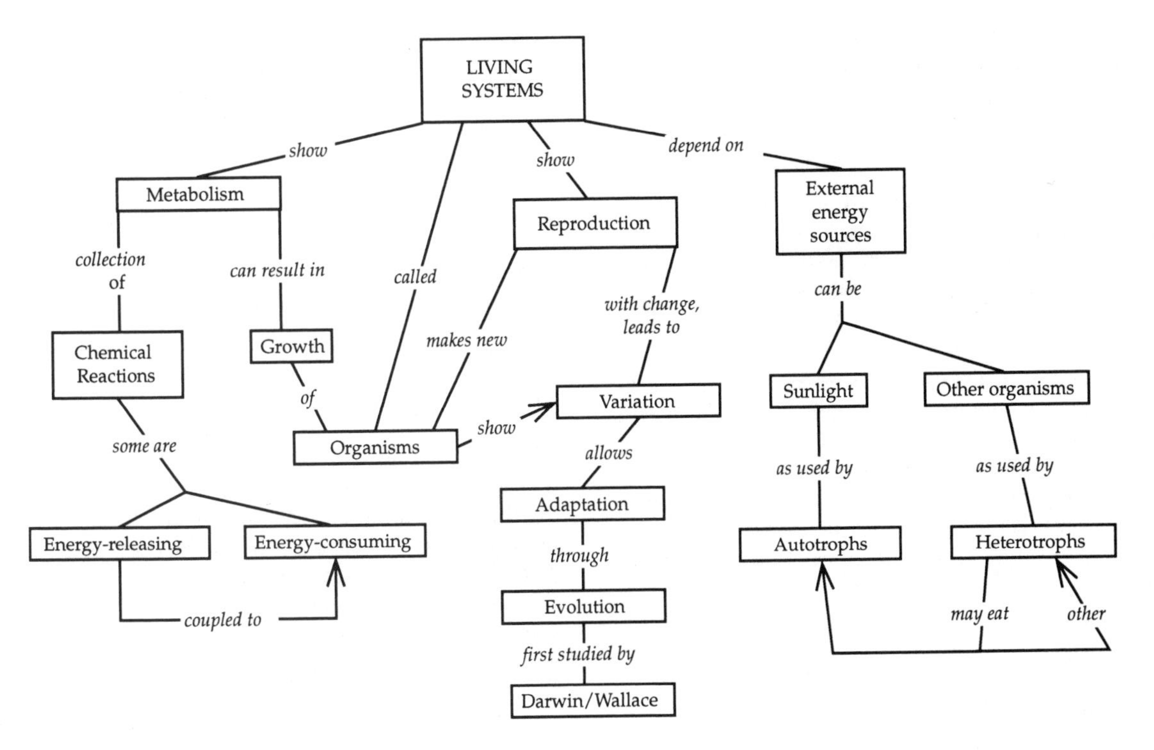

is being said about those concepts. One of the most important roles of a textbook is to provide the details that support the concepts presented in a lecture or laboratory setting. You should therefore rely on the book to provide much of the detail, rather than spending all of your time during lecture trying to write down things that are already available in your books and running the risk of missing most of what is being said. Think of the lecture as another way to prepare for the detailed reading of the textbook.

3. *Leave space in your notes* to include questions for the lecturer about the material just covered, to insert comments regarding relevant illustrations, and to fill in important or clarifying details as you read about the topic in the textbook. You should also strive to relate material from one section of a lecture to material in other sections, or to material presented in earlier lectures. If you get in the habit of asking "how" and "why" questions about the material you are currently studying, this will help you understand the connections between concepts and principles throughout the entire course.
4. *Develop a shorthand code* for common terms and phrases to increase your note-taking speed. Remember that your notes need not make sense to anyone but you, so feel free to use whatever abbreviations make sense in order to minimize the time spent writing while maximizing the time you devote to listening.

Preparing for and Taking Examinations

Unfortunately, examinations are a fact of life in the classroom. They come in many forms, ranging from oral examinations to essay and multiple choice formats. Although potentially stressful, examinations can also provide an excellent means of assessing your understanding of biology.

There is no better way to prepare for an exam than to practice answering exam questions. Your instructor may provide old examinations with which to practice, but even if this is not the case, you can still gain practice by creating your own exam questions. The idea here is not to "outguess" the instructor, but rather to try and make connections between linked concepts by organizing ideas in a logical, related fashion.

Studying for an exam should be an ongoing process which begins immediately after the previous exam. Do not expect to master a complex subject such as biology by cramming for many hours just prior to a test. Careful study spread over a period of many days or weeks prior to the exam not only makes the task more manageable, but also fosters long-term understanding of concepts.

If your test will have essay questions, practice writing the essays as part of your studying. The mechanics of organizing an answer in your head and actually writing it down on paper will help promote conceptual understanding. During the exam, make a brief outline of your essay before writing the answer; this will help insure that you cover all aspects of the topic.

Many biology examinations, especially those given in large lecture classes, are of necessity done using a multiple choice format. Students are often frustrated by these examinations because they fail to apply their understanding of a topic when answering the question. Assuming that you have spent the time needed to master the conceptual basis of a topic, the following strategy will greatly improve your chances of successfully answering multiple choice questions:

1. *Read the question carefully.* It is impossible to answer a question correctly if you don't understand what's being asked, so take the time to thoroughly read the premise

statement of the question. For example, are you to choose the true or false statement, the exception or the generalization? If necessary, underline key words that help you focus on the main point of the question.

2. *Answer the question in your own way before looking at the answers.* After you understand the question, cover up the possible answers and attempt to think of all relevant concepts and examples related to that question. This puts the question in terms that you already understand, which makes it easier for you to evaluate the possible answers offered by your instructor.
3. *Systematically evaluate each possible answer.* Uncover the answers and carefully evaluate each possibility based on what you have just reviewed about the topic. Assuming that you really do understand the material, you will likely be surprised at how many of the items you thought of will be present in the possible answers.
4. *Don't overthink the solution.* If you have followed steps 1–3 above, then you should be able to identify one best answer without much effort. As long as you have carefully read the question and truly understand the material, more often than not, second-guessing will produce the wrong answer. If you do decide to change your answer, ask yourself why the alternative choice is a better one.

Post-Exam Review

You can often learn a great deal about biology, about your study habits, and about how to successfully take a test, from a thorough review of your exam. Go back and reconsider the questions you got wrong, trying to understand where your thinking went astray. Did you read the question carefully? Did you consider all possible answers? Did you simply spend too little time thinking carefully about each answer? Did you have an inaccurate understanding of the material? Were your notes accurate? Make it a point to return to the textbook or study guide and correct any faulty understanding you brought into the exam, since biology courses frequently build upon material presented during earlier portions of the course.

Seeking Help

The field of biology is vast. There will be an ever-growing body of information that builds as you progress through your biology course, and a time may come when you need assistance to clarify your understanding beyond that provided by this study guide. Realize that there is no dishonor in not understanding everything you read in a textbook or hear in lecture, and do not let yourself become overwhelmed or intimidated. If you are having trouble, please seek help from your instructor, tutors, fellow students, or other available resources.

ACKNOWLEDGEMENTS

We wish to acknowledge the assistance of several colleagues and the forbearance of our families and friends during the preparation of this study guide. Ideas for several of the questions included in this book came from examinations prepared for the introductory biology courses at Cornell University and Clemson University by Kraig Adler, Carl Hopkins, Antonie Blackler, and William Surver. We also thank the authors of *Life: The Science of Biology* for their careful reading of the study guide and for the many suggestions they provided during its development. Any remaining errors, mis-statements, or omissions are our own. Special thanks go out to the staff at Sinauer Associates for assistance during the developmental, editorial, and production phases of this book. We want to especially acknowledge the editorial assistance of Kerry Falvey.

The authors and publisher encourage and appreciate any suggestions that you may have for ways to make this study guide more useful. Good luck in your study of biology, and may you find it as interesting and exciting as we do!

Jon C. Glase
Ithaca, New York

Jerry A. Waldvogel
Clemson, South Carolina

CONTENTS

1

The Science of Biology

CHAPTER LEARNING OBJECTIVES—after studying this chapter you should be able to:

- ❑ Describe three processes that characterize living things.
- ❑ Differentiate between autotrophs and heterotrophs.
- ❑ Define the term "homeostasis" and explain the importance of homeostasis to metabolism.
- ❑ Explain what is meant when we say that an organism is adapted to its environment and explain how adaptations arise.
- ❑ Describe the features and relationships of the following common levels in the hierarchical organization of biology: molecule, cell, tissue, organ, organism, species, population, community, biome, biosphere.
- ❑ Relate the concept of emergent properties to the hierarchical organization of biology.
- ❑ Describe the four stages of the hypothetico-deductive approach.
- ❑ Describe the relationship between a hypothesis and a null hypothesis and a hypothesis and a theory.
- ❑ Describe the strengths and weaknesses of laboratory and field experimentation.
- ❑ Describe the relationship between surface area and volume as the size of an object increases.
- ❑ Explain how metabolism and exchange of energy and materials with the environment vary with the surface area and volume of an organism.
- ❑ Explain the relationship between the size of supporting appendages and body size.
- ❑ Explain how naturally occurring radioactive materials and the arrangement of rock layers can be used to obtain both absolute and relative ages of evolutionary events.
- ❑ Describe differences in the time scales used to study physiology, population biology, microevolution, and macroevolution.
- ❑ Explain the organizing principles of biology concerned with (1) the laws of chemistry and physics, (2) organisms as energy-capturing/transforming systems, (3) the transmission of genetic information, (4) the cell theory, and (5) the role of natural selection in the evolution of adaptation.
- ❑ Describe some of the arguments advanced by Buffon against the special creation of organisms.
- ❑ Explain the theory of evolutionary change developed by Lamarck.
- ❑ List the underlying ideas that Darwin used to develop his theory of evolution.
- ❑ Explain why variation within a species and the potential for exponential growth shown by many populations are important to Darwin's theory.
- ❑ Describe natural selection and how it differs from artificial selection.
- ❑ Define paradigm and explain the Darwinian paradigm.
- ❑ Describe a hypothesis for the origin of eukaryotic cells from prokaryotic cells.
- ❑ Name and describe the distinguishing features of the six kingdoms of life.

CHARACTERISTICS OF LIVING ORGANISMS • THE HIERARCHICAL ORGANIZATION OF BIOLOGY • THE METHODS OF SCIENCE (pages 1–9)

Key Concepts

1. Living things (organisms) are characterized by three processes: metabolism, regulated growth, and reproduction.
2. All of the energy-consuming and energy-releasing chemical reactions occurring within living systems are called *metabolism*.
3. *Autotrophs* synthesize their own organic molecules using energy from inorganic sources, either sunlight or simple chemicals.
4. *Heterotrophs* obtain energy from complex chemicals produced either directly or indirectly by autotrophs.
5. In response to changes in the external environment, organisms vary the rates of their chemical reactions in order to maintain relatively constant internal conditions that are suitable for metabolism. Maintenance of a steady internal state is called *homeostasis.*
6. Through their metabolic activities, all living systems show *regulated growth*—a carefully controlled increase in the number of molecules of which they are composed.
7. During *reproduction,* organisms produce copies of themselves, which can show considerable variation. Modes of reproduction vary widely among organisms.

8. Reproduction with change produces *adaptations* that enhance an organism's survival and reproductive success in a particular environment.
9. A theory to explain adaptation was first proposed by *Charles Darwin* and *Alfred Russel Wallace* in the 1850's.
10. The objects that biologists study, from molecules to ecosystems, can be organized into a hierarchy in which each level consists of the objects at the next lower level.
11. *Molecules* are the smallest objects studied by biologists. The most important large molecules found in living systems are *proteins, nucleic acids, lipids,* and *carbohydrates.*
12. Molecules aggregate to form complex structures, like membranes, that have unique properties of their own.
13. The *cell* is the simplest unit capable of independent existence and reproduction. All organisms are cells or are composed of cells.
14. Structurally similar cells can associate to form *tissues,* and different tissues can aggregate to form *organs.*
15. The *organism* is the smallest unit of life. Organisms can be composed of a single cell (*unicellular*) or many cells (*multicellular*).
16. A group of structurally and functionally similar organisms that can interbreed with each other but not with other organisms is called a *species.*
17. Individuals of the same species living in a defined area are a *population.*
18. *An ecological community* consists of populations of different species living together in a particular environment.
19. Major types of ecological communities are *biomes.* All of the world's biomes form the *biosphere.*
20. *Emergent properties* are the new features of a system that result from the interaction of the system's components.
21. *Science* is both a collection of discovered knowledge about the universe and a process that can be used to discover new knowledge.
22. To discover new truths, scientists employ a cyclical process called the *hypothetico-deductive method* consisting of four steps: (1) making *observations,* (2) forming a *hypothesis,* (3) using the hypothesis to make *predictions,* (4) testing the predictions.
23. A hypothesis is a tentative answer to a question that usually predicts a specific effect. The corresponding *null hypothesis* states that the proposed effect is absent.
24. Assumptions and prior knowledge are usually involved in the formulation of a hypothesis.
25. In *experimentation,* extraneous factors are kept constant so that the effects of a single factor or group of related factors can be examined more clearly.
26. Greater control of environmental variables is possible in laboratory experimentation, but it is sometimes difficult to apply the results to nature. Field experimentation is less controlled, but the results may be easier to interpret.
27. A hypothesis cannot be proved or disproved with absolute certainty.
28. A *theory* is a hypothesis that has been extensively tested.

Questions (for answers and explanations, see page 46)

1. Which of the following is *not* a characteristic of all living systems?
   a. Regulated growth
   b. Autotrophism
   c. Metabolism
   d. Reproduction
   e. Adaptation

2. Which of the following is *not* a step in the hypothetico-deductive method of science?
   a. Stating hypotheses
   b. Making observations
   c. Establishing a conclusion with absolute certainty
   d. Deriving predictions from hypotheses
   e. Testing hypotheses

3. In the space provided before each of the following objects, provide a number that indicates the place of each object in a hierarchy from the least to most inclusive category. In the space following each object, specify its category.

   ____ A sperm ____________
   ____ A mouse ____________
   ____ Glucose ____________
   ____ A lung ____________
   ____ The Earth ____________
   ____ A woodland ____________
   ____ *Peromyscus leucopus* ____________

4. Biological membranes are composed of molecules called proteins and phospholipids. Which of the following statements best describes an emergent property at the level of biological membranes?
   a. All cells are bounded by membranes.
   b. Membranes regulate what chemicals can pass into or out of the cell.
   c. Phospholipids are composed of phosphorus and lipids.
   d. Phospholipids and proteins are found in all cells.
   e. Cells contain membrane-bounded structures called organelles.

5. In reference to the studies done on palatability of cryptic and conspicuous caterpillars described in the text, classify each of the following as either an *observation,* a *hypothesis,* a *null hypothesis,* a *prediction,* a *result,* or an *assumption.*
   a. Birds are visually oriented predators: ____________
   b. Conspicuous caterpillars tend to occur in groups: ____________
   c. Bird 1 consumed five conspicuous and one cryptic caterpillars: ____________
   d. This species of conspicuous caterpillar should not be eaten by blue jays: ____________
   e. Blue jays should eat equal numbers of conspicuous and cryptic caterpillars: ____________

6. What do laboratory and field experiments have in common and how do they differ?

WHY PEOPLE DO SCIENCE • SIZE SCALES • TIME SCALES • MAJOR ORGANIZING CONCEPTS IN BIOLOGY • EVOLUTIONARY CONCEPTS • LIFE'S SIX KINGDOMS • SCIENCE AND RELIGION (pages 9–15)

Key Concepts

1. The volume of an object approximates the cube of its linear dimension; e.g., the volume of a cube with an edge of 10 μm equals $(10\ \mu m)^3$ or $1{,}000\ \mu m^3$.
2. The surface area of an object approximates the square of its linear dimension; e.g., the surface area of a cube with an edge of 10 μm equals $6(10\ \mu m)^2$ or $600\ \mu m^2$.
3. The rate of metabolism of an organism is a function of its *volume*, while the rate of heat and material exchange between an organism and the environment is a function of the *surface area* of the organism.
4. As an object increases in size, its volume increases more rapidly than its surface area; as an object decreases in size, its volume decreases more rapidly than its surface area.
5. As a cell increases in size, its metabolic need for exchange of heat and materials with the environment outstrips its ability to provide this exchange because of its limited surface area.
6. *Multicellularity* and the evolution of *specialized cells* overcomes some of the size limitations imposed on unicellular organisms by surface-to-volume ratios.
7. Large organisms have lower metabolic rates than small organisms because their large volume (mass) produces more heat than a smaller organism. However, the proportionally smaller surface area of a large organism can dissipate less heat than can that of a smaller organism.
8. Large organisms tend to have proportionally larger supporting appendages than small organisms because body weight increases in proportion to body volume, whereas the strength of appendages is proportional to their cross-sectional area.
9. Biologist are concerned with size because it influences where and how an organism lives, what it can eat, and its abundance.
10. The age of evolutionary events can be determined relatively by studying the location of fossils within known rock strata, or absolutely by measuring the decay of naturally occurring radioactive elements.
11. Studies of physiological and biochemical processes are expressed in units ranging from milliseconds to a year. Studies of aging may span an organism's life time.
12. Studies of populations are expressed in units ranging from hours to many years.
13. Microevolutionary time, expressed in units of thousands of years, is appropriate for studying changes in the genetic makeup of populations.
14. Macroevolutionary time, expressed in units of millions of years, is appropriate for studying long-term changes in the evolution of groups of organisms.
15. All phenomena associated with living systems can be explained with the laws of *chemistry* and *physics*.
16. Organisms are living systems adapted for capturing *energy* from the environment and converting it into biologically useful forms.
17. Living systems contain *genetic information* that is passed from one generation to the next and directs the production of new individuals.
18. The *cell theory* states that all living systems are composed of cells, and that all cells are derived from preexisting cells.
19. *Evolution* by *natural selection* produces organisms that are better adapted to their environment.
20. *Buffon* advanced the following arguments against the special creation of organisms: all mammals share the same limb bones, although the bones are modified for the functions they perform, and some mammals have limbs with functionless toes.
21. *Jean Baptiste de Lamarck* argued that due to use or disuse structures become modified, and that these modifications are passed on to future generations.
22. Charles Darwin's theory of evolution (also independently developed by Alfred Wallace) is based on the ideas that Earth is very old, all organisms are related by common descent, diversity results from speciation, evolution produces gradual changes in populations, and that evolution occurs by natural selection.
23. Evolution by natural selection is based on two observations: that considerable variation exists among individuals of the same species and that most species have high reproductive rates.
24. The major inference of Darwin's theory is that slight variation among individuals can lead to differential reproductive success, in which the more fit individuals leave more offspring than the less fit individuals.
25. Differential reproductive success by individuals better adapted for their environment is called *natural selection.*
26. *Artificial selection* results from differential reproductive success as directed by humans and not the environment (as in natural selection).
27. The general world view, or *paradigm,* of biology holds that evolution is by natural selection, and that evolutionary change is not goal-directed.
28. In a common scheme for classifying organisms, all species are grouped into *six kingdoms.*
29. The kingdoms *Archaebacteria* and *Eubacteria* include all the single-celled *prokaryotic* organisms. Prokaryotic cells lack true nuclei and some other organelles.
30. *Eukaryotic* cells were formed when certain prokaryotes began permanently living inside other prokaryotes. Members of the other four kingdoms are or consist of eukaryotic cells.
31. The kingdom *Protista* generally includes single-celled, eukaryotic organisms.
32. The kingdom *Fungi* includes multicellular, heterotrophic organisms called molds, mushrooms, yeasts, etc., that obtain their nutrients by *absorption.*

33. The kingdom *Plantae* includes multicellular, autotrophic organisms called plants that obtain their nutrients by *photosynthesis.*

34. The kingdom *Animalia* includes multicellular, heterotrophic organisms called animals that obtain their nutrients by *ingestion.*

Questions (for answers and explanations, see page 46)

1. From the list of animals listed below, select the one expected to have the greatest total daily food intake.

   ____

   From the list of animals listed below, select the one expected to have the greatest daily food intake per kg of body weight. ________________

   horse — elephant — human — ostrich — hummingbird

2. If the length of a cube-shaped cell is doubled, then cell volume should increase by about
   a. 2 times.
   b. 4 times.
   c. 8 times.
   d. 16 times.
   e. 32 times.

3. Which of the following would *not* help an organism overcome problems related to surface area-to-volume relationships?
   a. Increasing surface area by developing infoldings
   b. Increasing surface area by developing appendages
   c. Maintaining the same size while becoming unicellular
   d. Developing internal gas-filled cavities
   e. Maintaining the same size while becoming multicellular with specialized cells

4. In describing the evolution of warm-bloodedness in mammals and birds, one would be mostly using
   a. a physiological time scale.
   b. a population time scale.
   c. a microevolutionary time scale.
   d. a macroevolutionary time scale.

5. Which of the following is *not* a major organizing principle of biology?
   a. All living systems obey the laws of chemistry and physics.
   b. All living systems consist of cells.
   c. The genetic code consists of information specifying the construction of new individuals.
   d. Evolution occurs by the inheritance of structures modified by use or disuse.
   e. Living systems capture and transform energy into useful forms.

6. Which of the following was *not* a major hypothesis incorporated by the theory of evolution by natural selection advanced by Charles Darwin?
   a. Earth is ancient.
   b. Evolution proceeds by the accumulation of gradual changes within populations.
   c. Offspring inherit genes from their parents.
   d. Biological diversity is produced by speciation.
   e. All present-day organisms have a common ancestor in the past.

7. Which of the following kingdoms consists of organisms that are mostly unicellular and have eukaryotic cells?
   a. Eubacteria
   b. Protista
   c. Fungi
   d. Plantae
   e. Animalia

8. Which of the following kingdoms consists of organisms that are mostly multicellular, photosynthetic autotrophs?
   a. Eubacteria
   b. Protista
   c. Fungi
   d. Plantae
   e. Animalia

9. Which of the following kingdoms consists of organisms that are mostly multicellular, absorptive heterotrophs?
   a. Archaebacteria
   b. Protista
   c. Fungi
   d. Plantae
   e. Animalia

10. Describe natural selection and how it differs from artificial selection.

11. Describe a hypothesis for the evolution of the eukaryotic cell.

2

Small Molecules

CHAPTER LEARNING OBJECTIVES—after studying this chapter you should be able to:

- ❑ Describe the structure of the atom, in terms of the locations of protons, electrons, and neutrons and the distribution of mass and charge.
- ❑ Relate an element's atomic number, mass number, and chemical properties to the number of protons, electrons, and neutrons in its atoms.
- ❑ Write the symbols, atomic numbers, and mass numbers of the biologically important elements using the conventions described in the text.
- ❑ Explain how isotopes of an element differ in terms of atomic number and mass number and how an element's atomic weight is determined.
- ❑ Describe how an element's isotopes can be separated and how the liquid scintillation and autoradiography methods are used to detect radioisotopes.
- ❑ Describe the concept of half-life and its use in radioisotope aging of biological materials.
- ❑ Describe the distribution of elections in an atom's orbitals and shells and how this distribution determines the chemical properties of atoms.
- ❑ Differentiate between molecules and ions and covalent and ionic bonds.
- ❑ Use the covalent bonding capabilities of biologically important elements to determine the types of chemical bonds they will form.
- ❑ Differentiate between compounds and elemental substances.
- ❑ Relate molecular and structural formulas for biologically important compounds.
- ❑ Determine molecular weight and understand its relationship to the concept of molarity.
- ❑ Relate equations showing biologically important chemical reactions to changes in chemical bonding between reactants and products.
- ❑ Explain what a calorie and kilocalorie are and how they are used to represent chemical bond energy.
- ❑ Describe the biologically important properties of water.
- ❑ Understand the concept of pH, relate it to the $[H^+]$ and $[OH^-]$, and given the pH of an aqueous system, be able to determine if it is acidic, basic, or neutral.
- ❑ Identify an acid and a base given a reversible chemical reaction that affects the pH of an aqueous system.
- ❑ Explain what a buffer is and how a buffer can maintain a relatively constant pH within a buffering range.
- ❑ Explain the concept of polarity and be able to identify compounds as polar or nonpolar.
- ❑ Describe the biological consequences of water's polar nature and its ability to form hydrogen bonds with other molecules.
- ❑ Explain the similarities and differences between van der Waals and hydrophobic interactions and the biological importance of these two forces.
- ❑ Recognize the distinguishing features of the following organic compounds: saturated and unsaturated hydrocarbons, alcohol, aldehyde and ketone, organic acid, organic base, and amino acid.
- ❑ Recognize the distinguishing features of the following functional groups: hydroxyl, carbonyl, carboxyl, amino, phosphate, and sulfhydryl.
- ❑ Explain the terms "isomer" and "optical isomer" and be able to identify an asymmetric carbon atom.

ATOMS • ELEMENTS (pages 19–22)

Key Concepts

1. *Atoms* are made of the elemental particles *protons, neutrons,* and *electrons.*
2. Protons and neutrons are found in the *nucleus* of the atom and represent most of the mass of the atom.
3. Electrons are found in *orbitals* surrounding the nucleus.
4. Protons and neutrons have the same mass; each is equal to one atomic mass unit (a.m.u) or *dalton.*
5. Protons have a charge of +1 unit, electrons have a charge of –1 unit, and neutrons are electrically neutral.
6. Atoms are electrically neutral because they contain the same number of electrons and protons.
7. A *chemical element* cannot be changed by chemical means into a simpler substance.
8. An element's chemical properties are determined by the number of electrons it contains.
9. The *atomic number* of an element is the number of electrons or protons present in its atoms.

10. All of the atoms of a chemical element have the same atomic number and the atomic number differs for each element.
11. The *mass number* is equal to the sum of the protons and neutrons in the nucleus of the atom; the mass of the electrons is negligible and is ignored.
12. Common notation for representing elements uses a one- or two-letter symbol with, optionally, the atomic number shown as a subscript and the mass number as a superscript preceding the symbol.

$$^{\textit{mass number}}_{\textit{atomic number}}\text{SYMBOL}$$

as in $^{12}_{6}C$ for carbon and $^{40}_{20}Ca$ for calcium.

13. Atoms with the same atomic number but a different mass number are *isotopes* of an element. Specifically, the isotopes of an element differ in the number of neutrons present.
14. Isotopes are usually shown with only the mass number preceding the symbol, as in ^{12}C and ^{14}C for two of carbon's isotopes.
15. The *atomic weight* of an element is the average of the mass numbers for all isotopes of an element weighted according to the natural occurrence of each isotope.
16. *Radioisotopes* undergo radioactive decay to become another element by giving off energy or elementary particles. Radioisotopes are used as markers in biological research. Isotopes of this type include ^{3}H, ^{14}C, ^{32}P.
17. *Liquid scintillation* and *autoradiography* are two common methods for detecting radioactivity in biological samples.
18. Each radioisotope loses one-half of its radioactivity in a constant time period called its *half-life*. The regularity of decay of a radioisotope allows it to be used in determining the age of biological materials.
19. *Stable isotopes* are nonradioactive but can be detected by their mass difference using an instrument called a *mass spectrometer*. Biologically important stable isotopes include ^{2}H and ^{18}O.

Questions (for answers and explanations, see page 47)

1. Which of the following statements about atomic structure is *not* true?
   a. Almost all the mass of an atom resides in the nucleus.
   b. Electrons are in orbitals that surround the nucleus.
   c. Protons and neutrons have approximately the same mass.
   d. All atoms contain electrons, protons, and neutrons.
   e. Protons and electrons have equal and opposite charge.
2. Which of the following statements about elements is *not* true?
   a. Elements cannot be changed by chemical means into simpler substances.
   b. The mass number of an element equals the number of protons plus neutrons.
   c. The atomic number of an element equals the number of electrons or protons.
   d. All atoms of an element have the same atomic number.
   e. All atoms of an element have the same mass number.
3. The mass number minus the atomic number
   a. is equal to the number of electrons in the atom.
   b. is equal to the number of protons in the atom.
   c. is equal to the number of neutrons in the atom.
   d. is equal to the atomic weight of the atom.
   e. is the same for all isotopes of an element.
4. There are two isotopes of a hypothetical element X: ^{30}X and ^{32}X. If the atomic weight of X is 30.5, what is the natural proportion of ^{30}X to ^{32}X in the world?
   a. 100:1
   b. 50:50
   c. 75:25
   d. 25:75
   e. 1:100
5. Isotopes of the same element can be physically separated based on
   a. electric charge differences.
   b. mass differences.
   c. radioactivity.
   d. atomic number.
   e. differences in chemical properties.

Activities (for answers and explanations, see page 47)

- Write the symbols for the most common isotopes of the following elements, showing the atomic number and mass number, using the notation described in key concept 12.

Hydrogen	**Carbon**
Nitrogen	**Oxygen**
Sulfur	**Phosphorus**

- Write the names and symbols for the less common, but biologically important, isotopes of hydrogen, carbon, and oxygen.

Hydrogen

Carbon

Oxygen

- Write the numbers of electrons, protons, and neutrons found in atoms of each of the following elements or isotopes.

	Electrons	Protons	Neutrons
$^{1}_{1}H$			
$^{14}_{6}C$			
$^{14}_{7}N$			
$^{65}_{30}Zn$			

THE BEHAVIOR OF ELECTRONS • CHEMICAL BONDS (pages 22–26)

Key Concepts

1. *Chemical reactions* occur when electrons are exchanged or shared differently between atoms.
2. An *electron orbital* is a region of space where an electron is located at least 90% of the time. A maximum of two electrons can occupy an orbital, and the two electrons spin in opposite directions.
3. Orbitals are grouped into *shells* that surround the nucleus and are designated by the letters K, L, M, N, O, P, and Q from the innermost to the outermost shell.
4. The *K shell* has a single spherical orbital (*s* orbital) holding a maximum of two electrons.
5. The *L shell* has four orbitals, a spherical *s* orbital and three dumbbell-shaped *p* orbitals oriented in the x, y, and z axes relative to the center of the nucleus. The L shell accommodates a maximum of eight electrons (two per orbital).
6. Successive shells (M–Q) are located at increasing distance from the nucleus and hold differing numbers of orbitals.
7. Atoms are most stable when each shell has a complete complement of electrons.
8. The energy level of electrons increases with their distance from the nucleus. Therefore, electrons in outer shells have more energy than electrons in more inner shells. Electron shells are filled from the innermost shell outward.
9. The outermost shell of an atom determines the chemical reactivity of the atom.
10. An atom with its outermost shell completely filled is most stable and, thus, least reactive with other atoms. Elements with completely filled shells are called the *inert elements* and include helium (He), neon (Ne), argon (A), and krypton (Kr).
11. A *molecule* consists of two or more atoms linked together by chemical bonds (attractive forces that hold the atoms together).
12. A *covalent chemical bond* results when two atoms share electrons in order to complete their outermost shell.
13. For an atom that can form covalent bonds, the number of vacant positions in its outer shell is the number of covalent bonds that it can form.
14. The *covalent bonding capabilities* of several biologically important elements include:

Hydrogen ($_1$H)	1
Oxygen ($_8$O)	2
Nitrogen ($_7$N)	3
Carbon ($_6$C)	4
Sulfur ($_{16}$S)	5 or 2

15. A single bond results if two atoms share a pair of electrons, a double bond results when four electrons are shared, and a triple bond results when six electrons are shared by two atoms.
16. Atoms that complete their outer shells by sharing electrons are stable because the attraction of the shared electrons for the nuclei of the atoms exactly balances the mutual repulsion of the nuclei.
17. *Ions* are atoms or groups of atoms that have gained or lost electrons and, as a result, are electrically charged. Positively charged ions are *cations;* negatively charged ions are *anions.*
18. Some covalently bonded molecules form ions when they dissolve in water.
19. When a molecule ionizes it forms positively and negatively charged ions whose total charge is zero, as in
$HCl \rightarrow H^+ + Cl^-$.
20. Some ions have multiple charges, as in Ca^{2+}, Fe^{3+}, and Al^{3+}. Groups of atoms can be ions, as in NH_4^+, SO_4^{2-}, and PO_4^{3-}.
21. Oppositely charged ions can be bound together to form *ionic compounds* such as common salt (NaCl).
22. An *ionic bond,* resulting from the electric attraction between oppositely charged ions, is less than one-tenth as strong as a covalent bond.

Questions (for answers and explanations, see page 47)

1. Sodium has an atomic number of 11. How many electrons are located in its M shell?
a. 1
b. 2
c. 7
d. 8
e. 0

2. Neon has an atomic number of 10. It is chemically ________ and has ____ completely filled orbitals in its L shell.
a. reactive, 2
b. reactive, 3
c. reactive, 4
d. inert, 2
e. inert, 3
f. inert, 4

3. If carbon can form the following stable molecule with Z, Z=C=Z, which of the following could be the atomic number of Z?
a. 2
b. 6
c. 8
d. 10
e. 11

4. The element with which of the following atomic numbers would be *most* stable?
a. 1
b. 3
c. 12
d. 15
e. 18

5. Given that the chemical properties of an element are determined by the number of electrons in its outer shell, which of the following pairs of elements would have the most similar chemical properties?
a. Carbon ($_6$C) and Nitrogen ($_7$N)
b. Carbon ($_6$C) and Oxygen ($_8$O)
c. Helium ($_2$He) and Fluorine ($_9$F)
d. Fluorine ($_9$F) and Chlorine ($_{17}$Cl)
e. Chlorine ($_{17}$Cl) and Sodium ($_{11}$Na)

Activities (for answers and explanations, see page 47)

- In the following figures, place electrons in the appropriate shells based on the atomic numbers of the elements carbon ($_6C$), nitrogen ($_7N$), and sodium ($_{11}Na$).

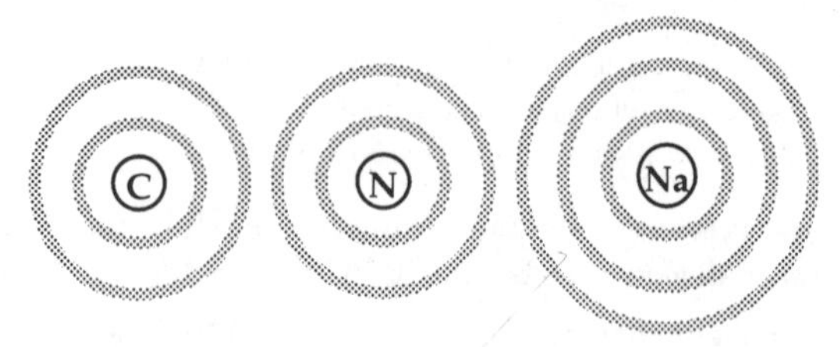

- In the following figures, place electrons in the appropriate shells based on the atomic numbers of the elements chlorine ($_{17}Cl$), argon ($_{18}A$), and oxygen ($_8O$).

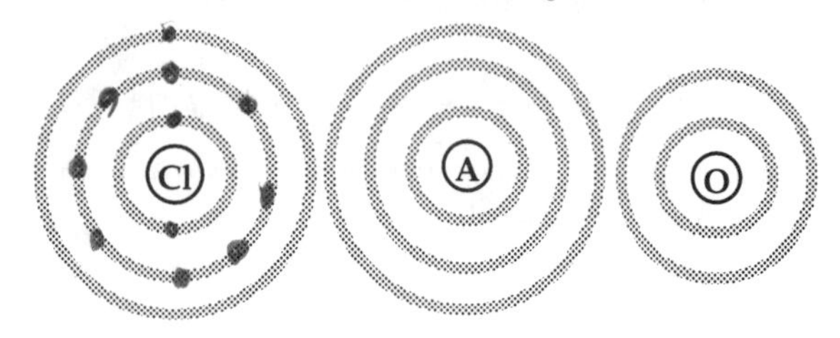

- In the following figures, place electrons in the appropriate locations based on the covalent bonding shown for each molecule.

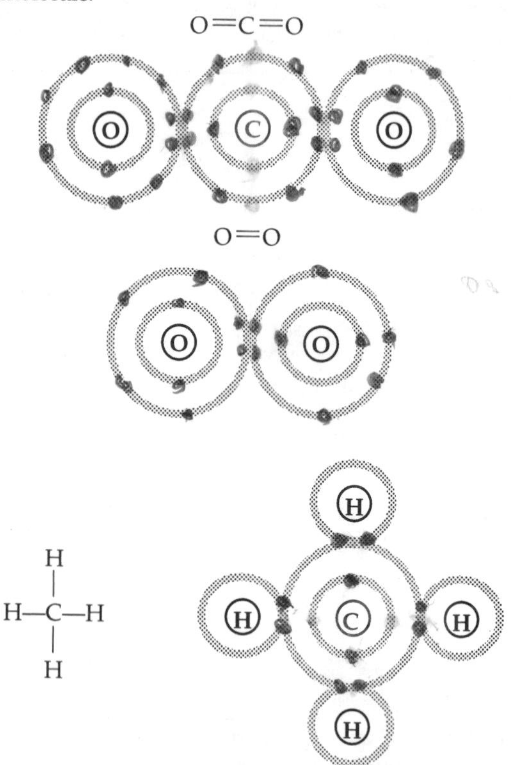

MOLECULES • CHEMICAL REACTIONS • WATER • ACIDS, BASES, AND pH • BUFFERS (pages 26–30)

Key Concepts

1. *Compounds* are substances whose molecules have two or more different atoms linked together by a chemical bond; *elemental substances* are composed of only one type of atom.
2. The *molecular formula* shows the symbols for each atom followed by the numbers of each atom written as a subscript, as in H_2O. *Structural formulas* show how the atoms are linked together.
3. The *molecular weight* of a compound is the sum of the atomic weights of the atoms in the molecule.
4. A gram molecular weight of a substance (its molecular weight in grams) is called a *mole*. If one mole is dissolved in one liter of water, the resulting solution is a 1.0-*molar* solution.
5. A mole of any substance contains the same number of molecules, called *Avogadro's number*, or 6.023×10^{23} molecules per mole.
6. A *chemical reaction*, represented by an arrow →, occurs when atoms change their pattern of electron sharing with other atoms.
7. Chemical reactions usually result in an exchange of energy with the environment. Some reactions release energy, while other reactions absorb it.
8. A *calorie* (cal) is the amount of heat required to raise 1 gram of water from 14.5 to 15.5°C. A *kilocalorie* (kcal) equals 1,000 cal.
9. Calories can be used as a measure of the energy associated with a chemical bond. In a chemical reaction, the bond energy of the reactants is usually different than the bond energy of the products.
10. More substances can dissolve in water than in any other solvent.
11. Most biologically important chemical reactions take place in water-based or *aqueous* solutions.
12. Ice floats because it is less dense than liquid water. This prevents bodies of water from freezing solid.
13. Water uses significant heat as it changes from liquid to gas (*evaporation*). Evaporation of sweat is an efficient cooling process for some organisms.
14. It requires a relatively large amount of heat to raise the temperature of water. This property causes the temperature of bodies of water to remain relatively constant or to change only slowly.
15. About 1 in every 550 million water molecules ionizes into a hydrogen ion (H^+) and a hydroxide ion (OH^-), $H_2O \rightarrow H^+ + OH^-$.
16. Pure water has equal concentrations of hydrogen and hydroxide ions and is called *neutral*. A solution with more hydrogen ions than hydroxide ions is called *acidic*. A solution with more hydroxide ions than hydrogen ions is called *basic* or *alkaline*.
17. A compound that releases H^+ in water is called an *acid*. Carbonic acid is an example: $H_2CO_3 \rightarrow H^+ + HCO_3^-$. Adding an acid to water makes the solution *acidic*.
18. A compound that accepts H^+ is called a *base*. The bicarbonate ion is an example of a base: $HCO_3^- + H^+ \rightarrow H_2CO_3$. Adding a base to water makes the solution *basic* or *alkaline*.
19. Acids and bases occur in pairs. H_2CO_3 is an acid; bicarbonate ion, HCO_3^-, is a base. $HCO_3^- + H^+ \rightleftharpoons H_2CO_3$ is an example of a *reversible reaction*, as indicated by double arrows.

20. The acidity or basicity of aqueous systems is expressed as *pH*. pH is the negative logarithm of the hydrogen ion concentration (pH = $-\log_{10}[H^+]$, where $[H^+]$ = molar concentration of hydrogen ions).

21. pH is a logarithmic scale. Each pH unit corresponds to a hydrogen ion concentration ten times greater than or less than the next nearest unit. Pure water has a pH of 7. Acidic solutions have pH values < 7; basic solutions have pH values > 7.

22. A *buffer* is a mixture of a weak acid (an acid that does not ionize completely) and its corresponding base.

23. Within a specific *buffering range*, buffers maintain a relatively constant pH by binding with or releasing H^+. If H^+ is added to a buffered system, the base binds with the additional H^+; if H^+ is removed from a buffered system, the weak acid ionizes to release additional H^+.

Questions (for answers and explanations, see page 47)

1. The molecular weight of glucose is 180. If you added 180 grams of glucose to 0.5 liter of water, what would be the molarity of the resulting solution?
   a. 18
   b. 1
   c. 9
   d. 0.5
   e. 2

2. One milliliter (ml) of water weighs one gram (g), so one liter (l) weighs one kilogram (kg). Given that the molecular weight of water is 18, how many moles of water are there in one liter of pure water?
   a. 1
   b. 100
   c. 56
   d. 18
   e. 1.8

3. How many kilocalories are required to raise 250 grams of water from 14.5 to 15.5° C?
   a. 0.25
   b. 4
   c. 0.4
   d. 2.5
   e. 1

4. In the reaction $CH_3COOH \rightleftharpoons H^+ + CH_3COO^-$, CH_3COO^- is an example of
   a. an acid.
   b. a base.
   c. a buffer.
   d. an ionic compound.
   e. a reversible reaction.

5. Cola has a pH of 3; blood plasma has a pH of 7. The hydrogen ion concentration of cola is ____ than the hydrogen ion concentration of blood plasma.
   a. four times greater
   b. four times lesser
   c. 400 times greater
   d. 400 times lesser
   e. 10,000 times greater

6. If solution A has a pH of 2 and solution B has a pH of 8, which of the following statements is true?
   a. A is basic and B is acidic.
   b. A is a base and B is an acid.
   c. A is acidic and B is basic.
   d. A is an acid and B is a base.
   e. A has a greater $[OH^-]$ than B.

Activities (for answers and explanations, see page 48)

- Add electrons to the following diagram showing the ionization of hydrogen chloride: HCL → $H^+ + Cl^-$.

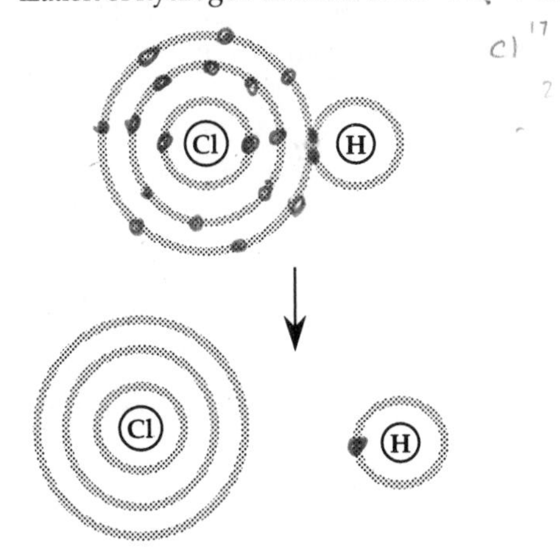

- Complete the missing parts of Table 1.

Table 1. The relationship between pH and the hydrogen ion concentration $[H^+]$ and hydroxide ion concentration $[OH^-]$ of water.

pH	[H⁺]	[OH⁻]
0	1.0	
1	0.1	
2	0.01	
3	0.001	
	10^{-4}	10^{-10}
	10^{-5}	10^{-9}
	10^{-6}	10^{-8}
	10^{-7}	10^{-7}
8		10^{-6}
9		10^{-5}
10		
11		
12		
13		
14		1.0

POLARITY • SOME SIMPLE ORGANIC COMPOUNDS AND FUNCTIONAL GROUPS (pages 30–36)

Key Concepts

1. *Polar* molecules are composed of atoms that share electrons unequally. The unequal sharing of electrons creates molecules with a positive and a negative pole.

2. *Nonpolar* molecules are composed of atoms that share electrons equally or in which the distribution of positive and negative poles is balanced within the molecule.
3. Water is a polar molecule because the shared electrons between the hydrogens and oxygen are more closely associated with the oxygen atom. Also, the hydrogen atoms are separated by an angle of 104.5° and, thus, are not aligned on exactly opposite sides of the oxygen.
4. The polarity of water explains its excellence as a solvent.
5. A *hydrogen bond* results from the attraction of the positive pole of a covalently bound hydrogen in one molecule and the negative pole associated with an atom in the same or a different molecule. A hydrogen bond is about one-tenth as strong as a covalent bond.
6. Hydrogen bonds form between water molecules, as each hydrogen atom is attracted to the negative pole of the oxygen atom in an adjacent water molecule.
7. Hydrogen bonding between water molecules explains the high surface tension of liquid water.
8. Hydrogen bonding of water to other molecules is the basis for the tendency of water to wet and climb up surfaces (called *capillary action*).
9. The attraction of the electron cloud of an atom in one nonpolar molecule for the positively charged nucleus in another nonpolar molecule is called a *van der Waals interaction*. Van der Waals interactions are about one-quarter to one-third as strong as a hydrogen bond.
10. The clumping of nonpolar molecules in the presence of water in called *hydrophobic interaction*. Hydrophobic interaction occurs to minimize the contact between nonpolar molecules and water. Hydrophobic interaction is an attractive force between nonpolar molecules equal in strength to the van der Waals interaction.
11. Both van der Waals interactions and hydrophobic interactions contribute to the tendency for nonpolar molecules to clump together.
12. *Hydrocarbons* are compounds containing only hydrogen and carbon. Hydrocarbons are nonpolar molecules and as a class are immiscible (not mixable) with water.
13. *Saturated hydrocarbons* contain only single covalent bonds between carbon atoms. *Unsaturated hydrocarbons* contain some double covalent bonds between carbon atoms.
14. An *alcohol* results when a *hydroxyl* group (—OH) replaces one hydrogen in a hydrocarbon. The —OH end of an alcohol is polar and, as a result, some alcohols are soluble in water.
15. In addition to a hydroxyl group, *sugars* also have a *carbonyl* group (a carbon with a double-bonded oxygen). If the carbonyl group includes a hydrogen, it is an *aldehyde* sugar; if it does not, it is a *ketone* sugar.
16. A compound containing a *carboxyl* group (—COOH) is an example of an *organic acid*, since the carboxyl group can ionize to become —COO⁻ and release a hydrogen ion.
17. A compound containing an *amino* group ($—NH_2$) is an example of an *organic base*, since the amino group can bind with a hydrogen ion to form the $—NH_3^+$ ion.
18. *Amino acids* have both an amino group and a carboxyl group attached to a central carbon called the *alpha (α) carbon*. Also attached to the α carbon are a hydrogen atom and a variable side chain.
19. Each of the 20 amino acids found in proteins has a different side chain which determines its chemical characteristics.
20. Amino acids can be joined together into chains to form the important class of organic molecules called *proteins*.
21. If a carbon atom is bound to four different atoms or groups it becomes an *asymmetric carbon*. The α carbon of an amino acid is an asymmetric carbon.
22. The *sulfhydryl* group, —SH, is important in protein structure and the *phosphate* group, $—OPO_3^{-2}$, is involved in energy transfer within the cell.
23. *Isomers* have the same molecular formula, but differ in their structural formula.
24. Each of the two different, mirror-image arrangements of the four groups around an asymmetric carbon is called an *optical isomer*. An asymmetric carbon's two optical isomers cannot be superimposed.
25. The two optical isomers of an amino acid are called the L- and D- forms. On Earth, only L-amino acids are commonly found in the proteins of living things.

Questions (for answers and explanations, see page 48)

1. Which of the following molecules is *not* polar?
   a. C_2H_5OH
   b. $H_3C—CH_3$
   c. H_2O
   d. CH_2NH_3COOH
   e. NH_3

2. Which of the following shows the *correct* arrangement of two water molecules and the hydrogen bond(s) linking them together?

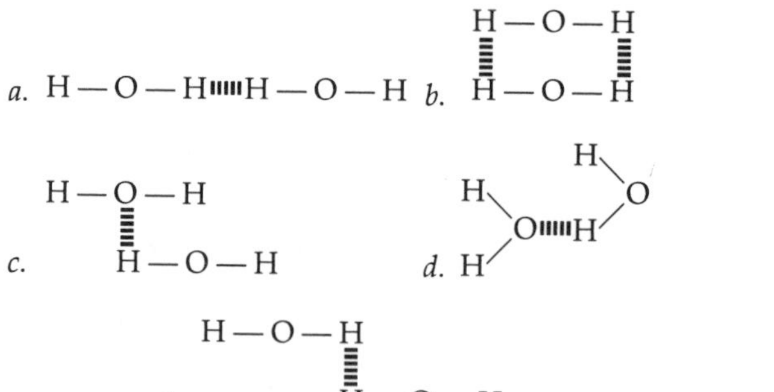

   a. H—O—H····H—O—H

   b.
   H—O—H
   ⋮ ⋮
   H—O—H

   c.
   H—O—H
   ⋮
   H—O—H

   d.
   H
   O
   H
   O····H
   H

   e.
   H—O—H
   ⋮
   H—O—H

3. Which of the following properties of water is *most* important in determining its excellence as a solvent?
   a. The pH of water is 7.
   b. Water has a high heat of evaporation.
   c. Water has a high surface tension.
   d. Water has a high heat capacity.
   e. The water molecule is polar.

4. Which of the following statements about optical isomers is *not* true?

a. Optical isomers have the same molecular weight.
b. Optical isomers must have an asymmetric carbon.
c. Optical isomers only occur in the L- form.
d. Optical isomers cannot be superimposed.
e. Optical isomers must have identical groups attached to the asymmetric carbon.

5. Which of the following molecules would you expect to be *least* soluble in water?

a. C_2H_5OH
b. $H_3C—CH_3$
c. NH_4OH
d. CH_2NH_2COOH
e. NH_3

6. Which of the following molecules would be expected to form van der Waals or hydrophobic bonds with another molecule?

a. NH_3
b. C_2H_5OH
c. NH_4OH
d. CH_2NH_2COOH
e. $H_3C(CH_2)_4CH_3$

7. Match the functional groups listed on the left with the correct structural formulas shown on the right.

______	Amino	*a.* —OH
______	Phosphate	*b.* —CHO
______	Hydroxyl	c. —SH
______	Carboxyl	*d.* —COOH
______	Carbonyl	*e.* $—NH_2$
______	Sulfhydryl	*f.* $—OPO_3^{-2}$

Activities (for answers and explanations, see page 48)

- Label each of the following as either an organic acid, an organic base, an amino acid, a saturated hydrocarbon, an unsaturated hydrocarbon, or an alcohol.

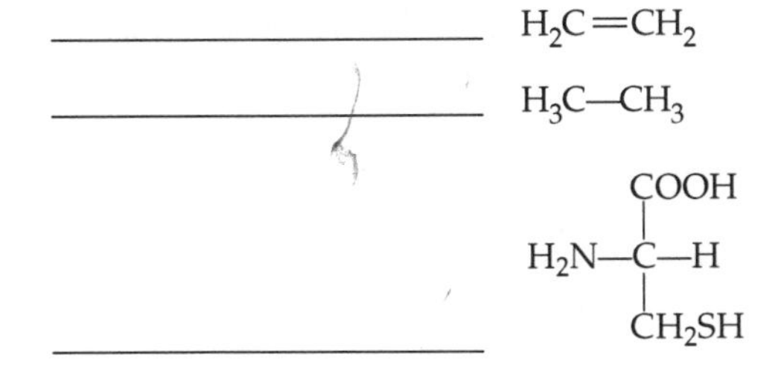

______ $H_2N—C(=O)—NH_2$

______ $H_3C—C(=O)—CH_2—COOH$

______ $H_3C—CH_2—OH$

- On the following diagram, circle and label the amino group, carboxyl group, α carbon, and side chain of the amino acid serine.

COOH
$H_2N—C—H$
CH_2OH

Integrative Questions (for answers and explanations, see page 48)

1. Which of the following types of bonds is strongest?
a. Hydrogen
b. Covalent
c. Van der Waals
d. Hydrophobic
e. Ionic

2. Consider the following element,

$$^{40}_{20}X$$

If an ion of this element has an electric charge of +2, how many electrons will there be in the outermost electron shell of the ion?
a. 8
b. 2
c. 4
d. 0
e. 1

3. Consider a hypothetical ion in which the number of protons is X, the number of neutrons is Y, and the number of electrons is Z. The atomic number of the ion would be the same as
a. X.
b. Y.
c. Z.
d. X or Z.
e. X + Y.

3

Large Molecules

CHAPTER LEARNING OBJECTIVES—after studying this chapter you should be able to:

- ❑ Describe a feature that characterizes all lipids and the two major biological functions of this class of macromolecules.
- ❑ Recognize a fatty acid and determine if it is saturated, unsaturated, or polyunsaturated.
- ❑ Recognize a triglyceride and describe the effects that fatty acid characteristics have on the properties of triglycerides.
- ❑ Recognize a phospholipid and describe how its structure determines its function in biological membranes.
- ❑ Differentiate between the terms "hydrophilic" and "hydrophobic" and recognize hydrophilic and hydrophobic regions of macromolecules.
- ❑ Describe the characteristics and roles of carotenoids and steroids.
- ❑ Describe the features of condensation reactions and how these reactions produce polymers from monomers.
- ❑ Explain the relationship between monosaccharides, disaccharides, oligosaccharides, and polysaccharides.
- ❑ Describe the transformation of glucose from straight chain to ring form, explain the numbering system used in referring to the carbons of a monosaccharide, and differentiate between α– and β-glucose.
- ❑ Recognize monosaccharides as pentoses or hexoses and list some of the common pentose and hexose isomers.
- ❑ Describe the formation of a disaccharide from two monosaccharides and differentiate between α and β linkages.
- ❑ Explain the principal features and biological functions of cellulose, starch, and glycogen.
- ❑ Explain the principal features and biological functions of some common sugar phosphates and amino sugars.
- ❑ Describe the classification of amino acids by the chemical properties of their R groups and the effect of R group properties on the location of specific amino acids within proteins.
- ❑ Identify the N-terminus and C-terminus of a polypeptide.
- ❑ Explain the formation of the following bonds within a polypeptide: peptide linkage, disulfide bridge, hydrogen bond.
- ❑ Determine the number of unique proteins that could be formed given a specific polypeptide length.
- ❑ Explain how protein primary structure determines higher levels of organization.
- ❑ Describe the three common types of protein secondary structure and give an example of each.
- ❑ Describe the organization of a protein such as hemoglobin relative to its primary, secondary, tertiary, and quaternary structure.
- ❑ Describe the main biological functions of proteins and how their primary structure is coded for by DNA.
- ❑ Describe the structure of a nucleotide and a nucleoside and their polymerization to form DNA and RNA.
- ❑ Describe differences in the structures of DNA and RNA.
- ❑ Classify the five nitrogenous bases as purines or pyrimidines and indicate their occurrence in DNA and RNA.
- ❑ Explain the basis for complementary base pairing and its importance in the transfer of information from DNA to RNA and RNA to protein.
- ❑ Describe the structure of the following macromolecules: glycoprotein, glycolipid, lipoprotein, and nucleoprotein.

LIPIDS • FROM MONOMERS TO POLYMERS (pages 39–45)

Key Concepts

1. As a class, *lipids* are insoluble in water but readily soluble in certain organic solvents. A mixture of water and lipids forms two distinct layers.
2. Because of the natural tendency of lipids to form a layer in water, lipids are a major constituent of biological *membranes*. The movement of materials through the lipid-containing membranes of cells and organelles depends largely on their lipid solubility.
3. Energy storage is another major function of cellular lipids.
4. *Fatty acids* are carboxylic acids with long hydrocarbon chains. *Saturated fatty acids* have no double bonds between carbon atoms in the hydrocarbon chain; *unsaturated fatty acids* have some double bonds. *Polyunsaturated fatty acids* have more than one double bond.
5. *Triglycerides* (also called simple lipids) contain three fatty acid molecules chemically bonded to a molecule of *glycerol*. The fatty acids in a triglyceride need not be the same in either the length of the chain or their saturated/unsaturated condition.

6. Triglycerides with short or unsaturated hydrocarbon chains are usually liquid at room temperature and are called *oils*. Triglycerides with longer, saturated hydrocarbon chains are usually solid at room temperatures and are called *fats*.

7. *Phospholipids* result when one of the fatty acids of a triglyceride is replaced by a phosphorus-containing compound. The phosphorus-containing end of the molecule is electrically charged and, thus, *hydrophilic*. The hydrocarbon end of the molecule is *hydrophobic*.

8. Phospholipids are a major component of membranes and are oriented within the membrane to form a *bilayer*. In the bilayer, the phospholipids are arranged with the hydrophobic, hydrocarbon "tails" forming the center of the bilayer and the phosphorus-containing, hydrophilic "heads" to the outside of the bilayer.

9 The lipids called *carotenoids* are light-absorbing pigments found in both plants and animals. β-carotene and its breakdown product vitamin A are examples of important carotenoids.

10 *Steroids* are lipids with a specific four-ring structure. Many steroids, such as testosterone, estrogen, and the cortisones, are the chemical messenger molecules called hormones; other steroids, such as *cholesterol*, are important components of membranes.

11. Most lipids can be synthesized by animals. Humans cannot produce certain lipids (three types of unsaturated fatty acids and the vitamins A, D, E, and K), and these lipids must be obtained in our diet.

12. Macromolecules are *polymers* composed of simple subunits called *monomers*. Macromolecules have molecular weights in excess of 1,000 daltons.

13. Oligomers are composed of only several monomers.

14. Polymerization involves *condensation* (also called *dehydration*) reactions of the type $A—H + B—OH \rightarrow A—B + H_2O$, where A and B are the *reactant* monomers.

15. During a condensation reaction a water molecule is released for each subunit added to the polymer; the hydroxyl group (OH) and a hydrogen atom (H) are contributed by the reactants. Condensation reactions are energy-requiring reactions.

Questions (for answers and explanations, see page 49)

1. Which of the following classes of lipids includes important light-capturing molecules?
   a. Phospholipids
   b. Carotenoids
   c. Fats
   d. Triglycerides
   e. Steroids

2. Which of the following pairs of lipid classes includes molecules that are important components of biological membranes?
   a. Fats and oils
   b. Steroids and phospholipids
   c. Triglycerides and phospholipids
   d. Carotenoids and triglycerides
   e. Carotenoids and fatty acids

Consider the following triglycerides (*A* and *B*) in answering questions 3–6.

A:
$CH_3(CH_2)_{16}—C(=O)—O—CH_2$
$CH_3(CH_2)_{16}—C(=O)—O—CH$
$CH_3(CH_2)_{16}—C(=O)—O—CH_2$

B:
$CH_3(CH_2)_3—C(=O)—O—CH_2$
$CH_3–HC=CH–CH_2—C(=O)—O—CH_2$
$CH_3–HC=CH–CH_2—C(=O)—O—CH_2$

3. In *B*, circle the remnant of the glycerol portion of the triglyceride.

4. Which triglyceride (*A* or *B*) is probably a solid at room temperature? _____ Explain your answer.

5. Which triglyceride (*A* or *B*) is probably derived from a plant? _____ Explain your answer.

6. How many water molecules will result during the formation of triglyceride *B* from glycerol and three fatty acids?
   _____

7. What type of lipid is the following molecule?

   (H_3C, OH, H_3C, O — four-ring steroid structure)

   a. Phospholipid
   b. Carotenoid
   c. Fat
   d. Triglyceride
   e. Steroid

Activities (for answers and explanations, see page 49)

- Circle and label the hydrophilic and the hydrophobic ends of the following phospholipid. Assuming that this phospholipid is part of a membrane, draw an arrow to indicate the direction to the outside of the membrane.

$CH_3–CH_2–(CH_2)_{14}–CH_2–C(=O)–O–CH_2$
$CH_3–CH_2–(CH_2)_6–CH_2–C(=O)–O–CH$
$H_2–C–O–P(=O)(O^-)–O^-$

- Circle the carboxylic end of the space-filling model of a fatty acid shown below.

  How many carbons are included in this fatty acid? ______

  Is this fatty acid saturated or unsaturated? ____________

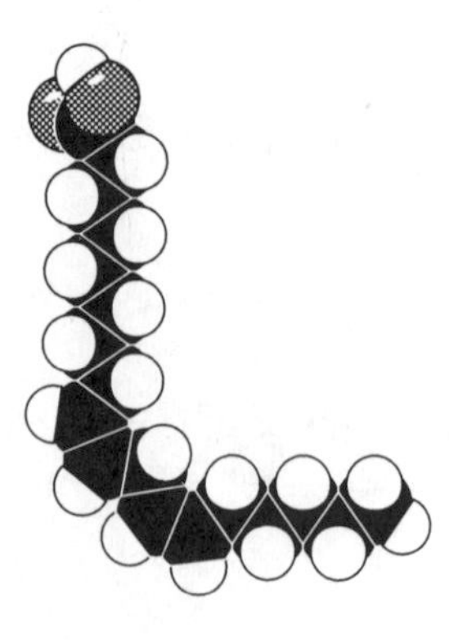

CARBOHYDRATES (pages 45–47)

Key Concepts

1. Carbohydrates consist of *monosaccharides*, *oligosaccharides*, and *polysaccharides*. Monosaccharides are simple sugars; oligosaccharides result when a few monosaccharides are linked together by condensation reactions; polysaccharides are polymers of many simple sugars.
2. Monosaccharides can exist in equilibrium in a straight chain form and a ring form.
3. Most carbohydrates share the general formula $C_nH_{2n}O_n$—there are twice as many hydrogen atoms as oxygen atoms.
4. *Isomers* are molecules with the same types and numbers of atoms, but the atoms are arranged in different ways. Two common isomers of the simple sugar *glucose* (α- and β-glucose) differ in the placement of the —H and —OH groups on carbon 1 (C-1).

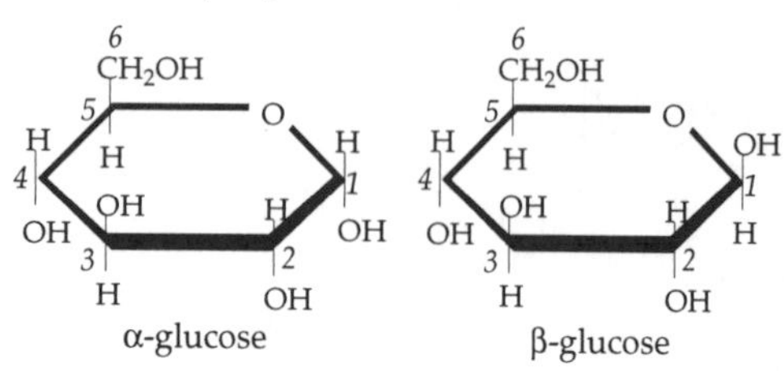

5. All isomers of glucose have the same formula ($C_6H_{12}O_6$) and include the sugars fructose, mannose, and galactose. *Hexoses* include all six-carbon sugars.
6. *Ribose* and *deoxyribose* are five-carbon sugars (called *pentoses*) found in the backbone of RNA and DNA, respectively. Deoxyribose has one less oxygen atom than ribose.
7. *Polysaccharides* are composed of many simple sugars covalently coupled together by condensation reactions.
8. A *disaccharide* results when a condensation reaction joins together two monosaccharides. If the monosaccharide providing C-1 of the linkage is an alpha sugar, than the linkage is an α linkage; if it is a beta sugar, then it is a β linkage.
9. *Maltose* results when C-1 of α-glucose is linked to another glucose. *Cellobiose* has the same formula as maltose ($C_{12}H_{22}O_{11}$), but involves a β-glucose linked to another glucose.
10. *Cellulose* is a polysaccharide with only β linkages between glucose molecules.
11. *Starch* (*amylose*) consists of hundreds to thousands of α-glucose molecules all joined C-1 to C-4. Unlike cellulose, starch can also form branches between C-6 of one glucose and C-1 of another.
12. *Glycogen* is identical to starch (α-glucose joined C-1 to C-4, with C-1 to C-6 branches), except it is more branched.
13. Starch and glycogen are important energy storage molecules in plants and animals, respectively.
14. Cellulose is the main structural building molecule used by plants.
15. *Derivative carbohydrates* contain other elements than carbon, hydrogen, and oxygen. *Sugar phosphates* like fructose 1,6-bisphosphate are important intermediates in photosynthesis and respiration. *Amino sugars* have an amino group in place of one —OH. *Galactosamine* is a major constituent of cartilage; *chitin* is derived from the amino sugar glucosamine.

Questions (for answers and explanations, see page 49)

1. Complete the following statements about the carbohydrate shown below.

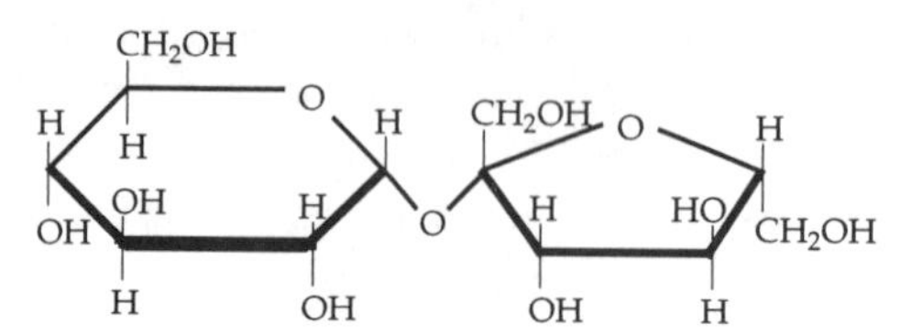

   a. The carbohydrate shown is an example of a ________________.

   b. The two monosaccharides forming this carbohydrate are both ______________ sugars.

   c. The two monosaccharides forming this carbohydrate are joined together by a ______ linkage.

   d. Prior to the condensation reaction that formed this carbohydrate, the two monosaccharides had the chemical formula ____________ .

   e. Based on *d*, we can call these two monosaccharides__________ .

2. Which of the following statements differentiates between starch and cellulose?
   a. Starch is composed of sucrose subunits, cellulose is formed from glucose subunits.
   b. Starch subunits are joined by β linkages; cellulose subunits are joined by α linkages.
   c. In addition to C-1 to C-4 linkages, starch also has C-6 to C-1 linkages; cellulose has only C-1 to C-4 linkages.
   d. Starch is found in animals, cellulose is found in plants.

3. Which of the following compounds is *not* classified as a derivative carbohydrate?
   a. Sugar phosphates
   b. Amino sugars
   c. Glucosamine

d. Glycogen
e. Fructose 1, 6-bisphosphate

4. Circle all of the following statements that correctly compare ribose and deoxyribose sugar.
   a. They are both pentose sugars.
   b. They are isomers.
   c. Ribose has one more oxygen that deoxyribose.
   d. They are both found in DNA.
   e. They have the same number of carbons.

PROTEINS • LEVELS OF PROTEIN STRUCTURE (pages 48–55)

Key Concepts

1. Proteins are important structural components of living things; functional proteins include the immunoglobulins, enzymes, and contractile proteins.
2. The twenty different *amino acids* are the monomers joined together to form proteins. The chemical properties of amino acids are determined by the nature of their side chains, called *R groups.*
3. Five amino acids have *hydrophobic R groups* composed predominantly of hydrocarbons. Protein chain segments with these amino acids are found in the interior of the protein when in an aqueous, hydrophilic environment. Other protein chain segments with these amino acids are found on the surface of the protein when in a hydrophobic environment.
4. Seven other amino acids have *hydrophilic R groups.* Protein chain segments with these amino acids behave the opposite of those with hydrophobic R groups.
5. The remaining eight amino acids have *polar, but uncharged R groups* and can occur either on the surface or in the interior of proteins.
6. *Cysteine* (cys) is a sulfur-containing amino acid. Two cysteines can form a covalent bond, called a *disulfide bridge,* between the sulfur atoms of their R groups.
7. A *peptide linkage* results when a condensation reaction joins the carboxyl group of one amino acid to the amino group of a second amino acid.
8. *Polypeptides* are chains of many amino acids joined by peptide bonds. A polypeptide has an amino end (N–terminus) and a carboxyl end (C–terminus). Proteins can consist of one or more polypeptides.
9. Within a protein, hydrogen bonding occurs between nearby monomers because the C=O oxygen has a partial negative charge and the N—H hydrogen has a partial positive charge.
10. The sequence of amino acids in a polypeptide is the *primary structure* of the protein. Because each amino acid in a polypeptide can be one of twenty different types, the number of possible polypeptides with *n* amino acids is 20^n.
11. The sequence of monomers (called nucleotides) in DNA determines the primary structure of the protein coded for by the DNA.
12. Twisting and folding of polypeptide chains may result in consistent structural configurations collectively called *secondary structure.* The three types of secondary structure are the α helix, the β-pleated sheet, and the triple helix. A protein may have one or more regions with a secondary structure, or none at all.
13. The *α-helix* configuration results when a polypeptide twists so that hydrogen bonds can form between atoms in amino acids four monomers apart. A group of fibrous proteins called *keratins* and the contractile protein *myosin* show an α helical secondary structure.
14. In the *β-pleated sheet* configuration, several different polypeptide strands (or different regions of one strand) are bonded together by hydrogen bonds to form a sheet of strands. From the side the sheet has a zigzag, pleated appearance with R groups projecting above or below the sheet. This type of conformation is shown in the protein *silk.*
15. The protein *collagen* consists of subunits that show the *triple helix* secondary structure, in which three polypeptide chains twist around each other and are held together by hydrogen bonding. Collagen is a strong, stretchable protein found in tendons, cartilage, and the skin of vertebrates.
16. The overall, three-dimensional shape of the whole protein molecule is its *tertiary structure.* Myoglobin is an important oxygen-binding protein with a tertiary structure. The oxygen in myoglobin is bound to an iron-containing *heme* group.
17. Proteins that consist of two or more polypeptides have a higher-order structure called the *quaternary structure.* For example, *hemoglobin* consists of four globular, myoglobin-like molecules of two types held together by a combination of ionic, hydrophobic, and hydrogen bonds. The relative arrangement of these subunits is hemoglobin's quaternary structure.
18. The sequence of amino acids in a protein (the primary structure) determines bonding potentials between different regions of the protein. Thus, the primary structure determines all higher-order structural levels.

Activity (for answers and explanations, see page 49)

- In the following polypeptide, label the N–terminus, C–terminus, and all the peptide linkages.

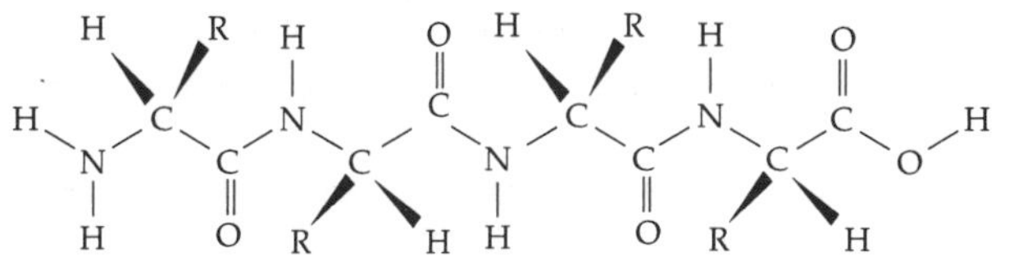

```
   H   R  H    O  H   R  H    O
H   C   N    C   C   N    C   H
  N   C    C   N    C   C   O
  H   O  R   H H    O  R   H
```

How many water molecules were released in the polymerization of this polypeptide? ________

Questions (for answers and explanations, see page 49)

1. Primary protein structure determines which of the following additional levels of protein organization?
   a. Secondary
   b. Tertiary
   c. Secondary and tertiary
   d. Quaternary
   e. Secondary, tertiary, and quaternary

2. Which of the following statements correctly categorizes the R groups (side chains) in the dipeptide shown below?

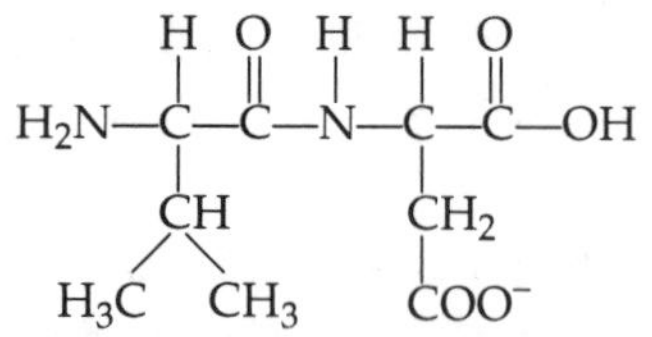

```
      H  O  H  H  O
      |  ‖  |  |  ‖
H2N—C—C—N—C—C—OH
      |        |
      CH       CH2
     /  \      |
  H3C   CH3    COO−
```

   a. Both R groups are hydrophobic.
   b. Both R groups are hydrophilic.
   c. The left R group is hydrophobic; the right is hydrophilic.
   d. The left R group is hydrophilic; the right is hydrophobic.
   e. None of these statements is correct.

3. How many different polypeptides *four* amino acids long could be made from *five* different kinds of amino acids?

   ____

4. Which of the following types of protein secondary structure are maintained by hydrogen bonding?
   a. α helix
   b. Triple helix
   c. β-pleated sheet
   d. α helix and triple helix
   e. All three

5. What is the *highest* structural organization found in *all* enzymes?
   a. Primary
   b. Secondary
   c. Tertiary
   d. Quaternary

NUCLEIC ACIDS • GLYCOLIPIDS, GLYCOPROTEINS, LIPOPROTEINS, AND NUCLEOPROTEINS (pages 55–59)

Key Concepts

1. Nucleic acids are polymers of *nucleotides* involved in controlling the production of proteins. *Deoxyribonucleic acid (DNA)* codes for proteins; *ribonucleic acid (RNA)* transfers this information to protein-making machinery within the cell.
2. *Nucleotides* consist of a pentose sugar (deoxyribose in DNA, ribose in RNA), a phosphate group, and a nitrogenous base. The phosphate group is attached to carbon 5 of the sugar, the nitrogenous base to the sugar's carbon 1.
3. A *nucleoside* consists of a pentose sugar and a nitrogenous base without the phosphate group.
4. The nature of the nitrogenous base determines the kind of nucleotide or nucleoside. Four bases are found in DNA: *adenine, cytosine, guanine,* and *thymine.* Adenine, cytosine, and guanine are also found in RNA, but instead of thymine, RNA has the base *uracil.*
5. In a polynucleotide chain, the backbone is formed of alternating sugar and phosphate groups, with the nitrogenous bases attached to the sugars.
6. Nucleotides are joined together by *phosphodiester linkages* between the phosphate group of one nucleotide and carbon 3 of the next nucleotide.
7. DNA consists of two antiparallel chains of nucleotides bound together by hydrogen bonding between the nitrogenous bases on the different chains. The DNA molecule is twisted into a double helix.
8. Most RNA molecules are single-stranded.
9. Adenine and guanine are double-ringed molecules called *purines;* uracil, cytosine, and thymine are single-ringed molecules called *pyrimidines.*
10. There is just enough space between two polynucleotide chains for a pyrimidine to pair with a purine, so that all base pairs are the same size (three rings). These hydrogen bonding restrictions are the basis for the *principle of complementary base pairing.*
11. For optimum hydrogen bonding to occur, adenine can only pair with thymine (two hydrogen bonds) and cytosine can only pair with guanine (three hydrogen bonds).
12. In RNA, adenine pairs with uracil, but when RNA binds to DNA, adenine from RNA binds to the thymine from DNA.
13. The sequence of bases in one strand of DNA is the information that specifies the sequence of amino acids in proteins.
14. The base-pairing rules ensure that a DNA sequence specifies the RNA sequence and RNA transmits this information to the cytoplasm where it is used to manufacture proteins.
15. All three major classes of macromolecules can covalently bond together to produce glycolipids, glycoproteins, lipoproteins, and nucleoproteins.
16. Glycolipids usually have their carbohydrate portion directed to the outside of the cell membrane. Many nucleoproteins are involved with regulation of DNA. Lipoproteins move hydrophobic lipids through the hydrophilic tissues of the body.

Questions (for answers and explanations, see page 50)

1. Which of the following choices is to DNA and RNA what amino acids are to proteins?
   a. Sugar, phosphate, nitrogenous base
   b. Nucleosides
   c. Nitrogenous bases
   d. Nucleotides
   e. Adenine, cytosine, thymine, guanine, uracil

2. Which of the following choices is to DNA and RNA what amino acid R groups are to proteins?
   a. Sugar, phosphate, nitrogenous base
   b. Nucleosides
   c. Ribose or deoxyribose
   d. Nucleotides
   e. Adenine, cytosine, thymine, guanine, uracil

3. Which of the following is *not* a difference between RNA and DNA?
   a. Adenine pairs with different bases in RNA and DNA.
   b. Purines are only found in RNA; pyrimidines are restricted to DNA.
   c. Only DNA is normally found in a double helix.
   d. Ribose is found in RNA; deoxyribose is found in DNA.
   e. DNA encodes the information for making proteins; RNA carries out these instructions.

Activities (for answers and explanations, see page 50)

- Use the base-pairing rules for DNA–RNA to label the complementary strand of RNA (right) to the single strand of DNA (left) shown below, where: C = cytosine, G = guanine, A = adenine, T = thymine, and U = uracil.

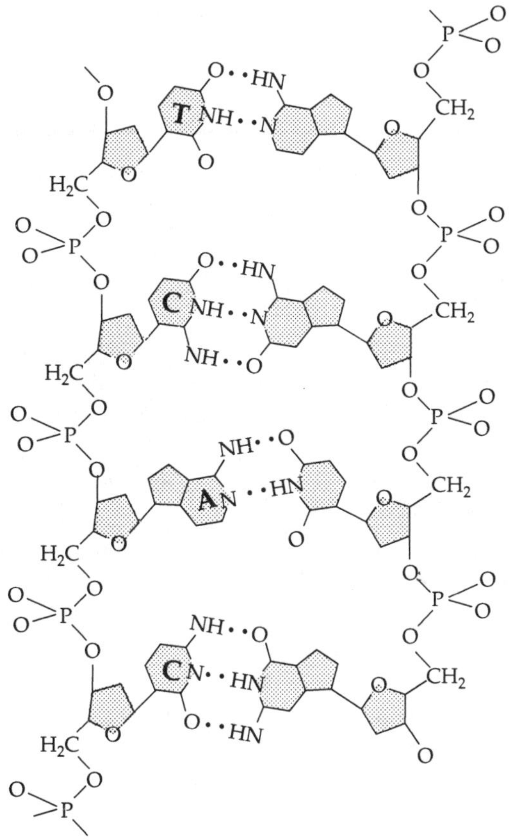

- Circle and label an example of a *nucleotide* and a *nucleoside* in the figure showing DNA–RNA.

- Based on the orientation of the sugar molecules, label the four ends of the molecule as 3' or 5' in the figure showing DNA–RNA.

Integrative Questions (for answers and explanations, see page 50)

1. Which of the following did *not* result from polymerization?
   a. RNA
   b. An enzyme
   c. Cellulose
   d. Cholesterol
   e. Myoglobin

2. What would you expect to be *true* of the R groups of amino acids located on the surface of protein molecules, found within the interior of biological membranes?
   a. The R groups should be hydrophobic.
   b. The R groups should be hydrophilic.
   c. The R groups should be polar.
   d. The R groups should be able to form disulfide.

3. Use numbers to represent the normal flow of information within a cell.

   _____ RNA nucleotide sequence

   _____ Primary structure of a protein

   ____ Folding of a protein into a three-dimensional shape

   _____ DNA nucleotide sequence

   _____ An enzyme catalyzes a specific reaction

4

Organization of the Cell

CHAPTER LEARNING OBJECTIVES—after studying this chapter you should be able to:

- ❑ State the two tenets of the cell theory.
- ❑ Describe the distinguishing features of prokaryotic and eukaryotic cells and know the kingdoms of the organisms composed of these two cell types.
- ❑ Describe prokaryotic cell structural features such as the nucleoid, cytoplasm, cytosol, ribosomes, cell walls, and several types of cell projections and membrane inclusions.
- ❑ Explain the term "resolution" and know the relative resolutions of the light and electron microscopes.
- ❑ Distinguish between scanning and transmission electron microscopy.
- ❑ Identify the major bacterial shapes.
- ❑ Name and describe the functions of all major eukaryotic organelles.
- ❑ Describe the characteristic features of the nucleus.
- ❑ Describe the characteristic features of ribosomes.
- ❑ Describe the characteristic features of the mitochondrion.
- ❑ Describe the characteristic features of the chloroplast.
- ❑ Explain the endosymbiotic theory and describe the evidence supporting this theory.
- ❑ Name the components of the endomembrane system and explain their interacting functions.
- ❑ Describe differences in the activities of the rough and smooth endoplasmic reticulum.
- ❑ Describe the structure of the Golgi apparatus and the movement of vesicles through the Golgi.
- ❑ Differentiate between endocytosis, exocytosis, pinocytosis, and phagocytosis.
- ❑ Describe the formation of lysosomes and their involvement in intracellular digestion.
- ❑ Describe the characteristic features of the several common types of microbodies and vacuoles.
- ❑ Describe the structural features and characteristic uses of microfilaments, intermediate filaments, and microtubules in the cell cytoskeleton.
- ❑ Distinguish between cilia and flagella and explain their structural similarities and differences.
- ❑ Explain what a basal body is and its relationship to a centriole.
- ❑ Describe the relationship between cell walls and plasmodesmata.
- ❑ Explain the techniques of cell fractionation and distinguish between equilibrium and differential centrifugation.

CELLS AND THE CELL THEORY • COMMON CHARACTERISTICS OF CELLS • PROKARYOTIC CELLS • PROBING THE SUBCELLULAR WORLD: MICROSCOPY (pages 61–68)

Key Concepts

1. According to the *cell theory*, the cell is the basic unit of life and all cells arise from other cells.
2. Most cells are tiny, depend on external energy sources, selectively regulate exchange of material with the environment, and use information in their DNA to regulate their chemistry.
3. *Prokaryotic cells* are characterized by the absence of a true nucleus and other membrane-bounded organelles. Organisms in the kingdoms Eubacteria and Archaebacteria consist of prokaryotic cells and are called *prokaryotes*.
4. *Eukaryotic cells* have true nuclei and membrane-bounded organelles. Organisms in the other four kingdoms are composed of eukaryotic cells and are called *eukaryotes*.
5. Important metric units used in studying cells are the *micrometer*, μm (1 μm = 1×10^{-6} m), and the *nanometer*, nm (1 nm = 1×10^{-9} m).
6. Cells have a *plasma membrane* responsible for regulating exchange of materials with the environment.
7. A prokaryotic cell has at least one *nucleoid* region containing DNA.
8. The remainder of the prokaryotic cell, called the *cytoplasm*, is a complex solution containing organic and inorganic molecules and a collection of many *ribosomes*.
9. The *cytosol* is the fluid portion of the cytoplasm.
10. Prokaryotic ribosomes each consist of three molecules of RNA and about 50 different protein molecules bound together into a roughly spherical structure with a diameter of 15–20 nm. Ribosomes coordinate the synthesis of proteins within the cell.
11. Many prokaryotic cells also have an external *cell wall* consisting of *peptidoglycan* (a polymer of amino sugars) and an

outer layer of polysaccharides called a *capsule*. The cell wall provides structural support and the capsule may protect the cell from dehydration or destruction by other cells.

12. In the photosynthetic prokaryotes, like the cyanobacteria, infolding of the plasma membrane produces an internal membrane system containing the molecules used in photosynthesis.
13. *Mesosomes* are other membrane systems involved in a variety of functions that also result from infolding of the plasma membrane.
14. Prokaryotic cell projections include *flagella* (sing. *flagellum*), whose function is propulsion, and *pili* (sing. *pilium*), used for adhesion to other cells.
15. *Bacteria* are prokaryotes. Bacterial species can occur as individual cells, short chains, or in colonies. Common bacterial shapes include the spherical *cocci*, rod-shaped *bacilli*, and corkscrew-shaped *spirilla* and *spirochetes*.
16. *Resolution* or resolving power is a measure of the ability to see two objects as separate. Objects closer than about 0.1 mm (100 μm) cannot be seen as separate objects by the unaided eye.
17. Microscopes are instruments designed to improve the resolution of the human eye. The *light microscope* has a resolving power of about 200 nm (0.2 μm, 2×10^{-4} mm). The *electron microscope* uses magnetic lenses and provides a resolution of about 2 nm under ideal conditions.
18. In *transmission electron microscopy*, high-energy electrons pass through a thin section of material. *Scanning electron microscopy* provides a three-dimensional view of the cell by bouncing electrons off the exposed surface.

Questions (for answers and explanations, see page 50)

1. Scanning electron microscopy provides a resolution of about 10 nm. If the resolution of the unaided human eye is 0.1 mm, how much is our resolution increased by this technology?
   a. 10^6 times
   b. 10^5 times
   c. 10^4 times
   d. 10^3 times
   e. 10^2 times

2. Match the appropriate letters for characteristics listed on the right to the cell structures listed on the left.

______	Nucleoid	*a.* Fluid portion of cytoplasm
______	Cytosol	*b.* Locomotory structure
______	Capsule	c. Membranous structure
______	Flagella	*d.* Contains the cell's DNA
______	Pili	*e.* Cell adherence
______	Mesosome	*f.* Can have protective function

3. What type of bacterial cell shape is shown in the following photomicrograph, and what type of microscopy was used to obtain the photomicrograph?
   a. Coccus, scanning electron microscopy
   b. Bacillus, transmission electron microscopy
   c. Spirillum, scanning electron microscopy
   d. Coccus, transmission electron microscopy
   e. Coccus, light microscopy

Activity (for answers and explanations, see page 50)

- Use the following terms to label the photomicrograph of the bacterial cell shown below: (A) peptidoglycan, (B) plasma membrane, (C) nucleoid, (D) capsule, (E) ribosomes, (F) cytoplasm.

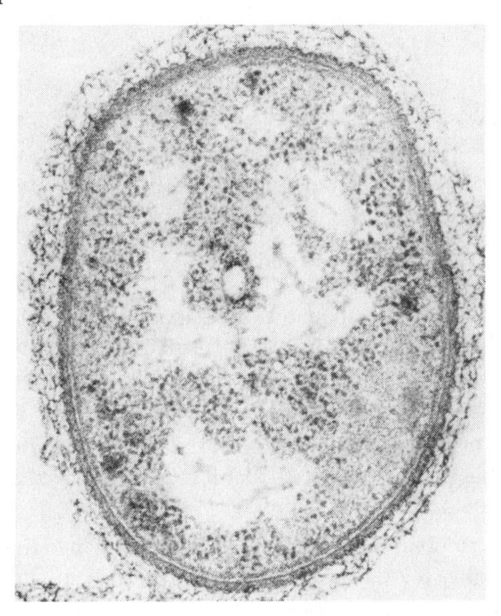

THE EUKARYOTIC CELL • INFORMATION-PROCESSING ORGANELLES • ENERGY-PROCESSING ORGANELLES (pages 68–77)

Key Concepts

1. The membrane-bounded systems that are unique to eukaryotic cells are included in the collection of intracellular structures called *organelles*. Although not membranous, ribosomes are also considered organelles.
2. Various membrane–bounded molecules carry out important functions seen in eukaryotic cells, such as transport of materials between the cell and the environment and between different organelles, interactions between cells, and energy transformation.
3. In eukaryotic cells, the DNA resides in the *nucleus*, the largest organelle found within the cell. The nucleus is surrounded by a double phospholipid bilayer membrane called the *nuclear envelope*.
4. The nuclear envelope is perforated by *nuclear pores* formed in areas where the inner and outer membranes merge. Eight protein granules are arranged around each pore. RNA and some proteins can move through the nuclear pores.
5. The outer membrane of the nuclear envelope is continuous with the endoplasmic reticulum of the cytoplasm and, like the endoplasmic reticulum, the outer surface of this membrane may be embedded with *ribosomes*.
6. The complex of nuclear DNA and proteins found in eukaryotes is called *chromatin*. Chromatin usually exists as a diffuse collection of threads, but when the nucleus is about to divide chromatin condenses to form *chromosomes*.
7. The *nucleoli* (sing. *nucleolus*) are also found in the nucleus. Ribosomal RNA synthesis and assembly with the protein components of the ribosomes occurs in the nucleoli.

8. The fluid contained within the nucleus is called the *nucleoplasm.*
9. Grafting experiments with species of the giant single-celled alga *Acetabularia* have been important in demonstrating that the nucleus controls the activities of the cytoplasm.
10. The ribosomes in both prokaryotic and eukaryotic cells are the sites of protein synthesis.
11. Ribosomes in eukaryotic cells are found free in the cytoplasm, attached to the endoplasmic reticulum, and within mitochondria and plastids.
12. Prokaryotic and eukaryotic ribosomes are similar in design and function. Both consist of a large and a small subunit; each is composed of a unique type of RNA and about 50 kinds of proteins.
13. Cellular respiration, the oxidization of food molecules and transformation of the resulting energy into a chemical form useful to the cell, occurs mainly in the eukaryotic organelle called the *mitochondrion.*
14. *Oxidation* is the removal of electrons. In mitochondria, energy from electrons that has been removed from food molecules is used to produce ATP.
15. Mitochondria consist of a smooth outer membrane and a highly folded inner membrane. Folds of the inner membrane are called *cristae* and the region enclosed by the inner membrane is called the *mitochondrial matrix.* The matrix includes ribosomes and some DNA.
16. *Plastids* are another class of eukaryotic organelles with double membranes. All plastids are derived from *proplastids;* major types include the *chloroplast,* the *chromoplast,* and the *leucoplast.*
17. A chloroplast is surrounded by two smooth lipid bilayer membranes. Within the chloroplast are found *grana,* stacks of circular sacs called *thylakoids.* Each thylakoid has a single membrane in which arrays of chlorophyll molecules are embedded.
18. The fluid interior of the chloroplast is called the *stroma.* The stroma contains some DNA and ribosomes.
19. Chromoplasts contain various types of carotenoids and give color to fruits, flowers, and vegetables. *Leucoplasts* store starch and lipids.
20. The *endosymbiotic theory* states that eukaryotic cells and their mitochondria, plastids, and flagella originated when certain prokaryotes began living within other prokaryotes as endosymbionts; aerobic endosymbionts became mitochondria, photosynthetic endosymbionts became chloroplasts, and certain motile endosymbionts became flagella.
21. Evidence for the endosymbiotic theory includes the presence of double membranes surrounding mitochondria and plastids, a similarity between present-day prokaryotic ribosomes and ribosomes found in eukaryotic organelles, functional similarity of the prokaryotic plasma membrane and the inner mitochondrial membrane, and the existence of a number of examples of endosymbiosis among living species.

***Questions** (for answers and explanations, see page 50)*

1. Label the indicated structures in the following diagram of a eukaryotic organelle. This organelle is a ______________.

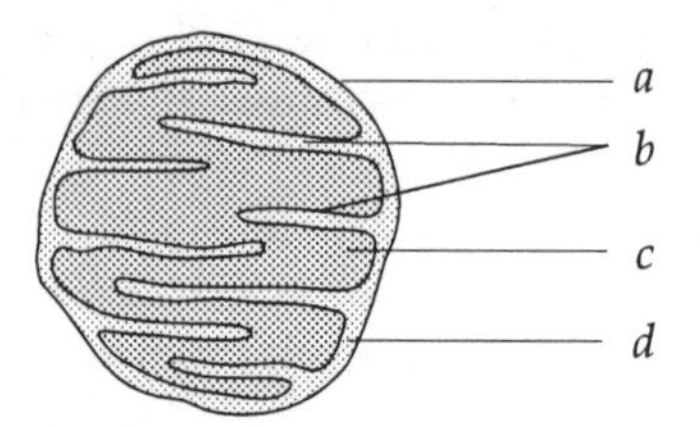

2. According to the endosymbiotic theory, which structure(s) in the preceding diagram would have been derived from the original host prokaryote? ____________________
3. Match the appropriate letters for characteristics listed on the right to the chloroplast components listed on the left.

______	Stroma	*a.* In mitochondria, gives rise to cristae
______	Inner membrane	*b.* Chlorophyll located here
______	Granum	c. Similar to mitochondrial matrix
______	Thylakoid	*d.* Stack of membranous sacs

4. Which of the following structures is common to both prokaryotic and eukaryotic cells?

   a. Nucleoplasm
   b. Chromatin
   c. Ribosomes
   d. Mitochondria
   e. Proplastids

5. The grafting experiments with *Acetabularia* showed that
   a. the cell's activities are controlled mostly by chemicals in the cytoplasm.
   b. the cell's activities are controlled both by the cytoplasm and the nucleus.
   c. the cell's activities are ultimately controlled by the nucleus.
   d. *Acetabularia* needs a nucleus to survive.
   e. the species donating the stalk determined the type of cap that would form.

THE ENDOMEMBRANE SYSTEM • OTHER ORGANELLES (pages 78–83)

Key Concepts

1. The *endomembrane system* includes membranes of the endoplasmic reticulum, the Golgi apparatus, vesicles, lysosomes, and microbodies like peroxisomes.
2. The *endoplasmic reticulum* or *ER* is a system of interconnected tubules and sacs that branches throughout the cell. The ER is continuous with the outer nuclear membrane.
3. *Rough ER* has ribosomes attached to its outer membrane surface; *smooth ER* lacks ribosomes.
4. The ribosomes of the rough ER produce proteins for export from the cytosol to be used outside the cell or within the endomembrane system.
5. Proteins produced by rough ER ribosomes have a special sequence of amino acids called a *signal sequence* that directs them into the interior of the ER where they may be modified. Smooth ER is also involved in protein modification.

6. Free ribosomes produce proteins that are used in the cytosol or within mitochondria or chloroplasts.
7. Cells that are specialized for protein synthesis, such as glandular cells, typically have an extensive endoplasmic reticulum.
8. The *Golgi apparatus* is a series of flattened sacs, with the sacs nearest the nucleus called the *cis* region. The sacs closest to the cell membrane are called the *trans* region. The middle sacs constitute the *medial* region.
9. The *trans* region of the Golgi apparatus receives *vesicles* containing protein from the ER. Small vesicles moving between the sacs of the medial region always move in a *cis* to *trans* direction.
10. Vesicles pinch off from the *trans* region, move through the cytosol, and merge with another membrane (either the plasma membrane or that of an organelle).
11. The Golgi apparatus chemically modifies proteins passing through it by adding specific address tags so the proteins are directed to the correct locations.
12. *Exocytosis* results as material in vesicles is moved to the outside of a cell when the vesicle fuses with the plasma membrane.
13. *Endocytosis* results when a portion of the plasma membrane infolds to form a new vesicle containing material that had previously been outside the cell.
14. *Pinocytosis* is a form of endocytosis is which only a small vesicle is formed, usually engulfing liquid; a very large vesicle is formed in *phagocytosis* and usually engulfs particulate material or another cell.
15. Vesicles formed by endocytosis may fuse with an organelle called a *lysosome* inside the cell. Lysosomes have a single membrane and contain enzymes that can digest all major classes of macromolecules.
16. Primary lysosomes are formed by the Golgi apparatus, become secondary lysosomes after they fuse with a food vacuole, and then fuse with the plasma membrane to release waste products to the outside of the cell after digestion is complete.
17. *Microbodies* are small organelles that pinch off directly from the rough ER. They include *peroxisomes*, which protect the cell from reactions producing peroxides, and *glyoxysomes*, sites for conversion of lipids to carbohydrates in some plant cells.
18. *Vacuoles* are commonly found in protists and plant cells. The large vacuole in plant cells serves as a place for deposition of waste materials and its turgidity gives nonwoody plants structural support.
19. Specialized vacuoles in protists include the *food vacuoles* formed by endocytosis and the *contractile vacuoles* used to bail out excess water that enters the cell by osmosis.

Activity (for answers and explanations, see page 51)

- In the diagram of intracellular digestion that follows, use the letters shown to label the following structures: (A) plasma membrane, (B) *cis* region of Golgi, (C) medial region of Golgi, (D) *trans* region of Golgi, (E) primary lysosome, (F) secondary lysosome, (G) vesicle formed by endocytosis, (H) site of intracellular digestion, and (I) waste products.

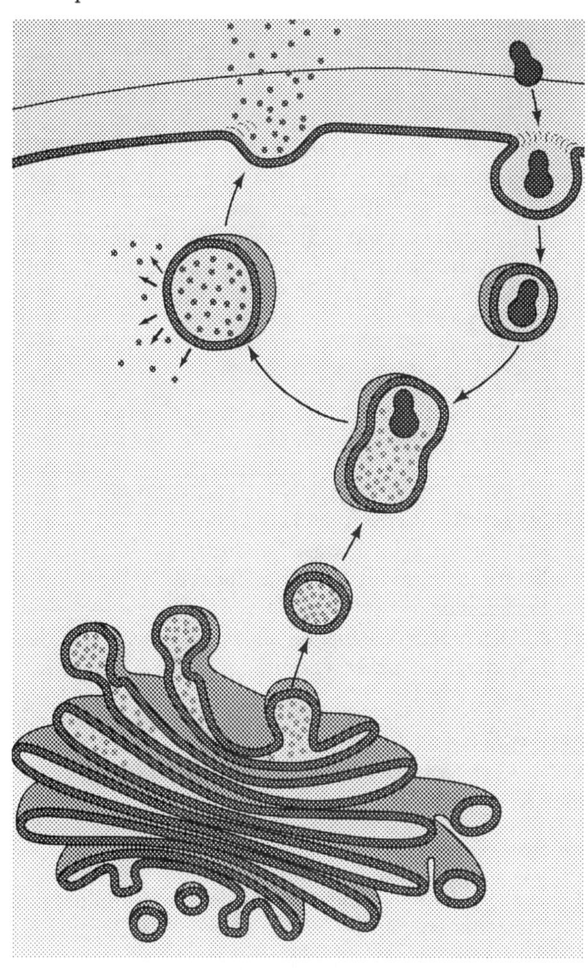

Questions (for answers and explanations, see page 51)

1. Match the appropriate letters for characteristics listed on the right to the cell structures listed on the left.

______	Glyoxysome	*a.* Produces proteins used within cytosol
______	Peroxisome	*b.* Formed by endocytosis
______	Food vacuole	c. Found only in plants
______	Free ribosome	*d.* Isolates dangerous reactions
______	Golgi apparatus	*e.* Endomembrane system component

2. Which of the following is the correct sequence of structures tracing the path of a protein from its site of synthesis to its point of export from the cell?
   a. Mitochondrion, *cis* region of Golgi apparatus, *trans* region of Golgi apparatus, vesicle
   b. Smooth ER, *cis* region of Golgi apparatus, *trans* region of Golgi apparatus, vesicle
   c. Smooth ER, vesicle
   d. Rough ER, *cis* region of Golgi apparatus, *trans* region of Golgi apparatus, vesicle
   e. Rough ER, vesicle, *cis* region of Golgi apparatus, *trans* region of Golgi apparatus, vesicle

3. Which, if any, of the following is *not* a function of the endoplasmic reticulum?
   a. Synthesis of proteins exported from cell
   b. Synthesis of proteins used in the cytosol
   c. Modification of proteins
   d. Lipid synthesis
   e. Transport of proteins

4. Which of the following is the *correct* sequence of activities associated with intracellular digestion?
   (1) Digestion of food (2) Exocytosis (3) Formation of secondary lysosome (4) Endocytosis (5) Formation of food vacuole
   a. 4, 5, 3, 1, 2
   b. 2, 5, 3, 1, 4
   c. 5, 4, 1, 3, 2
   d. 5, 3, 1, 4, 2
   e. 1, 2, 3, 4, 5

5. Which, if any, of the following is *not* a function of the types of vacuoles discussed in these sections?
   a. Storage of cellular waste products
   b. Provide turgidity to support the cell
   c. Digestion of food
   d. Attachment of signal sequence amino acids to proteins
   e. Removal of excess water

THE CYTOSKELETON • THE OUTER "SKELETON" — THE CELL WALL • EUKARYOTES, PROKARYOTES, AND VIRUSES • FRACTIONATING THE EUKARYOTIC CELL: ISOLATING ORGANELLES (pages 83–89)

Key Concepts

1. The *cytoskeleton* of the cell consists of microtubules, microfilaments, and intermediate filaments all located within the cytosol.
2. *Microfilaments* are composed of actin and other proteins. Microfilaments, arranged singly or in bundles, are involved in causing cell shape change, as in contraction of muscle cells, or movement of organelles within the cell, as in *cytoplasmic streaming*.
3. *Intermediate filaments* consist of fibrous proteins and act like tiny ropes to anchor in place cellular structures, such as the nucleus.
4. *Microtubules* are long, hollow tubes formed by subunits composed of the protein *tubulin*, a dimer consisting of two polypeptide subunits (α–tubulin and β–tubulin). The wall of the microtubule includes thirteen rows of tubulin dimers.
5. Microtubules change length by rapid addition or removal of tubulin subunits from one end (the "+" end) of the microtubule.
6. Microtubules are involved in the control of the construction of the cell wall in plants and are associated with the parts of the cell that are changing shape. Microtubules act as tracks along which molecular motors, such as the protein kinesin, move vesicles.
7. Microtubules are important components of the cellular organelles called *cilia* and *flagella*. Nine fused pairs of microtubules (called *doublets*) form the periphery of the cilium or flagellum and two unfused microtubules form the center (called a *9 + 2 arrangement*). A plasma membrane surrounds the cilium or flagellum.
8. Cilia are short and numerous; flagella are long and single or occur in pairs. Both move by bending that is caused by the sliding of enclosed microtubules.
9. *Dynein* molecules on one microtubule undergo a change in tertiary structure with energy from ATP and cause the movement of another microtubule to which they are bound.
10. Prokaryotic flagella are much smaller and do not have the 9 + 2 arrangement of microtubules, as seen in the flagella of eukaryotes.
11. At the base of each eukaryotic flagellum or cilium is a structure called a *basal body*. The basal body gives rise to the cilium or flagellum and is structurally attached to it.
12. Basal bodies have nine sets of three microtubules. Two of each set are continuous with the paired peripheral microtubules found in the cilium or flagellum. Basal bodies lack central microtubules.
13. A *centriole* is structurally like an isolated basal body; it may organize the microtubules involved in cellular division.
14. All plants, fungi, and some protists have a semirigid cell wall composed mostly of polysaccharides. Cell walls provide structural support and limit uptake of water.
15. *Plasmodesmata* are the cytoplasmic connections between adjacent plant cells that extend through holes in their cell walls.
16. Viruses consist of a nucleic acid core and a protein coat. They lack ribosomes and exist as nonliving parasites which are able to take over the metabolic machinery of living cells in order to make new viruses.
17. *Cell fractionation* techniques include mechanically rupturing the cell to disperse its components, followed by *centrifugation* to separate the components into different classes based on their size and density.
18. In *equilibrium centrifugation* (also called *density–gradient centrifugation*), the cell mixture is spun at a constant speed and allowed to move through a medium that differs in density from top to bottom. This causes components to reach an equilibrium position where their density equals that of the medium.
19. Varying centrifuge rotor speeds (and, thus, centrifugal force) are used to separate the components of a mixture in *differential centrifugation*.

Questions (for answers and explanations, see page 51)

1. Which of the following statements about equilibrium and differential centrifugation is true?
   a. Both equilibrium and differential centrifugation use a constant, high rotor speed.
   b. Both equilibrium and differential centrifugation use a centrifugation medium of fixed density.
   c. The rotor speed is varied in differential centrifugation.
   d. The density of the centrifugation medium varies in differential centrifugation.
   e. Differential centrifugation is used with intact cells; in equilibrium centrifugation, the cells must be ruptured.

2. Which of the following statements about the cross section of a flagellum shown below is *not* true?

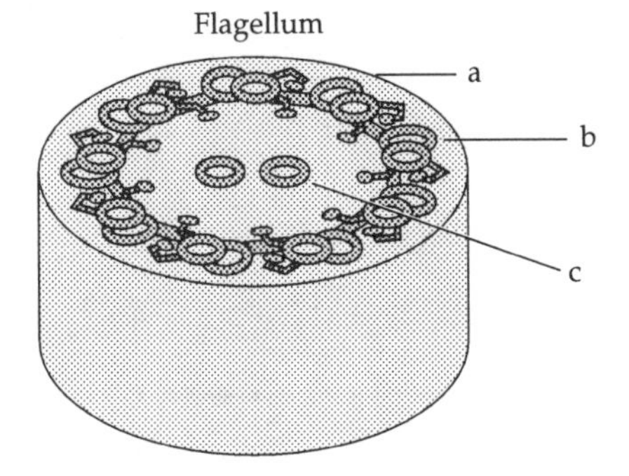

a. Structure b is a doublet.
b. Structure c is one of an unfused pair of microtubules.
c. Structure a is the plasma membrane.
d. The same arrangement of structures a, b, and c would be found in cilia of eukaryotic cells.
e. The same arrangement of structures a, b, and c is found in the flagella of prokaryotic cells.

3. How would you modify the diagram shown in question 2 to show the arrangement of microtubules seen in a *basal body*?

a. Add a single microtubule to all pairs.
b. Make all paired microtubules single.
c. Remove the central pair and add a single microtubule to each of the doublets.
d. Remove the central microtubule pair.
e. Add a second plasma membrane.

4. Match appropriate letters for the list on the right to the cytoskeletal elements listed on the left.

______	Microtubules	*a.* Actin
______	Microfilaments	*b.* Fibrous protein
______	Intermediate filaments	*c.* Tubulin
		d. Change cell shape
		e. Anchoring organelles
		f. Move cellular appendages

Integrative Activity (for answers and explanations, see page 51)

On the transmission electron micrograph shown below, label the following structures: (1) nuclear envelope, (2) nuclear pore, (3) nucleolus, (4) mitochondrion, (5) rough endoplasmic reticulum, (6) Golgi apparatus, (7) cytoplasm, (8) chromatin.

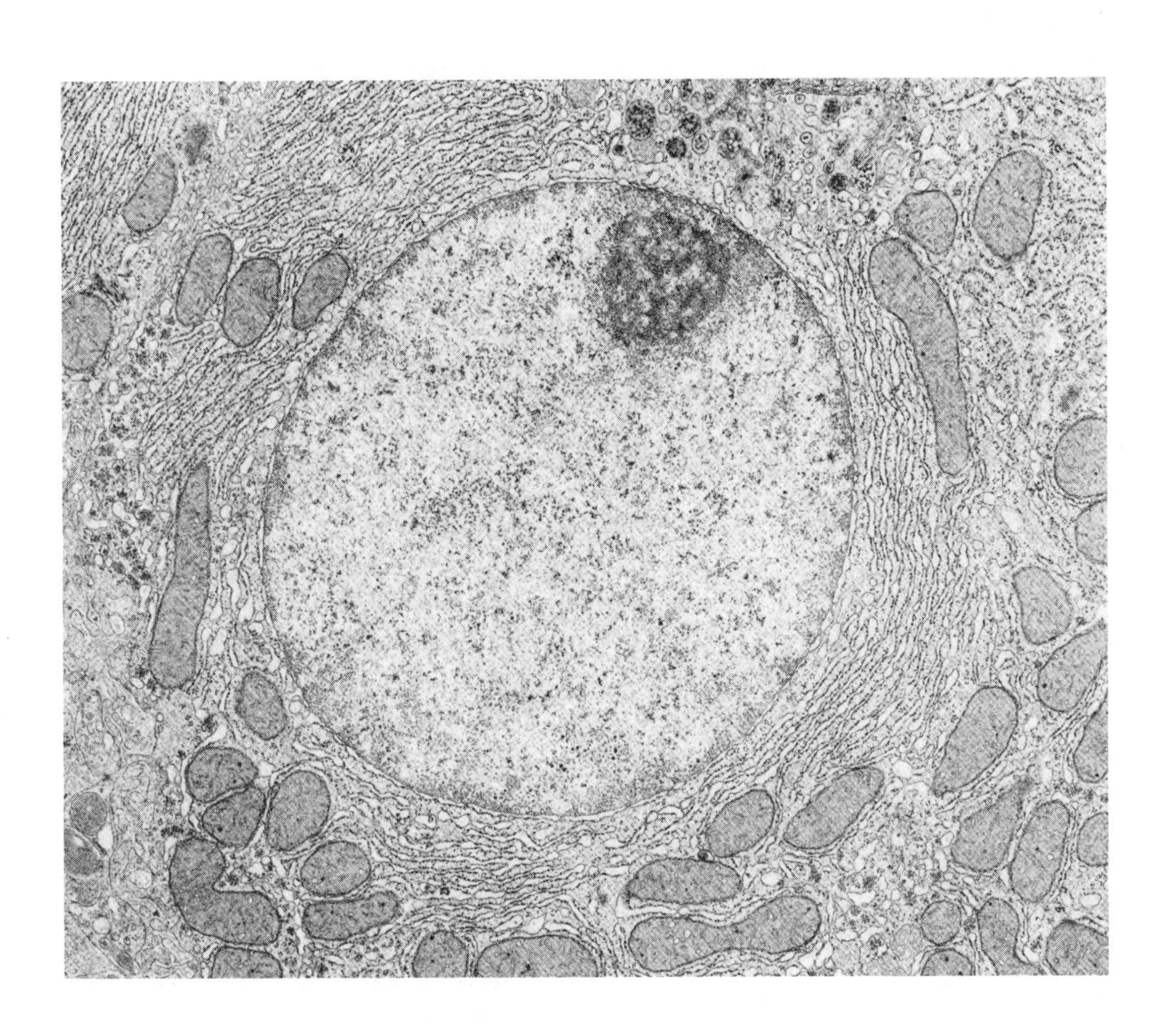

5

Membranes

CHAPTER LEARNING OBJECTIVES—after studying this chapter you should be able to:

- ❑ Describe the structure of a typical biological membrane and explain how that structure relates to the major roles of membranes.
- ❑ Describe the main types of lipids found in membranes and the functions they perform.
- ❑ Describe the main types of proteins found in membranes and the functions they perform.
- ❑ Differentiate between intrinsic and extrinsic proteins and explain what properties of a protein determine whether it will be intrinsic or extrinsic.
- ❑ Differentiate between glycoproteins and glycolipids and describe their primary functions.
- ❑ Describe how cell fusion and the electron microscopy techniques of freeze fracture, freeze etching, and shadowcasting have contributed to our current understanding of biological membranes.
- ❑ Differentiate between desmosomes, tight junctions, and gap junctions, and describe their primary functions.
- ❑ Describe the phenomenon of diffusion and the physical factors that affect the rate of diffusion.
- ❑ Differentiate between simple diffusion, facilitated diffusion, and active transport.
- ❑ With regard to a plot of a solute's transport rate against its concentration, explain how the concept of saturation can be used to differentiate between simple diffusion, facilitated diffusion, and active transport.
- ❑ Describe the role of lipid solubility in determining a solute's transport rate.
- ❑ Differentiate between membrane transport proteins that are uniports and those that are coupled transport systems.
- ❑ Differentiate between coupled transport systems that are symports and those that are antiports.
- ❑ Differentiate between primary active transport and secondary active transport and give an example of each.
- ❑ Describe the phenomenon of osmosis.
- ❑ Describe the concept of osmotic potential and relate it to the terms "isotonic," "hypertonic," and "hypotonic."
- ❑ Explain the relative importance of osmotic potential and pressure potential in determining the osmotic movement of water in cells with walls and cells without walls.
- ❑ Describe the importance of membranes to certain pathways of respiration and photosynthesis, to the establishment of electric and pH gradients, and to the cell's interaction with hormones, growth factors, antibodies, viruses, etc.
- ❑ Describe receptor-mediated endocytosis and the formation of a coated pit and coated vesicle.
- ❑ Explain what a ligand is and how ligands interact with membrane proteins.
- ❑ Discuss the importance of cell adhesion molecules to the formation of tissues.
- ❑ Describe the formation and assembly of the major membrane components and the typical flow of membranes within the cell.

MEMBRANE STRUCTURE AND COMPOSITION • MICROSCOPIC VIEWS OF BIOLOGICAL MEMBRANES • WHERE ANIMAL CELLS MEET (pages 92–100)

Key Concepts

1. Biological *membranes* typically consist of a bilayer of lipids, principally *phospholipids,* in which are embedded protein molecules. Carbohydrates may be attached to the membrane proteins or lipids.
2. The lipid bilayer is a barrier to movement of materials through the membrane and gives the membrane selective permeability and structural integrity.
3. Some membrane proteins transport specific materials across the membrane; other proteins are receptors for chemicals outside the cell or act as membrane-bound enzymes.
4. Carbohydrates are usually found attached to the outside of the membrane and help the cell recognize environmental chemicals.
5. The membrane's *phospholipid bilayer* has a hydrophobic interior composed of phospholipid fatty acid chains, and a hydrophilic exterior where the polar phosphate groups of the phospholipids are located.
6. Phospholipid and protein molecules can move around within their layer of the membrane, but, because of the hydrophobic membrane interior, do not usually transfer from one side of the membrane to the other.

7. *Cholesterol* is an important lipid component of membranes, but its abundance can vary widely from cell to cell. Cholesterol influences the fluidity of membranes.
8. *Intrinsic proteins* extend into the lipid bilayer, with some actually extending from one side of the membrane to the other; *extrinsic proteins* are bound to the hydrophilic groups of the inner or outer membrane layer.
9. The presence of specific membrane-bound proteins gives different membranes their unique functional characteristics.
10. *Cell fusion experiments* involving cells from different species have shown that many proteins freely migrate within the membrane; other proteins seem to be anchored by microfilaments.
11. The tertiary structure of a protein determines whether it will be intrinsic or extrinsic and, if intrinsic, whether it will extend just into or entirely through the membrane.
12. Many proteins with extensive hydrophobic α-helical regions, but hydrophilic ends, extend completely through the phospholipid bilayer.
13. Carbohydrates bound to membrane lipids are called *glycolipids.* Glycolipids may be involved in cell communication.
14. *Glycoproteins* consist of a membrane protein and an attached oligosaccharide, usually 15 or fewer monosaccharides in length.
15. The great diversity of glycoproteins results from the specific sequence of the nine kinds of monosaccharides, as well as the pattern of branching in the oligosaccharide chains.
16. Because of their role as extracellular recognition sites, all membrane carbohydrates (glycolipids and glycoproteins) are located on the outside of the plasma membrane.
17. In preparing cells for the electron microscope, surface views can be obtained by breaking frozen tissue, a process called *freeze fracture*. Fracturing tends to occur along or within membranes.
18. Freeze-fractured samples can be further enhanced by *freeze etching* (evaporating water from the exposed surface under a vacuum) and shadowing the specimen with platinum sprayed from an angle (called *shadowcasting*).
19. Animals cells may be joined together into tissues by an *extracellular matrix* of fibrous proteins, especially *collagen.*
20. Individual animal cells may be linked together by three types of junctions: tight junctions, desmosomes, and gap junctions.
21. *Tight junctions* result when adjacent cells share common membrane proteins which locally fuse the membranes of the cells and prevent extracellular material from "leaking" between the bound cells. Tight junctions are especially important in epithelial layers.
22. *Desmosomes* attach adjacent cells together via *keratin* fibers (an intermediate filament) between *plaques* located within the cytoplasm of the two cells. Cells in the epithelium are frequently bound by numerous desmosomes.
23. A *gap junction* of 2.7 nm between the plasma membranes of two cells results where intrinsic protein complexes called *connexons* are linked together. Materials can pass between the two cells through a strand of cytoplasm located within a central channel in the connexon.

Questions (for answers and explanations, see page 51)

1. Match the appropriate letters for the characteristics listed on the right with the membrane components listed on the left.

____	Lipid bilayer	*a.* Cellular communication
____	Membrane proteins	*b.* Permeability barrier
____	Glycolipids	*c.* Membrane transport
____	Glycoproteins	*d.* Accelerate reactions
		e. Molecule recognition

2. Which of the following is *not* a characteristic of biological membranes?
   a. Fluidity
   b. Components symmetrically distributed
   c. Membrane components can move about
   d. Structure reflects function
   e. Thickness varies
3. What would be the expected structural features of an intrinsic protein that spans the entire cell membrane?
   a. Entirely hydrophobic
   b. Entirely hydrophilic
   c. Hydrophobic ends, hydrophilic middle
   d. Hydrophilic ends, hydrophobic middle
   e. Bound to carbohydrate (a glycoprotein)
4. Which of the following statements about membrane carbohydrates is *not* true?
   a. They are either glycoproteins or glycolipids.
   b. They are very diverse.
   c. They are attached to either side of the membrane.
   d. They are all extrinsic to the membrane.
   e. They do not exceed 15 monosaccharides in length.
5. Add the appropriate letters for characteristics listed on the right to the junction types listed on the left.

____	Desmosome	*a.* Joins cells together
____	Tight junction	*b.* Prevents leakage
____	Gap junction	*c.* "Spot weld"
		d. Has connexons
		e. Cell communication
		f. Fused membranes

DIFFUSION • CROSSING THE MEMBRANE BARRIER (pages 100–107)

Key Concepts

1. *Diffusion* is the random movement of molecules from regions of greater to lesser concentration. Diffusion results in a uniform distribution of molecules within a system at equilibrium.
2. The rate of diffusion increases with temperature and the steepness of the concentration gradient. Size and electric charge of the diffusing molecules also influence their rate of diffusion.
3. In *simple diffusion* through a membrane, the substance must either pass directly through the lipid bilayer or through a protein *channel* in the membrane. Usually only small, nonpolar molecules can move through the lipid bilayer; some ions diffuse through channels.

4. In *facilitated diffusion*, the molecule combines with a membrane-bound *carrier protein*, which allows it to pass through the membrane.
5. Whereas simple and facilitated diffusion can only move substances down their concentration gradients, *active transport* mechanisms expend energy to move substances from regions of lesser to greater concentration (up concentration gradients).
6. If we plot the rate of solute transport against solute concentration, the rate due to simple diffusion always increases directly with solute concentration, whereas in both facilitated diffusion and active transport, the rate will eventually *saturate* when all of the transport proteins are fully engaged.
7. For substances that must move through the hydrophobic lipid bilayer, there is a direct relationship between their *lipid solubility* and their ability to pass through the membrane.
8. Ions like K^+ and Cl^-, which move through membranes at rates faster than their lipid solubilities warrant, do so via special channels located in membrane transport proteins.
9. Many channels created by membrane transport proteins are *gated* and open or close due to chemical or electrical signals.
10. Membrane transport proteins are intrinsic proteins that span the membrane and are specialized for transporting specific molecules or ions.
11. Membrane transport proteins can be *uniports* (transporting a single solute) or *coupled transport systems* (transporting two or more solutes). Coupled transport systems can be *symports* (transporting solutes in the same direction) or *antiports* (transporting solutes in opposite directions).
12. Most biochemical molecules are too large or hydrophilic to pass through membranes by simple diffusion and depend on specific membrane proteins for transport.
13. Facilitated diffusion carrier proteins allow solutes to move in both directions, with the net movement of solute being down the concentration gradient.
14. Active transport carrier proteins can only move their solutes in one direction, usually by expending energy to move solutes against their concentration gradients.
15. *Primary active transport* uses the energy released from the conversion of ATP to ADP to move ions across membranes against a concentration gradient.
16. In the *sodium-potassium pump*, an example of primary active transport found in all animal cells, an intrinsic protein catalyzes the breakdown of ATP and couples the movement of two K^+ ions into the cell with the release of three Na^+ ions. Thus, the sodium-potassium pump is an antiport.
17. *Secondary active transport* uses the energy released when ions diffuse back down a concentration gradient established by primary active transport. The uptake of glucose, coupled to Na^+ ion movement back into the cell, is an example of secondary active transport involving a symport mechanism.
18. The *osmotic potential* of a solution is always negative and its magnitude varies directly with the solute concentration of the solution. Pure water has an osmotic potential of zero.
19. *Osmosis* is the diffusion of water through a membrane. In animal cells, the net movement of water will be from the side with a less negative osmotic potential to the side with a more negative water potential.
20. Solutions with equal osmotic potentials are *isotonic*; a *hypotonic* solution has a less negative osmotic potential than another solution; a *hypertonic* solution has a more negative osmotic potential than another solution.
21. The osmotic movement of water in cells lacking walls (like those of animals) depends only on osmotic potential differences between the cells.
22. In cells with cell walls bacteria, plants, fungi, and some protists), pressure can build up within the cell and prevent further influx of water. Thus, in cells with walls, the net movement of water depends on the cell's osmotic potential and an opposing *pressure potential*.

Questions (for answers and explanations, see page 51)

1. Which of the following substances is most likely to permeate a membrane by diffusing *directly* through the lipid bilayer ?
   a. K^+
   b. Glucose
   c. Cl^-
   d. NH_3^+
   e. O_2

2. For a substance passing through a membrane by simple diffusion, select the curve in the following figure that best shows the relationship between its concentration outside the membrane and its rate of movement through the membrane.

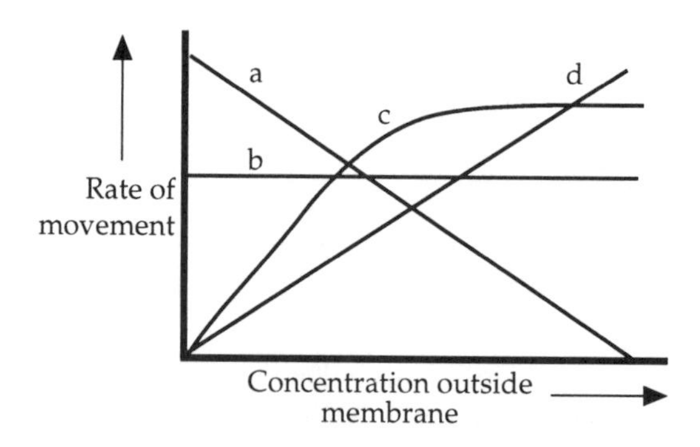

3. For a substance passing through a membrane by *facilitated diffusion*, which curve in the previous figure best shows the relationship between its concentration outside the membrane and its rate of movement through the membrane?

   Curve _____

4. Which of the following statements correctly characterizes the diagram shown at the top of the next page?
   a. Primary active transport involving a symport is shown on the right; secondary active transport involving an antiport is shown on the left.
   b. Primary active transport involving a symport is shown on the left; secondary active transport involving an antiport is shown on the right.

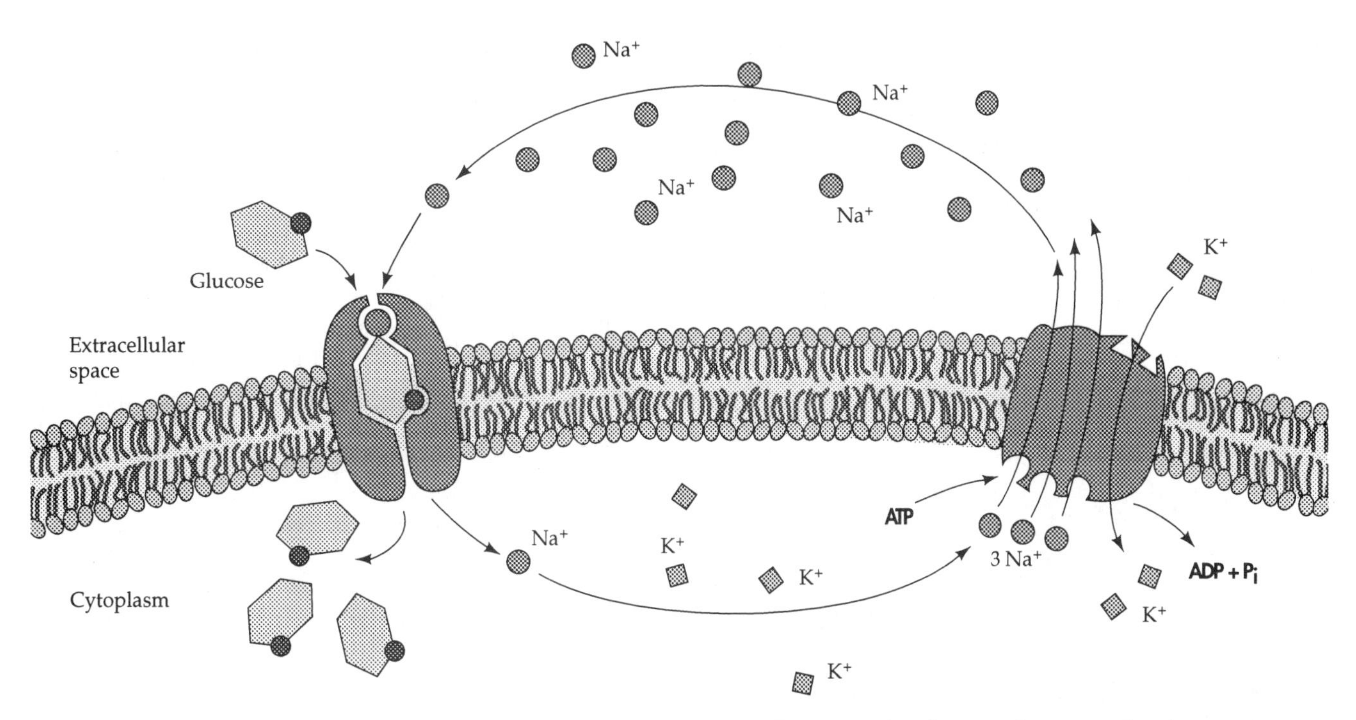

c. Primary active transport involving an antiport is shown on the right; secondary active transport involving a symport is shown on the left.
d. Primary active transport involving an antiport is shown on the left; secondary active transport involving a symport is shown on the right.
e. Both are examples of primary active transport; the antiport is on the right, the symport is on the left.

5. An *animal cell* is put into a bathing solution and you observe that it gains additional water. Choose *all* of the following statements that are true.
a. The cell was isotonic to the bathing solution.
b. The cell was hypotonic to the bathing solution.
c. The cell was hypertonic to the bathing solution.
d. The bathing solution was hypotonic to the cell.
e. The bathing solution was hypertonic to the cell.

6. If the cell in question 5 is a *plant cell,* which of the following statements is likely to be true at osmotic equilibrium?
a. The cell has a greater pressure potential and a more negative osmotic potential than the bathing solution.
b. The cell has a greater pressure potential and a less negative osmotic potential than the bathing solution.
c. The cell has a lesser pressure potential and a more negative osmotic potential than the bathing solution.
d. The cell has a lesser pressure potential and a less negative osmotic potential than the bathing solution.
e. At equilibrium, the cell and bathing solution would be the same for these two measurements.

MORE ACTIVITIES OF MEMBRANES • MEMBRANE INTEGRITY UNDER STRESS • MEMBRANE FORMATION AND CONTINUITY (pages 107–111)

Key Concepts

1. Important molecules in certain chemical pathways in cells (e.g., some parts of photosynthesis and respiration) are spatially arranged within membranes so that the reactions can take place in an orderly fashion.
2. By moving ions across membranes, certain membrane proteins can establish significant electric and pH differences across the membrane. Movement of charged particles back across the membrane can be used to perform work.
3. Antibodies, viruses, hormones, growth factors, and chemical messengers produced by the nervous system (like acetylcholine), interact with cells by binding to specific proteins and carbohydrates on the outer membrane surface.
4. In *receptor-mediated endocytosis,* membrane-bound receptor proteins bind specific, extracellular macromolecules to form a *coated pit* which invaginates to form a *coated vesicle.* Coated vesicles are stabilized by the fibrous protein *clathrin.*
5. Extracellular molecules that bind to receptor proteins or carbohydrates are called *ligands.*
6. Ligands have their effect on the cell by changing the tertiary structure of an intrinsic membrane protein. The inner portion of the intrinsic protein then effects a change within the cell.
7. *Cell adhesion molecules* are membrane proteins that bind similar cells together to form *tissues.*
8. In some cells, membrane proteins form microfibrils that provide the membrane with structural strength. *Spectrin* and *ankyrin* perform this function in red blood cells.
9. The phospholipid components of membranes are produced within the rough ER and then distributed to membranes throughout the cell.
10. Membrane proteins formed in ribosomes are deposited in the rough endoplasmic reticulum, and from there are moved to the Golgi apparatus. Carbohydrates may be added to these proteins within the ER or Golgi apparatus.
11. Golgi-derived vesicles move the completed membrane proteins to the plasma membrane where they are incorporated by exocytosis.

12. Formation and absorption of vesicles causes a steady flow of membranes from the ER to the Golgi apparatus (from *cis* to *trans* regions) to the plasma membrane.
13. Endocytosis moves membrane from the plasma membrane back into the cell interior.

Activity (for answers and explanations, see page 52)

- In the diagram (below) add arrows to show membrane flow within cells.

  Where in this diagram are carbohydrates added to newly synthesized membrane proteins?

  ____

  Label these areas.

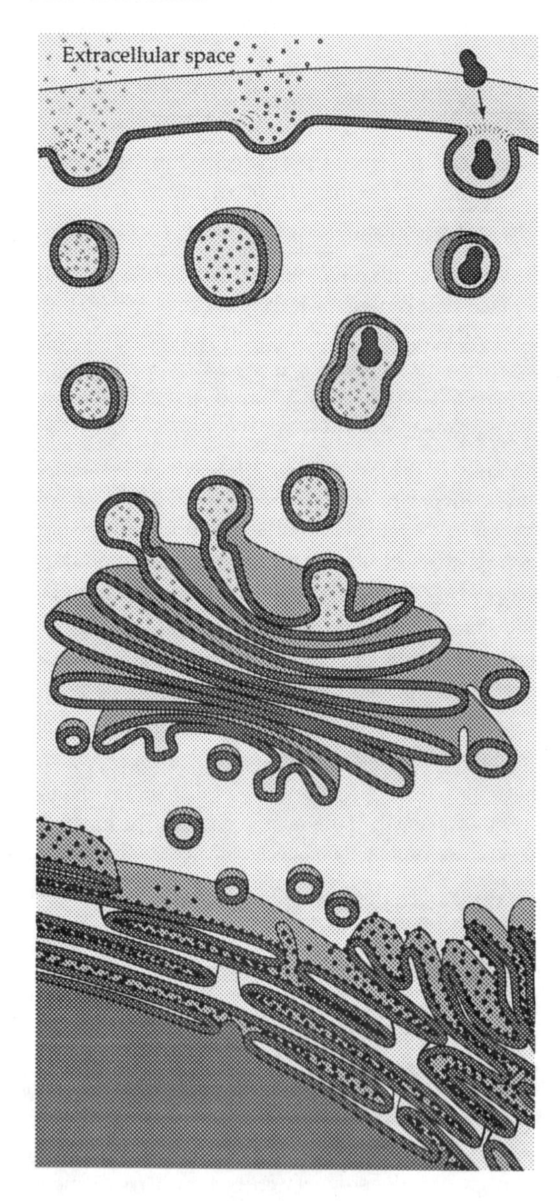

Questions (for answers and explanations, see page 52)

1. Which of the following is *not* a ligand?
   a. Hormone
   b. Membrane receptor protein
   c. Antibody
   d. Virus
   e. Nervous system messengers

2. Match the appropriate letters for the descriptions listed on the right to the molecules listed on the left.

_____	Clathrin	*a.* Cell adhesion molecule
_____	Spectrin	*b.* Strengthens membrane
_____	Acetylcholine	*c.* Coated pit protein
_____	Ankyrin	*d.* Ligand

3. Order the following terms in a way that depicts how membranes are likely to move through the cell: (1) ER (2) nuclear envelope (3) Golgi apparatus (4) vesicle (5) plasma membrane. [*Terms can be used twice.*]
   a. 1, 2, 3, 4, 5
   b. 1, 4, 2, 4, 3, 4, 5
   c. 2, 3, 1, 4, 5
   d. 2, 1, 4, 3, 4, 5
   e. 5, 4, 3, 4, 1, 2

Integrative Questions/Activities (for answers and explanations, see page 52)

1. Circle *all* of the following activities in which cell membranes are normally involved.
   a. Communication between different cells
   b. Tissue formation
   c. Establishment of pH gradients
   d. Osmosis
   e. Nutrient procurement
   f. Waste removal
   g. Movement
   h. Control of chemical reactions

2. Label the diagram of the cell membrane shown below with the numbers accompanying the following labels: (1) hydrophilic portion of phospholipid, (2) hydrophobic portion of phospholipid, (3) intrinsic protein, (4) extrinsic protein, (5) glycolipid, (6) glycoprotein, (7) cytoplasm. How do you know which side of the membrane is the cell's interior?

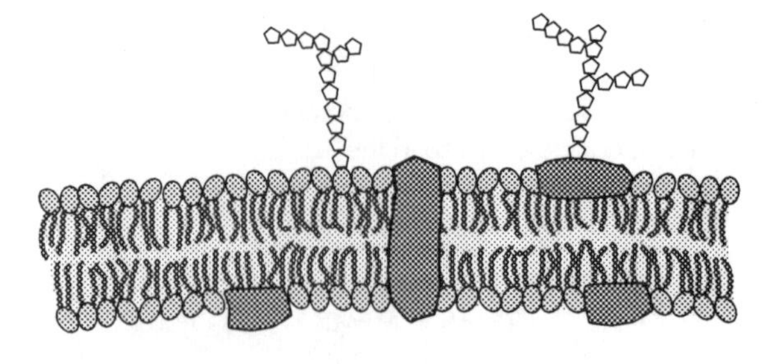

6

Energy, Enzymes, and Catalysis

CHAPTER LEARNING OBJECTIVES—after studying this chapter you should be able to:

- ❑ Define work and state the first law of thermodynamics relative to your definition.
- ❑ Define entropy and free energy and state the second law of thermodynamics relative to your definitions.
- ❑ Explain how the equilibrium constant can be used to determine if a reaction is spontaneous or not.
- ❑ Relate the change in free energy (ΔG) of a chemical reaction to the equilibrium constant of the reaction, and explain how this information can be used to determine if the reaction is endergonic or exergonic.
- ❑ Explain the equation relating ΔG to changes in heat and entropy of a reaction.
- ❑ Explain the concept of activation energy in terms of the formation of transition species and describe its relationship to the rate constant for the reaction.
- ❑ Explain how a catalyst accelerates a chemical reaction.
- ❑ State the generalized formula for an enzyme-catalyzed reaction and define the terms "substrate" and "enzyme–substrate complex."
- ❑ Explain enzyme specificity, the interaction of enzyme and substrate at the enzyme's active site, and the concept of induced fit.
- ❑ Explain the effect of an enzyme on the ΔG, the equilibrium constant, and the activation energy of a reaction.
- ❑ Explain what an enzyme kinetic curve is and be able to draw a typical curve showing the effect of substrate concentration on reaction rate.
- ❑ Describe the actions of the principal protein-digesting enzymes and explain the basis of their specificity.
- ❑ Define the terms "prosthetic group" and "coenzyme."
- ❑ Define what an inhibitor is and differentiate between irreversible, competitive, and noncompetitive inhibitors.
- ❑ Describe the structure of a simple allosteric enzyme and explain how its catalytic and regulatory subunits interact with substrate and effector molecule to determine the equilibrium between active and inactive forms of the enzyme.
- ❑ Explain how enzyme kinetic curves can be used to differentiate between competitive and noncompetitive inhibitors and between single subunit and allosteric enzymes.
- ❑ Explain the concept of negative feedback and understand how negative feedback is accomplished in metabolic pathways.
- ❑ Describe what is meant by "concerted feedback inhibition."
- ❑ Explain the effects of pH and temperature on enzymes and what is meant by the "denaturation" of an enzyme.
- ❑ Define the terms "isozyme," "ribozyme," and "abzyme."

ENERGY AND THE LAWS OF THERMODYNAMICS • CHEMICAL EQUILIBRIUM • REACTION RATES (pages 115–121)

Key Concepts

1. *Metabolism* is the total of all chemical reactions occurring within an organism. Catabolic reactions degrade complex molecules into simpler molecules; anabolic reactions build up complex molecules from simpler ones.
2. *Energy*, defined as the ability to do work, can exist in a variety of interconvertible forms, such as heat, light, electrical, mechanical, chemical, nuclear, etc.
3. The *first law of thermodynamics* states that in a closed system, energy can neither be created nor destroyed.
4. *Entropy* is a measure of the disorder in a system.
5. The *second law of thermodynamics* states that in a closed system, the amount of entropy increases with time.
6. *Chemical equilibrium* exists when the rate of the forward reaction equals the rate of the reverse reaction.
7. At chemical equilibrium, the *equilibrium constant*, K_{eq}, equals the ratio of the concentrations of the products and reactants, or...

$$K_{eq} = [\text{products}]/[\text{reactants}]$$

8. If $K_{eq} = 1$, then the reaction goes halfway to completion; if $K_{eq} < 1$, then the reverse reaction dominates; if $K_{eq} > 1$, then the forward reaction dominates. In *spontaneous reactions*, $K_{eq} > 1$.
9. *Free energy*, G, is energy that can be used to do work.

K_{eq} = equilibrium constant
ΔG = Change in free energy

10. The change in free energy of a reaction, ΔG, is directly related to the K_{eq} of the reaction: with $K_{eq} > 1$, ΔG is negative and the reaction is *exergonic* (it releases energy); with $K_{eq} < 1$, ΔG is positive and the reaction will be *endergonic* (it absorbs free energy).
11. The free energy of a reaction is determined by the amount of heat it gives off or absorbs, and by differences in the orderliness of reactants and products.
12. According to the second law of thermodynamics, in any closed system free energy decreases and entropy increases.
13. The relationship of free energy to changes in heat and entropy is $\Delta G = \Delta H - T\Delta S$, where ΔH is the change in heat, T is the absolute temperature, and ΔS is the change in entropy. Units for this equation are usually kilocalories (kcal).
14. Even if the free energy of the reactants is greater than the products (ΔG for the reaction is negative), some *activation energy* is usually required to get the reaction started.
15. Activation energy helps change reactants into *transition-state species* before they can be converted to products. Transition-state species have greater free energy than either reactants or products.
16. The rate of a reaction equals the concentration of reactant times a rate constant which is based on the amount of required activation energy. For $A \xrightarrow{k} B$, the rate of formation of B can be expressed as $r_B = k[A]$, where k reflects the size of the activation energy "hump." Rate constants vary with temperature.

Questions (for answers and explanations, see page 52)

1. Which of the following statements about spontaneous reactions is *not* true?
   a. In spontaneous reactions, the rate of the forward reaction exceeds the rate of the back reaction.
   b. Spontaneous reactions reach equilibrium cuickly.
   c. The change in free energy, ΔG, of spontaneous reactions is negative.
   d. Spontaneous reactions release energy.
   e. In spontaneous reactions, the concentration of products at equilibrium is greater than the concentration of reactants.

2. Which expression in the equation for change in free energy, $\Delta G = \Delta H - T\Delta S$, reflects changes in the orderliness of reactants and products?
   a. ΔG
   b. ΔS
   c. T
   d. ΔH
   e. $T\Delta S$

3. The reaction fructose 6-phosphate → glucose 6-phosphate has an equilibrium constant, K_{eq}, of 2.0. Which of the following conclusions about this reaction is true?
   a. The reaction is exergonic and ΔG is positive.
   b. The reaction is exergonic and ΔG is negative.
   c. The reaction is endergonic and ΔG is positive.
   d. The reaction is endergonic and ΔG is negative.
   e. None of these conclusions can be made based on a knowledge of K_{eq}.

4. Which of the following choices correctly identifies the activation energy and ΔG for the reaction graphed below?

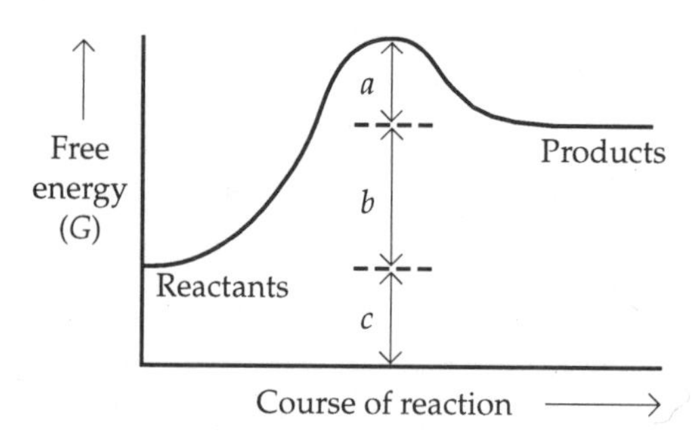

   a. Activation energy = a; $\Delta G = a + b$
   b. Activation energy = a; $\Delta G = b$
   c. Activation energy = $a + b$; $\Delta G = a + b + c$
   d. Activation energy = a; $\Delta G = a + b + c$
   e. Activation energy = b; $\Delta G = b + c$

5. Which of the following conclusions is true for the reaction graphed above?
   a. The reaction is exergonic and K_{eq} is greater than 1.
   b. The reaction is endergonic and K_{eq} is greater than 1.
   c. The reaction is exergonic and K_{eq} is less than 1.
   d. The reaction is endergonic and K_{eq} is less than 1.
   e. More information is needed.

6. Which of the following statements about transition-state species is *not* true?
   a. The free energy of transition-state species is always greater than that of reactants.
   b. The free energy of transition-state species is always greater than that of products.
   c. The ΔG of the reaction is unaffected by the quantity of energy required to create transition-state species.
   d. Transition-state species are present in exergonic reactions, but not endergonic reactions.
   e. The activation energy equals the quantity of energy required to create transition-state species.

THE HIGHLY SPECIFIC CATALYSIS OF ENZYMES • MOLECULAR STRUCTURE OF ENZYMES (pages 121–126)

Key Concepts

1. A *catalyst* increases the rate of a reaction by decreasing the activation energy for the reaction.
2. Biological catalysts are globular proteins called *enzymes*.
3. Reactants in enzyme-catalyzed reactions are called *substrates*.
4. The generalized form of an enzyme-catalyzed reaction is $E + S \rightleftharpoons E{\bullet}S \rightleftharpoons E + P$, where E•S is the *enzyme–substrate complex*.
5. Enzyme specificity results because the tertiary structure of the enzyme only allows chemically distinct types of substrates to attach to an area of the enzyme called its *active site*.
6. Substrate binding to the active site of the enzyme can involve hydrogen, hydrophobic, ionic, or covalent bonds.
7. Typically, the enzyme undergoes a conformational change when in the enzyme–substrate complex. The change in

shape, called an *induced fit* between substrate and enzyme, improves the conversion of reactant to product.

8. ΔG and K_{eq} for the reaction remain unchanged by the enzyme; only the activation energy is lowered. Thus, enzymes speed up both forward and back reactions equally.
9. As with active transport and facilitated diffusion, enzyme kinetic curves plotting reaction rate against substrate concentration show *saturation* at high substrate levels. Substrate saturation occurs when all enzyme molecules are involved in the enzyme–substrate complex.
10. *Coupled reactions* occur on the same enzyme. Typically, an endergonic reaction is coupled with an exergonic reaction, so that the energy released by the exergonic reaction allows the endergonic reaction to proceed.
11. *X-ray crystallography* is used to obtain diffraction patterns from crystals of proteins and nucleic acids. Electron density maps can be produced from these patterns and analysis of the maps using computers permits one to deduce the structure of the molecule.
12. The digestive enzymes carboxypeptidase, chymotrypsin, trypsin, and elastase each hydrolyze peptide bonds in specific locations on the protein being digested.
13. The specificity of protein-digesting enzymes depends on interactions between side groups of the amino acids making up the protein undergoing digestion and the active site of the enzyme.
14. Many enzymes possess a tightly bound nonprotein portion called a *prosthetic group*, consisting of either a metal ion or a *coenzyme*.
15. Metal ion prosthetic groups aid enzymes by binding the substrate or removing electrons from the substrate.
16. Coenzymes are complex organic compounds. In animals some coenzymes are derived from vitamins.
17. Coenzymes enhance the catalytic action of enzymes by either accepting or donating electrons, transferring phosphate groups, or altering the structure of the substrate in some other way.

Activities (for answers and explanations, see page 52)

- In the following graph, draw a curve showing the expected relationship between substrate concentration and the velocity of an enzyme-catalyzed reaction. Properly label the graph axes and mark areas on the curve where free enzyme molecules exist and where all enzyme is in the enzyme–substrate complex.

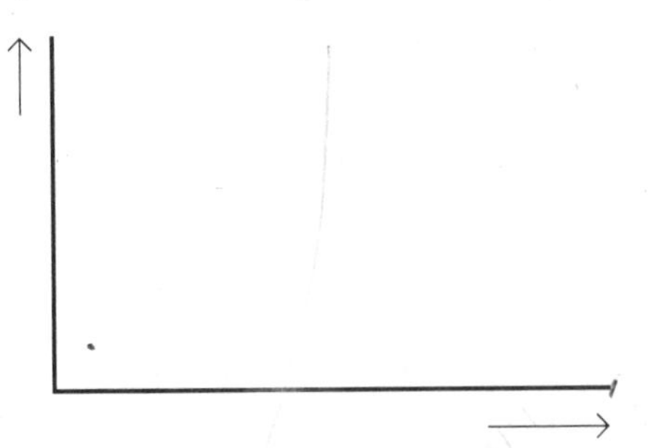

- Label the following graph to show the free energy change (ΔG) for the reaction, the activation energy without enzyme (E_a^{-e}), the activation energy with enzyme (E_a^{e}), the free energy associated with reactants (G_r), and the free energy associated with products (G_p).

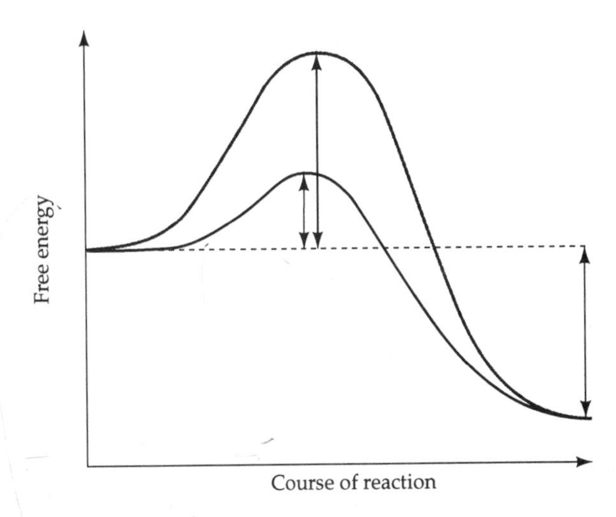

Questions (for answers and explanations, see page 53)

1. Which of the following statements about the interaction of enzyme and substrate is *not* true?
   a. The conformation of the enzyme may change as it binds to substrate.
   b. The enzyme can only form hydrogen bonds with the substrate.
   c. The enzyme is unchanged at the end of the reaction.
   d. The substrate only interacts with a small portion of the enzyme molecule.
   e. The conformation of the enzyme determines its specificity for substrate.
2. Which of the following statements about enzyme-catalyzed reactions is *true*?
   a. Reactions are accelerated by changing ΔG for the reaction.
   b. Reactions are accelerated by changing K_{eq} for the reaction.
   c. Reactions are accelerated by changing the activation energy of the reaction.
   d. Substrate can only become product through the action of the enzyme.
   e. Only the forward reaction is accelerated by the enzyme.
3. Which of the following statements about coupled reactions is *not* true?
   a. Usually an exergonic reaction is coupled to an endergonic reaction.
   b. Usually a single enzyme catalyzes both reactions.
   c. The product of one reaction is substrate for the other reaction.
   d. Energy released by the endergonic reaction allows the exergonic reaction to proceed.
   e. Active transport of molecules through membranes is an example of reaction coupling.

4. Which of the following statements about prosthetic groups and coenzymes is *not* true?
   a. Prosthetic groups interact chemically only with the enzyme and not the substrate.
   b. A coenzyme is a type of prosthetic group.
   c. Some coenzymes are free to move from enzyme to enzyme.
   d. Metal ions can function as prosthetic groups.
   e. Some coenzymes are vitamins.

REGULATION OF ENZYME ACTIVITY • ENZYMES, RIBOZYMES, AND ABZYMES (pages 126–133)

Key Concepts

1. Substances that interact with enzymes to reduce the catalytic activity of the enzyme are called *inhibitors.* Naturally occurring inhibitors control metabolism.
2. *Irreversible inhibitors* permanently modify the enzyme so that it can no longer act as a catalyst. The nerve gas DFP is an irreversible inhibitor of the enzyme acetylcholinesterase.
3. *Competitive inhibitors* resemble the enzyme's natural substrate and compete with it for the active site. Competitive inhibition can be overcome with additional substrate.
4. *Noncompetitive inhibitors* bind to the enzyme at a location other than the active site. Noncompetitive inhibitors change the conformation of the enzyme and alter the active site to make the enzyme less effective as a catalyst. Noncompetitive inhibition cannot be overcome with additional substrate.
5. Enzymes consisting of two or more subunits have a quaternary structure and may be subject to *allosteric control.*
6. *Allosteric enzymes* have two different types of subunits: a *catalytic subunit* with an active site and a *regulatory subunit* with an allosteric site. Substrate binds to the active site; *effector* molecules, either inhibitors or activators, bind to the allosteric site.
7. Allosteric enzymes can exist in two different forms which are in equilibrium with each other: an *active form,* in which the active sites can bind substrate but the allosteric sites cannot bind an inhibitor, and an inactive form, in which the active sites cannot bind substrate but the allosteric sites can bind an inhibitor.
8. Addition of substrate shifts the equilibrium between the active and inactive forms so that more active forms exist, and the reaction rate rapidly increases.
9. Addition of an effector that acts as an inhibitor shifts the equilibrium in favor of the inactive form, and the reaction rate rapidly decreases.
10. An *allosteric activator* is an effector that binds to the regulatory subunit of the active form and prevents it from converting to the inactive form.
11. Allosteric enzymes show more complex sigmoidal (S-shaped) kinetic curves when reaction rate is plotted against substrate concentration.
12. *Negative feedback* on metabolic pathways can occur when one or more products of the pathway are allosteric inhibitors of an enzyme catalyzing an earlier step in the pathway.
13. In a branched pathway, inhibition by a single end product of the *first committed step* in its portion of the pathway will not affect the other side of the pathway.
14. *Concerted feedback inhibition* results if the two (or more) end products of a branched pathway act together as allosteric regulators of the enzyme that catalyzes the first committed step for the whole pathway.
15. pH affects the ionization of carboxyl, amino, and other groups within the enzyme, which determines the shape of the enzyme and how it interacts with substrate.
16. Enzymes work best at specific pH values and may be *denatured* at extreme pH values.
17. Although enzyme activity increases with a moderate increase in temperature, at high temperatures enzymes become denatured.
18. *Isozymes* are enzymes that catalyze the same reaction but have slightly different structures; consequently, each isozyme requires different conditions for optimum performance as a catalyst.
19. Other classes of macromolecules can also act as catalysts; catalytic RNAs are called *ribozymes;* catalytic antibodies are *abzymes.*

Questions (for answers and explanations, see page 53)

1. The following enzyme kinetics graph shows the results of running a series of enzyme-catalyzed reactions at different substrate concentrations, either without an inhibitor or with an inhibitor.

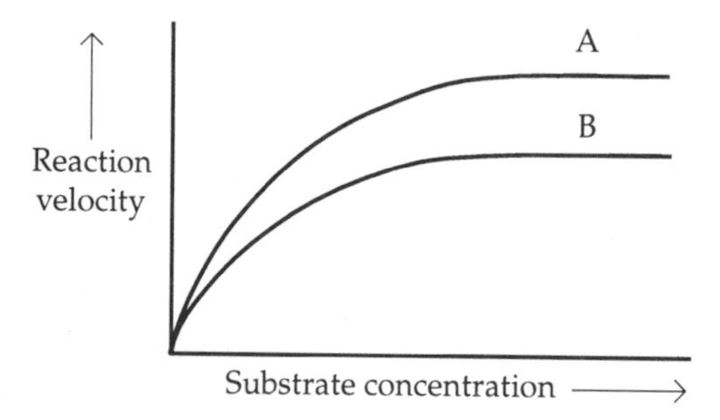

   Which of the following statements is a *correct* interpretation of this graph?
   a. Curve A is with the inhibitor, curve B is without the inhibitor, and the inhibitor probably resembles the normal substrate.
   b. Curve B is with the inhibitor, curve A is without the inhibitor, and the inhibitor probably resembles the normal substrate.
   c. Curve A is with the inhibitor, curve B is without the inhibitor, and the inhibitor probably does not resemble the normal substrate.
   d. Curve B is with the inhibitor, curve A is without the inhibitor, and the inhibitor probably does not resemble the normal substrate.
   e. Curve B is with the inhibitor, curve A is without the inhibitor, and the inhibitor probably resembles the enzyme.

2. Which of the following statements about an allosteric enzyme is *not* true?
   a. At high substrate, but low inhibitor concentrations, most of the enzyme is in the active form.
   b. At high substrate, but low inhibitor concentrations, most of the allosteric sites are empty.
   c. At high inhibitor, but low substrate concentrations, most of the enzyme is in the inactive form.
   d. At high inhibitor, but low substrate concentrations, most of the active sites are occupied.
   e. At high inhibitor, but low substrate concentrations, most of the allosteric sites are occupied.

3. Which of the following choices correctly shows the condition of the carboxyl and amino groups of an enzyme in an acidic environment?
   a. —COOH and NH_2
   b. —COOH and NH_3^+
   c. —COO^- and NH_2
   d. —COO^- and NH_3^+

Consult the following metabolic pathway to answer the next two questions:

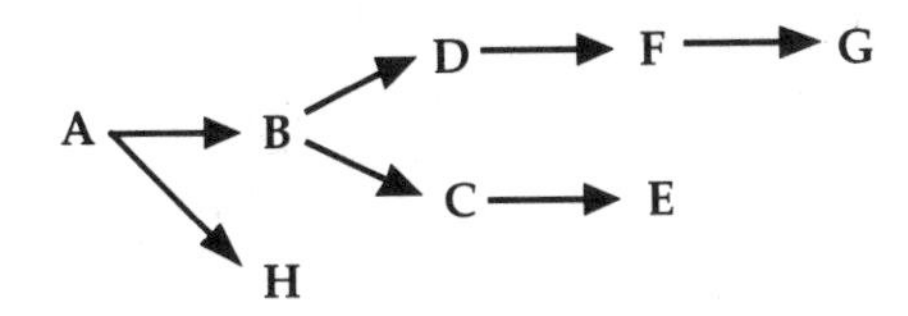

4. For substance G, the first committed step of the pathway is the
   a. A to B step.
   b. B to D step.
   c. B to C step.
   d. D to F step.
   e. F to G step.

5. Concerted feedback inhibition of G and E would be directed at the enzyme catalyzing the
   a. A to B step.
   b. B to D step.
   c. B to C step.
   d. D to F step.
   e. F to G step.

6. Which of the following statements is *not* true?
   a. Enzymes catalyzing the same reaction in different tissues of the body may actually be isozymes.
   b. Proteins were the first macromolecules to evolve a catalytic function.
   c. Catalytic RNAs are called ribozymes.
   d. Catalytic antibodies are called abzymes.
   e. When an enzyme denatures, it loses it normal conformation.

Activities (for answers and explanations, see page 53)

- In the diagram below showing allosteric regulation, label the following items: active site, allosteric site, substrate, inhibitor.

- Answer the following questions about the diagram shown below:
  a. If this enzyme catalyzed the first committed step in a metabolic pathway, what type of substance might act as an inhibitor?

  b. Which form of the enzyme (*a–f*) would be most prevalent if acted on by the type of substance described in *a* above? _____

  c. Could this enzyme be subject to concerted feedback inhibition? Explain your answer.

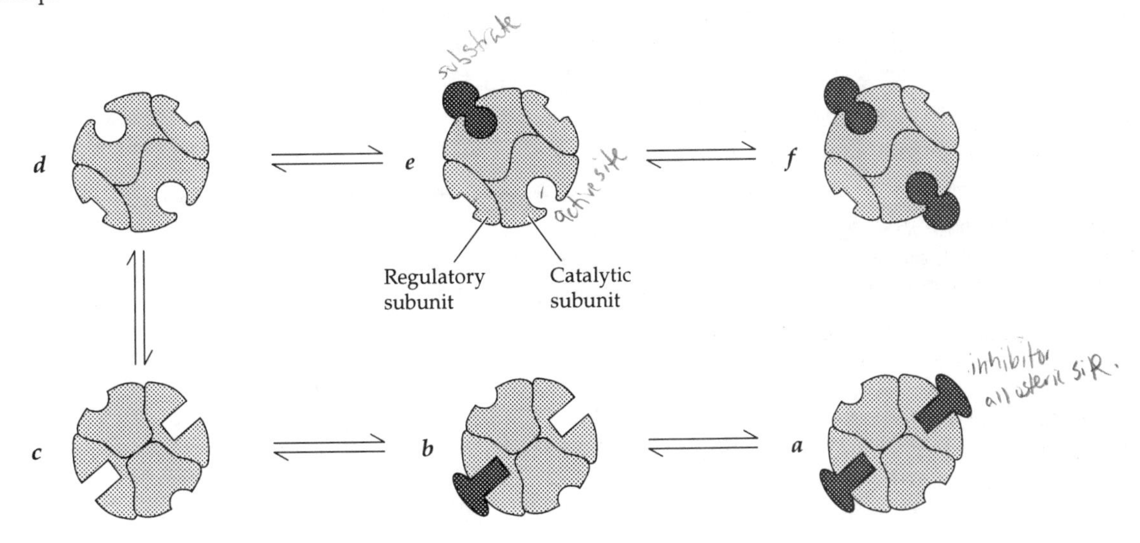

7

Pathways that Release Energy in Cells

CHAPTER LEARNING OBJECTIVES—after studying this chapter you should be able to:

- ❑ Describe the relationship between aerobic and anaerobic conditions and the occurrence of glycolysis, fermentation, and cellular respiration.
- ❑ Relate the structure of adenosine triphosphate (ATP) to its role as short-term energy storage molecule.
- ❑ Identify the oxidizing and reducing agent in a typical redox reaction and explain which portion of the reaction is endergonic and which portion is exergonic.
- ❑ Diagram the redox reaction involving nicotinamide adenine dinucleotide (NAD).
- ❑ Differentiate between ATP production via substrate-level phosphorylation and via the chemiosmotic mechanism (oxidative phosphorylation).
- ❑ Explain the involvement of proton pumps, ATP synthase, and the proton-motive force in the chemiosmotic mechanism.
- ❑ Describe the major inputs, outputs, and interconnections of glycolysis, the citric acid cycle, and the respiratory chain.
- ❑ Identify the locations of the reactions of glycolysis, the citric acid cycle, and the respiratory chain within the cell.
- ❑ Identify the reactions of glycolysis that involve "pump priming," those that involve oxidation–reduction agents, those that are endergonic, and those that are exergonic.
- ❑ Describe the overall reaction of the citric acid cycle and identify the location of reactants and products in the steps of the cycle.
- ❑ Describe the operation of the respiratory chain and its involvement in the production of ATP via the chemiosmotic mechanism.
- ❑ Describe the formation of water by cytochrome oxidase at the end of the respiratory chain.
- ❑ Explain how respiratory uncouplers can discharge the proton-motive force built up by the respiratory chain.
- ❑ Explain how respiratory control depends on the ratio of ADP/ATP in the cell and on the activity of ATP synthase.
- ❑ Describe the two most common forms of fermentation and explain what fermentation accomplishes.
- ❑ Trace the production of NADH + H^+ and ATP in the three major pathways of cellular respiration and determine the overall efficiency of the process.
- ❑ Explain the interconnections between the reactions of cellular respiration and other important metabolic pathways, especially those involving carbohydrates, fats, and amino acids.
- ❑ Describe the Pasteur effect and understand how it results.
- ❑ Explain the principle of allosteric control and its importance in the regulation of the reactions of cellular respiration.
- ❑ Identify the major positive and negative feedback control points in the reactions of glycolysis and the citric acid cycle and the allosteric activators and inhibitors involved.

GLYCOLYSIS, CELLULAR RESPIRATION, AND FERMENTATION • THE TRANSFER OF HYDROGEN ATOMS AND ELECTRONS • HOW DO CELLS PRODUCE ATP? • THE RELEASE OF ENERGY FROM GLUCOSE (pages 136–144)

Key Concepts

1. *Photosynthesis*, *glycolysis*, *fermentation*, and *cellular respiration* are the major processes of energy acquisition in cells.
2. The extraction of energy from food begins with glycolysis.
3. Under *aerobic* conditions (with oxygen present), cellular respiration follows glycolysis; under *anaerobic* conditions (with oxygen absent), fermentation follows glycolysis.
4. *Adenosine triphosphate* (*ATP*) is a short-term energy storage molecule; carbohydrates and fats are long-term energy storage molecules.
5. The exergonic conversion of adenosine triphosphate (ATP) to adenosine diphosphate (ADP) plus inorganic phosphate (P_i) is a hydrolysis reaction with ΔG equal to about –12 kcal of free energy per mole of ATP under conditions typical of most cells.
6. Expressing ATP as A–P~P~P indicates that the last two phosphate-to-phosphate bonds in ATP are high-energy bonds (A is adenine; R, the sugar ribose; P is inorganic phosphate or HPO_4^{2-}).
7. In ATP, high-energy bonds result because several negative charges are held in close proximity in the last two phosphates. Hydrolyzing these phosphate groups reduces the free energy by spreading out the negative charges.
8. *Bioluminescence*, the production of light by an organism, is an example of an ATP requiring process.

9. An *oxidation–reduction reaction*, or *redox reaction*, involves transfer of electrons or hydrogen atoms between chemical groups.
10. An atom or molecule is *reduced* when it gains an electron or hydrogen atom. An atom or molecule is *oxidized* when it loses an electron or hydrogen atom.
11. In a redox reaction, the reactant losing an electron or hydrogen atom is the *reducing agent;* the reactant accepting the electron or hydrogen atom is the *oxidizing agent.*
12. Redox reactions always involve both oxidation and reduction simultaneously; one reactant is reduced while the other reactant is oxidized.
13. The oxidative portion of a redox reaction is exergonic, with large negative ΔG values; the reductive portion of a redox reaction is endergonic, with large positive ΔG values. ΔG of the overall reaction will be negative if the reaction is spontaneous.
14. *Nicotinamide adenine dinucleotide (NAD)* is a reducing/oxidizing agent that exists in two different forms which can bind or release hydrogens in the reaction

$$NAD^+ + 2(H) \rightarrow NADH + H^+$$

NAD^+ is the oxidizing agent, and NADH is the reducing agent.
15. Another important electron carrier in cells is *flavin adenine dinucleotide (FAD).*
16. NAD and FAD transfer hydrogens and free energy during redox reactions.
17. *Substrate-level phosphorylation* produces ATP by the enzyme-catalyzed transfer of phosphate groups from donor molecules to ADP.
18. ATP production by the *chemiosmotic mechanism* involves *proton pumps,* a membrane that is relatively impermeable to protons, and membrane proton complexes called *ATP synthases.*
19. The proton pumps move H^+ across the membrane to produce a proton concentration gradient or *proton-motive force;* as protons diffuse through the ATP synthase, ATP is generated.
20. Glycolysis is the oldest and most universal of the energy-releasing pathways; all organisms have the glycolytic pathway. The reactions of glycolysis occur in the cytosol.
21. Beginning with a molecule of glucose, the major products of glycolysis are a net yield of two ATP, pyruvate, and two electrons carried by NAD as $NADH + H^+$.
22. Cellular respiration includes the *citric acid cycle* (also called the Krebs or tricarboxylic acid cycle) and the *respiratory chain.*
23. The reactions of the citric acid cycle, catalyzed by enzymes present in the mitochondrial matrix, move pyruvate through a series of reactions in which its carbons and hydrogens are removed. Products of the citric acid cycle are CO_2 (a waste product) and reduced NAD and FAD.
24. The respiratory chain transfers electrons from reduced NAD and FAD through a series of redox reactions that involve a chemiosmotic mechanism. Water is formed when oxygen serves as the final electron acceptor.
25. Chemiosmotic ATP production via the respiratory chain is called *oxidative phosphorylation.* Carriers of the respiratory chain are embedded in the inner mitochondrial membranes called the *cristae.*
26. Fermentation is an alternate pathway that can be used in the absence of oxygen for the oxidation of reduced NAD and other electron transfer agents

Activity (for answers and explanations, see page 53)

- Label the adenine, ribose, and phosphate components of the ATP molecule shown below and indicate the portion of the molecule that constitutes adenosine, AMP, and ADP. Add the missing bonds using a straight line for a normal covalent bond and a wavy line for a high-energy covalent bond.

Questions (for answers and explanations, see page 53)

1. In the conversion of ATP to AMP, how many water molecules will be used as reactants or produced as products?

   a. Three will be used as reactants.
   b. Two will be used as reactants.
   c. One will be used as a reactant.
   d. Two will be produced as products.
   e. One will be produced as product.

2. Which one of the following statements *incorrectly* characterizes the pair of coupled reactions shown below?

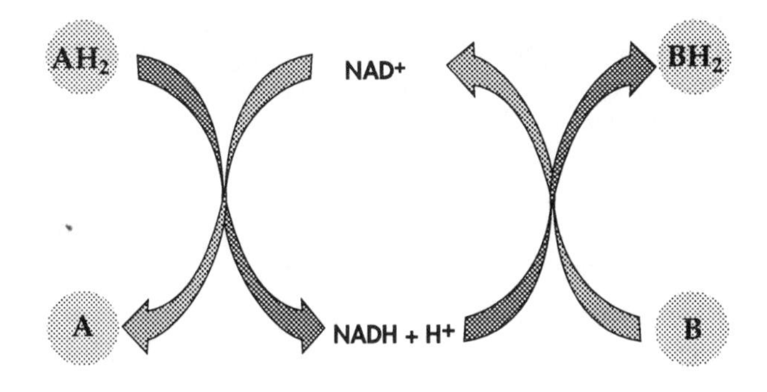

   a. AH_2 is oxidized.
   b. NAD^+ is a reducing agent.
   c. The reaction producing BH_2 and NAD^+ is exergonic.
   d. $NADH + H^+$ is a reducing agent.
   e. B is an oxidizing agent.

3. In eukaryotic cells with access to oxygen, the correct sequence showing movement of *carbon* atoms from glucose would include
   a. pyruvate, CO_2, citric acid cycle intermediates.
   b. glucose 6-phosphate, pyruvate, lactic acid.
   c. pyruvate, citric acid cycle intermediates, CO_2.
   d. glucose 6-phosphate, pyruvate, respiratory chain, CO_2.
   e. pyruvate, respiratory chain.

4. In eukaryotic cells with access to oxygen, the correct sequence showing movement of *hydrogen* atoms from glucose would include
   a. glucose 6-phosphate, coenzyme A, NADH + H^+, respiratory chain, H_2O.
   b. glucose 6-phosphate, NADH + H^+, respiratory chain, H_2O.
   c. glucose 6-phosphate, pyruvate, NADH + H^+, citric acid cycle intermediates, respiratory chain, H_2O.
   d. glucose 6-phosphate, NADH + H^+, H_2O, respiratory chain.
   e. glucose 6-phosphate, pyruvate, citric acid cycle intermediates, respiratory chain, H_2O.

GLYCOLYSIS • THE BEGINNING OF CELLULAR RESPIRATION: THE CITRIC ACID CYCLE • CONTINUATION OF CELLULAR RESPIRATION: THE RESPIRATORY CHAIN (page 144–154)

Key Concepts

1. As a result of the ten reactions of glycolysis (which take place in the cytoplasm), one molecule of glucose is converted into two molecules of pyruvate. The net production of energy carriers is two NADH + H^+ and two ATP per molecule of glucose.
2. The first five steps of glycolysis are endergonic and result in the addition of two phosphates from ATP (steps 1 and 3), the conversion of glucose to fructose (step 2), formation of two three-carbon sugars from fructose (step 4), and the production of two molecules of glyceraldehyde 3-phosphate (G3P) (step 5).
3. In step 6 of glycolysis, G3P undergoes three changes: it is oxidized with the addition of an oxygen atom, an inorganic phosphate group is added, and NAD removes two hydrogens to form two NADH + H^+ per molecule of G3P.
4. In both steps 7 and 10 of glycolysis, one ATP is produced as phosphate groups are removed from three carbon compounds. Since two ATP were used in steps 1 and 3, the net production of ATP is two per molecule of glucose.
5. The *citric acid cycle* is a cyclic series of reactions catalyzed by enzymes present within the liquid matrix of the mitochondrion. Pyruvate becomes oxidized by releasing its carbons as CO_2 and then passing its hydrogens on to NAD and FAD.
6. The overall reaction of the citric acid cycle is
   $(C_3H_4O_3) + 3\,H_2O + 5\text{ carrier} \rightarrow 3\,CO_2 + 5\text{ carrier} \bullet 2H.$
7. Of the five carriers involved in the citric acid cycle, four are NADH + H^+ and one is $FADH_2$.
8. The first reaction of the citric acid cycle, catalyzed by the *pyruvate dehydrogenase complex*, involves the removal of CO_2 (decarboxylation) and hydrogens (dehydrogenation) by NAD. This in turn converts pyruvate to acetate, which then combines with *coenzyme A (CoA)* to form *acetyl coenzyme A (acetyl CoA)*.
9. Acetyl coenzyme A enters the cyclic part of the citric acid cycle by combining with *oxaloacetate* to form *citrate*, releasing coenzyme A unchanged.
10. Each of the following steps in the cycle involves the reduction of NAD^+: isocitrate to α-ketoglutarate, α-ketoglutarate to succinyl CoA, and malate to oxaloacetate.
11. Decarboxylation occurs during steps 4 and 5: isocitrate to α-ketoglutarate and α-ketoglutarate to succinyl CoA.
12. Step 6, succinyl CoA to succinate, produces guanosine triphosphate (GTP). GTP is used to produce ATP.
13. Step 7, succinate to fumarate, also yields $FADH_2$ (the reduced form of FAD).
14. NADH + H^+ and $FADH_2$ pass their hydrogens to the *respiratory chain*, a series of oxidation–reduction agents built into the inner membrane of the mitochondrion.
15. The respiratory chain includes three large protein complexes (*NADH-Q reductase, cytochrome reductase, cytochrome oxidase*), a mobile carrier molecule (*ubiquinone*) that can move within the membrane, and an extrinsic protein (*cytochrome c*) lying in the space between the inner and outer membrane.
16. Cytochromes in the respiratory chain have heme groups with a central iron atom that is Fe^{3+} when oxidized or Fe^{2+} when reduced.
17. Each pair of hydrogens that is passed from NADH + H^+ to oxygen by way of the respiratory chain results in three new ATP; each pair of hydrogens that is passed from $FADH_2$ to oxygen by way of the respiratory chain results in two new ATP.
18. At various points within the respiratory chain, hydrogen atoms are separated into protons and electrons. Electrons continue to be moved by respiratory chain carriers; H^+ are deposited outside of the inner mitochondrial membrane.
19. The *chemiosmotic mechanism* for production of ATP depends on the proton gradient between the inside and outside of the inner mitochondrial membrane.
20. ATP is formed when hydrogen ions pass through a channel-like *ATP synthase* protein as they diffuse from the outside to the inside of the inner membrane.
21. The final acceptor in the respiratory chain is oxygen; cytochrome oxidase passes two electrons to two protons and an oxygen atom to form H_2O.
22. *Respiratory uncouplers* are substances that uncouple ATP production from respiratory metabolism by allowing protons to diffuse back across the membrane discharging the proton-motive force.
23. The supply of ATP–ADP in the cell provides *respiratory control* by regulating the rate at which protons can move through ATP synthase.

***Questions** (for answers and explanations, see page 53)*

The first four questions are concerned with the ten reactions of glycolysis, which are summarized below.

(1) glucose + ATP → glucose 6-phosphate + ADP

(2) glucose 6-phosphate → fructose 6-phosphate

(3) fructose 6-phosphate + ATP →
fructose 1,6-bisphosphate + ADP

(4) fructose 1,6-bisphosphate →
glyceraldehyde 3-phosphate + dihydroxyacetone phosphate

(5) dihydroxyacetone phosphate → glyceraldehyde 3-phosphate

(6) 2 glyceraldehyde 3-phosphate + 2 P_i + 2 NAD →
2 1,3-bisphosphoglycerate + 2 NADH + H^+

(7) 2 bisphosphoglycerate + 2 ADP →
2 3-phosphoglycerate + 2 ATP

(8) 2 3-phosphoglycerate → 2 2-phosphoglycerate

(9) 2 2-phosphoglycerate → 2 phosphoenolpyruvate + 2 H_2O

(10) 2 phosphoenolpyruvate + 2 ADP → 2 pyruvate + 2 ATP

1. Of the reactions shown, list those that are exergonic:
6, 7, 9, 10 ______

2. Of the reactions shown, list those that involve oxidation–reduction agents: ___6___

3. Of the reactions shown, list those that are described in the text as "pump-priming": ___1, 3, 6, 10___

4. In which reaction of glycolysis is a 6-carbon compound converted in to two 3-carbon compounds? ___4___

5. Which of the following are *not* products of the citric acid cycle?
a. Coenzyme A
b. CO_2
c. NADH + H^+
d. $FADH_2$
e. ATP

6. Which of the following compounds are *not* cycled in the citric acid cycle?
a. α-ketoglutarate
b. Citrate
c. Acetyl coenzyme A
d. Oxaloacetate
e. Fumarate

7. Which of the following is *not* a correct similarity between the reactions pyruvate → acetyl coenzyme A and α-ketoglutarate → succinyl CoA?
a. Both reactions are redox reactions.
b. Both reactions yield NADH + H^+.
c. Both reactions yield CO_2.
d. Both reactions are endergonic.
e. The enzymes catalyzing both reactions are located on the inner mitochondrial membrane.

8. Which of the following *correctly* lists the first and last components in the respiratory chain?
a. NADH-Q reductase, cytochrome oxidase
b. Q (ubiquinone), cytochrome oxidase
c. Succinate-Q reductase, cytochrome *c*
d. Cytochrome reductase
e. NADH-Q reductase, cytochrome reductase

9. Which of the following generalizations about the respiratory chain is *not* true?
a. Except for Q (ubiquinone), all respiratory components are proteins.
b. Some respiratory components move protons from the inside to the outside of the inner mitochondrial membrane.
c. Some components of the respiratory chain are chemically similar to hemoglobin.
d. The final acceptor of electrons and protons is oxygen.
e. Because of where its hydrogens enter the respiratory chain, the carrier $FADH_2$ produces only 2 ATP.

***Activities** (for answers and explanations, see page 54)*

- In the following transmission electron micrograph showing a portion of a mitochondrion and adjacent cytosol, add labeled arrows pointing to locations where (A) the reactions of glycolysis occur, (B) the pyruvate dehydrogenase complex and certain other enzymes of the citric acid cycle occur, (C) the remaining citric acid cycle reactions occur, and (D) the electron carriers of the respiratory chain are located. For the moment, ignore the arrows labeled 1 and 2.

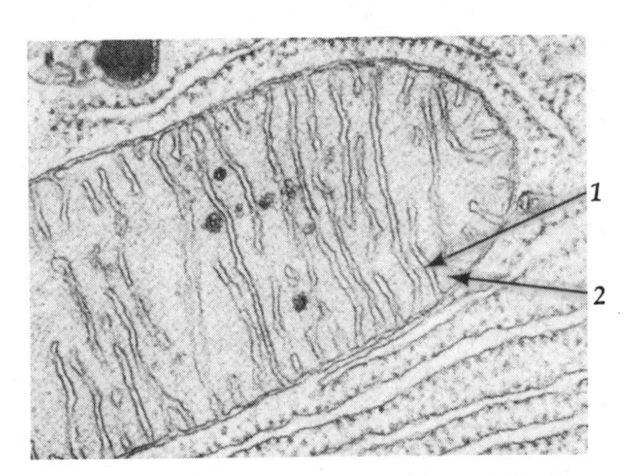

- Complete the following sentence to state what can be said about the pH and ATP production in the areas of the mitochondrion indicated by arrows 1 and 2 in the previous transmission electron micrograph.

The pH in area __2__ is greater than area __1__, and ATP is formed in area __1__, but not in area __2__.

FERMENTATION • COMPARATIVE ENERGY YIELDS • CONNECTIONS WITH OTHER PATHWAYS • FEEDBACK REGULATION (pages 155–160)

Key Concepts

1. In *fermentation*, pyruvate produced by glycolysis is reduced by NADH + H^+ to form either *lactic acid* or, via acetaldehyde, *ethanol*. The oxidized NAD^+ produced by fermentation can be used again in glycolysis.

2. Some bacteria carry on cellular respiration but use oxidizing agents other than oxygen, (for example, nitrate, NO_3^-), as the final acceptor of electrons in the respiratory chain.

3. *Fermentation* yields a net of 2 ATP per molecule of glucose. Glycolysis followed by cellular respiration yields a net of 36 ATP per molecule of glucose.
4. Glycolysis produces 2 ATP and 2 NADH + H^+. Movement of NADH + H^+ into the mitochondrion uses ATP, so net production of ATP from glycolysis is 6 ATP (2 ATP per NADH + H^+).
5. The citric acid cycle produces 2 ATP directly, 8 NADH + H^+, and 2 $FADH_2$, resulting in 30 ATP (3 ATP per NADH + H^+ and 2 ATP per $FADH_2$).
6. The efficiency of glycolysis plus cellular respiration is about 63%, since each ATP contains about 12 kcal for a total of 432 kcal (36 ATP x 12 kcal) as compared with the 686 kcal available per mole of glucose.
7. After monosaccharides are converted into glucose, and polysaccharides such as starch and glycogen are hydrolyzed into glucose, they can enter glycolysis.
8. Upon digestion, fats become glycerol and fatty acids. Glycerol enters glycolysis after conversion to glyceraldehyde 3-phosphate and fatty acids enter the citric acid cycle after they are converted into acetyl coenzyme A. When the cell has an adequate ATP supply, these reactions run in reverse to form fat.
9. Pyruvate and citric acid cycle intermediates can be used as building blocks for important amino acids.
10. In addition to its involvement in fat synthesis, acetyl coenzyme A can also be used to produce pigments, plant growth substances, and steroid hormones.
11. Because aerobic respiration is 18 times more efficient than glycolysis and fermentation, much less glucose is needed to meet the energy requirements if the cell has access to oxygen.
12. The observation that the rate of glycolysis in many cells increases when oxygen is not available and decreases when oxygen is available is called the *Pasteur effect*.
13. Because NADH-Q reductase (the first component of the respiratory chain) has a greater affinity for NADH than the enzyme that catalyses the reduction of pyruvate into lactose, the presence of oxygen will immediately convert a cell from anaerobic to aerobic respiration.
14. Regulation of the reactions of glycolysis, the citric acid cycle, and the respiratory chain depends on the *allosteric control* of the enzymes catalyzing these reactions.
15. In allosteric control, the intermediates of a biochemical pathway provide positive and negative feedback by inhibiting or enhancing the efficiency of enzymes catalyzing key steps in the pathway.
16. Examples of negative feedback of respiratory pathways include inhibition of phosphofructokinase, the enzyme catalyzing the third reaction of glycolysis by fructose 1, 6-bisphosphate (product of the reaction), and by citrate from the citric acid cycle.
17. Examples of positive feedback of respiratory pathways include the allosteric activation by acetyl coenzyme A of the reaction that converts pyruvate directly into oxaloacetate and the stimulatory effect of citrate on the conversion of acetyl coenzyme A into fatty acids.
18. Many reactions producing ATP and NADH + H^+ are inhibited by high concentrations of these molecules, but stimulated by high concentrations of ADP and NAD. For example, the rate of the third reaction of glycolysis is increased by ADP, but decreased by ATP. In the citric acid cycle, the rate of conversion of isocitrate to α-ketoglutarate is increased by ADP and NAD, but decreased by ATP and NADH + H^+.
19. ATP and NADH + H^+ are also allosteric inhibitors of the reaction that combines acetyl coenzyme A with oxaloacetate to form citric acid.

Questions (for answers and explanations, see page 54)

(Note: in several of the following questions, you may wish to refer to the accompanying figure on page 39 showing feedback regulation of glycolysis and the citric acid cycle.)

1. Of the 36 ATP produced from the complete aerobic respiration of one molecule of glucose, how many result from the chemiosmotic mechanism?
   a. 24
   b. 28
   c. 30
   d. 32
   e. 36

2. Acetyl coenzyme A is not usually involved in the synthesis of:
   a. lactate.
   b. citrate.
   c. fatty acids.
   d. steroids.
   e. pigments.

3. Which of the following molecules will result in the greatest energy yield if linked to the respiratory chain?
   a. NAD^+ produced through lactic acid fermentation
   b. NADH + H^+ produced in glycolysis
   c. NADH + H^+ produced in the citric acid cycle
   d. $FADH_2$ produced in the citric acid cycle
   e. $FADH_2$ produced in glycolysis

4. Which one of the following mechanisms most directly causes the phenomenon called the Pasteur effect?
   a. Allosteric activation by acetyl coenzyme A of the reaction converting pyruvate into oxaloacetate
   b. Allosteric inhibition by ATP and citrate of the enzyme (phosphofructokinase) catalyzing the step fructose 6-phosphate → fructose 1, 6-bisphosphate
   c. Allosteric activation by citrate of fatty acid synthesis to form acetyl coenzyme A
   d. Allosteric inhibition by ATP of the step isocitrate → α-ketoglutarate
   e. Allosteric activation by NAD^+ of the step isocitrate → α-ketoglutarate

Activity (for answers and explanations, see page 54)

- In the following figure, circle allosteric control points (both positive and negative) that would be important in regulating glycolysis and respiration in a cell that had been undergoing fermentation but which is now starting to respire aerobically.

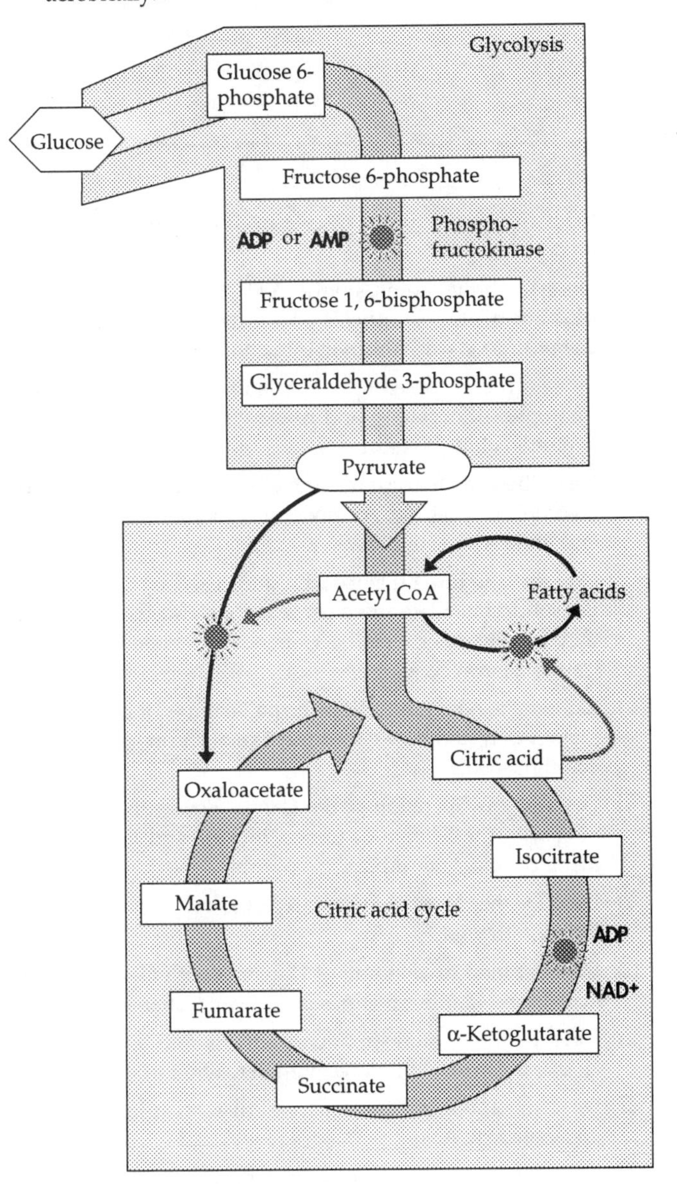

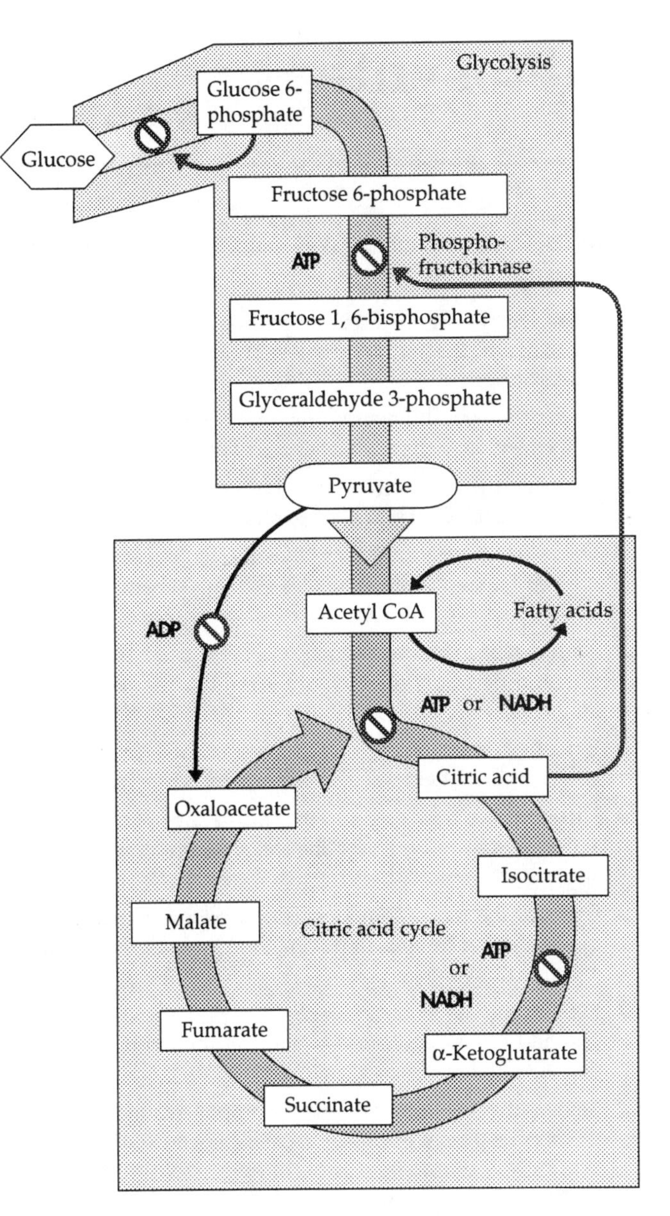

8

Photosynthesis

CHAPTER LEARNING OBJECTIVES—after studying this chapter you should be able to:

- ❑ Describe the importance of photosynthesis to life on Earth using the terms "photosynthetic autotroph" and "heterotroph."
- ❑ Given the overall equation for photosynthesis, trace the movement of atoms from reactants to products.
- ❑ Describe the role of nicotinamide adenine dinucleotide phosphate (NADPH) in photosynthesis.
- ❑ Describe the overall activities of the light reactions and dark reactions and the materials exchanged between these two pathways of photosynthesis.
- ❑ With regard to electromagnetic radiation, describe the relationship between velocity, wavelength, frequency, and light energy.
- ❑ Explain how the color of a pigment relates to the light wavelength that its molecules absorb, reflect, or transmit.
- ❑ Differentiate between an absorption spectrum and an action spectrum and describe the absorption and action spectrums of chlorophyll.
- ❑ Describe the major components of a photosynthetic energy-absorbing antenna and the energy movement that occurs within one.
- ❑ Explain the sequence of events following absorption of a photon by a molecule and relate these events to the phenomenon of fluorescence.
- ❑ Review the operation and major components of noncyclic photophosphorylation by describing the sequence of events following absorption of energized photons by antennae of photosystems I and II.
- ❑ Review the operation and major components of cyclic photophosphorylation by describing the sequence of events following absorption of an energized photon by an antenna of photosystem II.
- ❑ Describe how chemiosmotic ATP formation is accomplished in photosynthesis and how it differs from the chemiosmotic mechanism in mitochondria.
- ❑ Explain the proposed appearance of cyclic and noncyclic photophosphorylation and photosystems I and II in the evolution of life on Earth.
- ❑ Describe which contemporary autotrophs are able to accomplish cyclic and/or noncyclic photophosphorylation and under what conditions.
- ❑ Describe the three major techniques used by Calvin and Benson to elucidate the dark reactions and the approach they used to determine the sequence of compounds in the Calvin-Benson cycle.
- ❑ Explain the operation and principal features of paper chromatography and how two-dimensional chromatography is accomplished.
- ❑ Describe the operation of the Calvin–Benson cycle beginning with the fixation of carbon dioxide and list all of its major components, inputs and outputs.
- ❑ Describe the C_4 pathway and explain how it connects with the C_3 pathway.
- ❑ Contrast the leaf anatomy of C_4 plants with C_3 plants and explain the adaptive features of C_4 plant leaf anatomy.
- ❑ Describe the photorespiration pathway and explain how C_4 plants can keep photorespiration to a minimum.
- ❑ Describe the adaptive features of Crassulacean acid metabolism (CAM).
- ❑ Describe the cycling of fixed carbon, CO_2, and O_2 within a plant cell and between autotrophs and heterotrophs.

SUNLIGHT AND LIFE ON EARTH • EARLY STUDIES OF PHOTOSYNTHESIS • THE PATHWAYS OF PHOTOSYNTHESIS • LIGHT AND PIGMENTS (pages 162–169)

Key Concepts

1. *Autotrophs* are organisms that can use an energy source and simple raw materials to make reduced carbon compounds.
2. *Photosynthetic autotrophs* use the energy of visible light to reduce CO_2 with hydrogens supplied by water to produce carbohydrates, oxygen, and new water.
3. *Heterotrophs* are organisms that depend on preformed organic compounds in order to obtain needed free energy for life processes. Typically, heterotrophs consume autotrophs or other heterotrophs.
4. In studies using *heavy oxygen (^{18}O)*, it was shown that in the overall reaction of photosynthesis the oxygen gas produced comes from water:

$$6\,CO_2 + 12\,H_2O \rightarrow C_6H_{12}O_6 + 6\,O_2 + 6\,H_2O$$

5. The oxidation-reduction agent used in photosynthesis is *nicotinamide adenine dinucleotide phosphate (NADP)*, which can exist in a reduced or oxidized form,

$$NADP^+ + 2(H) \rightarrow NADPH + H^+$$

6. Photosynthesis consists of two pathways: the *light reactions* or *photophosphorylation*, which use light energy to produce ATP and NADPH + H^+, and the *dark reactions* or *Calvin–Benson cycle*, which fix CO_2 and convert it to sugar using ATP and NADPH + H^+ produced in the light reactions.

7. In the dark, the Calvin–Benson cycle stops as soon as it depletes its supply of NADPH + H^+ and ATP; if CO_2 availability limits the Calvin–Benson cycle, photophosphorylation will stop when it runs out of $NADP^+$ and ADP. Thus, the two pathways are interdependent.

8. Light is a form of energy that can be described either as particles called *photons* or *quanta*, or in terms of its wave-like characteristics. Wavelength is the distance from peak to peak, and is usually measured in nanometers (nm).

9. Visible light (to humans) is *electromagnetic radiation* with wavelengths from 400 nm (violet) to 700 nm (red). Wavelengths from 100 to 400 nm are ultraviolet; over 700 nm, infrared and radio.

10. The frequency of light (ν, Greek letter *nu*) is its speed (c) divided by its wavelength (λ, Greek letter *lambda*), or $\nu = c / \lambda$. The speed of light is a constant (3.0×10^{10} cm per sec, or 186,000 miles per sec). Frequency is sometimes expressed in hertz (cycles per second).

11. The energy (E) of a photon is directly proportional to its frequency (ν), such that $E = h\nu$, where h is Planck's constant. Since $\nu = c / \lambda$ and $E = hc / \lambda$, light energy is directly dependent on wavelength, with light of shorter wavelength having more energy than light of longer wavelength.

12. Brightness or light *intensity* is a measure of the quantity of energy striking a defined surface area. It is expressed in energy units such as calories or as photons per centimeter per second.

13. When a molecule absorbs a photon, one of its electrons moves to a more outer orbital, raising the molecule from a *ground state* to an *excited state*. The difference between the two energy levels equals the energy of the absorbed photon.

14. *Pigments* are molecules that absorb light within the visible region of the electromagnetic spectrum. The colors of pigments are determined by the wavelengths not absorbed (i.e., the wavelengths transmitted or reflected).

15. Molecules can only absorb photons of particular energy levels as evidenced by the *absorption spectrum* of a molecule, in which we plot light absorption against wavelength.

16. The rounded peaks of absorption spectra indicate that energy states (ground or excited) exist as families of sublevels and a molecule can move between sublevels by absorbing or releasing heat energy.

17. An *action spectrum* is produced if we plot a particular light-dependent activity, such as photosynthesis, as a function of wavelength.

18. The action spectrum for photosynthesis shows that light wavelengths in the blue and red regions of the spectrum are more effective than green, yellow, or orange light. Light wavelengths in the blue and red regions are best absorbed by chlorophyll, as shown by its absorption spectrum.

19. *Chlorophyll a* and *chlorophyll b* are two important photosynthetic pigments. Both have a complex ringed structure called a *chlorin*, with a central magnesium atom and a hydrocarbon chain. They differ only slightly in structure.

20. Photosynthetic accessory pigments include *carotenoids*, like β-carotene, and the *phycobilins*, like phycocyanin and phycoerythrin. Accessory pigments absorb mostly in the region between the peaks for chlorophyll *a* and *b* and pass their energy on to chlorophyll *a*.

21. An energy-absorbing *antenna* consists of chlorophyll *a* and chlorophyll *b* and any accessory pigments. Excitation is passed between pigment molecules within the antenna moving toward those pigments that absorb light with longer wavelengths and lower energies.

Questions (for answers and explanations, see page 54)

1. Examine the following absorption spectrum for an unknown pigment molecule.

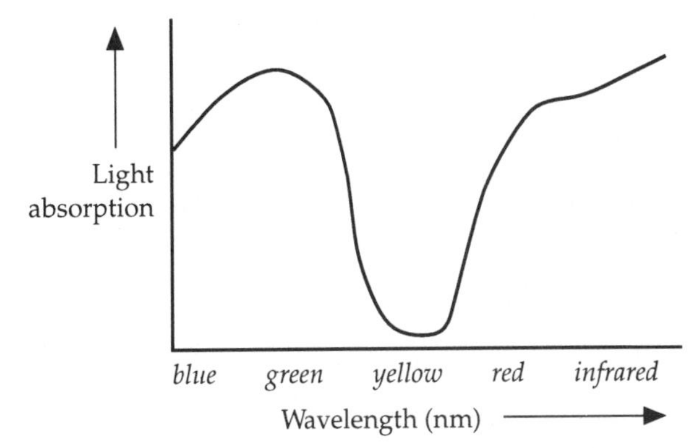

What color would this substance likely appear to you?
a. Blue
b. Green
c. Yellow
d. Blue-green
e. Infrared

2. Which one of the following statements about photophosphorylation and the Calvin–Benson cycle is *not* true?
a. The products of photophosphorylation are ATP, NADPH + H^+, and O_2.
b. The products of the Calvin–Benson cycle are CO_2 and sugar.
c. Photophosphorylation requires chlorophyll *a*.
d. ATP and NADPH + H^+ produced in photophosphorylation are used in the Calvin–Benson cycle.
e. Photophosphorylation is activated by light.

3. Light beam A has a frequency of 1,000 hertz; light beam B has a frequency of 900 hertz. Which of the following comparisons is true?
   a. Light beam A has a longer wavelength than B, and greater energy.
   b. Light beam A has a longer wavelength than B, and lesser energy.
   c. Light beam B has a longer wavelength than A, and greater energy.
   d. Light beam B has a longer wavelength than A, and lesser energy.
   e. More information is required to make these comparisons.

4. The rounded peaks seen in the action spectrum for photosynthesis result because
   a. the energy levels of pigment molecules consist of sublevels and molecules can move between sublevels by radiating heat.
   b. there are a variety of different pigment molecules that can trap light for photosynthesis.
   c. there is substantial heat energy transfer within an energy-absorbing antenna.
   d. light intensity varies with wavelength.
   e. light of shorter wavelength has more energy.

PHOTOPHOSPHORYLATION • THE CALVIN–BENSON CYCLE (pages 169–181)

Key Concepts

1. *Fluorescence* is the re-emission of light as a boosted electron falls back to its original, lower-energy orbital. In photosynthesis, the excited electron is used in a redox reaction, passing on its energy to the molecule it reduces.

2. In photophosphorylation, the excited chlorophyll of the reaction center (Chl*) is oxidized by an oxidizing agent (A) in the following general reaction. The oxidized chlorophyll (Chl^+) has transferred an electron and its energy to A.

$$Chl^* + A \rightarrow Chl^+ + A^-$$

3. *Noncyclic photophosphorylation* involves two different systems of chlorophyll molecules called *photosystem I* and *photosystem II.*

4. From photosystem I, pairs of excited electrons are passed to

$$NADP^+ + 2\,e^- + 2\,H^+ \rightarrow NADPH + H^+$$

5. The reaction center of the photosystem I antenna is a chlorophyll *a* molecule called P_{700} (its absorption peak is 700 nm).

6. P_{700} passes a pair of electrons to oxidized ferredoxin (Fd_{ox}), which becomes reduced ferredoxin (Fd_{red}). Fd_{red} reduces $NADP^+$ to form $NADPH + H^+$.

7. Electrons from photosystem II pass through a series of electron carriers to oxidized chlorophylls of photosystem I. The sequence of carriers is pheophytin-I, plastoquinone (PQ), a cytochrome complex , and plastocyanin (PC).

8. The reaction center of the photosystem II antenna is a chlorophyll *a* molecule called P_{680} (its absorption peak is 680 nm).

9. Water ionizes to form H^+ and OH^-. Two OH^- react to form O_2, two electrons, and two H^+. The electrons are used to replenish electrons in photosystem II.

10. The overall reaction for noncyclic photophosphorylation is H_2O + 4 photons + $NADP^+$ + ADP + $P_i \rightarrow NADPH + H^+ + ATP + \frac{1}{2}O_2$. Photosystems I and II each absorb two of the four photons.

11. *Cyclic photophosphorylation* involves photosystem I. P_{700} absorbs a photon and is raised to an excited state to become the reducing agent $P_{700}{}^+$. Excited electrons from two different $P_{700}{}^+$ are used to reduce oxidized *ferredoxin* (Fd_{ox}) to form reduced ferredoxin (Fd_{red}),

$$2\,e^- + Fd_{ox} \rightarrow Fd_{red}$$

12. Pairs of electrons from Fd_{red} are passed through a sequence of redox reactions involving the oxidation-reduction agents plastoquinone (PQ), a cytochrome complex (Cyt), and finally, plastocyanin (PC).

13. One of the reactions in the sequence is sufficiently exergonic to generate ATP from ADP, plus inorganic phosphate. Electrons are then returned to oxidized $P_{700}{}^+$.

14. The electron carriers of photophosphorylation are located within the *thylakoid* membrane of the chloroplast.

15. Protons resulting from the ionization of water are pumped across the thylakoid membrane into the interior of the thylakoid.

16. *Chemiosmotic ATP formation* occurs when protons diffuse back out of the thylakoid into the stroma through proteins which are ATP synthases.

17. The model for chemiosmotic ATP formation during photophosphorylation was tested by varying the pH within the thylakoids of isolated chloroplasts. If the pH was lower inside the thylakoid than in the surrounding medium, ATP would be formed in the dark.

18. Noncyclic photophosphorylation evolved from cyclic photophosphorylation and photosystem I predates photosystem II.

19. Some photosynthetic bacteria have only cyclic photophosphorylation. Cyanobacteria, algae, and plants mostly use noncyclic photophosphorylation, but can switch to cyclic photophosphorylation when the ratio of $NADPH + H^+$ to $NADP^+$ in their chloroplasts is low.

20. Elucidation of the Calvin–Benson cycle depended on three techniques: labeling CO_2 with carbon-14 to get $^{14}CO_2$, *paper partition chromatography* for separating the intermediates in the pathway, and *autoradiography* for locating the colorless intermediates on the chromatogram.

21. In paper chromatography, movement of components in the mixture being separated depends on their relative affinity for the *stationary phase* (water bound to the paper) and the *mobile phase* (the solvent).

22. The distance moved by a substance in a chromatographic system divided by the distance moved by the solvent

is called that substance's front ratio (R_F). The R_F value is constant for a substance under defined conditions.

23. In *two-dimensional chromatography*, a mixture is separated in one direction with one solvent system, the paper is rotated 90°, and the spots are run again with a different solvent system.

24. The method used by Calvin and Benson to determine the sequence of compounds in the Calvin–Benson cycle was to sample through time from a population of photosynthesizing algal cells exposed to $^{14}CO_2$ and identify the labeled organic compounds formed under different conditions.

25. Fixation of CO_2 by RuBP (ribulose bisphosphate) is catalyzed by the enzyme *RuBP carboxylase* (*rubisco*). The resulting six-carbon compound immediately breaks down to form two molecules of 3PG (3-phosphoglycerate).

26. 3PG is reduced by NADPH + H^+ to form a three-carbon sugar phosphate (G3P), in a reaction requiring ATP. The "*sugar shuffle*" is a complex series of reactions that regenerates RuBP from G3P, while some G3P is directed into the synthesis of glucose and other organic compounds.

27. The Calvin–Benson cycle must "turn" six times to produce a molecule of glucose. This requires six CO_2, 12 NADPH + 12 H^+, and 18 ATP from the light reactions.

28. Whereas the components of photophosphorylation are built into the thylakoid membrane, the enzymes of the Calvin–Benson cycle are found in the stroma of the chloroplast, except for rubisco, which is also part of the thylakoid membrane.

Questions (for answers and explanations, see page 54)

1. Select the correct path that an electron could take during the complete process of photosynthesis.
   a. CO_2 → ribulose bisphosphate → G3P → glucose
   b. H_2O → photosystem I → photosystem II → NADPH + H^+ → G3P
   c. photosystem II → H_2O → photosystem I → NADPH + H^+ → glucose
   d. CO_2 → photosystem I → NADPH + H^+ → ribulose bisphosphate → G3P
   e. H_2O → photosystem II → photosystem I → NADPH + H^+ → G3P

2. Choose *all* of the following materials that are *not* cycled between the light reactions and the Calvin–Benson cycle.
   a. Water
   b. NADPH$^+$
   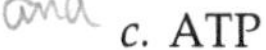
   c. ATP
   d. Oxygen
   e. NADPH + H^+

3. The following graph shows changes in the concentration of substances A, B, and C during a "lollipop" experiment in which the CO_2 supply was *eliminated* at time 30.

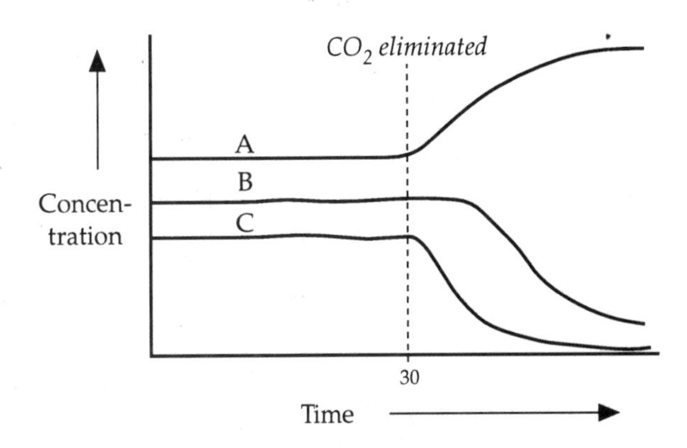

Based on these data, which of the following shows the correct pathway for the substances involved?
   a. A + CO_2 → B → C
   b. C + CO_2 → B → A
   c. A + CO_2 → C → B
   d. B + CO_2 → C → A
   e. A → B → C + CO_2

4. Two groups of chloroplasts (A and B) are kept in the dark and allowed to equilibrate with a solution of pH 6. Group A is then transferred to a solution of pH 7, and group B to a solution of pH 5. Which of the following would be true?
   a. Chloroplasts from both groups produce ATP.
   b. Chloroplasts from neither group produce ATP.
   c. Both groups produce ATP, but group A chloroplasts produce more ATP than group B chloroplasts.
   d. Both groups produce ATP, but group B chloroplasts produce more ATP than group A chloroplasts.
   e. Only group A chloroplasts produce ATP.

Activities (for answers and explanations, see page 54)

- A mixture of compounds was separated using two-dimensional chromatography, first with solvent A, then with solvent B. The following chromatogram results. Circle the substance that has the greatest total R_F for *both* solvents.

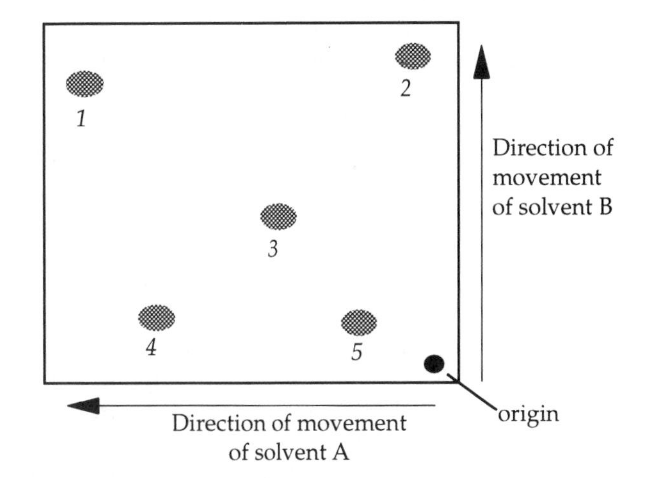

- In the following transmission electron micrograph showing a portion of a chloroplast and adjacent cytosol, label the arrows pointing to (A) the chloroplast outer membrane, (B) a thylakoid membrane, (C) the stroma, (D) the cytosol, (E) the location of photosystem I and photosystem II components, and (F) the location of the enzymes of carbon dioxide fixation.

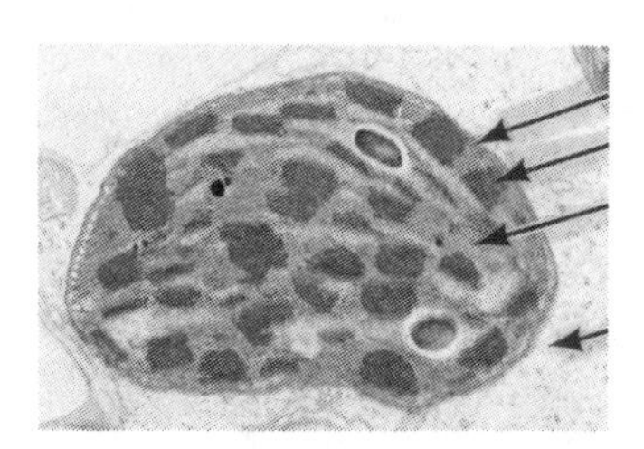

PHOTORESPIRATION • ALTERNATE MODES OF CARBON DIOXIDE FIXATION • PHOTOSYNTHESIS AND CELLULAR RESPIRATION (pages 181–185)

Key Concepts

1. *Photorespiration* occurs when rubisco catalyzes the reaction of RuBP with oxygen to form, in part, the two-carbon compound *glycolate.*
2. Glycolate leaves the chloroplast and enters organelles called microbodies where it is oxidized eventually to form CO_2. Photorespiration is a light-dependent process.
3. Photorespiration is a wasteful process because it uses RuBP, but produces no ATP. Photorespiration results because O_2 and CO_2 compete as substrates for rubisco.
4. *C_4 plants,* like sugar cane and corn, have an additional pathway for capturing CO_2 involving the enzyme *PEP carboxylase,* which has a much greater affinity for CO_2 than does rubisco.
5. PEP carboxylase catalyzes a reaction combining CO_2 with phosphoenolpyruvate (PEP) to form the four-carbon compound oxaloacetate. This pathway is called *C_4 photosynthesis* because the first stable compounds formed have four carbons instead of three, as in the Calvin–Benson cycle, which is also called *C_3 photosynthesis.*
6. C_4 plants have a unique leaf anatomy, in which vascular bundles are surrounded by an outer *mesophyll layer* and an inner *bundle sheath layer*. The layers are composed of different types of cells, and each cell has a distinct type of chloroplast.
7. Cells of the mesophyll layer have the C_4 pathway only and produce oxaloacetate and other four-carbon acids. These four-carbon acids diffuse into the bundle sheath layer from the mesophyll.
8. Bundle sheath cells have the C_3 pathway only. The four-carbon acids from the mesophyll layer are decarboxylated to form CO_2 and three-carbon compounds that cycle back to the mesophyll cells. The CO_2 is incorporated into the C_3 pathway by rubisco.
9. The C_4 pathway and the unique anatomy of C_4 plants permit them to pump CO_2 from an area where it is in low concentration (the intercellular spaces) into the bundle sheath cells, thereby creating a high CO_2 concentration in these cells. This enables rubisco and the C_3 pathway to function normally and adapts C_4 plants for life in arid areas because they can keep their stomata closed more of the time.
10. Some plants, including many in the family Crassulaceae, keep their stomata closed all day to conserve water and carry on C_3 photosynthesis using CO_2 derived from four-carbon acids that were formed from CO_2 at night, using PEP carboxylase. This mode of carbon fixation is called *Crassulacean acid metabolism.*
11. The C_4 pathway may be an adaptation for limiting photorespiration in arid environments, and C_4 plants show little photorespiration.
12. Photosynthetic autotrophs, such as the green plants and photosynthetic bacteria and protists, cycle fixed carbon, CO_2, and O_2 with heterotrophic organisms.
13. Photosynthetic autotrophs simultaneously carry on photosynthesis and respiration and cycle fixed carbon, CO_2 and O_2 between chloroplasts and the sites of glycolysis and respiration in the cytosol and mitochondria.

Questions (for answers and explanations, see page 55)

1. Which one of the following reactions is *least likely* to occur within the cells of a C_4 plant?
   a. The Calvin–Benson cycle
   b. Photorespiration
   c. CO_2 + RuBP → 2 3PG
   d. Photophosphorylation
   e. CO_2 + PEP → oxaloacetate

2. Which one of the following categories does *not* differ between a C_3 and C_4 plant?
   a. Initial CO_2 acceptor
   b. Extent of photorespiration
   c. Enzyme catalyzing reaction that fixes CO_2
   d. Presence of Calvin–Benson cycle
   e. Leaf anatomy

3. Which of the following features is *not* characteristic of some plants in the family Crassulaceae?
   a. Stomata open only at night
   b. Trap CO_2 with PEP carboxylase
   c. PEP produced photosynthetically during day
   d. Calvin–Benson cycle during day only
   e. CO_2 captured by rubisco from four-carbon acids

4. Which of the following processes would *not* be inhibited if a plant cell were placed in the dark?
   a. Cellular respiration
   b. Cyclic photophosphorylation
   c. Noncyclic photophosphorylation
   d. Photorespiration
   e. The Calvin–Benson cycle

Integrative Questions (for answers and explanations, see page 55)

1. Which one of the following statements about NAD and NADPH is *not* true?
   a. Both compounds are oxidation–reduction agents.
   b. NAD participates in respiratory reactions; NADPH participates in photosynthetic reactions.
   c. NADPH is restricted to photosynthetic autotrophs; NAD is restricted to heterotrophs.
   d. Both compounds exist in two different molecular forms.
   e. Both compounds function as electron carriers.

2. Which of the following statements about chemiosmotic ATP production in chloroplasts and mitochondria is *not* true?
   a. Both depend on membrane proteins that are ATP synthases.
   b. In both, ATP is generated by passive diffusion of H^+.
   c. Both depend on a proton gradient across a membrane.
   d. Both depend on the impermeability of cell membranes to charged particles.
   e. The electron carriers are the same in both.

Integrative Activity (for answers and explanations, see page 55)

- In the diagram of noncyclic photophosphorylation shown below, label the appropriate parts with the following labels: (1) stroma, (2) thylakoid interior, (3) photosystem I, (4) photosystem II, (5) cytochrome complex, (6) ATP synthase, (7) NADP reductase, (8) location of dark reactions.

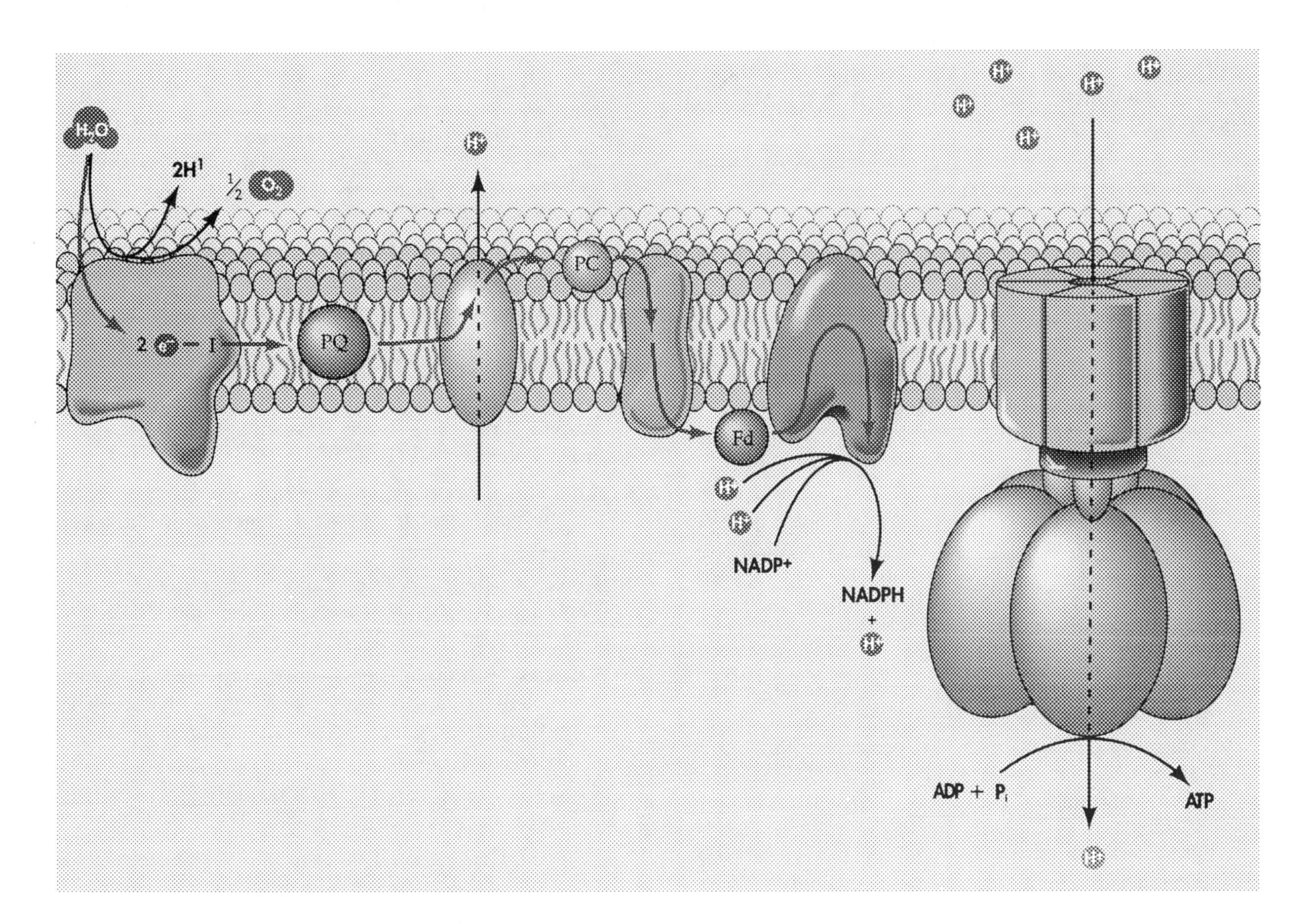

Answers and Explanations

Chapters 1–8

CHAPTER 1 – THE SCIENCE OF BIOLOGY

Note : The page numbers listed with the section titles refer to the Study Guide; numbers in parentheses for each question refer to the relevant key concepts.

CHARACTERISTICS OF LIVING ORGANISMS • THE HIERARCHICAL ORGANIZATION OF BIOLOGY • THE METHODS OF SCIENCE (pages 1–2)

1. Choice *b*. Only a subset of all organisms (the photosynthetic and chemosynthetic organisms) is autotrophic. (1–8)
2. Choice *c*. Absolute certainty is not possible in science. (22–28)
3. (10–19)

2	A sperm	**cell**
4	A mouse	**organism**
1	Glucose	**molecule**
3	A lung	**organ**
7	The Earth	**biosphere**
6	A woodland	**ecological community**
5	*Peromyscus leucopus*	**species**

4. Choice *b*. Emergent properties are the new features of a system that result from the interaction of the system's components. The molecules of which membranes are composed, such as proteins and phospholipids, result in the regulation that membranes exert over what enters or leaves cells. (20)
5. (21–24)
   a. Birds are visually oriented predators: ***assumption***. This is not tested but is assumed true based on past studies.
   b. Conspicuous caterpillars tend to occur in groups: ***observation***. Observations like this led to the development of the hypothesis that conspicuous caterpillars are not palatable.
   c. Bird 1 consumed five conspicuous and one cryptic caterpillars: ***result***. This is data collected in the study.
   d. This species of conspicuous caterpillar should not be eaten by blue jays: ***prediction***. If the data shows this to be true, the hypothesis is validated.
   e. Blue jays should eat equal numbers of conspicuous and cryptic caterpillars: ***null hypothesis***. The null hypothesis always states that the factor being studied has no effect.
6. Both use an experimental approach in which extraneous factors are kept constant so that a single variable can be tested. It is easier to control extraneous factors in the laboratory than in the field, but the results of field experiments are easier to relate to nature than laboratory experiments. (25–26)

WHY PEOPLE DO SCIENCE • SIZE SCALES • TIME SCALES • MAJOR ORGANIZING CONCEPTS OF BIOLOGY • EVOLUTIONARY CONCEPTS • LIFE'S SIX KINGDOMS • SCIENCE AND RELIGION (pages 3–4)

1. Greatest total daily food intake—***elephant***.

   Greatest daily food intake per kg of body weight—***hummingbird***. Because the hummingbird has the greatest surface area to volume ratio, it will lose more heat to the environment, have a higher metabolism, and require more food per gram of body weight than the other, larger organisms. (1–8)
2. Choice *c*. Volume increases with the cube of the linear dimension, so if the length doubles, say from 2 to 4, than the volume would change from $(2)^3 = 8$ to $(4)^3 = 64$, producing an increase in volume of 64/8 or 8. (1, 2)
3. Choice *c*. Choices *a*, *b*, and *e* would increase surface area, while maintaining or reducing cell volume. Choice *d* would help reduce metabolism-producing mass while keeping surface area constant. Becoming unicellular would make matters worse. (1–8)
4. Choice ***d***. A macroevolutionary time scale is appropriate for studying adaptations seen in larger groups of organisms. (10–14)
5. Choice ***d***. The theory of evolution first proposed by Jean Baptiste de Lamarck was historically important but is now known to be incorrect. (15–19, 21)
6. Choice *c*. Although Darwin recognized that offspring resemble their parents because they inherit characteristics from them, the concept of the gene was not developed until much later. (22–25)
7. Choice ***b***. The kingdom Protista includes organisms that are unicellular and have eukaryotic cells. (28–34)
8. Choice ***d***. The kingdom Plantae includes organisms that are mostly multicellular, photosynthetic autotrophs. (28–34)
9. Choice *c*. The kingdom Fungi includes organisms that are mostly multicellular, absorptive heterotrophs. (28–34)
10. Natural selection is differential reproductive success by individuals better adapted for their natural environment. Artificial selection is differential reproduction as directed by humans and not the natural environment. (25–26)
11. Eukaryotic cells arose when prokaryotic cells that had been living inside other prokaryotic cells became permanent residents. (29–30)

CHAPTER 2 – SMALL MOLECULES

Note : The page numbers listed with the section titles refer to the Study Guide; numbers in parentheses for each question refer to the relevant key concepts.

ATOMS • ELEMENTS (pages 5–6)

1. Choice *d*. Hydrogen, the most abundant element in the universe, consists of atoms with one proton, one electron, and no neutrons. (1–6)
2. Choice *e*. Isotopes of an element differ in the number of neutrons, so isotopes of an element have different mass numbers. (6, 9–11, 14)
3. Choice *c*. Choices *a* and *b* are only true for elements in which the number of neutrons and protons is the same. (9–11, 13, 14)
4. Choice *c*. With mass numbers of ${}^{30}X$ and ${}^{32}X$ and an atomic weight of 30.5, the proportions of the two isotopes must be 75:25; $(30 \times .75) + (32 \times .25) = 30.5$. (12–15)
5. Choice *b* is correct; isotopes of an element differ in mass. Other choices are incorrect: all isotopes are electrically neutral (*a*), have the same atomic number (*d*), identical chemical properties (*e*), and only some isotopes are radioactive (*c*). (13–19)

- Hydrogen ${}^{1}_{1}\mathbf{H}$ Carbon ${}^{12}_{6}\mathbf{C}$

  Nitrogen ${}^{14}_{7}\mathbf{N}$ Oxygen ${}^{16}_{8}\mathbf{O}$

  Sulfur ${}^{32}_{16}\mathbf{S}$ Phosphorus ${}^{30}_{15}\mathbf{P}$

- Deuterium ${}^{2}_{1}\mathbf{H}$ Tritium ${}^{3}_{1}\mathbf{H}$

  Carbon–14 ${}^{14}_{6}\mathbf{C}$

  Oxygen–18 ${}^{18}_{8}\mathbf{O}$

-

	Electrons	Protons	Neutrons
${}^{1}_{1}\mathbf{H}$	1	1	0
${}^{14}_{6}\mathbf{C}$	6	6	8
${}^{14}_{7}\mathbf{N}$	7	7	7
${}^{65}_{30}\mathbf{Zn}$	30	30	35

THE BEHAVIOR OF ELECTRONS • CHEMICAL BONDS (pages 7–8)

1. Choice *a*. Recall that the atomic number is equal to the number of electrons in the atom and that orbitals are filled from innermost to outermost, with the K shell holding a maximum of two electrons and the L shell holding a maximum of eight electrons. The electron distribution of sodium is 2 in the K shell, 8 in the L shell, and 1 in the M shell. (3–5, 7)
2. Choice *f*. Neon is inert with a completely filled outer (L) electron shell. Recall that the L shell can accommodate 8 electrons organized into four orbitals with 2 electrons per orbital, so all four orbitals are completely filled. (3–10)
3. Choice *c*. Since carbon is able to form a double bond with each Z atom, Z must need two electrons to complete its outer shell. An atomic number of 8 would result in 2 electrons in the K shell and 6 in the L shell. (3–15)
4. Choice *e*. The inert elements have their outermost shells completely filled and are least reactive. These include helium (${}_{2}He$), neon (${}_{8}Ne$), and argon (${}_{18}A$). All other choices are elements with only partially filled outer shells. (3–15)
5. Choice *d*. Since chemical reactivity is determined by the number of electrons in the outermost shell, you must determine which elements have the same number of electrons in the outermost shell; chlorine and fluorine each have seven electrons in their outermost shell (3–15).

- (2–7)

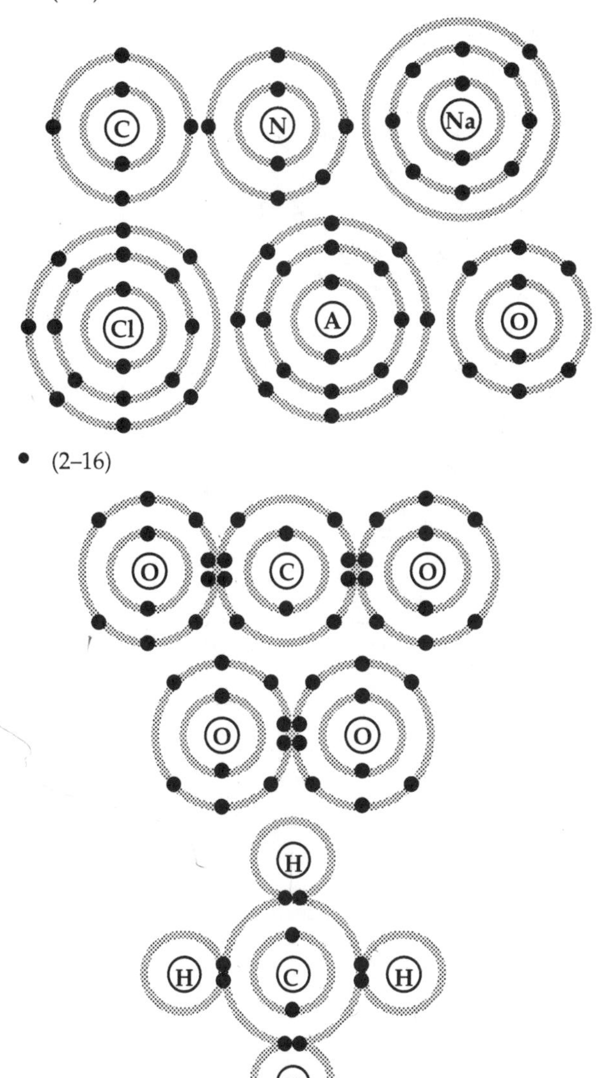

- (2–16)

MOLECULES • CHEMICAL REACTIONS • WATER • ACIDS, BASES, pH, AND BUFFERS (pages 8–9)

1. Choice *e*. One mole in one liter is a 1.0-molar solution. Adding one mole to 0.5 liter produces a 2.0-molar solution. (3–5)

2. Choice *c*. A 1.0-molar solution contains one gram molecular weight of a substance. The gram molecular weight of water is 18 grams. A liter contains 1,000g/18g/mole = 56 mole. (3–5)
3. Choice *a*. One kilocalorie is the heat needed to raise 1.0 kilogram(1,000 g) from 14.5°C to 15.5°C, so 0.25 kilocalorie would be needed to increase 250 g by the same amount. (8)
4. Choice *b*. Since CH_3COO^- combines with H^+ to form CH_2COOH, it is a base. (16–19)
5. Choice *e*. The pH scale is logarithmic, so each pH unit is ten times greater or lesser than the next nearest unit. Therefore, the $[H^+]$ of Cola is $10 \times 10 \times 10 \times 10$ (10^4) greater than blood plasma! (20, 21)
6. Choice *c*. Recall that "acid" and "base" refer to the compounds or ions that are donating the H^+ and OH^-, while "acidic" and "basic" refer to solutions. (16, 20, 21)

- (17–19)

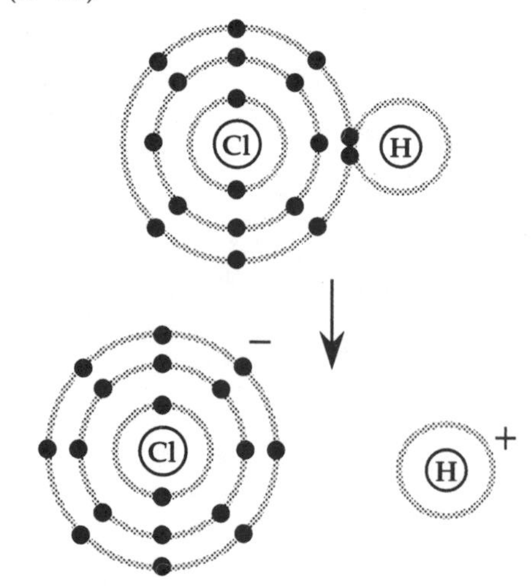

- (15–21)

pH	$[H^+]$	$[OH^-]$
0	1.0	10^{-14}
1	0.1	10^{-13}
2	0.01	10^{-12}
3	0.001	10^{-11}
4	10^{-4}	10^{-10}
5	10^{-5}	10^{-9}
6	10^{-6}	10^{-8}
7	10^{-7}	10^{-7}
8	10^{-8}	10^{-6}
9	10^{-9}	10^{-5}
10	10^{-10}	10^{-4}
11	10^{-11}	0.001
12	10^{-12}	0.01
13	10^{-13}	0.1
14	10^{-14}	1.0

POLARITY • SOME SIMPLE ORGANIC COMPOUNDS AND FUNCTIONAL GROUPS (pages 9–11)

1. Choice *b*. Other molecules shown have hydrogens bound to atoms with greater electron affinity or have a structure in which the distribution of positive and negative poles is not balanced. (1–3)
2. Choice *d*. Hydrogen bonding occurs between hydrogen and oxygen and the two hydrogens are oriented at an angle of 104.5° relative to oxygen. (3–6)
3. Choice *e*. The polarity of the water molecule allows it to bind to ions and polar molecules, preventing solute from clumping and coming out of solution. The properties listed in choices *b*, *c*, and *d* also result from the polar nature of the water molecule. (3–8)
4. Choice *c*. Optical isomers can occur in both L- and D-forms, although only the L-form of amino acids are found in biological systems. (23–25)
5. Choice *b*. Because water is polar, it cannot readily mix with or dissolve nonpolar substances. Only molecule *b* is nonpolar. (4, 12–17)
6. Choice *e*. Van der Waals and hydrophobic bonds will only form between nonpolar molecules like the hydrocarbon shown in *e*. (9–17)
7. Amino: ***e*** (14–17)
   Phosphate: ***f***
   Hydroxyl: ***a***
   Carboxyl: ***d***
   Carbonyl: ***b***
   Sulfhydryl: ***c***

- (12–18)

unsaturated hydrocarbon	$H_2C{=}CH_2$
saturated hydrocarbon	$H_3C{-}CH_3$
amino acid	$H_2N{-}C(COOH)(CH_2SH){-}H$ (COOH above C, CH_2SH below C)
organic base	$H_2N{-}C({=}O){-}NH_2$
organic acid	$H_3C{-}C({=}O){-}CH_2{-}COOH$
alcohol	$H_3C{-}CH_2{-}OH$

- (15–18)

carboxyl — COOH
α-carbon — C
amino — H_2N
side chain — CH_2OH

$H_2N{-}C{-}H$, with COOH above C and CH_2OH below C

Integrative Questions

1. Choice *b*. Van der Waals and hydrophobic bonds are equal in strength and are the weakest of the bonds. Covalent bonds are strongest, while ionic and hydrogen bonds are about equal and intermediate in strength to the others.

2. Choice *a*. Element X, with an atomic number of 20, would normally have two electrons in its outer shell (2–8–8–2). When in the ionic form X^{2+}, it has lost two electrons and has a full outer shell with 8 electrons.
3. Choice *a*. In a neutral atom, the number of protons and electrons is the same and the atomic number is that number. Ions have gained or lost electrons, so the atomic number equals the number of protons.

CHAPTER 3 – LARGE MOLECULES

Note : The page numbers listed with the section titles refer to the Study Guide; numbers in parentheses for each question refer to the relevant key concepts.

LIPIDS • FROM MONOMERS TO POLYMERS (pages 12–13)

1. Choice *b*. (9)
2. Choice *b*. Steroids, especially cholesterol, and phospholipids are important lipid components of biological membranes. (7, 8, 10)
3. (4–5)

$$CH_3(CH_2)_3\text{—}\overset{O}{\overset{\|}{C}}\text{—}O\text{—}CH_2$$
$$CH_3\text{-}HC\text{=}CH\text{-}CH_2\text{—}\overset{O}{\overset{\|}{C}}\text{—}O\text{—}CH_2$$
$$CH_3\text{-}HC\text{=}CH\text{-}CH_2\text{—}\overset{O}{\overset{\|}{C}}\text{—}O\text{—}CH_2$$

4. Triglyceride *A* is probably solid at room temperature; its fatty acid chains are saturated (no double bonds) and relatively long, both characteristic of solid, animal-derived triglycerides. (4–6)
5. Triglyceride ***B*** is probably derived from a plant; its fatty acid chains are unsaturated (double bonds) and relatively short, both characteristic of liquid, plant-derived triglycerides. (4–6)
6. ***Three*** water molecules will result. A water molecule results for each of the three fatty acids added to glycerol by a condensation reaction. (5, 14, 15)
7. Choice *e*. The interlocking, four-ring structure is characteristic of steroids. (4–10)

- (9, 10)

hydrophobic

$$CH_3\text{–}CH_2\text{–}(CH_2)_{14}\text{–}CH_2\text{–}\overset{O}{\overset{\|}{C}}\text{–}O\text{–}CH_2$$
$$CH_3\text{–}CH_2\text{–}(CH_2)_6\text{–}CH_2\text{–}\overset{O}{\overset{\|}{C}}\text{–}O\text{–}CH \longrightarrow$$
$$H_2\text{–}C\text{–}O\text{–}\overset{O^-}{\overset{\|}{P}}\text{–}O^-$$

hydrophilic

- (4–6)

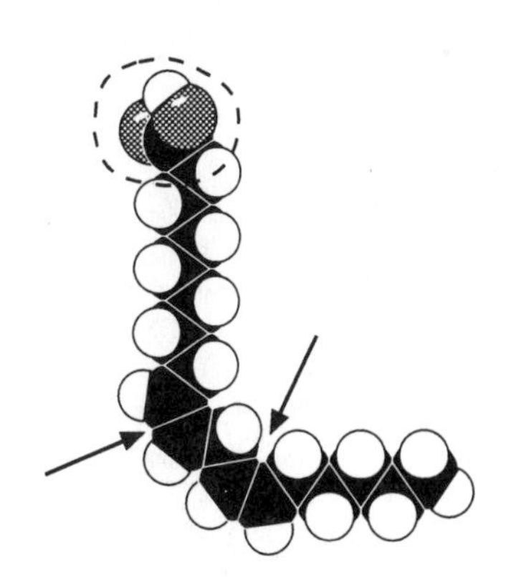

18 carbons
unsaturated (actually polyunsaturated with two double bonds as indicated above by arrows)

CARBOHYDRATES (pages 14–15)

1. Disaccharide, hexose, α, $C_6H_{12}O_6$, isomers (2–4, 7)
2. Choice *c*. Starch is joined by α linkages and cellulose is joined by β linkages, so choice *b* is false. In addition, starch can also form C-6 to C-1 branches between its glucose monomers. (9, 10)
3. Choice *d*. Derivative carbohydrates contain other elements than carbon, hydrogen, and oxygen; glycogen is a polymer of glucose. (11, 14)
4. Choices *a, c, e*. (5)

PROTEINS • LEVELS OF PROTEIN STRUCTURE (pages 15–16)

- (2, 7, 8)

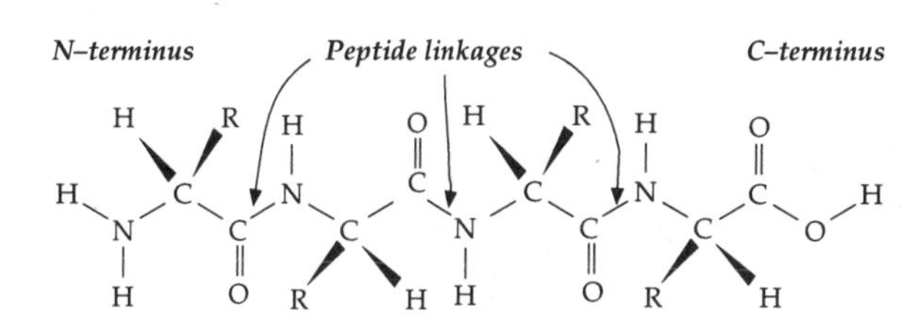

3 water molecules—one for each peptide bond formed.

1. Choice *e*. Because the sequence of amino acids in a protein determines the arrangement of R groups and bonding potentials between different regions of the protein, the primary structure determines all higher organizational levels. (10–17)
2. Choice *c*. The left R group is nonpolar; the right R group is polar. Nonpolar groups are hydrophobic; polar groups are hydrophilic. (2–8)
3. $m^n = 5^4 =$ ***625*** different polypeptides, where *n* is the length of the chain and *m* is the number of different kinds of subunits. (10)
4. Choice *e*. All three secondary types result from hydrogen bonding between or within polypeptide chains. (12–15)
5. Choice *c*. Although some enzymes consist of several globular polypeptide subunits joined together in a quaternary structure, the highest structural level seen in all enzymes is tertiary. (16, 17)

NUCLEIC ACIDS • GLYCOLIPIDS, GLYCOPROTEINS, LIPOPROTEINS, AND NUCLEOPROTEINS (pages 16–17)

1. Choice *d*. Nucleotides are the subunits of which DNA and RNA are composed, just as proteins are polymers of the subunits called amino acids. (1–5)
2. Choice *e*. Just as R groups give each amino acid its unique chemical features, the nitrogenous bases are unique for each nucleotide. (1–5)
3. Choice *b*. Purines and pyrimidines are found in both DNA and RNA. However, the pyrimidine uracil is restricted to RNA and the pyrimidine thymine is restricted to DNA. (7–12)

• (1–10)

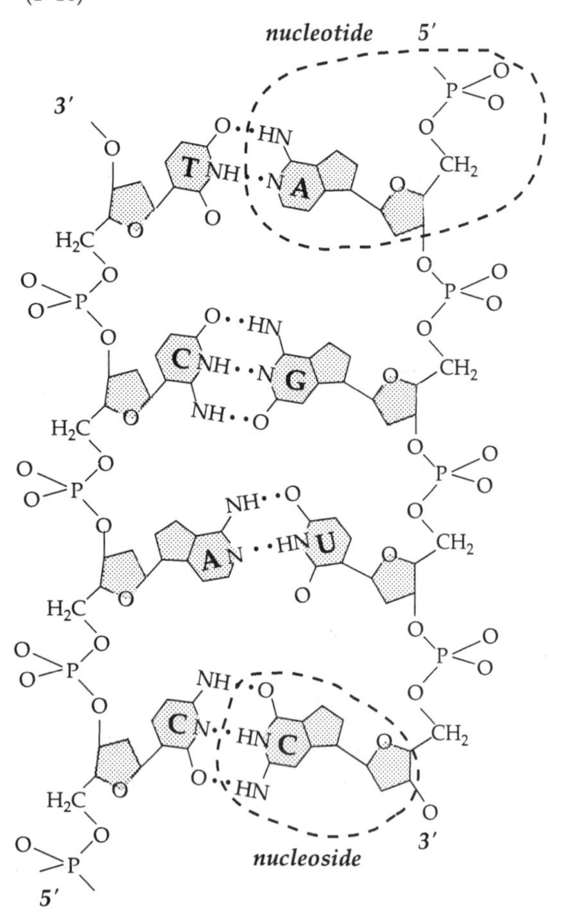

Integrative Questions

1. Choice *d*. Lipids are a class of large molecules that are not polymers of simple subunits. Cholesterol is a lipid.
2. Choice *a*. Since the interior of membranes consists of the hydrophobic hydrocarbon tails of phospholipids, the R groups of amino acids located on the surface of protein molecules found within membranes would also be expected to be hydrophobic.
3.
   2 RNA nucleotide sequence
   3 Primary structure of a protein
   4 Folding of a protein into a three-dimensional shape
   1 DNA nucleotide sequence
   5 An enzyme catalyzes a specific reaction

CHAPTER 4 – ORGANIZATION OF THE CELL

Note : The page numbers listed with the section titles refer to the Study Guide; numbers in parentheses for each question refer to the relevant key concepts.

CELLS AND THE CELL THEORY • COMMON CHARACTERISTICS OF CELLS • PROKARYOTIC CELLS • PROBING THE SUBCELLULAR WORLD: MICROSCOPY (pages 18–19)

1. Choice *c*. Recall that 1 nm = 10^{-9} m or 10^{-6} mm, so 10 nm = 10^{-5} mm. If our unaided resolving power is 0.1 mm or 10^{-1} mm, then the increase in resolving power with the scanning electron microscope is $10^{-5} - 10^{-1} = 10^{-4}$. (5, 16–17)
2. Nucleoid: ***d***
   Cytosol: ***a***
   Capsule: ***f***
   Flagella: ***b***
   Pili: ***e***
   Mesosome: ***c***
   (6–14)
3. Choice *d*. (15–18)

• (3, 6–14)

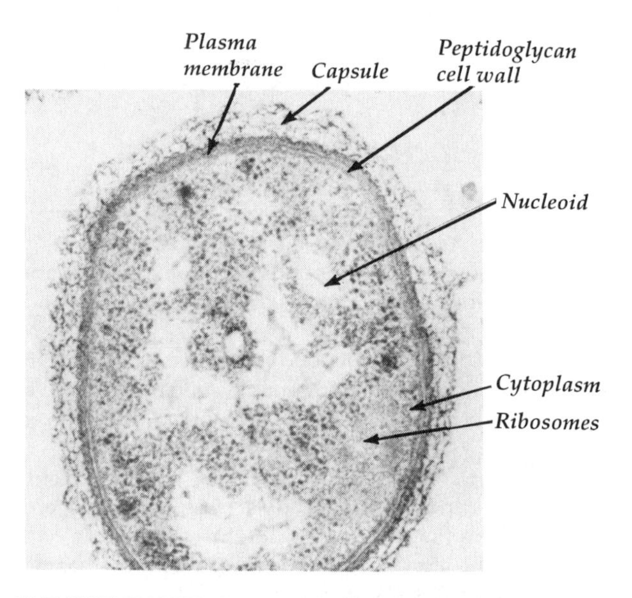

THE EUKARYOTIC CELL • INFORMATION-PROCESSING ORGANELLES • ENERGY-PROCESSING ORGANELLES (pages 19–20)

1. Mitochondrion (15–19)

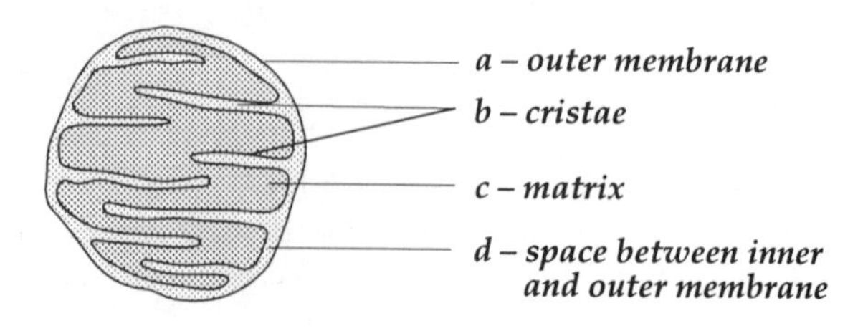

2. Item *a*. The outer membrane of the organelle would have been derived from the plasma membrane of the host. (20, 21)

3. Stroma: *c*
   Inner membrane: *a*
   Granum: *d*
   Thylakoid: *b*
   (15–17)
4. Choice *c*. Prokaryotes lack true nuclei and double membrane organelles, but have ribosomes. (1-8, 10-13)
5. Choice *c*. So, in choice *e*, it was the species donating the base where the nucleus is located that determined the type of cap that would form. (9)

THE ENDOMEMBRANE SYSTEM • OTHER ORGANELLES (pages 20–22)

- (8–16)

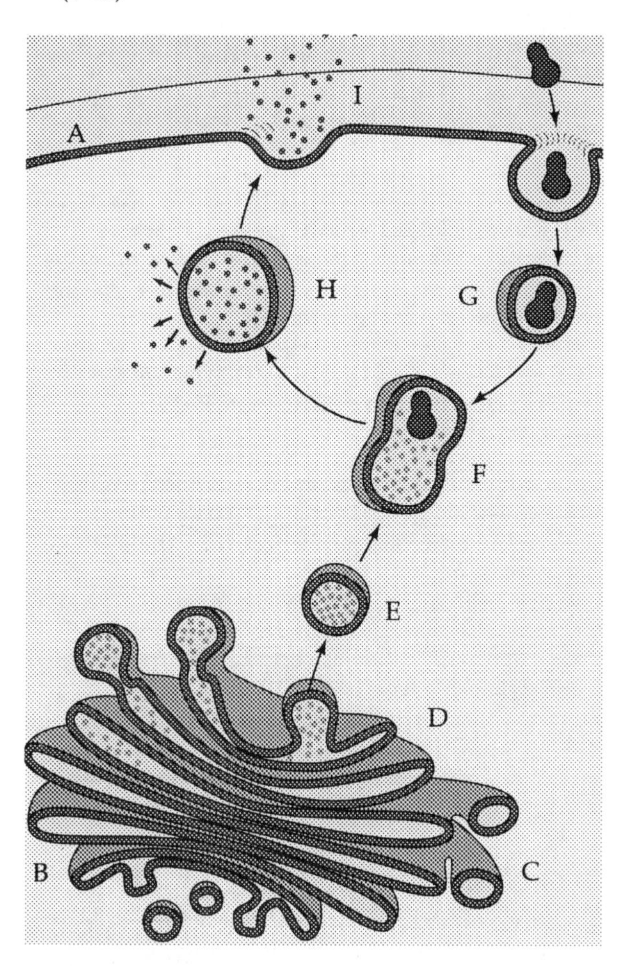

1. Glyoxysome: *c*
   Peroxisome: *d*
   Food vacuole: *b*
   Free ribosome: *a*
   Golgi apparatus: **e**
   (1–11, 17–19)
2. Choice *e*. Exported proteins are formed by ribosomes attached to ER, are moved to the *cis* region of the Golgi apparatus by a vesicle, depart from the *trans* region of the Golgi apparatus in another vesicle, and from there are transported to the outside of the cell. (1–14)
3. Choice *b*. Isolated ribosomes manufacture proteins for local use within the cell. (2–6)
4. Choice *a*. Endocytosis forms the food vacuole, fusion of the primary lysosome with the food vacuole produces the secondary lysosome, digestion of food occurs, and waste is expelled from the cell by exocytosis. (12–16)
5. Choice *d*. Attachment of signal sequence amino acids to newly synthesized proteins prior to their entry into rough ER does not involve vacuoles. (5, 18, 19)

THE CYTOSKELETON • THE OUTER "SKELETON" — THE CELL WALL • EUKARYOTES, PROKARYOTES, AND VIRUSES • FRACTIONATING THE EUKARYOTIC CELL: ISOLATING ORGANELLES (pages 22–23)

1. Choice *c*. Differential refers to the variation in the rotor speed and, thus, centrifugal force. (17–19)
2. Choice *e*. Prokaryotic flagella do not have the 9+2 arrangement of microtubules. (4–12)
3. Choice *c*. Basal bodies lack central microtubules and have 9 sets of three peripheral microtubules. (4–12)
4. (1–6)
   Microtubules: *c, f*
   Microfilaments: *a, d*
   Intermediate filaments: *b, e*

Integrative Activity

- Please see page 71 in the textbook for the correct labeling of this photomicrograph.

CHAPTER 5 – MEMBRANES

Note : The page numbers listed with the section titles refer to the Study Guide; numbers in parentheses for each question refer to the relevant key concepts.

MEMBRANE STRUCTURE AND COMPOSITION • MICROSCOPIC VIEWS OF BIOLOGICAL MEMBRANES • WHERE ANIMAL CELLS MEET (pages 24–25)

1. Lipid bilayer: *b* (1, 2, 5)
   Membrane proteins: *a, c, d, e* (3, 8, 9)
   Glycolipids: *a* (4, 13)
   Glycoproteins: *e* (4, 14, 15)
2. Choice *b*. Membrane components, especially proteins, glycolipids, and glycoproteins, are distributed asymmetrically. (1, 3, 4–6, 8, 9)
3. Choice *d*. This type of tertiary structure allows intrinsic proteins to fit into the lipid bilayer, with its hydrophobic interior and hydrophilic exterior. (8, 11, 12)
4. Choice *c*. Since most membrane carbohydrates are involved in interacting with the environment, most are attached to the outside of the membrane only. (13–16)
5. Desmosome: *a, c* (20, 22)
   Tight junction: *a, b, f* (20, 21)
   Gap junction: *a, d, e* (20, 23)

DIFFUSION • CROSSING THE MEMBRANE BARRIER (pages 25–27)

1. Choice *e*. Usually only small, nonpolar molecules can move through the lipid bilayer of the membrane by simple diffusion. Because a molecule of oxygen has a balanced electron distribution, it is nonpolar. (1–3)
2. Choice *d*. The rate of simple diffusion of a solute always increases directly with solute concentration. (1–3, 6)

3. Curve *c*. In both facilitated diffusion and active transport, the rate of movement will eventually saturate when all of the transport proteins are fully engaged. (4, 6)
4. Choice *c*. The sodium-potassium pump (see page 27) depends on ATP so it is primary active transport and since Na^+ and K^+ move in opposite directions it is an antiport. Since Na^+ is passively diffusing down a concentration gradient established by active transport and glucose is accompanying it (left), the process is secondary active transport involving a symport. (11, 15–17)
5. Choices *c, d*. Because the cell absorbs water, we know that its osmotic potential was more negative than the osmotic potential of the environment. (18–20)
6. Choice *a*. Because the cell has absorbed some water and developed an opposing pressure potential, its osmotic potential will still be more negative than the environment. (18, 20–22)

MORE ACTIVITIES OF MEMBRANES • MEMBRANE INTEGRITY UNDER STRESS • MEMBRANE FORMATION AND CONTINUITY (pages 27–28)

- Carbohydrates are added to newly synthesized membrane proteins in the rough ER and the Golgi apparatus. (9–13)

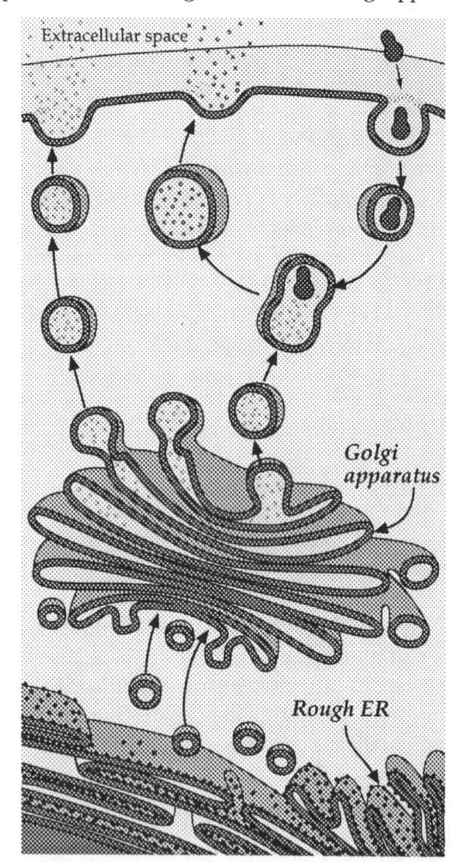

1. Choice *b*. A ligand is an extracellular molecule that binds to a membrane receptor protein. (3, 5, 6)
2. Clathrin: *c* (4)
   Spectrin: *b* (8)
   Acetylcholine: *d* (3, 5)
   Ankyrin: *b* (8)
3. Choice *d*. The typical flow of membrane within the cell is from nuclear envelope to rough ER, from rough ER to Golgi apparatus and from Golgi apparatus to plasma membrane. Membrane movement between nuclear envelope, Golgi apparatus, and ER is via vesicles. (11–13)

Integrative Questions/Activities

1. Cell membranes are involved in all these activities!
2. Glycolipids and glycoproteins are on the external surface of the membrane.

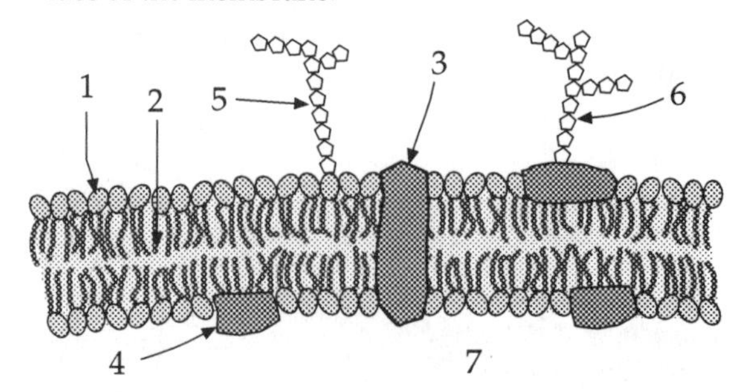

CHAPTER 6 – ENERGY, ENZYMES, AND CATALYSIS
Note : The page numbers listed with the section titles refer to the Study Guide; numbers in parentheses for each question refer to the relevant key concepts.

ENERGY AND THE LAWS OF THERMODYNAMICS • CHEMICAL EQUILIBRIUM • REACTION RATE (pages 29–30)

1. Choice *b*. The rate of a reaction is independent of whether it is spontaneous or not. (6–11)
2. Choice *b*. The entropy (disorder) of the system is expressed by *S*. (13)
3. Choice *b*. With $K_{eq} > 1$, the reaction is exergonic and ΔG is negative. (6–13)
4. Choice *b*. b is the difference between the free energy of reactants and products or ΔG. (13–15)
5. Choice *d*. If ΔG is positive, the reaction is endergonic and $K_{eq} < 1$. (7–10)
6. Choice *d*. Transition-state species are part of both forward and back, exergonic and endergonic reactions. (10, 14–16)

THE HIGHLY SPECIFIC CATALYSIS OF ENZYMES • MOLECULAR STRUCTURE OF ENZYMES (pages 30–32)

- (1–4, 9)

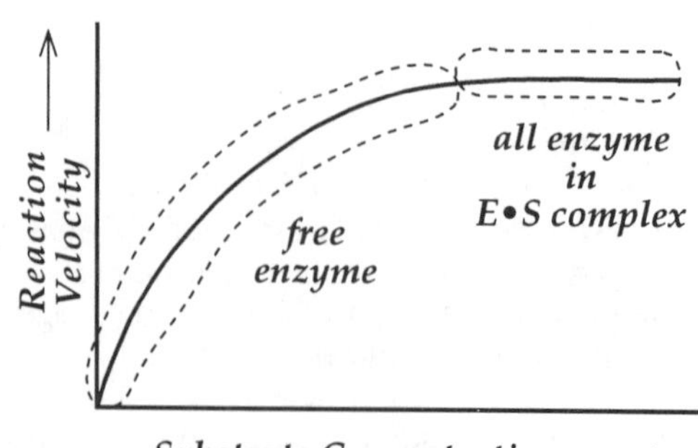

• (1–4, 8)

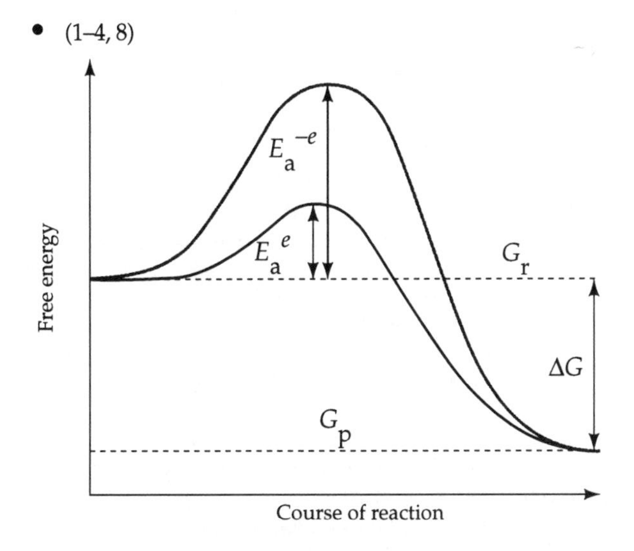

1. Choice *b*. Substrate binding to an enzyme can involve all types of chemical bonds. (2–7)
2. Choice *c*. Only the activation energy is lower. Also, in *d*, recall that a catalyst can only accelerate a reaction that would occur in the absence of the catalyst. (1–8)
3. Choice *d*. Energy released by the exergonic reaction allows the endergonic reaction to proceed. (10)
4. Choice *a*. Prosthetic groups have their main effect by interacting chemically with substrate. (14–17)

REGULATION OF ENZYME ACTIVITY • ENZYMES, RIBOZYMES, AND ABZYMES (pages 32–33)

1. Choice *d*. This is a graph typical of noncompetitive inhibition, with curve B resulting from reactions with the inhibitor. A noncompetitive inhibitor does not normally resemble the substrate. (1–4)
2. Choice *d*. At high inhibitor, but low substrate concentrations, most of the allosteric sites are occupied and, consequently, most of the active sites are empty. (5–12)
3. Choice *b*. In an acidic environment, carboxyl groups would not ionize, but amino groups would ionize. (15–16)
4. Choice *b*. That step closes down only the branch leading to G. (13, 14)
5. Choice *a*. That step controls the production of both G and E. (13, 14)
6. Choice *b*. Evidence now indicates that catalytic RNAs (ribozymes) were the first catalytic macromolecule. (18, 19)

• (5–14)

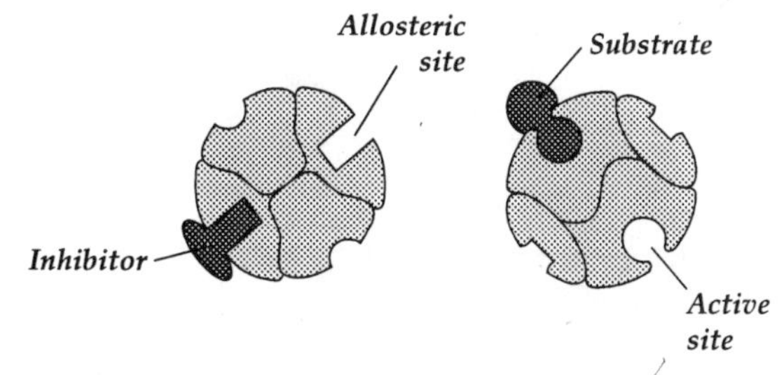

a. The end product of that branch of the pathway.
b. Form *a*
c. No. Concerted feedback inhibition would require an enzyme with at least two different allosteric sites; one for each of the two or more inhibitors.

CHAPTER 7 – PATHWAYS THAT RELEASE ENERGY IN CELLS

Note : The page numbers listed with the section titles refer to the Study Guide; numbers in parentheses for each question refer to the relevant key concepts.

GLYCOLYSIS, CELLULAR RESPIRATION, AND FERMENTATION • THE TRANSFER OF HYDROGEN ATOMS AND ELECTRONS • HOW DO CELLS PRODUCE ATP? • THE RELEASE OF ENERGY FROM GLUCOSE (pages 34–36)

• (6)

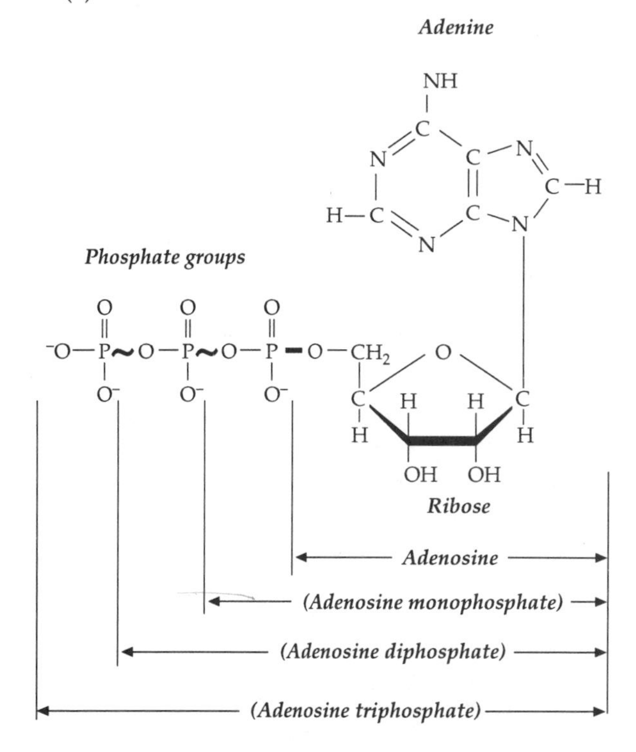

1. Choice *b*. One water molecule is used to break each bond. (4–7)
2. Choice *b*. NAD^+ is an *oxidizing* agent. See Figure 7.6 in the textbook. (9–14)
3. Choice *c*. Carbons move from pyruvate into the citric acid cycle and are eventually released as CO_2. (20–23)
4. Choice *b*. Hydrogens are moved from glycolysis intermediates, like glucose 6-phosphate, by NADH to the respiratory chain and eventually incorporated into water. (20–24)

GLYCOLYSIS • THE BEGINNING OF CELLULAR RESPIRATION: THE CITRIC ACID CYCLE • CONTINUATION OF CELLULAR RESPIRATION: THE RESPIRATORY CHAIN (pages 36–37)

1. ***6, 7, 9, 10***. (1–4)
2. ***6***. (1–4)
3. ***1, 3, 6, 10***. In reactions 1 and 3, the ADP formed can be used elsewhere. (1–4)

4. *4*. (1–4)
5. Choice *a*. Coenzyme A is a permanent part of the citric acid cycle. (5–14)
6. Choice *c*. Acetyl coenzyme A is the compound derived from pyruvate that enters the cycle. (5–14)
7. Choice *d*. Both reactions are *exergonic*. (5–14)
8. Choice *a*. NADH + H^+ passes hydrogens to NADH-Q reductase and cytochrome oxidase passes electrons to molecular oxygen. (15–21)
9. Choice *b*. All carrier-based movement of protons is from the outside to the inside of the membrane. (15–21)

- (1, 5, 14, 15)

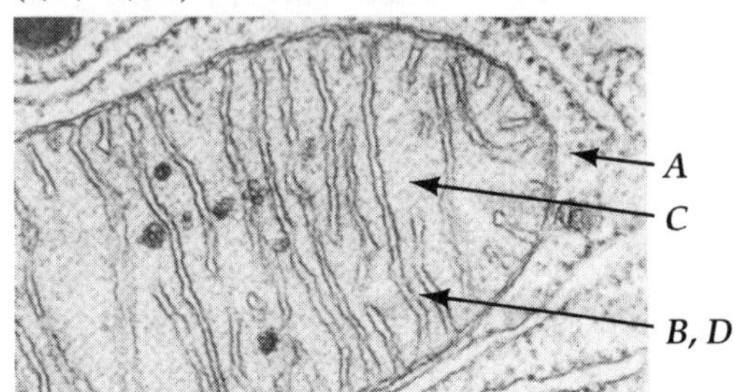

- The pH in area *2* is greater than area *1* and ATP is formed in area *1*, but not in area *2*. Hydrogen ions are moved from the inside to the outside of the membrane, so the pH is lower on the outside of the membrane; ATP is formed on the inside of the membrane as H^+ diffuse back. (18–20)

FERMENTATION • COMPARATIVE ENERGY YIELDS • CONNECTIONS WITH OTHER PATHWAYS • FEEDBACK REGULATION (page 37–39)

1. Choice *d*. The remaining four ATP are produced via substrate-level phosphorylation (a net of two during glycolysis and two from the citric acid cycle) and not via the chemiosmotic mechanism. (3–6)
2. Choice *a*. In certain organisms, lactate results from fermentation. (1, 10)
3. Choice *c*. NADH + H^+ produced in glycolysis results in only two ATP, as does $FADH_2$.
4. Choice *b*. All other choices are valid allosteric control mechanisms but do not regulate the rate of glycolysis directly, which is what the Pasteur effect describes. (12, 13)
5. See Figure 7.23 on page 159 in the textbook. Because of the greater efficiency of aerobic respiration, less glucose needs to be processed, so the rate of glycolysis in the cell should decrease. Allosteric inhibition of the enzymes catalyzing the following reactions would accomplish this: glucose + ATP → glucose 6-phosphate + ADP by glucose 6-phosphate, fructose 6-phosphate + ATP → fructose 1,6-diphosphate + ADP by ATP and citrate. (14–19)

CHAPTER 8 – PHOTOSYNTHESIS

Note : The page numbers listed with the section titles refer to the Study Guide; numbers in parentheses for each question refer to the relevant key concepts.

SUNLIGHT AND LIFE ON EARTH • EARLY STUDIES OF PHOTOSYNTHESIS • THE PATHWAYS OF PHOTOSYNTHESIS • LIGHT AND PIGMENTS (pages 40–42)

1. Choice *c*. The color of a pigment is indicated by the portion of the absorption spectrum where little light is absorbed. For the absorption spectrum shown, minimum absorbance occurs in the yellow portion of the spectrum. (8, 9, 14–16)
2. Choice *b*. CO_2 is a reactant in the Calvin-Benson cycle, not a product. All other statements are true. (4, 6, 7)
3. Choice *d*. Frequency and wavelength are inversely related and light of longer wavelength has less energy than light of shorter wavelength. (8–11)
4. Choice *b*. Choice *a* is true for absorption spectra, but does not explain the rounded peaks in the action spectrum of photosynthesis. (15–18)

PHOTOPHOSPHORYLATION • THE CALVIN-BENSON CYCLE (pages 42–44)

1. Choice *e*. Electrons move from water to photosystem II to photosystem I and are carried to G3P in the Calvin-Benson cycle by NADP + H^+. (3–10, 25–27)
2. Choice *a* and *d*. Only $NADP^+$, NADPH + H^+, and ATP are cycled. Water and oxygen are reactants and products, respectively. (10, 25–27)
3. Choice *c*. Since the concentration of A increases when CO_2 is withheld, A must combine with CO_2. C is first to decrease, so A + CO_2 forms C first. B, which starts to decrease last, must be formed at a later step. (24)
4. Choice *e*. Recall that H^+ are pumped into the interior of the chloroplast thylakoids and that ATP formation occurs when H^+ diffuse back out of the thylakoids. In thylakoids of group A chloroplasts, H^+ will be diffusing from inside to outside (the pH is greater outside, so the $[H^+]$ is less) and ATP *will* form; in group B thylakoids, H^+ diffusion will be from outside to inside and *no* ATP will form. (14–17)

- Substance 1 moved the greatest distance with both solvents; thus, its total R_F is greatest. Substance 2 moved the greatest distance with solvent B, but little distance with solvent A. (20–23)
- (14, 28)

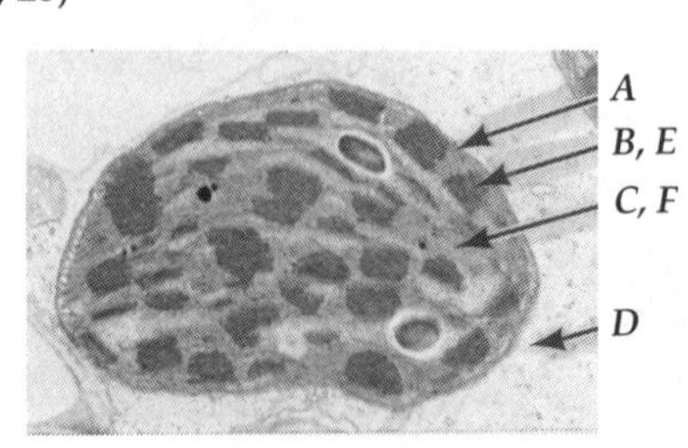

PHOTORESPIRATION • ALTERNATE MODES OF CARBON DIOXIDE FIXATION • PHOTOSYNTHESIS AND CELLULAR RESPIRATION (pages 44–45)

1. Choice *b*. All of the reactions listed occur in a C_4 plant except for the reactions of photorespiration. Because of their unique anatomy and the presence of the enzyme PEP carboxylase, C_4 plants do not photorespire. (1–9)
2. Choice *d*. Both C_3 and C_4 plants have the Calvin-Benson cycle; C_4 plants have an additional CO_2 fixation pathway in their mesophyll cells called the C_4 cycle and the Calvin-Benson cycle occurs only in their bundle sheath cells. (1–9)
3. Choice *c*. In some plants in the family Crassulaceae, PEP is produced from CO_2 only at night using PEP carboxylase. The CO_2 is then passed on to the Calvin-Benson cycle during the day. (10)
4. Choice *a*. Cellular respiration continues in the dark and is only limited by glucose and oxygen. (1–3, 12, 13)

Integrative Questions

1. Choice *c*. Since photosynthetic autotrophs have mitochondria and respire just like heterotrophs, plants use NAD as well as NADP.
2. Choice *e*. The electron carriers in the respiratory chain differ from those in photophosphorylation. All other statements are true.

Integrative Activity

- (See the figure below. Also, for more information see Figure 8.17 in your text book)

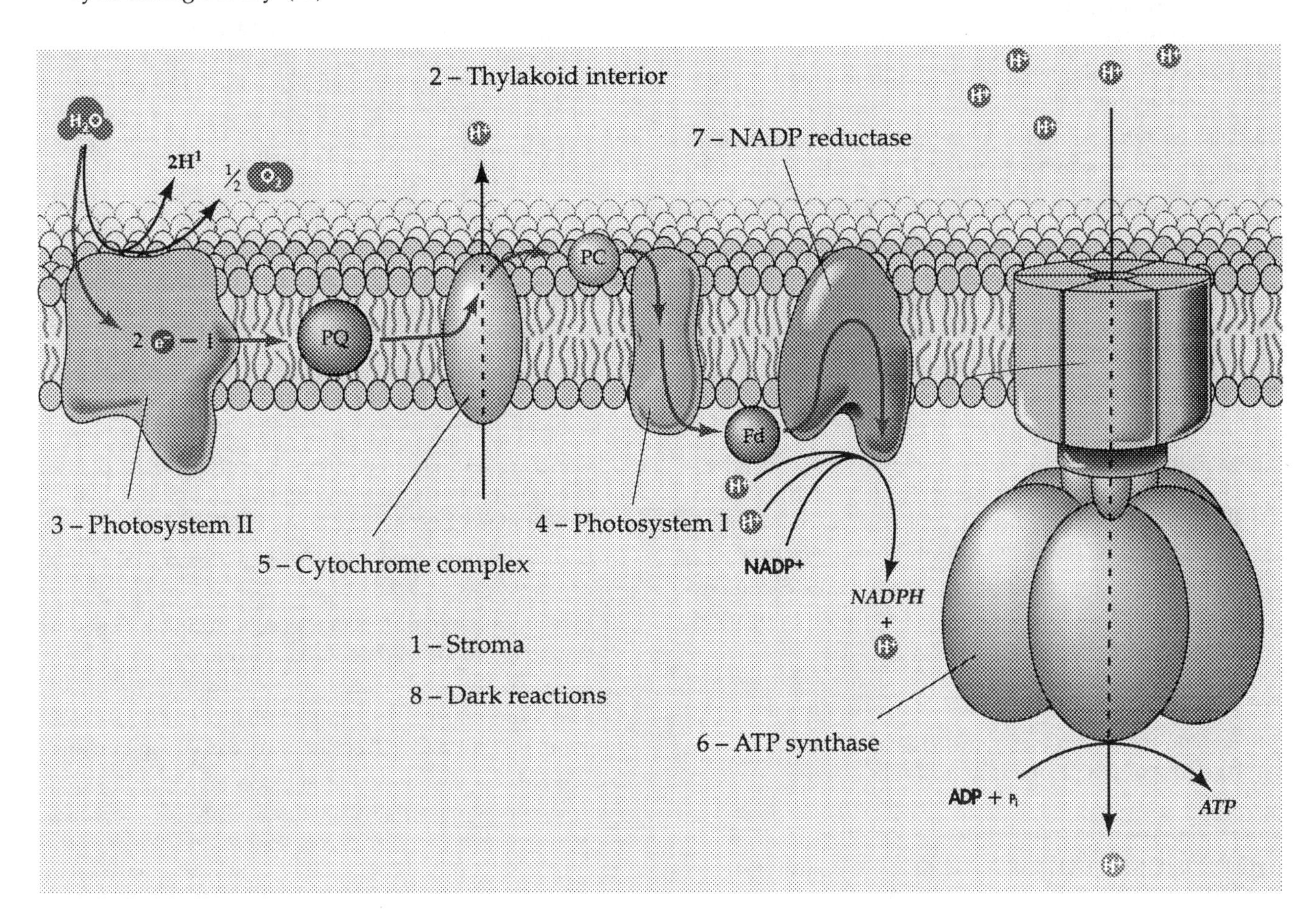

9

Chromosomes and Cell Division

CHAPTER LEARNING OBJECTIVES—after studying this chapter you should be able to:

- ❑ Differentiate between the two main steps involved in any type of cell division.
- ❑ State the fundamental genetic differences between mitosis and meiosis.
- ❑ Discuss why mitosis is useful for overall growth and repair, while meiosis is necessary for sexual reproduction.
- ❑ Describe the component parts of a chromosome and indicate when they are present during mitosis.
- ❑ Clearly differentiate between the terms "chromatin," "chromosome," "chromatid," and "centromere."
- ❑ Discuss the role of histone proteins in determining the shape of nucleosomes and chromosomes.
- ❑ Diagram the generalized life cycle of a cell and discuss the events that occur in the S, G1, G2, and M phases.
- ❑ Explain the difference between interphase and cell division, indicate the relative length of time cells spend in these phases, and describe the molecular mechanism that determines the transition from one phase to the next.
- ❑ Name the stages of mitosis and describe the conditions of chromosomes and other important cellular structures during each stage.
- ❑ Identify the roles of mitotic spindles, centrosomes, asters, kinetochores, and microtubules in nuclear division.
- ❑ Explain the purpose of cytokinesis and describe how it differs between plant and animal cells.
- ❑ Differentiate between asexual and sexual reproduction, and discuss the relative genetic/evolutionary advantages or disadvantages of each.
- ❑ Relate the terms "haploid" and "diploid" to the terms "gamete," "zygote," "egg," and "sperm," and discuss the role of fertilization in determining the genetic makeup of an organism.
- ❑ Explain what is meant by a "karyotype" and discuss how karyotypes are determined.
- ❑ Describe the importance of the reduction division in meiosis.
- ❑ Explain what is meant by a cell's ploidy level, and describe how ploidy level changes (if at all) during mitosis and meiosis.
- ❑ Identify the stages of meiosis and diagram the conditions of chromosomes and other important cellular structures during each stage.
- ❑ Compare and contrast meiosis with mitosis, placing special emphasis on why meiosis has two divisions while mitosis has only one.
- ❑ Explain when synapsis occurs in meiosis and discuss its genetic consequences.
- ❑ Explain how nondisjunction events during meiosis can lead to the condition known as aneuploidy.
- ❑ Relate the terms "trisomy" and "monosomy" to chromosomal abnormalities, identify mechanisms for how these occur, and cite a specific example of trisomy in humans.
- ❑ Describe the fundamental differences between prokaryotic fission and the cell division mechanisms of eukaryotes.

THE DIVISIONS OF EUKARYOTIC CELLS • EUKARYOTIC CHROMOSOMES AND CHROMATIN • THE CELL CYCLE (pages 191–196)

Key Concepts

1. Nucleic acids contain the hereditary information that controls cell activity. When cells divide, these nucleic acids must be properly allocated to new daughter cells.
2. *Mitosis* is a type of nuclear division that produces genetically identical copies of the parent nucleus in two daughter cells. The overall growth and repair of organisms is achieved through mitosis.
3. *Meiosis* is nuclear division where the amount of DNA in the nuclei of four daughter cells is precisely reduced by one-half. Meiosis promotes genetic variability between parents and offspring in sexually reproducing species.
4. Eukaryotic cell division is a two-step process involving the separation of nuclear material followed by division of the cytoplasm.
5. Most of a cell's life is spent in *interphase*, the period between successive cell divisions.
6. DNA is the primary nucleic acid that contains the genetic information of cells. It is a large, double-stranded, linear molecule that combines with certain proteins to form structures called *chromosomes*. In eukaryotes, chromosomes reside in the nucleus.

7. DNA molecules replicate prior to cell division. After replication, each chromosome consists of two *chromatids* joined together by a *centromere*, which functions in chromosome movement.

8. Chromosomes contain five kinds of *histone* proteins which, combined with DNA, make up *chromatin*. A portion of an individual DNA molecule coils around a complex of four histones. The histone–DNA complex is then locked together by the fifth histone (designated H1) to form a spool or *nucleosome*. Many repeating nucleosomes connected by a DNA thread yield a beaded chromatin chain.

9. Chromatin undergoes changes in form during cell division. While in interphase, chromatin exists as a long string spread throughout the nucleus. During mitosis and meiosis, the beaded chromatin becomes increasingly coiled and condensed into an individual chromosome. This condensed form facilitates chromosome movement.

10. The *cell cycle* consists of nuclear division (*M phase*) and interphase, with the latter comprising the vast majority of the cycle. Some cells never leave interphase, but most have at least some probability of dividing.

11. Interphase is divided into three subphases. *G1* is a gap period that follows immediately after mitosis; cells that will not divide again linger in this phase. *S* is a phase characterized by the synthesis of new DNA and histones in preparation for nuclear division. *G2* represents a second gap period prior to mitosis.

12. The transitions from G1 to S and from G2 to M are regulated by the activity of a protein complex called MPF (maturation-promoting factor). An MPF complex contains two types of proteins: cdc2 and a type of *cyclin*. In the presence of appropriate enzymes, the MPF complex is activated and initiates either the S or M phase, depending on which type of cyclin is involved with the MPF complex at the time.

13. Activated MPF functions as an enzyme to phosphorylate, and thereby activate, other enzymes. These enzymes then add phosphates to still other enzymes in a *cascade* of activation which multiplies the original effect of MPF. The final products of this cascade are ultimately responsible for the cellular effects that occur during the S and M phases.

Questions (for answers and explanations, see page 114)

1. If a cell has 44 chromosomes at the end of mitosis, how many will it have at the end of the next S phase?
   a. 4
   b. 11
   c. 22
   d. 44
   e. 88

2. Suppose you examine under the microscope a thin section of tissue removed from the actively growing portion of a plant shoot. Which of the following will be true of the tissue section?
   a. Most of the cells will be in one of the three stages of interphase.
   b. All the cells will be undergoing a reduction in the amount of DNA present.
   c. The chromatin in most of the cells will already have condensed into clearly recognizable chromosomes.
   d. No histones or DNA will be present in the interphase cells.
   e. A large number of cells will have stopped dividing and will have become fixed in G2 of the cell cycle.

3. Most new histone molecules are manufactured during the __________ phase of the cell cycle.
   a. G1
   b. S
   c. G2
   d. mitosis
   e. G1 and G2

4. Which of the following statements about chromosomes is *not* true?
   a. Chromosomes observed just prior to nuclear division in mitosis will have two chromatids each.
   b. Chromosomes are made of DNA and several types of histone proteins.
   c. The shape of chromosomes and chromatin stays remarkably constant during the cell cycle.
   d. New cells produced by meiosis have fewer chromosomes than the parent cell.
   e. The centromere connects together chromatids within a chromosome.

5. During which of the following phases of the cell cycle is chromatin present in cells?
   a. G1
   b. S
   c. G2
   d. Mitosis
   e. All of the above

6. Which of the following statements about the structure or function of maturation-promoting factor (MPF) in cells is *not* true?
   a. MPF is a complex of two proteins: cdc2 and H1 histone.
   b. Cyclin proteins determine whether the cell enters the S or M phase of the cell cycle.
   c. Active forms of MPF catalyze the addition of phosphates to other enzymes.
   d. The repeated phosphorylation of many different enzymes initiated by active MPF is called a cascade.
   e. Active MPF causes the activation of cellular enzymes that degrade cyclin, which in turn leads to the inactivation of MPF.

MITOSIS (pages 196–202)

Key Concepts

1. Mitotic division yields two new nuclei that are genetically identical to the parent nucleus. Although mitosis is a continuous process, it can be divided into several phases.

2. *Interphase* is marked by the condensation of chromatin and the appearance of a pair of *centrosomes* near the nuclear envelope. In many organisms, each centrosome also contains a pair of centrioles oriented at right angles to one another. One centriole (the daughter) is smaller than the other (the parent).
3. During *prophase*, centrosomes move to opposite ends of the cell where they act as *mitotic centers*, organizing the microtubules necessary for chromosome movement.
4. Details of microtubule arrangement differ with cell type. All cells have *polar microtubules* running from one mitotic center to the midline of the cell; these form the two halves of the *spindle*. Some cells also have microtubules that radiate away from the spindle in patterns called *asters*.
5. Chromatin structure begins to change during prophase. Individual chromosomes condense, revealing that each chromosome consists of paired chromatids.
6. Late in prophase *kinetochores* develop near each centromere. They produce *kinetochore microtubules*, which interact with polar microtubules in the spindle.
7. *Prometaphase* occurs when chromosomes actually begin moving. Kinetochore microtubules make contact with microtubules from each pole of the spindle. The tension of these opposing forces moves the chromosomes toward the cell midplane, or *equatorial plate*. Disintegration of the nuclear envelope also occurs at this stage due to the action of one kind of MPF.
8. *Metaphase* is a brief phase that commences when all kinetochores are aligned along the equatorial plate. Chromatin condensation continues throughout metaphase, and the centromeres divide.
9. *Anaphase* is marked by the separation of chromatids. The action of polar microtubules somehow pulls the chromatids of each chromosome toward opposite ends of the spindle, creating two identical sets of *daughter chromosomes*. In a manner probably similar to the dynein-mediated mechanism of movement in eukaryotic cilia and flagella, the spindle poles also move apart during anaphase, further separating the two sets of daughter chromosomes. Surprisingly little ATP energy is required for these chromosome movements.
10. The new daughter chromosomes each contain one double-stranded molecule of DNA. Although each still has a centromere, the chromatids characteristic of prophase chromosomes are no longer present.
11. *Telophase* begins when the movement of daughter chromosomes stops as they reach the spindle poles. The chromatin starts to uncoil and assumes the dispersed state that characterizes interphase, marking the end of mitosis.
12. *Cytokinesis* "packages" the new, genetically identical nuclei produced by mitosis into unique cells. In animals, actin and myosin microfilaments contract, causing the center portion of the cell membrane to divide the cytoplasm.
13. In plants, the presence of a cell wall prevents direct constriction of the plasma membrane. Instead, membrane vesicles group along the equatorial plate and, with the help of microtubules, form a cell plate that provides the foundation for a new cell wall.
14. Although genetically identical, the new cells may not necessarily share cytoplasmic structures equally. However, a sufficient quantity of every organelle is passed on to each new cell so as to ensure proper metabolic function.
15. Once thought to actually help organize the mitotic centers, centrioles (where present) are now believed to be inherited simply to ensure that each daughter cell will possess its own copies of them.

Activities ***(for answers and explanations, see page 114)***

- Consider the ovals below to represent animal cells from a species with *four* chromosomes. Diagram what these cells and their chromosomes would look like during *both* metaphase and anaphase of mitosis.

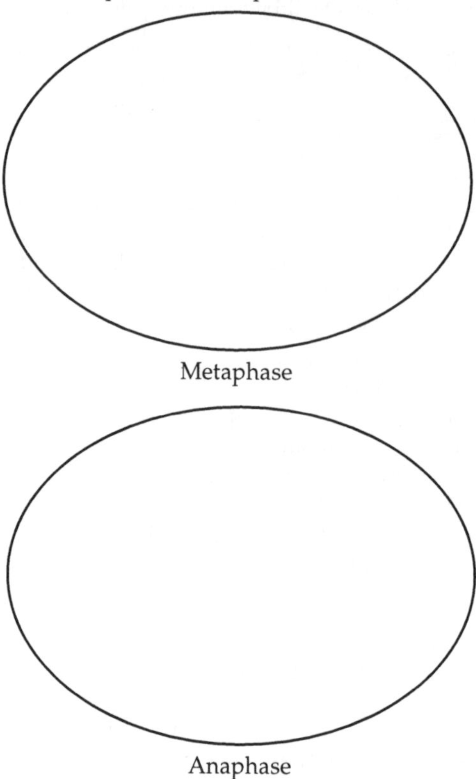

Questions ***(for answers and explanations, see page 114)***

1. Polar microtubules are an important part of the spindle apparatus used in cell division. During which of the following phases of mitosis would you *not* expect to find well-formed polar microtubules?
   a. Interphase
   b. Prophase
   c. Prometaphase
   d. Metaphase
   e. Anaphase
2. During which of the following phases of mitosis do chromosomes contain distinct chromatids?
   a. G2 and prophase
   b. Prophase and prometaphase

c. Prophase, prometaphase, and metaphase
d. Metaphase and anaphase
e. Anaphase and telophase

3. Match the letter of the appropriate phase(s) of mitosis with all relevant characteristics of cell division.

a. Interphase	C	Chromosomes align on equatorial plate
b. Prophase	e	Chromatin redisperses after cell division
c. Metaphase	b	Centrosomes migrate to opposite ends of cell
d. Anaphase	a-e	Double-stranded DNA present in cell
e. Telophase	b-e	Centromeres present on chromosomes
	a	Cytokinesis occurs

4. All but one of the following statements about mitosis is true. Select the *exception*.
   a. Cytokinesis in plants involves the production of a cell plate.
   b. Chromatids separate during anaphase.
   c. Anaphase represents the first time when chromosomes move during cell division.
   d. The completion of telophase signals the end of mitosis.
   e. Kinetochore microtubules interact with polar microtubules of the spindle.

5. Daughter chromosomes first appear during ________ of mitosis.
   a. interphase
   b. prophase
   c. metaphase
   d. anaphase
   e. telophase

6. In cells which have them, asters are present during ________ of mitosis.
   a. prophase
   b. prometaphase
   c. metaphase
   d. anaphase
   e. All of the above

7. Which of the following would you expect to occur after telophase for both a plant and an animal cell?
   a. Separation of chromatids
   b. Chromosome movement toward spindle poles
   c. Cytokinesis
   d. Movement of spindle poles
   e. Polar microtubule formation

8. Which of the following would you expect to be *absent* from the newly formed daughter cells of an animal cell that has just completed cytokinesis?
   a. Centrioles
   b. Chromosomes
   c. Mitochondria
   d. Cytoplasm
   e. None of the above

SEX AND REPRODUCTION • THE KARYOTYPE • MEIOSIS • PLOIDY, MITOSIS, AND MEIOSIS • CELL DIVISION IN PROKARYOTES (pages 202–211)

Key Concepts

1. The formation of new individuals by mitotic cell division is called *asexual reproduction*. It is an easy, rapid, and effective way to reproduce.
2. A drawback to asexual reproduction is the lack of genetic diversity among the resulting offspring (i.e., all progeny are genetically identical, or *clones*). Genetic uniformity can be a disadvantage in changing environments.
3. *Sexual reproduction* promotes genetic diversity by recombining hereditary material from two genetically different individuals. Diversity promotes evolutionary opportunity through novel genetic combinations.
4. To reproduce sexually, specialized cells called *gametes* must be produced. Gametes are *haploid*, containing only one set of chromosomes, whereas body cells are typically *diploid*, containing two sets of chromosomes.
5. The number of chromosomes in each set is denoted by n; haploid cells thus have n chromosomes, diploid cells have $2n$ chromosomes.
6. Haploid female gametes (*eggs*) unite with haploid male gametes (*sperm*) during *fertilization* to form a diploid *zygote*. The zygote then undergoes repeated mitotic divisions to become a multicellular organism.
7. *Meiosis* is the cell division process that produces haploid gametes (animals) or *spores* (plants and some fungi). It involves the *reduction division* of a single diploid nucleus into four haploid nuclei.
8. Organisms alternate between haploid and diploid phases of their life cycle by successive meiosis and fertilization events. Many organisms have only a multicellular diploid phase in their life cycle, others have only a multicellular haploid phase. Some species, particularly plants, have both.
9. During metaphase of mitosis, it is possible to count the numbers and types of chromosomes present in a cell. Such *karyotypes* show that diploid cells have *homologous pairs* of chromosomes, one member of the pair coming from each parent.
10. Only a single member of each chromosome pair is present in a haploid cell. Fertilization restores the homologous pairs by combining two haploid sets of chromosomes in the diploid zygote. We can thus refer to cells by their *ploidy level*, which is usually either n or $2n$. However, in some cases *triploid* ($3n$), *tetraploid* ($4n$), or higher-order *polyploid* nuclei are formed.
11. The phases of meiosis are the same as in mitosis, except that in meiosis each phase occurs twice. There are also distinctly different chromosomal arrangements during the phases of meiosis.
12. *Meiosis I* reduces the cell's ploidy level by one-half through separation of homologous chromosome pairs. *Meiosis II* separates the chromatids of chromosomes in these haploid nuclei.

13. Homologous pairs of chromosomes are able to separate correctly because of an important pairing process known as *synapsis*, which lasts from prophase through the end of metaphase in meiosis I.

14. During *prophase I*, synapsed pairs of chromosomes form crosses among the arms of their chromatids. These *chiasmata* represent areas where genetic material is exchanged between chromatids.

15. *Prometaphase I* occurs with the breakdown of the nuclear envelope, spindle formation, and kinetochore attachment to the polar microtubules. Unlike mitosis, only a single kinetochore is present in each chromosome.

16. During *metaphase I*, the synapsed homologous pairs of chromosomes move to the equatorial plate. They separate (still with paired chromatids attached by the centromere) during *anaphase I*. Note how this process differs from chromosome alignment and separation during metaphase and anaphase of mitosis.

17. *Telophase I* follows anaphase I in some species, with features similar to those observed during telophase in mitosis.

18. *Interkinesis*, which is somewhat like interphase of mitosis, follows telophase I. However, no DNA synthesis occurs because each chromosome is still composed of two chromatids.

19. *Meiosis II* resembles mitosis in that chromatids separate at their centromeres during metaphase to produce daughter chromosomes. Four haploid nuclei result.

20. Genetic diversity among the products of meiosis occurs for two reasons. First, crossing over during synapsis in prophase I recombines genetic material. Second, chance alone determines in which cell a particular member of a chromosome pair will end up during anaphase I.

21. Division errors occasionally take place during meiosis. *Nondisjunction* occurs when a chromosome pair moves to the same pole during anaphase I, or when sister chromatids fail to separate during anaphase II. The resultant cells are *aneuploid*, having an abnormally low or high number of chromosomes.

22. *Trisomy* (three copies of the same chromosome) is one consequence of nondisjunction. In humans, trisomy of chromosome 21 leads to the developmental abnormality known as Down syndrome.

23. *Translocation* is another division error that produces aneuploidy. Large portions of one chromosome are moved to another, leading to the functional equivalent of a nondisjunction trisomy. Translocation of chromosome 21 in humans can also lead to Down syndrome.

24. Trisomies (or *monosomies*, where only one copy of a chromosome is present) are fairly common in humans. However, most are lethal to the fetus during early development and usually result in natural termination of the pregnancy before birth.

25. The ploidy of a cell places limits on whether meiosis can even occur. Because chromosomes must undergo synapsis, only cells with an even ploidy ($2n$, $4n$, etc.) can produce viable gametes or spores. Mitosis is not similarly constrained by ploidy level.

26. Prokaryotic cells, such as the human intestinal bacterium *Escherichia coli*, have a single, circular chromosome attached to the cell membrane. This chromosome contains DNA and some loosely bound protein components.

27. Bacteria divide by *fission*, in which the circular chromosome is first copied and then attached to an adjacent part of the cell membrane. The new chromosome separates from its parent via cell elongation, at the end of which a new cell wall forms to complete the division process.

Questions (for answers and explanations, see page 114)

1. Which of the following statements is *not* true?
   a. Diploid cells contain two sets of chromosomes.
   b. Meiosis I represents a reduction division.
   c. Karyotypes indicate the numbers and kinds of chromosomes in a cell.
   d. Gametes are necessary for successful reproduction.
   e. Crossing over during meiosis is one factor contributing to increased genetic diversity.

2. A diploid plant cell has 36 chromosomes at the start of meiosis. How many *chromatids* does this cell have?
   a. 9
   b. 18
   c. 36
   d. 54
   e. 72

3. A diploid animal cell has 40 chromosomes at the start of meiosis. How many *homologous pairs of chromosomes* does this cell have?
   a. 10
   b. 20
   c. 40
   d. 60
   e. 80

4. A plant cell with a diploid number of 18 undergoes meiosis. How many *centromeres* do each of the four daughter cells have at the end of meiosis II?
   a. 9
   b. 18
   c. 27
   d. 36
   e. Cannot determine from information given

5. An important feature of metaphase in meiosis I is that
   a. chromatids separate to produce the final products of meiosis.
   b. kinetochore microtubules do not make attachments with the spindle.
   c. cytoplasmic division begins in preparation for chromosome separation.
   d. synapsed chromosome pairs line up along the equatorial plate.
   e. homologous pairs of chromosomes move to opposite poles of the spindle.

6. Synapsed chromosomes are present during which of the following phases of meiosis?
   a. Prophase I only
   b. Prophase I and II
   c. Prophase I through metaphase I

d. Prophase II through metaphase II
e. Metaphase II only

7. ___________ can be the result of a nondisjunction event that occurs during meiosis.
a. Haploidy
b. Karyotypy
c. Diploidy
d. Interkinesis
e. Aneuploidy

8. Which of the following is usually thought to result from an isolated error in cell division?
a. Ploidy reduction
b. Translocation
c. Synapsis
d. Interkinesis
e. Fission

Activities (for answers and explanations, see page 115)

- Consider the oval below to represent an animal cell from a species whose diploid number equals *four*. Diagram how this cell and its chromosomes would appear during *metaphase I* of meiosis.

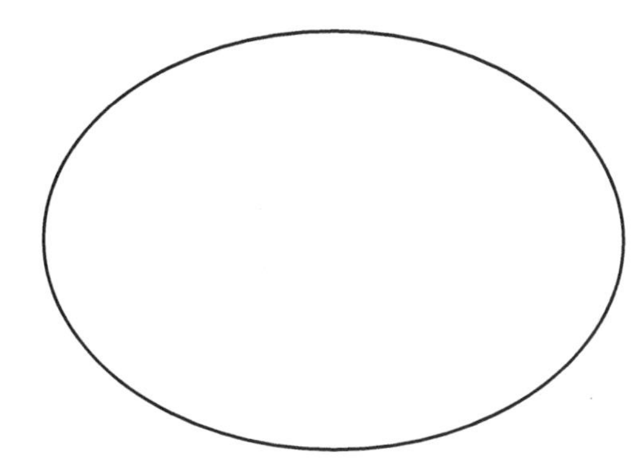

- Now diagram one of the immediate progeny of this same cell and show its chromosomes as they would appear during *metaphase II* of meiosis.

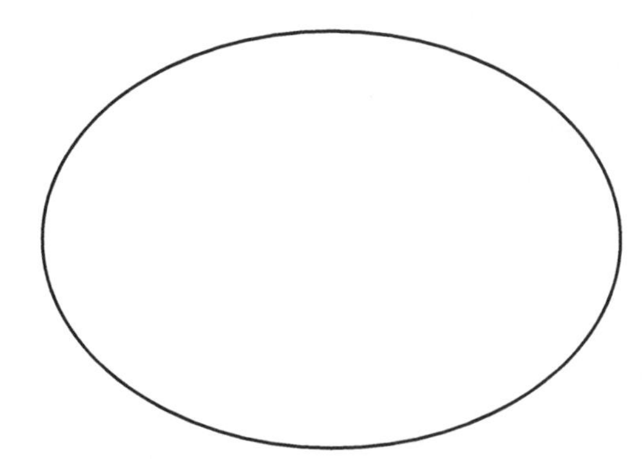

- Use the following diagram and the labels to complete the generalized life cycle of a organism showing both multicellular haploid and diploid stages as well as sexual reproduction: *multicellular diploid organism, multicellular haploid organism, meiosis, fertilization, zygote, gametes, haploid cells, germ cells.*

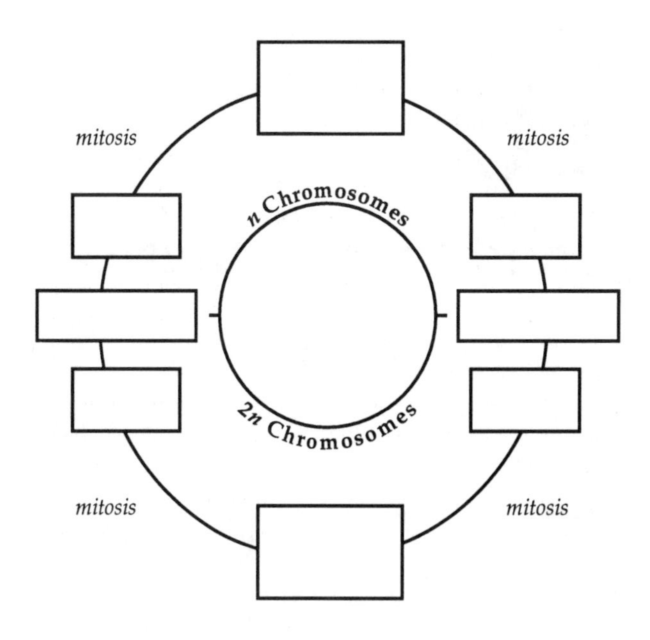

Integrative Questions (for answers and explanations, see page 115)

1. Anaphase of meiosis II is similar to anaphase of mitosis in that both involve
a. reduction divisions.
b. the production of aneuploid cells.
c. separation of chromatids.
d. condensation of chromatin.
e. diploid cells.

2. You are given a microscope slide on which is mounted a cell from an organism whose diploid number is 18. This cell has 19 chromosomes, all of which appear to lack obvious chromatids. Which of the following is most likely a true statement about this cell?
a. The cell is haploid and would probably have gone on to become a sperm cell.
b. It had completed telophase of mitosis, and had probably suffered a nondisjunction event somewhere in its genetic past.
c. If allowed to continue dividing, this cell would have undergone meiosis II.
d. The cell had just completed interphase and was about to enter mitosis.
e. It had just completed telophase of meiosis I, and was about to enter interkinesis.

3. All but one of the following would be present in animal cells isolated from both prophase of mitosis and metaphase of meiosis II. Select the *exception*.
a. Kinetochore microtubules
b. Polar microtubules
c. Chromatids
d. Diploid number of chromosomes
e. Centromeres

10

Mendelian Genetics and Beyond

CHAPTER LEARNING OBJECTIVES—after studying this chapter you should be able to:

- ❑ Explain how Mendel's quantitative approach to the study of inheritance allowed him to better document the ways that traits are passed between generations.
- ❑ Define and apply the following terms: "hybrid," "reciprocal cross," "monohybrid cross," "dihybrid cross."
- ❑ Identify the parental and the first or second filial generation in any genetic cross.
- ❑ Discuss the criteria used to determine whether a trait is dominant or recessive.
- ❑ Differentiate between genotypes, phenotypes, genes, alleles, and gene loci, and relate these terms to Mendel's particulate theory of inheritance.
- ❑ Describe which phenotypes result from homozygous or heterozygous genotypes, given a simple dominant–recessive trait.
- ❑ Use the Punnett-square method to determine the outcome of genetic crosses.
- ❑ Explain the chromosomal basis for Mendel's law of segregation, and discuss how this law predicts the observed phenotypic and genotypic ratios of a monohybrid cross.
- ❑ Discuss the importance of test crosses in determining the genotype of an individual expressing a dominant phenotype.
- ❑ Explain the chromosomal basis for Mendel's law of independent assortment, and discuss how this law predicts the observed phenotypic and genotypic ratios of a dihybrid cross.
- ❑ Define the term "recombinant phenotype," and explain the origin of this phenotype in a dihybrid cross.
- ❑ Compare and contrast the patterns of inheritance shown for incomplete dominance, codominance, and multiple allele systems, providing examples of each.
- ❑ Discuss the concept of pleiotropy and provide an example of its effects.
- ❑ Discuss the difference between wild-type and mutant phenotypes, and explain how they can be used to understand inheritance patterns.
- ❑ Explain how the concept of gene linkage alters the typical pattern of Mendelian inheritance, and use the correct symbolism to represent linkage groups.
- ❑ Differentiate between the mechanisms of sex determination in various organisms, and discuss how sex-linked inheritance differs from patterns of autosomal inheritance.
- ❑ Describe how crossing over relates to genetic recombination, how recombination frequencies are calculated, and how genetic markers are used to help generate gene maps of chromosomes.
- ❑ Explain how cytogenetics can be used to study the effects of chromosomal inversions and deletions in polytene chromosomes.
- ❑ Define the term "epistasis," and discuss the role of complementary genes in this phenomenon.
- ❑ Discuss the difference between continuous and discontinuous variation, and explain the role of polygenes in determining continuous variation.
- ❑ Distinguish between the concepts of penetrance and expressivity as they relate to the genotype and phenotype of an organism.
- ❑ Explain the cytological basis of maternal inheritance, and differentiate between it and Mendelian inheritance.

MENDEL'S DISCOVERIES (pages 214–222)

Key Concepts

1. Gregor Mendel was the first to carefully quantify inheritance patterns in organisms (1866). However, his work on garden peas went unnoticed for nearly 40 years.
2. Prior to Mendel, the botanist Kölreuter showed that cross-pollination yielded *hybrids* which, instead of being intermediate in form to their parents, often exhibited traits much like one or the other parent. *Reciprocal crosses,* where both possible mating combinations of parental sexes and traits are arranged, always gave identical results.
3. Mendel's quantitative approach identified several easily measured traits in true-breeding pea plants (those whose progeny always resemble the parents). Suitable traits were distinct and possessed contrasting alternative forms. Such careful attention to quantification was uncommon in the science of Mendel's day.

4. The original male and female plants in Mendel's crosses are referred to as the parental generation (*P*). Their immediate progeny belong to the first filial, or F_1, generation. The offspring of these individuals belong to the F_2 generation, and so on.
5. In Mendel's *monohybrid cross,* the parents differ for a single trait only. The form of this trait that shows up in the F_1 progeny is said to be *dominant* over the other, *recessive* form.
6. When F_1 individuals from a monohybrid cross are allowed to interbreed, the "lost" recessive trait reappears in some of the F_2 individuals. The ratio of dominants to recessives is always approximately 3:1 in the F_2 generation.
7. Mendel's findings disproved the blending hypothesis of inheritance, suggesting instead that hereditary information travels in discrete units. This *particulate theory* of inheritance has since been verified by our understanding of *genes* and other important cellular structures in the hereditary transfer system.
8. The *genotype* of an individual represents its actual genetic makeup. Different forms of genes (*alleles*) are found at the gene's location, or *locus,* on the chromosome. Alleles combine in pairs to form the diploid genotype. The *phenotype* is the physical expression of an individual's genotype (e.g., blue eyes, dented seeds, etc.).
9. When gametes are formed, only one allele for each trait in the parent's genotype is included in any given egg or sperm. Mendel' law of *segregation* recognizes that alleles separate during gamete formation, restoring the diploid genotype in the zygote at the time of fertilization.
10. The dominant allele for a gene is usually abbreviated with an uppercase letter, the recessive allele with a lowercase. *Homozygous* individuals have genotypes with two copies of the same allele, either dominant or recessive. *Heterozygous* individuals possess two different alleles for a particular gene.
11. For example, if the allele for spherical seeds in pea plants (*S*) is dominant to the allele for dented seeds (*s*), both the homozygous dominant genotype *SS* and heterozygous *Ss* will produce plants with the same phenotype (spherical seeds), while the homozygous recessive genotype *ss* will yield a plant with dented seeds.
12. The F_1 individuals from a monohybrid cross between true-breeding dominant and recessive parents are always heterozygotes. When the F_1 individuals form gametes, the dominant and recessive alleles segregate into separate gametes, which form with equal probability.
13. This known proportion of gamete types makes it possible to calculate the predicted outcomes of an F_1 cross. A Punnett square, in which gamete probabilities are plotted in tabular form to facilitate the determination of F_2 genotypes, is often used in this process.
14. The Punnett square method verifies the 3:1 ratio of dominant to recessive phenotypes found in the F_2 generation. This phenotypic ratio results because two of the three individuals with the dominant phenotype are heterozygous for the gene.
15. Individuals expressing a dominant phenotype can be either homozygous dominant or heterozygous. Which genotype they possess is determined using a *test cross,* in which the unknown dominant phenotype is crossed with an individual homozygous recessive for the trait. If all offspring have dominant phenotypes, then the test individual must be homozygous. If half the offspring have recessive phenotypes, then the test individual is heterozygous.
16. *Dihybrid crosses* involve cases where individuals that differ for two traits are mated. Inheritance patterns from such crosses support Mendel's law of *independent assortment,* which states that alleles for different traits assort independently of one another during gamete formation.
17. As in a monohybrid cross, the F_1 progeny of a dihybrid cross all have the same genotype and phenotype. In the F_2 generation, however, there are now nine unique genotypes and a phenotypic ratio of 9:3:3:1. Because they are genetically independent events, we can mathematically treat dihybrid crosses as the probabilistic interaction of two separate monohybrid crosses.
18. Some F_2 progeny of a dihybrid cross show phenotypes unlike either parent. These *recombinant phenotypes* result from the formation of novel genotypes due to the action of independent assortment.
19. Independent assortment of genes is not a universal hereditary law; it applies only to genes residing on different chromosomes, as was the case in all of Mendel's studies.

Activities (for answers and explanations, see page 115)

- Indicate how many different kinds of gametes can be made by individuals possessing the following diploid genotypes:

  AABBCCDDEE

  AaBbCcDdEe

  AABbCCDDEE

- Explain the fundamental differences between Mendel's laws of segregation and independent assortment.

- Complete a Punnett square for a dihybrid cross involving two individuals that are heterozygous for both gene loci.

Questions (for answers and explanations, see page 115)

1. An individual has the genotype *PpQQrr*. Which of the following is true about this individual?
   a. It is the result of a cross between two *PPqqRR* individuals.
   b. The *P* gene is dominant to both the *Q* and *R* genes.
   c. The individual is homozygous recessive for one trait.
   d. It must be a fungus because the genotype is haploid.
   e. Gametes of the genotype *PQr* are not possible from this individual.

2. If all the F_2 progeny of a cross are homozygous recessive for the trait in question, what does this tell you about the two individuals mated together in the original parental generation?
   a. They were also homozygous recessive for this trait.
   b. One was homozygous recessive, the other homozygous dominant.
   c. Both were heterozygotes.
   d. Both were homozygous dominant.
   e. Neither was true-breeding.

3. Mendel's particulate theory of inheritance is supported by modern evidence about the genetic role of
   a. chromosomes.
   b. genes.
   c. alleles.
   d. gametes.
   e. All of the above

4. If allele *K* is dominant to *k*, then the offspring of a cross between *KK* and *kk* individuals will
   a. all be homozygous.
   b. display the same phenotype as the *KK* parent.
   c. display the same phenotype as the *kk* parent.
   d. have the same genotype as the *KK* parent.
   e. have the same genotype as the *kk* parent.

5. A cross between true-breeding short-haired and long-haired rats always produces short-haired individuals. If these F_1 animals are test crossed, half of the progeny are short-haired and the other half long-haired. This suggests that
   a. long hair is the dominant trait and that the F_1 progeny are all heterozygotes.
   b. long hair is the recessive trait and that the F_1 progeny are all homozygotes.
   c. short hair is the recessive trait and that the F_1 progeny are all homozygotes.
   d. short hair is the recessive trait and that the F_1 progeny are all heterozygotes.
   e. short hair is the dominant trait and that the F_1 progeny are all heterozygotes.

6. In cocker spaniels, black coat color (*B*) is dominant over red (*b*), and solid color (*S*) is dominant over spotted (*s*). If a red, spotted male is crossed with a black, solid female, and all the offspring from several litters express only the dominant traits, what is the most likely genotype for the female parent?
   a. *BBSS*
   b. *BbSS*
   c. *BbSs*
   d. *BBSs*
   e. None of these

7. In a species of ornamental fruit tree, curly leaves (*C*) are dominant to flat leaves (*c*), and solid leaf color (*L*) is dominant to striped (*l*). A cross between a tree that is heterozygous for curly, solid leaves and one that has flat, striped leaves will produce an F_1 generation with what percentage of individuals that are homozygous recessive for both traits?
   a. 0
   b. 25
   c. 50
   d. 75
   e. 100

8. An individual with genotype *NnOO* is crossed with one having the genotype *nnOo*. How many different genotypes will be represented in their progeny?
   a. One
   b. Two
   c. Three
   d. Four
   e. Cannot determine from information given

GENETICS AFTER MENDEL: ALLELES AND THEIR INTERACTIONS • FOCUS ON CHROMOSOMES (pages 222–233)

Key Concepts

1. For certain genes, heterozygous individuals exhibit phenotypes intermediate to either parent. This arises because of *incomplete dominance*, where the phenotypic effect of the single dominant allele in the heterozygote is qualitatively different from that of having two dominant alleles in a homozygote. Flower color in snapdragons is an example.

2. Incomplete dominance occurs more frequently than does complete dominance. It was fortuitous that all seven of Mendel's traits in peas showed complete dominance, which made his deciphering inheritance patterns an easier task.

3. Monohybrid crosses involving incomplete dominance produce phenotypic and genotypic ratios of 1:2:1 in the F_1 generation.
4. *Codominance* occurs when two alleles at a gene locus produce different phenotypic effects, both of which are visible in heterozygote individuals. The color and patterning of white clover leaves are an example.
5. *Pleiotropic* alleles have phenotypic effects on more than one trait. An example is the influence of a single gene on both coat coloration and crossed eyes in Siamese cats.
6. New alleles arise through *mutation*, the heritable change of nucleic acid material in the gene. The most common allelic form of the gene is called the *wild-type*; other forms are known as mutant alleles.
7. *Multiple alleles* may exist at a gene locus. An example is the gene that regulates the ABO blood group system in humans. In this case, three different alleles interact to produce six genotypes and four phenotypes. The Rh blood factor of primates also exhibits inheritance patterns consistent with multiple alleles.
8. In some dihybrid crosses, outcomes occur where homozygous recessive genotypes appear in the F_2 generation at levels in excess of those predicted by Mendel's laws. This is explained by *linkage*, where the genes involved occur on the same chromosome and do not assort independently during gamete formation.
9. Linked genes are symbolized by drawing a line over the appropriate allele symbols ($\overline{AB}$ $\overline{ab}$). A full set of linked genes on a chromosome constitutes a *linkage group*.
10. In Mendel's studies it did not matter which parental sex possessed a given form of a trait; reciprocal crosses always yielded similar results. For many other traits, however, parental origin has profound genetic consequences.
11. Most plants produce both male and female gametes on the same organism; they are *monoecious*. By contrast, most animals produce eggs and sperm inside separate individuals; they are *dioecious*.
12. Morphological sex in some dioecious species is determined by precise differences in the combination of X or *Y sex chromosomes*, which may occur singly or in pairs.
13. *Autosomes* are the chromosomes present in equal numbers in both sexes. The sex chromosomes are typically present in unequal numbers.
14. In mammals, the presence or absence of the Y chromosome determines sex; XX individuals are female and XY individuals are male. Atypical combinations of these sex chromosomes (e.g. XO or XXY) produce males or females with physical and/or mental abnormalities. In birds and some insects, the pattern of sex determination is reversed. To avoid confusion, females in these species are designated ZW and males ZZ.
15. In the fruit fly *Drosophila*, sex is determined by the ratio of the number of X chromosomes to autosome sets. In other insects, whether the individual is haploid or diploid determines its sex.
16. Some genes are located on the X and Y chromosomes. Depending on their sex, individuals may have only a single copy of each chromosome; thus they are *hemizygous* (having only one allele) for that trait. Patterns of *sex-linked inheritance* differ markedly from those of autosomal inheritance, where two allelic copies are present.
17. Many types of inheritance patterns, including those exhibiting sex linkage, can be studied by following gene loci whose alleles code for easily observable phenotypes or *markers*. An example is Morgan's work on the sex-linked inheritance of eye color in *Drosophila*.
18. *Carriers* of X-linked recessive traits will differentially pass the trait to their male and female offspring. In humans, where females are XX, a woman who carries an X-linked recessive allele in the heterozygous condition can pass it to either her sons or daughters, but it is only in her hemizygotic sons (XY) that the trait will be expressed. Hemophilia in humans is an example.
19. Because of its small size, there are fewer genes on the Y chromosome than on the larger X chromosome. Nevertheless, some of these Y chromosome genes are quite important in determining male characteristics. In humans, Y-linked inheritance is obviously limited to males.
20. Since Mendelian ratios are statistical averages, and not absolutes, it is essential to have large *sample sizes* in order to guard against the effects of chance deviations from predicted ratios. Slow-breeding species with long life spans (e.g., humans) therefore do not make especially good subjects for genetic study.
21. *Genetic maps* showing relative gene locations on the chromosome have been determined for some species, including corn plants (*Zea*), fruit flies (*Drosophila*), and the fungus *Neurospora*.
22. *Neurospora* is a particularly useful organism for genetic study because its haploid spores exhibit all alleles present. Moreover, the spores are packaged within a saclike ascus whose characteristic development allows precise quantification of the genetic recombination that takes place during meiosis.
23. By documenting phenotypic changes that occur relative to known markers on the chromosome, *recombination frequencies* can be calculated based on the amount of genetic exchange that takes place during *crossing over*, when homologous chromosomes pair and synapse during prophase I of meiosis.
24. Since the probability that crossing over will occur is directly proportional to the distance between linked genes, recombination frequencies yield a quantitative map of the relationships between gene loci on the chromosome.
25. *Cytogenetics* is the study of microscopic chromosome structure in relation to genetics. Giant *polytene chromosomes* from the salivary glands of *Drosophila* larvae have distinct banding patterns that make it possible to monitor changes in the sequence of genetic material. Common chromosomal changes that strongly influence gene sequence include *inversions* (reversal of linked gene sequences) and *deletions* (actual loss of genetic material).

Activity (for answers and explanations, see page 116)

- For each combination of sex chromosomes and animal species listed below, indicate whether the individual would be male or female, and whether or not they would be reproductively normal.

_____________ *ZZ* chicken
_____________ *X* human
_____________ *XXY* human
_____________ *XY* fruit fly
_____________ *XXY* fruit fly
_____________ haploid honeybee

Questions (for answers and explanations, see page 116)

1. In cases of incomplete dominance, the phenotype associated with the homozygous recessive genotype is
a. indistinguishable from the heterozygote phenotype.
b. only compatible with the homozygous dominant genotype as a mating partner.
c. different from both the homozygous dominant and heterozygous phenotypes.
d. a very rare condition which occurs much less frequently than predicted by Mendelian ratios.
e. unrecognizable because it looks just like the homozygous dominant phenotype.

2. Individuals in a species of deer with the genotype *BB* have dark brown skin, while those with the genotype *bb* have tan skin. Deer with the genotype *Bb* have mottled skin with mixes of brown and tan patches. This situation is probably an example of
a. incomplete dominance.
b. codominance.
c. simple dominance.
d. pleiotropy.
e. hemizygosity.

3. The ABO blood group system in humans is controlled by
a. multiple alleles.
b. Rh factors.
c. multiple genes.
d. incomplete dominance.
e. environmental influences on gene expression.

4. If a child belongs to blood type O, he or she could *not* possibly have been born to which of the following sets of parents?
a. Type A mother and type B father
b. Type O mother and type A father
c. Type AB mother and type O father
d. Type B mother and type O father
e. Type O mother and type O father

5. A man with blood type A, one of whose parents was blood type O, has a child by a woman with blood type AB. What is the probability that their baby will have blood type O?
a. 0%
b. 25%
c. 50%
d. 75%
e. 100%

6. In humans, hemizygous individuals for sex-linked traits are also known as
a. males.
b. females.
c. heterozygotes.
d. autosomal mutants.
e. wild types.

7. Red–green color blindness is an X-linked recessive trait in humans. A color-blind woman and a man with normal vision have a son. What is the probability that the son is color-blind?
a. 0%
b. 25%
c. 50%
d. 75%
e. 100%

8. If a human female expresses an X-linked recessive trait, then she must have inherited the trait from:
a. her mother.
b. her father.
c. both parents.
d. her maternal grandmother.
e. through a mutation.

9. Data useful for constructing genetic maps comes from which of the following sources?
a. Measurements of the rate of mutation at specific gene loci
b. Patterns of sex-linkage in the ABO blood group system of humans
c. The speed at which wild-type alleles segregate during gamete formation
d. Recombination frequencies that measure the amount of genetic change due to crossing over
e. Counts of the numbers of carriers of X-linked recessive traits in the F_1 generation

INTERACTIONS OF GENES WITH OTHER GENES AND WITH THE ENVIRONMENT • NON-MENDELIAN INHERITANCE (pages 233–237)

Key Concepts

1. *Epistasis* occurs when the action of one allelic form of a gene masks the expression of alleles at other gene loci. The epistatic gene usually affects production of substances manufactured early in the biochemical pathway that leads to the trait. An example is determination of coat color in mice.

2. *Complementary* genes are epistatic to each other; that is, the phenotypic expression of each one is dependent upon the influence of the other. Flower color in sweet peas is an example.

3. Mendel studied traits showing *discontinuous* variation; they had distinct, contrasting forms. However, many heritable traits such as human height or skin pigmentation show *continuous* variation, which is under the combined control of many genes (*polygenes*), often located on different chromosomes.

4. Genotype and environment also interact to affect an organism's phenotype. Nutrition, light, and temperature

can all influence the translation of genotype into phenotype. Dark pigments in the fur of Siamese cats represent an example of temperature influences on phenotype.

5. *Penetrance* measures the proportion of individuals in a population with a certain genotype that actually show the expected phenotype. The *expressivity* of a genotype is a measure of how much the phenotype emerges in any given individual with that genotype.
6. Differences between the influence of genetic and environmental factors are best studied using identical twins, where genetic variation is negligible and differences between individuals can usually only result from environmental influences.
7. Mendelian inheritance is based on the precise allocation of genetic material during meiosis. However, some DNA is also passed between generations by non-Mendelian mechanisms. One mechanism is *maternal inheritance*, where cytoplasmic DNA from mitochondria or chloroplasts is provided to the zygote through the egg cytoplasm. The variegated color pattern of stems in the four-o'clock plant is an example.

Questions (for answers and explanations, see page 116)

1. Two genes regulate coat color in a breed of show dog: one for pigment production and one for pigment deposition. If either gene is homozygous recessive, the animal is an albino. This situation represents an example of which of the following genetic conditions?
   a. Epistasis
   b. Complementary genes
   c. Discontinuous variation
   d. Expressivity
   e. Penetrance

2. Himalayan rabbits normally have an all-white coat except for black on the tail, legs, nose, and ears. If, however, you shave the hair off the back of one of these rabbits and put an ice bag on the skin, the new hair will grow in black. Which of the following represents a likely genetic explanation for this observation?
   a. Environmental influences on penetrance or expressivity
   b. Complementary gene interaction
   c. Masking of one gene by another in epistasis
   d. Simple dominance and recessiveness
   e. None of the above

3. Novel gene combinations produced by the events of meiosis are not a factor in maternal inheritance because
   a. genetic recombination is due entirely to the DNA found in the male sperm.
   b. Y-linked traits control gene recombination, and are expressed only in males.
   c. maternally inherited DNA does not mutate, so change never occurs.
   d. errors that occur in replication of this DNA counteract all new gene combinations.
   e. maternally inherited DNA comes from mitochondria and chloroplasts, and is not involved in meiosis.

4. Imagine a plant in which flower color is limited to three forms; red, white, and pink. By contrast, skin color in humans covers an enormous variety of colors and shades. This difference is best attributed to the fact that
   a. flower color is controlled by maternal inheritance and skin color is not.
   b. skin color is controlled by maternal inheritance and flower color is not.
   c. flower color involves only one gene whereas skin color involves polygenes.
   d. skin color involves only one gene whereas flower color involves polygenes.
   e. epistasis controls flower color and penetrance controls skin color.

Integrative Questions (for answers and explanations, see page 116)

1. If the linked gene sequence *ABCDEFGHIJK* becomes *ABHGFEDCIJK*, we would say that ________ had probably occurred.
   a. a deletion mutation
   b. an inversion mutation
   c. some polygenic event
   d. linkage reduction
   e. independent assortment

2. Which of the following statements is *not* true?
   a. Chromosomes have many genes, each of which may have multiple alleles.
   b. Incomplete dominance produces heterozygotes that are phenotypically different from either parent.
   c. The effects of one gene are masked by another in epistasis.
   d. The F_1 progeny of a cross between two homozygous dominant parents will all be heterozygotes.
   e. Wild-type alleles are the most common form of genes found under natural conditions.

3. Two separate crosses are performed using fruit flies. The first cross involves a dark-bodied male and a light-bodied female. The second involves a light-bodied male and a dark-bodied female. All of the offspring from these crosses are dark-bodied. This is an example of
   a. reciprocal crosses involving a likely autosomal trait.
   b. reciprocal crosses involving a likely sex-linked trait.
   c. test crosses involving a likely autosomal trait.
   d. test crosses involving a likely sex-linked trait.
   e. dihybrid crosses involving incomplete dominance.

4. Purple flowers are dominant to white flowers in pea plants, and tall plants are dominant to short plants. If a true-breeding, purple-flowered, tall plant is crossed with a true-breeding, white-flowered, short plant, what proportion of the offspring in the F_2 generation will be purple and short?
   a. 1/4
   b. 3/4
   c. 1/16
   d. 3/16
   e. 0/16

11

Nucleic Acids as the Genetic Material

CHAPTER LEARNING OBJECTIVES—after studying this chapter you should be able to:

- ❑ Describe the molecular basis of bacterial transformation and give a historical chronology of important experiments in the search for the so-called transforming principle.
- ❑ Explain the reasoning behind and the major conclusions from Hershey and Chase's experiments on T2 bacteriophage.
- ❑ Discuss the importance of x-ray crystallography in the study of DNA structure.
- ❑ Differentiate between purine and pyrimidine bases in DNA and tell which specific bases pair with one another.
- ❑ Explain what is meant by complementary base pairing in DNA and how this concept relates to Chargaff's rules.
- ❑ Describe our current understanding of the physical structure of a DNA molecule.
- ❑ Explain what is meant by the 5′ and 3′ ends of a polynucleotide, and use these ideas to describe the direction in which new strands of DNA are formed.
- ❑ Discuss the differences between normal DNA and Z-DNA.
- ❑ Compare and contrast the structures and functions of DNA and RNA.
- ❑ Distinguish among the three models for DNA replication, and provide evidence for why biologists believe the semiconservative model is correct.
- ❑ Describe in detail the molecular mechanism of DNA replication using terms such as "helicase," "polymerase," "replication origin," "replication fork," "lagging strand" and "leading strand," "Okazaki fragments," "ligases," etc.
- ❑ Discuss the various molecular mechanisms cells possess for dealing with errors in DNA replication.
- ❑ Explain the molecular relationship between genes and an organism's phenotype.
- ❑ Discuss the one-gene, one-polypeptide theory and describe how experimental evidence from work with the fungus *Neurospora* provides evidence for this idea.
- ❑ Differentiate between auxotrophs and prototrophs and explain their use in the study of molecular genetics.
- ❑ State the central dogma of molecular biology, and explain how retroviruses are exceptions to this dogma.
- ❑ Describe the roles of mRNA, tRNA, and rRNA in the process of protein synthesis.
- ❑ Compare and contrast in detail the mechanisms of transcription and translation with respect to the primary molecules involved, how each occurs, and where in the cell each takes place.
- ❑ Define the term "codon" and explain how codons provide information for the construction of polypeptides.
- ❑ Differentiate between a degenerate genetic code and one that is ambiguous.
- ❑ Use the terms "codon," "anticodon," "start codon," "stop codon," and "signal sequence" in describing the various stages of protein synthesis.
- ❑ Explain how tRNA is charged and how ribosomes associate with various RNA molecules during translation.
- ❑ Describe what biologists mean by the term "mutation," and explain why mutations are evolutionarily important events.
- ❑ Discuss some molecular mechanisms by which mutagens can produce mutations.
- ❑ Distinguish among the following types of mutations and describe their effects: point, chromosomal, missense, nonsense, and frame-shift.
- ❑ Further distinguish among the following types of chromosomal mutations and describe their effects: deletions, duplications, inversions, and translocations.
- ❑ Appreciate the rate at which natural mutations occur and how this affects evolution.

WHAT IS THE GENE? • NUCLEIC ACID STRUCTURE • REPLICATION OF THE DNA MOLECULE • PROOFREADING AND DNA REPAIR (pages 241–254)

Key Concepts

1. Early attempts to explain the chemical basis of heredity focused on proteins. This idea was first questioned in the 1920s when Griffith discovered the phenomenon of *transformation* in pneumococcus bacteria. His experiments showed that a living strain of nonvirulent bacteria could be converted into a virulent form by exposure of the first to heat-killed specimens of the second.
2. During the 1940s Avery and colleagues used Griffith's transformation technique along with highly purified cell extracts to clearly identify the *transforming principle* as DNA. However, their important work went unnoticed for several years.
3. In the 1950s Hershey and Chase provided definitive evidence that DNA is the genetic material. They raised T2 bacteriophage under two conditions: one with radioactive phosphorus (^{32}P), which labels the DNA, and the other with radioactive sulfur (^{35}S), which labels the viruses' external proteins. Labeled bacteriophage were then allowed to infect normal bacteria, and the locations of radioactive markers were determined.
4. Only the ^{32}P was found inside the infected bacteria; the ^{35}S was outside the cells. This finding that viruses transfer only their DNA from one generation to the next led to the conclusion that DNA (not protein) must be the hereditary material.
5. The next task was to determine the molecular structure of DNA. X-ray crystallographic work by Franklin and Wilkins, coupled with earlier biochemical analyses, indicated that DNA is a long polymer of nucleotides each of which consists of a deoxyribose sugar, a phosphate group, and a nitrogen-containing base.
6. Four nucleotides of DNA are known to exist, each differing only in their nitrogenous bases. Two are purines (*adenine* and *guanine*) and two are pyrimidines (*cytosine* and *thymine*). In 1950, Chargaff noted that for any DNA sample the amount of adenine always equals the amount of thymine (A = T), and the amount of guanine always equals the amount of cytosine (G = C). Thus, the total amount of purines always equals the total amount of pyrimidines (A + G = T + C).
7. Watson and Crick used these data in 1953 to build our current model of DNA structure. The model portrays DNA as a double-stranded *helix* (cylindrical spiral) of uniform diameter that twists to the right. It contains two *antiparallel* chains of nucleotides (i.e., polynucleotides that run in opposite directions).
8. The "backbone" of each nucleotide chain consists of repeated sugar–phosphate connections, with the nitrogenous bases oriented toward the center of the molecule. Hydrogen bonds form between the bases of one chain and those of the other. A system of *complementary base pairing* occurs in which G pairs with C, and A with T. The geometry of this bonding pattern produces major and minor grooves along the length of the helix.
9. DNA is formed by connecting the phosphate group attached to the 5′ carbon of a new nucleotide to the 3′ carbon of the preceding sugar molecule; the molecule is thus constructed in the 5′ → 3′ direction. This constraint on the formation of DNA strands explains why the two chains run in antiparallel directions.
10. The helix of normal DNA spirals to the right. A related nucleic acid called *Z-DNA* has a left helical twist and sugar–phosphate connections that are zigzagged; it thus possesses only a single helical groove. The function of Z-DNA remains unclear, but may have to do with interactions between DNA and regulatory proteins.
11. Another important nucleic acid utilizes ribose sugar instead of deoxyribose. Although *RNA* is structurally quite similar to DNA, it differs by usually having only a single strand of nucleotides and also by substituting *uracil* (U) for the base thymine. Like thymine in DNA, the uracil of RNA bonds only with its complementary base adenine.
12. DNA carries its hereditary information in the linear sequence of nucleotide bases. Because of the specific complementary base-pairing rules and double-stranded structure of the molecule, unwinding the helix provides an efficient mechanism for DNA replication. The helical model of DNA also provides mechanisms that account for its overall function and for gene mutation.
13. DNA will replicate outside the cell (i.e., in a test tube), given an appropriately complex mixture of nucleotide precursors and assembly enzymes.
14. Three possible DNA replication mechanisms were originally conceived. Conservative replication would produce one new double-stranded DNA while retaining intact the old double-stranded molecule. Dispersive replication would require fragmentation of the old DNA, followed by its reassembly along with new DNA. All available evidence supports a third model of *semiconservative replication*, where each old strand serves as a template for the manufacture of one new strand.
15. In 1957 Meselson and Stahl verified the semiconservative model by raising bacteria on a medium enriched with "heavy" nitrogen (^{15}N). This forced the bacteria to manufacture all heavy DNA.
16. When bacteria that have grown for several generations on ^{15}N are transferred to a medium of ^{14}N, each parental strand of DNA is used as a template to manufacture a new double-stranded DNA molecule for the next generation. The result is that one strand of the new DNA molecule has nothing but heavy ^{15}N bases (old DNA), and the other strand contains only light ^{14}N bases (new DNA). This difference in DNA types can be measured using density gradient centrifugation in a CsCl solution. With additional growth on a ^{14}N medium, the bacterial DNA becomes increasingly dominated by new ^{14}N strands.
17. DNA replication is only initiated when the double helix unwinds at a specific location, called the *origin of replication*, exposing the two antiparallel parent strands and forming a *replication fork*. DNA polymerase III attaches at the origin and begins making new DNA along the

leading strand through the continuous addition of nucleotides in the 5′ → 3′ direction.

18. Along the *lagging strand*, new DNA is also added in the 5′ → 3′ direction, but is not made continuously because the parent DNA is being exposed in a direction incompatible with continuous replication. In this case, new DNA is synthesized in a direction opposite to the movement of the replication fork by way of small *Okazaki fragments*, which are later connected together by *DNA ligase* to form functional daughter strands.

19. Many other polypeptides such as *helicases* (which cause the spiralled DNA to unwind), *single-stranded binding proteins* (which prevent the parent DNA strands from recombining before replication can occur), and *primases* (which provide primer strands for the production of Okazaki fragments), are also involved in coordinating activities at the replication fork.

20. DNA replication is normally rapid and accurate, with synthesis rates in excess of one thousand new base pairs per second, and error levels of only one mistake per billion or more bases utilized. In addition, DNA polymerases and other enzymes also carry out inspection or "proofreading" functions to detect, excise, and replace mispaired bases. The absence or malfunction of such inspection mechanisms greatly enhances an organism's susceptibility to DNA-repair defects. An example is development of the human skin disease xeroderma pigmentosum.

Activities (for answers and explanations, see page 117)

- What will be the sequence of nucleotide bases in a daughter strand of DNA made from the parent DNA strand shown below?

5′ – A T C C G T A A C G C A G G G C T T A – 3′

- Based on the results of Meselson and Stahl's experiments with heavy and light forms of DNA, explain why the following three density gradient centrifuge tubes and their contents provide support for the semiconservative model of DNA replication.

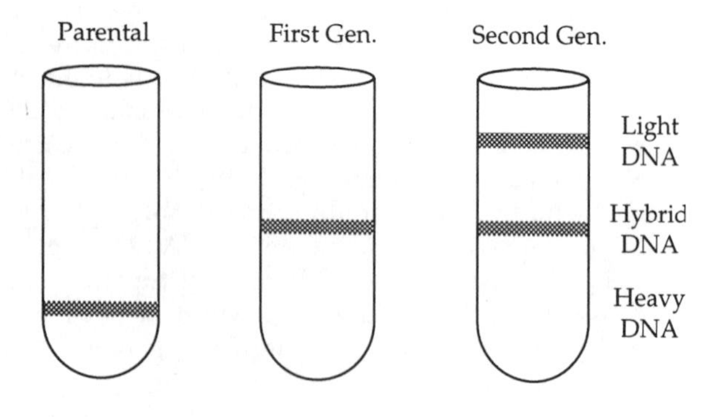

- Explain the conceptual relationship between Griffith's experiments on bacterial transformation and those of Hershey–Chase using T2 bacteriophage. What is the main conclusion regarding hereditary transfer that can be drawn from these two sets of experiments?

Questions (for answers and explanations, see page 117)

1. DNA contains ________ sugar and ________ as one of its nitrogenous bases, while RNA contains ________ sugar and ________ as one of its bases.
   a. ribose, thymine, deoxyribose, uracil
   b. deoxyribose, thymine, deoxyribose, uracil
   c. ribose, thymine, deoxyribose, thymine
   d. deoxyribose, thymine, ribose, thymine
   e. deoxyribose, thymine, ribose, uracil

2. Which of the following represents a bond between a purine and a pyrimidine (in that order)?
   a. C – T
   b. G – A
   c. G – C
   d. T – A
   e. C – A

3. The Hershey–Chase experiments, in which T2 bacteriophage were formed in the presence of the radioactive precursors of potential hereditary macromolecules, demonstrated that
   a. RNA, and not DNA, was the hereditary material in this virus.
   b. RNA, and not protein, was the hereditary material in this virus.
   c. DNA, and not protein, was the hereditary material in this virus.
   d. protein, and not RNA, was the hereditary material in all bacteria.
   e. protein, and not DNA, was the hereditary material in all bacteria.

4. Thirty percent of the bases in a sample of DNA extracted from eukaryotic cells is adenine. What percentage of cytosine is present in this DNA?
   a. 10
   b. 20
   c. 30
   d. 40
   e. 50

5. Which one of the following statements about DNA or its replication is *not* true?
   a. Okazaki fragments are the initiators of continuous DNA synthesis along the leading strand.

b. Replication forks represent areas of active DNA synthesis on the chromosome.
c. Error rates for DNA replication are often less than one in every billion base utilizations.
d. Ligases and binding proteins function in the vicinity of replication forks.
e. Z-DNA and normal DNA differ in the rotational directions of their polynucleotide helices.

6. In the Watson and Crick model of the DNA molecule
a. neither nitrogen nor phosphorus is present.
b. the width of the molecule changes regularly along its length.
c. the amount of cytosine always equals the amount of thymine.
d. a single helix of polynucleotides is held together exclusively by covalent bonds between bases.
e. the double helix of polynucleotides is held together largely by hydrogen bonds between bases.

7. Which of the following was *not* a technique used to determine the structure of DNA?
a. X-ray crystallography
b. Biochemical analysis of purine/pyrimidine ratios
c. Radioactive labeling of sulfur and phosphorus
d. Construction of three-dimensional molecular models
e. Biochemical analysis of generalized nucleotide structure

8. Consider a segment of nucleic acid with the nucleotide sequence AACGCAUCGG. Which of the following best describes this molecule?
a. It is Z-DNA.
b. It is normal DNA.
c. It is DNA made with ^{15}N.
d. It is RNA.
e. It is an Okazaki fragment.

WHAT DO GENES CONTROL? • FROM DNA TO PROTEIN • MUTATIONS • THE ORIGIN OF NEW GENES (pages 255–270)

Key Concepts

1. Phenotypic traits do not emerge directly from genes. Rather, genes code for the production of the macromolecules necessary for the phenotype to appear.
2. Garrod (1908) was the first to explore the mechanistic relationship between genes and phenotype. He reasoned that an enzyme defect was responsible for the inherited urinary disease alkaptonuria.
3. Beadle and Tatum extended Garrod's ideas in the 1940s using the fungus *Neurospora*. Wild-type *prototrophs* of *Neurospora* were grown on a *minimal medium*, which provided only the minimum amount of all basic nutrients needed for growth. Mutant *auxotrophs*, produced by irradiation, required one supplemental nutrient in order to grow.
4. By producing a series of auxotrophs, Beadle and Tatum showed that each mutant lacked only one enzyme necessary for the synthesis of a required nutrient, suggesting that numerous genes are involved in the control of common biosynthetic pathways.
5. Beadle and Tatum's findings, in combination with work on hemoglobin by Pauling, spawned the *one-gene, one-polypeptide theory*, which states that an individual gene controls the production of only one polypeptide. After its manufacture, the polypeptide usually functions in cell metabolism, often as an enzyme.
6. The *central dogma* of molecular biology states that DNA codes for the manufacture of RNA, which in turn codes for the production of protein. The reverse process is not possible, indicating a one-way flow of genetic and molecular information.
7. Crick proposed that "adaptor" molecules permit RNA and the specific amino acid components of proteins to associate properly during protein formation. We call these adaptor molecules transfer RNAs (*tRNA*).
8. Genetic information moves from the nucleus to the cytoplasm by way of messenger RNA (*mRNA*). Crick also proposed the idea of *translation*, the process by which amino acids are organized into specific proteins using the mRNA message.
9. Viruses (noncellular entities) are exceptions to the central dogma in that RNA replaces DNA as the genetic material. Certain of these viruses, called *retroviruses*, carry a *reverse transcriptase* enzyme that transcribes RNA into DNA. The DNA is then replicated by normal host cell mechanisms to make new viruses. The human immunodeficiency virus (HIV) that causes AIDS is an example.
10. Before translation can occur, the DNA genetic message must first become an mRNA message through the process of *transcription*. The enzyme *RNA polymerase* binds to one side of the double-stranded DNA (the *template strand*) and begins assembling a continuous sequence of complementary ribonucleotides in the 5′ → 3′ direction only. The other strand of DNA remains untranscribed. As the mRNA transcript is made, it peels away from the DNA template allowing the separated DNA strands to reassociate.
11. Gene transcription begins at well-defined *initiation sites* along the DNA molecule, and ends at specific termination sites. Initiation sites allow RNA polymerase to bind to the DNA and also determine which strand will serve as the template. In this way, only certain genes are transcribed at any given time.
12. All RNA is made by transcription. In addition to the tRNAs and mRNAs, ribosomal RNA (*rRNA*) is coded for by DNA. rRNA is a major structural component of the ribosomes, which are used during the translation of mRNA into polypeptide chains.
13. mRNA transcripts can be viewed as a series of three-letter genetic "words," or *codons*, with nucleotide bases serving as the "letters." During translation, it is the job of tRNA to associate the proper amino acids with each codon in the mRNA message.
14. tRNA molecules have a characteristic three-dimensional shape. Amino acids attach to one portion of the molecule. The *anticodon*, a complementary 3-nucleotide sequence that matches one particular mRNA codon, is found on another region of tRNA. Other portions of the tRNA molecule are involved with ribosome associations.

15. tRNA molecules become *charged* when one of a family of *activating enzymes* (e.g., aminoacyl-tRNA synthase) identifies an amino acid and attaches it to the correct tRNA.
16. Ribosomes are the sites of protein assembly in both prokaryotic and eukaryotic cells. Each ribosome consists of one heavy and one light subunit made of several different rRNAs and many proteins arranged in a precise pattern. Each ribosome has binding sites for two tRNAs and one mRNA molecule.
17. Translation cannot begin without formation of an *initiation complex*. This complex consists of the ribosomal light subunit bound to an appropriately-charged tRNA and the start codon of an mRNA message. The anticodon of a specific tRNA then forms hydrogen bonds with the initial mRNA codon. At this point, a heavy ribosome subunit attaches to complete the functional ribosome complex and translation begins.
18. Two RNA binding sites exist on each ribosome. The first mRNA codon and a charged tRNA bind to the P site (where the growing polypeptide forms). Other charged tRNAs bind to the A site (attachment location) in complementary accordance with the specific codons present in the mRNA sequence.
19. Once two tRNAs are in place on the ribosome, a peptide bond forms between the first and second amino acids. The first tRNA is then released from the P site. The second tRNA, bearing the dipeptide, shifts position from the A to P site. A third charged tRNA then enters the vacated A site. Translation of the polypeptide continues in this manner until an "end-chain" mRNA codon appears in the A site causing termination.
20. Multiple ribosomes often move sequentially down the length of an mRNA thread during translation, trailing their growing polypeptide chains behind. These *polysomes* are an efficient way of increasing the rate of protein synthesis.
21. Ribosomes are generalized "workbenches" for the translation process; they can therefore associate with any particular piece of mRNA and successfully translate the protein for which it codes.
22. The above description applies mainly to translation in prokaryotes. The process is somewhat more complex in eukaryotes, where several intervening steps occur between transcription and translation.
23. For proteins that remain suspended in the cytoplasm, translation takes place on "free" (unattached) ribosomes. Proteins intended for export from the cell, or for use in lysosomes and other membrane-bounded cell structures, are manufactured by ribosomes attached to the rough ER.
24. All protein synthesis begins on free ribosomes regardless of the protein's final destination. If specific amino acid chains known as *signal sequences* are present at the beginning of the polypeptide chain, then the protein is destined for export or incorporation into a membranous structure. These proteins and their ribosomes must attach to rough ER in order to complete translation.
25. The four possible nucleotides in DNA or RNA, taken three at a time, produce 64 possible codon combinations (e.g., UUU, UUC, UAC, etc.). Since there are only 20 amino acids, we say that the code is *degenerate*, meaning more than one codon codes for the same amino acid. This does not mean that the code is *ambiguous*, where one codon would code for more than one amino acid.
26. Some codons serve as translational punctuation marks. The *start codon* acts as an initiation signal for transcription, while the *stop codon* causes chain termination.
27. The genetic code is universal among living organisms on this planet. The only known exception involves minor alterations in mitochondrial DNA, the functional significance of which remains unknown.
28. Errors in replication, transcription, or translation are infrequent, but do occur. Heritable changes in genetic material are called *mutations*. They result from alterations in the nucleotide sequence of DNA, and provide the ultimate source of genetic variation for evolutionary change.
29. Genetic mutations are most easily detected if they have clear phenotypic effects. Some mutations are conditional and appear only under highly restrictive circumstances, such as with certain temperature-sensitive growth effects.
30. *Point mutations* affect the alleles of individual genes and may involve an alteration of only one nucleotide. *Chromosomal mutations* are more extensive, involving insertions, deletions, or inversions of many genes within entire chromosome segments.
31. *Missense mutations* result from a single base substitution that causes only one codon to change; they do not alter the mRNA reading frame for translation. The result is a mutant protein with only one erroneous amino acid. Such proteins are usually still functional, and may even confer an evolutionary advantage on the bearer.
32. *Nonsense mutations* have more dramatic effects, since base substitution produces a chain-terminator codon that causes protein translation to stop prematurely.
33. *Frame-shift mutations* involve actual insertions or deletions of nucleotide bases. The effect is to alter the reading frame for all subsequent codons, causing dramatic changes in protein structure and thus phenotype. Such changes often prove lethal, or are otherwise evolutionarily maladaptive.
34. Genetic changes can be induced by *mutagens*, substances or treatments that cause alterations of nucleic acid sequences. Some mutagens are chemical analogs of nucleotide bases. Others produce chemical changes in natural bases that lead to improper base pairing during DNA replication.
35. Chromosomal mutations can involve *deletions* (loss of genes), *duplications* (DNA copies not properly separated from the chromosome), *inversions* (reversed sequence of chromosome segment), or *translocations* (gene sequence retained but moved to another part of the chromosome, thus affecting synapsis during meiosis).

36. Mutation rates are typically less than 1 per 10^4 genes per DNA duplication (often as low as 1 per 10^9), and involve mostly point mutations. The vast majority are harmful to the organism and are selected against. Other mutations are neutral and have no selective effects. A very few mutations actually improve the organism's adaptation for its environment, and are thus favored evolutionarily.

Activities (for answers and explanations, see page 117)

- Using the universal genetic code depicted in Figure 11.25 of your textbook, determine the amino acid sequences that would result from the translation of each of the following mRNA molecules:

  Case 1: 5′ – A U G A C A U G U C G U G G G U G A – 3′

  Case 2: 5′ – U G A A C A U G U C G U G G G C G U – 3′

  Case 3: 3′ – G G U A C A U G A A C A G G G G U A – 5′

- Why must proteins be synthesized in a two-step process that first involves transcription, followed by translation?

Questions (for answers and explanations, see page 117)

1. In a species of *Neurospora*, it is known that ten genes have an important influence on the control of a biochemical pathway. How many auxotrophs related to this pathway can theoretically be made?
   a. One, since auxotrophs exhibit the most common genetic form of the trait.
   b. More than ten, since each gene that helps control the pathway could potentially mutate in many ways.
   c. Less than five, since two or more of these genes likely produce the same polypeptide product.
   d. None, since *Neurospora* is not a good organism for this kind of research.
   e. None, since autotroph strains cannot survive in the laboratory.

2. According to the one-gene, one-polypeptide theory
   a. all proteins are used as enzymes during DNA replication.
   b. each polypeptide codes for the production of a single nucleotide chain called a gene.
   c. genes can only be active in the presence of the polypeptide for whose production they code.
   d. each gene codes for production of a single polypeptide chain which can often influence phenotype.
   e. the polypeptide made by the gene incorporates itself into the chromosome next to the gene locus.

3. Which of the following represents a proper hierarchical organization of genetic information?
   a. Gene – nucleotide – DNA – chromosome
   b. Nucleotide – DNA – gene – chromosome
   c. Gene – DNA – chromosome – nucleotide
   d. Nucleotide – gene – DNA – chromosome
   e. DNA – gene – nucleotide – chromosome

4. When interpreting the universal genetic code during translation, the codons being read to produce the correct sequence of amino acids for a protein are found in molecules of
   a. normal DNA.
   b. Z-DNA.
   c. mRNA.
   d. tRNA.
   e. rRNA.

5. Which of the following is true about a DNA triplet code?
   a. Each DNA base codes for three proteins.
   b. Each gene codes for three proteins.
   c. Each triplet codes for three proteins.
   d. It takes three genes to code for one protein.
   e. Each amino acid in a protein is coded for by a set of three DNA bases.

6. Of the molecules listed below, which one is *not* directly involved in translation?
   a. DNA
   b. tRNA
   c. mRNA
   d. Ribosomes
   e. Enzymes

7. Transfer RNA is still sometimes referred to as an adaptor molecule, in the sense that Francis Crick used this label, because
   a. tRNA adapts to changing cell environments by controlling the rate of mRNA production.
   b. tRNA connects the mRNA message with appropriate amino acids for protein synthesis.
   c. transcription cannot be accomplished without tRNA initiators binding to the DNA template.
   d. the shape of the other RNAs is based on the kind of tRNA present in the cell.
   e. None of the above

8. Which of the following best represents the sequence of events that takes place as a free amino acid is incorporated into a developing polypeptide chain?
   a. Free in cell cytoplasm – binds with mRNA on ribosome – attaches to tRNA – forms peptide bond
   b. Binds with mRNA on ribosome – free in cell cytoplasm – attaches to tRNA – forms peptide bond
   c. Free in cell cytoplasm – attaches to tRNA – binds with mRNA on ribosome – forms peptide bond
   d. Attaches to tRNA – forms peptide bond – binds with mRNA on ribosome – free in cell cytoplasm
   e. Attaches to tRNA – free in cell cytoplasm – forms peptide bond – binds with mRNA on ribosome

9. Digestive enzymes destined for use in the food vacuole of an amoeba are made
   a. primarily on ribosomes attached to the rough ER.
   b. exclusively on free ribosomes in the cytoplasm.
   c. by first translating the appropriate genes into mRNA and then transcribing this into protein.
   d. using the enzyme reverse transcriptase.
   e. whenever polysomes are present within the cell.

10. Consider the short DNA sequence 3′ – C G A C A T A C G T G C – 5′. If this sequence suddenly mutated to become 3′ – C G A C A A A C G T G C – 5′, what kind of mutation would have to be involved?
   a. Insertion
   b. Deletion
   c. Inversion
   d. Substitution
   e. Chromosomal

Integrative Questions (for answers and explanations, see page 118)

1. Frame-shift mutations are particularly deleterious because
   a. entire segments of genetic material are lost from the chromosome.
   b. they can alter the kinds of amino acids coded for by large portions of a gene.
   c. they tend to affect only initiation codons, thus inhibiting the start of translation.
   d. they produce a situation where DNA can no longer replicate, leading to chromosomal abnormalities.
   e. they primarily affect genes that code proteins necessary in transcription.

2. If you raise several generations of bacterial cells on a nutrient medium in which all the cytosine is radioactively labeled, in which of the following molecules would you expect to find evidence of radioactive cytosine?
   a. mRNA
   b. Z-DNA
   c. Normal DNA
   d. tRNA
   e. All of the above

3. Which of the following is *not* a product of translation?
   a. mRNA
   b. DNA polymerase
   c. Certain components of ribosomal subunits
   d. Signal sequence peptides
   e. Reverse transcriptase

12

Molecular Genetics of Prokaryotes

CHAPTER LEARNING OBJECTIVES—after studying this chapter you should be able to:

- ❑ Provide reasons why prokaryotes and viruses are good models for the study of molecular genetics.
- ❑ Differentiate between asexual (clonal) reproduction and the process of conjugation in bacteria.
- ❑ Explain how bacterial colonies are produced under laboratory conditions, and describe how bacteriophages can be used to test for genetic mutations in bacteria.
- ❑ Discuss how auxotrophic and prototrophic strains of bacteria are produced, and how these strains can be used to study the molecular genetics of prokaryotes.
- ❑ Discuss methods employed by molecular geneticists to isolate and maintain specific mutant strains of bacteria.
- ❑ Explain the process of replica plating and describe its usefulness in the study of bacterial genetics.
- ❑ Define and use the terms "F factor," "F-pili," "Hfr mutant", and "F^+" and "F^-" mating types in bacteria.
- ❑ Explain how the technique of interrupted mating can be used to map the bacterial chromosome.
- ❑ Differentiate between sexduction and conjugation and describe the role of each in genetic transfer between bacteria.
- ❑ Discuss the difference between the lytic and lysogenic life cycles of bacteriophage, and explain the role of a prophage in the lysogenic cycle.
- ❑ Compare and contrast the process of restricted transduction with that of general transduction.
- ❑ Define and discuss the following terms: "episome," "plasmid," "replicon," and "R factor."
- ❑ Explain what is meant by the terms "transposable element" and "transposon," and relate these to the function and evolution of plasmids, F factors, and Hfr mutants.
- ❑ Discuss the difference between inducible and constitutive enzymes, and explain the role of inducers in the function of such enzymes.
- ❑ Describe the role of structural genes in prokaryotic cell metabolism, and discuss their relationship to polycistronic messengers.
- ❑ Discuss the basic structure and function of an operon using the following terms: "promoter," "repressor," "corepressor," "operator," "regulator gene," and "inducer."
- ❑ Differentiate between inducible and repressible operon systems, and explain the molecular mechanisms that regulate the activity of each.
- ❑ Compare and contrast the molecular control mechanisms of the *lac* operon with those of the tryptophan operon in *E. coli*.
- ❑ Describe the control mechanism for an operon regulated by enhanced promoter efficiency, and compare this mechanism to those of repressible and inducible operon systems.
- ❑ Discuss the differences between negative and positive transcriptional control of bacterial operons.

MUTATIONS IN BACTERIA AND BACTERIOPHAGES • BACTERIAL CONJUGATION • BACTERIOPHAGES (pages 273–283)

Key Concepts

1. Prokaryotic cells (bacteria) and viruses offer many advantages for the study of molecular genetics. They are small, easy to manipulate, and breed rapidly in large numbers. Most importantly, they have less than 1/1000 as much DNA per cell as do humans and other eukaryotes.
2. Although much of their reproduction is asexual, bacteria and viruses do engage in genetic recombination through mechanisms analogous to sex. Their genetic material can also mutate.
3. The bacterium *E. coli* is a representative prokaryote. Asexual reproduction yields offspring that are exact genetic copies (i.e., clones) of the original parent cell. Cloning can be maintained indefinitely under favorable conditions, whether in nature or the laboratory.
4. Bacterial mutation is demonstrated in experiments where T4 bacteriophages are introduced into a culture dish supporting a well-developed *E. coli* colony (or "lawn"). Shortly after the virus is introduced, small circular openings called *plaques* appear in the lawn, indicating areas where viruses have killed bacteria.
5. In a few plaques, small colonies of virus-resistant bacteria arise due to the clonal growth of one mutant bacterium. If this mutant is grown alone on another culture dish, intro-

duced bacteriophages will generate no plaques, indicating that the mutation for viral resistance is heritable.

6. Mutations may also occur in the bacteriophage, allowing it to better circumvent a host's defenses. This demonstrates the genetic basis for evolutionary change: mutations provide the raw material for phenotypic changes, and environmental selection pressures determine the viability of these changes.

7. The details of genetic recombination in bacteria were investigated by the Lederbergs and Tatum during studies of metabolic auxotrophs in *E. coli* K12. They observed that when certain auxotrophic combinations were grown together, occasionally wild-type (prototrophic) colonies would appear.

8. One explanation for this outcome is that the prototrophs arise by spontaneous mutation. However, mutation is too rare an event to explain the high rate at which prototrophs appear.

9. A second explanation is that prototrophs are transformed bacteria that have been changed by a transforming principle which was secreted into the growth medium by one of the auxotrophic strains. This possibility was also ruled out by Davis' U-tube experiments, where two auxotrophic strains growing in the same culture were prevented from touching one another by a filter. Even though the growth medium (where transforming DNA would exist) could freely move between both sides of the culture tube, no wild-type bacteria appeared.

10. A third explanation is *conjugation*, in which bacteria pair and mix their genetic material. This type of genetic recombination produces some new bacteria that are wild-type for the mutant genes in their parents.

11. New mutant strains of bacteria are created by exposing wild-type strains to procedures such as ultraviolet or X-ray irradiation and to mutagenic chemicals. The treated bacteria are then cultured on a complete nutrient medium to increase their numbers.

12. When the colony is of sufficient size, a sample is transferred to a new culture medium lacking the nutrient needed by the desired mutant. This new medium also contains the drug penicillin, which inhibits cell wall formation in actively growing bacteria.

13. Prototrophic bacteria flourish on this new medium, but by engaging in active growth they suffer penicillin-induced death. The desired mutants do not grow, and thus avoid the effects of the drug. Mutant bacteria may then be transferred to yet another medium that lacks penicillin but contains the necessary nutrient, where they grow.

14. *Replica plating* is a technique for identifying and isolating bacterial strains exhibiting particular phenotypes. Mixed populations of bacteria are grown on complete medium in a Petri dish, and then pressure-transferred to a velvet cloth. Many exact replicates of this sample are made by stamping the cloth on different growth media lacking specific nutrients. By keeping track of which colonies grow on which medium, recombinant forms can be isolated.

15. Conjugation in *E. coli* is a one-way process where genetic material flows from donor cell (male) to recipient cell (female). During conjugation, the female acquires a small extrachromosomal piece of DNA called a *fertility factor* (F), which transforms her into a male.

16. Male *E. coli* are thus F^+, and females F^-. One mating type can change into the other either by gaining or losing the fertility factor. Genes on the F factor control the presence of hairlike, tubular projections called *F-pili*, through which DNA is transferred between cells.

17. *Hfr* mutants are bacteria that act as high-frequency donors of genetic material. They yield a higher than normal percentage of recombinant offspring because their F factor is actually incorporated into the chromosome.

18. Gene location on the bacterial chromosome can be mapped using the technique of *interrupted mating*. Conjugating bacteria are interrupted at different times during the mating process, permitting different amounts of DNA to be transferred between cells. By comparing offspring characteristics with the known duration of conjugation, a gene map can be constructed. The map reveals that the bacterial chromosome is circular.

19. Interrupted mating studies reveal that in Hfr males the F factor is actually incorporated into the chromosome, and that the insertion point of this factor varies with different Hfr strains. The F factor also marks the point where the bacterial chromosome opens during conjugation. This opened end of the chromosome always leads the way into the female.

20. Genetic transfer between Hfr and F^- cells begins when one strand of the Hfr chromosome is nicked and starts moving into the F^- cell through the F-pilus. This stimulates DNA replication of the remaining Hfr chromosome. Once inside the F^- cell, the transferred Hfr DNA also replicates and becomes double-stranded. It must then incorporate into the F^- cell's chromosome by crossing over for recombination to occur.

21. *Sexduction* is a variation of the conjugation process. Here, the F factor actually separates from the chromosome, taking some chromosomal markers with it. The modified autonomous F factor (now called an F′ factor) is a circular DNA molecule capable of introducing new genetic material into the F^- cell.

22. Phage viruses can also undergo recombination. They too have a single, circular chromosome which, when injected into a host bacterium, may recombine with other phage chromosomes to produce novel phenotypes.

23. Bacteriophages often show a *lytic* life cycle in which the host cell ruptures shortly after new viruses are produced. Bacteriophage, can also exhibit a *lysogenic* life cycle, where viral DNA known as a *prophage* becomes incorporated into the host cell's DNA for several generations. Environmental changes later induce the host to synthesize new viruses, which leads to cell lysis.

24. Bacteriophages can also act as vectors to move bacterial DNA between cells. In *restricted transduction*, only those bacterial genes closest to where the prophage actually attaches to the bacterial chromosome are

transferred. Transducing phages cause infected bacteria to become lysogenic.

25. *General transduction* occurs when some portion of the bacterial chromosome, minus the prophage, gets incorporated into the phage coat. When the resulting particle (which lacks phage genes and is not infectious) attaches to a bacterium, that cell gains a segment of bacterial DNA, which may recombine with its own chromosome. Unlike restricted transduction, general transduction may move any portion of the bacterial chromosome.

Activities (for answers and explanations, see page 118)

- The cultures shown below represent the original master plate and one replica from an experiment in which a molecular geneticist is trying to produce new mutants of *E. coli*. In particular, she is looking for a mutant that requires the nutrient methionine. Based on what you see here, has the geneticist succeeded? How can you tell?

Normal medium

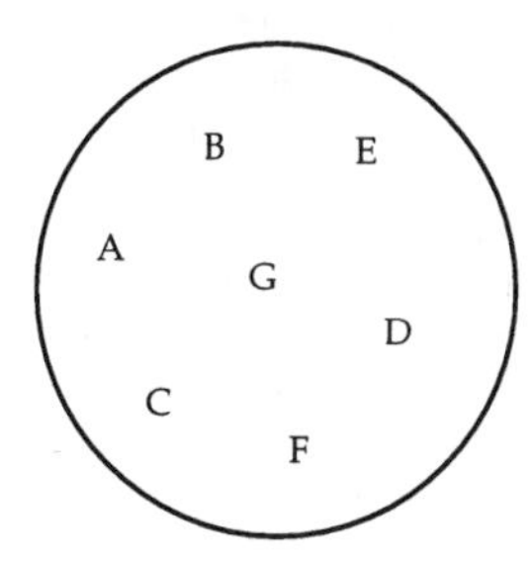

Lacks methionine

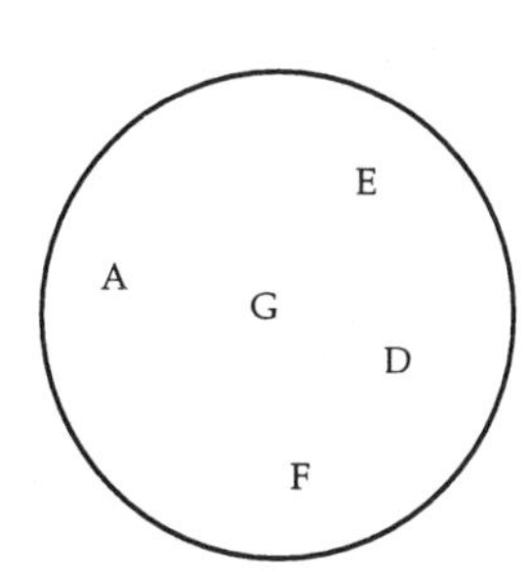

- In the same experiment the geneticist makes other replica plates, including one on a growth medium lacking threonine. The results are shown below. Is it possible to tell from these data if she has any mutant forms that might require *both* methionine and threonine? Explain your answer.

Normal medium

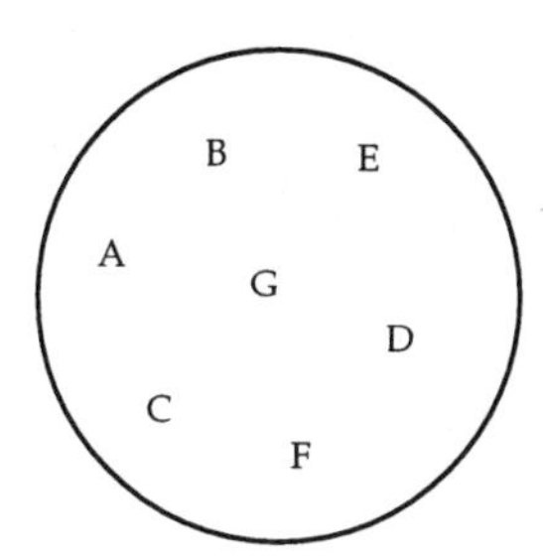

Lacks threonine

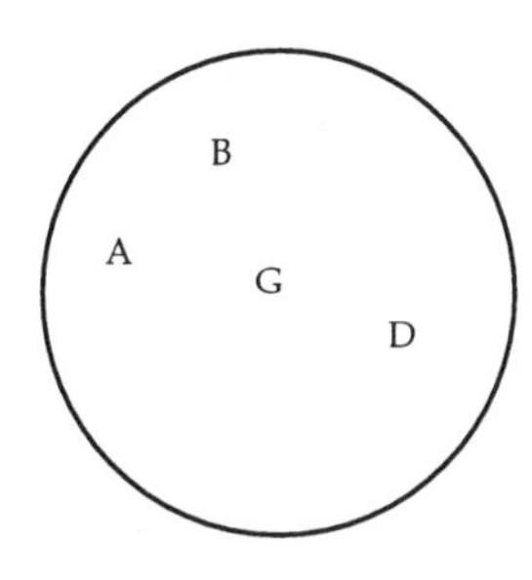

Questions (for answers and explanations, see page 118)

1. Which of the following is *not* a reason why prokaryotes and viruses make good subjects for the study of molecular genetics?
   a. Their DNA can mutate.
   b. They are small in size.
   c. They are easy to manipulate.
   d. They typically breed rapidly and in large numbers.
   e. They possess large amounts of DNA.

2. Clear plaques appear in a cultured bacterial lawn when it is exposed to an appropriate bacteriophage. This occurs because
   a. the viruses in the area of the plaques are killed by the bacteria, exposing the culture medium below.
   b. the bacteria in the area of the plaques are killed by the virus, exposing the culture medium below.
   c. bacteriophages secrete a clear fluid which covers up the bacterial lawn at the plaque locations.
   d. antiviral drugs are made by the bacteria in plaque areas, thus killing the virus and clearing space.
   e. None of the above

3. In the studies of Joshua and Esther Lederberg and Edward Tatum on sexual reproduction in bacteria, which of the following mechanisms is the basis for the appearance of prototrophs on the minimal medium plates?
   a Transforming substances that convert one phenotype into another
   b. Lysogenic induction
   c. Spontaneous mutation of appropriate genes
   d. DNA transfer via conjugation
   e. Interrupted mating

4. An interrupted mating experiment is performed using *E. coli*. Five phenotypic characters are monitored, and it is found that only those cells that engaged in conjugation for the longest time express character number five. What does this tell you about the gene for character five?
   a. It is a recessive trait that is expressed in ⅕ of all offspring.
   b. It controls the transcription of the other four genes.
   c. It is situated furthest away from the F factor on the chromosome.
   d. It regulates the ability of *E. coli* to engage in sexduction.
   e. Nothing useful

5. Which of the following is true of bacterial conjugation?
   a. Genetic exchange via conjugation always occurs from male to female mating types.
   b. Mating types can change from one form to another via conjugation.
   c. Conjugation by Hfr mutants tends to produce more recombinant offspring than non-Hfr matings.
   d. After conjugation, the original F^+ cell in a mating of two *E. coli* remains F^+.
   e. All of the above

6. Sexduction differs from regular bacterial conjugation in that
   a. sexduction only occurs between F^- bacteria.
   b. conjugation only occurs between Hfr bacteria.
   c. conjugation never involves the transfer of new genetic material to the recipient cell.
   d. sexduction permits genetic material from the donor chromosome to travel along with F factor DNA.
   e. sexduction occurs only in bacteriophages.

7. Which of the following is *true* of general transduction in bacteria?
   a. Bacterial DNA is incorporated into a phage coat.
   b. Prophages play an active role in the process.
   c. The infected cell is induced to become lysogenic.
   d. Only highly restricted portions of bacterial chromosomes can be moved this way.
   e. Replica plating is the best way of promoting the likelihood that this process will occur.

8. Which of the following statements about Hfr bacteria is *not* true?
   a. They produce higher than normal numbers of recombinant offspring.
   b. Their F factor is actually incorporated into the bacterial chromosome.
   c. The F factor marks the spot where the chromosome opens during conjugation.
   d. The leading end of the opened chromosome enters the female cell last.
   e. Interrupted mating can be used to study the genetics of Hfr mutants.

9. The patterns of mutations seen in bacteria and viruses demonstrate which of the following ideas regarding evolution?
   a. Mutations almost never occur in eukaryotes.
   b. Mutations are an important source of raw material for evolutionary change.
   c. Mutations are limited to *E. coli* K12 and bacteriophage T4.
   d. Bacterial and viral mutations only occur via conjugation.
   e. Mutations are limited exclusively to Hfr cells and bacteriophage.

EPISOMES AND PLASMIDS • TRANSPOSABLE ELEMENTS • CONTROL OF TRANSCRIPTION IN PROKARYOTES (pages 283–289)

Key Concepts

1. F factors and prophages are *episomes*, nonessential genetic elements that exist either independently of the host chromosome or incorporated into it. Neither arise from mutation; they are only acquired by infection.

2. *Plasmids* are another nonessential genetic element in bacteria that exist as self-replicating circles of DNA. They are not required for bacterial survival, nor are they ever incorporated into the bacterial chromosome.

3. *R factors* are an important class of plasmids that confer drug resistance in bacteria. Although they probably existed long before the advent of modern medicine, the increased drug resistance exhibited by microorganisms during recent decades is likely due to R factor actitivty.

4. Plasmids must exist as *replicons* in order to be maintained within a bacterial population. Each replicon has its own origin of replication; it can thus divide independently and in synchrony with the bacterial chromosome.

5. Another agent for gene transport is a chromosomal segment or type of plasmid DNA known as a *transposable element*. These independently replicating, mobile elements carry the genes for their own insertion into DNA molecules, often with dramatic disruptions of normal gene activity.

6. Large transposable elements (>5,000 base pairs) are called *transposons*. Transposons may include additional genes besides the transposable elements themselves.

7. Also known as "jumping genes," transposable elements actually move at very low frequencies. Transposition is accomplished by a complex and poorly understood mechanism.

8. Plasmid evolution and F factor development in Hfr males are probably both the result of transposable elements. For example, the same drug-resistance genes found in transposable elements also exist in R factor plasmids, making it likely that R factors gained their resistance from transposons.

9. Cells do not require all of the substances they are capable of synthesizing at all times. Gene activity must therefore be regulated. Substances that cause enzyme synthesis are called *inducers*, and the enzymes made are known as *inducible enzymes*. If enzymes are made continuously at fairly constant rates, then they are known as *constitutive enzymes*.

10. The genetic blueprint for enzyme structure is contained in *structural genes*. When multiple enzymes are involved in the same biochemical pathway, the structural genes coding for those enzymes are usually adjacent on the chromosome. These genes are often simultaneously transcribed to form a *polycistronic messenger*, an mRNA sequence that codes for the entire set of enzymes.

11. Polycistronic messengers lead to the manufacture of several polypeptide chains because they contain nucleic acid punctuation marks to indicate the end of one chain and the start of another. The same ribosome may therefore work down the entire polycistronic messenger, translating several different proteins in sequence.

12. Some genes need to be transcribed more often than others. *Promoters* are segments of DNA on which RNA polymerase binds to begin the transcription process. Each set of structural genes has its own promoter whose binding efficiency for RNA polymerase controls the frequency with which those particular enzymes are produced by the cell.

13. Genes must be able to halt transcription when particular enzymes are no longer needed; that is, they must be able to negatively influence transcription rate. An often-employed strategy is to use a physical obstacle to block mRNA synthesis. A specific protein known as a *repressor* molecule does this by binding to a second site on the

DNA called the *operator*. Such controllable units of transcription are known as *operons*.

14. An operon consists of a repressor binding site (the operator), and one or more structural genes. Regulation of the operon occurs through control of repressor activity. When the repressor binds to the operator, transcription of the structural genes is halted. However, when the repressor binds to certain inducers, the repressor changes shape (i.e., becomes inactive) and transcription takes place.

15. The *lac* operon of *E. coli* is an example of an inducible operon. Lactose serves as the inducer; it combines with (inactivates) the repressor to permit mRNA transcription for the enzymes needed in lactose metabolism. If lactose concentration drops, the repressor is reactivated and blocks mRNA synthesis. Thus, by using the lactose stimulus itself as an inducer, the availability of enzymes for lactose breakdown is keyed directly to the need for those enzymes.

16. *Regulatory genes* code for repressor proteins; they may be located near to, or at some distance from, the operon they control.

17. In their unregulated condition, inducible operons are continuously "on." They make their structural gene products unless turned off by the repressor. Some regions in the operon make proteins whose only function is to control other genes, while other regions (operators and promoters) do not code for protein synthesis and may never be transcribed.

18. It is also useful for prokaryotes to switch off some operons in the presence of high concentrations of substances that are ordinarily manufactured by the cell. These *repressible* operons turn off mRNA synthesis in response to sudden increases in the concentration of important molecules. The tryptophan operon of *E. coli* is an example.

19. In repressible systems such as the tryptophan operon, the operon cannot be shut down unless the repressor first combines with a *corepressor*, which is often the stimulus substance itself (e.g., tryptophan) or some chemical analogue. Once combined, the repressor/corepressor complex blocks the operator to inhibit gene transcription.

20. Subtle but important differences exist between these two kinds of operons. Inducible operons use an inducer to inactivate the repressor. This frees up the operator, allowing the promoter to bind RNA polymerase and initiate transcription. Repressible systems use a corepressor to activate the repressor molecule. The activated repressor then binds to the operator and blocks transcription.

21. Some operons containing genes for catabolic enzymes are not regulated by being turned on or off. Rather, they adjust the efficiency of promoter function. In such systems, *CRP* (cAMP receptor protein) first forms a complex with cAMP. The CRP–cAMP complex then attaches near the promoter site to facilitate the binding of RNA polymerase. In this way polymerase binding and transcription rate are improved by as much as 50-fold.

22. Because they prevent transcription via the action of repressor molecules, both inducible and repressible operon systems are examples of negative transcriptional control. By contrast, promoter operon systems are examples of positive transcriptional control because they actually enhance the rate of transcription.

Activities** **(for answers and explanations, see page 118)

- Consider a bacterial operon that controls the production of four enzymes needed in the synthesis of arginine by bacteria. It is known from chromosome mapping studies that the structural genes for these enzymes are all adjacent to one another. Assuming that this is a repressible enzyme system, make a schematic diagram like the ones in your textbook that shows how the components of the arginine operon might be arranged on the chromosome, and briefly describe how this operon would function.

- Refer to the discussion of R factors and drug resistance on page 284 of your textbook. Propose a hypothesis to explain why R factors might be more widespread in certain microorganisms today than they were in the past.

Questions (for answers and explanations, see page 118)

1. Episomes and plasmids are important players in the molecular genetics of prokaryotes. Which of the following is *neither* an episome nor a plasmid?
   a. F factor
   b. R factor
   c. F pilus
   d. Prophage
   e. All of the above are plasmids or episomes.

2. The reason that plasmids must be replicons in order to survive within a bacterial population is because
   a. replicons allow the plasmid to integrate into the host chromosome prior to DNA replication.
   b. plasmids use their replicons to extract themselves from the host chromosome after sexduction.
   c. bacteria will reject the plasmid during conjugation unless the replicon is present.
   d. the replicon allows the plasmid to divide independently of the host chromosome.
   e. polycistronic messengers identify the plasmid by its replicon.

3. Large segments of DNA that can make multiple copies of themselves and then insert these copies anywhere within the chromosomal material are known as
   a. R factors.
   b. transposons.
   c. episomes.
   d. regulator genes.
   e. promoters.

4. Which of the following statements about transposable elements is *not* true?
   a. Independent replication is an essential feature of their activity.
   b. They can be made of either chromosomal or plasmid DNA.
   c. They have probably had important influences on the evolution of certain plasmids.
   d. R factor genes probably have their origins in certain transposable elements.
   e. They have very high rates of movement within the chromosomal material of the cell.

5. The use of inducible enzymes is usually the cell's best option in situations where
   a. the substance upon which the enzyme acts is always present.
   b. constitutive enzymes can act as corepressors of the inducible enzyme.
   c. the substance upon which the enzyme acts is present only on an irregular basis.
   d. constitutive enzymes control the presence or absence of the inducible enzyme's target substance.
   e. plasmids are needed to act on a continuously available nutrient that is required for cell growth.

6. In a repressible operon, the repressor molecule binds with the ________ before influencing ________ by interaction with the operator.
   a. corepressor; gene transcription
   b. regulator; protein translation
   c inducer; conjugation
   d. promoter; gene recombination
   e. plasmid; episome transfer

7. Polycistronic messengers are
   a. manufactured from tRNA templates and used during transcription.
   b. a category of sex-determining factors that regulate bacterial recombination.
   c. not associated with regulation or functioning of prokaryotic operon systems.
   d. made of mRNA and code for several polypeptide chains whose activity is closely related.
   e. an important structural component of prokaryotic ribosomes.

8. With which of the following operon systems would you expect to find associated regulator genes?
   a. *Lac* operon
   b. Tryptophan operon
   c. Inducible operon
   d. Repressible operon
   e. All of the above

9. The concentration of enzyme X in a bacterium is continuously maintained at a relatively constant level, regardless of changes in cell conditions. Enzyme X is best considered
   a. an inducible enzyme.
   b. a repressible enzyme.
   c. a constitutive enzyme.
   d. a repressor protein.
   e. a corepressor protein.

10. In bacterial operons where the mechanism of regulation is by enhancement of promoter efficiency, which of the following is true?
    a. The structural genes usually code for catabolic enzymes.
    b. cAMP receptor protein is involved.
    c. A CRP–cAMP complex is formed.
    d. Transcription rates can sometimes be increased by 50-fold.
    e. All of the above

13

Molecular Genetics of Eukaryotes

CHAPTER LEARNING OBJECTIVES—after studying this chapter you should be able to:

- ❑ Describe the features of eukaryotic organisms that make the regulation of their gene expression more complex than that of prokaryotic organisms.
- ❑ Describe the process of DNA hybridization and how it led to the discovery of noncoding and repetitive DNA in eukaryotes.
- ❑ Identify introns and exons in the hybridization patterns formed with mRNA hybridized to the coding DNA.
- ❑ Describe the three classes of DNA defined by the rate of reannealing and the characteristics of each class.
- ❑ Discuss the structure of the telomere and how its structure solves the problem associated with the replication of a linear chromosome.
- ❑ Explain the three requirements for the construction of artificial chromosomes.
- ❑ Describe the importance of tranposable elements in the DNA of eukaryotes and to the endosymbiotic theory for the origin of mitochondria and chloroplasts.
- ❑ Describe what is meant by a gene family and how one might arise.
- ❑ Describe what is meant by a pseudogene and the two mechanisms that create pseudogenes.
- ❑ Explain the major steps in the processing of heterogenous nuclear RNA to form a mature, translation-ready RNA molecule.
- ❑ Describe the principal features of the RNA splicing process in eukaryotes.
- ❑ Describe several examples of the transcriptional control of gene expression called gene inactivation, including those examples involving euchromatin and heterochromatin, chromosome puffs, and DNA methylation.
- ❑ Discuss the existence of the Barr body relative to the concepts of euchromatin and heterochromatin.
- ❑ Describe important examples of gene duplication as a mechanism for amplifying gene product.
- ❑ Interpret electron micrographs showing transcription in the nucleoli of amphibian egg cells.
- ❑ Discuss the involvement of steroid hormones in selective gene transcription.
- ❑ Describe differences in transcriptional control and the involvement of promoters in prokaryotes and eukaryotes.
- ❑ List the activities of the three types of RNA polymerase.
- ❑ Explain the involvement of the TATA box, transcription factors, and RNA polymerase II in forming the transcription complex.
- ❑ Provide some examples of translational control involving longevity of mRNAs and modification with the G cap.
- ❑ Provide some examples of posttranslational control involving activation, inactivation, and insertion into the proper cellular environment.

EUKARYOTES AND EUKARYOTIC CELLS • HYBRIDIZATION OF NUCLEIC ACIDS • EUKARYOTIC GENE STRUCTURE • REPETITIVE DNA IN EUKARYOTES (pages 293–298)

Key Concepts

1. Eukaryotic cells are larger, contain numerous membrane-bounded organelles, and have much more DNA than prokaryotic cells.
2. Unlike prokaryotes, unicellular eukaryotes compartmentalize activities within different organelles and commonly employ sexual reproduction.
3. Multicellularity in eukaryotes led to division of labor among different cell types, but because genes must be turned on and off at appropriate times, regulation of gene expression is complex.
4. Eukaryotic genes contain stretches of noncoding DNA that must be removed during a processing step not found in prokaryotes.
5. *Nucleic acid hybridization* has been a powerful tool in the study of gene expression in eukaryotes. Nucleic acid hybridization depends on complementary base pairing.
6. When warmed, the two strands of a DNA molecule will separate as hydrogen bonds are broken to form single-stranded, *denatured* DNA. Cooling will allow complementary strands to *reanneal*. Typically, special enzymes are used to initially cut the DNA into shorter, more manageable fragments .

7. If an RNA transcript is mixed with a sample of denatured (single-stranded) DNA, the RNA transcript will hybridize with the DNA strand that codes for it.

8. The presence of noncoding DNA in eukaryotes was first discovered when cytoplasmic mRNA was allowed to hybridize with fragmented, denatured DNA from the same cell. Double-stranded DNA–mRNA coding sections were associated with thin, single-stranded DNA loops and thick, double-stranded DNA loops formed by the noncoding DNA.

9. The coding sections of a gene are called *exons*. The noncoding sections of a gene are called *introns*. Typical eukaryotic genes consist of several exons separated by intervening introns.

10. Products of introns are edited out after transcription, so that the base sequence of all the gene's exons determines the mature mRNA product of the gene.

11. Nucleic acid hybridization studies showed that not all eukaryotic DNA reanneals at the same rate, indicating that the number of copies of eukaryotic genes in the *genome* varies widely.

12. The class of DNA that anneals most slowly consists of *single-copy sequences*, which include the genes coding for most enzymes and structural proteins.

13. *Highly repetitively DNA* reanneals most quickly. This class of DNA may represent a third or more of the genome and much of it is located near the centromere, suggesting involvement in cell division. It is usually not transcribed.

14. *Moderately repetitive DNA* has intermediate reannealing rates and copy numbers of hundreds to thousands per genome. Included in this class of DNA are duplicate genes, such as those for rRNA, the ends of eukaryotic chromosomes called *telomeres*, and *transposable elements*.

15. Telomeres are sequences of moderately repetitive DNA located at the ends of eukaryotic chromosomes. The repeated DNA loops back on itself and the looped-back section can serve as a primer for completion of the lagging strand during DNA replication.

16. Artificial chromosomes consist of three essential components: an origin of replication where DNA polymerase can attach, a centromere (essential for proper movement of chromosomes during mitosis and meiosis), and telomeres, so that replication can be completed at the end of the chromosome.

17. *Barbara McClintock* first discovered transposable elements in maize. Unlike that of prokaryotes, a transposable element in eukaryotes requires an RNA intermediate before it can move to another location within the genome.

18. The cellular functions of the gene products of transposable elements have not been discovered.

19. By causing insertions, transpositions, deletions, and inversions, transposable elements are an important source of mutation in eukaryotes.

20. Some evidence suggests that transposable elements may have moved certain genes from the DNA of mitochondria and chloroplasts to the nuclear DNA where they now reside. The original genes in the mitochondria and chloroplasts may then have been lost.

Activities (for answers and explanations, see page 119)

1. The following diagram shows a section of DNA with two exons and one intron during hybridization with complementary mRNA. Label (1) the coding DNA strands, (2) the noncoding DNA strand, (3) the mRNA, (4) exon 1, (5) exon 2, and (6) the intron.

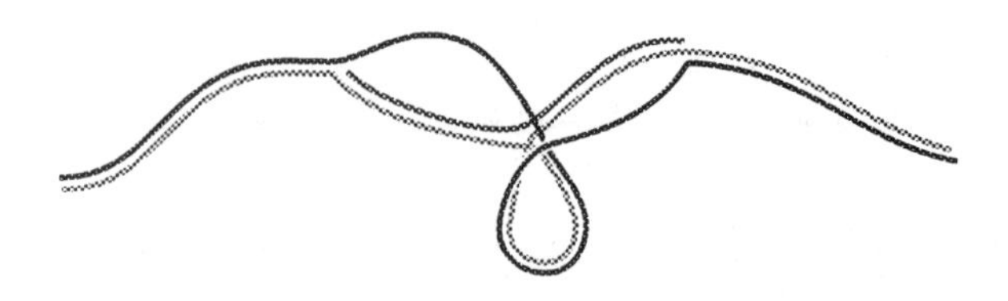

2. In the space provided below, diagram how the DNA–RNA shown above would appear if there was no intervening intron present.

Questions (for answers and explanations, see page 119)

1. Place numbers to the left of the following events to indicate their correct sequence in a typical application of mRNA–DNA nucleic acid hybridization.

   __5__ Add mRNA
   __2__ Warm the nucleic acid mixture
   __3__ Observe denaturation
   __6__ Cool the nucleic acid mixture
   __7__ Observe reannealing
   __1__ "Cut" the DNA into small fragments
   __4__ Immobilize DNA on nitrocellulose filter

2. Which of the following DNA types should have the *fastest* reannealing rate during nucleic acid hybridization?
   a. DNA coding for enzymes
   b. DNA coding for cellular structures
   c. DNA located near the centromeres
   d. DNA coding for ribosomal subunits (rRNA)
   e. DNA from the ends of chromosomes

3. Which one of the following features is *not* characteristic of the telomeres found in eukaryotic chromosomes?
   a. Serves as a primer for completion of the leading strand
   b. Found only in linear chromosomes
   c. Consists of moderately repetitive DNA
   d. Interacts with the last Okazaki fragment
   e. Form loops through complementary base pairing

4. Which one of the following statements about transposable elements is *not* true?
   a. Transposable elements in eukaryotes require an RNA intermediate before they can move to a different location.
   b. Transposable elements are an example of moderately repetitive DNA.
   c. Transposable elements may have been important in the endosymbiotic origin of mitochondria and chloroplasts.
   d. The gene products of many transposable elements have important cellular functions.
   e. Insertion of transposable elements are an important source of mutation for many genes.

5. Circle all of the following that need to be present in an artificial chromosome designed to be inserted into a yeast cell.
   a. Telomeres at each end
   b. A structural gene with introns and exons
   c. An origin of replication
   d. A centromere
   e. A transposable element

6. A gene that includes 1,200 nucleotides codes for a protein consisting of 250 amino acids. How many of the nucleotides reside in exons and how many reside in introns?

   _750_ Nucleotides in exons

   _450_ Nucleotides in introns

GENE DUPLICATION AND GENE FAMILIES • RNA PROCESSING IN EUKARYOTES • CONTROL OF GENE EXPRESSION IN EUKARYOTES (Transcriptional Control: Gene Inactivation) (pages 298–302)

Key Concepts

1. Duplicate gene copies can be formed by several processes, such as *unequal crossing over* and by the action of transposable elements.
2. Multiple copies of a gene can sustain different mutations thereby producing a *gene family*. The genes of a gene family may be clustered together on the same chromosome or may be on different chromosomes.
3. Human *hemoglobin* is a tetramer composed of four globin polypeptide subunits; two α *globins* and two β *globins*. A family of globin genes code for the various types of α and β globins that are found in different developmental stages of humans. Similarities in amino acid sequence indicate that all members of the globin gene family descended from a common ancestral gene.
4. Genes for α-like and β-like globins are organized into different clusters. Like many gene families, the different globin genes differ more in their introns than in their exons.
5. *Pseudogenes* are nonfunctional genes that are not expressed; they are usually gene family members that have been extensively modified and may have inactived promoter regions or exons with nonsense mutations. Some pseudogenes may have been functional in the past, but their functions are now performed by other gene family members.
6. Some pseudogenes (called processed pseudogenes) were derived by *reverse transcription* of mRNA. They lack introns and are not clustered with their gene family members.
7. The original product of transcription is called *heterogeneous nuclear RNA*, or *hnRNA*; it contains the RNA transcript of the gene's exons and any introns that were present.
8. The first step in RNA processing is the addition of a *cap* consisting of a modified form of *guanosine triphosphate* to the 5' end of the hnRNA. The cap is later necessary for binding to a ribosome.
9. Next, a chain of 100–200 adenine nucleotides, called a *poly A tail*, may be added to the 3' end. This step is called polyadenylation. The tail may protect the RNA from degradation.
10. Neither cap nor tail is coded for in the DNA. A significant fraction of hnRNA molecules in mammalian cells lack poly A tails.
11. Removal of the introns, called *RNA splicing*, is the next step in RNA processing. In the case of ovalbumin, over 75% of the nucleotides are removed as seven introns are spliced out and eight exons joined together.
12. The boundaries between introns and exons consist of short sections of DNA called *consensus sequences*, so called because the base sequences are very similar in many genes.
13. Small nuclear RNA molecules, or *snRNA*, within the nucleus have regions complementary to the consensus sequences. The snRNA first binds to proteins to form a small nuclear ribonucleoprotein particle or *snRNP*. Six different snRNP molecules form a *spliceosome*; some bind to the beginning and end of the intron, some bind within the intron.
14. snRNP molecules interact to form a loop or *lariat* of the intron transcript. The exons at the base of the lariat are spliced together and the excised intron is later degraded in the nucleus.
15. The RNA precursor of rRNA can catalyze the removal of its own intron—it is autocatalytic.
16. Once removed from the nucleus by carrier proteins, mature eukaryotic mRNA is relatively long-lived (unlike prokaryotic mRNA, which usually lasts for only a few minutes).
17. The stability and lifetime of mRNAs differ and this plays an important role in the development process.
18. Normal development depends on the regulation of transcription and translation within the cell.
19. Regulation of gene expression occurs at various levels involving transcription, translation, and posttranslation.
20. In some cases, gene expression depends on rearrangement of the DNA, as in *mating type determination* in yeast, where the rearrangement of alleles between active and inactive loci determines the mating type of a cell.
21. During interphase, DNA can exist as uncoiled *euchromatin* or coiled *heterochromatin*. Because of the coiled nature of heterochromatin, DNA of this type cannot participate in transcription. The genes included in heterochromatin are said to be *inactivated*.

22. Heterochromatin at the chromosome level is shown by the presence of an *inactivated X chromosome* or *Barr body* in the cells of female mammals. One or the other of the two X chromosomes condenses into heterochromatin so that only a single active X chromosome remains in each cell.

23. *Chromosome puffs*, most obvious in the giant, polytene chromosomes of insect salivary gland cells, are regions of individual chromosomes where DNA is in the euchromatin form and is being transcribed.

24. *DNA methylation*, the addition of methyl groups to cytosine nucleotides, inactivates the affected genes. In some vertebrates, globin genes remain methylated except in cells requiring globin for hemoglobin synthesis.

Questions (for answers and explanations, see page 119)

1. Which of the following statements about the genes in a gene family is *not* true?
   a. Gene families can include pseudogenes.
   b. Gene families can be clustered together on the same part of a chromosome.
   c. Some members of a gene family may never be transcribed.
   d. Genes in a gene family differ more in their exons than their introns.
   e. Members of a gene family may be functional at certain times, but nonfunctional at other times.

2. Place numbers to the left of the following events to show the correct sequence during RNA processing in eukaryotes.
   5 Removal of intron "lariat"
   3 Binding of snRNA to protein
   6 Splicing of exons
   4 Binding of snRNPs to RNA
   7 Degradation of intron
   1 Addition of G cap
   2 Polyadenylation

3. How many Barr bodies would you observe in the cells of an individual with Klinefelter syndrome and the genotype XXXXY?
   a. 0
   b. 1
   c. 2
   d. 3
   e. 4

4. Chromosome puffs consist of DNA
   a. in the form of heterochromatin, and the genes in the puff region are active.
   b. in the form of euchromatin, and the genes in the puff region are active.
   c. in the form of heterochromatin, and the genes in the puff region are inactive.
   d. in the form of euchromatin, and the genes in the puff region are inactive.
   e. that has undergone methylation.

5. Select *all* of the following that are found in both an hnRNA immediately after transcription and its mature mRNA.
   a. Introns
   b. Poly A tail
   c. Modified G cap
   d. Start and stop sequences
   e. Exons

6. Describe the importance of mRNA stability as a factor in the control of development and how it differs in prokaryotes and eukaryotes.

CONTROL OF GENE EXPRESSION IN EUKARYOTES (Transcriptional Control: Gene Amplification • Transcriptional Control: Selective Gene Transcription • Translational Control • Posttranslational Control) (pages 302–308)

Key Concepts

1. *Gene duplication* is an important mechanism for *amplifying* the amount of a needed gene product. Examples of gene amplification include the genes coding for the *histone* proteins (those proteins involved in the formation of *nucleosomes*—many organisms have hundreds of copies of these genes), and the genes coding for tRNAs and rRNAs.

2. Many copies of the genes coding for rRNA are arranged in a series—called a *tandemly repetitive region*—within the nucleolar region of the nucleus. In amphibian egg cells (oocytes), as many as a thousand nucleoli, each with tandemly repetitive rRNA genes, ensure adequate rRNA for the production of the ribosomes needed during the rapid development of the embryo.

3. Electron micrographs of transcription in the nucleolus of oocytes show the tandem arrangement of the copies of the rRNA genes. Each gene is separated from the next by a "spacer" that may function like a "loading zone" for the RNA polymerase proteins that transcribe the genes.

4. Most genes are not amplified; instead, their transcription is turned on and off as needed. This is referred to as selective *gene transcription*.

5. *Steroid hormones* regulate selective gene transcription in eukaryotes. For example, the insect hormone *ecdysone* causes certain chromosome puffs to appear or disappear in many insects.

6. Whereas transcriptional control in prokaryotes can be both positive and negative, regulation is almost always positive in eukaryotes—the genes are turned off until they are turned on by a protein.

7. *Promoters*—sequences of DNA to which RNA polymerase must bind to initiate transcription—are much more complex in eukaryotes than in prokaryotes. This allows more precise control over transcription.

8. Eukaryotes have three kinds of RNA polymerase: *RNA polymerase I* transcribes the DNA coding for rRNA; *RNA polymerase II* transcribes genes that produce mRNAs; *RNA polymerase III* transcribes genes that code for tRNAs and other small RNA molecules.

9. Various regulatory protein molecules called *transcription factors* must bind to the *promoter region* of the gene before RNA polymerase II can attach and begin transcription. Each transcription factor binds to a unique sequence of nucleotides.

10. The *TATA box* is an eight-base-pair sequence of thymine and adenine pairs located about 25 base pairs before the starting point for transcription in many genes. The transcription factor TFIID binds to the TATA box. The TATA box helps to specify the exact location of the starting point for transcription.

11. After two additional transcription factors (TFIIA and TFIIB) also bind to the promoter and RNA polymerase II attaches, a final transcription factor (TFIIE) binds and the *transcription complex* is complete.

12. Although the TATA box is common to many genes and is recognized by transcription factors found in many cells, other genes have sequences only recognized by the transcription factors found in certain cell types.

13. *Enhancers* are DNA sequences located to either side of a gene that stimulate specific promoters and enhance transcription of the gene.

14. One mechanism seen in eukaryotes for the control of translation during differentiation is to change the *longevity* of specific mRNAs. For example, in lactating mammals, the hormone prolactin increases the longevity of casein mRNA by a factor of 25.

15. Another translational control mechanism depends on the fact that mRNA must be capped with a modified G unit before it can be translated. In some species, the G cap on a given variety of mRNA will not be modified until translation of that kind of mRNA is required. Once modified, translation can begin.

16. Several mechanisms affecting translation of hemoglobin components are responsible for maintaining the proper balance of α-globin chains, β-globin chains, and the heme groups. For example, excess heme increases translation of the two globin mRNAs.

17. *Posttranslational control* regulates gene expression by affecting the gene product.

18. Posttranslational control can involve inactivation of a gene product by degradation, production of an inactive gene product that is later modified to be made active, as in the conversion of proinsulin to insulin, or the production of an inactive gene product that becomes active when it combines with other gene products, as in the combination of actin and tubulin to form microtubules and microfilaments.

19. Another posttranslational control mechanism found in proteins that must pass through membranes is the presence of *leader sequences* on their N-terminal ends. These sequences interact with a specific membrane receptor protein which transports the gene product through the membrane. After the protein arrives in the correct compartment, it is made active by proteases that remove the leader sequences.

Activities (for answers and explanations, see page 120)

- The following photomicrograph shows transcription of an rRNA gene in the nucleolus of an oocyte. Label (1) approximate location of an RNA polymerase protein, (2) the DNA molecule, (3) an RNA molecule, and (4) the location of a "spacer" region on the DNA. Use an arrow to show the direction of transcription.

  Is rRNA or mRNA being produced? rRNA

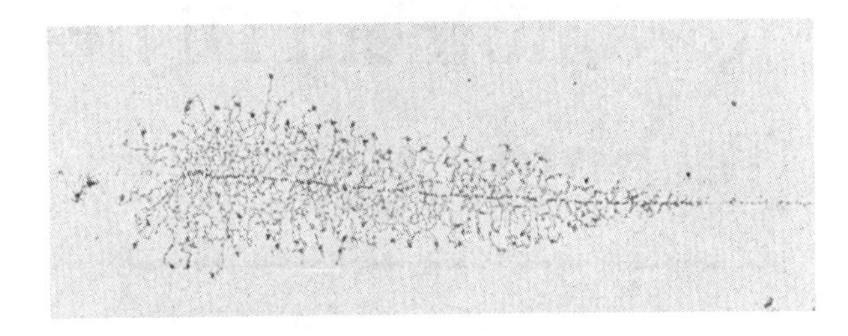

Questions (for answers and explanations, see page 120)

1. Which of the following statements contrasting transcriptional regulation in prokaryotes and eukaryotes is *not* true?
   a. Eukaryotes lack operons.
   b. Unlike prokaryotes, transcriptional control in eukaryotes can be both positive and negative.
   c. In eukaryotes, hormones can trigger selective gene transcription.
   d. The promoter region in eukaryotes is more complex than the promoter region of prokaryotes.
   e. Eukaryotes have three types of RNA polymerases; prokaryotes have only one type.

2. Which of the following RNA polymerases transcribe genes that code for proteins?
   a. RNA polymerase I
   b. RNA polymerase II
   c. RNA polymerase III
   d. RNA polymerase I and RNA polymerase III
   e. All three

3. The TATA box is
   a. a transcription factor.
   b. an enhancer.
   c. a binding site for TFIID.
   d. a leader sequence for a protein that must move through a membrane.
   e. a growth factor.

Integrative Activity

- Classify the following events associated with the control of gene expression as transcriptional, posttranscriptional, or posttranslational.

posttranscript — Increasing the longevity of an mRNA
posttranscript — Modification of the G cap
trans — Presence or absence of enhancers
trans — Methylation of a gene
trans — Presence or absence of the appropriate transcription factor
trans. — Gene amplification through duplication.
posttranslational — Conversion of proinsulin into insulin
trans — Conversion of heterochromatin into euchromatin
posttranslational — Modification of a gene product

14

Recombinant DNA Technology

CHAPTER LEARNING OBJECTIVES—after studying this chapter you should be able to:

- ❑ Describe what is meant by the terms "recombinant DNA" and "transgenic."
- ❑ Explain what restriction endonucleases are, what functions they perform in the bacteria that produce them, and how a bacterium is protected from the action of its own restriction endonucleases.
- ❑ Differentiate between restriction endonucleases that produce sticky-ended and blunt-ended fragments and discuss the utility of each in recombinant DNA studies.
- ❑ Describe what a cloning vector is and list the essential features of a cloning vector.
- ❑ Describe the steps involved in using restriction endonucleases and DNA ligase to insert foreign DNA into a plasmid.
- ❑ Evaluate the relative importance of plasmids and viruses as cloning vectors
- ❑ Discuss the various methods that have been used to insert DNA into cells, specifically transformation, transfection, use of microprojectiles, and electroporation.
- ❑ Explain why baker's yeast has become the most commonly used eukaryotic host cell for recombinant DNA studies.
- ❑ Describe how a plasmid with an antibiotic resistance gene and the β-galactosidase, *z* gene are used to select for transgenic cells.
- ❑ Explain how the *lac* operon was used to control the expression of the human somatostatin gene.
- ❑ Describe the shotgunning technique that is used to produce a DNA library.
- ❑ Describe the steps involved in producing a cDNA library.
- ❑ Explain how gene libraries and cDNA libraries differ and describe the special uses of each.
- ❑ Discuss how synthetic genes and probes are produced and describe their uses.
- ❑ Explain the principles underlying electrophoresis and the importance of size standards to this technique.
- ❑ Describe the following recombinant DNA techniques: Southern blotting, in situ hybridization, restriction mapping, and chromosome walking.
- ❑ Explain the DNA sequencing technique involving ddNTPs with attached fluorescent dyes.
- ❑ Explain the steps involved in the polymerase chain reaction, the principles underlying it, and its main uses.
- ❑ Contrast plant agricultural biotechnology with traditional plant breeding.
- ❑ Explain what is meant by the term "germ plasm" and what are its sources and storage media?
- ❑ Describe the *Ti* plasmid in *Agrobacterium tumefaciens* and how it is used as a cloning vector in plant biotechnology.
- ❑ Discuss several current uses for transgenic bacteria.
- ❑ Explain what the human genome project is and some of its potential benefits.

THE PILLARS OF RECOMBINANT DNA TECHNOLOGY • CLEAVING AND SPLICING DNA • CLONING GENES (pages 311–318)

Key Concepts

1. *Recombinant DNA* results when DNA from different sources (different species or natural versus synthetic) are combined into a single molecule.
2. Recombinant DNA technology includes techniques from the fields of microbial genetics, molecular biology, and biochemistry. DNA technology is contributing to basic research in many areas and has practical applications in medicine, agriculture, environmental science, etc.
3. Recombinant DNA technology uses naturally occurring enzymes that cleave, replicate, transcribe, and repair DNA to sequence, splice, locate, and identify DNA fragments in the laboratory. The complementary base pairing rules underlie most of the recombinant DNA techniques.
4. *Restriction endonucleases* are enzymes produced by bacteria to protect themselves against attack by *bacteriophages.* Each different enzyme recognizes and binds to a specific sequence of bases in DNA. The enzyme cleaves the DNA within this *recognition site.*
5. Once the polynucleotide chains have been broken by the restriction endonuclease, only the hydrogen bonds between the bases hold the DNA pieces together. Warm temperatures disrupt these bonds and the pieces separate.

6. Restriction endonucleases are named for the bacterial species in which they were first identified. *Eco*RI cuts within the sequence G–A–A–T–T–C; it was first discovered in *E. coli*. Several hundred different restriction endonucleases are now available to molecular biologists for cleaving DNA.
7. The bacterium is protected from attack by its own endonucleases because methyl groups have been added to bases within the recognition sites of its DNA by specific *methylase* enzymes.
8. Some restriction endonucleases cut the two DNA chains between different pairs of bases within the recognition site to produce *sticky ends*; blunt ends are made by enzymes that cut between the same base pairs.
9. Sticky ends can be rejoined by hydrogen bonding between complementary bases but the enzyme *DNA ligase* must join the polynucleotide chains together to effect a permanent union of the sticky ends.
10. A piece of foreign DNA can be inserted into a *plasmid* if both DNAs have recognition sites for the same restriction endonuclease. DNA ligase then joins the complementary sticky ends of the two DNA sources to create a plasmid containing the foreign DNA.
11. A *transgenic* cell or organism has a genome that includes some recombinant DNA.
12. *Cloning vectors*—usually plasmids or viruses—are used to move foreign DNA, either synthetic or natural, into a cell, thereby making the cell transgenic. In order to be replicated, foreign DNA must be part of a cloning vector.
13. To be useful as a cloning vector, a plasmid must have an *origin of replication*—it must be a *replicon*—so it can replicate with the host chromosome.
14 More is known about the molecular biology of baker's yeast, *Saccharomyces cerevisiae*, than any other eukaryotic organism. There are several reasons for the popularity of this species in research: it has a manageable amount of DNA; it can multiply rapidly, making gene cloning easier; foreign DNA can either be inserted into a plasmid, an artificial chromosome, or directly into one of the yeast's haploid set of 17 chromosomes.
15. A cloning vector should carry genes conferring properties on the host cell that will make it easier to identify the transgenic cells. Antibiotic resistance genes are particularly popular because antibiotics can be used to select for cells with resistance.
16. Ideally, a plasmid cloning vector should have a single recognition site for the restriction endonuclease so the plasmid is opened at a single point for insertion of the gene.
17. Although plasmids were the first vectors used, viruses are now more commonly employed. One constraint in using viral cloning vectors is that the amount of DNA inserted into the viral genome must fit within the viral coat. An advantage of some viruses is that they naturally infect host cells and insert their genes into their chromosomes.
18. *Transformation*, the uptake of foreign DNA by a cell, is a commonly used technique to insert a vector into a host bacterium. A high calcium ion concentration facilitates transformation.
19. *Transfection* is the uptake and expression of foreign DNA by a eukaryotic cell. Enzymatic removal of cell walls to create *protoplasts* may be necessary when transfecting plant or fungal cells.
20. The so-called "DNA particle gun" fires microprojectiles—tiny, DNA-coated tungsten particles—at plant cells.
21. Other methods for inserting DNA into cells include electroporation (use of pulses of high-voltage current) or direct injection of DNA into individual cells. Biological methods tend to yield higher transfection rates than nonbiological methods.
22. Selection with antibiotics can be useful for identifying transgenic cells that contain plasmid vectors with genes conferring resistance to specific antibiotics.
23. A common method for selecting transgenic cells utilizes a plasmid with an antibiotic resistance gene and the *E. coli z* gene—the *lac* operon gene that codes for β-galactosidase. A restriction endonuclease is used to cut the foreign DNA and also to cut within the plasmid's *z* gene. If the foreign DNA is inserted into the plasmid's *z* gene, the gene is nonfunctional. Cells transformed with any plasmid can grow on antibiotic-supplemented plates and those with a plasmid lacking the foreign DNA make β-galactosidase. If the artificial substrate XGal is provided to these cells, the colonies they produce are blue; cells transformed with the plasmid containing the foreign DNA produce white colonies.
24. Sometimes the expression of the foreign gene can be controlled by inserting it into an operon that can be turned on or off at will. For example, the human somatostatin gene was inserted into a structural gene of the *lac* operon of *E. coli*.

Questions (for answers and explanations, see page 120)

1. Which of the following statements about restriction endonucleases is *not* true?
   a. On the average, longer restriction endonuclease recognition sites (more base pairs) would be less common in DNA than shorter recognition sites (fewer base pairs).
   b. Methylated recognition sites are not recognized by restriction endonucleases.
   c. Restriction endonucleases catalyze the breakage of hydrogen bonds between the nitrogenous bases within the recognition site.
   d. Restriction endonucleases are named after the bacterial species in which they were first discovered.
   e. The blunt-ended fragments produced by some restriction endonucleases can rejoin with any other blunt-ended fragment.
2. Which of the following is *not* an important characteristic of a plasmid cloning vector?
   a. It should be a replicon.
   b. It should have many recognition sites for the restriction endonuclease used to isolate the foreign DNA.
   c. It should have a gene for antibiotic resistance.
   d. It should have a single origin of replication.
   e. It should have a gene that makes it possible to easily identify transgenic cells.

3. Which of the following techniques is *not* commonly used to create a transgenic cell?
   a. Transformation
   b. Lysing a protoplast in the presence of the vector
   c. Infection by a viral cloning vector
   d. Use of microprojectiles
   e. Use of electroporation

4. Order the following steps in the gene cloning process.
   _____ Encourage growth of the transgenic cells
   _____ Treat with ligase
   _____ Create fragments with restriction endonucleases
   _____ Transform cells
   _____ Use selection plating with antibiotics

5. In a study involving a plasmid with an antibiotic gene and the *z* gene, you use an appropriate restriction endonuclease to insert foreign DNA into the plasmid's *z* gene, as described in the textbook. You then transform bacteria with the plasmid mixture and plate the cells on an agar plate supplemented with antibiotic and Xgal. You count 55 blue colonies and 10 white colonies.

   Of these 65 colonies:
   How many consist of cells that were transformed with a plasmid? _______
   How many consist of cells that contain a plasmid *without* foreign DNA? _______
   How many consist of cells that contain a plasmid *with* foreign DNA? _______

   You spread an equal volume of the bacterial culture described above onto a plate with the antibiotic but *no* Xgal. Specifically, what do you expect to observe on this second plate?

6. Which of the following is *not* a reason why baker's yeast (*Saccharomyces cerevisiae*) is a popular organism for recombinant DNA studies?
   a. It is a eukaryote.
   b. There are several places where a foreign gene can reside within the cell.
   c. It can be used to clone genes rapidly.
   d. It is multicellular.
   e. It has only 3½ times as much DNA as *E. coli*.

Activities (for answers and explanations, see page 120)

- The following diagram shows the recognition sequences for the two restriction endonucleases **(a)** *Hpa*I and **(b)** *Hin*dIII. Complete the complementary ends for each fragment.

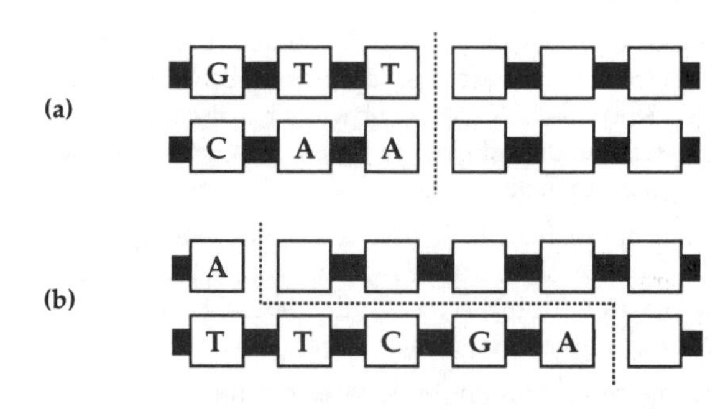

- Which of the two restriction endonucleases (*Hpa*I or *Hin*dIII) would be best for inserting foreign DNA into a plasmid? Explain your answer.

SOURCES OF GENES FOR CLONING • EXPLORING DNA ORGANIZATION (pages 318–326)

Key Concepts

1. A *gene library* is obtained by a technique called *shotgunning*. First a DNA source, such as a species' genome, is fragmented with restriction endonucleases. The collection of fragments is then inserted into plasmids or bacteriophage DNA and cloned into bacteria. Each clone, called a *volume*, contains a single fragment from the library.
2. *Complementary DNA*, called *cDNA*, is made from RNA, usually mRNA, with the enzyme *reverse transcriptase*. First a string of T residues (*oligo-dT*) is allowed to bind to the poly A tail on the 3' end of mRNA. The oligo-dT serves as a primer for reverse transcriptase.
3. If DNA precursors (2'–deoxyribonucleoside triphosphates) are available, reverse transcriptase produces a strand of cDNA complementary to the mRNA. The pH is then adjusted to favor separation of the mRNA–cDNA chain and degradation of the mRNA.
4. cDNA is made double-stranded using *DNA polymerase* and can then be incorporated into bacteriophage vectors to produce a *cDNA library* of the source DNA.
5. Because a cDNA library contains only *expressed DNA* (DNA representing mRNAs), comparing cDNA from different cells or tissues can provide insight into which genes are on or off during the differentiation process
6. Gene libraries include all of the DNA, including introns, promoters, and regulatory sequences like the TATA box.
7. Automated DNA synthesis equipment can be used to produce *synthetic DNA* for small genes like that for *somatostatin*. If the amino acid sequence of the desired protein is known, the base sequence of an artificial gene can be determined and initiation and termination codons added. Appropriate restriction endonuclease recognition sequences must also be added to the ends of the gene so it can be spliced into the cloning vector.

8. An optical device and lasers can be used to separate individual chromosomes by detecting slight differences in fluorescence of chromosomes stained with two different fluorescent dyes. Drops containing specific chromosomes are deflected into a collection tube.

9. In *electrophoresis*, negatively charged DNA fragments migrate through a porous gel in response to an electrical field. The rate of fragment movement is proportional to the size of the the fragments; smaller fragments move faster than larger fragments.

10. Several samples are run in different *lanes* on an electrophoresis gel. *Size standards* (DNA fragments of known size) are also usually run so that the sizes of all fragments can be estimated.

11. A *probe* is DNA or RNA with a base sequence complementary to a "target" DNA sequence. For example, mRNA can be used to locate the gene from which it was transcribed.

12. In *Southern blotting*, DNA fragments are first separated by electrophoresis. A basic pH is used to denature the double-stranded DNA into single strands and a sheet of nitrocellulose is used to blot the gel. Heat is used to fix in place the DNA that has been transferred to the sheet and a radioactive probe is poured over the sheet. Base pairing between the probe and fragments with a sequence complementary to the probe form double-stranded molecules and autoradiography is used to locate the fragments with the hybrid molecules.

13. Southern blotting is named after E. M. Southern. Similar techniques are used for detection of specific RNA sequences (Northern blotting) and protein sequences (Western blotting).

14. *In situ* ("in place") *hybridization* is similar to Southern blotting, except it is carried out directly on a microscope slide with metaphase chromosomes whose DNA has been denatured. Autoradiography identifies the specific region of the chromosome to which the probe has hybridized.

15. *Restriction mapping* is a technique that is used to locate restriction endonuclease recognition sequences on DNA by comparing fragment sizes obtained with different combinations of these enzymes. Fragment sizes are determined using electrophoresis.

16. *Chromosome walking* is used to order the volumes of a gene library. First a DNA source is cleaved with two different restriction endonucleases and the fragments from each enzyme are used to create two different libraries. Each clone in the library is a different volume.

17. A previously cloned gene is used as a probe to identify the volume in the first library with the fragment containing the gene. The second library is probed with this fragment in order to identify volumes in the second library with fragments that overlap it. The first library is now probed with the fragment from the second library and this procedure is repeated until all the volumes have been ordered.

18. The *DNA base sequencing* technique makes use of 2',3'-dideoxyribonucleoside triphosphates (*ddNTPs*) in addition to the normal DNA precursors 2'-deoxyribonucleoside triphosphates (*dNTPs*). Each of the four ddNTPs is bound to a fluorescent molecule that emits a different colored light.

19. The single-stranded DNA to be sequenced is mixed with DNA polymerase, a primer, the four dNTPs (dATP, dGTP, dCTP, and dTTP), and a small amount of each of the ddNTPs (ddATP, ddGTP, ddCTP, and ddTTP). Whenever a ddNTP is added to the growing complementary strand, the chain is terminated.

20. After synthesis is complete, the reaction mixture is separated into fragments on an electrophoresis gel. By progressing from the shortest to the longest fragments, the color of each band identifies the base at that position in the newly synthesized DNA. The complementary base pairing rules then allow us to deduce the sequence of bases in the original DNA.

Questions (for answers and explanations, see page 120)

1. Which of the following statements characterizing gene and cDNA libraries is *not* true?
   a. Gene libraries can include the entire genome of a species.
   b. Gene libraries are especially useful for determining which genes are active during differentiation.
   c. Gene libraries can be used to study regulatory sequences like the TATA box
   d. cDNA libraries contain only expressed DNA.
   e. Chromosome walking cannot be done with cDNA libraries.

2. Order the following steps in the process of creating a cDNA library.
   ____ Add oligo-dT primer to mRNA
   ____ Denature DNA–RNA hybrid molecules
   ____ Insert cDNA into DNA vector
   ____ Add DNA polymerase
   ____ Degrade mRNA
   ____ Add reverse transcriptase

3. Five DNA samples (a–e) were treated with a restriction endonuclease and the resulting fragments separated in the following electrophoresis gel. Which of the DNA samples *both* contained the greatest number of recognition sites for this enzyme and produced the smallest fragments?

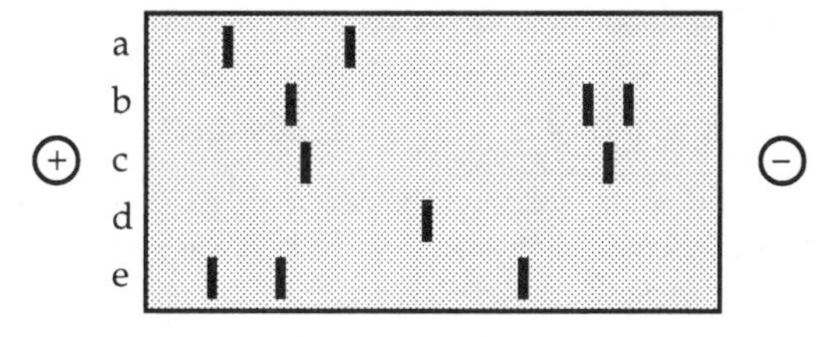

4. Which of the following techniques would you use to identify the location of a gene producing a specific mRNA on a chromosome?
   a. In situ hybridization
   b. DNA base sequencing
   c. Southern blotting
   d. Restriction mapping
   e. Shotgunning
   f. Chromosome walking

5. Which of the following techniques would you use to identify the molecular defects of a genetic disease?
   a. In situ hybridization
   b. DNA base sequencing
   c. Southern blotting
   d. Restriction mapping
   e. Shotgunning
   f. Chromosome walking

6. Match the molecular techniques and tools on the left with the appropriate characteristics on the right.

____ Shotgunning	*a* Makes cDNA from mRNA
____ Chromosome walking	*b.* Used to create a gene library
____ Southern blotting	*c.* Requires use of a probe
____ Reverse transcriptase	*d.* Used in electrophoresis
____ Size standards	*e.* May involve autoradiography
____ DNA base sequencing	*f.* Uses ddNTPs
____ In situ hybridization	

Activities (for answers and explanations, see page 121)

- The following diagram shows an electrophoresis gel produced by application of the Sanger sequencing method to a fragment of DNA 12 base pairs in length. The DNA was divided into four samples, and the specific ddNTP (A, T, C, G) added to each sample is shown across the top of the gel.

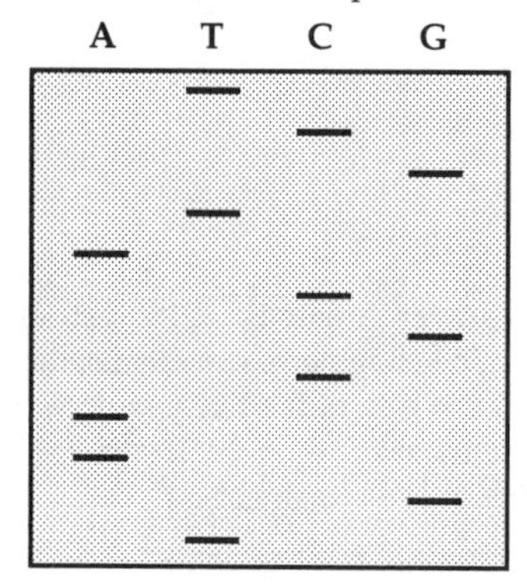

What is the sequence in the newly synthesized strand?

What is the sequence in the original DNA fragment?

- The following electrophoresis gel shows fragment sizes obtained from a plasmid treated with the restriction endonucleases x and y, both alone (lane 1 and 2) and in combination (lane 3). Fragment size is expressed in thousands of base pairs (bp). In the space provided next to the gel, draw a possible map of the plasmid, showing the location of the recognition sites for both restriction endonucleases and the distance between sites.

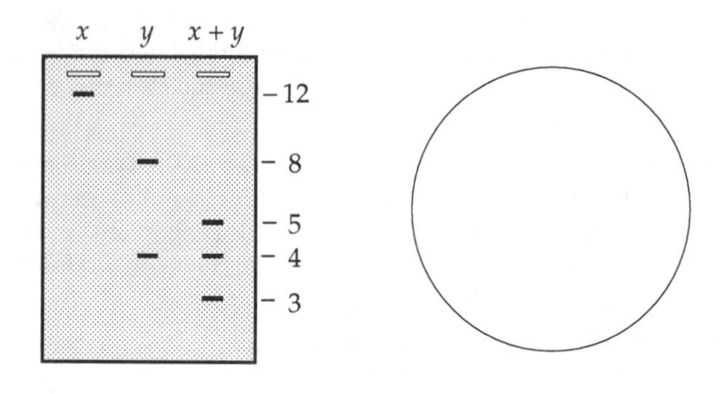

GENE COPIES BY THE BILLION • PROSPECTS (pages 326–332)

Key Concepts

1. The *polymerase chain reaction* (PCR) makes it possible to produce billions of copies of a gene in a cell-free environment in a few hours. First, two probes of several dozen bases in length must be made that are complementary to sequences at the 3' ends bounding the DNA chain to be cloned. The probes serve as primers to initiate synthesis of strands complementary to the target DNA.
2. By alternately heating and cooling the DNA in the presence of DNA polymerase, additional primer, and DNA precursors, DNA strands separate and hybridize with the primer, and synthesis of new strands continues. The amount of DNA doubles with each cycle.
3. Because of the high temperatures needed to denature the DNA (about 94°C), a temperature-resistant DNA polymerase from the bacterium *Thermus aquaticus* is used in the polymerase chain reaction.
4. The polymerase chain reaction is used to amplify DNA collected from crime scenes, in genetic screening (see Chapter 15), and in the study of prehistoric DNA.
5. Recombinant DNA technology is being used by the pharmaceutical industry to engineer strains of bacteria, yeast, plant and animal cells to make gene products that improve human health.
6. Guidelines established by the U.S. National Institutes of Health are designed to minimize the dangers of recombinant DNA research.
7. *Traditional plant breeding* involves the selection and breeding of strains of plants with desirable phenotypes. The limitations of traditional plant breeding are that it is space and time intensive, the entire genome of the parents is passed on to the progeny, and the availability of natural genetic strains is limited.
8. *Plant agricultural biotechnology* is much more economical of time and space, only the genes of interest need to be transferred from one plant to another, and genes can be transferred from any other living cell.
9. *Germ plasm* refers to the supply of genes that is available for both conventional plant breeding and recombinant DNA research. Seed repositories, plants in nature, and gene libraries can be storage media for germ plasm.

10. A popular cloning vector for plant biotechnology is the *Ti* (tumor inducing) *plasmid* in the bacterium *Agrobacterium tumefaciens*, an organism that causes *crown gall disease* in many plants. Diseased plants have large tumors.

11. Genes to be cloned are inserted into a *transposon* in the Ti plasmid, called T-DNA. Transformation is used to insert the plasmid into *A. tumefaciens* and the T-DNA transposon transfers copies of itself and any other genes present to the chromosomes of an infected plant. Tissue culture can be used to grow whole transgenic plants from tumor cells.

12. Transgenic bacteria are being designed using recombinant DNA technology to improve the nitrogen fixation process, reduce frost damage in plants, and metabolize oil spills and organic pollutants, such as sewage and dioxins.

13. Yeast and other microorganisms are being genetically manipulated to concentrate metals such as nickel, gold, and plutonium for recycling and pollution reduction.

14. *Genome projects* are efforts to map and sequence all of the genetic information of an organism. The human genome project will result in detailed information about the hundreds of hereditary diseases that afflict humans (see Chapter 15). For example, if the base sequence of a gene is known, its polypeptide product can be determined.

Questions (for answers and explanations, see page 121)

1. The first step in the use of the polymerase chain reaction to clone a gene is to:
   a. make cDNA for the gene of interest.
   b. produce probes that can serve as primers on the borders of the gene.
   c. add DNA polymerase.
   d. sequence the gene using the Sanger method.
   e. add reverse transcriptase.

2. Assume that it requires 5 minutes for one cycle of the polymerase chain reaction. If you start with one copy of a gene, how many minutes will be required to produce a thousand copies?
   a. 9
   b. 20
   c. 30
   d. 50
   e. 60

3. Select *all* of the following that are sources of germ plasm.
   a. Cells in tissue culture
   b. A gene library
   c. A living plant
   d. Stored seeds
   e. Recombinant DNA

4. Which of the following characteristics of the bacterium *Agrobacterium tumefaciens* is *not* a reason it is much used in plant recombinant DNA research?
   a. It readily infects a variety of plants.
   b. It has a large plasmid called Ti.
   c. Its plasmid contains the transposon T-DNA.
   d. The plasmid can serve as a cloning vector.
   e. Transformation can be used to insert the plasmid into the plant.

5. What are some of the major limitations of traditional plant breeding and how is recombinant DNA technology helping to address these limitations?

Integrative Questions (for answers and explanations, see page 121)

1. The enzyme needed to make cDNA from mRNA is
   a. DNA ligase.
   b. DNA polymerase.
   c. reverse transcriptase.
   d. a restriction endonuclease.
   e. RNA polymerase.

2. The enzyme that is used to combine nucleic acid restriction fragments is
   a. DNA ligase.
   b. DNA polymerase.
   c. reverse transcriptase.
   d. a restriction endonuclease.
   e. RNA polymerase.

3. Select all of the following techniques that depend on the rules of complementary base pairing.
   a. Estimating DNA fragment sizes with gel electrophoresis
   b. Production of a DNA probe
   c. Production of a cDNA library
   d. Polymerase chain reaction
   e. Automated separation of chromosomes
   f. In situ hybridization

4. Based on information provided in the textbook, describe the various recombinant DNA techniques that were used to induce *E. coli* to produce human somatostatin.

15

Genetic Disease and Modern Medicine

CHAPTER LEARNING OBJECTIVES—After studying this chapter you should be able to:

- ❑ Explain the mode of inheritance and general features of the following genetic diseases: cystic fibrosis, Tay-Sachs disease, phenylketonuria, sickle-cell anemia, Duchenne muscular dystrophy, fragile-X syndrome, hemophilia, and Huntington's disease.
- ❑ Describe several autosomal dominant mutations and explain why their symptoms are usually mild and nonlethal.
- ❑ Describe the features of several diseases, such as fragile-X syndrome and Huntington's disease, that suggested that their disease gene was changing over time.
- ❑ Explain what a triplet repeat is and describe the characteristics of triplet repeats seen in several genetic diseases.
- ❑ Discuss the multistep process involved in understanding and treating a genetic disease.
- ❑ Analyze a pedigree and determine the most likely pattern of inheritance (autosomal dominant, autosomal recessive, sex-linked) that it shows.
- ❑ Describe the technique of somatic-cell hybridization and how it is used to determine on which chromosome a gene resides.
- ❑ Describe the technique of chromosome-mediated gene transfer and how it is used to determine on which chromosome a gene resides.
- ❑ Explain what a restriction fragment length polymorphism (RFLP) is, the steps involved in studying RFLPs, and how a RFLP can be used as a genetic marker for a disease gene.
- ❑ Describe how various recombinant DNA techniques were used to sequence the cystic fibrosis gene.
- ❑ Explain the essential characteristics of an animal model and how knockout mice are used to test therapies for genetic disease.
- ❑ Discuss the advantages and disadvantages of gene therapy using somatic cells such as white blood cells or stem cells.
- ❑ Discuss the advantages and disadvantages of gene therapy using germ cells.
- ❑ Describe the major approaches to genetic screening.
- ❑ Explain why cancer is considered a genetic disease that is not inherited.
- ❑ Use the following terms to discuss the general characteristics of cancer: "transformation," "metastasis," "benign tumor," "malignant tumor," "carcinoma," "sarcoma," "lymphoma," and "leukemia."
- ❑ Describe the characteristics of carcinogens and list common types.
- ❑ Explain the involvement of proto-oncogenes, growth factors, and receptor proteins in the regulation of cell growth.
- ❑ Describe the principal mechanisms that can disrupt normal cell growth regulation and cause cancer.
- ❑ Discuss the involvement of tumor-suppressor genes, such as *p53*, in the development of cancer.
- ❑ Discuss the importance of retroviruses, oncogenic viruses, and transposable elements to cancer.
- ❑ Discuss various cancer treatments including the use of antisense nucleic acids.
- ❑ Describe the Ames test and its uses.

SOME INHERITED DISEASES • TRIPLET REPEATS AND THE FRAGILITY OF SOME HUMAN GENES • DEALING WITH GENETIC DISEASE: AN OVERVIEW (pages 334–337)

Key Concepts

1. The incidence of many genetic diseases varies with different racial groups because human populations were geographically isolated for long periods of time. (See the concept of *genetic drift* in Chapter 19.)
2. *Cystic fibrosis* results from an autosomal recessive allele for a gene that is responsible for production of a transmembrane chloride channel protein. Homozygous recessive individuals cannot export chloride and thus take on water osmotically, and produce thick extracellular mucus in the lungs and other organs.
3. Ashkenazic Jews have a much higher incidence of *Tay-Sachs disease*, caused by an autosomal recessive allele that results in the inability to produce an enzyme (hexosaminidase A) essential for normal development. Afflicted children usually die of nervous system degeneration by age five.
4. Individuals homozygous for the recessive allele causing *phenylketonuria* cannot metabolize phenylalanine. Severe mental retardation results unless individuals are placed on a diet low in phenylalanine early in life.

5. *Sickle-cell anemia*, common in populations derived from equatorial Africa and the Mediterranean region, is characterized by abnormal hemoglobin and defective red blood cells in individuals homozygous for the recessive, autosomal allele. Heterozygous carriers do not express the disease.

6. *Duchenne muscular dystrophy* results from an X-linked recessive allele that incorrectly codes a muscle cell membrane protein. Progressive muscular deterioration usually leads to death in the twenties.

7. *Fragile-X syndrome* is a common form of mental retardation caused by an X-linked allele that also produces a structural abnormality of the X chromosome. Whether mental retardation results depends on the number of triplet repeats (see concept 12, below) within the allele.

8. *Hemophilia* is an X-linked recessive trait that results from a mutant allele that improperly codes for a blood clotting protein. Transfusion with normal blood and provision of the clotting factor is the common treatment.

9. *Huntington's disease* results from an X-linked dominant allele. The symptoms, degeneration of the nervous system, do not begin until after age 35, but invariably lead to premature death. Presymptom diagnosis is now available for this disease.

10. Most autosomal dominant mutations that are expressed prior to reproductive age produce relatively mild, nonlethal symptoms. Examples include a form of dwarfism and the production of extra digits (polydactyly).

11. The observation that within a family fragile-X syndrome was more severe in successive generations suggested that the defective allele was changing over time.

12. A *triplet repeat* of the bases CGG occurs within the fragile-X gene. CGG repeats of 6 to 54 are seen in the normal allele. Carriers of the disease without symptoms have an allele (called a *premutation*) with 50 to 200 repeats. Mentally retarded individuals have an allele with 200 to 1,300 repeats.

13. In the fragile-X syndrome, the premutation allele is especially likely to accumulate additional repeats and become disease-producing when passed from a mother to her children.

14. Triplet repeats are common in the normal alleles of many genes of humans and other animals. Triplet repeats can be found both within and outside the protein-coding regions of genes.

15. Some other defective genes represented by a normal allele with few repeats, a premutation allele with more repeats, and a disease-causing allele with the most repeats include myotonic dystrophy, Kennedy's disease, and Huntington's disease. In some cases, such as Huntington's disease, expansion of triplet repeats is more common when the allele is transmitted by the father.

16. The first step involved in understanding and developing a treatment for a genetic disease involves identifying its symptoms.

17. *Epidemiologists* look for correlations between the occurrence of a disease and its incidence in specific groups of people or geographic locations.

18. Sickle-cell anemia is most common in areas with a high incidence of malaria. Individuals heterozygous for the sickle-cell allele are more resistant to malaria than homozygous individuals; this maintains the defective allele at a relatively high level in populations exposed to malaria.

19. *Pedigree analysis* of a disease in families can lead to an understanding of its mode of inheritance.

20. Using *recombinant DNA technologies* it is now possible to identify the chromosome that contains a disease gene, identify its locus on the chromosome, and clone the gene in order to study the molecular basis of the disease.

Questions (for answers and explanations, see page 121)

1. Which of the following statements about triplet repeats is *not* true?
   a. Triplet repeats are not restricted to disease genes.
   b. Expansion of triplet repeats appears to be more common in humans than in other organisms.
   c. Expansion of triplet repeats is always more likely when transmission of the disease gene is from the mother.
   d. The number of triplet repeats is intermediate in the premutation condition.
   e. Triplet repeats can occur within or outside the protein-coding region of the gene.

2. Which of the following features does *not* apply to the genetic disease sickle-cell anemia?
   a. Caused by a recessive, autosomal allele
   b. Under certain conditions, the heterozygous phenotype is advantageous.
   c. In different regions of the world, the incidence of the disease gene correlates with the prevalence of malaria.
   d. The mutant allele causes production of abnormal hemoglobin.
   e. Onset of the disease occurs in postreproductive years.

3. Match the following genetic diseases on the left with all of the appropriate characteristics on the right.

___ Cystic fibrosis	*a.* Common in Ashkenazic Jews
___ Tay-Sachs disease	*b.* Associated with malaria
___ Phenylketonuria	*c.* Blood clotting disorder
___ Sickle-cell anemia	*d.* Associated with triplet repeats
___ Duchenne muscular dystrophy	*e.* Defective transmembrane chloride channel protein
	f. Dominant trait
___ Fragile-X syndrome	*g.* Dietary treatment is effective
___ Hemophilia	*h.* Caused by X-linked allele
___ Huntington's disease	

4. Describe the expected features of a genetic disease whose symptoms have a prereproductive onset.

5. Use numbers to chronologically order the following steps in the study and treatment of a genetic disease.
   ______ Cloning the disease gene
   ______ Pedigree analysis
   ______ Comparison of the DNA of normal and afflicted individuals
   ______ Characterization of the symptoms
   ______ Epidemiological studies
   ______ Replacement of the defective allele
   ______ Isolation of the locus to a specific chromosome

6. Of the steps listed in question 5, which is considered to be the most controversial, and why?

FINDING A DEFECTIVE GENE: MAPPING HUMAN CHROMOSOMES • DEALING WITH A DEFECTIVE GENE (pages 337–344)

Key Concepts

1. *Pedigree analysis* is useful for determining the pattern of inheritance of a genetic disease (autosomal or sex-linked, dominant or recessive). Pedigree analysis of two or more genes can determine if they are linked and roughly estimate the map distances between linked genes.
2. Long generation time, few offspring, and a reliance on pedigree analysis alone have made mapping human genes difficult. New, technological approaches associated with the human genome project are helping to overcome these difficulties.
3. The *somatic-cell hybridization* technique relies on the ability of somatic (nonreproductive) cells of different species (such as mice and humans) to fuse and form hybrid cells with nuclei containing chromosomes of both species. During successive mitotic cell divisions, the chromosomes of one species (usually human) are lost.
4. Hybrid cells form colonies and each colony is tested for the presence of a specific human gene product and screened to determine which human chromosomes remain. Only colonies with a certain chromosome will express a specific human gene product and this verifies on which chromosome the gene resides.
5. In *chromosome-mediated gene transfer*, mouse cells exposed to human chromosomes take up small parts of some chromosomes. Only closely linked genes will be transferred to and, later, expressed together by the mouse cell and these data can be used to map the locations of the genes.
6. Restriction endonucleases bind to a specific sequence of bases in DNA called their recognition sequence and catalyze the breakage of the nucleotide chain within that sequence. Point mutations that occur within the recognition sequence of a restriction endonuclease can prevent the restriction enzyme from making its normal cut.
7. Mutations can cause individuals to differ in the occurrence of specific recognition sequences in their DNA. Using recombinant DNA techniques to study chromosome structure, these differences can be visualized as *restriction fragment length polymorphisms (RFLPs)* and are useful as markers for genetic disease.
8. Steps involved in studying RFLPs include digesting DNA with a specific restriction enzyme, separating the fragments into bands on an electrophoresis gel, Southern blotting to transfer the DNA band pattern to a filter, and applying a single-stranded DNA probe complementary to a region of DNA.
9. If the probe hybridizes to a region of DNA where a mutation has altered a recognition sequence for the restriction endonuclease used, individuals with and without the mutation can have different band patterns on the Southern blot. Also, since each individual has two chromosomes, this technique can be used to determine if an individual is homozygous or heterozygous for the mutation.
10. If pedigree analysis shows that a RFLP for a particular probe is always inherited with a genetic trait, then it must be tightly linked to the gene determining that trait. Determining the chromosome to which the RFLP-associated probe binds shows the location of the gene.
11. *Chromosome jumping* and chromosome walking can be used to further narrow the search for the gene's location. These techniques, and others, narrowed the locus for the cystic fibrosis gene to a specific region of chromosome 7 and this gene has now been sequenced.
12. The first step in developing an effective treatment for a defective gene involves identifying the protein encoded by the gene. *Animal models* are useful at this stage.
13. The ideal animal model should show the disease symptoms and should be useful in testing possible therapies. Transgenic rodents, particularly *knockout mice*, have become important models for many human genetic diseases.
14. In knockout mice, techniques from molecular biology are used to replace the functional allele with a disease allele. In some cases, the mice appear normal because another member of a gene family performs the functions for the "knocked out" allele, or an entirely different biochemical pathway may become involved.
15. *Gene therapy* usually involves removing somatic cells from a patient, transferring normal genes into them (sometimes using retroviruses to accomplish this), and then returning the genetically modified cells to the individual.
16. An early successful attempt at gene therapy involved a defective allele for the gene that codes for adenosine deaminase (ADA), an enzyme involved in the immune response. Normal ADA genes were transferred to some white blood cells removed from the patient; the genetically modified cells were then returned to the patient.
17. Because white blood cells cannot divide, patients require repeated gene therapy treatment. Genetic modification of stem cells (undifferentiated cells that give rise to various types of blood cells) eliminates this problem because stem cells continue to divide.

18. Genetic modification of germ cells (gametes) offers the possibility of permanently correcting a genetic defect, but the potential for misuse of this approach causes most scientists to argue against its use.

19. Cells from a developing fetus can be removed using amniocentesis and chorionic villi sampling and the cells can then be subjected to *genetic screening* for defects.

20. Genetic screening tests are based on detecting chromosome abnormalities, observing certain proteins, verifying the presence or absence of biochemical reactions, or examining the DNA directly by looking for specific RFLP patterns that are tightly linked to known disease genes.

21. Screening results can be used to devise treatment strategies, if possible, or to consider abortion in cases where no therapies are available for the genetic disease.

22. Many ethical issues, such as right-to-life and individual privacy, are associated with the practice of genetic screening.

Activities (for answers and explanations, see page 122)

- With reference to the symbols described in Figure 15.3 on page 338 in the textbook, determine the most likely mode of inheritance shown in the following pedigree.

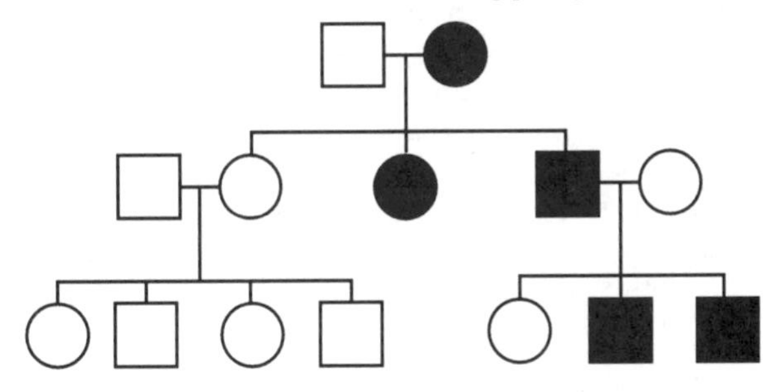

Most likely mode of inheritance:

What genetic disease described earlier would show this pedigree pattern?

- The following pedigree shows the children born to a woman and two different men. Using the symbols described in Figure 15.3 on page 338 in the textbook, determine the most likely mode of inheritance shown in this pedigree and indicate all known heterozygotes in the pedigree.

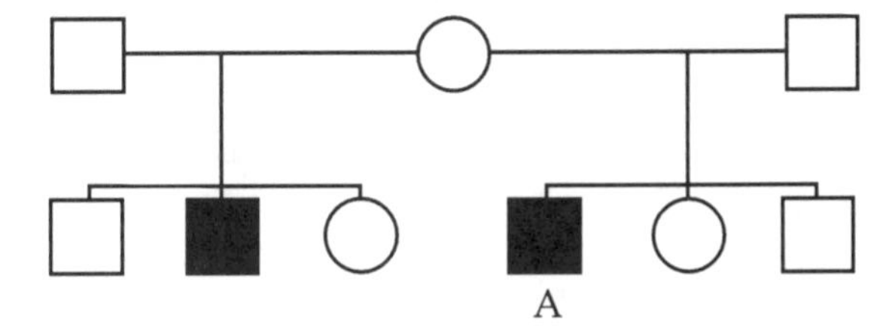

Most likely mode of inheritance:

What is the probability that male A and a normal female would have a normal daughter?

What is the probability that male A and a normal female would have a normal son?

- You are studying five human genes (*a*, *b*, *c*, *d*, *e*) that code for enzymes catalyzing different reactions. Using the somatic-cell hybridization technique you obtain five clones of mouse cells containing human chromosomes and whose cytoplasm has the appropriate enzymes to catalyze one or more of the five reactions. You then determine which human chromosomes are present in the cells of each clone. The following table shows the results of your analysis. Based on these data, indicate below which genes are on which chromosomes.

		Mouse Cell Clones				
		1	2	3	4	5
Chromosome #	6	−	+	−	−	−
	12	−	−	+	+	−
	19	+	−	−	−	+
	22	−	−	+	−	+
Genes	*a*	−	−	+	+	−
	b	−	+	−	−	−
	c	+	−	−	−	+
	d	−	+	−	−	−
	e	−	−	+	−	+

Gene *a*:

Gene *b*:

Gene *c*:

Gene *d*:

Gene *e*:

- Sickle-cell anemia results from a single point mutation in the β-globin gene. This point mutation also eliminates a recognition site for the restriction endonuclease *mst*II. Thus, whereas *mst*II cuts the normal β-globin gene at three sites, producing restriction fragments of 1150 and 200 base pairs (bp), *mst*II cuts the mutated β-globin gene at only two sites, producing a large restriction fragment of 1350 bp. The following pedigree shows the incidence of sickle-cell anemia in a family. Also shown are RFLP patterns for all individuals, obtained using a probe for the β-globin gene. Using symbols described in Figure 15.3 on page 338 in the textbook, complete the pedigree to indicate unaffected individuals, affected individuals, and carriers of the disease.

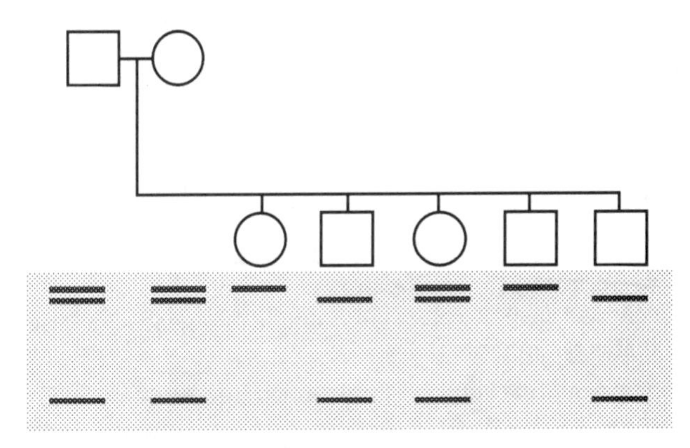

Questions (for answers and explanations, see page 122)

1. Select the following condition that is *most* important in utilizing a restriction fragment length polymorphism (RFLP) as a marker for a genetic disease.
   a. The availability of many pedigrees of families afflicted with the disease
   b. Establishing that the RFLP is tightly linked to the disease gene
   c. A knowledge of which chromosome contains the disease gene locus
   d. The existence of many triplet repeats within the coding portion of the gene
   e. An available animal model for the genetic disease

2. Which one the following approaches has been *least* important in the attempts to understand cystic fibrosis?
   a. Breeding experiments
   b. RFLP mapping
   c. Chromosome walking and jumping
   d. Pedigree analysis
   e. Nucleotide sequencing of the gene

3. Which one of the following statements about knockout mice and animal models is *not* true?
   a. A useful animal model must show disease symptoms.
   b. Knockout mice are especially useful for somatic-cell hybridization studies.
   c. Knockout mice are transgenic because human alleles have been inserted into their genome.
   d. Knockout mice may appear normal.
   e. New gene therapies can be tested on knockout mice.

4. Which one of the following statements about gene therapy is *not* true?
   a. Viruses may be useful vectors for getting normal alleles into a patient's cells.
   b. Gene therapy involving the transfer of normal alleles into a patient's cells has already been accomplished.
   c. Therapies using genetically engineered stem cells will not require repeated treatments.
   d. Genetic modification of germ cells is a common practice.
   e. Tissue culture is an important technique used in gene therapy.

5. What is the connection between the issue of abortion and the technology of genetic screening?

CANCER (pages 345–350)

Key Concepts

1. *Cancer* is a category of over 200 different diseases that result from genetic alteration of an individual's somatic cells. As such, cancer is an example of a genetic disease that is not inherited.
2. When normal cells are transformed into cancerous cells they lose control of their rate of division and they spread to other parts of the body. Spreading of cancer cells is called *metastasis*.
3. Rapidly dividing cells can form cell masses called *tumors*. Non-cancerous tumors, called *benign tumors*, do not metastasize, but remain localized. Cancerous tumors, called *malignant tumors*, undergo metastasis.
4. After spreading to surrounding tissue, cancerous cells can then either enter the lymphatic system or the bloodstream. Metastasis through the lymphatic system is slower than movement through the bloodstream because the lymph nodes retard the spread of the disease. Removal of affected lymph nodes and intervening ducts is a treatment for certain types of cancer.
5. *Carcinomas*, such as lung, breast, colon, and liver cancer, are cancers of skin or the gut lining and account for the majority of all cancers. *Sarcomas* are cancers of bone, blood vessels, and muscle. *Lymphomas* are cancers of the white blood cells and *leukemias* affect the red blood cells.
6. Cancer-causing agents, called *carcinogens,* include chemicals in the diet or the immediate environment, high energy radiation (ultraviolet light, X rays), and tumor-inducing viruses. Although many viruses have been shown to cause cancer in animals, only a few viruses are cancer-inducing in humans. The best way to reduce your risk of developing cancer is to avoid exposure to carcinogens.
7. Cancer may be a disease of old age because it depends on the accumulation of mutations within cells. Carcinogens react directly or indirectly with DNA and cause mutations, especially of genes involved in controlling normal growth.
8. *Growth factors* are a group of proteins that regulate cell division. Each growth factor binds to and activates a *receptor protein* embedded in the cell membrane of a particular target cell type. The activated receptor acts as an enzyme in reactions that enhance cell division.
9. *Proto-oncogenes* code for growth factors and receptor proteins. They are essential for normal growth. *Oncogenes* are the dominant, cancer-producing alleles that arise by mutation from the recessive proto-oncogene. Oncogenes lead to uncontrolled cell division.
10. Cancer can result from the mutation of a proto-oncogene into an oncogene, from the overproduction of proto-oncogenes by gene amplification, or by the insertion of a proto-oncogene into a chromosomal site near the promoter of a very active gene. The third mechanism results in continuous transcription of the proto-oncogene.
11. Tumor-suppressor genes encode proteins that inhibit cell division. Mutations that convert a dominant tumor-suppressor into a recessive allele can lead to the formation of tumors. A common tumor-suppressor gene, *p53*, codes for a protein that arrests cells before division so that their DNA can be repaired. Mutations of *p53* are common in many human cancers.
12. Transformation of a normal cell into a cancerous cell may require mutation of several tumor-suppressor genes and also conversion of a proto-oncogene into an oncogene.

13. A retrovirus is a single-stranded RNA virus that must reverse transcribe its RNA into DNA before it can be inserted into the host chromosome. The enzyme reverse transcriptase, coded for by one of its genes, accomplishes the reverse transcription.

14. In many birds and mammals, the insertion of the DNA reverse transcript of the retroviral RNA into a host chromosome transforms the host cell into a cancerous cell. In addition to three genes needed for their own reproduction, highly *oncogenic* (cancer-producing) retroviruses possess an oncogene with a strong promoter.

15. Retroviruses without oncogenes can also cause cancer if they insert close to a proto-oncogene in the host chromosome. The high rate of transcription of the viral genes would cause overproduction of the proto-oncogene product.

16. Retroviral oncogenes probably arose from a proto-oncogene RNA transcript that became part of the genome of a retrovirus by a process similar to transduction. Evidence for this is that the retroviral oncogene is similar in base sequence to the presumed "parental" proto-oncogene except that it lacks introns (introns would have been edited out of the RNA transcript).

17. Evidence from DNA sequencing studies suggests that some transposable elements arose from retroviruses that became restricted to the host cell, but were still able to change location on the chromosomes.

18. *Antisense nucleic acids* are single-stranded RNA or DNA polymers that can be used to inactivate mRNAs transcribed from harmful genes. They are designed to be complementary to a target RNA, form a duplex with it, and thus block its translation by ribosomes.

19. Viral transduction has been used to insert a gene encoding an antisense nucleic acid into a cell, enabling the cell to make its own antisense nucleic acids against both viral genes as well has cellular oncogenes. To be successful at fighting cancer, the antisense nucleic acid must be specific and not interfere with any proto-oncogenes.

20. Cancer treatments include surgery, radiotherapy (treatment with ionizing radiation), and chemotherapy. Radiotherapy breaks chromosomes and kills cells when they attempt to undergo mitosis. Chemotherapy involves treatment with drugs that, like radiotherapy, specifically affect dividing cells. Cells of the immune system are also susceptible to chemotherapy and radiotherapy because they are mitotically active.

21. The *Ames test* for cancer-causing agents is based on the observation that all chemical carcinogens are themselves mutagenic or are converted to mutagens in the liver. Liver cells, a mutant bacterial strain that requires a specific amino acid for growth, and the chemical being screened are combined. If bacteria arise that can grow without the amino acid, the chemical tested must be mutagenic, and is likely to be carcinogenic. Results with the Ames test correlate well with other, more expensive tests of tumor formation in mice and rats.

22. Prior to the development of modern medicine, human populations were limited by infectious diseases. Today noninfectious diseases, such as cancer and heart disease are more important. A major, recent exception to this is the spread of AIDS, acquired immunodeficiency syndrome.

Questions (for answers and explanations, see page 122)

1. Match the terms on the left with the appropriate characteristics on the right.

___ Benign	*a.* Cancer-causing agents
___ Metastasis	*b.* Cancer of red blood cells
___ Transformation	*c.* Nonmalignant
___ Carcinogens	*d.* Loss of mitotic control
___ Lymphoma	*e.* Spreading of transformed cells
___ Leukemia	*f.* Cancer of white blood cells

2. Which of the following events would *not* normally lead to cancer?
   a. Overproduction of growth factors due to gene amplification of proto-oncogenes
   b. Insertion of a retrovirus lacking an oncogene near the locus of a proto-oncogene
   c. Infection by a retrovirus carrying an oncogene
   d. Presence of a proto-oncogene "downstream" from a strong promoter
   e. A dominant mutation of a tumor-suppressor gene

3. Select all of the following ways that retroviral oncogenes differ from proto-oncogenes.
   a. No introns
   b. Shortened at N-terminal or C-terminal
   c. Part of a transposable element
   d. Point mutations have modified some bases
   e. Unable to produce effective growth factors

4. Which one of the following statements about the use of antisense nucleic acids as a cancer treatment is *not* true?
   a. To be effective, they must be specific to an oncogene.
   b. Their use depends on the principle of complementary base pairing.
   c. They form a duplex with the oncogene's RNA.
   d. They form a duplex and block transcription.
   e. Genes coding for antisense nucleic acids could be used.

5. Explain why chemotherapy and radiotherapy are both extensively used to combat cancer, and both also suppress the immune system.

6. Why are most forms of cancer diseases of old age?

7. What are the functions of the mutant bacterial strain and the liver cells in the Ames test?

16

Defenses against Disease

CHAPTER LEARNING OBJECTIVES—after studying this chapter you should be able to:

- ❑ Describe the most important nonspecific defenses that organisms use against pathogens that have and have not penetrated the host.
- ❑ Explain the major differences in how plants and animals react to infection.
- ❑ Differentiate between the terms "blood," "lymph," "plasma," and "serum."
- ❑ List the major types of white blood cells and describe the general roles that each type play in the nonspecific or specific immune response.
- ❑ Characterize the features of the cellular immune response and the humoral immune response.
- ❑ Differentiate between the terms "antigen," "antibody," "immunoglobulin," and "antigenic determinant."
- ❑ Characterize the roles of macrophages, T lymphocytes, and B lymphocytes in the cellular and humoral immune responses.
- ❑ Describe the basis for immunological memory and its involvement in the process of immunization.
- ❑ Describe the clonal selection theory.
- ❑ Explain the roles that clonal deletion and suppressor T cells may play in preventing the production of anti-self antibodies and the development of an autoimmune disease.
- ❑ Explain what immunological tolerance is and how it normally develops.
- ❑ Describe the events that occur during plasma cell development.
- ❑ List the steps in the production of monoclonal antibodies and describe the major uses of this technology
- ❑ Describe the structure of an immunoglobulin and identify and discuss the functions of its variable and constant regions.
- ❑ List the major types of immunoglobulins and describe their functions.
- ❑ Explain how the complement system functions and how it is activated.
- ❑ Describe the genetic mechanisms involved in the production of antibody diversity.
- ❑ Explain the basis for the phenomenon known as class switching.
- ❑ List the classes of the major histocompatibility complex (MHC) proteins and describe the functions of each class.
- ❑ Characterize the structure of the T-cell receptor and how it is able to bind to both processed antigen and MHC proteins.
- ❑ List the major antigen-presenting cells and describe how they present processed antigen.
- ❑ List and describe the roles of the major types of T-cells.
- ❑ Describe the normal flow of events that occurs in the activation and effector phases of the humoral response.
- ❑ Describe the normal flow of events that occurs in the activation and effector phases of the cellular response.
- ❑ Explain the events that take place during the maturation of T cells in the thymus gland.
- ❑ Describe the underlying genetic mechanisms controlling production of antibodies, MHC proteins, and T-cell receptors and similarities that exist in the genes coding for these molecules.
- ❑ Describe the involvement of the various interleukins in the cellular and humoral immune responses.
- ❑ List some autoimmune diseases and immune deficiency disorders.
- ❑ Describe the life cycle of the HIV virus.
- ❑ Describe the normal course of AIDS, the groups most at risk, what conditions cause its spread to uninfected individuals, and the prospects for a cure or development of a vaccine.

NONSPECIFIC DEFENSES AGAINST PATHOGENS • SPECIFIC DEFENSES: THE IMMUNE SYSTEM • RESPONSES OF THE IMMUNE SYSTEM (Immunological Memory and Immunization) (pages 353–359)

Key Concepts

1. *Pathogens* are disease-producing organisms, such as a bacterium, virus, protist, fungus, or animal parasite.
2. To be successful, a potential pathogen must arrive at the surface of the host, penetrate the body surface, proliferate within the host, and be transmitted to the next host.

3. Both the physical integrity of unbroken skin and the *normal flora* of bacteria and fungi that live on the skin protect against invasion by a pathogen to deeper layers. Frequently, the normal flora outcompete pathogens for nutrients or produce substances that are toxic to them.
4. The *mucous membranes* covering surfaces of the visual, digestive, respiratory, and urogenital systems provide several lines of defense against penetration by pathogens, including secretion of lysozyme (in tears, saliva, and nasal secretions), entrapment in mucus, removal by beating cilia, and the sneeze and cough reflexes. Low pH in the stomach and bile salts in the small intestine further discourage pathogens.
5. Nonspecific defenses against pathogens that have penetrated the host include competition for iron with host cells, antimicrobial molecules secreted by host cells, and ingestion via endocytosis by white blood cells called *phagocytes*. Other white blood cells called *natural killer cells* destroy cells infected with certain viruses and some tumor cells.
6. The *inflammation response* includes a localized redness, heat, and swelling. It results from dilation of blood vessels, leakage of blood plasma (including key defensive proteins), and invasion by phagocytes. Increased heat may also help to destroy pathogens.
7. *Interferons* are a class of antimicrobial molecules that provide generalized resistance to a variety of viruses, such as the influenza virus. Interferons are glycoproteins of about 160 amino acid subunits; they bind to cell surfaces and inhibit viral replication.
8. Whereas animals generally attempt to repair infected tissue, plants prevent the spread of pathogens by isolating the diseased area with nonliving tissue, such as wood in trees. Plants also produce substances that provide them with some nonspecific protection against fungi and some bacteria.
9. Major functions of the vertebrate immune system are to recognize, selectively eliminate, and remember foreign invaders.
10. The cells of the immune system are restricted to the *blood* and *lymph*. Blood consists of a cellular component, including red blood cells and white blood cells called *leukocytes*, and a fluid component called *plasma*. Plasma consists of water and simple and complex molecules; when blood clots, the liquid remaining is *serum*.
11. Lymph, a blood filtrate including water, solutes, and leukocytes, accumulates in extracellular spaces. It is returned to blood vessels near the heart by lymph ducts.
12. White blood cells are large, colorless cells with nuclei. They move by extending pseudopods and can squeeze between the cells that form capillaries. Two kinds of white blood cells are *lymphocytes* and *phagocytes*.
13. Phagocytes are part of the nonspecific defense system. Lymphocytes are the major cells of the immune system; they include *B cells* and *T cells*.
14. B and T lymphocytes originate in the bone marrow. T-cell precursors migrate to the *thymus gland*, where they develop into mature T cells. After B lymphocytes leave the bone marrow, they disperse thoughout the circulatory and lymph vessels.
15. Two forms of immune response are recognized: the *cellular immune response* is carried out by specialized cells against fungi, foreign tissue, multicellular parasites, and cells that are infected with viruses; the *humoral immune response* is carried out by protein molecules and is directed against bacteria and free viruses.
16. An *antigen* is a foreign cell or macromolecule that bears one or more specific sites, called *antigenic determinants*, that the immune systems attacks. A large antigen may contain several antigenic determinants.
17. The humoral immune response is based on the secretion of *antibodies* by B cells that have differentiated into *plasma cells*. Antibodies are highly specific; they recognize and bind to a single antigenic determinant. Antibodies can circulate free within the blood and lymph or may be part of a B-cell membrane.
18. *T lymphocytes*, the cells responsible for the cellular immune system, have *T-cell receptors* on their surfaces. T-cell receptors are glycoproteins that recognize and bind to antigenic determinants just like antibodies.
19. Both B and T lymphocytes are specialized to recognize a single antigenic determinant; phagocytes react nonspecifically with any nonself cell or molecule.
20. Originally, biologists thought that the body used a newly encountered antigenic determinant as a template for the production of the correct antibody. It is now known that a genetic mechanism generates a tremendous diversity of B and T cells capable of producing antibodies specific to any conceivable antigenic determinant.
21. *Immunological memory* refers to the more rapid and effective response to second and subsequent exposure to an antigen. The immune system remembers a previously encountered antigen, and production of antibodies and T cells specific to it is enhanced.
22. The success of *immunization* depends on immunological memory. *Vaccination* involves the injection of a small inoculum of a pathogen, a bacterium or virus, usually rendered harmless by heat treatment, or foreign protein from a pathogen. Subsequent exposure will elicit a massive response from the immune system.

Questions (for answers and explanations, see page 122)

1. Characterize the following items as adaptations for A) preventing pathogens from penetrating the body surface or B) combating pathogens that have done so.

_____ Production of mucus
_____ Secretion of lysozyme
_____ Normal flora
_____ Sneeze reflex
_____ Inflammation
_____ Low pH in stomach
_____ Interferons

2. Which one of the following is *not* a major function of the immune system?
 a. Recognition of self
 b. Production of the inflammation response
 c. Selective elimination of pathogens
 d. Immunological memory
 e. Recognition of antigens

3. Match the names on the left with the most appropriate characteristics on the right.

______	Plasma	*a.*	Foreign protein or cell
______	White blood cells	*b.*	Blood minus blood cells
______	Lymph	*c.*	Produced by plasma cell
______	Lymphocytes	*d.*	Includes phagocytes
______	Antigen	*e.*	B and T cells
______	Antibody	*f.*	Blood filtrate with white blood cells

4. Which of the following does *not* characterize the humoral immune response?
 a. Directed against circulating viruses and bacteria
 b. Based on activities of phagocytes
 c. Recognition of antigens is by specific antibodies
 d. Effector molecules bind to antigenic determinants
 e. Second exposure to antigen is more rapid and effective

5. Which of the following statements is *not* true?
 a. Antigens can lead to the production of more than one antibody.
 b. Antigens serve as templates for the production of antibodies specific to them.
 c. Only plasma cells make antibodies.
 d. B and T cells both originate in bone marrow.
 e. A protein molecule can have more than one antigenic determinant.

6. Examine the figure shown below and answer the following questions:

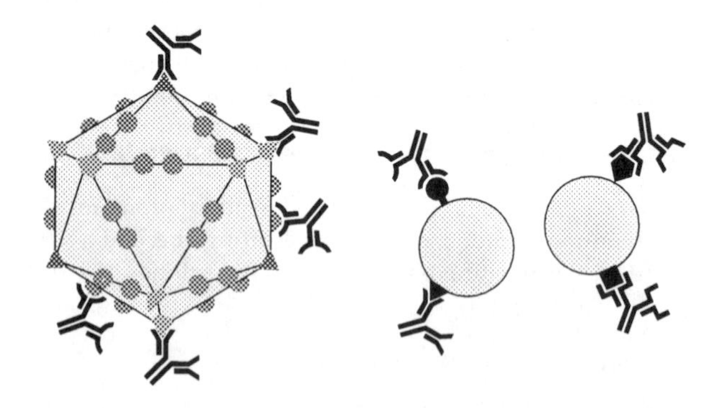

How many antigens are shown? _______

How many different antibodies are present? _______

How many different antigenic determinants are shown?

____

Activity (for answers and explanations, see page 123)

- Removal of the thymus gland in young mice usually results in most of the mice losing weight and dying. Mice that are thymectomized but maintained in a germ-free environment survive. Explain these observations.

RESPONSES OF THE IMMUNE SYSTEM (Clonal Selection and Its Consequences • Self, Nonself, and Tolerance • Development of Plasma Cells) (pages 359–361)

Key Concepts

1. Niels Jerne proposed that an antigen selects for a particular preexisting lymphocyte that is specific to it. The activated lymphocyte (either a T or a B cell) then gives rise to a large clone of cells that are all able to recognize the antigen. MacFarlane Burnet elaborated on Jerne's theory and named it the *clonal selection theory.*
2. A fundamental prediction of the clonal selection theory is that each B cell should produce a single antibody. This prediction has been verified by careful experimentation.
3. Activated lymphocytes produce two types of progeny: *effector cells* and *memory cells.* Effector cells are relatively short-lived antibody-secreting B cells or T cells with receptors that can bind antigenic determinants. Memory cells may circulate in the body for decades and respond rapidly upon a second exposure to the same antigen.
4. The *clonal deletion theory* states that immature anti-self lymphocytes are eliminated when they encounter their anti-self antigens and that no clones of anti-self lymphocytes ever develop.
5. Another proposed mechanism to prevent the production of antiself antibodies is that antiself lymphocytes are suppressed, probably by a class of T cells called *suppressor T cells.* Suppressor T cells are antigen-specific and suppress the activities of effector B and T cells. Each antiself lymphocyte would have a unique suppressor T cell suppressing its activity.
6. Based on the discovered by *Ray Owen* that some nonidentical twin cattle had blood cells of two different types, *MacFarlane Burnet* suggested that the blood may have been exchanged between the twins before the immune system had differentiated self from nonself and the blood cells from the other twin became recognized as self.
7. *Peter B. Medawar* demonstrated that a developmentally immature animal exposed to foreign antigens would develop *immunological tolerance* to those antigens, thereafter recognizing them as self. He injected highly inbred, newborn mice of one strain with lymphoid cells from another strain. Later, only the injected mice would accept skin grafts from the other strain as self tissue; uninjected controls rejected skin grafts as nonself.
8. Because lymphocytes are constantly being produced, continued exposure to self-antigen is required to main-

tain immunological tolerance. An *autoimmune disease* results if the immune system attacks a part of the body as nonself.

9. B cell activation begins when the antigenic determinant of an antigen binds to the antibody carried on the surface of a particular B cell. A type of T cell induces the B cell to divide and form a line of plasma cells and memory cells.
10. Plasma cell development involves an increase in the number of ribosomes and endoplasmic reticulum, and the development of an extensive Golgi complex. The cell is now ready to synthesize antibodies identical to those on the surface of the parent B cell.
11. Most antigens contain many different antigenic determinants. Lymphocytes produce large quantities of a single antibody, but cannot grow in tissue culture. *Cesar Milstein* and *Georges Köhler* fused single lymphocytes with myelomas (cancerous plasma cells) to produce hybrid cells called a *hybridomas*. Hybridomas grow well in tissue culture and produce large quantities of a single antibody called a *monoclonal antibody*.
12. The steps in the production of monoclonal antibodies are the injection of an animal with an antigen, isolation of lymphocytes from the animal's lymphoid tissue (like the spleen), fusion of lymphocytes with myeloma cells to produce hybridomas, growth of clones of hybridomas in tissue culture, and selection of the cell line producing the desired antibody.
13. Hybridomas grown in tissue culture secrete large quantities of a specific antibody. Hybridomas can be stored by freezing or injected into animals to give rise to tumors that produce a monoclonal antibody.
14. Monoclonal antibodies are important in studies of antibody chemistry and cell membranes, and for tissue typing and transplants. Medical applications include passive immunization (inoculation with antibody instead of antigen) and detection of specific cancers. Cancer treatments include the attachment of poisons to monoclonal antibodies specific for certain tumors.

Questions (for answers and explanations, see page 123)

1. Which of the following statements is *not* in agreement with the clonal selection theory?
   a. Antigens with more than one antigenic determinant can select for plasma cells each coding for more than one antibody.
   b. An antigen activates a preexisting lymphocyte with receptors for it.
   c. Exposure to an antigen leads to the production of both memory and effector cells.
   d. Exposure to an antigen leads to the proliferation of specific T cells and B cells.
   e. An antigen may lead to the production of more than one clone of B cells.

2. Which of the following statements about memory or effector cells is *not* true?
   a. Only effector cells carry out the attack on the antigen.
   b. Only memory cells are long-lived.
   c. Only effector cells are specific to a single antigenic determinant.
   d. Only memory cells can divide to produce both memory and effector cells.
   e. Effector cells include both B-cell and T-cell descendants.

3. Which of the following statements about immunological tolerance is *not* true?
   a. "Forbidden" antiself lymphocyte clones are either suppressed or eliminated.
   b. The development of immunological tolerance is independent of the age of the animal.
   c. Suppressor T cells may be important in the development of immunological tolerance.
   d. Continued exposure to nonself antigens is necessary to maintain immunological tolerance.
   e. The failure of immunological tolerance may lead to an autoimmune disease.

4. Which of the following does *not* occur as a plasma cell matures from a B cell?
   a. Increase in the number of ribosomes
   b. Synthesis of antibodies specific to the antigenic determinant-binding receptors on the surface of the parent B cell
   c. Increase in the amount of ER
   d. Development of an extensive Golgi complex
   e. Loss of the nucleus and ability to divide mitotically

5. Which of the following steps is *not* normally a part of the production of monoclonal antibody?
   a. Isolation of the antigens from the immunized animal
   b. Immunization with the antigen
   c. Tissue culture
   d. Cell fusion
   e. Formation of hybridomas

IMMUNOGLOBULINS: AGENTS OF THE HUMORAL RESPONSE • ANTIBODIES AND NONSPECIFIC DEFENSES WORKING TOGETHER • THE ORIGIN OF ANTIBODY DIVERSITY (pages 361–370)

Key Concepts

1. Antibodies are also called *immunoglobulins (Igs)*.
2. The structure of the most common immunoglobulin, first worked out by Gerald M. Edelman and Rodney M. Porter, is a tetramer, consisting of four polypeptides. Two identical light chains and two identical heavy chains are held together by disulfide bonds. Each of the halves of the Y-shaped molecule consists of one light and one heavy chain.
3. Each chain includes a *constant region*, in which the amino acid sequence does not differ much from one species to the next, and a *variable region*, in which the amino acid sequence does differ.
4. The variable regions of heavy and light chains are unique for each immunoglobulin. Each arm of the immunoglobulin molecule combines with a specific antigenic determinant.
5. Since the two halves of an immunoglobulin are identical, each can bind to an identical antigenic determinant, leading to the formation of large antibody–antigen complexes.

6. The constant regions of the molecule determine if it will be incorporated into a cell membrane or be secreted into the bloodstream. The constant regions of the heavy chains differ in the five classes of immunoglobulins.
7. *Immunoglobulin M (IgM)* is the first immunoglobulin produced by a plasma cell; the other four types arise by *class switching*. IgM consists of five immunoglobulin subunits and is the immunoglobulin produced in the primary immune response. It activates the complement system and promotes phagocytosis of antibody-coated cells. It has a μ heavy chain.
8. *IgG* (γ heavy chain) molecules consist of single subunits and are produced during the secondary immune response. Like IgM, IgG with bound antigens promotes phagocytosis by attaching to macrophages; they also activate the complement system.
9. The major function of *IgD* (δ heavy chain) is as a membrane receptor on B cells.
10. *IgE* (ε heavy chain) is the most common immunoglobulin in allergic reactions (such as hives, hay fever, eczema, and asthma) and in inflammation. IgE bound to antigenic determinants on allergens attaches to receptor sites on *mast cells* and the mast cell–IgE–allergen complex stimulates the release of *histamine*. IgE also helps fight worm parasites, like those causing *schistosomiasis*.
11. *IgA* (α heavy chain) is transported across epithelial membranes and added to body secretions such as tears, saliva, milk, and gastric secretions.
12. Twenty common antimicrobial blood proteins make up the nonspecific defense system called the *complement system*.
13. Once certain immunoglobulins have attached to an antigen, the first complement protein binds to the foreign cell surface. A cascade of reactions involving other complement proteins leads to the development of a *lytic complex*—a doughnutlike protein complex that causes the membrane to become leaky.
14. Two other complement-mediated responses are the promotion of endocytosis by attaching to and marking microorganisms for easier recognition by phagocytes and the attraction of phagocytes to sites of infection.
15. The genes specifying the four polypeptides of an immunoglobulin (two identical light and two identical heavy chains) are assembled from hundreds of separate DNA segments during the maturation of a B cell. The assembly of immunoglobulin genes from physically separated DNA segments, first demonstrated in 1976 by S. Tonegawa, is the basis of antibody diversity.
16. DNA segments coding for immunoglobulin light and heavy chains are on separate chromosomes. The light-chain variable region is determined by two families of DNA segments. The heavy-chain variable region is determined by three families of DNA segments. Since each family may have several hundred different forms and the heavy and light chains are independently determined, many millions of different immunoglobulins are possible.
17. Through a poorly understood mechanism, only one of each homologous chromosome producing the light and heavy chains is transcribed into mRNA, so only one light and one heavy chain are produced.
18. Mutation in the variable-region genes further increases diversity by at least 10–100 times.
19. The heavy-chain variable region has three families of DNA fragments. The mouse has over 100 different V (variable) fragments, 10 D (diversity) fragments, and 4 J (joining) fragments. Since each variable region is formed by one V, one D, and one J fragment, $100 \times 10 \times 4 = 4{,}000$ different heavy-chain variable regions can be produced before mutation further increases the diversity.
20. Assembly of the segments is accomplished with the deletion of the unused V, D, and J segments and the joining of the three used segments. The variable region of the heavy chain is then joined to one of eight C (constant) segments representing the constant region of the heavy chain. The other constant region segments are not deleted. The functional heavy-chain gene consists of one V, one D, one J, and one C region.
21. After transcription, editing of the primary RNA transcript of the functional gene removes any products of introns.
22. Initially all B cells produce IgM molecules because the gene coding for that type of heavy-chain constant region is adjacent to the variable-region gene. *Class switching* occurs if a later deletion removes constant-region segments and juxtaposes a different segment with the variable-region gene. Class switching causes a plasma cell to produce antibodies that still recognize the same antigen, but have a different function.

Activity(for answers and explanations, see page 123)

- In the following diagram of an immunoglobulin, label the light and heavy chains and the variable and constant regions of each chain. Circle areas on the molecule where antigen can bind.

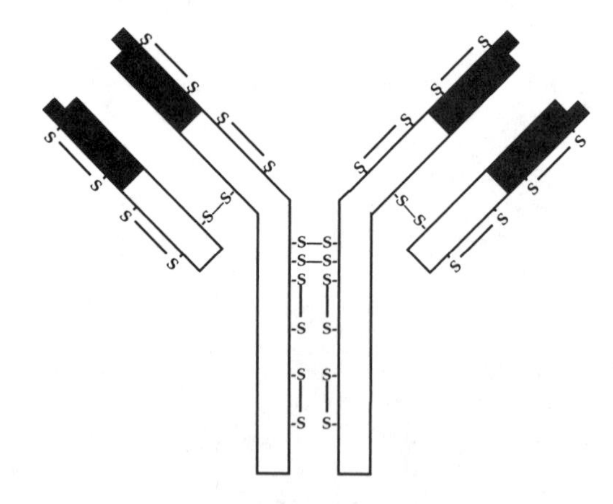

Questions (for answers and explanations, see page 123)

1. Which of the following statements about the structure of an immunoglobulin molecule is *not* true?
   a. The constant regions of heavy chains are identical for a given molecule.
   b. The variable regions of light chains differ for a given molecule.

c. The two antigen-binding sites are identical for a given molecule.
d. The constant regions of light chains are the same for the same type of immunoglobulin.
e. Precipitation of antigen would not occur if each immunoglobulin had only one antigen-binding site.

2. Select all the letters for characteristics listed on the right that apply to the immunoglobulin types listed on the left.

______	IgG	*a.* Primary immune response
______	IgM	*b.* Secondary immune response
______	IgA	*c.* Activates the complement system
______	IgD	*d.* Membrane receptor on B cells
______	IgE	*e.* Binds to mast cells
		f. Found in tears and saliva
		g. Combats schistosomiasis
		h. Can bind ten antigenic determinants

3. The complement system is involved in all but one of the following activities. Select the *exception*.
a. Elimination of foreign cells
b. Attraction of phagocytes to sites of infection
c. Formation of a lytic complex
d. Recognition of specific pathogens
e. Increasing the effectiveness of endocytosis

4. You discover a simple animal in which the variable regions of *both* the heavy and light immunoglobulin chains are assembled from three families of DNA segments and each family is represented by only three different segments. How many different antigenic determinants could this genetic system specify?
a. 9
b. 18
c. 27
d. 729
e. 10,512

5. Which of the following statements about the determination of class in immunoglobulins is *not* true?
a. The DNA segment for the constant-region gene that is adjacent to the assembled variable-region gene for the heavy chain determines the immunoglobulin class.
b. Class switching produces an immunoglobulin with a different class, but the same antigenic specificity.
c. Class switching can only occur once for a given immunoglobulin.
d. All immunoglobulins begin as members of the IgM class.
e. Deletion of DNA is the mechanism underlying class switching.

T CELLS: AGENTS OF BOTH RESPONSES • DISORDERS OF THE IMMUNE SYSTEM (pages 370–378)

Key Concepts

1. Various types of T cells are the effectors of the cellular immune response, but also coordinate the humoral immune response.

2. T-cell receptors are membrane-bound glycoproteins composed of two polypeptide chains (α and β) with variable and constant regions. T-cell receptor genes are composed of V, D, and J regions and are formed by the same joining process seen in B cells.

3. The genes determining the major histocompatability complex (MHC) proteins are located in a tight cluster of loci. Because the large number of possible alleles for each gene generates a huge number of possible genotypes, each individual of a species can differ in its cell surface MHC proteins. MHC loci fall into three classes.

4. Class I MHC loci code for proteins present on the surface of every cell of the body. Virus-infected cells present processed antigen on class I MHC proteins.

5. Class II MHC loci code for proteins found only on the surface of B cells, T cells, and macrophages. Interaction between these cells depends on recognition of self-class II MHC proteins and underlies the ability of these cells to recognize self versus nonself cells.

6. Class III MHC loci code for some of the complement proteins involved in the antibody-mediated lysis of foreign cells.

7. T-cell receptors recognize processed antigen fragments that are attached to class II MHC proteins on the surface of cells such as macrophages. One region of the T-cell receptor recognizes the antigen, another region recognizes the MHC protein.

8. T cells activated by antigen produce several distinct cell lines, including memory cells and effector cells such as *cytotoxic T cells* (T_C), *helper T cells* (T_H), and *suppressor T cells* (T_S). A given clone of T cells produces T_C, T_H, and T_S cells with receptors that only respond to the same antigenic determinant–MHC combination.

9. T_C cells eliminate virus-infected cells by causing the infected cell to lyse.

10. T_H cells bind to processed antigen with the appropriate MHC protein on B cells and stimulate the B cell to proliferate and differentiate into a clone of memory and antibody-secreting plasma cells.

11. T_S cells are regulatory cells that inhibit specific lines of B cells and T cells. Their activity may be important in turning off an immune response at the end of an infection and, perhaps, in immunological tolerance and clonal deletion.

12. Cells that can take up antigens, degrade them in their lysosomes, and present the antigen fragments as *processed antigen* in association with their MHC proteins are called *antigen-presenting cells*. Antigen-presenting cells include macrophages, B cells, and virus infected cells of any type.

13. The *humoral activation phase* begins when a macrophage presents processed antigen in association with class II MHC proteins anchored to its cell surface. Helper T cells with T-cell receptors that recognize the processed antigen and the class II MHC protein bind to them and are activated to proliferate by chemicals released by the macrophage.

14. The *humoral effector phase* begins when a B cell ingests an antigen that has bound to one of its membrane-anchored IgM receptors. The B cell displays processed antigen in association with its class II MHC proteins. When a T_H cell recognizes and binds to the processed antigen and the class II MHC protein, it releases chemicals ("helping signals") that stimulate the B cell to produce a clone of plasma cells that secrete antibody specific to the antigen.
15. The *cellular activation phase* begins when a virus-infected cell presents processed antigen in association with membrane-bound class I MHC proteins. Cytotoxic T cells with appropriate T-cell receptors recognize the processed antigen and the class I MHC protein and bind to the complex. The bound T_C cell is activated to proliferate to form a clone.
16. The *cellular effector phase* begins when an activated T_C cell recognizes processed antigen bound to class I MHC protein on a virus-infected cell. It binds to the antigen–MHC complex and releases chemicals ("lytic signals") that cause lysis of the infected cell.
17. During maturation in the thymus, T cells are tested to determine that they can recognize self MHC proteins; without this ability they could not participate in an immune response. Those T cells that pass the first test are also tested to determine that their T-cell receptors are not specific to any self antigens. Cells that fail either test are eliminated.
18. Base sequence similarities among the genes coding for MHC proteins, antibodies, and T-cell receptors suggest that all three systems may have evolved from an ancestral defense system.
19. MHC proteins of one individual can act as antigens to another individual during organ transplant. Unless transplant is done shortly after birth (before the immune system has matured), the MHC compatibility of host and donor must be carefully matched or immunosuppressant drugs used to inhibit B and T cells.
20. *Interleukins* are protein signals passed among macrophages, T cells, and B cells. Interleukin-1 (*IL-1*) is released by macrophages and activates T cells.
21. Interleukin-2 (*IL-2*) and IL-2 receptors are produced by activated T cells. The receptors become part of the activated T cells' membranes and those of other T cells specific for the same antigenic determinant. Binding of IL-2 to the receptor causes proliferation of the cell.
22. IL-2 also stimulates proliferation of B cells and activation of natural killer cells. Two immunosuppressant drugs commonly used during tissue transplants function by inhibiting IL-2 synthesis.
23. *Allergies* represent an over-response of the immune system to an antigen, such as pollen or bee sting toxin.
24. *Autoimmune diseases* (including rheumatic fever, rheumatoid arthritis, ulcerative colitis, myasthenia gravis, and perhaps multiple sclerosis) result when the immune recognition of self fails and one or more "forbidden clones" of B or T cells proliferate and attack components of the body.
25. Examples of *immune deficiency disorders* include cases where B cells never form or cannot produce plasma cells. In both cases, antibodies are never produced.
26. *Acquired immune deficiency syndrome* or *AIDS* devastates the immune system because the virus that causes AIDS, *HIV* (human immunodeficiency virus), attacks the T_H cells. The T_H cells play a key role in both the cellular and humoral immune responses.
27. The HIV virus is a *retrovirus* that inserts a reverse transcript of its RNA genetic material into the host genome. The central core of the virus includes a protein coat (p24 capsule protein) enclosing two identical RNA molecules and enzymes including the enzyme *reverse transcriptase*, used to make DNA from the RNA genome. An envelope, derived from the host cell plasma membrane, surrounds the core.
28. The HIV virus envelope is studded with protein (gp120) that can couple with a membrane protein called *CD4* that is found only on T_H cells. After coupling, the viral core is released into the cell, reverse transcriptase makes double-stranded DNA from the RNA genetic material, and the viral DNA is spliced into a host chromosome. The HIV enzyme *integrase* catalyzes the splicing.
29. The integrated viral DNA may remain latent within the host chromosome for a decade or more. Upon activation, viral DNA is transcribed into HIV RNA, translation produces viral enzymes and structural proteins, and assembly of new virus takes place.
30. As the virus buds from the infected cell, it acquires an envelope of modified cell membrane. During rapid growth, lysis of the T_H cell occurs. Reduction in the T_H cell population cripples the immune system and most people with AIDS die of opportunistic diseases. A milder form of the disease, *AIDS-related complex*, usually progresses into full-blown AIDS.

Activities ***(for answers and explanations, see page 123)***

- Explain why the high mutation rate of the AIDS gene coding for the antigenic properties of the envelope protein is complicating efforts to develop a vaccine for AIDS.

- List ways that the HIV-I virus can and cannot be transmitted.

- The following figure shows the effector phase of the humoral immune response. Label all important structures.

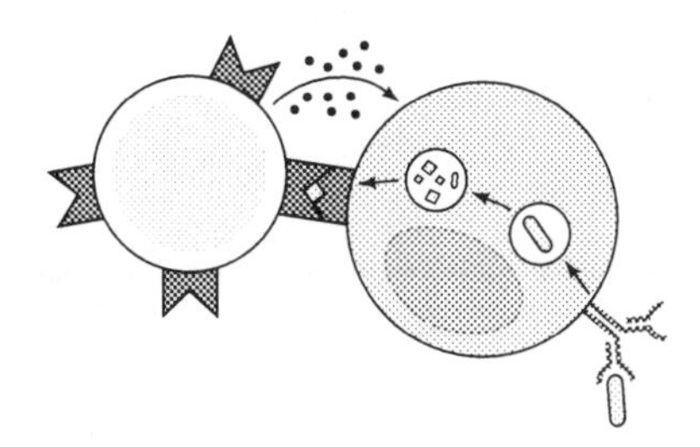

- The following figure shows the effector phase of the cellular immune response. Label all important structures.

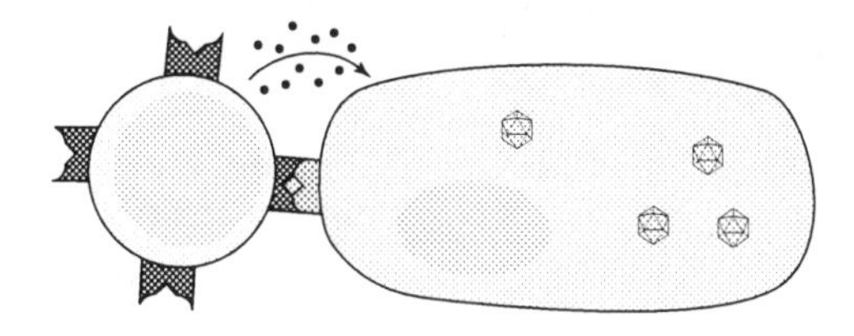

Questions (for answers and explanations, see page 123)

1. What do immunoglobulins, T-cell receptors, and MHC proteins share in common?
a. The base sequences of their genes are similar.
b. They are restricted to cells of the immune system.
c. They are all membrane-bound proteins.
d. Their genes are assembled through the same V–D–J joining process.

2. In which of the following activities are major histocompatibility complex proteins *not* involved?
a. Binding of T_H cells to macrophages
b. Binding of antigenic determinant to B cell
c. Complement-induced lysis of foreign cells
d. Binding of T_C cell to virus-infected cell
e. Transplant rejection

3. Which of the following statements contrasting T-cell receptors and immunoglobulins is *not* true?
a. Unlike immunoglobulins, T-cell receptors can only recognize processed antigen.
b. The genetic system underlying T-cell receptor diversity is different than the system specifying immunoglobulins.
c. T-cell receptors are more specific than immunoglobulins; they have a dual specificity for antigen and MHC protein.
d. Both T-cell receptors and immunoglobulins consist of multiple polypeptide chains held together with disulfide chains.
e. Like immunoglobulins, T-cell receptors have constant regions and variable regions.

4. To reduce the risk of rejection of a tissue transplant, the immune system can be suppressed by administration of drugs that inhibit production of interleukin-2 (IL-2). Which one of the following activities would *not* be affected by this drug?
a. Secretion of antibodies by B cells
b. Activation of T cells by macrophages
c. Rapid growth of T cell clones
d. Activation of natural killer cells
e. Rapid growth of B cell clones

5. Select all of the following that are *not* autoimmune diseases.
a. AIDS
b. Rheumatic fever
c. Allergies
d. Rheumatoid arthritis
e. Multiple sclerosis

Integrative Questions (for answers and explanations, see page 124)

1. Characterize each of the following as part of the A) nonspecific defense system or the B) immune system.

_____ Natural killer cells

_____ Macrophages

_____ Inflammation

_____ T_H cells

_____ T_S cells

_____ Plasma cells

_____ Interleukins

_____ Complement proteins

_____ IgM

_____ Interferons

2. Number the following events in the humoral immune response to show their correct sequence.

_____ Binding of T_H cell to processed antigen-class II MHC on B cell

_____ Proliferation of T_H cell clone

_____ Proliferation of effector and memory cells

_____ Binding of antigen to a surface receptor on B cell

_____ Stimulation of B cell by interleukin-2

_____ Maturation of plasma cell

_____ Binding of T_H cell to processed antigen-class II MHC on macrophage

17

Animal Development

CHAPTER LEARNING OBJECTIVES—after studying this chapter you should be able to:

- ❑ Describe the effects of egg yolk concentration on the early cleavage patterns seen in animals .
- ❑ Differentiate between mosaic and regulative development.
- ❑ Describe changes in cell number and size during the early cleavage stages.
- ❑ Describe the developmental sequence from the fertilized egg through the gastrula stage for the sea urchin, frog, and chicken, using the following terms: "blastula," "blastomere," "blastocoel," "blastoderm," "blastopore," "archenteron," "gastrulation," "dorsal lip," "primitive streak," "gray crescent," and "yolk plug."
- ❑ Show the derivation of the three primary germ layers and list the structures they form in the adult.
- ❑ Explain what kinds of animals are included in the protostomes and the deuterostomes and describe an important developmental difference seen in these two groups.
- ❑ Describe important features seen in the early development of mammals.
- ❑ Describe the process of neurulation seen in vertebrates, including formation of the notochord and somites.
- ❑ Explain how the anterior–posterior and the bilateral axes are determined in the early embryo.
- ❑ Explain what is meant by growth and discuss the main types of growth curves seen during animal development.
- ❑ Explain what metamorphosis is and differentiate between complete and gradual metamorphosis, using the terms "larva," "pupa," "instar," and "imaginal disc."
- ❑ Discuss the roles of determination, differentiation, and pattern formation in development.
- ❑ Describe the significance of the nuclear transfer experiments performed by Robert Briggs and Thomas King and also by John Gurdon.
- ❑ Discuss the importance of the experiments involving the transplantation of imaginal discs in *Drosophila* fruit flies and the phenomenon of transdetermination.
- ❑ Explain the significance of the studies done by Sven Hörstadius and Hans Spemann on polarity in frog eggs and embryos
- ❑ Describe the importance of segregation of cytoplasmic determinants in the eggs of *Drosophila* and in the nematode worm *Caenorhabditis elegans.*
- ❑ Explain how the studies of Spemann and Hilde Mangold on the dorsal lip of the frog embryo contribute to our understanding of embryonic induction.
- ❑ Describe the inductive control of the development of the vertebrate eye.
- ❑ Discuss induction at the cellular level with respect to vulva formation in *Caenorhabditis elegans.*
- ❑ Describe the work done by Lewis Wolpert on the role of morphogens in embryonic chick wing development.
- ❑ Describe the sequence of events that leads to pattern formation in *Drosophila* larvae, including the involvement of segmentation, gap, pair-rule, segment-polarity, and homeotic genes, and explain how these genes have their effects.
- ❑ Explain what the homeobox is and discuss its relevance to the development of many animals.

THE STUDY OF ANIMAL DEVELOPMENT • CLEAVAGE • GASTRULATION (pages 381–388)

Key Concepts

1. *Development* is the progressive change in the structure and function of an individual as it passes through the stages of its life cycle.
2. The embryo stage begins with the zygote and continues until the organism becomes independent of the nutrients provided by the mother.
3. Mitotic cell division within the embryo is called *cleavage.* In early embryos, cleavage is synchronized and the total number of daughter cells, or *blastomeres,* doubles with each division cycle.
4. In eggs with uniformly distributed yolk, such as the sea urchin egg, cleavage is complete and the resulting blastomeres are all about the same size at the eight-cell stage.
5. Frog eggs have yolk concentrated in the end of the egg called the *vegetal pole.* The end of the egg with relatively little yolk is the *animal pole* after fertilization. The *gray crescent* is a pigmented area that forms opposite the site of sperm entry.

6. In the frog, the first two cleavages form four equal-sized cells. The third cleavage is transverse and closer to the animal pole, forming blastomeres of unequal size at the eight-cell stage.
7. In large eggs with much yolk, like those of birds, yolk-free cytoplasm is restricted to a small part of one end; cleavage is incomplete and the embryo develops on top of the yolk.
8. In animals with *mosaic development*, each blastomere is necessary to form a portion of the adult body. Other animals, such as vertebrates and their relatives, show *regulative development*: the loss of certain early cells is compensated for by the remaining cells and development forms a normal adult.
9. In most animals, cleavage is rapid and results in an increase in cell number, but no growth in the volume of the embryo. Consequently, cells become smaller and smaller and the ratio of nuclear to cytoplasmic volume increases.
10. Cleavage ends with the formation of the *blastula*—a hollow structure with a fluid-filled central cavity called the *blastocoel*. The cell layer surrounding the blastocoel is called the *blastoderm*.
11. Zygotes from small eggs with uniformly distributed yolk tend to develop into blastulas with spherical blastocoels and a blastoderm composed of a single layer of cells.
12. Zygotes from large, yolk-filled eggs develop into disc-shaped blastulas with flattened blastocoels enclosed by a thin blastoderm above and a thicker layer below.
13. The distribution of yolk and other materials within the blastomeres reflects their original distribution in the egg.
14. *Gastrulation* results in the formation of a two- or three-layered embryo called a *gastrula*. During gastrulation, the rate of cell division is less than during cleavage and the total volume of the embryo changes little, but massive movement of cells may occur by processes such as *invagination*.
15. During gastrulation, masses of surface cells move into the interior to form the gut tube or *archenteron*. In the gastrula, the outer germ layer is the *ectoderm*, the gut tube consists of the germ layer called *endoderm*, and the *mesoderm* develops between the endoderm and ectoderm. The opening of the archenteron is the *blastopore*.
16. Ectoderm gives rise to the following: epidermis of the skin and epidermal derivatives like hair, feathers, and scales; the lining of the oral and nasal cavities; sweat, oil, and milk glands; the nervous system.
17. The inner lining of the digestive tract, the respiratory passages, and most of the major internal organs are derived from endoderm.
18. Mesoderm forms the skeleton, muscles, circulatory system, heart, blood, gonads, kidneys, and the coverings of internal organs.
19. In the sea urchin, mesoderm forms from two groups of cells that migrate or *ingress* into the blastocoel. Prior to gastrulation, *primary mesenchyme* arises from cells derived from the blastopore; *secondary mesenchyme* forms from cells that separate from the tip of the archenteron during gastrulation.
20. In *deuterostomes*, animals such as sea urchins, other echinoderms, and vertebrates, the anus forms from the blastopore, and the mouth forms secondarily. In the *protostomes*, animals such as segmented worms, insects, and mollusks, the blastopore gives rise to the mouth, and the anus forms later.
21. In frogs, invagination begins at a point just below the gray crescent called the *dorsal lip* of the blastopore. Sheets of surface cells, first from the animal hemisphere and then from the vegetal hemisphere, stream inward to form the archenteron as the dorsal lip expands to become a complete circle. The inward movement of a sheet of cells is called *involution*.
22. A *yolk plug*, composed of vegetal hemisphere cells displaced by the expanding archenteron, protrudes from the blastopore and mesoderm is formed by both migration and multiplication of cells. Mesoderm at the roof of the archenteron differentiates to form a stiff, supportive rod called the *notochord*.
23. In birds, involution of sheets of cells from the surface of the disc-shaped blastula creates an elongated blastopore called the *primitive streak*. Ingression of these cells forms the endoderm and mesoderm of the gastrula.
24. Although mammal eggs have little yolk, their early development shows some of the same features seen in birds, such as the presence of a primitive streak during gastrulation.
25. The mammalian blastula, or *blastocyst*, consists of an outer layer of cells with tight junctions and an inner cell mass enclosed by a cell layer called the trophoblast.
26 After the blastocyst implants in the uterus, the trophoblast gives rise to the placenta, part of the inner cell mass develops into extraembryonic membranes (see Chapter 27), and a disc-shaped portion of the inner mass becomes the embryo.

Questions (for answers and explanations, see page 124)

1. Match the structures on the left with the appropriate descriptors on the right.

____	Blastocoel	*a.* A hollow sphere of cells
____	Blastocyst	*b.* Created by ingression
____	Blastoderm	*c.* Yolk-rich end of egg
____	Blastomeres	*d.* Sheet of cells
____	Blastopore	*e.* Cells from early cleavage
____	Blastula	*f.* Forms opposite sperm penetration
____	Gray crescent	*g.* Infolding
____	Mesenchyme	*h.* Blastula cavity
____	Invagination	*i.* Mammalian blastula
____	Vegetal pole	*j.* Becomes mouth in protostomes

2. Select the correct curve describing changes in nuclear volume to cytoplasmic volume during the cleavage stage of development and the correct statement about changes in the size of blastomeres during that stage.

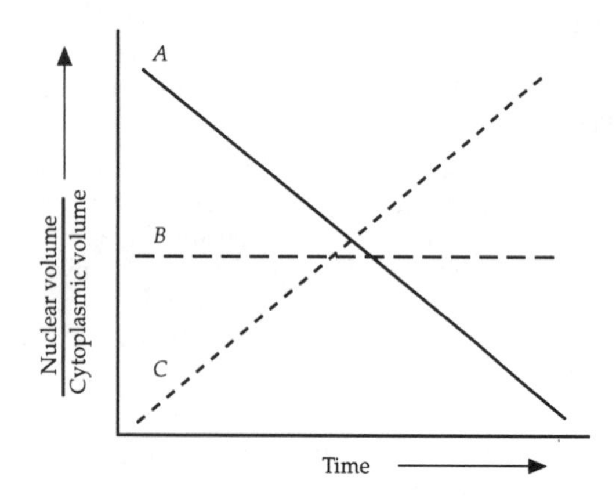

a. Curve *A*; blastomeres decrease in size during this stage
b. Curve *B*; blastomeres decrease in size during this stage
c. Curve *C*; blastomeres decrease in size during this stage
d. Curve *A*; blastomeres decrease in size during this stage
e. Curve *C*; blastomeres increase in size during this stage

3. Which one of the following structures is *not* formed as a result of gastrulation in frogs?
a. Yolk plug
b. Archenteron
c. Anus
d. Gray crescent
e. Endoderm

4. For each of the following organisms, select from the choices listed on the right the structure or location where invagination begins during formation of the gastrula.

_____	Sea urchin	*a.* Primitive streak
_____	Frog	*b.* Blastopore
_____	Bird	*c.* Archenteron
		d. Yolk plug
		e. Dorsal lip

5. You stain cells at the *ventral* lip of a frog blastopore with an intense dye. These cells will most likely become part of the
a. gut tube.
b. nervous system.
c. notochord.
d. ectoderm.
e. yolk plug.

6. Which of the following statements about the mammalian blastocyst is *not* true?
a. The trophoblast gives rise to the embryo proper.
b. Some of the cells of the blastocyst give rise to non-embryonic structures.
c. The blastocyst implants in the mother's uterus.
d. Early mammalian development is slow.
e. Tight junctions are common between the outer layer of cells in the blastocyst.

Activities ***(for answers and explanations, see page 124)***

- In the following diagram of a frog blastula, label the blastoderm, blastocoel, and the vegetal and animal poles.

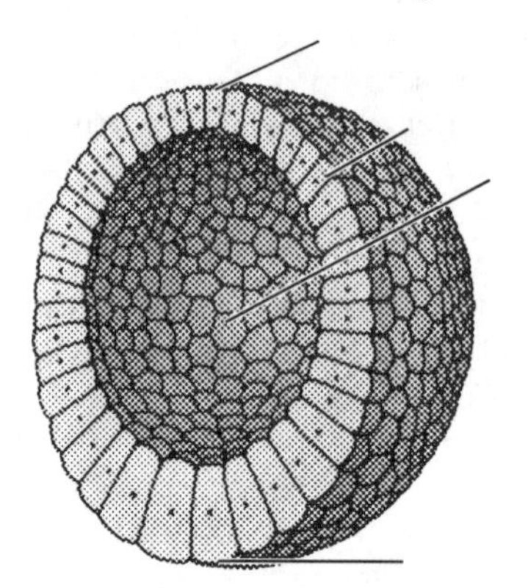

- In the following diagram of a sea urchin gastrula, label the ectoderm, endoderm, primary mesenchyme, secondary mesenchyme, blastopore, blastocoel, and archenteron.

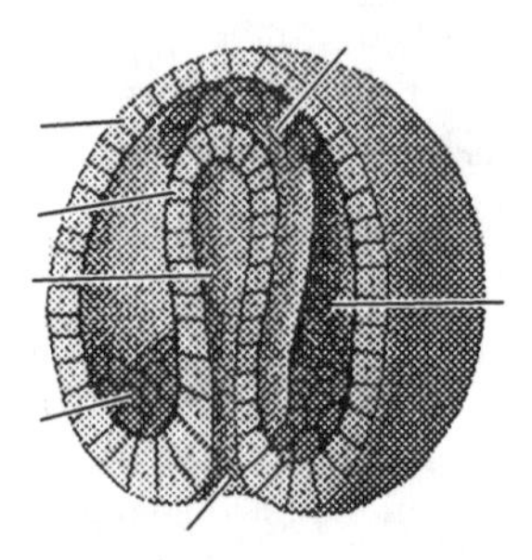

NEURULATION • LATER STAGES OF DEVELOPMENT • LOOKING CLOSER AT DEVELOPMENT • DIFFERENTIATION • DETERMINATION BY CYTOPLASMIC SEGREGATION (pages 388–397)

Key Concepts

1. Formation of the vertebrate nervous system includes *neurulation*, a common morphogenic process. An internal tubular structure forms from a sheet of initially external ectoderm. Whereas the location of the blastopore determines the anterior–posterior axis, neurulation forms the bilateral axis.
2. A *neural plate* of thickened ectoderm forms on the dorsal surface of the gastrula. A neural groove forms and the edges of the plate become neural folds. The plate rolls up to form the *neural tube*, which becomes detached from and covered by the neural folds. The spinal cord is derived from the neural tube, with the most anterior portion becoming the brain.
3. The notochord forms from mesoderm lying immediately below the neural tube. Lateral blocks of mesoderm form somites; some somite cells develop into muscles and others into the vertebrae making up the backbone.

4. Growth—a irreversible increase in size—characterizes the development of most animals. Many animals show an S-shaped growth curve: an early period of little growth is followed by a long period of rapid growth. Growth slows again as the individual approaches old age.
5. Larvae are sexually immature forms of an animal that are able to feed independently. The extensive rearrangement of structures that takes place when an animal passes from the larval to the adult stage is called *metamorphosis*. In addition to rearrangement of tissues, programmed cell death is an important part of metamorphosis.
6. In *complete metamorphosis*, insect larvae pass through stages called *instars*, undergoing a series of molts until the final instar enters a resting stage and transforms into a *pupa*. After radical reorganization, the adult emerges from the pupa.
7. Insects that pass through a series of molts as they gradually change from the larval to the adult form show *gradual metamorphosis*. These instars are known as nymphs or naiads in aquatic insects.
8. In a typical insect larva, clusters of undifferentiated cells called *imaginal discs* are responsible for transforming the larva into an adult during metamorphosis. Each disc will give rise to a distinct part of the adult insect.
9. Developmental changes occur through three processes: *determination* causes the developmental fate of a cell to become fixed; the changes in the biochemistry, structure, and function of cells as their fate becomes determined is *differentiation*; the organization of tissues to form composite structures is *pattern formation*.
10. The zygote is *totipotent*; it is able to give rise to every cell in the adult. When the prospective fate of cellular descendants of the zygote becomes determined, the cell is *differentiated*.
11. *Robert Briggs* and *Thomas King* transferred nuclei from the cells of embryos of various stages into enucleated eggs. If the nuclei were from blastula cells, development of the activated donor egg was normal. Nuclei from the cells of older embryos did not develop normally.
12. *John Gurdon* was able to achieve successful development in nuclei derived from tadpoles and adult frogs by performing serial transplants: nuclei were repeatedly transplanted into enucleated eggs, the eggs allowed to develop to the blastula stage, and nuclei from these blastula cells transplanted again. Occasionally, a normal tadpole would be formed.
13. Nuclear transplant experiments convinced biologists that differentiation does not result from the inactivation or loss of genetic material.
14. An imaginal disc transplanted to a different site in another larva will form its distinctive organ wherever it is placed in the host insect. The cells of discs transplanted to adults continue to divide but give rise to undifferentiated cells. If such a disc is later transplanted to a larva it will usually form its distinctive body part.
15. Occasionally imaginal discs transplanted from larvae to adults and back to larvae will *transdetermine*—they will give rise to unexpected organs—indicating that all genetic information is present in imaginal disc cells.
16. There may be a *polarity* in the distribution of yolk and other substances in the egg, and after fertilization this polarity may change. Cytoplasmic differences in the blastomeres of an embryo can result from polarity in the zygote.
17. Sven Hörstadius showed that if sea urchin embryos at the eight-cell stage were split in two, only those divisions that separated cells from the animal and vegetal poles equally produced normal embryos. Since nuclear transfer experiments have shown that the nuclei in these cells are still totipotent, these results must be due to cytoplasmic differences in the blastomeres.
18. Further studies by Hörstadius with sea urchin eggs showed that an unequal distribution of materials exists in the eggs, such that only eggs divided into halves with animal and vegetal portions have the potential to develop into normal larvae.
19. *Hans Spemann* found that the gray crescent—a pigmented region of the frog egg formed at the time of fertilization—contains materials essential for normal development. If the zygote is made to divide such that only one of the new cells receives gray crescent material, only the cell that did will develop normally.
20. Polarity in *Drosophila melanogaster* is based on the distribution of *cytoplasmic determinants*—more than a dozen mRNA and protein species produced by the mother's genes and distributed to the egg, where they determine the dorsoventral (top–bottom) and the anteroposterior (head–rear) axes of the embryo.
21. Female *Drosophila* with mutant alleles for any of the genes coding for cytoplasmic determinants produce abnormal larvae. For example, females homologous for the *bicoid* allele produce larvae with no head and no thorax. Females homologous for the *nanos* allele produce larvae with missing abdominal segments.
22. If eggs from females homozygous for the *bicoid* or *nanos* genes receive cytoplasm from the region of a normal egg where the mutation has its effect, they will develop normally. Also, removal of cytoplasm from specific regions of a normal egg can produce larvae similar in appearance to a *bicoid* or *nanos* larva.
23. Cytoplasmic determinants called *germ-line granules* determine which cells will eventually produce sperm and eggs in the nematode *Caenorhabditis elegans*. At each cell division during development, microfilaments are responsible for moving the granules into the correct cell.

Questions (for answers and explanations, see page 124)

1. Whereas the location of the ___________ determines the anterior–posterior axis of the embryo, the ___________ determines the embryo's bilateral symmetry.
   a. blastopore, neural plate
   b. sperm penetration, blastopore
   c. blastopore, gray crescent
   d. vegetal pole, notochord
   e. primitive streak, imaginal disc

2. An imaginal wing disc from a *Drosophila* larva is transplanted into the head region of an adult. If these disc cells are later transplanted into a second larva, they will most likely give rise to
   a. a wing.
   b. a head.

c. undifferentiated tissue.
d. cells that die.
e. a wing or head, depending on where in the larva they are placed.

3. Which of the following statements about complete and gradual metamorphosis is *not* true?
a. A final pupa stage occurs in complete metamorphosis, but not in gradual metamorphosis.
b. In complete metamorphosis instars are called larvae, in gradual metamorphosis instars are called either nymphs or naiads.
c. Imaginal disc cells of insects showing complete metamorphosis are fully determined; imaginal disc cells in insects with gradual metamorphosis are not determined.
d. Both types of metamorphosis involve a series of molts.
e. Only in gradual metamorphosis do the instars resemble the adults.

4. Select the following choice to create a statement that is *not* true. When a cellular descendent of the zygote can no longer give rise to all the cells of a multicellular organism, we can say that
a. its developmental potential is greatest.
b. it is no longer totipotent.
c. it is differentiated.
d. it is determined.
e. its developmental fate is achieved.

5. The work of Briggs and King, and later John Gurdon, on the transplantation of nuclei into enucleated eggs in the frog showed that
a. sperm are not necessary for development in the frog.
b. while differentiation is reversible in plants, it is irreversible in animals.
c. gene loss is not the cause of differentiation.
d. early gastrula cells are fully determined.
e. older nuclei are more likely to cause normal development in host eggs.

6. Which of the following observations *does not* suggest that a polarity of cytoplasmic substances in eggs or embryos is important in normal development?
a. Bisection of the eight-celled sea urchin embryo only leads to normal development if each half receives cells from both animal and vegetal hemispheres.
b. Gray crescent material is essential for normal development.
c. Sea urchin eggs divided into halves that receive only animal pole egg cytoplasm cannot undergo gastrulation.
d. Germ-line granules in *Caenorhabditis elegans* are only synthesized in the posterior end of the worm, where sex organs later develop.
e. Eggs from *Drosophila* females homozygous for a cytoplasmic determinant gene can be made to develop normally if incubated with appropriate cytoplasm from a normal egg.

7. If an egg from a homozygous *bicoid* female is incubated with cytoplasm from the anterior region of the egg of a homozygous *nanos* female, the resulting embryo will likely
a. be normal.
b. be *bicoid.*
c. be *nanos.*
d. be both *bicoid* and *nanos.*
e. die.

Activities (for answers and explanations, see page 125)

- Label the listed structures in the following longitudinal diagram of a late frog gastrula: blastopore, area where brain will form, yolk plug, notochord, neural plate, archenteron, ectoderm, mesoderm, and endoderm. Also, indicate the (A) anterior, (P) posterior, (D) dorsal, and (V) ventral ends of the embryo.

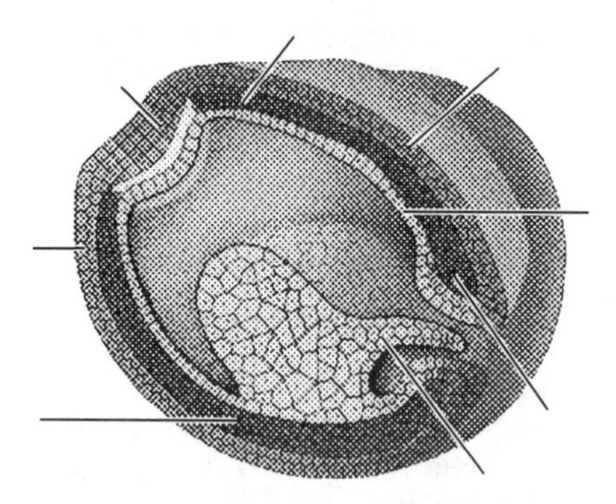

- Draw an idealized growth curve for a fast growing animal such as a chicken

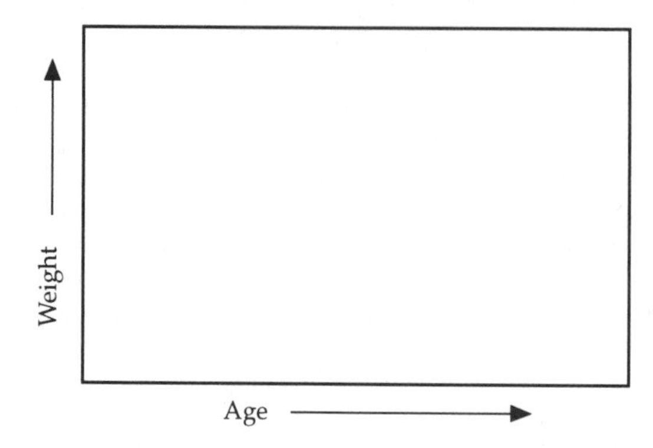

- Draw an idealized growth curve for an insect with gradual metamorphosis and four instars.

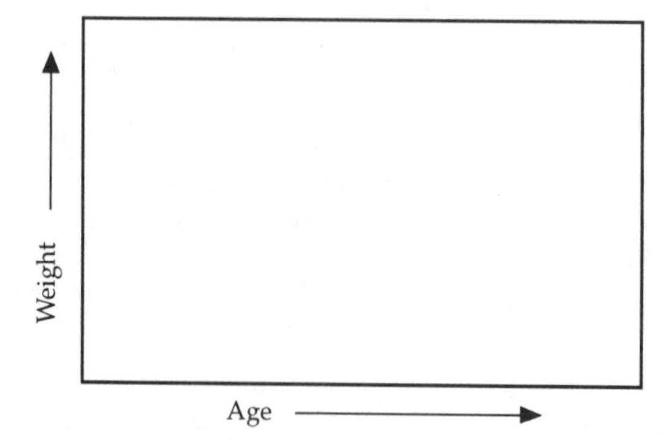

DETERMINATION BY EMBRYONIC INDUCTION • PATTERN FORMATION • YOU'VE SEEN THE PIECES; NOW LET'S BUILD A FLY • DEVELOPMENTAL BIOLOGY AND EVOLUTION (pages 397–405)

Key Concepts

1. Experiments by *Hilde Mangold* and *Hans Spemann* on the dorsal lip of the blastopore showed that the fate of tissues is frequently determined by interaction with other tissues in the embryo by a process called *embryonic induction.*
2. Mangold and Spemann found that a piece of the dorsal lip transplanted to the ectoderm of a late gastrula newt embryo induced the formation of an extra neural plate. Spemann called the dorsal lip of the blastopore the *embryonic organizer* because the formation of the neural tube establishes anteroposterior and dorsoventral axes of the embryo.
3. Formation of the lens of the eye is another example of induction. *Optic vesicles* from the forebrain induce the formation of a *lens placode* in overlying ectoderm. The lens placode invaginates, detaches from the ectoderm, and forms the lens. A signal released by the optic vesicle is essential for lens formation.
4. A transplanted optic vesicle will only induce lens formation if the overlying ectoderm had earlier been made *competent* by developing mesenchyme. The lens also interacts with the optic vesicle so that the *optic cup* that forms from the vesicle is properly sized to fit the lens. The lens also induces ectoderm overlying it to become the *cornea* of the eye.
5. There is much yet to be learned about the chemical nature of *inducers.* Some are diffusible proteins, such as a growth factor discovered in frog gastrulas, but some inducers are insoluble, extracellular proteins.
6. The complete development of the nematode, or roundworm, *Caenorhabditis elegans* from zygote to the 959-celled adult has been studied in sufficient detail so that the fate of each cell is known. This hermaphrodite, which has both sperm- and egg-producing organs, completes its development in about three and a half days.
7. Work on *Caenorhabditis elegans* has shown that development is controlled by chemical switches that determine which developmental path cells will take.
8. In the case of vulva formation in *C. elegans,* the primary switch is an inducer produced by the anchor cell. Cells receiving enough of this inducer produce a second inducer. Only cells receiving enough of the secondary inducer become vulva cells. Inducers act by turning on specific genes in the cells receiving the inducers.
9. *Pattern formation*—the organization of tissues to form composite structures such as organs—is regulated, in part, by *positional information,* the spatial relationship of one structure to another. *Lewis Wolpert* developed a theory of positional information based on his studies of limb bud formation in the chick.
10. Wolpert's theory of positional information postulates the existence of gradients of *morphogens*—diffusible chemicals that establish concentration gradients. Cells exposed to different concentrations of the morphogen develop along different lines.
11. The *zone of polarizing activity* (*ZPA*), a region on the posterior margin of the wing bud, determines pattern formation along the anteroposterior axis. Grafting of an additional, donor ZPA to different positions on a wing bud leads to abnormal pattern formation, suggesting that distance from the ZPA determines what wing structures will develop.
12. The ZPA produces an apparent morphogen called *retinoic acid.* A concentration gradient of retinoic acid may determine the anteroposterior axis of wing development.
13. Research on the imaginal discs of *Drosophila melanogaster* larvae has provided important insights into the process of pattern formation. The *Drosophila* larva consists of 13 segments—one head segment, three thoracic segments, eight abdominal segments, and a genital segment.
14. Development of the *Drosophila* larva depends on the sequential activation of key genes. The genes that produce the cytoplasmic determinants are active first and establish the anteroposterior and dorsoventral axes of the larva.
15. Three classes of *segmentation genes* in *Drosophila* larva become active after the cytoplasmic determinant genes. First the *gap genes* become active, then the *pair-rule* genes, and last, the *segment-polarity* genes. Each class of segmentation genes determines finer and finer details of the larval segmentation pattern.
16. *Homeotic genes* specify what specific structures will develop on each segment. *Homeotic mutations* result in the development of an inappropriate body part on a segment. The *antennapedia* mutation causes legs to grow in place of antennae; the *opthalmoptera* mutation results in wings where eyes should be.
17. Loci of the homeotic genes fall into several tight clusters. The best known cluster is the *bithorax complex*—eight or more genes that control development of the abdomen and posterior thorax of the fly. The *antennapedia complex* controls the development of the head and anterior thorax.
18. Mutations of one or more genes in the bithorax and antennapedia complexes produce larvae with the normal number of segments, since the segmentation genes determine the number and polarity of segments, but the way the segments differentiate is abnormal. For example, some mutations of the bithorax complex produce adults with the third thoracic segment like the second—a fly with two pairs of wings, but no halteres.
19. Recombinant DNA techniques have shown that a sequence of DNA called the *homeobox* is common to the homeotic and segmentation genes and to some genes of all animals with segmentation, including humans.
20. The homeobox, a sequence of DNA about 180 base pairs in length, codes for a 60-amino acid region, called the *homeodomain,* that is part of the protein produced by the gene in which the homeobox resides. Proteins containing this 60-amino acid sequence remain within the nucleus and bind to DNA, regulating the transcription of other genes.

21. The presence of the homeobox suggests that both the bithorax and antennapedia complexes evolved from an ancestral gene that gave rise to what is now a widespread developmental control system in many organisms.
22. Differences in the concentration of mRNAs for different morphogens in eggs of *Drosophila* produce morphogen concentration gradients in the larvae. Concentration of the *bicoid* protein morphogen is greatest in the anterior end, lowest in the posterior end; the *nanos* protein concentration gradient runs in the opposite direction.
23. Nanos and bicoid are transcription factors that enhance or repress the gap genes. The gap genes code for transcription factors that control the pair-rule genes. The pair-rule genes code for transcription factors that control the segment-polarity genes. The interacting gradients of the various transcription factors set up the segmentation pattern of the larva.
24. Different homeotic genes are expressed in different regions of the body and these six genes are arranged in the same order on a single chromosome. Each homeotic gene has a homeobox and codes for a transcription factor.
25. Both mice and humans have homeobox genes arranged in clusters on different chromosomes. Within each cluster, the genes are arranged in the same sequence as the anterior-posterior order of their expression.
26. Through macroevolutionary time, the appearance of homeotic genes may have converted an annelid-like animal composed of identical segments into an insect-like animal composed of specialized and diversified segments.

Questions (for answers and explanations, see page 125)

1. Why did Hans Spemann call the dorsal lip of the blastopore the embryonic organizer?
   a. It is the point where gastrulation begins.
   b. It becomes part of the notochord.
   c. It becomes part of the nervous system.
   d. It leads to the the establishment of the embryonic axes.
   e. It induces formation of an extra neural fold if transplanted to another embryo.

2. Pattern formation in insects depends on the sequential activation of key groups of genes. Select the gene group that is normally *first* to be activated in the fruit fly, *Drosophila melanogaster*?
   a. Gap genes
   b. Segment-polarity genes
   c. Homeotic genes
   d. Segmentation genes
   e Pair-rule genes

3. Select the gene group in *Drosophila melanogaster* that causes differentiation of the segments?
   a. Gap genes
   b. Segment-polarity genes
   c. Homeotic genes
   d. Segmentation genes
   e Pair-rule genes

4. The homeobox is
   a. a gene involved in pattern formation in *Drosophila melanogaster.*
   b. part of a protein that remains in the nucleus and regulates transcription of other genes.
   c. a DNA sequence common to the genes of many animals with a segmented body plan.
   d. important in determining pattern formation in the chick wing bud.
   e. a group of genes subject to homeotic mutations.

5. Complete the following statements about pattern formation in *Drosophila melanogaster.*

   Within the *Drosophila* larva, the concentration gradients of the morphogens ____________ and ____________ are established by the unequal distribution of their mRNAs within the egg.

   The proteins that are coded for by the segmentation and homeotic genes are ________________.

   The ____________ genes are arranged on their chromosome in the same order as the order of their expression from anterior to posterior in the larva.

6. Grafting an extra ZPA to the anterior margin of a wing bud would likely result in
   a. a normal wing.
   b. mirror-image duplication of the wing with two complete sets of wing bones.
   c. mirror-image duplication of the wing with two sets of lower wing bones.
   d. mirror-image duplication of the wing with two sets of upper wing bones.
   e. an abnormal wing with no wing bones.

7. Which of the following statements about vulva development in the roundworm *Caenorhabditis elegans* is *not* true?
   a. The anchor cell produces the primary inducer.
   b. Only in the epidermal precursor cells is gene 3 turned "on".
   c. Only primary precursor cells produce the secondary inducer.
   d. Proximity to the anchor cell determines the fate of cells relative to vulva formation.
   e. If the anchor cell is destroyed, a neighboring cell will produce the primary precursor.

Activity (for answers and explanations, see page 125)

- Beginning below with the simple diagram showing the initial stage in induction of the vertebrate eye, draw the stage where a complete lens has been formed. Label the following structures on these drawings: *optic vesicle, optic cup, lens placode,* and *cornea.*

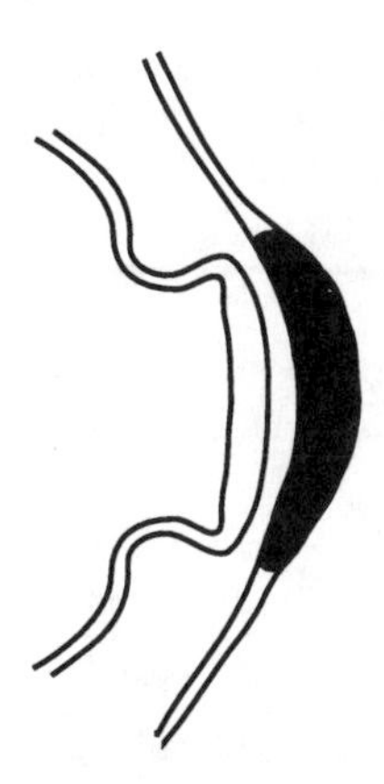

Integrative Questions (for answers and explanations, see page 125)

1. Match the following scientists with the key studies they conducted.

_____ W. Gehring	*a.* Wing formation in the chick
_____ J. Holtfreter	*b.* Polarity in sea urchin eggs
_____ L. Wolpert	*c.* Homeobox
_____ S. Hörstadius	*d.* Cellular affinities in gastrulas
_____ R. Briggs, T. King	*e.* Embryonic organizer in the frog
_____ H. Spemann	*f.* Nuclei transfer in sea urchins

2. Number the following processes to show their position in the normal sequence of animal development.

_____ Neurulation

_____ Cleavage

_____ Fertilization

_____ Invagination at the blastopore

_____ Blastula formation

_____ Mesoderm formation

3. Match the organisms on the right with the developmental descriptors on the left.

_____ Imaginal discs	*a.* Chick
_____ Spherical blastocoel	*b.* Sea urchin
_____ Anus forms first	*c.* Frog
_____ Complete metamorphosis	*d.* Fruit fly
_____ Gray crescent	
_____ Primitive streak	

Answers and Explanations

Chapters 9–17

CHAPTER 9 – CHROMOSOMES AND CELL DIVISION

Note : The page numbers listed with the section titles refer to the Study Guide; numbers in parentheses for each question refer to the relevant key concepts.

EUKARYOTIC CHROMOSOMES AND CHROMATIN • THE CELL CYCLE (pages 56–57)

1. Choice *d*. The number of chromosomes does not change between the end of mitosis and the beginning of the next cell division. During the S phase, new DNA and chromatids are synthesized, but no new chromosomes appear. (2, 7, 9, 11)
2. Choice *a*. Since most of the cell cycle involves the various stages of interphase, most cells from this tissue sample will be in interphase. (5, 10)
3. Choice *b*. Although other kinds of proteins are synthesized throughout interphase, histones and DNA are manufactured primarily during the S phase. (11)
4. Choice *c*. Chromatin and chromosomes form change dramatically during the cell cycle. (9)
5. Choice *e*. Chromatin is present at all times during the cell cycle; only its form changes. (8–9)
6. Choice *a*. MPF is made of the proteins cdc2 and cyclin. Histone proteins are involved in the formation of nucleosomes. (8, 12–13)

MITOSIS (pages 57–59)

- (5–10)

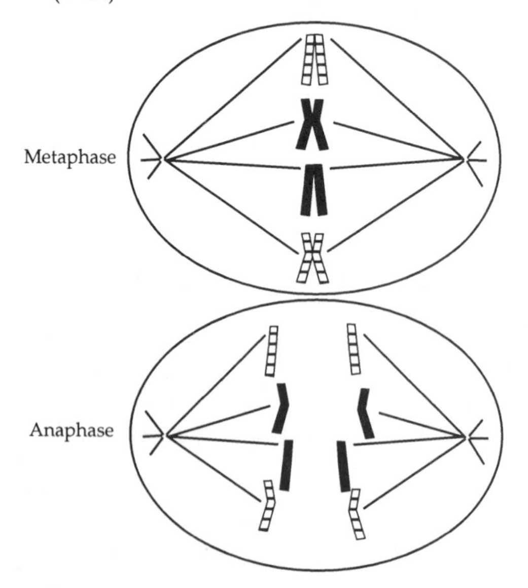

1. Choice *a*. Polar microtubules are present whenever any trace of the spindle is present, which includes all phases except interphase. (3–9, 11)
2. Choice *c*. Chromatids are present from prophase until they separate at the beginning of anaphase. (2–11)
3. *c*: Chromosomes align on equatorial plate (8)
   e: Chromatin redisperses after cell division (11)
   b: Centrosomes migrate to opposite ends of cell (3)
   a–e: Double-stranded DNA present in cell (2–11)
   b–e: Centromeres present on chromosomes (3–11)
   a: Cytokinesis occurs (12)
4. Choice *c*. Chromosomes begin moving during prometaphase. (7)
5. Choice *d*. (9)
6. Choice *e*. Asters are present whenever the spindle is present; this includes every phase except interphase. (2–9, 11)
7. Choice *c*. Cytokinesis occurs after telophase for both plants and animals. (11–13)
8. Choice *e*. In mitosis, sufficient quantities of all important cell structures are passed along to each daughter cell, although not necessarily in equal proportions. (14-15)

SEX AND REPRODUCTION • THE KARYOTYPE • MEIOSIS • PLOIDY, MITOSIS, AND MEIOSIS • CELL DIVISION IN PROKARYOTES (pages 59–61)

1. Choice *d*. Reproduction can involve mitosis only (i.e., it can be asexual). Gametes (or spores) are formed through meiosis and are used in sexual reproduction. (3–6)
2. Choice *e*. The cell is diploid at this stage, so it has all of its chromosomes. Since the chromosomes have yet to divide, there are twice as many chromatids as chromosomes. (13–16)
3. Choice *b*. This cell is diploid, which by definition means that the number of homologous pairs of chromosomes equals one-half the diploid number. (9, 12)
4. Choice *a*. Cells are haploid at the end of meiosis II. The number of centromeres equals the number of chromosomes. (12, 19)
5. Choice *d*. (16)
6. Choice *c*. Homologous pairs of chromosomes remain in synapsis from interphase through metaphase of meiosis I. (13–16)
7. Choice *e*. Nondisjunction causes an improper number of chromosomes to end up in daughter cells, a condition known as aneuploidy. (21)
8. Choice *b*. Translocation is a replication/division error

where large segments of one chromosome are incorrectly moved onto another. (23)

- (12, 16)

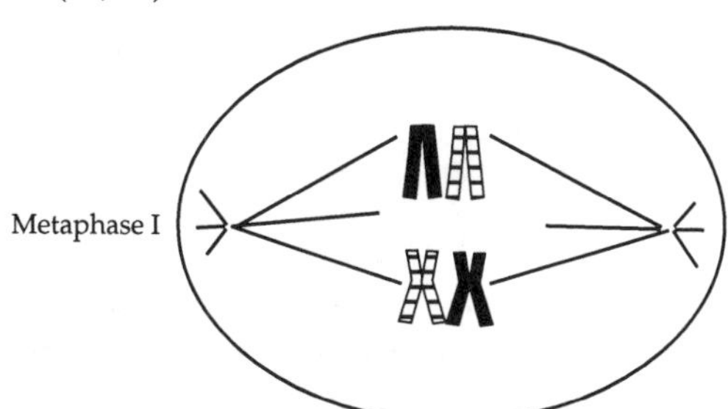

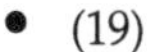

- (19)

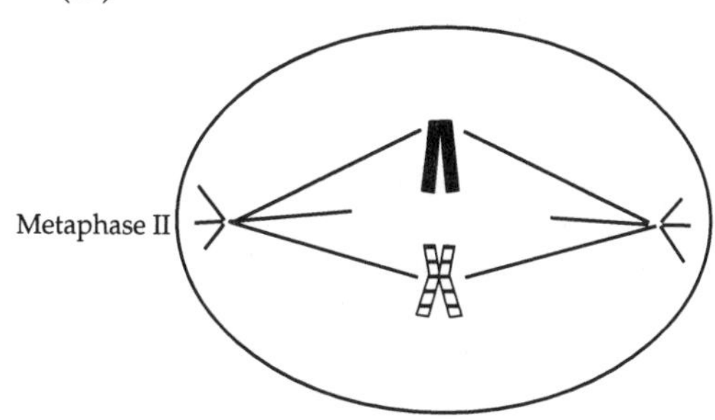

- (3–6, 8)

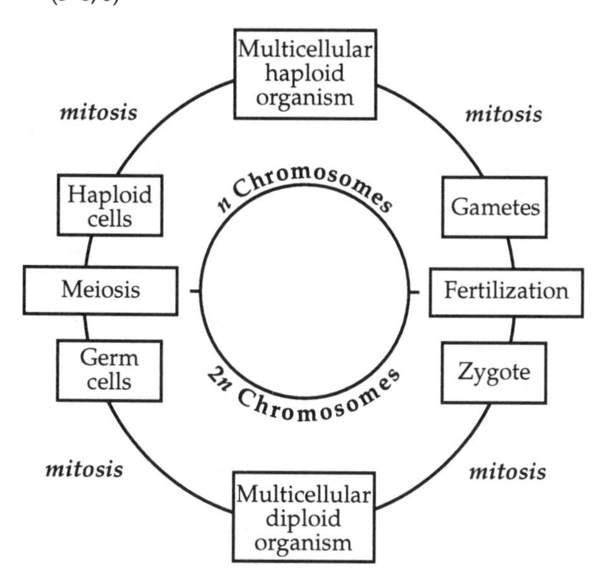

Integrative Questions

1. Choice *c*. In anaphase of mitosis, the chromatids of diploid cells separate. In anaphase of meiosis II, the chromatids of haploid cells separate.
2. Choice *b*. Since the cell has 19 chromosomes, one more than the normal diploid number, we assume it has just completed telophase and cytokinesis of mitosis. The one extra chromosome probably resulted from a nondisjunction event.
3. Choice *d*. Only cells from prophase of mitosis are diploid; those from metaphase of meiosis II would be haploid.

CHAPTER 10 – MENDELIAN GENETICS AND BEYOND

Note : The page numbers listed with the section titles refer to the Study Guide; numbers in parentheses for each question refer to the relevant key concepts.

MENDEL'S DISCOVERIES (pages 62–64)

- All five genes are homozygous dominant in the first genotype, so only ***one*** kind of gamete can be made (*ABCDE*). In the second case all five genes are heterozygous, which means that there are two alleles for each gene, and thus a total of **32 *(i.e., 2^5)*** different allele combinations possible in the gametes. The third genotype is heterozygous for only one gene locus, so only ***two (2^1)*** kinds of gametes are possible (*ABCDE* and *AbCDE*). (8–10)
- Mendel's law of segregation states that alleles at the same diploid gene locus separate from one another during gamete formation. The law of independent assortment states that alleles from one gene assort into gametes independently of all other genes, unless the genes are members of a linkage group (i.e., on the same chromosome). Segregation thus concerns the separation of alleles for the same gene during meiosis, while assortment has to do with the interaction between alleles from genes on different chromosomes. (9, 16)
- We are told that both individuals are heterozygotes, which we can symbolize as *AaBb*. To set up the Punnett square, the four possible gametes that each individual can make are assigned to rows (sperm) and columns (eggs). The F_1 genotypes are then assembled in each cell of the square. (13, 14, 16, 17)

EGGS (columns) / *SPERM* (rows)

	AB	*Ab*	*aB*	*ab*
AB	*AABB*	*AABb*	*AaBB*	*AaBb*
Ab	*AABb*	*AAbb*	*AaBb*	*Aabb*
aB	*AaBB*	*AaBb*	*aaBB*	*aaBb*
ab	*AaBb*	*Aabb*	*aaBb*	*aabb*

1. Choice *c*. This individual could not have come from PPqqRR parents because the allele combinations are inappropriate. Moreover, there is no evidence that *P* is dominant over the other genes. Finally, this diploid genotype could indeed make a *PQr* gamete. (5, 9–11)
2. Choice *a*. The way to maintain absolutely true-breeding characteristics throughout several generations is to mate only those individuals that are homozygous for the same form of the desired trait. Thus, all parents in this lineage had to be homozygous recessive. (3, 12–13)

3. Choice *e*. The functions of all these components in the hereditary transfer system are consistent with Mendel's particulate theory. (7)
4. Choice *b*. All offspring from this cross will be heterozygotes (*Kk*) and will show the dominant phenotype. (5, 10–12)
5. Choice *e*. True breeding indicates that the parents are both homozygotes, probably with short hair being dominant. Since the test cross (mating with homozygous recessive individual) produces one-half dominant phenotypes, the F_1 individuals must be heterozygotes. (5, 10–12, 15)
6. Choice *a*. We know that the male has to be homozygous recessive (*bbss*). The female must have at least one dominant allele for each trait (*B—S—*). In essence, this is simply a test cross. Since all the subsequent offspring show only the dominant phenotype, the female's genotype is probably *BBSS*. (5, 10, 15, 16)
7. Choice *b*. The cross is *CcLl* x *ccll*. Four F_1 genotypes will result, of which 25% will be *ccll*. (9, 13, 16, 17)
8. Choice *d*. Each parent can make two kinds of gametes (*NO* or *nO* for one, and *nO* or *no* for the other) which can then combine to form four different genotypes in the offspring. (9, 13, 16, 17)

GENETICS AFTER MENDEL: ALLELES AND THEIR INTERACTIONS • FOCUS ON CHROMOSOMES (pages 64–66)

- (12, 14–15)
  normal male: ZZ chicken
  abnormal female: X human
  abnormal male: XXY human
  normal male: XY fruit fly
  normal female: XXY fruit fly
  normal male: haploid honeybee

1. Choice *c*. Incomplete dominance is characterized by a unique phenotype for all three genotypes. (1)
2. Choice *b*. In codominance, neither trait is recessive nor dominant. Both alleles have equal but different effects on the phenotype of the heterozygote. (4)
3. Choice *a*. There is only one gene locus for the ABO component of human blood type. However, there are multiple alleles for this gene, which combine to produce five different genotypes. (7)
4. Choice *c*. A type O child must have the genotype *ii*, meaning that each parent needs at least one *i* allele to contribute. A woman with blood type AB cannot make eggs with the *i* allele. (7)
5. Choice *a*. Again, the mother is type AB and so cannot make eggs with the *i* allele. Even though the father has the *i* allele to contribute (by way of his mother), none of his sons by this woman can have type O blood. (7)
6. Choice *a*. Human males have an XY combination of sex chromosomes, whereas females are XX. Males are thus hemizygous for traits on either of the sex chromosomes. (12, 14–16)
7. Choice *e*. Since women carry two alleles for X-linked traits, this woman must be homozygous recessive because she expresses the color-blind phenotype. The man is hemizygous, so he must have a single copy of the dominant allele to produce his normal phenotype. But the father will always donate a Y chromosome when producing a son, so the son's X chromosome must come from his homozygous mother, who has only the recessive allele to contribute. All sons from this couple will therefore be color-blind. (14–16)
8. Choice *c*. For females to express an X-linked recessive phenotype they must have the homozygous recessive genotype. Since one X chromosome in a female comes from each of her parents, each parent must have contributed to their daughter's homozygous condition. (14–16)
9. Choice *d*. Useful gene mapping information must have something to say about the relative distances between genes. Of the things listed here, only recombination frequencies provide such information. (16, 20–22)

INTERACTIONS OF GENES WITH OTHER GENES AND WITH THE ENVIRONMENT • NON-MENDELIAN INHERITANCE (pages 66–67)

1. Choice *a*. This is an example of epistasis, where the effect of one gene masks the other. If no pigment is produced, it doesn't matter that the alleles for proper pigment deposition are present. Likewise, the animal will be colorless even if pigment is made, as long as the recessive alleles for deposition are present. (1)
2. Choice *a*. The products manufactured by the gene that controls pigment deposition are temperature sensitive, showing how environment can influence penetrance or expressivity of a phenotype. (4–5)
3. Choice *e*. Maternal inheritance refers to the passage of extranuclear DNA from mother to offspring by way of cytoplasmic sources, usually the mitochondria or chloroplasts. The DNA in these structures does not undergo meiosis during cell division, and so genetic recombination is not an issue. (7)
4. Choice *c*. Skin color is a likely polygenically controlled trait in humans, whereas flower color in our hypothetical plant involves only a single gene showing incomplete dominance. (3)

Integrative Questions

1. Choice *b*. The sequence of genes from C through H has been inverted within the linkage group.
2. Choice *d*. This statement is false because homozygous dominants can only produce gametes containing the dominant allele, which cannot then combine with one another to form heterozygous offspring.
3. Choice *a*. These crosses are reciprocals of one another with respect to both sex and trait. Since the offspring are the same regardless of which parent possesses which trait (i.e., the trait is gender-independent), the inheritance pattern is likely to be autosomal.
4. Choice *d*. The information about both parents being true-breeding tells us that they are homozygous for their respective traits. Independent assortment allows us to treat this dihybrid cross as two monohybrid crosses occurring simultaneously. Thus, the chance of getting purple flowers in the F_2 generation is 3/4 and the chance of getting short plants is 1/4. The chance of both phenotypes occurring together is the product of the two independent probabilities, or 3/16.

CHAPTER 11 – NUCLEIC ACIDS AS THE GENETIC MATERIAL

Note : The page numbers listed with the section titles refer to the Study Guide; numbers in parentheses for each question refer to the relevant key concepts.

WHAT IS THE GENE? • NUCLEIC ACID STRUCTURE • REPLICATION OF THE DNA MOLECULE • PROOFREADING AND DNA REPAIR (pages 69–71)

- (6, 9, 14)
  (parent strand)
  5′ – A T C C G T A A C G C A G G G C T T A – 3′
  3′ – T A G G C A T T G C G T C C C G A A T – 5′
  (daughter strand)

- If parental DNA is made exclusively with ^{15}N, as was the case in Meselson and Stahl's experiments, it can only form one heavy band during centrifugation. The first generation of DNA raised on ^{14}N will, however, form a different centrifugation band due to its hybrid DNA, which is made from a ^{15}N template strand and a ^{14}N replicate strand. In the second and subsequent generations, the hybrid band and a new light DNA band will appear. Some new DNA will still be made using the old ^{15}N templates (hybrid band), but most new DNA will now be made by ^{14}N templates (light band). (14–16)

- Griffith's experiments with pneumococci showed that bacterial transformation occurs, but did not provide an actual mechanism for this process. Hershey and Chase used radioactive-specific labeling of nucleic acids and proteins to conclusively demonstrate that DNA is the transforming principal. The main conclusion from the combined experiments is that DNA is the hereditary material that ultimately controls the characteristics of organisms. (1–4)

1. Choice *e*. The "D" of DNA stands for deoxyribose, and the "R" in RNA for ribose. Thymine is a nucleotide unique to DNA, while uracil is found only in RNA. (5–6, 11)
2. Choice *c*. Guanine and cytosine are the only combination shown that fits both requirements. (6)
3. Choice *c*. These experiments provided definitive evidence that DNA, not protein, was the hereditary material for T2 bacteriophage (and by extension, other organisms). (3–4)
4. Choice *b*. If 30 percent of the bases are adenine, then according to Chargaff's rules 30 percent must also be thymine, accounting for a total of 60 percent of the DNA in this particular sample. That leaves 40 percent to be divided equally among the guanine and cytosine, or 20 percent for each. (6)
5. Choice *a*. Okazaki fragments form in 100–200 nucleotide segments along the lagging strand of DNA, and are then connected together to make a functional strand. They cannot form continuous strands of DNA because the lagging strand has the wrong sugar–phosphate backbone orientation. (17–19)
6. Choice *e*. The Watson–Crick model contains nitrogen in the bases and phosphorous in the phosphate groups. It has a constant cylindrical shape due to the double helix, and the amount of any given purine always equals the amount of its complementary pyrimidine. The double helix is held together mainly by hydrogen bonds between the bases of opposing strands. (6–9)
7. Choice *c*. Although radioactive labeling of phosphorus and sulfur in the Hershey–Chase experiments showed that DNA is the hereditary material, it did not provide information about the actual structure of DNA. (3–10)
8. Choice *d*. The presence of the nucleotide base uracil suggests that this molecule is RNA. (11)

WHAT DO GENES CONTROL? • FROM DNA TO PROTEIN • MUTATIONS • THE ORIGIN OF NEW GENES (pages 71–74)

- (25–26)
  Case 1: ***methionine – threonine – cysteine – arginine – glycine***
  Case 2: ***nothing***, because there is no initiator codon to start translation
  Case 3: ***methionine – alanine – threonine – serine – threonine – tryptophan***
  (Note: during translation mRNA codons are read in the 5′ to 3′ direction)

- Genetic information is inherited in the form of DNA (or RNA in the case of retroviruses). Because this material stays in the nucleus, a messenger nucleic acid (mRNA) must be transcribed to get the genetic message out into the cytoplasm where protein synthesis occurs. Once in the cytoplasm, the mRNA message must then be translated into the amino acid language of proteins. (6–8)

1. Choice *b*. Genes code for the production of a single polypeptide, so any one gene can control only one part of a pathway. Thus, assuming that these are the only genes involved, it is theoretically possible to have as few as ten different auxotroph strains, although each gene can mutate in many different ways. (1–4)
2. Choice *d*. The one-gene, one-polypeptide theory says that any give gene can produce only a single polypeptide, which then usually plays a role in shaping the organism's phenotype. (1, 4)
3. Choice *b*. Genetic information starts with nucleotides (the "letters" in the triplet codons). The nucleotides are organized into DNA segments, which are grouped into genes. Many genes are contained on each chromosome. (6, 10)
4. Choice *c*. The universal genetic code as it relates to translation is based on the nucleotide triplets found in mRNA, the blueprint for protein manufacture. (8, 13)
5. Choice *e*. One gene leads to one polypeptide, with the triplet bases in the DNA sequence corresponding to specific amino acids in the polypeptide. (5, 13, 25)
6. Choice *a*. Translation involves turning an mRNA message into a polypeptide. DNA does not play a direct role in this process, although it does in transcription. (6, 10, 17–19)
7. Choice *b*. Adaptors allow things to coexist or work together. tRNA is an adaptor that connects the nucleic acid language of DNA and mRNA with the amino acid language of proteins. (7, 13)
8. Choice *c*. Free amino acids in the cell are attached to an appropriate tRNA through the action of activating

enzymes. The charged tRNA then combines with mRNA on the ribosome and forms a dipeptide bond with an adjacent amino acid in the growing polypeptide sequence. (13–18)

9. Choice *a*. Digestive enzymes would most likely be packaged in membranous vesicles in order to avoid self-digestion. This suggests that these particular proteins would be manufactured on ribosomes attached to the rough ER. (23–24)
10. Choice *d*. Careful examination of the two sequences reveals a substitution of the sixth nucleotide base, thymine, with an adenine. No other detectable mutations have occurred. (28–33)

Integrative Questions

1. Choice *b*. Frame-shift mutations result from insertions or deletions of bases in the DNA sequence. This causes the triplet reading frame to shift so that all subsequent codons are altered. It is the equivalent of moving all the spaces in this line of text one character to the left or right. Now the "words" defined by that convention of punctuation no longer convey the same message.
2. Choice *e*. Cytosine should appear in all these molecules, since both DNA and RNA use this pyrimidine in their base sequences.
3. Choice *a*. Translation makes proteins, not nucleic acids, using mRNA as the blueprint for this process. mRNA is therefore not a direct product of translation, whereas all of the other molecules listed contain at least some protein made directly through translation.

CHAPTER 12 – MOLECULAR GENETICS OF PROKARYOTES

Note : The page numbers listed with the section titles refer to the Study Guide; numbers in parentheses for each question refer to the relevant key concepts.

MUTATIONS IN BACTERIA AND BACTERIOPHAGES • BACTERIAL CONJUGATION • BACTERIOPHAGES (pages 75–78)

- Each letter represents a colony of bacteria. On the normal medium, all forms grow. However, colonies B and C do not develop on the medium lacking methionine, indicating that they are mutants for the synthesis of this substance. (11–14)
- In this case, colonies C and E do not develop, showing that they are auxotrophic for threonine. Colony C consists of bacteria mutant for both methionine and threonine. (11–14)

1. Choice *e*. Bacteria and viruses have relatively small amounts of DNA compared to eukaryotes. (1)
2. Choice *b*. The lawn in a bacterial culture represents large numbers of individual bacteria. When phage are applied, they kill all (or virtually all) of the bacteria near them, thus exposing the medium below. (4)
3. Choice *d*. All other mechanisms could be ruled out, except for DNA transfer via conjugation. (8–10, 18–19, 23)
4. Choice *c*. The F factor is where the circular chromosome breaks during conjugation. The DNA replicate formed starts entering the recipient cell in a linear sequence beginning with the F factor. If character five only appears in cells that mated the longest, this means that this character must be located farthest away from the F factor on the chromosome. (15–20)
5. Choice *e*. All of these statements are true. (15–20)
6. Choice *d*. Sexduction is a variation on conjugation in which the F factor physically separates from the chromosome, taking a small amount of chromosomal DNA on either side along to the recipient cell. (21)
7. Choice *a*. General transduction occurs when bacterial DNA is inserted into a phage coat, which can associate with another bacterium and thus pass along the bacterial DNA without the need for viral infection. (24–25)
8. Choice *d*. The opened end of the broken chromosome always enters the female cell first for Hfr cells. (19)
9. Choice *b*. Advantageous mutations can be naturally selected and result in heritable phenotypic changes in organisms, including bacteria and viruses. Thus, mutations are the raw material of evolution. (6)

EPISOMES AND PLASMIDS • TRANSPOSABLE ELEMENTS • CONTROL OF TRANSCRIPTION IN PROKARYOTES (pages 78–80)

- In this operon, an inactive repressor (R) cannot normally block the transcription of the structural genes (1–4). All four enzymes needed for the synthesis of arginine are therefore made continuously. Arginine acts as the co-repressor (co-R) in this system. When present it binds with R to form an active R/co-R complex that settles down at the operator (O) to block transcription of the structural genes. When arginine is removed, the repression ends and the genes are once again transcribed. (13–20)

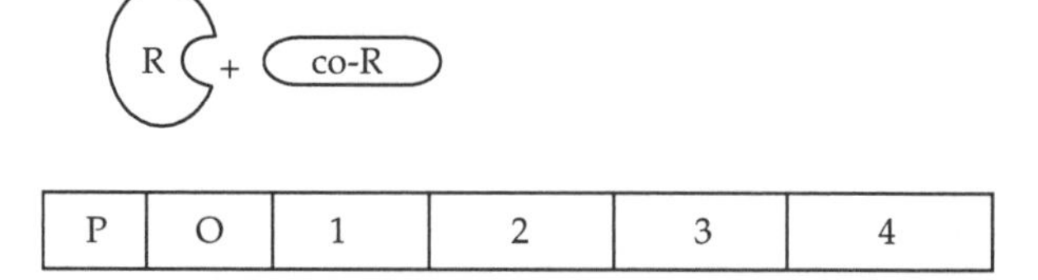

- The most likely hypothesis is that as the use of new antibiotics increased with the advent of modern medicine, bacteria were subjected to intense selection pressures. Mutant forms that were resistant to these drugs appeared in the bacterial population, and the genes for drug resistance were transmitted via R factors. Since drug resistance was selectively advantageous, the frequency of R factors increased in these strains of bacteria. (3)

1. Choice *c*. The F pilus is a hairlike, tubular projection through which genetic material is exchanged during conjugation. All others in the list are indeed plasmids and episomes. (1–3)
2. Choice *d*. Replicons are necessary for independent plasmid replication, since the plasmid remains distinct from the host cell chromosome and must replicate in synchrony with the cell in order to remain in the bacterial population. (4)

3. Choice *b*. See the definition of a transposon. (5–6)
4. Choice *e*. Although transposable elements do move around within the chromosome, the rate at which they move is rather low. (5–7)
5. Choice *c*. Inducible enzymes have the advantage of only being made when the substance on which they are intended to act is actually present. This mechanism is particularly efficient when the substance is available to the cell only irregularly. (9, 15)
6. Choice *a*. In repressible operon systems, the repressor is normally incapable of preventing gene transcription, which goes on routinely in the absence of an external stimulus. The repressor becomes active when combined with a corepressor. The complex then proceeds to block transcription. (18, 19)
7. Choice *d*. Polycistronic messengers are needed for simultaneous translation of the multiple enzymes coded for by the structural genes of an operon. Through the careful use of nucleic acid punctuation signals, a single ribosome may translate several proteins from this long, single strand of mRNA. (10, 11)
8. Choice *e*. All four types (of which a and b are specific examples of c and d) contain regulatory genes that code for repressor molecules. (16, 20)
9. Choice *c*. See the definition of a constitutive enzyme. (9)
10. Choice *e*. All of these structures, events, or features occur in the regulatory mechanism of enhanced promoter efficiency. (21–22)

CHAPTER 13 – MOLECULAR GENETICS OF EUKARYOTES

Note : The page numbers listed with the section titles refer to the Study Guide; numbers in parentheses for each question refer to the relevant key concepts.

EUKARYOTES AND EUKARYOTIC CELLS • HYBRIDIZATION OF NUCLEIC ACIDS • EUKARYOTIC GENE STRUCTURE • REPETITIVE DNA IN EUKARYOTES (pages 81–83)

- (8–10)

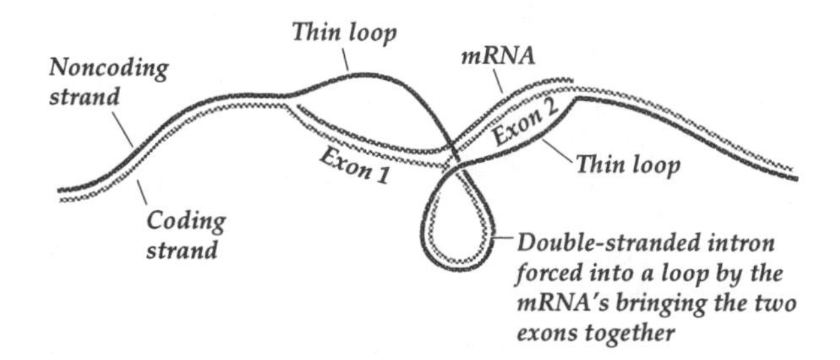

- (8–10)

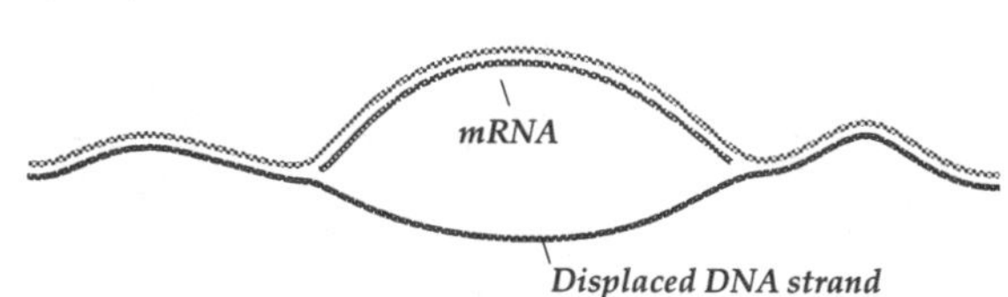

1. (5–7)
   5: Add mRNA
   2: Warm the nucleic acid mixture
   3: Observe denaturation
   6: Cool the nucleic acid mixture
   7: Observe reannealing
   1: "Cut" the DNA into small fragments
   4: Immobilize nucleic acids on nitrocellulose filter
2. Choice *c*. The DNA type that is most abundant will reanneal most rapidly. This includes the type known as highly repetitive DNA, typically located near the centromeres. (11–14)
3. Choice *a*. The telomere serves as a primer for completion of the *lagging* strand. (14–15)
4. Choice *d*. Although the gene products of some transposable elements have been characterized, their functions are unknown. (17–20)
5. Choices ***a,c, d***. (16)
6. (4, 9)
   750 nucleotides in exons
   450 nucleotides in exons
   Since each amino acid in the protein is coded for by three nucleotides, 750 nucleotides are included in the exons and the remaining 450 nucleotides must be in introns.

GENE DUPLICATION AND GENE FAMILIES • RNA PROCESSING IN EUKARYOTES • CONTROL OF GENE EXPRESSION IN EUKARYOTES (Transcriptional Control: Gene Inactivation) (pages 83–84)

1. Choice *d*. Actually, genes in a gene family differ more in intron sequences than in exon sequences. Since introns are spacer regions of DNA, variation in introns is less detrimental to the functioning of a gene than variation in exons, which actually code for gene product. (1–5)
2. (7–14)
   5: Removal of intron "lariat"
   3: Binding of snRNA to protein
   6: Splicing of exons
   4: Binding of snRNPs to RNA
   7: Degradation of intron
   1: Addition of GTP cap
   2: Polyadenylation
3. Choice *d*. Since each cell needs one functional X chromosome, and no more, three of the X chromosomes in each cell of this individual will be inactivated, so three Barr bodies will be evident. (21, 22)
4. Choice *b*. DNA that is uncoiled and nonstaining is euchromatin. Because euchromatin is uncoiled, the genes located there are actively producing mRNA. Puffs are regions of euchromatin. (21, 23)
5. Choices *d, e*. The start and stop sequences and exons are found in both snRNA immediately after transcription and its mature mRNA. The introns are removed and a modified G cap and a poly A tail are added during processing. (1–8)
6. Eukaryotic mRNA is relatively long-lived, unlike mRNA in prokaryotes, and its stability determines how much protein is produced. Eukaryotic mRNAs vary widely in their stability.

CONTROL OF GENE EXPRESSION IN EUKARYOTES (Transcriptional Control: Gene Amplification • Transcriptional Control: Selective Gene Transcription • Translational Control • Posttranslational Control) (pages 84–85)

- *rRNA*. The nucleolus represents the chromosomal region that specializes in production of the RNA subunits for ribosomes, so rRNA is being produced. (1–3)

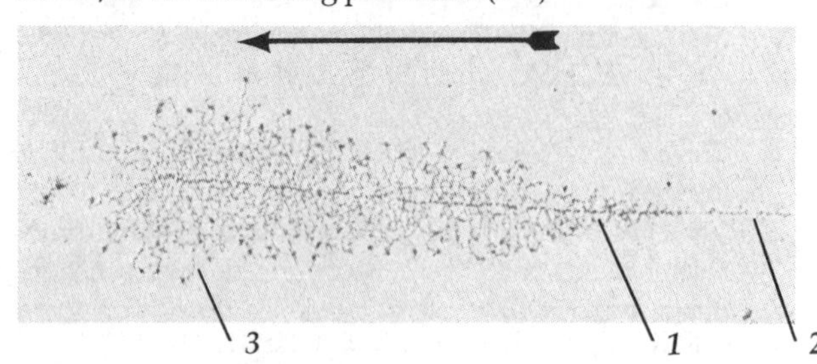

1. Choice *b*. Prokaryotes have both positive and negative regulation of their operons; eukaryotes lack operons and only have positive regulation (their genes are "off" until turned "on" by regulatory proteins).(5–9)
2. Choice *b*. Only RNA polymerase II catalyzes transcription of genes that code proteins; RNA polymerase I and III catalyze transcription of genes coding for either rRNA (I) or tRNAs (III). (8)
3. Choice *c*. The TATA box is a binding site on DNA for the transcription factor TFIID. (9–12)

Integrative Activity

posttranscriptional	Increasing the longevity of the mRNA coding for a gene product
posttranscriptional	Modification of the G cap
transcriptional	Presence or absence of enhancers
transcriptional	Methylation of the gene
transcriptional	Presence or absence of the appropriate transcription factor
transcriptional	Gene amplification through duplication
posttranslational	Conversion of proinsulin into insulin
transcriptional	Conversion of heterochromatin into euchromatin
posttranslational	Modification of a gene product

CHAPTER 14 – RECOMBINANT DNA TECHNOLOGY

Note : The page numbers listed with the section titles refer to the Study Guide; numbers in parentheses for each question refer to the relevant key concepts.

THE PILLARS OF RECOMBINANT DNA TECHNOLOGY • CLEAVING AND SPLICING DNA • CLONING GENES (pages 86–88)

1. Choice *c*. Restriction endonucleases break the phosphodiester bonds within the backbone of the DNA molecule; hydrogen bonds between the bases are not broken, but cannot persist at warm temperatures. Choice *a* is true, because the distribution of the four bases in DNA is random, so longer recognition sequences will be less common than shorter sequences. (4–9)
2. Choice *b*. The ideal plasmid would have a single recognition site to open the plasmid for insertion of the new DNA. (10–13)
3. Choice *b*. Lysing the protoplast would kill the cell. All other techniques are used to insert foreign DNA into cells. (17–21)
4. (3, 4, 9, 18, 22)
   4: Encourage growth of the transgenic cells
   2: Treat with ligase
   1: Create fragments with restriction endonucleases
   3: Transform cells
   5: Use selection plating with antibiotics
5. (23)
   65. Only cells with the plasmid can grow on an antibiotic supplemented plate, so all of the colonies resulted from transformed cells .
   55. The 55 blue colonies resulted from cells containing a plasmid with a functional z gene that can make β-galactosidase.
   10. The 10 white colonies resulted from cells containing a plasmid with foreign DNA inserted into the z gene. Since the z gene can not make β-galactosidase, these cells can not convert the substrate XGal into a blue product. Thus, these cells appear white.
   About 65 colonies should grow on the plate, but all colonies will be white.
6. Choice *d*. Yeast is a *unicellular* fungus. (14)

- Frequently the recognition sequences for restriction endonucleases are six base pairs long and palindromic, that is, the two sequences are the same when read in the opposite direction. (4–9)

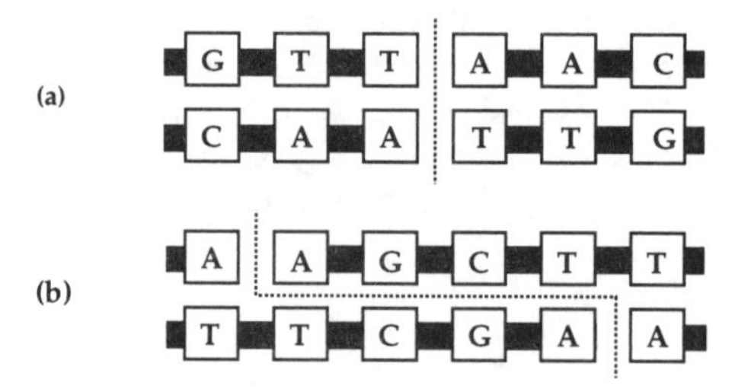

- Because *Hpa*I produces a blunt-ended cut, its fragments will rejoin with any other fragment. *Hin*dIII produces so-called "sticky-ended" fragments that can only rejoin with a fragment that has a complementary end. This allows precise control over which fragments will produce recombinant molecules.

SOURCES OF GENES FOR CLONING • EXPLORING DNA ORGANIZATION (pages 88–90)

1. Choice *b*. cDNA libraries would be most useful for determining which genes are active during differentiation because they are normally obtained from mRNA. (1–6)
2. (2–4)
   1: Add oligo-dT primer to mRNA
   3: Denature DNA–RNA hybrid molecules
   6: Insert cDNA into DNA vector
   5: Add DNA polymerase
   4: Degrade mRNA
   2: Add reverse transcriptase

3. Choice *e*. The sample with the greatest number of recognition sites would produce the largest number of fragments. B and e both have three fragments. Since DNA prepared for electrophoresis is negatively charged, it will move toward the positive pole and the smallest fragments will move the greatest distance. (9, 10)
4. Choice *a*. In situ hybridization can be used to hybridize a DNA probe made from mRNA to the DNA of intact chromosomes. (11–20)
5. Choice *b*. DNA base sequencing could be used to compare the DNA sequence of a mutant allele causing a genetic disease with the normal allele for the gene. (11–20)
6. (1–20)
   b: Shotgunning
   c, e: Chromosome walking
   e: Southern blotting
   a: Reverse transcriptase
   d: Size standards
   e, f: DNA base sequencing
   c, e: In situ hybridization

- (18–20)
  New strand: T–C–G–T–A–C–G–C–A–A–G–T
  Original strand: A–G–C–A–T–G–C–G–T–T–C–A

- (9, 10, 15)

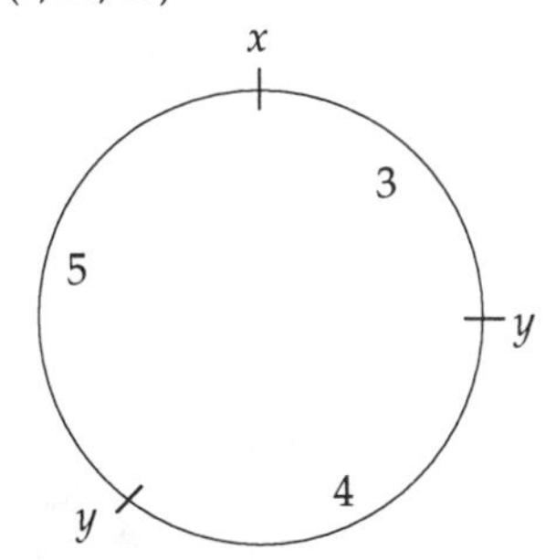

GENE COPIES BY THE BILLION • PROSPECTS (pages 90–91)

1. Choice *b*. Probes with base sequences complementary to the sequences at the boundaries of the gene must be constructed to act as primers for the DNA polymerase. (1–3)
2. Choice *d*. Since the number of DNA molecules doubles with each cycle, it will require ten 5-minute cycles, or 50 minutes, to get 1024 copies. (1–3)
3. Choices *a, b, c, d, e*. All choices are valid sources of plant germ plasm. (9)
4. Choice *e*. Transformation is used to insert the plasmid into the bacterium. (10, 11)
5. Traditional plant breeding is a slow process because it depends on the completion of a plant's normal life cycle. Also, seed production from the selected strains requires a significant land area. Since recombinant DNA techniques are directed at the cellular level, the time and space constraints are reduced considerably. A traditional genetic cross between two plants involves all the the genes, not just the genes of interest, and only plants that are capable of interbreeding can be used. Recombinant DNA technology allows transfer of individual genes, not whole genomes, and gene transfer can involve any organism, not just those that are closely related. (7–8)

Integrative Questions

1. Choice *c*.
2. Choice *a*.
3. Choices *b, c, d, f*.
4. Major techniques for synthesizing the somatostatin gene were based on the amino acid sequence of its product, insertion of the gene into a plasmid using restriction endonucleases, transformation of *E. coli* with the plasmid, and selection of transgenic bacteria. Control of expression of the somatostatin gene was accomplished by using a plasmid containing the *lac* operon and inserting the gene into the *z* gene of the operon. Somatostatin production by *E. coli* could then be induced by simply adding lactose to the culture.

CHAPTER 15–GENETIC DISEASE AND MODERN MEDICINE

Note : The page numbers listed with the section titles refer to the Study Guide; numbers in parentheses for each question refer to the relevant key concepts.

SOME INHERITED DISEASES • TRIPLET REPEATS AND THE FRAGILITY OF SOME HUMAN GENES • DEALING WITH GENETIC DISEASE: AN OVERVIEW (pages 92–94)

1. Choice *c*. Although it is true that the fragile-X premutation allele is more likely to accumulate additional repeats when transmitted by the mother, this is not true for all diseases linked to triplet repeats. All other statements are true. (7, 11–15)
2. Choice *e*. Onset of the disease is early in life. (5, 18)
3. (2–9)
   e Cystic fibrosis
   a Tay-Sachs disease
   g Phenylketonuria
   b Sickle-cell anemia
   h Duchenne muscular dystrophy
   d, h Fragile-X syndrome
   c Hemophilia
   d, f, h Huntington's disease
4. Genetic diseases that act early in life could be caused by dominant or recessive mutations. Those caused by dominant alleles would be expected to produce relatively mild symptoms, because dominant alleles with lethal or highly deleterious effects would have been eliminated by natural selection. (10)
5. (16–20)
   5 Cloning the disease gene
   3 Pedigree analysis
   6 Comparison of the DNA of normal and afflicted individuals
   1 Characterization of the symptoms
   2 Epidemiological studies
   7 Replacement of the defective allele
   4 Isolation of the locus to a specific chromosome
6. ***Step*** *7*, replacement of the defective allele with a normal allele, is the most controversial because many people think that science should not alter the human genome.

FINDING A DEFECTIVE GENE: MAPPING HUMAN CHROMOSOMES • DEALING WITH A DEFECTIVE GENE (pages 94–96)

- ***Autosomal dominant***. The trait is passed to about one-half of the offspring of both sexes. The female in generation 1 must be heterozygous, otherwise all generation 2 offspring would show the disease. ***Huntington's disease*** is caused by an autosomal dominant allele. (1, 2)
- ***Sex-linked recessive***. The only known heterozygote is the original female; the second generation females could be carriers or homozygous for the dominant allele. The probability that male A and a normal female would have a normal daughter is ***100%***, although there is a 50% chance that she would be a carrier for the disease. The probability that male A and a normal female would have a normal son is ***100%***. (1, 2)
- Note that genes *b* and *d* are linked. (3–5)
  Gene *a*: chromosome ***12*** Gene *b*: chromosome ***6***
  Gene *c*: chromosome ***19*** Gene *d*: chromosome ***6***
  Gene *e*: chromosome ***22***
- (6–10)

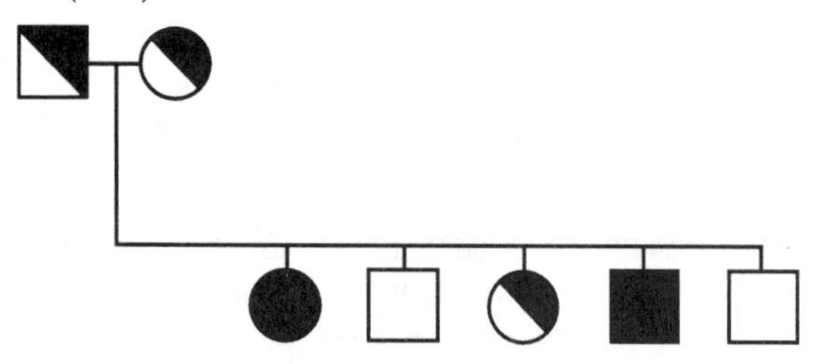

Remember that individuals have two chromosomes and thus will show bands for both alleles. Carriers would have a RFLP with three bands (1350, 1150, and 200 bp).

1. Choice ***b***. Without tight linkage, a RFLP is useless. (6–10)
2. Choice ***a***. OK with fruit flies, but not with humans! (2, 11)
3. Choice ***b***. Except that they are rodents, knockout mice have little to do with somatic-cell hybridization. (3, 13–14)
4. Choice ***d***. Germ cells have not been modified. (15–18)
5. Genetic screening can be used to detect defects in the fetus and abortion may be considered if gene therapy is not available to correct the problem. (21–22)

CANCER (pages 96–97)

1. (1–6)
   c: Benign
   e: Metastasis
   d: Transformation
   a: Carcinogens
   f: Lymphoma
   b: Leukemia
2. Choice ***e***. *Recessive* mutations of tumor-suppressor genes normally lead to cancer. (8–15)
3. Choices ***a, b, d***. (15–17)
4. Choice ***d***. They form a duplex and block *translation*. (18)
5. Chemotherapy and radiotherapy both damage actively dividing cells, such as cancer cells. Since most of the cells of the immune system are also mitotically active, they are also adversely affected by these treatments. (20)
6. Transformation of a normal cell into a cancerous cell may require mutation of several tumor-suppressor genes and also conversion of a proto-oncogene into an oncogene. The accumulation of mutations requires time. (7, 12)
7. Liver cells convert carcinogens into mutagens. The mutant bacteria develop different nutritional requirements as they are changed genetically by the mutagens. (21)

CHAPTER 16 – DEFENSES AGAINST DISEASE

Note : The page numbers listed with the section titles refer to the Study Guide; numbers in parentheses for each question refer to the relevant key concepts.

NONSPECIFIC DEFENSES AGAINST PATHOGENS • SPECIFIC DEFENSES: THE IMMUNE SYSTEM • RESPONSES OF THE IMMUNE SYSTEM (Immunological Memory and Immunization) (pages 98–100)

1. (1–9)
   A: Production of mucus
   A: Secretion of lysozyme
   A: Normal flora
   A: Sneeze reflex
   B: Inflamation
   A: Low pH in stomach
   B: Interferons
2. Choice ***b***. The inflammation response is a nonspecific response that is primarily hormonal in nature. (6, 9)
3. (10–13, 16, 17)
   b: Plasma
   d: White Blood cells
   f: Lymph
   e: Lymphocytes
   a: Antigen
   c: Antibody
4. Choice ***b***. Although the T cells facilitate the humoral immune response, the B cells that have differentiated into plasma cells secrete the immunoglobulins. (10, 16)
5. Choice ***b***. It is now known that plasma cells specific for a huge diversity of antibodies pre-exist in the body and are selected for by the antigen. In choice *a*, an antigen with more than one antigenic determinant can lead to the production of more than one antibody, as stated in *e*. (15–17, 20)
6. The figure shows **3** different antigens. The antigens share **4** different antigenic determinants. Each antigenic determinant has its own antibody, so there are **4** different antibodies. (16–17, 20)

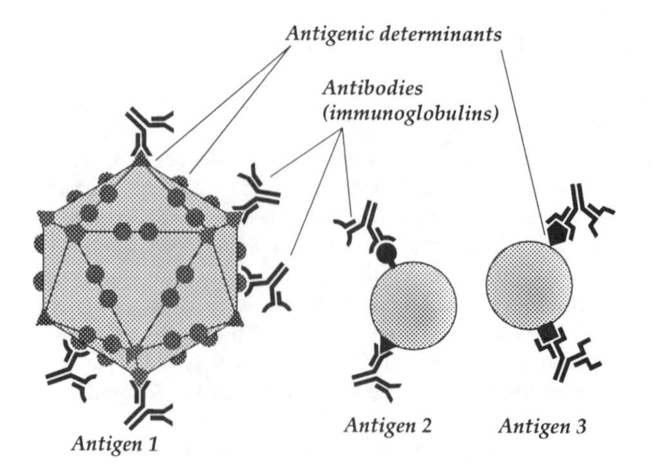

- Since T cells mature in the thymus gland, removing this gland would seriously cripple the immune system. Only animals kept in isolation from pathogens could survive. (13, 14, 19, 20))

RESPONSES OF THE IMMUNE SYSTEM (Clonal Selection and Its Consequences • Self, Nonself, and Tolerance • Development of Plasma Cells) (pages 100–101)

1. Choice *a*. A fundamental postulate of the clonal selection theory is that a plasma cell can make only one immunoglobulin. In choice *b*, an antigen with more than one antigenic determinant can lead to the production of more than one clone of B cells. (1–3, 11)
2. Choice *c*. Both effector cells *and* memory cells must be able to recognize the antigen. (1–3)
3. Choice *b*. Normally, the development of immunological tolerance occurs in a very young animal, although continued exposure to self-antigens is required throughout the life of the animal. (4–8)
4. Choice *e*. Plasma cells have nuclei and are capable of continued cell division. (9, 10)
5. Choice *a*. *Lymphoid tissue*, not antigens, is isolated from the animal immunized to the antigen for which the monoclonal antibodies are desired. (11–14)

IMMUNOGLOBULINS: AGENTS OF THE HUMORAL RESPONSE • ANTIBODIES AND NONSPECIFIC DEFENSES WORKING TOGETHER • THE ORIGIN OF ANTIBODY DIVERSITY (pages 101–103)

- (1–6)

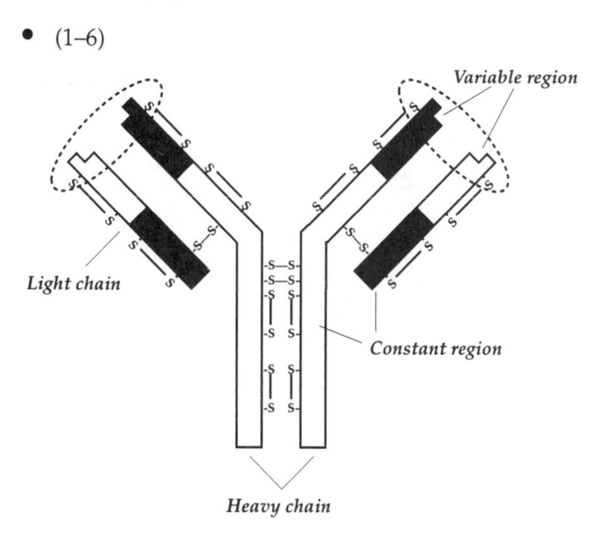

1. Choice *b*. For a given molecule, the two light chains are identical in amino acid sequence, as are the two heavy chains (the two halves of the "Y" are identical), so their variable and constant regions are also identical. (1–6)
2. (7–11))
   b, c: IgG
   a, c, h: IgM
   f: IgA
   d: IgD
   e, g: IgE
3. Choice *d*. The complement system is part of the nonspecific defenses of the body; it attacks any cell marked by agents of the immune system. (12–14)
4. Choice *d*. Since the variable regions of both chains will be assembled by joining only one of the three different segments of each family, there will be 3 × 3 × 3 or 27 different variable regions for both light and heavy chains. Since the heavy and light chains are independently assembled, there will be 27 × 27 or 729 different combinations, each able to specify a different antigenic determinant. (15–21)
5. Choice *c*. An antibody always starts in the IgM class and class-switches to other classes. Class switching can be repeated if there are remaining constant region genes present on the chromosome. (22)

T CELLS: AGENTS OF BOTH RESPONSES • DISORDERS OF THE IMMUNE SYSTEM (pages 103–105)

- Developing a vaccine means isolating an antigen, like the an HIV envelope protein, and exposing animals to it so they can develop immunoglobulins specific to the antigen. If the antigenic determinants of the envelope proteins are constantly changed by mutation of the gene that codes for them, then previously vaccinated individuals will have immunoglobulins unable to recognize the changed antigen. (27–30)
- The HIV virus is transmitted by exchange of blood (transfusion, contaminated hypodermic needles), various forms of sexual intercourse, and through the placenta to the fetus. HIV cannot be transmitted by insect vectors or by kissing and casual contact. (27–30)
- (13–14)

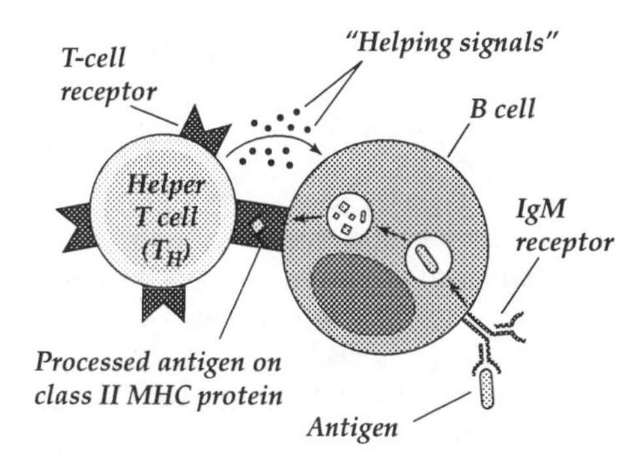

- (15–16)

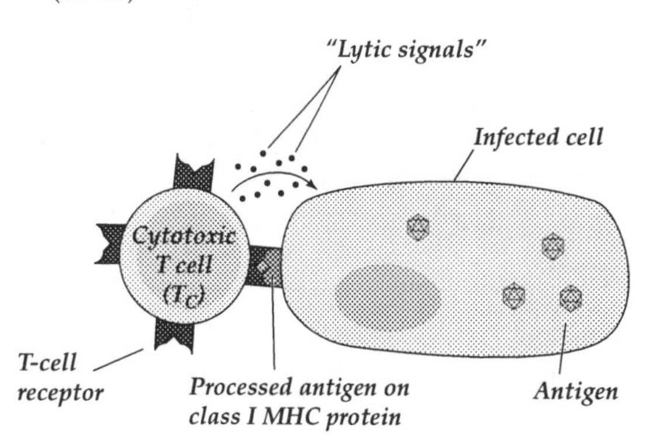

1. Choice *a*. They are all proteins (T-cell receptors are glycoproteins) and the genes that code for them have similar base sequences, suggesting evolution from a common ancestral gene. Choice *b* does not apply to MHC proteins

(they are found on all body cells), choice *c* does not apply to immunoglobulins (immunoglobulins can occur free in the blood/lymph), and choice *d* does not apply to the genetic determination of MHC proteins (large numbers of multiple alleles produce their diversity). (2–7, 18)

2. Choice ***b***. Antigenic determinants are bound by the membrane-embedded immunoglobulins of the B cell. In choice *c*, remember that some of the complement proteins involved in cell lysis are class III MHC proteins.(3–7, 19)
3. Choice ***b***. T-cell receptor genes are composed of V, D, and J regions and are formed by the same joining process seen in B cells. (2, 7)
4. Choice ***b***. Activation of the T cell by the macrophage is caused by interleukin-1 released by the macrophage. (20–22)
5. Choices ***a, c***. AIDS is an immune deficiency disease and allergies involve overreaction of the immune system to a harmless antigen. (24–26)

Integrative Questions

1. *A*: Natural killer cells
   A: Macrophages
   A: Inflammation
   B: T_H cells
   B: T_S cells
   B: Plasma cells
   B: Interleukins
   A: Complement proteins
   B: IgM
   A: Interferons
2. Note: step 3, binding of antigen to a surface receptor on B cell could have occurred earlier
   4: Binding of T_H cell to processed antigen-class II MHC on B cell
   2: Proliferation of T_H cell clone
   6: Proliferation of effector and memory cells
   3: Binding of antigen to a surface receptor on B cell
   5: Stimulation of B cell by interleukin-2
   7: Maturation of plasma cells
   1: Binding of T_H cell to processed antigen-class II MHC on macrophage

CHAPTER 17 – ANIMAL DEVELOPMENT

Note : The page numbers listed with the section titles refer to the Study Guide; numbers in parentheses for each question refer to the relevant key concepts.

THE STUDY OF ANIMAL DEVELOPMENT • CLEAVAGE • GASTRULATION (pages 106–108)

1. (3–5, 10–15, 19–21, 25)
   h: Blastocoel
   i: Blastocyst
   d: Blastoderm
   e: Blastomeres
   j: Blastopore
   a: Blastula
   f: Gray crescent
   b: Mesenchyme
   g: Invagination
   c: Vegetal pole
2. Choice *c*. During cleavage, the embryo remains about the same size, even though there is a large increase in cell number. Consequently, cells (blastomeres) become smaller and the ratio of nuclear volume to cytoplasmic volume increases. (9)
3. Choice ***d***. The gray crescent is a pigmented area that forms at fertilization opposite the point of sperm penetration. (5, 15–23)
4. (16, 14, 15, 21–23)
   b: Sea urchin
   e: Frog
   a: Bird
5. Choice ***e***. In the frog, cells immediately below the blastopore (the ventral lip) become part of the yolk plug. (21, 22)
6. Choice ***a***. In the mammalian blastocyst, the trophoblast forms the fetal part of the placenta. A disc-shaped portion of the inner cell mass surrounded by the trophoblast becomes the embryo. (24–26)

- (3–10).

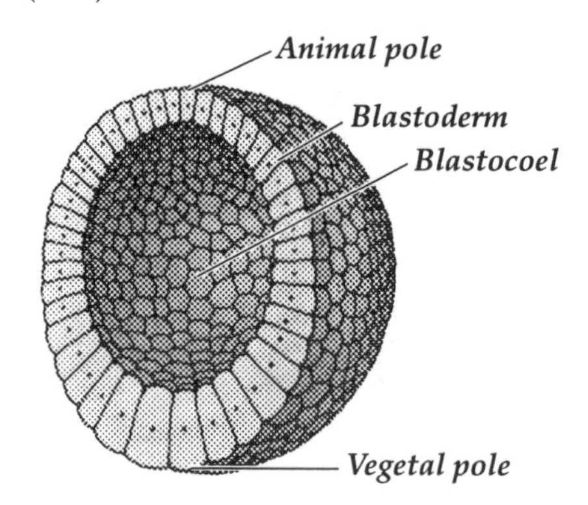

- (3–10).

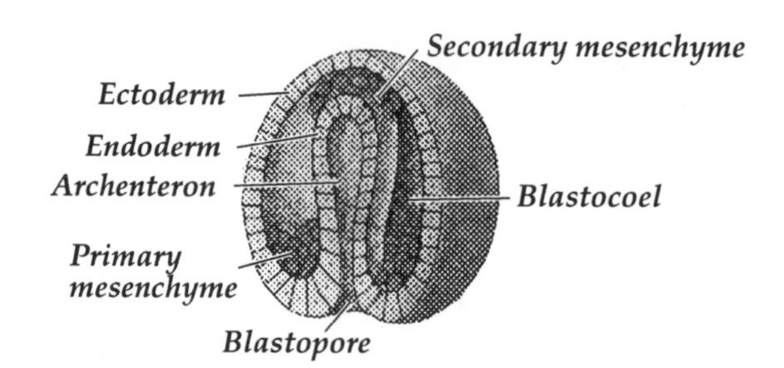

NEUROLATION • LATER STAGES OF DEVELOPMENT • LOOKING CLOSER AT DEVELOPMENT • DIFFERENTIATION • DETERMINATION BY CYTOPLASMIC SEGREGATION (pages 108–110)

1. Choice *a*. (1, 2)
2. Choice *a*. Normally, imaginal discs retain their ability to produce the distinctive structures they would have produced if undisturbed. (8, 14, 15)
3. Choice *c*. Imaginal disc cells from insects showing *both* types of metamorphosis are fully determined. (5–10)
4. Choice *a*. At this point, the developmental potential of the cell is less than the zygote; it is no longer totipotent. (9, 10)
5. Choice *c*. Their studies convincingly demonstrated that differentiation is not caused by loss or permanent inactivation of the genetic material. (11–13)
6. Choice *d*. Germ-line granules in *Caenorhabditis elegans* are uniformly distributed in the egg, but differential move-

ment of granules during cell division leads to their segregation in the posterior end of the worm. (16–23)

7. Choice *a*. The *nanos* female is mutant for a gene coding for a cytoplasmic determinant affecting the abdominal segments of the larva. Because it has a normal allele for the *bicoid* gene, cytoplasm from the anterior region of its egg would have the normal cytoplasmic determinant that the *bicoid* female cannot produce. Consequently, this cytoplasm should permit normal development of the *bicoid* female egg. (20–22)

• (1, 2)

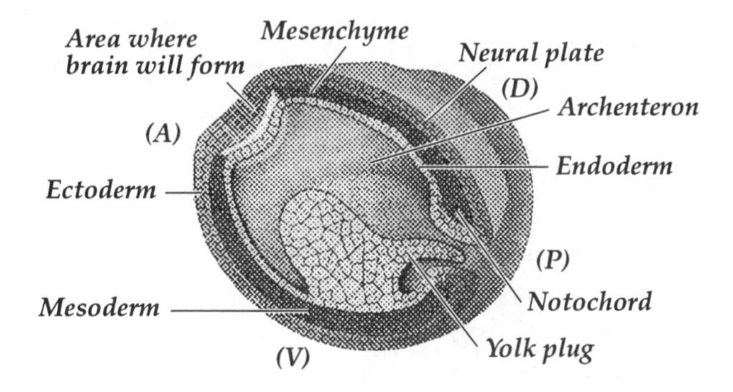

• (4–7)

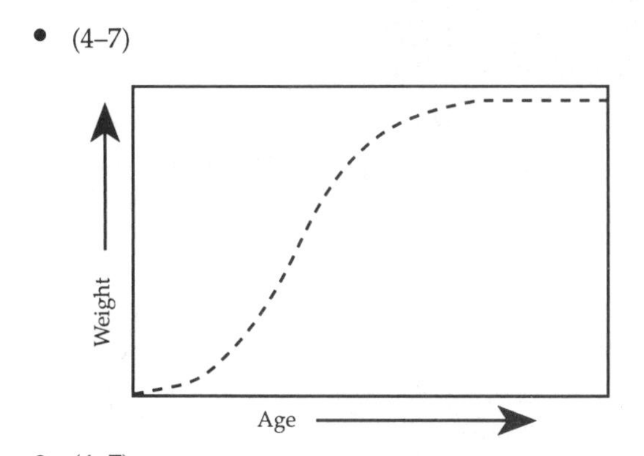

• (4–7)

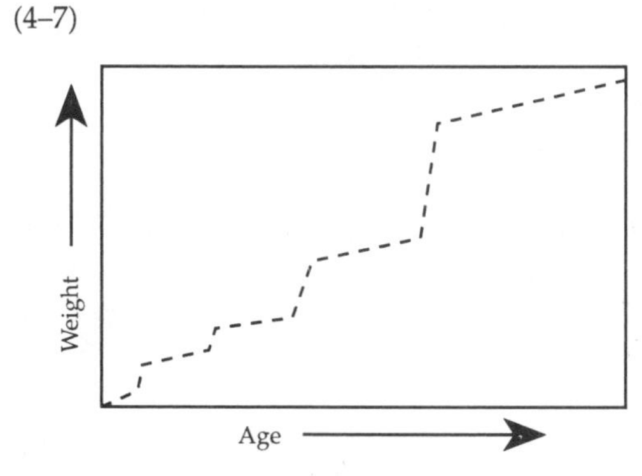

DETERMINATION BY EMBRYONIC INDUCTION • PATTERN FORMATION • YOU'VE SEEN THE PIECES; NOW LET'S BUILD A FLY • DEVELOPMENTAL BIOLOGY AND EVOLUTION (pages 111–113)

1. Choice *d*. This structure determines the principal axes of the embryo. (1, 2)
2. Choice *a*. The gap genes are activated first. They organize large areas along the anteroposterior axis. (13–16)
3. Choice *c*. The homeotic genes are activated last and they determine what specific structures will develop on each segment. (13–16)
4. Choice *c*. The homeobox is a sequence of DNA common to the genes of a host of organisms with segmented body plans. Choice *b* is the definition of the homeodomain, the polypeptide segment coded for by the homeobox. (19–21)
5. (22–25)
   Within the *Drosophila* larva, the concentration gradients of the morphogens ***bicoid protein*** and ***nanos protein*** is established by the unequal distribution of their mRNAs within the egg.
   The proteins that are coded for by the segmentation and homeotic genes are ***transcription factors***.
   The ***homeotic*** genes are arranged on their chromosome in the same order as the order of their expression from anterior to posterior in the larva.
6. Choice *d*. A ZPA grafted to the anterior margin should produce a mirror-image duplication of the entire wing. (9–12)
7. Choice *e*. If the anchor cell is destroyed, no vulva will be formed. Only the anchor cell can produce the primary inducer. (7–9)

• (3, 4)

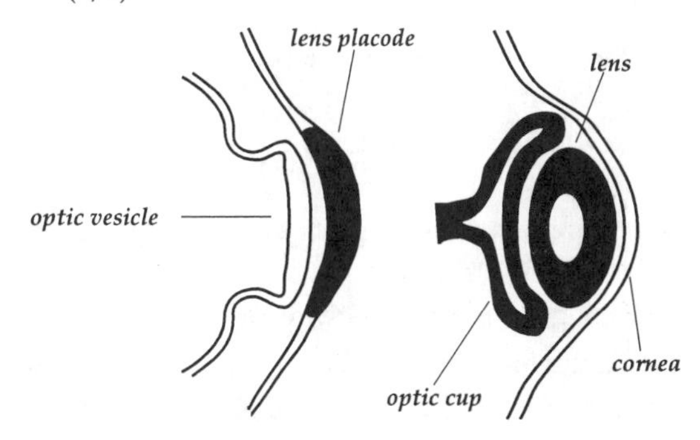

Integrative Questions

1. *c*: W. Gehring
   d: J. Holtfreter
   a: L. Wolpert
   b: S. Hörstadius
   f: R. Briggs, T. King
   e: H. Spemann
2. *6*: Neurulation
   2: Cleavage
   1: Fertilization
   4: Invagination at the blastopore
   3: Blastula formation
   5: Mesoderm formation
3. *d*: Imaginal discs
   b: Spherical blastocoel
   a, b, c: Anus forms first
   c, d: Complete metamorphosis
   c: Gray crescent
   a: Primitive streak

18

The Origin of Life on Earth

CHAPTER LEARNING OBJECTIVES—after studying this chapter you should be able to:

- ❑ Discuss the significance of the work of Redi and Spallanzani on spontaneous generation.
- ❑ Describe the work of Pasteur on spontaneous generation and explain why his studies were more convincing than earlier experiments.
- ❑ Explain how the principle of superposition can be used to determine the relative ages of fossils.
- ❑ Describe the use of radioactive isotopes in obtaining absolute ages for fossils.
- ❑ Explain how to get an estimate for the age of the genetic code.
- ❑ Describe the major events leading to the formation of Earth and its early atmosphere.
- ❑ List the major gases in Earth's early reducing atmosphere.
- ❑ Describe Earth's history as a 30-day month and identify the approximate location of the major events in the origin of life on this calender.
- ❑ Describe the work done by Stanley Miller on prebiotic synthesis of molecules, and discuss the importance of these studies.
- ❑ Explain how polymerization of simple organic molecules could have produced larger, more complex molecules, such as pyrimidine and purine bases and monosaccharides.
- ❑ Discuss the importance of phosphorylation to the formation of the first nucleic acids.
- ❑ Describe what a ribozyme is and give several examples of ribozymes described in the text.
- ❑ Explain the evidence for the notion that the first genetic code was based on RNA.
- ❑ Describe how the evolution of RNA molecules might have improved the catalytic efficiency of ribozymes.
- ❑ Discuss the properties of coacervates that suggest that they may have evolved into cells.
- ❑ Explain the conditions that would have favored the replacement of RNA by DNA as the molecule of heredity.
- ❑ Describe the insights into the metabolism of early organisms provided by a study of the metabolism of living bacteria.
- ❑ Discuss the characteristic features of the anaerobic photosynthetic bacteria, the purple and green sulfur bacteria, and the nonsulfur bacteria.
- ❑ Discuss the characteristic features of the first aerobic photosynthetic organisms.
- ❑ Explain the planetary changes that followed the liberation of free oxygen by cyanobacteria.
- ❑ Explain why new life is not being formed on Earth today.

IS LIFE EVOLVING FROM NONLIFE TODAY? • WHEN DID LIFE ON EARTH ARISE? • HOW DID LIFE EVOLVE FROM NONLIFE? (Conditions on Earth at the Time of Life's Origin • An Overview of Earth's History • Beginning the Sequence of Events: The Synthesis of Organic Compounds • Continuing the Sequence: More Polymerization) (pages 411–419)

Key Concepts

1. Studies done on maggots by *Francesco Redi* and on microorganisms by *Lazzaro Spallanzani* cast doubt on the validity of *spontaneous generation*—the direct formation of organisms from nonliving matter.
2. Studies done by *Louis Pasteur* in 1862 convinced most scientists that organisms can only arise from other organisms.
3. *Fossils* are the remains of organisms that lived in the past.
4. Fossils within different layers, or *strata,* of rock can be relatively aged based on the *principle of superposition*: younger sedimentary rocks are formed on top of older sedimentary rocks.
5. Changes in the appearance of organisms in successively younger strata show evolutionary change through time.
6. Strata from different locations with similar kinds of fossils can be assumed to be of the same age.
7. Absolute dating of the history of life depends on decay of *radioactive isotopes* of elements like uranium, thorium, rubidium, and potassium.
8. The constant rate of decay of radioactive isotopes into stable isotopes as electrons are lost provides a *radiometric clock.*
9. Rocks can be aged by comparing the proportions of radioactive and stable isotopes present in the rocks.

10. Isotopes with slow decay rates are best for aging ancient events; isotopes with rapid decay rates are best for aging more recent events.
11. Similarity in the base sequences of tRNA of different organisms suggests that the genetic code originated before the separation of the eubacteria and the archaebacteria, or about 3.7 billion years ago based on the fossil record.
12. The universe began with a "big bang" about 10 to 20 billion years ago and has been expanding since that time. Our solar system coalesced from cold dust particles in a small region of the Milky Way galaxy about 5 billion years ago.
13. During the formation of Earth, much of the hydrogen and noble (inert) gases, like neon and xenon, were lost.
14. Earth's early atmosphere probably consisted mostly of carbon dioxide (CO_2), water vapor (H_2O), hydrogen (H_2), nitrogen (N_2), ammonia (NH_3), and methane (CH_4).
15. Unlike the present-day atmosphere, the early atmosphere was a *reducing* atmosphere, because all of the original free oxygen had combined with other elements.
16. Earth's history is divided into four *eons*: the *Hadean eon*, the *Archean eon*, the *Proterozoic eon*, and the *Phanerozoic eon*.
17. The Phanerozoic eon is divided into three *eras* (the Paleozoic era, the Mesozoic era, and the Cenozoic era) and eras are divided into *periods*.
18. Absolute dating of eons, eras, and periods is done using radioactive isotopes; large changes in the fossil record are the basis for establishing boundaries between the units.
19. The first three eons occupy the first 26 days of an arbitrary 30-day month representing Earth's history (each day = 150 million years).
20. The oldest fossils date back to about day 7 of the calender—about 3.6 billion years ago, during the Archean eon.
21. From day 7 to day 20 of the 30-day calender, prokaryotes flourished.
22. On day 21, about 2.5 billion years ago, the first eukaryotic cells appeared in the fossil record.
23. On day 27, the beginning of the Phanerozoic eon, 600 million years ago (mya), most of the major groups of organisms were present in the ancient seas.
24. In 1951, *Stanley Miller* demonstrated that amino acids and other organic compounds formed if a reducing atmosphere of hydrogen, ammonia, methane, and water vapor was exposed to electric discharge within a spark chamber.
25. In these studies, simple organic compounds were formed under a variety of different conditions, provided no oxygen was present and an energy source was available.
26. *Polymerization* of small organic molecules via condensation reactions can form larger, more complex molecules. Polymerization of hydrogen cyanide and other molecules can form the nucleic acid bases. Polymerization of formaldehyde can generate sugars.
27. *Phosphorylation* (the addition of a phosphate group) produces stable, yet reactive polymers. Nucleic acids may have formed as phosphorylated polymers of nucleic acid bases and three- and four-carbon sugars.
28. Primitive nucleic acids, based on three- and four-carbon sugars, may have acted as catalysts to polymerize amino acids into proteins, and then self-replicated by complementary base pairing. Nucleic acids based on pentose sugars formed later by an unknown mechanism.

Questions (for answers and explanations, see page 144)

1. Which of the following statements about the studies on spontaneous generation done by Spallanzani and Pasteur is *not* true?
   a. Both individuals showed that life would not appear in sterilized containers that were kept isolated from the outside world.
   b. Both individuals used heat to sterilize the broth.
   c. Both individuals studied microbes.
   d. Because their flasks were closed, both individuals were criticized for not controlling for the possibility that sterilization made the air unfit for life.
   e. Both individuals did their studies during a time when spontaneous generation was considered valid.

2. Which of the following gases was probably *least* abundant in Earth's early atmosphere? __________

   Which of the following gases was probably *most* abundant in Earth's early atmosphere? __________

   CO_2 CO N_2 H_2 NH_3 H_2S O_2

3. Which of the following statements about experiments to synthesize organic compounds under simulated early-Earth conditions is *not* true?
   a. Organic compounds were formed in atmospheres containing different proportions of methane, ammonia, and hydrogen.
   b. Organic compounds were formed in atmospheres containing a variety of gases, provided oxygen was absent.
   c. A variety of energy sources could be used to promote the synthesis of organic compounds.
   d. A variety of organic compounds were formed in these experiments, including the amino acids found in most organism.
   e. These studies showed that thermal energy released by deep sea vents was probably the most important energy source for prebiotic synthesis of organic compounds.

4. A radioactive isotope with a very *fast* decay rate would be best for aging rocks from which of the following time divisions?
   a. Hadean eon
   b. Proterozoic eon
   c. Paleozoic era
   d. Mesozoic era
   e. Cenozoic era

5. Which of the following molecules was probably formed *last* during the synthesis of pre-life biochemicals?
   a. Hydrogen cyanide
   b. Methane
   c. Amino acids
   d. Tetramers of HCN
   e. Adenine

6. On about what day of the arbitrary 30-day calender representing Earth's history would each of the following events occur?
Oldest prokaryotic fossils: day _____

Oldest eukaryotic fossils: day _____

First land plants: day _____

Reptiles dominant: day _____

Rise of mammals: day _____

7. Which, if any, of the following approaches has *not* been used to estimate the age of the origin of life on Earth?
a. Application of the principle of superposition
b. Comparisons of fossils
c. Ratios of radioactive isotopes
d. A radiometric clock
e. Similarities in the base sequences of different organisms

HOW DID LIFE EVOLVE FROM NONLIFE? (RNA: The First Biological Catalyst • From Ribozymes to Cells • The Evolution of DNA) • METABOLISM OF EARLY ORGANISMS • HOW PROBABLE WAS THE EVOLUTION OF LIFE? (pages 419–423)

1. A central problem that slowed research on the origin of life was the need to explain how catalysis was provided to allow replication of early nucleic acids before protein synthesis had evolved.

2. In the late 1970's, *Manfred Eigen* showed that RNA is autocatalytic—it could catalyze the polymerization of new RNA. Other studies revealed that some RNA can also catalyzes the processing of pre-tRNA.

3. Recent studies of the protist *Tetrahymena thermophilia* have shown that RNA can catalyze the excision and splicing of its own intron. Also, in the *Tetrahymena* ribosome, RNA rather than ribosomal protein is the catalyst of protein synthesis.

4. *Ribozymes* are RNAs with catalytic activity. Ribozymes have been discovered in many organisms.

5. Many scientists now believe that the first genetic code was based on RNA that catalyzed its own replication as well as other reactions. RNA's enzymatic functions were assumed by proteins after the evolution of protein synthesis.

6. Evolution of RNAs that were more efficient catalysts for replication involved competition between different RNAs for RNA monomers. RNAs with base sequences that allow the molecule to fold into shapes that are more catalytically efficient could be favored by natural selection if they also unfold and replicate.

7. Ribozymes have "evolved" from random-sequence RNA under laboratory conditions.

8. The Russian scientist *Alexander Oparin* believed that the first cells evolved from *coacervates*—drops of protein and polysaccharide with some water that form spontaneously in an aqueous solution with low concentrations of these substances.

9. Coacervates can become coated with a membranelike lipid layer and will differentially exchange materials with the aqueous environment.

10. Coacervates with enzyme molecules can absorb substrate and release products. Oparin was able to create coacervates with chlorophyll that could absorb oxidized dye from the environment, reduce it using light energy, and release the reduced dye.

11. Competition for limited organic molecules might favor coacervates with better controlled chemical reactions, perhaps involving enzymes to make them more stable.

12. DNA is most stable in the hydrophobic environment provided within cells. DNA probably evolved as the molecule of inheritance after RNA-based life became cellular.

13. Like some present-day bacteria, the earliest organisms were probably *obligate anaerobes*, obtaining energy in an oxygen-free environment. The earliest photosynthesizers were also obligate anaerobes and did not generate oxygen.

14. Present-day bacteria may provide clues about the stages in the evolution of photosynthetic pathways in early autotrophs.

15. Present-day anaerobic photosynthesizers include the *green sulfur bacteria*, the *purple sulfur bacteria*, and the *purple nonsulfur bacteria*.

16. The anaerobic photosynthesizers contain a type of chlorophyll called *bacteriochlorophyll.* They also contain red and yellow carotenoids that assist chlorophyll in capturing light energy.

17. All anaerobic photosynthesizers have their photosynthetic pigments and electron transport components embedded in *thylakoid* membrane complexes.

18. Unlike plants and the cyanobacteria, which use water as the source of hydrogen to reduce CO_2 into carbohydrates and liberate oxygen, the anaerobic photosynthesizers use other compounds as hydrogen donors.

19. The sulfur bacteria, both green and purple, use H_2S and release sulfur; the purple nonsulfur bacteria use a variety of organic compounds as hydrogen donors, including ethanol, lactic acid, pyruvic acid, and molecular hydrogen.

20. Anaerobic bacteria were key organisms in Earth's early history, and today they continue to be important in cycling many important elements.

21 About one billion years ago, *cyanobacteria* evolved that released oxygen as a product of photosynthesis, as evidenced by the first appearance of oxidized minerals in rocks from that period.

22. As molecular oxygen (O_2) accumulated in the atmosphere, its presence favored those species that could tolerate and use it.

23 Two major planetary changes caused by early photosynthesizers were 1) the increase in atmospheric oxygen and 2) the decrease in atmospheric CO_2 and the deposition of CO_2 in marine sediments.

24. The prevalence of oxygen on present-day Earth prevents the formation of the simple biological molecules that would be necessary for the origin of new life forms.

Questions (for answers and explanations, see page 144)

1. Which one of the following statements about the earliest catalysts is most probably *not* true?

a. The first catalysts were proteins that lacked the specificity shown by enzymes.
b. The earliest catalysts were nonprotein cofactors.
c. The earliest catalysts were restricted to membranes.
d. RNA had important catalytic functions before the evolution of enzymes.
e. DNA had important catalytic functions before the evolution of enzymes.

2. Which of the following statements about the importance of coacervates in the evolution of cells is *not* true?
a. Under appropriate conditions, coacervates form spontaneously.
b. Early coacervates developed a system of nucleic acid replication after they became true cells.
c. Oparin was able to create coacervates that showed some of the properties of chloroplasts.
d. Coacervates can modify the chemistry of the solution in which they reside.
e. Coacervates are clumps of protein and polysaccharide in an aqueous solution.

3. The earliest organisms were probably
a. chemosynthetic.
b. photosynthetic.
c. obligate anaerobes.
d. obligate aerobes.
e. like present-day cyanobacteria.

4. Select *all* of the following ways that present-day cyanobacteria and anaerobic photosynthesizers (purple and green sulfur bacteria and nonsulfur bacteria) are similar?
a. Type of chlorophyll used
b. Source for hydrogens used in photosynthesis
c. Use of CO_2 as carbon source for photosynthesis
d. Production of O_2 as a waste product of photosynthesis
e. Energy source for photosynthesis

5. The earliest photosynthetic organisms were probably *least* like present-day
a. eukaryotic photosynthesizers.
b. cyanobacteria.
c. purple sulfur bacteria.
d. purple nonsulfur bacteria.
e. green sulfur bacteria.

6. Which of the following is *not* a planetary effect caused by the evolution of photosynthesis and increased oxygen production?
a. Development of an oxidizing atmosphere
b. Reduction in amount of ultraviolet light reaching the Earth's surface
c. Increase in the atmospheric CO_2 concentration
d. Predominance of aerobic organisms
e. Increased deposition of oxidized minerals in the ocean

7. Which of the following was *not* presented in the text as evidence that catalysis by RNA proceeded catalysis by enzymes?
a. RNA can excise and splice its own intron.
b. RNA is only stable in hydrophobic environments.
c. Ribozymes have been discovered in many different organisms.
d. RNA can catalyze the polymerization of RNA.
e. Ribozymes have "evolved" from random-sequence RNA under laboratory conditions.

8. Which of the following statements about the anaerobic photosynthesizers (purple and green sulfur bacteria and nonsulfur bacteria) is *not* true?
a. All use a form of bacteriochlorophyll.
b. All have carotenoid pigments that assist chlorophyll in light capture.
c. All use inorganic compounds as a hydrogen donors for reducing CO_2.
d. All have their photosynthetic pigments embedded in thylakoid membrane systems.
e. None liberate oxygen.

9. Discuss why life does not evolve from nonliving matter on Earth today.

Integrative Questions (for answers and explanations, see page 144)

1. Place numbers next to the following events to reflect the sequence of their occurrence in Earth's history.

____ Coacervates become cells

____ Photosynthesis without oxygen production appears

____ DNA assumes central role in genetic code

____ Cyanobacteria appear

____ Polymerization of monomers

____ Ribozymes appear

____ Organic compounds formed in a reducing atmosphere

2. Match the letter for the name of the scientist on the left with the phrase on the right characterizing his work.

a. Oparin	____	First to study spontaneous generation
b. Eigen	____	Work lead to acceptance of aphorism "*omne vivum e vivo*"
c. Pasteur	____	Estimated that Earth was created in 4004 B.C.
d. Redi	____	Based on heat measurements, said Earth began 10–20 billion years ago
e. Kelvin	____	Studied ribozymes
f. Usher	____	Studied coacervates

19

The Mechanisms of Evolution

CHAPTER LEARNING OBJECTIVES—after studying this chapter you should be able to:

❒ Define the term "biological evolution" and describe the major agents of evolutionary change.

❒ Relate the concepts of a geographic population and a deme, or Mendelian population.

❒ Differentiate between genetically and environmentally based phenotypic variation and give examples of each.

❒ Discuss the concept of a population gene pool.

❒ Determine allele frequencies given the numbers of population members with specific genotypes.

❒ Describe what factors determine the genetic structure of a population.

❒ Discuss what is meant by a population polymorphic for a heterozygous gene locus

❒ Discuss how gel electrophoresis has been used to study the extent of genetic variability in populations.

❒ Use the Hardy–Weinberg formula to show the relationship between allele frequencies and genotype frequencies during genetic equilibrium.

❒ Discuss the conditions necessary for Hardy–Weinberg genetic equilibrium and identify which conditions are usually met or not met in most populations.

❒ Use the Hardy–Weinberg formula to calculate either the allele or genotype frequencies, given the other.

❒ Describe the major role of mutation in biological evolution.

❒ Differentiate between the two types of genetic drift, called bottlenecks and founder effects.

❒ Discuss how nonrandom mating such as assortive mating and self-fertilization can change genotype but not allele frequencies.

❒ Differentiate between natural selection and artificial selection and give examples of each.

❒ Describe the type of expected graph showing the proportions of phenotypes in a population for a trait determined by polygenic inheritance.

❒ Differentiate between the types of selection known as stabilizing, disruptive, and directional selection.

❒ Explain the relationship between inclusive fitness, individual fitness, and kin selection.

❒ Describe the major evolutionary phenomena that reduce genetic variation within a population.

❒ Describe the major evolutionary phenomena that increase genetic variation within a population.

❒ Explain how studies of cytochrome *c* have contributed to our understanding of neutral and adaptive changes at the molecular level.

❒ Differentiate between microevolutionary and macroevolutionary change.

WHAT IS EVOLUTION? • THE STRUCTURE OF POPULATIONS • THE HARDY–WEINBERG RULE (pages 426–431)

Key Concepts

1. A population undergoes *biological evolution* when it changes genetically over time.
2. *Genotype* is the genetic constitution of an individual relative to a *heritable* trait. Genetic change in populations occurs when individuals with different genotypes within the population survive and reproduce at different rates.
3. *Phenotype* is the observable expression of a genotype. Agents of evolutionary change act on genetically based phenotypic variation of population members.
4. The environment can influence the expression of the genotype—that is, some portion of phenotypic variation is environmentally based.
5. A *population* is a geographically isolated group of organisms of the same species.
6. Most populations possess a high level of genetic variation among their members.
7. Plant and animal breeders have produced new varieties of organisms with desirable traits by artificially selecting from the high levels of genetic variation found in most species.
8. A *deme* (or *Mendelian population* or *subpopulation*) is a local population of potentially interbreeding individuals. A large, *geographic population* may consist of numerous demes.
9. Genetic variation is determined by the number of different alleles for the genes of a species found in population members.

10. The *gene pool* of a population consists of all the alleles for all the genes present in the species.

11. *Allele frequencies* are the proportions of the total number of alleles for a gene represented by each different allele. Allele frequencies are expressed on a 0 to 1.0 basis.

12. The *genetic structure* of a population is determined by the genotype frequencies and allele frequencies in the population.

13. Populations can have the same allele frequencies for a gene (the same gene pool), but because the alleles can be distributed differently in the genotypes, they can have different genetic structures.

14. A gene locus is *heterozygous* if two or more alleles are present for the gene. A population is *polymorphic* for a heterozygous gene locus if the frequency of the most common allele is not greater than 95%.

15. *Gel electrophoresis* can be used to visualize differences in proteins and nucleic acids in different individuals. In gel electrophoresis, molecules move through a gel under the influence of an electric field; separation is based on molecular size and charge differences in the molecules.

16. Gel electrophoresis studies have shown that most populations show considerable variation in proteins and DNA structure. Vertebrates have proportionally fewer heterozygous loci than invertebrates; inbred populations have proportionally fewer heterozygous loci than outbred populations.

17. *Genetic equilibrium* within a population occurs when allele and genotype frequencies do not change over time.

18. *G. H. Hardy* and *W. Weinberg* independently showed in 1908 that under certain conditions a population will remain in genetic equilibrium.

19. Important Hardy–Weinberg conditions are that the population must be large, all individuals must be equal in terms of reproduction and survival, and mating must be random (independent of genotype).

20. The *Hardy–Weinberg formula,* $p^2 + 2pq + q^2$, where p is the frequency of the first allele and q is the frequency of the second allele, applies the principle of independent probabilities to determine the expected genotype frequencies. The expected frequencies of the two homozygous genotypes are given by p^2 and q^2, and $2pq$ is the expected frequency of the heterozygous genotype.

21. A population under Hardy–Weinberg conditions will show an equilibrium between allele and genotype frequencies.

22 Very few real populations meet the Hardy–Weinberg conditions.

23. The main use of the Hardy–Weinberg rule is to determine if evolutionary change is occurring; if a population is evolving, its genotype frequencies differ from those predicted by the Hardy–Weinberg rule.

Questions (for answers and explanations, see page 145)

1. Evolution occurs at the level of
   a. the individual genotype.
   b. the individual phenotype.
   c. environmentally based phenotypic variation.
   d. the population.
   e. the species.

2. Which choice would form a statement about demes that is *not* true? A deme (Mendelian population or subpopulation) must
   a. consist of members of the same species.
   b. be geographically isolated from other populations.
   c. have members that are capable of interbreeding.
   d. show genetic variation.
   e. have a gene pool.

3. In comparing two populations of the same species, the population with the greatest genetic variation would have (select all correct answers):
   a. the greatest number of genes.
   b. the greatest number of alleles per gene.
   c. the greatest number of population members.
   d. the greatest number of different genotypes.
   e. the largest gene pool.

4. The Hardy–Weinberg expression is an expansion of a binomial or

   $$(p + q)^2 = p^2 + 2pq + q^2$$

   In this equation, the left-hand side represents ________ frequencies and the right-hand side represents ________ frequencies.

5. The ability to taste the chemical PTC (phenylthiocarbamide) is determined in humans by a dominant allele *T*, with tasters having the genotypes *Tt* or *TT* and nontasters, *tt*. If you discover that 36% of the members of a population cannot taste PTC, then according to the Hardy–Weinberg rule, the frequency of the *T* allele should be
   a. 0.2
   b. 0.4
   c. 0.6
   d. 0.64
   e. 0.8

6. A gene in humans has two alleles, *M* and *N*, that code for different surface proteins on red blood cells. If you know that the frequency of allele *M* is 0.3, according to the Hardy–Weinberg rule, the frequency of the genotype *MN* in the population should be
   a. 0.21
   b. 0.3
   c. 0.42
   d. 0.49
   e. 0.7

7. If the frequency of allele *b* in a gene pool is 0.2, according to the Hardy–Weinberg rule, the expected frequency of the genotype *bbb* in a triploid ($3n$) plant species would be
   a. 0.008
   b. 0.04
   c. 0.08
   d. 0.2
   e. 0.4

8. Which one of the following would be expected to have the smallest proportion of heterozygous gene loci?
   a. An individual from an inbred population of vertebrates
   b. An inbred population of vertebrates
   c. An individual from an outbred population of vertebrates
   d. An inbred population of invertebrates
   e. An outbred population of invertebrates

9. Discuss the main use of the Hardy–Weinberg rule in evolutionary biology.

CHANGING THE GENETIC STRUCTURE OF POPULATIONS (pages 432–437)

Key Concepts

1. Evolutionary agents are forces that alter the genetic structure of populations. They can change both allele and genotype frequencies.
2. Major evolutionary agents are mutation, gene flow, genetic drift, nonrandom mating, and natural selection.
3. Although mutation rates are low (usually one mutation in a million zygotes per generation), mutation creates genetic variation by producing new alleles, and also replaces alleles that are being eliminated by other evolutionary agents.
4. Although mutation is present in all populations, its rate is so low that mutation *per se* does not disturb Hardy–Weinberg genetic equilibrium.
5. *Gene flow* results when individuals migrate from one population to another and breed in the new location.
6. Gene flow between different subpopulations can keep the gene pools of the subpopulations from diverging due to genetic drift or selection.
7. A population undergoes *genetic drift* when allele frequencies change randomly in a population. Genetic drift is most prevalent in small populations.
8. Because genetic drift is random, deleterious alleles may become more frequent in the gene pool, or advantageous alleles may be entirely lost.
9. A population *bottleneck* is a form of genetic drift brought on when a population is reduced to a very small size by some natural or human-produced catastrophe.
10. If a small number of individuals (founders) that are not genetically representative of the larger population from which they come founds a new population, the new population may have allele frequencies that are quite different than the parent population. This is called the *founder effect.*
11. Nonrandom mating, called *assortative mating,* occurs when the phenotype of individuals influences mate selection. Assortative mating is common in vertebrates.
12. *Self-fertilization,* common in many plants, is also a form of nonrandom mating.
13. Assortative mating and self-fertilization both lead to more homozygous genotypes than predicted by the Hardy–Weinberg rule, but do not alter the expected allele frequencies.
14. *Natural selection* occurs when individuals differing in genotype reproduce and survive differentially. Natural selection leads to populations whose members are better adapted to the environment.
15. In *artificial selection,* animal and plant breeders are the selective agents; in natural selection, the environment is the selective agent.
16. In traits determined by more than one gene, *sexual recombination* increases the *evolutionary potential* of populations by producing new combinations of alleles.
17. For traits determined by multiple genes, a bell-shaped curve usually results if you plot some measurement of the trait on the y-axis and the proportion of the population with that measurement on the x-axis.
18. *Stabilizing selection* results if individuals with extreme forms of a trait contribute fewer offspring to the next generation than individuals that are intermediate for the trait. Stabilizing selection leads to reduced genetic variation.
19. Enhanced reproduction and survival of individuals from one of the two ends of the distribution curve can result in *directional selection.*
20. *Disruptive selection* occurs if individuals from both extremes of the distribution are reproductively superior to the intermediate forms.
21. Artificial selection of populations of *Drosophila* fruit flies with low and high bristle number has shown that significant disruptive selection can occur in just 35 generations.
22. An African finch called the black-bellied seed cracker (*Pyrenestes ostrinus*) shows a dramatic example of disruptive selection for individuals with small and large bills.
23. The reproductive contribution of a genotype or phenotype to the next generation, relative to all other genotypes or phenotypes, is called the organism's *fitness.*
24. *Individual fitness* results from direct reproduction by an individual.
25. *Kin selection* results from behavior of an individual that promotes the survival and reproduction of relatives who are genetically similar.

26. The *inclusive fitness* of an individual is the sum of its individual fitness and any kin selection component that may result due to the organism's behavior.

27. Individual fitness is more important than kin selection in solitary species. Kin selection is most important in highly social species.

Questions (for answers and explanations, see page 145)

1. The following graph shows the range of variation among population members for a trait determined by multiple genes.

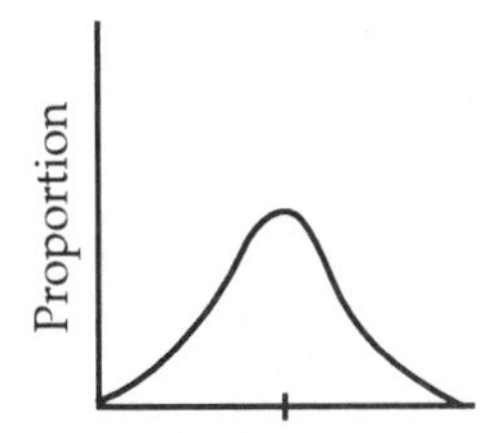

If this population is subject to *stabilizing selection* for several generations, which of the distributions (*a-d*) is most likely to result?

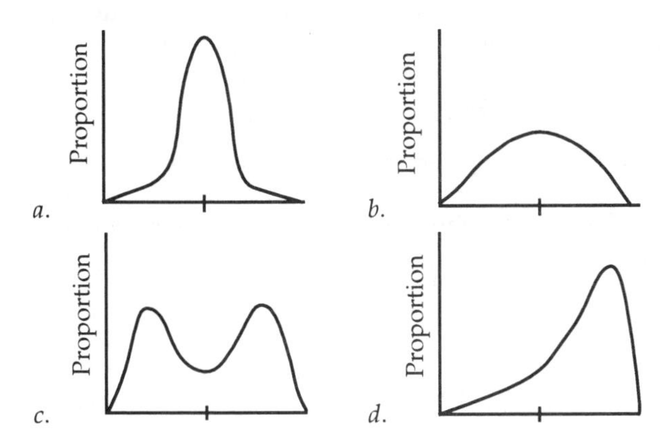

2. A small, isolated population would most likely be subject over time to
a. assortative mating.
b. a founder effect.
c. genetic drift.
d. gene flow.
e. natural selection.

3. Allele frequencies for genes determining traits such as blood groups are *least* likely to be changed by
a. assortative mating.
b. the founder effect.
c. genetic drift.
d. gene flow.
e. mutation.

4. Select *all* of the following evolutionary agents that produce *nonrandom* changes in the genetic structure of a population.
a. Self-fertilization
b. The founder effect
c. Bottlenecks
d. Mutation
e. Natural selection

5. The following data were collected in a study in which dark and light moths of the peppered moth (*Biston betularia*) were released and later recaptured in several different areas (*a*, *b*, *c*, and *d*). Based on these data, in which area is the *light* phenotype most advantageous?

Area	Moth Type	Released	Recaptured
a	dark	125	25
	light	200	40
b	dark	50	20
	light	200	40
c	dark	250	40
	light	425	125
d	dark	125	75
	light	200	25

6. Peppered moths are subject to predation by birds who eat them as the moths rest on bark surfaces during the day. Assuming that moths that are more cryptic in coloration (look more like their background) will be preyed on less, what can you say about the color of bark surfaces in area *d*?

7. A bird species in which females help sisters feed their nestlings is an example of
a. kin selection.
b. fitness
c. inclusive fitness.
d. artificial selection.
e. nonrandom mating.

8. Describe the relationship between inclusive fitness, individual fitness, and kin selection.

GENETIC VARIATION AND EVOLUTION • INTERPRETING LONG-TERM EVOLUTION (pages 437–443)

Key Concepts

1. Genetic drift, stabilizing selection, and directional selection tend to reduce genetic variation within populations.
2. Some genetic variation in populations is correlated with geography.

3. *Clines* are geographic changes in phenotypes and underlying genotypes within a species.
4. *Step clines* are created when geographically based phenotypic variation changes abruptly in a widely distributed species.
5. *Phenotypic polymorphism,* the presence of two or more phenotypes for a given characteristic within a population, shows that a population is polymorphic for a certain gene locus, and reveals what natural selective forces may be at work.
6. *Balanced polymorphism* exists in a population when two or more alleles exist at stable frequencies within the population
7. Preservation of genetic variation within a population can involve natural selection, favoring different traits in different areas, or can involve natural selection favoring different traits in the same area, but at different times.
8. Some genetic variation is not adaptive, but neutral. On the molecular level, any change that does not influence the functioning of the molecule would be neutral.
9. Cytochrome *c*, an important protein component of the electron transport chain of mitochondria, consists of 104 amino acids and has a molecular weight of 12,400 daltons.
10. Many of the positions in the cytochrome *c* molecule are occupied by the same amino acids in all species. These positions probably are important in determining the overall shape and functioning of the molecule.
11. Amino acids that *do* vary in cytochrome *c* molecules from different species may either be adaptive or selectively neutral.
12. The functionally important regions of molecules can be discovered by determining which regions *do not* vary from species to species.
13. Amino acid substitutions that are selectively neutral should accumulate at a constant rate over time due to mutation.
14. Microevolution is short-term change in the gene pool of a population.
15. Macroevolution refers to long-term change that cannot be explained completely by the agents of microevolution.
16. Macroevolutionary change may be dependent on events that occur infrequently and that are not likely to be observed in short-term microevolutionary studies.

Questions (for answers and explanations, see page 145)

1. You use electrophoresis to examine the amino acid sequence of a protein in different kinds of organisms and discover that 25% of the amino acid positions are occupied by the same amino acids in all the organisms. Which of the following statements about these invariant positions is *not* true?
   a. No mutations produced variation at these positions.
   b. The amino acids at these positions are probably involved in determining the conformation of this protein.
   c. Changes at these positions would probably adversely affect the functioning of this protein.
   d. Natural selection helped to conserve the amino acids occupying these positions.
   e. Amino acid substitutions at these positions are not selectively neutral.
2. In areas of Africa where malaria is prevalent, many human populations exist in which the frequencies of the phenotypes normal, sickle-cell trait, and severe sickle-cell anemia are constant. This is an example of
   a. the founder effect.
   b. balanced polymorphism.
   c. transitional polymorphism.
   d. a cline.
   e. natural selection.
3. Which of the following statements about clinal variation is *not* true?
   a. Species showing clinal variation usually are widely distributed.
   b. Clinal variation usually involves external morphology.
   c. Some important environmental factor usually varies geographically over the range of the species.
   d. Clinal variation can be gradual or occur in steps.
   e. Clinal variation is the result of natural selection.
4. Which of the following is *not* an example of microevolution?
   a. The founder effect
   b. Genetic drift
   c. Balanced polymorphism
   d. Speciation
   e. A cline
5. Contrast microevolution with macroevolution and provide an example of each.
6. Select all of the following that would tend to promote genetic variation in populations.
   a. Stabilizing selection
   b. Genetic drift
   c. Clinal variation
   d. Selection that varies seasonally
   e. Directional selection

20

Species and Their Formation

CHAPTER LEARNING OBJECTIVES—after studying this chapter you should be able to:

- ❑ Define what a biological species is and explain each of the definition's criteria.
- ❑ Explain what is meant by the "cohesiveness of species" and discuss how gene flow, genetic drift, and adaptation to local conditions affect it.
- ❑ Differentiate between vertical evolution, or anagenesis, and speciation.
- ❑ Use concepts of population genetics from Chapter 19 to describe the condition that underlies all modes of speciation.
- ❑ Explain the allopatric, or geographic, mode of speciation and illustrate its features by referring to the evolution of Darwin's finches.
- ❑ Discuss the parapatric mode of speciation, the necessary conditions that promote it, and the groups of organisms that are most subject to it.
- ❑ Discuss the sympatric mode of speciation caused by polyploidy and the groups of organisms that are most subject to it.
- ❑ Explain how sympatric speciation can occur in animals through nonpolyploid means and give an example.
- ❑ Describe what is meant by "reproductive isolating mechanisms," and differentiate between those that are prezygotic and postzygotic in their action.
- ❑ Describe the several possible outcomes that can result when two populations that have been genetically isolated from each other come into contact again.
- ❑ Discuss what is known about the genetic similarities of sympatric species based on studies of the Hawaiian *Drosophila*.
- ❑ Explain the influence of behavioral complexity on the time required for speciation.
- ❑ Explain what is known about the influence of ecology, especially diet, on the time required for speciation, based on studies of African mammals.
- ❑ Describe what is meant by an "evolutionary radiation" and explain why evolutionary radiations are most prominent on island archipelagos.
- ❑ Explain what is meant by the term "endemic" and explain the observation that many endemic species live on islands.
- ❑ Explain what is known about the evolutionary radiations on islands based on studies of Hawaiian silverswords and tarweeds.
- ❑ Discuss the view that most population genetic change takes place during periods of speciation that are followed by intervals of genetic stability called stasis.

WHAT ARE SPECIES? • THE COHESIVENESS OF SPECIES • HOW DO NEW SPECIES ARISE? (pages 446–452)

Key Concepts

1. The *biological species concept*, developed by *Ernst Mayr* in 1940, defines a species as groups of natural populations of actually or potentially interbreeding organisms that are reproductively isolated from other such groups. The groups are collections of local populations, or demes.
2. Because organisms of the same species share a common gene pool, species are key evolutionary units.
3. The potential to interbreed means that species members would produce viable offspring if a natural barrier that prevents them from doing so were removed—it excludes organisms that are able to interbreed only in captivity.
4. *Gene flow* promotes the cohesiveness of species by preventing populations from diverging genetically due to genetic drift or adaptations to local conditions.
5. *Vertical evolution,* or *anagenesis,* results when a single lineage changes through time to the extent that its descendants are given a different species name.
6. *Speciation* is the temporal process whereby one species gives rise to two species.
7. Separation of the gene pool of the ancestral species into two isolated pools is a necessary part of all speciation modes. Evolutionary agents can cause changes in allele and genotype frequencies in the isolated gene pools.
8. *Geographic speciation* (also called *allopatric speciation*) involves the establishment of a physical barrier within the range of a species, which creates two reproductively isolated populations.

9. In geographic speciation, the physical barrier establishes two separate gene pools; genetic drift may proceed separately in the two pools or, if environmental conditions are different, natural selection may operate differently in the two pools.
10. In geographic speciation, founder effect may have been involved when the two pools were initially established.
11. In general, the greater the degree of reproductive isolation within a group of species, the greater the degree of genetic divergence.
12. The 14 species of *Galapagos*, or *Darwin's finches*, arose by allopatric speciation from a single ancestral species that arrived from the South American mainland.
13. The scale of the barrier required to create reproductive isolation varies with the mobility and lifestyle of the organism.
14. The area required for speciation varies with the mobility and lifestyle of the organism; birds need more space than small mammals, and animals like land snails can form new species in adjacent valleys.
15. When speciation occurs without geographic isolation due to an environmental discontinuity, it is called *parapatric speciation*.
16. Parapatric speciation is most common in plants adapted for different soils. Because their hybrids are less fit, reproductive isolating mechanisms quickly evolve.
17. Like parapatric speciation, *sympatric speciation* also occurs without geographic isolation, but also without any environmental discontinuity.
18. Among plants, and some animals, sympatric speciation results from *polyploidy*—an increase in the chromosome (ploidy or n) number.
19. Polyploidy arises in two different ways: through duplication of chromosomes of a single species (usually due to nondisjunction of chromosomes during meiosis) and from combining chromosomes in hybrids of different species.
20. If polyploids can mate with other polyploids or be self-fertile, polyploidy can create a new species in one generation.
21. Animal species that arose by polyploidy are either *parthenogenetic* (self-fertilizing), or initially depend on mating with siblings to maintain the new species.
22. Sympatric speciation among animals (especially insects) occurs when individuals within the same area become specialized for specific resources; for example, herbivores or parasites specializing in different species of host.
23. Sympatric speciation is common among plants via polyploidy, but uncommon in animals.
24. Geographic speciation is most common among animals, but it may be difficult to distinguish between hybrid zones formed by geographic isolation with subsequent range overlap and parapatric speciation.

Questions (for answers and explanations, see page 145)

1. It is difficult to apply the biological species concept to groups of organisms that (select all correct answers):
   a. are asexual.
   b. are widespread.
   c. produce hybrids only in captivity.
   d. exist only in the fossil record.
   e. show little morphological diversity.
2. Which of the following statements about geographic speciation is *not* true?
   a. Geographic speciation is also called allopatric speciation.
   b. Geographic speciation only occurs in species that are widely distributed.
   c. Geographic speciation always involves a physical barrier that interrupts gene flow.
   d. Geographic speciation can sometimes involve genetic drift.
   e. Geographic speciation can sometimes involve founder effect.
3. You look at several polymorphic loci in land snails removed from the areas A–E of four adjacent city blocks, as shown in the following map.

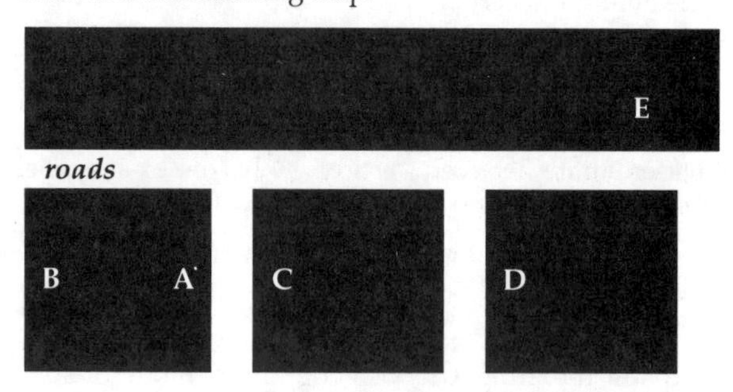

   Snails from area A should be genetically most like individuals from area _____ and least like individuals from area _____.
4. The activities of a mining company results in deposition of a new soil type within the range of a widespread plant species. Which of the following phenomena is likely to occur as a result of this?
   a. Geographic speciation
   b. Sympatric speciation
   c. Parapatric speciation
   d. Allopatric speciation
   e. Anagenesis
5. By means of sympatric speciation due to nondisjunction, how many different tetraploid ($4n$) species could be formed from three diploid ($2n$) species?
   a. 0
   b. 3
   c. 4
   d. 6
   e. 9
6. Which type of speciation is most common among flowering plants?
   a. Geographic speciation
   b. Sympatric speciation
   c. Parapatric speciation
   d. Allopatric speciation
   e. Anagenesis
7. Match the evolutionary concepts on the left with the most appropriate descriptions on the right.

_____ Allopatric speciation *a.* Associated with the fossil record

_____ Anagenesis *b.* Sometimes seen in specialized parasites

_____ Sympatric speciation *c.* Can be associated with soil differences

_____ Parapatric speciation *d.* Most common speciation mode

8. Discuss the conditions on the Galapagos Islands that led to the evolution of the birds known as Darwin's finches.

REPRODUCTIVE ISOLATING MECHANISMS • HYBRID ZONES • HOW MUCH DO SPECIES DIFFER GENETICALLY? • HOW LONG DOES SPECIATION TAKE? • THE SIGNIFICANCE OF SPECIATION (pages 452–459)

Key Concepts

1. *Reproductive isolating mechanisms* are organismal traits that prevent hybrid formation between previously separated populations.
2. Isolating mechanisms are adaptations that prevent interbreeding between genetically different individuals in cases where hybrids would be less fit.
3. Differences in breeding seasons, courtship behavior, and reproductive physiology (including differences in courtship pheromones) are examples of *prezygotic isolating mechanisms* that reduce the chances that hybrid zygotes are produced.
4. *Postzygotic isolating mechanisms* reduce the number of offspring produced by matings between genetically dissimilar individuals because the hybrids are less viable or sterile.
5. If few changes have accumulated in the gene pools of isolated populations, when the barrier is removed interbreeding may occur and the two populations may again become one.
6. Stable *hybrid zones* can form between two species and only in these areas are hybrids common. Hybrid zones sometimes occur in areas where the environment is intermediate for some important habitat variable and hybrids are only well adapted in that area.
7. Interbreeding between snow geese and blue geese and the formation of a hybrid zone in the Canadian Arctic may present biologists with an opportunity to study the development of isolating mechanisms.
8. Studies on species of Hawaiian *Drosophila* show that sympatric species—species living in the same area—can be morphologically distinct, yet genetically very similar.
9. There is a wide range in the time required for speciation in different organisms: some plants can speciate in one season by polyploidy. American and European sycamores can still interbreed after 20 million years of geographic isolation.
10. Increased behavioral complexity, especially in the degree of mate selection, tends to reduce time required for speciation.
11. Within groups of organisms, some lineages give rise to more species than others.
12. Studies of large hoofed mammals of Africa show that the diet of these animals is correlated with the rate of speciation, with lineages of grazers and browsers showing higher speciation rates than omnivores and anteaters.
13. We expect organisms with ecological roles that cause their populations to fragment and be more sensitive to environmental change to show the highest speciation rates.
14. Evolution may be most rapid during periods of speciation, followed by a period of *stasis* after speciation is complete.
15. *Evolutionary radiations* occur when a lineage of organisms gives rise to a large number of new species.
16. Because islands offer geographic isolation and ecological opportunities due to reduced competition with other organisms, the best examples of evolutionary radiations have occurred on islands.
17. The Hawaiian islands are the most isolated archipelago in the world. The islands also vary in age.
18. Over 90% of all plant species on the Hawaiian islands are *endemic*—found nowhere else.
19. The group of Hawaiian plants called the *silverswords* and *tarweeds* provide an example of an evolutionary radiation. The 28 species of silverswords are much more diverse and widespread than are their mainland relatives.
20. The silversword radiation shows that major diversification can be accomplished by allele differences in a small number of genes.
21. The silversword radiation occurred because the original ancestors encountered few competing species and could diversify ecologically.
22. The cichlid fish of Lake Victoria are the result of an adaptive radiation that gave rise to more than 300 species from a single parental species that colonized Lake Victoria about 200,000 years ago.
23. It is theorized that major population genetic changes take place at the time of speciation due to genetic drift and natural selection. Long intervals of *stasis* characterized by genetic stability separate periods of active speciation.

Questions (for answers and explanations, see page 145)

1. Classify each of the following as an example of a **pre**zygotic isolating mechanism, a **post**zygotic isolating mechanism, or **not** an isolating mechanism.

_____ Non-overlapping species ranges

_____ Non-overlapping breeding seasons

_____ Production of sterile offspring

_____ Differences in courtship pheromones

_____ Incompatible gametes

_____ Poorly adapted offspring

2. Which of the following factors would *not* be expected to increase the rate of speciation in a group of organisms?
   a. A species range consisting of fragmented populations
   b. A diet consisting of food items whose abundance varies widely
   c. High birth rates
   d. Increased behavioral complexity
   e. A high degree of mate selection

3. Which of the following is *not* a suggested reason for the evolutionary radiation of silverswords and tarweeds on the Hawaiian archipelago?
   a. Water is an effective barrier for many organisms.
   b. Because islands are small compared to mainland areas, you would expect more species to develop there.
   c. Frequently, competition is reduced on islands.
   d. The Hawaiian islands differ in age from west to east.
   e. More ecological opportunities exist on islands.

4. Studies on cichlid fish of Lake Victoria, East Africa, show that
   a. a long period of time was required for the evolution of the 300 species found there.
   b. the lake must have been fragmented into many smaller lakes during the period of speciation leading to these fish.
   c. their most striking differences have to do with diet.
   d. this group originated from several dozen ancestral species.
   e. DNA evidence has shown that sympatric speciation probably led to this evolutionary radiation.

5. Explain what is known about the influence of ecology, especially diet, on the time required for speciation, based on studies of African mammals.

6. Studies of species of Hawaiian *Drosophila* show that
   a. sympatric speciation via polyploidy is common in insects.
   b. few genes need be involved in establishing reproductive isolation.
   c. matings between siblings can sometimes lead to new species.
   d. Species with non-overlapping ranges can be morphologically very similar.
   e. Geographic isolation is not always necessary for the establishment of reproductive isolation.

7. Discuss what is known about the evolutionary radiations on islands, based on studies of Hawaiian silverswords and tarweeds.

Integrative Questions (for answers and explanations, see page 146)

1. Select *all* of the following statements about speciation that are *not* true?
   a. Founder effect can be involved in speciation.
   b. The time required for speciation can vary for different types of organisms.
   c. Speciation always involves interruption of gene flow between different groups of organisms.
   d. The rate of speciation can vary for different groups of organisms.
   e. Speciation can occur in a single generation.
   f. Only plants can speciate by polyploidy.
   g. Speciation is a temporal process.

2. Which of the following concepts in population genetics is *least* involved in explaining the formation of new species by geographic isolation?
   a. Gene flow
   b. Genetic drift
   c. Founder effect
   d. Selection pressures
   e. Stabilizing selection
   f. Geographic variation

3. Select all of the following observations suggesting that two overlapping populations that had been geographically isolated have *not* diverged into distinct species.
   a. A stable hybrid zone exists where their ranges overlap.
   b. Matings between members of the two populations produce viable hybrids.
   c. Electrophoresis shows that the two populations share a common gene pool.
   d. Interbreeding is common between members of the two populations.
   e. Distinct prezygotic isolating mechanisms have evolved.

4. If you examine geographic variation in a widely distributed species, you would expect to see the least morphological variation in
   a. mountainous areas.
   b. island archipelagos.
   c. environmentally homogeneous regions.
   d. environmentally heterogeneous regions.
   e. areas with numerous large rivers.

21 Systematics and Reconstructing Phylogenies

CHAPTER LEARNING OBJECTIVES—after studying this chapter you should be able to:

- ❑ Explain what systematics is and discuss the major objectives of this discipline of biology.
- ❑ Explain what taxonomy is and discuss its relationship to systematics.
- ❑ Explain some of the past and present uses of biological classification schemes.
- ❑ Discuss the limitations of using common names to identify organisms.
- ❑ Employ the conventions used in the Linnaean nomenclature system for constructing binomial names.
- ❑ Identify the major hierarchical taxa within the Linnaean classification system.
- ❑ Explain why reclassification is a major activity of systematists today.
- ❑ Explain what cladistic systematics is and discuss its major objectives.
- ❑ Define and relate the following terms used in cladistic systematics: "clade," "cladogram," "ancestral traits," and "derived traits."
- ❑ Differentiate between homologous traits and homoplastic traits.
- ❑ Explain the roles of convergent and parallel evolution in creating homoplasy.
- ❑ Differentiate between general homologous traits and special homologous traits.
- ❑ Discuss what an outgroup is and explain the relationship of the outgroup to general and special homologous traits.
- ❑ Explain the provisional assumptions employed in the construction of a cladogram.
- ❑ Construct and interpret cladograms using the conventions presented in the textbook.
- ❑ Define what a fossil is and explain the major role of the fossil record in cladistic systematics.
- ❑ Discuss the application of the principle of parsimony in cladistic systematics.
- ❑ Discuss the strengths and limitations of using developmental traits to infer phylogenies.
- ❑ Explain why shared behavioral characteristics are often special homologous traits
- ❑ Discuss how biochemical differences in proteins and nucleic acids can be used to infer phylogenies.
- ❑ Explain why DNA from eukaryotic organelles has been especially useful in systematics.
- ❑ Discuss some of the features of chloroplast DNA and give an example of its use in inferring phylogenies.
- ❑ Explain the steps involved in the DNA hybridization technique and the rationale for its use in systematics.
- ❑ Describe the study of evolution in T7 bacteriophage presented in the textbook and the implications that the study has for cladistic systematics.

THE IMPORTANCE OF CLASSIFICATIONS • THE HIERARCHY OF THE LINNAEAN SYSTEM (pages 462–464)

Key Concepts

1. *Systematics* is the scientific study of the evolution of biological diversity.
2. The underlying objective of biological classification is to reveal the evolutionary relationships—the *phylogeny*—among living and fossil organisms.
3. The goal of systematics is to devise taxonomic classification schemes that express the evolutionary relationships among organisms.
4. Some classification schemes used in the past were concerned with the usefulness of organisms for humans or with revealing divine laws.
5. Biological classification systems help us place organisms into easily remembered groups that share common characteristics.
6. Biological classification systems allow us to make predictions about the characteristics of new organisms once they have been placed within a group.
7. Another objective of biological classification is to provide a system for naming different species in a uniform, unambiguous way.

8. Common names for organisms often vary regionally and do not reflect evolutionary relationships.
9. A *taxon* (plural, *taxa*) is a group of organisms treated as a unit within a classification system.
10. *Taxonomy* is a branch of systematics specifically concerned with classifying organisms.
11. The *Linnaean system*, originated by *Carolus Linnaeus* (1707–1778), forms the basis of modern classification systems.
12. In the Linnaean system, each species is assigned a two-part, or *binomial name*, composed of a genus name and a species name. This two name system is called *binomial nomenclature*.
13. A *genus* (plural, *genera*) is a group of related species. The species name is unique within the genus.
14. In some cases, the name of the scientist who named the organism is included after the binomial, as in *Homo sapiens* Linnaeus.
15. Both genus and species names are italicized; the genus name is capitalized, the species name is not.
16. To refer to more than one species within a genus, follow the genus name with "spp."; to refer to an unknown species within a known genus, follow the genus name with "sp." Within a paragraph, after the genus name is given once, it can then be abbreviated to the first letter of the genus name followed by a period in further references.
17. Taxonomic classification schemes are hierarchical; each higher taxon includes all lower taxa, and organisms in higher taxa have fewer characteristics in common than do those in lower taxa.
18. The number of species included in a taxon depends on their relative degrees of similarity and dissimilarity.
19. Higher taxa contain organisms that presumably shared a common ancestor in the less recent past than organisms in lower taxonomic groups, although this information is usually not available.
20. Related genera are grouped into *families*. Family names are based on a name of one of the included genera and end in "*-idae*" for animals or "*-aceae*" for plants.
21. Related families are grouped into *orders*, orders are grouped into *classes*, related classes are grouped into either *phyla* (animals and protists), or *divisions* (plants, bacteria, and fungi), and phyla and divisions are grouped into *kingdoms*.

Questions (for answers and explanations, see page 146)

1. The *most* important attribute of a biological classification scheme is that it
   a. allows a more systematic procedure for naming organisms.
   b. reflects the evolutionary relationships among organisms.
   c. helps us remember organisms and their traits.
   d. improves our ability to make predictions about the morphology and behavior of organisms.
   e. avoids the ambiguity created by using common names.

2. Suppose you are writing a scientific paper about a unicellular green alga called *Chlamydomonas reinhardi*. What would be the proper way to refer to this species after you had used the full binomial earlier in the same paragraph?
   a. Chlamydomonas reinhardi
   b. Chlamydomonas sp.
   c. Chlamydomonas spp.
   d. C. reinhardi.
   e. C. sp.

3. Organisms in a higher taxon are _____ similar, usually have diverged from a common ancestor _____ recently, and include _____ species than organisms in a lower, included taxon.
   a. less; less; fewer
   b. less; more; fewer
   c. less; less; more
   d. more; more; fewer
   e. more; more; more

4. Number, from lower to higher, the following taxonomic categories as you would use them to classify a new species of salamander. If a listed taxon should not be used, leave it blank.

   _____ Order
   _____ Species
   _____ Family
   _____ Kingdom
   _____ Division
   _____ Genus
   _____ Phylum
   _____ Class

CLASSIFICATION SYSTEMS AND EVOLUTION (pages 465–473)

Key Concepts

1. Classification systems should be judged based on how well they achieve their objectives: reflecting evolutionary history is the major objective of today's biological classification systems.
2. Biological classification by Linnaeus and his followers was based exclusively on morphology. The process of reclassifying organisms as we learn more about their phylogeny goes on today.
3. *Cladistic systematics* attempts to classify organisms based exclusively on their evolutionary history.
4. In cladistics, all of the species derived from a common ancestor form a *clade*, and members of a clade are all descended from a common ancestor.
5. A *cladogram* is a phylogenetic tree showing the relationships among different clades. Contemporary organisms are shown across the top of the cladogram; the common ancestor of all the clades being represented is at the base of the cladogram.
6. *Ancestral traits* are those traits that are shared by a group of species and that were also present in their common ancestor.

7. A *derived trait* has been modified from an ancestral trait due to evolution within a group.
8. Cladograms are based on the identification of ancestral and derived traits.
9. Structures derived from a common ancestor are called *homologous* traits. Homologous traits come from ancestral traits (i.e. they are derived traits), but may be modified by divergence.
10. *General homologous traits* are shared by many organisms because they appeared only once and in the distant past.
11. *Special homologous traits* are shared by only a few organisms because they appeared only recently.
12. *Homoplasy* is the presence of similar structures in different lineages that were not present in the ancestor common to the lineages.
13. Homoplasy can result from convergent evolution or parallel evolution.
14. In *convergent evolution*, formerly dissimilar structures become more alike because the different lineages are subject to similar selection pressures.
15. In *parallel evolution*, different lineages evolve similar modifications of an ancestral trait.
16. Constructing cladograms that accurately reflect the phylogeny of a lineage requires distinguishing ancestral from derived traits and homologies from homoplasies.
17. An *outgroup* is a taxon that is closely related to the group whose phylogeny is being constructed. General homologous traits are found within some species of the group being studied and within the outgroup; specific homologies are only found within members of the studied group.
18. A method for devising cladograms developed by Willi Hennig provisionally assumes that a trait shared by two species is homologous unless other evidence indicates that the trait is homoplastic.
19. Other assumptions used in constructing cladograms are that derived traits appear only once, and that no derived traits are lost during the evolution of the lineage.
20. Using Hennig's system, species are clustered according to the number of special homologous traits they share. If numerous traits are considered, most will be homologous if the species share a recent common ancestor.
21. In a cladogram, the taxon with no derived traits is the outgroup. The number of derived traits that two species share is a reflection of how recently they had a common ancestor.
22. In constructing a cladogram, contradictory data suggest that derived traits may have appeared multiple times or been lost in the evolution of the lineage.
23. *Fossils* are any evidence of organisms that lived in the past, including remains of the organism or its impressions, such as footprints. The fossil record helps reveal ancestral traits.
24. *Parsimony* is employed in arranging the taxa in a cladogram, such that the number of changes in traits is minimized. A parsimonious cladogram is likely to be more accurate.

Activities (for answers and explanations, see page 147)

- Based on the following cladogram showing the phylogeny of five species (*A–E*) relative to five traits (*1–5*), fill in the table shown below using "1" to indicate the presence of a derived trait, and "0" to indicate the presence of an ancestral trait.

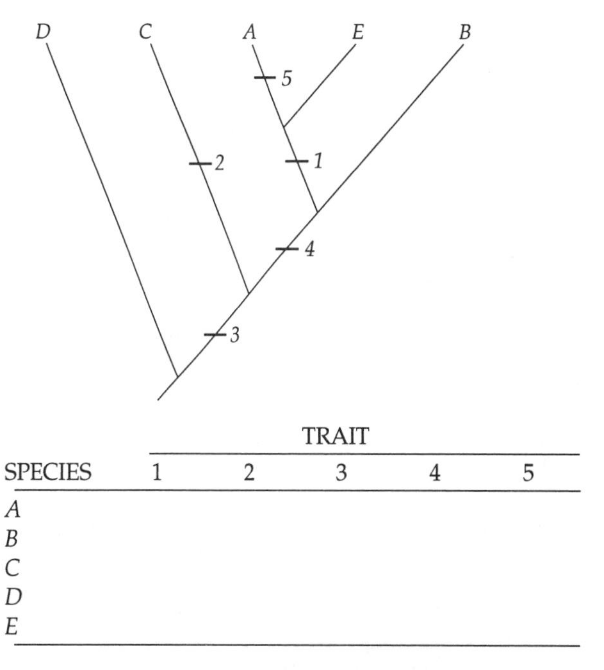

	TRAIT				
SPECIES	1	2	3	4	5
A					
B					
C					
D					
E					

- In the cladogram shown in activity 1, which species is considered the outgroup? _____ Explain your answer.

- Based on the following table showing the ancestral and derived traits of five species (*A–E*), construct a cladogram the represents the phylogeny of this group using conventions presented in the textbook.

	TRAIT				
SPECIES	1	2	3	4	5
A	1	1	1	0	0
B	0	0	0	0	0
C	1	1	0	1	0
D	0	1	0	0	0
E	1	1	0	1	1

The ancestral state of each trait is indicated by a 0 and the derived state is indicated by a 1.

- Discuss the application of parsimony in the construction of cladograms.

Questions (for answers and explanations, see page 147)

1. Which of the following would *not* be expected to result in homoplasy?
   a. Convergent evolution
   b. The independent evolution of similar structures in different lineages
   c. Parallel evolution
   d. The inheritance of ancestral traits
   e. Selection for traits that perform similar functions

2. Which of the following choices completes a statement about clades that is *not* true? A clade
   a. includes all organisms in one branch of a cladogram.
   b. is the basic taxon of cladistic systematics.
   c. includes only living species.
   d. shares a common ancestor.
   e. shares homologous structures only.

3. Choose a letter to make an *incorrect* statement. In a cladogram, the outgroup would
   a. have a branch point nearest the bottom.
   b. have all ancestral traits.
   c. not share any derived traits with other species.
   d. be determined based on evidence from the fossil record.
   e. share the least recent common ancestor with the group.

4. Which of the following is *not* an assumption used in the construction of cladograms?
   a. Shared traits are provisionally considered to be homologous.
   b. Special homologous traits are not found within the outgroup.
   c. General homologous traits are only found within the outgroup.
   d. Derived traits appear only once in the evolution of the lineage.
   e. Once derived traits appear, they can not disappear.

TRAITS USED IN RECONSTRUCTING PHYLOGENIES • EVALUATING DIFFERENT METHODS • THE FUTURE OF SYSTEMATICS • OTHER USES OF PHYLOGENIES (pages 473–477)

Key Concepts

1. Because the sizes and shapes of body parts, called *gross morphology*, are more likely to become part of the fossil record and are easy to measure and quantify, they are widely used by systematists.
2. Developmental stages of organisms may provide useful evidence about their phylogenies, as shown by the the similarities between sea squirts and vertebrates.
3. Larval traits in some species are highly modified by evolution and are not useful for inferring phylogenies.
4. Because behavior is easily modified by evolution, behavioral characteristics shared between different organisms tend to be special homologous traits.
5. *Konrad Lorenz* showed that many morphologically distinct species of ducks show very similar courtship behavior.
6. Biochemical differences in molecules of organisms, especially DNA, RNA, and proteins, is valuable information for systematists.
7. Amino acid sequencing techniques allow direct comparison of the primary structures of homologous proteins, such as cytochrome *c,* in different species. Amino acid differences in the primary structure reflect genetic differences between the taxa.
8. A comparison of the *base sequences* of nucleic acids is also an important aspect of determining the evolutionary relationships of different species.
9. Studies of the fine structure of DNA have been greatly facilitated with the invention of polymerase chain reaction technology in 1985. Amplification of DNA from fossil organisms is now commonly done.
10. DNA from the organelles of eukaryotes, such as chloroplast DNA (cpDNA) and mitochondrial DNA, has been especially useful for systematists because it is more evolutionary stable than nuclear DNA.
11. cpDNA is a circular, double-stranded molecule with a size of 151-kilobases. It includes about 100 genes that code for transfer RNAs, ribosomal subunits, and some proteins involved in photosynthesis.
12. cpDNA has been used to study the phylogenetic relationships between members of the sunflower family (Asteraceae), one of the largest families of flowering plants. The fossil record has not been useful for understanding the relationships within the Asteraceae because the family underwent a very rapid evolutionary radiation.
13. *DNA hybridization* results when DNAs from two different species that have been denatured into single-stranded molecules by heating are cooled and allowed to form interspecific double helices.
14. DNA hybridization can be used to compare DNAs even if their exact sequences are not known.
15. The stability of hybrid DNA molecules depends on the similarity of their base sequences.
16. The amount of heat required to separate hybrid DNAs relates directly to the similarity of the molecules and is a measure of the relatedness of the species from which the DNAs were obtained.
17. Nucleic acid hybridization studies have helped settle debates about the phylogenies of such organisms as humans, chimpanzees, and the giant panda.
18. Systematists now use cladistic techniques to combine molecular, morphological, and behavioral data. This approach can lead to more powerful conclusions.
19. Five different cladistic methods were recently used to construct phylogenies for nine different genetic strains of T7

bacteriophage viruses whose evolutionary radiation in the laboratory had been carefully documented. The cladogram based on the principle of parsimony most closely duplicated the true phylogeny.

Questions (for answers and explanations, see page 147)

1. Which of the following statements about traits used in reconstructing phylogenies is *not* true?
   a. Gross morphology is the only type of systematic evidence obtainable from the fossil record.
   b. Most shared behavioral characteristics are special homologous traits.
   c. The studies done by Konrad Lorenz on duck courtship showed that morphologically distinct species can have very similar behavior.
   d. The presence of a notochord in sea squirt larvae reveals their close relationship with other chordates
   e. Larval characteristics can be highly modified by natural selection.

2. Which of the following statements about DNA hybridization is *not* true?
   a. DNA hybridization is not dependent on a precise knowledge of DNA base sequences.
   b. DNA hybridization is not useful if the species being compared are too closely related.
   c. The amount of heat required to denature the hybrid DNA is inversely related to the degree of mismatching in the DNA from the two sources.
   d. The amount of heat required to denature the hybrid DNA is directly related to the genetic relatedness of the two species.
   e. DNA hybridization helped biologists realize that the giant panda is a bear.

3. DNA hybridization has shown that the percentage difference in human and chimpanzee DNA is approximately
   a. less than 1%.
   b. 1–2%.
   c. 10–20%.
   d. 40–50%.
   e. greater than 50%.

4. Which of the following statements about chloroplast DNA (cpDNA) is *not* true?
   a. cpDNA is very similar in most flowering plant species.
   b. cpDNA is a circular, double-stranded molecule.
   c. cpDNA includes about 100 genes.
   d. A single 22-kilobase inversion in cpDNA is an ancestral trait.
   e. Studies show that cpDNA is subject to a high degree of mutation.

5. The major conclusion of the laboratory study on an evolutionary radiation of T7 bacteriophage was that
   a. none of the cladistic methods were very accurate.
   b. the principle of parsimony yielded a cladogram best fitting the data.
   c. nonliving systems like viruses cannot be studied using cladistic systematics.
   d. mutation rates in viruses is too low to produce significant evolutionary changes.
   e. the simplest cladogram is always the most accurate.

6. Discuss the implications the following statement has for systematics: "DNA is the genetic material for all eukaryotes."

Answers and Explanations

Chapters 18–21

CHAPTER 18 – THE ORIGIN OF LIFE ON EARTH

Note : The page numbers listed with the section titles refer to the Study Guide; numbers in parentheses for each question refer to the relevant key concepts.

IS LIFE EVOLVING FROM NONLIFE TODAY? • WHEN DID LIFE ON EARTH ARISE? • HOW DID LIFE EVOLVE FROM NONLIFE? (Conditions on Earth at the Time of Life's Origin • An Overview of Earth's History • Beginning the Sequence of Events: The Synthesis of Organic Compounds • Continuing the Sequence: More Polymerization) (pages 126–128)

1. Choice *d*. Some critics argued that the air in Spallanzani's closed containers was damaged by boiling; Pasteur avoided this criticism by using swan-necked flasks that left the containers open to the air but avoided microbial contamination. (1, 2)
2. Oxygen (O_2) was probably the least abundant and carbon dioxide (CO_2) was probably the most abundant gas in Earth's early atmosphere. (14, 15)
3. Choice *e*. Although it is hypothesized that thermal energy may have been a more important energy source than ultraviolet light and electric discharge (because of their damaging effects on life), these studies did not deal with this issue directly. (24, 25)
4. Choice *e*. Rocks are aged using radioactive isotopes by comparing the proportion of radioactive and stable forms of the element in the rock. An isotope with a rapid decay rate would be best for more recent times since in older rocks most of the element would be in the stable form and the accuracy would be reduced. (5–10, 16–23)
5. Choice *e*. Hydrogen cyanide (HCN) and methane were constituents of the early atmosphere. Some amino acids were formed directly in Miller's experiments and presumably in the early atmosphere. The synthesis of adenine is more complex, requiring the polymerization of HCN and rearrangement of its HCN tetramers. (14, 24–28)
6. (16–23)
   Oldest prokaryotic fossils: day *7*
   Oldest eukaryotic fossils: day *21*
   First land plants: day *28*
   Reptiles dominant: day *29*
   Rise of mammals: day *29*
7. All of these approaches have been used to estimate the age of the origin of life on Earth. (3–11)

HOW DID LIFE EVOLVE FROM NONLIFE? (RNA: The First Biological Catalyst • From Ribozymes to Cells • The Evolution of DNA) • METABOLISM OF EARLY ORGANISMS • HOW PROBABLE WAS THE EVOLUTION OF LIFE? (pages 128–129)

1. Choice *e*. Unlike RNA, there is no evidence that DNA has ever had a catalytic function. (1–7, 12)
2. Choice *b*. The theory explaining the evolution of the first cells from coacervate droplets assumes that replication via nucleic acids appeared later. (8–12)
3. Choice *c*. Very little oxygen was present in the atmosphere of the early Earth and until organisms evolved adaptations to use it, oxygen was poisonous . (13)
4. Choices *c* and *e*. Except for the use of CO_2 as a carbon source and sunlight as an energy source, these groups differ in all other ways listed. (14–19)
5. Choice *a*. The earliest photosynthetic organisms were prokaryotic and, therefore, most like the bacteria listed in choices *b–e* and least like eukaryotic photosynthesizers. (14–19)
6. Choice *c*. With the evolution of noncyclic photophosphorylation, photosynthetic organisms became very successful, resulting in a *reduction* in CO_2 within the atmosphere as more became fixed in organic form or deposited in ocean bottoms. (21–24)
7. Choice *b*. *DNA* is only stable in hydrophobic environments. (2–7, 12)
8. The purple nonsulfur bacteria use a variety of organic compounds as hydrogen donors. (15–20)
9. The present-day oxidative atmosphere prevents the accumulation of abiotically formed simple organic compounds. Even if these materials could accumulate, they would be consumed by already living organisms. (24)

Integrative Questions

1. *4* Coacervates become cells
   6 Photosynthesis without oxygen production appears
   5 DNA assumes central role in genetic code
   7 Cyanobacteria appear
   2 Polymerization of monomers
   3 Ribozymes appear
   1 Organic compounds formed in a reducing atmosphere
2. ***Oparin:*** Studied coacervates
   Eigen: Studied ribozymes
   Pasteur: Work lead to acceptance of aphorism "*omne vivum e vivo*"
   Redi: First to study spontaneous generation
   Kelvin: Based on heat measurements, said Earth began 10–20 billion years ago
   Usher: Estimated that Earth was created in 4004 B.C.

CHAPTER 19 – THE MECHANISMS OF EVOLUTION

Note : The page numbers listed with the section titles refer to the Study Guide; numbers in parentheses for each question refer to the relevant key concepts.

WHAT IS EVOLUTION? • THE STRUCTURE OF POPULATIONS • THE HARDY–WEINBERG RULE (pages 130–132)

1. Choice ***d***. Evolution is defined as changes in the genetic structure of a population over time, so evolution occurs at the level of the population. (1–4)
2. Choice ***d***. Although most populations show genetic variation, this condition is not part of the definition of a deme. (8)
3. Choices ***b, d***. Genetic variation has to do with the number of different alleles per gene, which would also result in the greatest number of different genotypes. Regarding choice *a*, recall that all species members have the same number of genes. Population size, *per se*, has little to do with genetic variation, so choices *c* and *e* are also incorrect. (2, 3, 10–13)
4. ***allele*** frequencies, ***genotype*** frequencies (18–20)
5. Choice ***b***. Recall that q^2 = the frequency of the *tt* genotype, so $\sqrt{q^2}$ = the frequency of the *t* allele or $\sqrt{0.36} = 0.6$ and $1 - 0.6 = 0.4$ is the frequency of the *T* allele. (18–20)
6. Choice ***c***. Since $p = 0.3$ and, therefore, $q = 1 - 0.3 = 0.7$, the frequency of the *MN* genotype is $2pq$ or $2(0.3)(0.7) = 0.42$. (18–20)
7. Choice ***a***. The probability of one allele *b* in a genotype is equal to its frequency, or 0.2, so the probability of 3 *b* alleles in a genotype (*bbb*) would be $(0.2)^3$ or 0.008. (18–20)
8. Choice ***a***. Vertebrates have proportionally fewer heterozygous gene loci than invertebrates, inbred populations have proportionally fewer than outbred populations, and individuals have proportionally fewer than populations. (14–16)
9. Most populations do not meet the Hardy–Weinberg conditions and the main use of the rule is to determine if evolutionary change is occurring within a population. (22, 23)

CHANGING THE GENETIC STRUCTURE OF POPULATIONS (pages 132–133)

1. Distribution ***a***. Stabilizing selection results when individuals that are intermediate in phenotype make a larger contribution to future generations than individuals of more extreme phenotype. This leads to reduced variation for the trait and causes the curve to be higher and narrower. Curve *b* shows greater variation, curve *c* would result from disruptive selection, and curve *d*, from directional selection. (17–22)
2. Choice ***c***. Genetic drift is always most prevalent in small populations. The founder effect may be important in the act of creating a small population, but after it is established, random variation due to small size is called genetic drift. (1–14)
3. Choice ***a***. Since assortative mating involves mate selection based on phenotype, a mostly invisible phenotype like blood group is not likely to be involved. (1–14)
4. Choices ***a, e***. Self-fertilization leads to increased numbers of homozygous individuals and natural selection is also nonrandom. (1–14)
5. Area ***c***. Relatively more light moths are recaptured in this area than dark moths, indicating that this phenotype has greater fitness. (14)
6. Dark moths seem to survive best in area *d*. If the increased survival of the moths is due to cryptic coloration then the bark surfaces in that area should also be dark. (14)
7. Choice ***a***. Helping raise young that are related through a sister contributes to the bird's inclusive fitness through kin selection. (23–27)
8. Inclusive fitness is a measure of the reproductive success of an individual relative to all other individuals. Inclusive fitness results from an individual's direct reproduction (individual fitness) and any behavior shown by the individual that promotes reproduction and survival of relatives (kin selection). (23–27)

GENETIC VARIATION AND EVOLUTION • INTERPRETING LONG-TERM EVOLUTION (page 133–134)

1. Choice ***a***. Mutations leading to amino acid substitutions are probably equally frequent for all positions, but those mutations causing changes in critical parts of the protein are eliminated by natural selection. (8–13)
2. Choice ***b***. Balanced polymorphism is the existence of two or more alleles at stable frequencies within the population over time. If phenotypes are constant through time, then the underlying alleles will also be constant. (5, 6)
3. Choice ***b***. As shown by the example in the text of clinal variation in cyanide production in the European clover *Trifolium repens*, clines can involve any type of phenotypic feature of the organism, including biochemical characteristics. (2–4)
4. Choice ***d***. All except speciation are the short-term changes in population gene pools characteristic of microevolution. (14–16)
5. Microevolution involves agents such as gene flow, genetic drift, and selection that cause changes in allele frequencies of population gene pools. Macroevolution refers to long-term changes in groups of organisms due to unpredictable events like continental drift and climate change. (14–16)
6. Choices ***c*** and ***d***. Genetic drift, directional selection, and stabilizing selection tend to reduce genetic variation in populations. (1–4, 7)

CHAPTER 20 – SPECIES AND THEIR FORMATION

Note : The page numbers listed with the section titles refer to the Study Guide; numbers in parentheses for each question refer to the relevant key concepts.

WHAT ARE SPECIES? • THE COHESIVENESS OF SPECIES • HOW DO NEW SPECIES ARISE? (pages 135–137)

1. Choices ***a, d***. The key criterion of a biological species is that its members be reproductively isolated from other such groups. This criterion is impossible to evaluate in asexual and fossil species. (1–3)
2. Choice ***b***. A wide distribution is not requisite for geographic speciation. (7–14)

3. Snails from area A would be genetically most like individuals from area *B* and least like individuals from area *D*. Snails from area A are in the same block of habitat with area B snails, but must cross two barriers (roads) to interbreed with area D snails. They need only cross one road to interbreed with area C or E snails. (11, 13, 14)
4. Choice *c*. Differences in soil type and other environmental discontinuities frequently lead to parapatric speciation. (16–17)
5. Choice ***b***. Nondisjunction of homologous chromosomes during meiosis I can lead to a doubling of the chromosome number within a single species; polyploid individuals would have cells with four copies of each chromosome. Each of the three species could form a different tetraploid species. (17–24))
6. Choice ***b***. Sympatric speciation is most common among flowering plants. It has been estimated that over one-half of all flowering plant species are polyploid. (18)
7. (8–23)
   d: Allopatric speciation
   a: Anagenesis
   b: Sympatric speciation
   c: Parapatric speciation
8. The relatively great distance between the Galapagos Islands and the South American mainland and also between each of the islands in the archipelago insured that once emigrants had arrived on an island, they would be genetically isolated for a substantial period of time. Also, because the islands differ greatly in climate and vegetation, the resident birds were subject to different selection pressures. This in combination with reduced gene flow between the islands led to a rapid evolutionary radiation of finches. (11, 12)

REPRODUCTIVE ISOLATING MECHANISMS • HYBRID ZONES • HOW MUCH DO SPECIES DIFFER GENETICALLY? • HOW LONG DOES SPECIATION TAKE? • THE SIGNIFICANCE OF SPECIATION (pages 137–138)

1. (1–4)
   not Non-overlapping species ranges
   pre Non-overlapping breeding seasons
   post Production of sterile offspring
   pre Differences in courtship pheromones
   pre Incompatible gametes
   post Poorly adapted offspring
2. Choice *c*. Birth rates *per se* do not seem to effect the rate of speciation in organisms. All other factors have been shown to increase speciation rates in the lineages of some organisms. (9–14)
3. Choice *b*. Actually, biogeographers have found that larger islands tend to have more species than smaller islands, so you might expect the reverse effect. (19–21)
4. Choice *c*. All other statements are false. (22)
5. Studies of the African savanna mammals show that grazers and browsers have undergone more rapid speciation than omnivores and anteaters. Climate change has a more direct effect on the availability of food for grazers and browsers, causing frequent fragmentation of populations of these species. Climate change influences the food supply of omnivores and anteaters less and their population ranges are more continuous. Thus, diet indirectly affects the degree of gene flow in these mammals and this results in different speciation rates. (11–13)
6. Choice *b*. Studies of the Hawaiian *Drosophila* show that these species have only slight genetic differences, despite being morphologically diverse. (8)
7. Studies of the silverswords and tarweeds of the Hawaiian archipelago show that taxa that evolve on islands frequently show great morphological diversity because of the reduced competition that emigrants encounter on islands. Thus, Hawaiian silverswords and tarweeds have evolved tree- and shrub-like species because there were few resident tree and shrub species there with which they had to compete. (15–21)

Integrative Questions

1. Choice *f*. There is evidence that some parthenogenetic animals have undergone sympatric speciation by polyploidy. All other statements are true.
2. Choice *e*. Geographic isolation would tend to reduce gene flow and small isolated populations would be subject to founder effect or genetic drift. Substantial clinal variation in the parent population would facilitate speciation after geographic isolation occurred. Selection pressures might be different in the isolated populations leading to genetic divergence. Stabilizing selection would reduce the amount of variation in the parent population and this would tend to inhibit speciation after geographic isolation occurred.
3. Choice *c*. If the two populations share a common gene pool, they are still a single species. Limited interbreeding, production of viable hybrids, and establishment of a hybrid zone do not necessarily mean that speciation is not complete.
4. Choice *c*. Areas where environmental change is least rapid (environmentally homogeneous areas) and gene flow greatest are areas where the least geographic (clinal) variation would be expected.

CHAPTER 21 – SYSTEMATICS AND RECONSTRUCTING PHYLOGENIES

Note : The page numbers listed with the section titles refer to the Study Guide; numbers in parentheses for each question refer to the relevant key concepts.

THE IMPORTANCE OF CLASSIFICATIONS • THE HIERARCHY OF THE LINNAEAN SYSTEM (pages 139–140)

1. Choice *b*. Although all of the statements listed are important attributes of biological classification schemes, the most important attribute is that biological classification reflect evolutionary relationships. (2–8)
2. Choice *d*. After a scientific name is referenced once in a paragraph, the genus is abbreviated, but the species name is given in full. (11–16)
3. Choice *c*. Organisms in a higher taxon are *less* similar, have diverged from a common ancestor *less* recently, and include *more* species than organisms in a lower, included taxon. (17–19)

4. (20–21)
4 Order
1 Species
3 Family
7 Kingdom
Division (plants, fungi, bacteria only)
2 Genus
6 Phylum
5 Class

CLASSIFICATION SYSTEMS AND EVOLUTION (pages 140–142)

- (16–22)

	TRAIT				
SPECIES	1	2	3	4	5
A	1	0	1	1	1
B	0	0	1	1	0
C	0	1	1	0	0
D	0	0	0	0	0
E	1	0	1	1	1

- Species ***D***. The outgroup has the ancestral form for all traits. (16–22)
- (16–22)

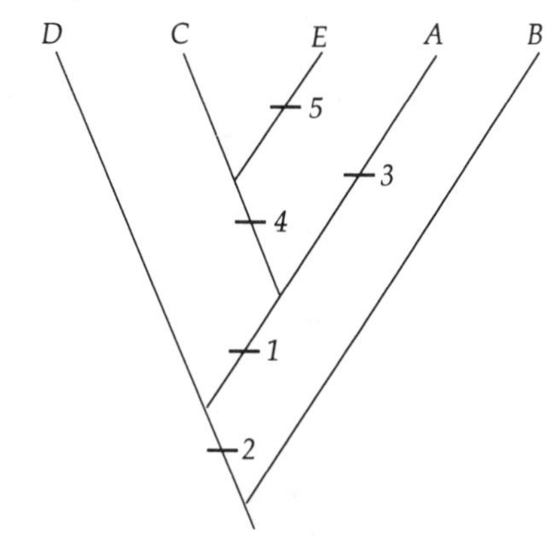

- In the construction of a cladogram, you initially assume that derived traits appear only once and never disappear. Given a set of traits for a group of species, sometimes these restrictions must be relaxed in order to produce a cladogram for the group. Parsimony involves arranging the species so that you minimize the number of required reversals and multiple origins. Generally, the simplest explanation is most likely to be correct. (24)

1. Choice ***d***. Homoplasy is the appearance of similar structures in different lineages that were not present in the common ancestor. (12–15)
2. Choice ***c***. A clade can include living species or species known only from the fossil record. (4)
3. Choice ***d***. The outgroup includes the taxon with all ancestral traits for those traits being considered. The fossil record is used for deciding what are ancestral traits, but it is not used directly in the selection of the outgroup. (17–23)
4. Choice ***c***. General homologous traits are found in *both* the outgroup and the group being classified. (10, 11, 16)

TRAITS USED IN RECONSTRUCTING PHYLOGENIES • EVALUATING DIFFERENT METHODS • THE FUTURE OF SYSTEMATICS • OTHER USES OF PHYLOGENIES (pages 142–143)

1. Choice ***a***. With the availability of polymerase chain reaction technology, DNA from fossils can now be amplified, sequenced, and compared. All other statements are true. (1–5, 9)
2. Choice ***b***. Actually, DNA hybridization is only useful if the species are closely related. If the DNA sequences are too dissimilar, strand hybridization will not take place. (13–17)
3. Choice ***b***. See Table 21.3 on page 475 in the textbook. (13–17)
4. Choice ***e***. Actually, cpDNA is a very stable molecule that has changed very little in flowering plants. This has made the changes that have occurred very useful systematic data. (10–12)
5. Choice ***b***. All other statements are false. (19)
6. Some of the implications of this statement are that DNA evolved as the genetic material before eukaryotes had diverged from prokaryotes, that DNA is an ancestral and general homologous trait, and that all surviving eukaryotes use DNA as the genetic material.

22

Bacteria and Viruses

CHAPTER LEARNING OBJECTIVES—after studying this chapter you should be able to:

- ❑ Name the two kingdoms of bacteria and identify the major phyla within each kingdom.
- ❑ Explain the importance of peptidoglycan, lipids, and rRNA as taxonomic criteria for assigning specific bacteria to a kingdom or phylum.
- ❑ Discuss the metabolic similarities and differences among bacteria that are designated as obligate aerobes, obligate anaerobes, and facultative anaerobes.
- ❑ Define the following modes of bacterial nutrition: photoautotrophy, photoheterotrophy, chemoautotrophy, and chemoheterotrophy.
- ❑ Name the principal photosynthetic pigments used by photoautotrophic bacteria.
- ❑ Describe the process of nitrogen fixation by bacteria, and discuss its importance in nature.
- ❑ Explain the Gram-staining method, describe the functional difference between gram-positive and gram-negative bacteria, and discuss how this technique can be used as a medical diagnostic tool.
- ❑ Identify the kingdom to which the mycoplasmas belong and discuss their general characteristics.
- ❑ Differentiate between the bacteria known as thermoacidophiles, methanogens, and strict halophiles.
- ❑ Discuss the basic structural features of prokaryotic cells, the difference between prokaryotic and eukaryotic flagella, and the function of the internal membranes found in some bacteria.
- ❑ Explain the difference between the terms "bacillus," "coccus," and "spiral," or "curved," bacteria.
- ❑ Describe the roles of fission, transformation, conjugation, transduction, and endospores in bacterial reproduction.
- ❑ Name Koch's four postulates and explain their use in associating a disease with a particular microogranism.
- ❑ Differentiate between the concepts of invasiveness and toxigenicity when discussing bacterial infection.
- ❑ Associate specific bacterial diseases discussed in your textbook with the appropriate phyla of those bacteria.
- ❑ Describe the anatomical and physiological specializations of cyanobacteria that make them nutritionally independent, and discuss the form of reproduction in this phylum.
- ❑ Explain why viruses are considered to be acellular obligate intracellular parasites, and list the main taxonomic criteria used to classify viruses.
- ❑ Distinguish between a virus, a virion, a viral capsid, a viroid, and a scrapie-associated fibril.
- ❑ Identify where in a virus the nucleic acid material is typically found, and where the protein is located.
- ❑ Discuss the different ways that viruses enter plant, animal, and bacterial cells, and explain the role of vectors in these processes.
- ❑ Compare and contrast the lytic life cycle of a bacteriophage with a lysogenic life cycle.
- ❑ Define the term "arbovirus," and explain how arboviruses are transmitted from a vector to a host.

GENERAL BIOLOGY OF THE BACTERIA (pages 483–489)

Key Concepts

1. The two kingdoms of bacteria include an incredibly abundant, diverse, and highly successful array of microscopic organisms exhibiting prokaryotic cell structure; their evolutionary origins date back over 3.5 billion years.
2. Some species of bacteria are *obligate anaerobes* and can live only in oxygen-free environments. Others require oxygen for survival and are designated *obligate aerobes.* Bacteria that are able to shift their metabolism between aerobic and anaerobic mechanisms are the *facultative anaerobes.*
3. Some bacteria perform respiratory electron transport in the absence of oxygen by using inorganic ions such as nitrate, nitrite, or sulfate to accept electrons. This action helps to complete the nitrogen and sulfur cycles in nature.
4. Four main categories of nutrition are recognized in bacteria: photoautotrophy, photoheterotrophy, chemoautotrophy, and chemoheterotrophy.
5. *Photoautotrophs* use light energy to fix CO_2 into organic carbon compounds. Two types of photosynthesis are known to occur in bacteria. The cyanobacteria use *chlorophyll a* as their key photosynthetic pigment in the

process of noncyclic photophosphorylation, which produces oxygen as a by-product. Other species use *bacteriochlorophyll* and generate pure sulfur instead of oxygen. Because of differences in the absorption spectra between these two photopigments, different species of bacteria can grow in diverse habitats spanning a broad range of light quality and intensity.

6. *Photoheterotrophs* also use light energy, but obtain carbon from organic compounds produced by other organisms. Also known as purple nonsulfur bacteria, they typically "feed" on alcohols, fatty acids, and carbohydrates.
7. *Chemoautotrophs* gather energy by oxidizing inorganic substances and then use it to reduce CO_2. Included are species that oxidize ammonia and nitrite ions (the nitrifiers important to plants), as well as those capable of oxidizing hydrogen gas, hydrogen sulfide, and sulfur. Oceanic ecosystems based on chemoautotrophy exist at great depths (~2,500 m) where no light reaches. Here, chemoautotrophic bacteria oxidize hydrogen sulfide emanating from thermal vents in the ocean floor to obtain energy which they use to fix carbon, thus creating food for other purely heterotrophic organisms.
8. *Chemoheterotrophs* obtain both energy and carbon from one or more preexisting organic compounds. The vast majority of bacteria are chemoheterotrophic.
9. *Nitrogen fixation*, the conversion of atmospheric nitrogen into chemical forms usable by cells, is a vital function performed only by certain bacteria, many of which live in symbiotic association with organisms that require this nitrogen.
10. Because of their astounding diversity and ancient lineages, bacteria represent a difficult challenge for taxonomists to classify into biologically meaningful groups. At one time bacteria were lumped into a single kingdom, the Monera.
11. Your textbook adopts Woese's recent classification of bacteria, which recognizes two kingdoms: the *Archaebacteria* and *Eubacteria*. However, other than the use of these two kingdoms, bacterial classification is still based on the standard method described in *Bergey's Manual* which groups species so as to facilitate identification of unknown organisms.
12. The Eubacteria are divided into three phyla based on the nature of their cell walls. These are the *gram-negative* and *gram-positive* bacteria (typical bacterial cell walls which respond differently to specific staining methods), and the *mycoplasmas* (lack cell walls). The Archaebacteria, whose cell walls are chemically unrelated to those of the Eubacteria, are subdivided into three phyla based on habitat. These are the *thermoacidophiles*, *methanogens*, and *strict halophiles*.
13. Bacteria show prokaryotic cell structure. Prokaryotes differ from eukaryotes in terms of the molecular organization of their genetic material and its replication mechanisms. Prokaryotes also lack the typical membrane-bounded cytoplasmic organelles of eukaryotes. Some bacteria do, however, show elaborate internal membrane structure used in respiration and photosynthesis. Many possess prominent *mesosomes* which often associate with the cell wall or DNA.
14. A number of prokaryotes move by means of *flagella*, which may occur singly or in tufts along the cell's exterior. The design of prokaryotic flagella differs significantly from that of eukaryotes. Bacterial flagella consist of a single fiber of the protein flagellin which rotates about its base, rather than beating from side to side as in eukaryotes.
15. Eubacteria are classified based on structural characteristics of their cell wall and its response to the chemical dye known as *Gram stain*. Those with large amounts of the polymerized amino sugar peptidoglycan in their cell walls readily take up this stain and appear purple under the microscope; they are called *gram-positive*. Other Eubacteria contain much less peptidoglycan, and thus do not absorb the stain. These are called *gram-negative*, and appear pink or red after counterstaining.
16. The Gram stain response of Eubacteria represents a useful diagnostic tool in medicine, since chemical treatments against bacterial infection are often aimed at destroying the cell wall of the pathogen.
17. Eubacteria are also classified by their shape. Three types are recognized: spherical (*coccus*), rod (*bacillus*), and spiral or curved forms. Each may exist individually, or in two- and three-dimensional arrays of adhering cells known as filaments.
18. Most bacteria reproduce asexually by fission, a form of *vegetative* reproduction. However, sexual processes such as transformation, conjugation, and transduction also allow for exchange of genetic material between individuals.
19. Some bacteria endure harsh conditions by replacing the parent cell with a nonreproductive resting structure called an *endospore*, which can remain dormant for long periods of time. The endospore germinates to become a metabolically active bacterium whenever favorable conditions return.
20. *Koch's postulates* state that, in order to definitively associate a disease with some particular microorganism:
    a. The microorganism must always be found in diseased individuals.
    b. Microorganisms taken from the host must be grown in pure culture.
    c. Samples of this culture must produce disease when injected into a new, healthy host.
    d. A newly infected host must yield a new pure culture of the suspected microorganism.
21. The consequences of infection for a host depend on factors related to a pathogen's *invasiveness* (ability to multiply within the host) as well as its *toxigenicity* (ability to produce chemical substances injurious to the host). Some bacterial pathogens have low invasiveness but high toxigenicity (e.g., diphtheria bacterium), while others have high invasiveness and low toxigenicity (e.g., anthrax bacterium).

22. Despite the importance of pathogenic bacteria, only a small minority of bacterial species are agents of disease. Most bacteria play very beneficial roles such as aiding animal digestion, processing nitrogen, and acting as agents of decomposition in the environment.

Activity (for answers and explanations, see page 205)

- Briefly explain how each of the following is related to the general biology of bacteria.
  1. Invasiveness and toxigenicity:
  2. Facultative anaerobism:
  3. Chemoautotrophy:
  4. Differential uptake of Gram stain:
  5. Coccus, bacillus, spiral, or curved:

Questions (for answers and explanations, see page 205)

1. Which of the following is *not* a necessary condition to satisfy Koch's postulates?
   a. Ability to grow a pure culture of the isolated microorganism
   b. Cultured samples must be able to produce disease symptoms in newly infected host.
   c. Isolated microorganisms must respond positively to gram-staining procedures.
   d. Newly infected hosts must yield the same suspect microorganism in pure culture.
   e. The suspected microorganism must always be found in diseased hosts.

2. After using the Gram-staining procedure on a sample of bacteria from your toothbrush, you observe a cluster of tiny, rod-shaped purple cells under the microscope. These are most likely
   a. gram-positive bacilli.
   b. gram-negative bacilli.
   c. gram-positive cocci.
   d. gram-negative cocci.
   e. spiral bacteria insensitive to Gram stain.

3. An important distinction between eukaryotic and prokaryotic flagella is that
   a. the overall design of eukaryotic and prokaryotic flagella is quite different.
   b. prokaryotic flagella consist of a single fiber of flagellin, while eukaryotic flagella are made of other proteins.
   c. the mechanism of motion in the two types of flagella is completely different.
   d. All of the above

4. Which of the following statements about bacterial metabolism is *not* true?
   a. Cyanobacteria utilize bacteriochlorophyll as their key photosynthetic pigment.
   b. Bacteria that use chlorophyll *a* for photosynthesis also produce oxygen as a by-product.
   c. Chemosynthetic bacteria sometimes form the basis of deep water oceanic food chains.
   d. In chemoautotrophic bacteria, energy to fix carbon comes from oxidizing inorganic substances.
   e. Most bacteria are chemoheterotrophic.

5. Match the following categories of bacterial nutrition with the correct biochemical mechanism for that category.

____ Photoautotrophy	*a.* Energy and carbon from preexisting organic compounds
____ Photoheterotrophy	*b.* Light used as energy source, CO_2 is the carbon source
____ Chemoautotrophy	*c.* Carbon comes from CO_2, energy from oxidizing inorganic materials
____ Chemoheterotrophy	*d.* Preformed organics from other organisms, energy from light

6. All but one of the following is a true statement about the structure of prokaryotic and eukaryotic cells. Select the *exception.*
   a. The molecular organization and shape of the chromosomal material in eukaryotes is different than that of prokaryotes.
   b. Unlike eukaryotes, prokaryotes lack a membrane-bounded nucleus.
   c. While there are extensive internal membrane systems within eukaryotes, no such membranes exist in prokaryotes.
   d. Some prokaryotes have mesosomes associated with their DNA or cell walls.
   e. Photosynthesis based on the pigment chlorophyll *a* occurs in both eukaryotes and prokaryotes.

7. Which of the following criteria is the primary one used to subdivide the prokaryotes belonging to the kingdom Archaebacteria into different phyla?
   a. Habitat type
   b. Whether they are aerobic or anaerobic
   c. Presence or absence of mesosomes
   d. Presence or absence of flagella
   e. Shape and gram-stain response

PHYLA OF THE ARCHAEBACTERIA • PHYLA OF THE EUBACTERIA (pages 489–497)

Key Concepts

1. Members of the kingdom Archaebacteria are recognized as related taxonomic groups based on three criteria. All exhibit a definitive lack of peptidoglycan in their cell walls, they have lipids of distinctive composition, and they possess great similarity in the base sequences of their ribosomal RNA.

2. The Archaebacteria can be further subdivided by habitat and lifestyle into the *thermoacidophiles* (heat- and acid-loving bacteria), the *methanogens* (methane-producing bacteria), and the *strict halophiles* (salt-loving bacteria).
3. Thermoacidophilic bacteria such as those of the genus *Sulfolobus* are common in hot sulfur springs, where they endure temperatures as high as 75°C. Although the pH in such habitats is typically 2–3, *Sulfolobus* can tolerate pH as low as 0.9.
4. The ten species of methanogens are obligate anaerobes that use methane production as a key step in metabolism. They are responsible for producing all the methane gas released annually into our atmosphere (~2 billion tons). Habitats for methanogens range from hot volcanic vents to the digestive tracts of grazing mammals.
5. Strict halophiles live in aquatic and terrestrial habitats nearly saturated with salt, where pH values can be as high as 11.5. Many species contain pink carotenoid pigments, which make them visible when present in large concentrations.
6. The great majority of Eubacteria (nearly three-fourths of all species) belong to the *gram-negative* phylum. Included are the gliding bacteria, spirochetes, curved and spiral bacteria, gram-negative rods, gram-negative cocci, rickettsias, chlamydias, and cyanobacteria.
7. *Gliding bacteria* (e.g., members of the genus *Beggiatoa*) are rod-shaped or filamentous, and locomote via a gliding motion whose mechanism remains unknown. Individual bacteria aggregate to form *fruiting bodies;* thick-walled, desiccation-resistant spores or cysts are produced inside.
8. *Spirochetes* are coiled, cylindrical cells that possess unique structures called *axial filaments*. These filaments are composed of flagella positioned between the cell wall and an outer envelope, and are thought to aid the bacterium's locomotion. Many species are parasitic (e.g., *Treponema pallidum*, which causes syphilis in humans).
9. *Curved* and *spiral bacteria* are a loosely affiliated group of gram-negative species separated from the spirochetes by the absence of axial filaments. Some are aquatic and free-living, others are parasitic and cause disease in animals and other bacteria.
10. *Gram-negative rods* are a highly diverse group related primarily by the nature of their cell walls. Included is *Escherichia coli*, an important resident of the human digestive system. Also counted among this group are the causative agents of human plague, dysentery, cholera, and salmonella food poisoning. The tumors characteristic of *crown gall* in plants result from infection by a gram-negative bacterium whose plasmids are used in recombinant DNA studies.
11. *Gram-negative cocci* include the nitrifiers which use oxides of nitrogen as terminal electron acceptors in cellular respiration. Also included is *Neisseria gonorrhoeae*, the causative agent of the human veneral disease gonorrhea.
12. The *rickettsias* and *chlamydias* are extremely small intracellular parasites once thought to be viruses. They can reproduce only within the cells of other organisms. Many are agents of serious human diseases such as Rocky Mountain spotted fever, typhus, certain types of venereal disease, and some forms of pneumonia. Some are transmitted to humans through the bite of arthropod hosts.
13. The *cyanobacteria* can perform photosynthesis using chlorophyll *a* in much the same way that plants do, and many have elaborate internal photosynthetic membrane systems. Some carry out fermentation, and others fix nitrogen. They are a closely related group, as evidenced by studies of rRNA sequences.
14. Cyanobacteria are nutritionally independent; they can be photosynthetic, rely on fermentation, or act as nitrogen fixers. Although prokaryotes, they contain an elaborate system of internal membranes. They may exist colonially or live free as single cells. Filamentous forms show at least three cell types: vegetative cells, spores, and *heterocysts* (which serve as break points during reproduction and also as deposition sites for the oxygen-sensitive nitrogenase enzymes needed in nitrogen fixation). Fission is the only form of reproduction in cyanobacteria.
15. The phylum of *gram-positive bacteria* has three subgroups: the gram-positive rods, gram-positive cocci, and the actinomycetes.
16. *Gram-positive rods* fall into two groups: those that make highly resistant endospores, and those that do not. The first group includes the causative agent of anthrax, as well as species that produce potent toxins responsible for tetanus and botulism food poisoning in humans. The second group includes members of the genus *Lactobacillus*, a lactic acid producer.
17. *Gram-positive cocci* are abundant and common both outside and inside the human body. The genus *Staphylococcus* contains species causing respiratory, intestinal, and skin diseases, as well as ones that produce toxins responsible for toxic shock syndrome. *Streptococcus* species can infect virtually every organ system in the human body. One member of this genus was used by Avery et al. for the first experimental demonstration that DNA is the hereditary material.
18. *Actinomycetes* possess a highly branched mat of filaments similar to that found in fungi, with which they were once classified. Some reproduce by forming spores, others by fission. The genus *Streptomyces* produces natural chemical defenses, including streptomycin and other antibiotics.
19. Members of the Eubacterial phylum *mycoplasmas* entirely lack cell walls. They are the smallest cellular organisms known, and have less than half the DNA typical of other prokaryotes. This may represent the minimum level of genetic material necessary to maintain cellular existence. Virtually all mycoplasma species are parasites of animal or plant cells. However, since they lack cell walls, traditional antibiotics are ineffective at fighting their attack.

Activities (for answers and explanations, see page 205)

- For each of the bacterial types listed below, name the kingdom and phylum to which they belong, and cite one specific taxonomic characteristic of that phylum.

  a. Mycoplasmas

  b. Gram-positive rods

  c. Cyanobacteria

  d. Methanogens

  e. Gram-negative cocci

  f. Thermoacidophiles

- Briefly describe what is meant by each of the following terms when used with reference to bacteria:

  a. Axial filament

  b. Antibiotic

  c. Fruiting body

  d. Heterocyst

Questions (for answers and explanations, see page 205)

1. Bacterial species capable of producing toxic substances or disease in other organisms belong to which of the following groups?
   a. Gram-positive cocci
   b. Chlamydias
   c. Curved bacteria
   d. Spirochetes
   e. All of the above

2. All of the following bacterial groups belong to the gram-negative phylum *except* the
   a. methanogens.
   b. spirochetes.
   c. rickettsias.
   d. cyanobacteria.
   e. gliding bacteria.

3. Members of the phylum mycoplasmas are uniquely characterized by
   a. being among the largest of all known bacteria.
   b. having more DNA than 80% of all other bacteria.
   c. being the only bacterial group without parasitic forms.
   d. entirely lacking cell walls.
   e. None of the above

4. Which of the following statements correctly applies to methanogen bacteria?
   a. They do not typically cause disease.
   b. They are prokaryotes.
   c. They possess cell walls.
   d. They are representatives of the Archaebacteria.
   e. All of the above

5. *Escherichia coli*, a common bacterial resident in the human gut, is a member of which kingdom and phylum of prokaryotes?
   a. Archaebacteria; gram-negative rods
   b. Archaebacteria; thermoacidophiles
   c. Eubacteria; gram-positive cocci
   d. Eubacteria; gram-negative rods
   e. Eubacteria; rickettsias

6. Bacteria of the genus *Streptomyces* are particularly notable because they
   a. are important members of the phylum cyanobacteria.
   b. form colonial aggregations and produce fruiting bodies.
   c. locomote by means of flagella which contain axial filaments.
   d. are the causative agents of the veneral disease gonorrhea.
   e. produce antibiotics as chemical defenses.

VIRUSES (pages 497–500)

Key Concepts

1. *Viruses* are even smaller than bacteria, and have only become well understood in the past 50 years. Although structurally simple, they are neither primitive nor ancient. Viruses most likely evolved from the host plants, animals, and bacteria they infect.
2. Viruses are classified as *acellular*. They are not cells, nor do they consist of cells, and they do not metabolize energy.
3. Viruses are *obligate intracellular parasites*. They never arise directly from preexisting viruses, but depend instead on their specific host cells for reproduction and development.
4. When outside the host cell, viruses are known as *virions*. Each virion consists of a central core of DNA or RNA surrounded by a *capsid*, which contains one or more types of protein. The assembly of capsid proteins gives virions a characteristic shape which is used to classify viruses.
5. Some viruses may also possess membrane coatings derived from the host cell, or protein "tails" used to attach the virus to its host.
6. Viruses reproduce by taking over the host cell's metabolism, with the viral nucleic acid directing the host's biochemistry to make new viruses.

7. Animal viruses enter the host cell by endocytosis, and most often emerge by budding through virus-modified regions of the plasma membrane.
8. Plant viruses often enter their host cells by way of an animal *vector* (an intermediate carrier of the virus) usually through feeding or reproductive activity. Viruses may also penetrate the plant cell wall as a result of mechanical damage caused by weather or abrasion.
9. Bacteriophages are specialized bacterial viruses equipped with tail assemblies which can inject viral DNA through bacterial cell walls. Virions escape from plant and bacterial cells by lysing the host cell.
10. Some bacteriophages exhibit lytic life cycles, where host cells rapidly lyse as new viruses form. Others show lysogenic life cycles, where host and viral DNA (a provirus) replicate together for many cell generations.
11. Virion recognition of host cells remains poorly understood, but likely involves specific interactions between viral proteins and the host cell's membrane or cell wall.
12. Viral classification is based on many characteristics including nucleic acid composition (DNA or RNA, single- or double-stranded, one or more molecules, linear or circular), overall shape of the virus, presence or absence of a membranous envelope around the virion, capsid symmetry, and capsid size.
13. Three capsid shapes are recognized: *helical* (coiled), *icosahedral* (regular solid with 20 faces), and *binal* (polyhedral, many-faceted head with helical tail).
14. Susceptibility to viral infection varies within groups of host organisms. Flowering plants are more susceptible than cone-bearing plants, ferns, algae, or fungi. In animals, vertebrates and arthropods are most susceptible.
15. *Arboviruses* are transmitted to mammalian hosts by the bite of insect vectors (e.g., mosquitoes), which are themselves not severely affected by the virus.
16. *Viroids* are infectious agents that consist only of nucleic acid; i.e., they lack the normal viral protein coat. Each viroid is made of a relatively short segment of RNA (270–380 nucleotides) folded into double-stranded rods.
17. Viroids have so far been isolated only from plant cells, where they cause a variety of diseases. Viroids may be passed between plants by physical contact, or through infected pollen and ovules. Since viroids apparently are not translated into proteins, how they cause disease remains unknown. Evidence suggests that their origin may be by evolution from transposable elements.
18. Prions, also known as *scrapie-associated fibrils*, are viruslike entities consisting entirely of protein with no trace of a nucleic acid component. Their mechanism of action is unknown, but they may act as proviruses. Scrapie-associated fibrils have been found in association with central nervous system diseases such as scrapie. Several human nervous system diseases, including kuru, Creutzfeldt-Jakob disease, and Alzheimer's disease, may also be caused by these fibrils.

Activities (for answers and explanations, see page 206)

- What exactly do biologists mean when they describe viruses as "acellular" and "obligate intracellular parasites"?

- Discuss the difference between a virus and a virion, and describe the role of the capsid in viral structure.

Questions (for answers and explanations, see page 206)

1. Viruses are considered "nonliving" for the primary reason that they lack
   a. the ability to self-replicate.
   b. cell walls.
   c. DNA or RNA.
   d. membrane-bounded organelles.
   e. a covering of proteins and lipids.

2. Which statement about viruses or their relatives is *not* true?
   a. Viruses are usually highly specific about the host cells they infect.
   b. Viroids consist only of proteins and lack any trace of nucleic acid.
   c. The nucleic acid of some viruses is DNA, and in others is RNA.
   d. During the lysogenic cycle, viral DNA incorporates itself into the host cell's genetic material.
   e. Scrapie-associated fibrils are viruslike entities that consist only of proteins.

3. Which of the following is *not* a criterion normally used to classify viruses?
   a. Symmetry of the protein capsid
   b. Whether or not the virus is an intracellular parasite
   c. Overall shape of the virus
   d. Characteristics of the nucleic acid core

4. Which of the following statements about viroids is *not* true?
   a. Viroids are found only in plants.
   b. RNA is the only nucleic acid found in viroids.
   c. Like arboviruses, viroids are transmitted by insect bites.
   d. Plant reproduction may be a way that viroids are spread.
   e. Unlike most viruses, viroids lack a capsid.

5. Which of the following would be the most likely vector for an arbovirus?
   a. Bacterium
   b. Mammal
   c. Plant
   d. Bird
   e. Insect

6. Which of the following human diseases is believed to be linked with a scrapie-associated fibril?
   a. Alzheimer's disease
   b. Kuru
   c. Creutzfeldt-Jakob disease
   d. All of the above

Integrative Questions (for answers and explanations, see page 206)

1. In what ways is it accurate to say that the bacteria are "primitive" organisms, and in what ways is this a misleading statement?

2. In terms of overall size, the largest virus is ____________ the smallest bacterium.

3. A bacteriophage virus infects a gram-positive rod bacterium. How many kingdoms of organisms are involved in this interaction?

4. Below is a list of potential habitats in which bacteria may be found. For each habitat, name a likely bacterial group that could reside there.
   a. Hot spring ____________
   b. Intestines of a cow ____________
   c. Intracellular parasite ____________
   d. Skin surface ____________
   e. Alkaline lake ____________

23

Protists

CHAPTER LEARNING OBJECTIVES—after studying this chapter you should be able to:

- ❑ State the general characteristics that separate the kingdom Protista from all other kingdoms.
- ❑ Identify the main features of the animallike, funguslike, and plantlike protists and state their evolutionary relationships to the other eukaryotic kingdoms.
- ❑ Describe the diversity of structure, locomotion, reproduction, and modes of nutrition in protists.
- ❑ Define endosymbiosis and explain the types of endosymbiotic relationships in which protists are involved.
- ❑ Explain the concept of alternation of generations, and differentiate between isomorphic and heteromorphic forms of these life cycles.
- ❑ Define and correctly apply the following terms with respect to protists: "gamete," "spore," "zygote," "gametophyte," "sporophyte," and "sporocyte."
- ❑ Discuss the distinguishing characteristics of the following protozoan phyla: Zoomastigophora, Rhizopoda, Actinopoda, Foraminifera, Apicomplexa, and Ciliophora.
- ❑ Describe the basic life cycle of the human malarial parasite *Plasmodium.*
- ❑ Identify the structures and functions of a typical ciliate body, and explain the method of locomotion in these protists.
- ❑ Discuss the distinguishing characteristics of the following funguslike protistan phyla: Myxomycota, Acrasiomycota, Protomycota, and Oomycota.
- ❑ Differentiate between a plasmodium and a pseudoplasmodium, and apply these terms to the concept of a coenocytic body plan in the funguslike protists.
- ❑ Define and correctly apply the following terms to the funguslike protists: "cytoplasmic streaming," "saprobic," "mycelium," "hyphae," and "rhizoids."
- ❑ Discuss the distinguishing characteristics of the following algal phyla: Pyrrophyta, Chrysophyta, Euglenophyta, Phaeophyta, Rhodophyta, and Chlorophyta.
- ❑ Relate the photosynthetic activity of algae to the total amount of photosynthesis that occurs on Earth.
- ❑ Identify the kinds of algal protists whose bodies have contributed to various forms of sedimentary rock formations.
- ❑ Explain the asexual and sexual reproduction of diatoms.
- ❑ Describe the purpose of the phaeophyte structure known as a holdfast, and state how this and other structures have allowed the brown algae to inhabit high energy habitats.
- ❑ Correctly apply the concepts of isomorphic and heteromorphic alternation of generations to the various forms of algae.
- ❑ Discuss the role of chromatic adaptation in the red algae, and explain its effects on the color of these protists.
- ❑ Describe the difference between isogamous and anisogamous reproduction in algae.
- ❑ Differentiate between haplontic and diplontic heteromorphic alternation of generations in the green algae.
- ❑ Discuss the evolutionary trends observed in the transition from simple photosynthetic protists to the more complex forms of algae.

PROTISTA AND THE OTHER EUKARYOTIC KINGDOMS • GENERAL BIOLOGY OF THE PROTISTS (pages 503–507)

Key Concepts

1. The kingdom *Protista* includes only eukaryotic organisms. Most are unicellular or colonial, but some are multicellular and achieve large sizes (e.g., giant kelps).
2. The protist kingdom is a grab bag of organisms, formed by exclusion; the best general definition for protists is all eukaryotes not included among the plants, fungi, or animals.
3. All the other eukaryotic kingdoms are believed to have evolved from protists. The three major groups of protists reflect these evolutionary relationships: the *protozoans* (animallike), the *algae* (plantlike), and the *funguslike protists.* The taxonomic line between each protist group and its related eukaryotic kingdom is often indistinct.

4. There are three important differences between fungi and the funguslike protists. Fungi do not possess flagella nor do they have visibly different male and female gametes, whereas funguslike protists do, and fungal cell walls contain chitin, which is absent from the cell walls of most funguslike protists.
5. The main difference between plantlike protists (algae) and the plants is reproductive; plant embryos develop inside protective tissues formed by the parent plant, algae do not.
6. Animals are the eukaryotes defined as multicellular ingestive heterotrophs. Animallike protists (protozoans) are best characterized as unicellular ingestive heterotrophs.
7. Most protists are aquatic and include both freshwater and marine species, as well as blood parasites. Others live in moist organic matter or damp soil.
8. As with bacteria, protists exhibit considerable metabolic diversity. Some are autotrophs, others absorptive heterotrophs, and still others ingestive heterotrophs. Some can switch between autotrophic and heterotrophic modes.
9. Means of locomotion are also diverse among protists, ranging from completely nonmotile to amoeboid movement and sophisticated ciliary or flagellar motion.
10. Freshwater protozoans have a more negative osmotic potential than their hypotonic environment, and must therefore constantly pump out water gained through osmosis. These protists use specialized membranous vesicles called *contractile vacuoles* to cope with excess water. Experiments show that the rate at which the contractile vacuole operates changes in accordance with the rate of osmotic water flow into the cell.
11. Food vacuoles are another type of vesicle found in protists. These vacuoles form around engulfed food particles; digestion then occurs within the vesicle. Smaller vesicles then develop that increase the surface area for absorption of digested food into the cytoplasm.
12. The exact nature of the protist cell surface varies among taxonomic groups. Some groups have a highly flexible plasma membrane only, while others have additional coverings such as cell walls and various forms of secreted or accreted shells. Many protists also possess specialized organelles that make them sensitive to environmental stimuli such as light, chemicals, temperature, and touch.
13. *Endosymbiosis*, where one organism lives within the cells of another, is quite common among protists. Such relationships often, but not always, lead to mutualistic benefits of metabolism and protection for the protist and/or its partner. In protistan endosymbiosis, the partner is almost always a bacterium or another protist.
14. Reproduction takes many forms among the protists. Asexual reproduction can employ binary fission, multiple fission, budding, or spore formation. Sexual reproduction may result in either a haploid or diploid multicellular generation, or both. Some protists practice a form of genetic recombination not directly related to reproduction.
15. *Alternation of generations* occurs in the life cycle of some protists and yields both multicellular diploid ($2n$) and multicellular haploid (n) stages. Mature diploid individuals produce spores, while mature haploid individuals produce gametes. If the stages are morphologically and genetically different from one another, the system is called *heteromorphic alternation of generations*. When the stages do not differ morphologically (even though they are genetically different), the system is known as *isomorphic alternation of generations*.
16. In multicellular plants, fungi, and protists, gametes are generally not produced directly by meiosis, as they are in animals. Instead, the multicellular haploid *gametophyte* generation makes gametes by mitosis and cytokinesis.
17. Fusion of the gametes produces a diploid *zygote*, which then undergoes mitotic divisions to produce the diploid *sporophyte* generation. Sporophytes produce haploid spores through meiosis, which occurs in specialized cells called *sporocytes*.

Activities ***(for answers and explanations, see page 206)***

- In the space provided below, construct a simple diagram that depicts the supposed evolutionary relationships between the kingdoms Archaebacteria, Eubacteria, Protista, Fungi, Plantae, and Animalia.

- Discuss the importance of the contractile vacuole to freshwater protists, and predict the change in rate of vacuole contraction that would occur after a protist is moved from an isotonic solution to a hypertonic solution.

Questions ***(for answers and explanations, see page 206)***

1. Which one of the following characteristics is *not* present in at least some members of the kingdom Protista?
   a. Prokaryotic cells
   b. Absorptive heterotrophy
   c. Endosymbiosis
   d. Multicellularity
   e. Alternation of generations

2. The main difference(s) between animallike protists and animals is that
   a. animals are prokaryotes and protists are eukaryotes.
   b. animals are photoautotrophs and protists are chemoautotrophs.
   c. animals are multicellular ingestive heterotrophs and protists are unicellular ingestive heterotrophs.
   d. animals never exhibit mutualistic interactions, whereas many protists do.
   e. All of the above

3. Which of the following is true of endosymbiotic protists?
   a. The endosymbiotic relationship only forms with plants.
   b. The partner in the relationship is typically a bacterium or another protist.
   c. Very few protists engage in such relationships.
   d. Endosymbiotic protists usually reside on the exterior of their partner.
   e. The endosymbiotic relationship only lasts during the partner's breeding season.

4. Suppose you observe a freshwater protist resting in a sample of pond water for five minutes and count the number of times its contractile vacuole expels water from the cell. You then put this same protist into a solution of unknown solute concentration for the same amount of time and count the number of contractions, which is now twice what it was before. Which of the following *best* describes the nature of the second solution?
   a. Pond water
   b. Salt water
   c. Distilled water
   d. A 1:1 mixture of pond water and salt water

5. Which of the following modes of reproduction can be found in at least some protists?
   a. Binary fission
   b. Sexual reproduction
   c. Budding
   d. Spore formation
   e. All of the above

6. Which of the following statements about the kingdom Protista is *not* true?
   a. Some protists locomote by means of ciliar or flagellar action.
   b. The sporophyte stage in some protists produces spores by mitosis.
   c. The sporophyte stage in some protists is formed by mitotic division of the zygote.
   d. Some protists switch between autotrophic and heterotrophic modes of metabolism.
   e. Both isomorphic and heteromorphic alternation of generations is known among the protists.

PROTOZOANS (pages 507–513)

Key Concepts

1. Protozoans are the unicellular protists that ingest their food primarily by endocytosis.

2. The phylum *Zoomastigophora* represents the largest (>10,000 species) and probably most ancient group of protozoans. All possess at least one flagellum, and hence have the common name "zooflagellates." They reproduce asexually by mitosis and cytokinesis.

3. Some zooflagellates are free-living predators that eat other protists, while others are specialized as internal parasites of animals such as insects and humans.

4. The anatomical structure of the zooflagellate group Choanoflagellida closely resembles a characteristic cell type found in sponges. The Choanoflagellida are therefore thought to be the evolutionary ancestors of this primitive animal group.

5. Some zoomastigophorans are important human pathogens. A few are carried by insect vectors, including members of the genus *Trypanosoma*, which cause a lethal sleeping sickness in Africa. Others (e.g., *Giardia*) contaminate water supplies and produce intestinal disorders. Still others (e.g., *Trichomonas*) are transmitted by sexual contact and cause venereal disease.

6. The phylum *Rhizopoda* includes the amoebas and their relatives, noted for the ability to form flexible extensions of their body mass, called pseudopods, which serve both in locomotion and feeding.

7. All Rhizopoda are animallike and exist as predators, parasites, or scavengers. They assimilate food by phagocytosis after surrounding it with pseudopods. Many are aquatic and highly adapted for life as bottom-dwellers, where a rich food supply is present.

8. Amoebas are not primitive organisms; their apparent simplicity is secondarily derived. They may have evolutionary connections to the zoomastigophorans, as evidenced by living forms that possess both pseudopods and flagella (*Mastigamoeba*), and those that retain flagellated and amoeboid stages in the life cycle (*Naegleria*).

9. Members of the phylum *Actinopoda* are all aquatic and have very thin pseudopods reinforced with microtubules. These structures greatly increase the surface area for feeding and chemical exchange with the environment, and also help keep the organism afloat.

10. Radiolarians are a group of drifting marine actinopods. Their silica-based shells are often quite large and elaborate. Over hundreds of millions of years these shells have accumulated to form significant sedimentary deposits in tropical oceans. Heliozoans are the freshwater actinopods; they differ mainly from the radiolarians with regard to internal shell structure.

11. The phylum *Foraminifera* consists of drifting marine protozoans with porous calcium carbonate shells. Threadlike nets of sticky pseudopods emerge from these pores and are used to capture planktonic food items. Foraminiferans shed their old shells after mitotic reproduction. Massive numbers of discarded shells have contributed over millions of years to extensive limestone deposits on the ocean floor and to sand beaches.

12. The protozoan phylum *Apicomplexa* is so named because its members produce spores whose apical end contains many organelles. They are exclusively parasitic, have an amoeboid body form (although are unrelated to the Rhizopoda), and lack a contractile vacuole. Osmotic cell expansion is limited by rigid cell walls.

13. Human malarial parasites of the genus *Plasmodium* are apicomplexans. They exhibit complex life cycles involving multiple hosts (humans and female *Anopheles* mosquitoes), through which they cycle during different developmental stages.
14. *Plasmodium* parasites enter humans through the bite of an infected female mosquito. The parasites take up residence in the liver and lymph system, and then change form before reentering the bloodstream to attack red blood cells. When bitten by another uninfected mosquito, the parasites are transferred from the human back to the insect and the life cycle is completed. This cycle is best interrupted by eliminating breeding sites for the mosquito host.
15. The ciliates, members of the phylum *Ciliophora*, are named for the presence of hairlike cilia on their cells. The phylum represents a highly diverse and ecologically important group of protozoans.
16. Ciliates are all heterotrophic and have highly specialized body forms. Two types of nuclei are found in the ciliate cell: a large *macronucleus* whose DNA controls cell function, and up to 80 *micronuclei* that play an essential role in genetic recombination.
17. *Paramecium* is a representative ciliate genus. These organisms possess a *pellicle*, which is an elaborate cell covering made from the outer plasma membrane and many associated alveoli, through which individual cilia protrude. The pellicle also contains needlelike defensive organelles called trichocysts, which can be rapidly expelled.
18. Although some of the cilia in paramecia are sensory in function, most aid in locomotion. The beating motion of these cilia can be directed to provide precise locomotory control, allowing the organism to move forwards, backwards, and to alter its direction in response to changing environmental conditions.
19. Reproduction in paramecia is usually by simple cell division. However, they also engage in an elaborate, nonreproductive sexual behavior called *conjugation*, where two cells reciprocally exchange genetic material derived from their micronuclei. After conjugation, the genetically recharged individuals separate to undergo asexual cell division.
20. Some ciliates (along with other complex protists) exhibit such extreme specialization that they almost appear to function like multicellular animals. Specialized ciliate organelles include *cirri* (fused cilia) which permit highly coordinated walking movements, almost like animal appendages. *Neurofibrils* are nerve-like fibers that control the movements of cirri. *Myonemes* act like muscles within the cytoplasm to cause rapid movement of the entire cell or specific cell structures.
21. Finally, some very highly specialized ciliates which inhabit the digestive systems of hoofed mammals possess not only cirri, neurofibrils, and myonemes, but also structures analogous to skeletons and gut systems.

Activities (for answers and explanations, see page 206)

- Correctly associate the letters of each of the ciliate structures and functions listed below with the unlabeled lines on the following diagram of *Paramecium*.

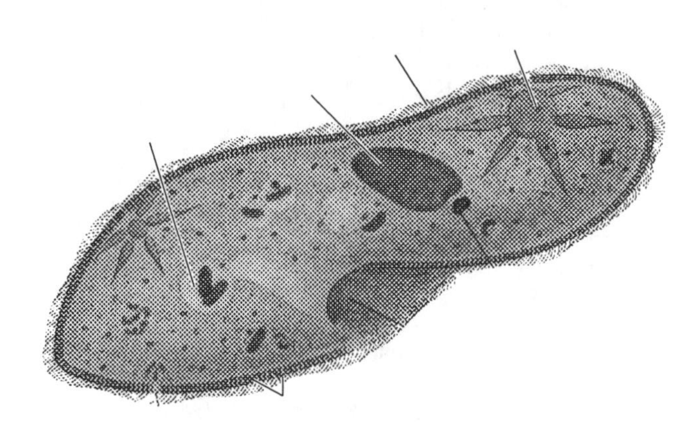

Structures	Functions
a. Food vacuole	*f.* Regulates water balance of cell
b. Pellicle	*g.* Nutrient digestion
c. Macronucleus	*h.* Cell covering to which cilia attach
d. Contractile vacuole	*i.* Defensive structures
e. Trichocysts	*j.* Contains DNA that controls cell functions

- Besides flagella, what other locomotory-related structures are found in protozoans?

Questions (for answers and explanations, see page 207)

1. Which one of the following statements about protozoans is *not* true?
   a. All protozoans are capable of ingesting food by endocytosis.
   b. Apicomplexans are the only protozoan group without parasitic representatives.
   c. Foraminiferans and radiolarians are shelled protozoans.
   d. Some ciliates have organelles that permit coordinated "walking" movements.
   e. Although they appear structurally simple, amoebas are not considered primitive organisms.

2. All of the following are members of the phylum Zoomastigophora, *except*
   a. Giardia.
   b. Trypanosoma.
   c. Trichomonas.
   d. Choanoflagellida.
   e. Plasmodium.

3. The many species of actinopods, including foraminiferans, heliozoans, and radiolarians, all share which of the following characteristics?
   a. They are photosynthetic.
   b. They use pseudopods in locomotion or feeding.
   c. They have calcium carbonate shells.
   d. They are parasites.
   e. They are flagellates.

4. Trichocysts are defensive structures found in the protozoan phylum known as
   a. Zoomastigophora.
   b. Rhizopoda.
   c. Apicomplexa.
   d. Ciliophora.
   e. All of the above

5. Microscopic examination of a ciliate could conceivably reveal all but one of the following structures. Select the *exception.*
   a. Pseudopod
   b. Cirrus
   c. Trichocyst
   d. Pellicle
   e. Myoneme

6. Although not closely related, members of the phyla Rhizopoda and Apicomplexa share a similar __________ body form.
   a. shelled
   b. ciliated
   c. flagellated
   d. amoeboid
   e. None of the above

FUNGUSLIKE PROTISTS (pages 513–517)

Key Concepts

1. Unlike true fungi, all funguslike protists possess flagella during some stage of their life cycle. They also lack the dikaryotic stage typical of true fungi, and most have no chitin in their cell walls.
2. The phylum *Myxomycota* includes the acellular slime molds, while the phylum *Acrasiomycota* is home to the cellular slime molds. At one time these superficially similar groups were lumped into a single phylum. In both phyla, motile independent cells undergo striking organizational changes and aggregate to form fruiting bodies, which then produce spores. For ease of movement, slime molds prefer moist, cool habitats like forests.
3. The vegetative state of an acellular slime mold is called the *plasmodium*, a wall-less mass of cytoplasmic strands containing numerous diploid nuclei. This exemplifies the *coenocyte* body plan, in which many nuclei are surrounded by a single plasma membrane.
4. Myxomycetes move by means of *cytoplasmic streaming*. In this type of locomotion one region of the outer cytoplasm becomes more fluid, causing the plasmodium to bulge in that area. Contractile proteins and microfilaments help control these streaming movements. Overall growth of the plasmodium can be almost indefinite under favorable conditions of food, moisture, and pH.
5. If conditions become unsuitable for continued growth, the plasmodium may either form a resistant structure called a *sclerotium* (which can revert back into a plasmodium upon the return of favorable conditions), or it may transform into a branched, spore-producing fruiting body called a *sporangiophore*.
6. Sporangiophores emerge from piled masses of the plasmodium. Diploid nuclei within each fruiting body divide meiotically and collect within *sporangia* located at the tip of the sporangiophore. The haploid nuclei become surrounded by cell walls to form resistant spores.
7. These haploid spores are eventually shed and germinate, at which time they become flagellated *swarm cells*, which function as gametes. These can either assume a resting state (cysts) or fuse to create a diploid zygote that divides mitotically without wall formation to become a new, coenocytic plasmodium.
8. Unlike the acellular slime molds, the vegetative state of cellular slime molds (phylum Acrasiomycota) is amoeboid. These *myxamoebas*, each containing a single haploid nucleus, engulf food particles and can exist indefinitely under favorable conditions. They reproduce asexually by mitosis and fission.
9. When conditions become unfavorable, myxamoebas aggregate to form fruiting bodies in response to cAMP signals. The independent cells gather into a structure called a *pseudoplasmodium*, from which the fruiting bodies emerge, but each myxamoeba retains its individual identity. A pseudoplasmodium is therefore not coenocytic.
10. Haploid spores form within thick-walled cells on these fruiting bodies. The spores then germinate and produce new myxamoebas to complete this asexual reproductive cycle. Sexual reproduction also occurs in cellular slime molds. This process involves the direct fusion of two myxamoebas into a structure that later germinates to form new myxamoebas.
11. The phylum *Protomycota* contains the chytrids and hypochytrids. Both are aquatic, and have either parasitic or *saprobic* (feeding on dead organic matter) lifestyles. Many are unicellular; others develop into massive chains of filamentous branching cells called *mycelia* whose cell walls consist mainly of chitin. Hypochytrids employ only asexual reproduction, while chytrids can reproduce either asexually or sexually.
12. The chytrid genus *Allomyces* exhibits isomorphic alternation of generations. The diploid stage produces *zoospores*, which are diploid flagellated cells that arise by mitosis. Haploid zoospores are also produced by meiosis. Each type of zoospore mitotically gives rise to an adult organism of that same ploidy.
13. Mature haploid chytrids produce gametes in specialized *gametangia* located at the tips of the filaments (*hyphae*) that make up the mycelium.
14. Certain hypochytrids (e.g., *Rhizidiomyces*) parasitize their host by developing a massive network of *rhizoids*, rootlike structures that invade the host's body and help procure nutrients. Zoospores then form in sporangia and are released to infect a new host.

15. The phylum *Oomycota*, commonly known as water molds, is frequently the target of parasitism by hypochytrids. As their name suggests, water molds (e.g. *Saprolegnia*) are aquatic and saprobic, and are often found feeding on dead animals. They are coenocytic, have flagellated reproductive cells, are diploid through most of the life cycle, and have cellulose cell walls.

16. Terrestrial species of Oomycetes are also known as the downy mildews. One member of the genus *Phytophthora* causes late potato blight, the disease that was responsible for the Irish potato famine of the mid-1840s.

Activities (for answers and explanations, see page 207)

- What do biologists mean when they refer to a funguslike protist as being coenocytic? How does the vegetative state of an acellular slime mold (the plasmodium) exemplify this concept? Why is the pseudoplasmodium of a cellular slime mold not considered coenocytic?

- Explain the concept of alternation of generations as it applies to both the morphology and the genetics of funguslike protists.

Questions (for answers and explanations, see page 207)

1. A major difference between the *vegetative* states of acellular and cellular slime molds is that
   a. acellular slime molds have haploid nuclei while cellular slime molds are diploid.
   b. acellular slime molds produce fruiting bodies and cellular slime molds do not.
   c. acellular slime molds use cAMP as a chemical cue for aggregation while cellular slime molds do not.
   d. acellular slime molds form a large, coenocytic mass of cytoplasm, whereas cellular slime molds usually exist as individual myxamoebas.

2. One characteristic of the chytrid genus *Allomyces* is isomorphic alternation of generations, which means that the haploid and diploid generations
   a. are indistinguishable morphologically.
   b. are radically different morphologically.
   c. only appear every other generation.
   d. alternate between saprobic and parasitic lifestyles.
   e. None of the above

3. Which of the following are *not* members of the protist phylum Oomycota?
   a. Downy mildews
   b. Hypochytrids
   c. Water molds
   d. *Phytophthora infestans*

4. All but one of the following processes or events occurs in at least one representative of the funguslike protists. Select the *exception.*
   a. Engulfing particulate foods
   b. Zoospore production via mitosis from diploid cells
   c. Sporangiophore development from the plasmodium stage
   d. Rhizoid development that penetrates a parasitized host's body
   e. Sexual reproduction by means of conjugation

5. Which of the following characteristics or structures of the funguslike protists is *incorrectly* associated with the phylum in which it occurs?
   a. Myxamoebas....................................Acrasiomycota
   b. Sclerotium......................................Myxomycota
   c. Gametangia....................................Myxomycota
   d. Hyphae...Protomycota
   e. Hypochytrid parasitism................Oomycota

6. Some hypochytrid protists improve their chances of acquiring nutrients with the aid of
   a. rootlike structures called rhizoids.
   b. a plasmodium.
   c. a pseudoplasmodium.
   d. a coenocyte body plan.
   e. large numbers of zoospores.

ALGAE (pages 517–526)

Key Concepts

1. All algae are eukaryotic and photosynthetic. As a group, they are responsible for over half of the total photosynthesis that occurs on Earth.

2. Unlike true plants, the zygotic stage in the algal life cycle is an independent cell(s) that receives no parental protection during development.

3. Extreme diversity is the rule for algal growth forms, which can be unicellular and relatively simple, or multicellular and very complex.

4. Algal life cycles are also highly varied. However, all phyla except for the Rhodophyta exhibit flagellated motile cells during at least one stage in the life cycle.

5. Biochemical characteristics are often used when classifying algae. The specific kinds of polysaccharide compounds used for storage, and the substances found in cell walls are important in this regard.

6. The phylum *Pyrrophyta* is dominated by the dinoflagellates, unicellular algae whose unique mixture of accessory and photosynthetic pigments give them a golden-brown color. They are of enormous importance in

oceanic photosynthesis, and are common endosymbionts with corals.

7. Dinoflagellates have a peculiar anatomy, with two prominent grooves that run along the cell surface at right angles to one another. Each groove contains a flagellum that helps propel the alga through the water.

8. Enormous population explosions of dinoflagellates (e.g. *Gonyaulax*) cause "red tides," with a resultant accumulation of potent nerve toxins that may kill fish and other vertebrates, or accumulate in shellfish. Many species are also bioluminescent.

9. The phylum *Chrysophyta* contains the diatoms and their relatives. Most are single-celled and contain carotenoid pigments that give them a yellow or brownish color. Their principal storage products are the carbohydrate chrysolaminarin and several oils.

10. Most diatoms have radially symmetrical silicon shells that fit together like the two halves of a box. Each species' shell is so unique that it is the only taxonomic character used to identify diatoms.

11. When reproducing asexually, each half of the diatom shell serves as the top half of the new organism. To avoid the ever-decreasing cell size that would result if this process continued indefinitely, sexual reproduction takes place. Gametes form, shed their cell walls, and fuse to make a zygote, which then grows substantially before a new cell wall is produced.

12. Like other marine protists with mineralized shells, diatoms have contributed to massive oceanic sedimentary rock formations. Diatomaceous earth, obtained from such rocks, has many industrial uses.

13. The phylum *Euglenophyta*, sometimes grouped with the zoomastigophorans, contains unicellular flagellates. Although they lack the cell walls characteristic of algae, their photosynthetic abilities and mode of reproduction justify their inclusion with the other algal phyla.

14. Typical representatives of the Euglenophyta are members of the genus *Euglena*. These common freshwater protists have a complex body plan, multiple flagella, and reproduce asexually by mitosis and cytokinesis. They are nutritionally flexible, being able to survive on photosynthetic activity while in the light and heterotrophy when in the dark.

15. Brown algae make up the phylum *Phaeophyta*. They earn their name from the combination of a yellow-brown carotenoid pigment called fucoxanthin and the green of chlorophyll. They are always multicellular and have branched filaments or leaflike growths called *thalli*.

16. Phaeophyta include relatively small organisms such as *Ectocarpus* as well as the giant kelps of the genus *Macrocystis*, the largest of all algal species. Some are free-floating on the ocean's surface (e.g. *Sargassum*), while others are glued to the substrate by means of *holdfasts*. These holdfasts, coupled with specialized body forms resulting from *heteromorphy* within a single species, allow some brown algae to survive in high energy habitats such as coastal intertidal zones.

17. Considerable tissue differentiation is observed in the larger brown algae. In addition to stalks and blades that are analogous to the stems and leaves of plants, photosynthetic and food-conducting filaments resembling those of plants are also present in some species of Phaeophytes.

18. Both isomorphic and heteromorphic alternation of generations are seen in the brown algae. In the isomorphic forms, sporophyte and gametophyte are morphologically indistinguishable; an example is the brown alga *Ectocarpus*. In the heteromorphic species (e.g. *Laminaria*), distinct anatomical differences are evident between stages in the life cycle. In *Fucus* this is taken to the extreme, with the multicellular haploid phase being eliminated altogether.

19. The cell walls of brown algae contain large amounts of a sticky sugar polymer called alginic acid. In nature it holds cells together and provides the glue used by holdfasts; commercially it is used as an emulsifier.

20. Members of the phylum *Rhodophyta* (red algae) are mostly multicellular and contain substantial amounts of the photosynthetically active red pigment phycoerythrin. In addition they have chlorophyll, phycocyanin, and certain other accessory pigments.

21. The quantity of phycoerythrin in a red alga varies depending on environmental light conditions. This happens through the process known as *chromatic adaptation*. While the amount of chlorophyll is the same in both surface and deeper-water species, the quantity of red pigment is reduced if the alga grows near the surface. In deeper water, where mostly blue-green light penetrates, phycoerythrin is more photosynthetically useful and therefore accumulates in the alga. The "redness" of the alga thus depends on the ratio of green chlorophyll to red phycoerythrin.

22. The use of floridean starch as a photosynthetic storage product, and the lack of motile flagellated cells in any stage of the life cycle are characteristics unique to the Rhodophyta. Agar, the substance used in laboratory cell and tissue cultures, is also derived from a mucilaginous polysaccharide made by certain red algae.

23. Some red algae can enhance the formation of coral reefs by secreting calcium carbonate in and around their cell walls. Moreover, some marine red algae have recently been found to parasitize other red algae by injecting their nuclei into the host cell cytoplasm. The host's metabolism is then diverted to express the parasite's genes.

24. The *Chlorophyta* (green algae) constitute the remaining phylum of plantlike protists. They are the only protists with the full complement of photosynthetic pigments characteristic of true plants, of which they are thought to be the evolutionary ancestors. Chlorophylls *a* and *b* dominate, along with carotenoids and xanthophylls. Starch is the principle photosynthetic storage product.

25. Virtually every imaginable shape and body construction is seen within the green algae, from simple unicellular to colonial to advanced multicellular. Both uninucleate and multinucleate forms are known. Life cycles are also diverse and sometimes complex.

26. In sea lettuce (*Ulva lactuca*), an isomorphic life cycle is observed. However, the male and female gametes produced by the haploid gametophyte are of distinctly different sizes (i.e., they are *anisogamous*, in contrast to the *isogamous* gametes of other algae and protists).
27. Other algae have heteromorphic life cycles. One variation on this theme is the *haplontic* life cycle (e.g. *Ulothrix*), in which the only diploid stage is the zygote itself (i.e., there is no multicellular diploid sporophyte). Zygotes act as sporocytes and directly undergo meiosis to produce spores.
28. Still other algae are *diplontic* (like animals) and show no multicellular haploid stage. In these algae, meiosis produces gametes directly.
29. Important evolutionary patterns in the algae include a trend towards multicellularity along with a trend from isogamy towards anisogamy. Oogamy is an extreme example of the latter shown by the giant green alga *Oedogonium*. Here the female gamete is large and totally immobile, while the male gametes are small, free-swimming, and flagellated.

Activities (for answers and explanations, see page 207)

- The algae encompass an enormous range of sizes, structural and metabolic complexities, and ecological lifestyles. List the characteristics that unify this otherwise diverse assemblage of organisms.

- Explain why, from a structural point of view, sexual reproduction is periodically necessary in diatoms.

- Review Figure 23.27 in your textbook (life cycle of *Ulva*) and then explain why this life cycle is considered to be isomorphic alternation of generations involving anisogamy.

Questions (for answers and explanations, see page 207)

1. Select all of the following algal phyla that show *both* isomorphic and heteromorphic alternation of generations.
   a. Chlorophyta
   b. Rhodophyta
   c. Phaeophyta
   d. Chrysophyta
   e. Pyrrophyta
2. Through the process known as chromatic adaptation
   a. brown algae regulate the manufacture of alginic acid depending on variations in wavelength of light.
   b. the quantity of phycoerythrin found in red algae can vary depending on light conditions.
   c. the quantity of chlorophyll *a* found in brown algae can vary depending on light conditions.
   d. red algae regulate the deposition rate of floridean starch depending on light conditions.
   e. diatoms make new silica shells as a function of light conditions.
3. Match each algal characteristic in the second column with an appropriate phylum from the first column.

____	Rhodophyta	*a.* Large amounts of phycoerythrin present
____	Chlorophyta	*b.* Made of two boxes fitted tightly together
____	Phaeophyta	*c.* Common endosymbionts with corals
____	Chrysophyta	*d.* Cell walls contain alginic acid
____	Pyrrophyta	*e.* Often glued to substrate via holdfasts
		f. Chlorophyll *a* the dominant pigment

4. Red tides often produce deadly toxins that result in major fish kills. This phenomenon results from enormous population increases by species from which of the following algal groups?
   a. Chlorophyta
   b. Rhodophyta
   c. Phaeophyta
   d. Chrysophyta
   e. Pyrrophyta
5. Which of the following molecules, structures, or characteristics can be found in *all* representatives of the Chlorophyta?
   a. Chlorophyll *a*
   b. Anisogamy
   c. Haplontic life cycle
   d. Diplontic life cycle
   e. Oogamy

6. Thalli, holdfasts, and alginic acid are all structures or substances characteristic of which algal phylum?
   a. Chlorophyta
   b. Rhodophyta
   c. Phaeophyta
   d. Chrysophyta
   e. Pyrrophyta

Integrative Questions (for answers and explanations, see page 207)

1. Your biology professor rushes into the classroom one day and announces that she has discovered a previously unknown organism, which she claims is a member of the kingdom Protista. Which of the following sets of characteristics would unequivocally place this new creature in with the other protists?
   a. Contains chlorophyll *a*, multicellular, photosynthetic, flagellated
   b. Diploid, multicellular, eukaryotic, switches between heterotrophy and autotrophy
   c. Eukaryotic, shows traits which exclude it from animal, plant, or fungal kingdoms
   d. Prokaryotic, possesses cilia, heterotrophic ingestive feeding
   e. All of the above

2. Consider the list of biological characteristics shown below. During a lifetime of studying all kinds of protists, Dr. Amoeba says that the only one of these characteristics that he has *never* observed in protists is
   a. mitosis.
   b. nitrogen fixation.
   c. endosymbiosis.
   d. blood parasitism.
   e. autotrophy.

3. A protist exhibiting heteromorphic alternation of generations with a haplontic life cycle would *not* have
   a. a multicellular haploid stage.
   b. a multicellular diploid stage.
   c. any gametes.
   d. any spores
   e. All of the above

4. A nonparasitic amoeba is discovered crawling on a freshwater green alga. Which two phyla of protists are involved in this situation?
   a. Ciliophora and Rhodophyta
   b. Foraminifera and Chrysophyta
   c. Apicomplexa and Pyrrophyta
   d. Actinopoda and Phaeophyta
   e. Rhizopoda and Chlorophyta

24

Fungi

CHAPTER LEARNING OBJECTIVES—after studying this chapter you should be able to:

- ❑ Cite the main taxonomic characteristics used to distinguish the fungi from all other kingdoms.
- ❑ Correctly assign representative fungi to their appropriate classes (Zygomycetes, Ascomycetes, Basidiomycetes, or Deuteromycetes), and provide the basic structural and functional characteristics of these classes.
- ❑ Discuss the term "absorptive heterotrophy" as it applies to saprobic fungi, and appreciate the ecological importance of fungi as decomposers in the environment.
- ❑ Define and provide examples of facultative and obligate parasitism in the fungi.
- ❑ Provide several examples of mutualistic relationships in fungi, describe the organism with which the fungus associates, and cite the benefits of these relationships to each organism.
- ❑ Understand the basic anatomy of a multicellular fungus, and correctly use the terms "hyphae," "mycelia," "rhizoids," and "haustoria."
- ❑ Describe four different mechanisms for asexual reproduction in fungi.
- ❑ Discuss the various stages in the life cycle of sexually reproducing fungi, and clearly distinguish between haploid and diploid conditions, mating types, dikaryotic cells, heterokaryotic cells, fruiting bodies, asci, and spores.
- ❑ Understand how conidia differ from other kinds of spores which are produced by fungi.
- ❑ Differentiate between euascomycete and hemiascomycete fungi, and explain the role of ascospores and perithecia in their life cycles.
- ❑ Discuss how yeast is an atypical fungus with regards to its anatomy, metabolic activity, and reproduction.
- ❑ Describe the basic anatomy of the basidiomycete fruiting body and explain the process of spore formation within this structure.
- ❑ Cite examples of how deuteromycete fungi are relevant to the lives of humans.
- ❑ Explain how a lichen differs structurally from a true fungus, and describe the ways that lichens reproduce.
- ❑ Discuss the reasons why lichens are able to survive in habitats considered too harsh for other fungi and plants, and describe how lichens contribute to soil formation.

GENERAL BIOLOGY OF THE FUNGI (pages 528–536)

Key Concepts

1. The kingdom Fungi encompasses all eukaryotic, heterotrophic organisms with absorptive nutrition. Most are multicellular.
2. Fungi play a crucial ecological role as decomposers of dead organic matter, helping return nutrients to various biogeochemical cycles.
3. Fungi inhabit all environments and attack virtually all other eukaryotes, including other fungi. Their life cycles are diverse and often complex.
4. The lifestyles of fungi fall into one of three categories: *saprobes* (live on dead matter), *parasites* (attack living matter), or *mutualists* (live in a mutually beneficial symbiotic association with other organisms).
5. All fungal species form spores and their cells never possess flagella, a feature that distinguishes them from the funguslike protists. Sexual reproduction is by *conjugation*.
6. The fungal cell wall is made of the polysaccharide *chitin*, also found in some funguslike protists and arthropod exoskeletons.
7. A single division *Eumycota* encompasses all four fungal classes: the *Zygomycetes* (conjugating fungi), the *Ascomycetes* (sac fungi), the *Basidiomycetes* (club fungi), and the *Deuteromycetes* (imperfect fungi).
8. Individual fungal cells form filaments called *hyphae*, which grow rapidly to form a larger body, the *mycelium*. Saprobic fungi use specialized *rhizoids* to anchor themselves to the substrate on which they feed. Parasitic fungi have modified hyphae called *haustoria* that penetrate living cells of their host's body to procure nutrients.
9. During reproduction, many fungi generate elaborate fruiting bodies, which produce and disseminate spores. Examples include puffballs, mushrooms, and brackets.
10. The fungal mycelium has an enormous surface-to-volume ratio, making it easy to absorb food and insuring that all cells within the mycelium are near the food source.

11. Fungi are tolerant of extreme environmental conditions, including strongly hypertonic surroundings and low or high temperatures.

12. Saprobic fungi, along with bacteria, are the major decomposers of the biosphere, a function made possible by their absorptive nutrition. Their preferred carbon sources are sugars, and they obtain nitrogen from proteins or the breakdown products of protein. Essential B vitamins and minerals are also obtained from their food.

13. Parasitic fungi may be either *facultative* or *obligate* parasites. Obligate parasites can grow only on their specific host organisms, which are usually plants.

14. Some parasitic fungi have *predatory* adaptations, such as sticky traps or constricting rings, which they use to capture prey. These prey are often tiny nematode worms. Once prey is captured, haustoria penetrate the prey to obtain nutrients.

15. *Lichens* are mutualistic associations of a fungus with either a unicellular alga or a cyanobacterium. The fungus provides protection and support, while the protist or bacterial partner provides photosynthetic products.

16. Other fungi form mutualistic associations with the roots of certain plants via structures called *mycorrhizae*. These structures cause the plant's root hairs to swell, which increases the surface area for absorption of nutrients into the roots. They also help the plant retain water. By helping the plant with its nutrient acquisition, the fungus also ensures food for itself. Seed germination in some plants (e.g., orchids) requires the presence of mycorrhizae.

17. Insects are sometimes involved in mutualistic interactions with fungi. Certain leaf-cutting ants use plant leaves or flowers as a substrate for fungal growth in their colony. These fungal "farms" are meticulously tended by the ants, who use the fungus as food. Scale insects use the coverings of another fungus species for protection from drying and to avoid predators. In the process, they provide food for the nonlethal parasitic fungus.

18. Asexual reproduction in fungi can occur in four ways. Haploid spores may be produced in specialized sporangia. Naked spores, or *conidia*, may also be formed outside of sporangia at the tips of individual hyphae. Unicellular fungi carry out simple cell division, which may produce two equivalent-sized cells or a tiny cell called a bud. Finally, some fungi reproduce by simple breakage of the mycelium, which then forms a new fungus.

19. Sexual reproduction in some fungi (e.g., the bread mold *Neurospora crassa*) begins when hyphae of opposite *mating types* come into contact. Their hyphal cell walls are digested, and eventually the two haploid nuclei reside in a single hyphal cell, a condition that is both *dikaryotic* ($n + n$) and *heterokaryotic* (because the two nuclei are different). The heterokaryotic cell then typically develops into a heterokaryotic mycelium.

20. Within the heterokaryotic mycelium specialized fruiting structures called *asci* develop. Here two haploid nuclei fuse to form a diploid zygote, which then undergoes meiosis to produce haploid spores. The spores become modified, depending on the type of fungus, and germinate to produce new hyphae.

21. Three important features of sexual reproduction in fungi are 1) the absence of motile gamete cells, 2) a liberation from liquid water habitats for reproduction, and 3) no true diploid tissue at any point in the life cycle (i.e., all cells except the zygote are either haploid or dikaryotic). Indeed, dikaryosis is perhaps the most significant genetic peculiarity of the fungi.

22. *Puccinia graminis*, a fungus that causes black stem rust in wheat, demonstrates three important points about the life cycles of fungal parasites: they are complex, they show extensive dikaryosis, and they often utilize multiple hosts.

23. *P. graminis* produces numerous spore types (*uredospores, teliospores, basidiospores, pycniospores,* and *aeciospores*) during its life cycle, all of which are well adapted for the peculiar ecological requirements of each stage. The spores include haploid, diploid, and dikaryotic types. Figure 24.7 in your textbook provides a detailed life cycle for *P. graminis* showing the different spore types and their utilization of multiple host plants.

Questions (for answers and explanations, see page 207)

1. Which of the following cell types would you expect to *never* find in a fungus?
   a. Haploid
   b. Diploid
   c. Chitinous
   d. Flagellated
   e. Eukaryotic

2. Fungal cell filaments called ___________ combine to form the ___________.
   a. rhizoids; mycelium
   b. hyphae; mycelium
   c. haustoria; hyphae
   d. rhizoids; fruiting body
   e. haustoria; lichen

3. All fungi are absorptive heterotrophs. Which of the following is an adaptation shown by many fungi that greatly aids this mode of nutrient procurement?
   a. Dikaryosis
   b. Conjugation
   c. Complex life cycle
   d. Chitinous cell wall
   e. Large surface-to-volume ratio

4. Assume that two normal hyphal cells of different fungal mating types unite. After a period of time the cell walls between these cells will dissolve, producing a
   a. dikaryotic cell.
   b. fruiting body.
   c. zygote.
   d. dikaryotic cell which is also heterokaryotic.
   e. mycelium.

5. Which of the following statements about the fungal structures known as conidia is *not* true?
   a. Conidia appear at the tips of hyphal cells.
   b. Conidia are formed within sporangia.
   c. Conidia are haploid.
   d. Formation of conidia is a type of asexual reproduction in fungi.
   e. All of the above

6. In the fungal structure known as an ascus
   a. haploid nuclei fuse to form a diploid zygote.
   b. mitosis produces diploid spores.
   c. budding takes place to yield new hyphae.
   d. sticky traps used in predation are formed.
   e. None of the above

Activities (for answers and explanations, see page 208)

- Complete the diagram below which shows the generalized life cycle of a fungus by adding the following terms in their appropriate locations: *spores, dikaryotic hyphae, meiosis, mitosis, fertilization.* Also, circle and label the portions of the life cycle corresponding to the haploid (n), diploid ($2n$), and the dikaryotic ($n + n$) stages.

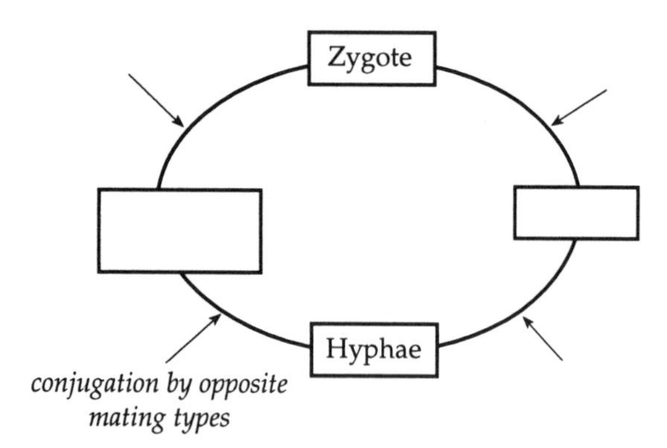

- Explain the functional differences between the following fungal structures: hyphae, mycelia, rhizoids, and haustoria.

PHYLUM EUMYCOTA • LICHENS (pages 536–542)

Key Concepts

1. *Zygomycetes* (the conjugating fungi) number some 600 species. They possess hyphae that lack cross walls, have no motile cells, and produce no fleshy fruiting body. The only diploid cell in the entire life cycle is the zygote.
2. Many similarities exist between the zygomycetes and the actinomycete bacteria. However, since the two groups are evolutionarily unrelated, this is an example of *convergent evolution* (where similarity is the result of nonhomologous structure).
3. The black bread mold *Rhizopus stolonifer* is a zygomycete that can reproduce vegetatively or engage in sexual reproduction through conjugation. Gaseous pheromones direct the conjugation process. Spores formed sexually are known as *zygospores*.
4. About 30,000 species comprise the class *Ascomycetes* (sac fungi). All produce a unique saclike structure called an ascus, inside which meiosis occurs to produce *ascospores*. These spores then germinate to re-form the mycelial stage.
5. Unlike zygomycetes, ascomycete hyphae do have cross walls, but these walls contain perforations that permit cytoplasmic contact among cells.
6. Ascomycete species that form fruiting bodies called *perithecia* are known as the *Euascomycetes*. Species that do not produce perithecia are called *Hemiascomycetes*.
7. Hemiascomycetes are generally very small, with many species being unicellular. The yeasts (such as *Saccharomyces*) are among the best known, having major commercial importance in the baking and brewing industries because of their ability to generate ethanol and carbon dioxide during alcohol fermentation. Yeasts usually reproduce asexually, either by fission or *budding*. Conjugation takes place only occasionally in yeasts, and there is no dikaryon stage.
8. The genus *Neurospora* (common bread mold) is an example of an euascomycete fungus. Euascomycetes are multicellular with heterokaryotic hyphae. Sexual reproduction involves the passing of nuclei from one mycelium to another. The donor nuclei then divide independently of the host nuclei for some time. The nuclei only undergo fusion at the actual time of ascus formation.
9. Euascomycetes can be serious parasites of higher plants (e.g., chestnut blight and Dutch elm disease), as well as gourmet delicacies (morels and truffles). This group also includes the cup fungi.
10. Asexual reproduction in euascomycetes is by means of conidia, which develop by the millions on the tips of specialized hyphae and give the molds their characteristic coloration.
11. The 25,000 species of club fungi comprise the class *Basidiomycetes*. They have prominent fruiting structures (e.g., mushrooms) where spore formation occurs. Some are commercially important as foods, others are serious plant pathogens (e.g., brackets, smuts, and rusts).
12. Basidiomycete hyphae are completely walled off (*septate*). Dikaryotic hyphae undergo nuclear fusion and meiosis in a fruiting structure called a *basidium*, where haploid *basidiospores* are produced by mitosis. These spores are scattered and germinate to reestablish the haploid mycelium.
13. The fleshy fruiting structure of basidomycetes is topped by a cap (the pileus) with gills on its underside. Basidia develop between these gills and discharge their spores for dispersal, most often by wind or water.
14. The class *Deuteromycetes* (*imperfect fungi*) is a leftover grouping of some 25,000 species that do not produce fruiting structures. As such, they cannot be placed in the other three fungal classes, which are distinguished from one another primarily by reproductive criteria. Included in the Deuteromycetes are species that cause athlete's foot and ringworm.

15. Some chemical by-products of Deuteromycetes are of major importance to humans. Members of the genus *Penicillium* produce antibiotics against bacteria which humans have usurped for our own medical advantage. Other species are responsible for the characteristic flavors of certain cheeses, soy sauce, and the alcoholic beverage known as saki.
16. The approximately 20,000 "species" of lichens are not really true organisms. Instead, they are mutualistic associations between two organisms: a fungus and either a unicellular alga or a cyanobacterium.
17. The combination of tough protection and structural support provided by the fungus, and the photosynthetic nutrition provided by the alga or bacterium, allows lichens to survive in extremely harsh environments. However, because they cannot eliminate the toxins associated with pollution, they do not typically occur near industrialized cities.
18. The fungal component of lichens is most often an ascomycete, but other classes of fungi are also known to form lichens. Some species, such as the reindeer "moss" *Cladonia subtenuis*, are important food items for large mammals. Lichens come in a variety of colors and forms, including crustlike (crustose), leafy (foliose), and shrubby (fruticose).
19. Reproduction of lichens is by simple fragmentation of the vegetative body (*thallus*), or by specialized structures known as *soredia*. These soredia consist of a few photosynthetic cells surrounded by fungal hyphae, and are dispersed by wind or water to new locations where they develop into another lichen.
20. The fungal component of a lichen may produce its own characteristic fruiting bodies, but when spores are released from these structures they will lead only to the production of a new fungus, since no algal or bacterial cells are present.
21. Lichens are often the first colonists to appear on bare rock. By acidifying their environment during growth, lichens help break down the rock and hasten the formation of soils. Because of their incredible resistance to cold and dry habitats, lichen species can outnumber plant species by as much as 100-fold in harsh environments like Antarctica.

Questions ***(for answers and explanations, see page 208)***

1. Select *all* of the following fungal groups that contian members likely to appear on your dinner plate as an *intended* part of the meal!
   a. Basidiomycetes
   b. Zygomycetes
   c. Ascomycetes
   d. Deuteromycetes
   e. None of the above

2. Athlete's foot, ringworm, and penicillin all have a common link to which of the following fungal groups?
   a. Basidiomycetes
   b. Zygomycetes
   c. Ascomycetes
   d. Deuteromycetes
   e. None of the above

3. A unknown specimen of fungus is brought to you for identification. Upon close examination, you discover that its hyphae are completely septate, and it possesses gills on the underside of the pileus. To which fungal group does it most likely belong?
   a. Basidiomycetes
   b. Zygomycetes
   c. Ascomycetes
   d. Deuteromycetes
   e. None of the above

4. A process known as karyogamy, where haploid nuclei fuse, occurs in certain groups of fungi during sexual reproduction. Which of the following statements about this process is true?
   a. It occurs only in the hemiascomycetes, such as yeast.
   b. The rate of growth of the fungal mycelium is maximum at this time.
   c. It takes place just prior to ascus formation in euascomycetes.
   d. It denotes the point where hyphal cell walls dissolve and the remaining cell becomes heterokaryotic.
   e. The characteristic flavor of some cheeses and alcoholic beverages is the direct result of this process.

5. If you subjected the soredium of a lichen to microscopic and chemical analysis, which of the following would you *never* expect to find?
   a. Chlorophyll *a* in the algal component
   b. Chitin in the cell walls of the fungal component
   c. Eukaryotic cells
   d. A fungal fruiting structure actively producing spores
   e. An ascomycete fungus

6. Although there are over 20,000 known "species" of lichens, unlike other members of the kingdom Fungi, none of them are true organisms. This is because lichens
   a. are actually a mutualistic association between two true organisms.
   b. are really viruses, which are not considered organisms.
   c. have no means of self-reproduction.
   d. cannot live in harsh conditions.
   e. do not contain nucleic acids.

7. Which of the following taxonomic classifications best describes the fungus *Neurospora*?
   a. Euascomycete
   b. Hemiascomycete
   c. Basidiomycete
   d. Zygomycete
   e. Deuteromycete

8. Common bread molds such as *Neurospora* and the fungi known as truffles share which of the following traits?
   a. They are both ascomycetes.
   b. They both produce ascospores.
   c. Their spores will germinate to form a mycelium.
   d. Meiosis occurs inside an ascus to yield spores.
   e. All of the above

Activities (for answers and explanations, see page 208)

- For a variety of reasons, yeasts are considered somewhat peculiar fungi. List a few characteristics of the genus *Saccharomyces* that are atypical of the fungal kingdom as a whole.

- Refer to Figure 24.13(*c*) in your textbook, which shows a photograph of the bracket fungus *Polyporus sulphureus*. Explain the purpose of this bracket structure, and identify where in the basidiomycete life cycle the bracket occurs.

- Folklore has it that one can determine directions in the woods by looking to see on which side of tree trunks the lichens grow. Suggest a possible environmental reason why this observation may have biological validity.

Integrative Questions (for answers and explanations, see page 208)

1. Which of the following is true of all fungi?
a. They are multicellular
b. They are heterotrophic and eukaryotic
c. They obtain nutrients by absorption
d. Both *a* and *b*
e. Both *b* and *c*

2. You find a small, flattened organism that looks like a leafy plant growing on an otherwise bare rock surface. Microscopic analysis shows numerous threadlike, eukaryotic cells surrounding what are clearly bacterial cells. The best description for this organism is a(n)
a. mushroom.
b. alga.
c. foliose lichen.
d. fruticose lichen.
e. club fungus.

3. A saprobic fungus could utilize which of the following objects as a food source?
a. Spoiled cheese
b. A dead fly
c. Overripe fruit
d. Animal feces
e. All of the above

4. The fungal part of a lichen and members of the genus *Neurospora* share which of the following characteristics?
a. Haploid hyphae
b. Autotrophic nutrition
c. Asexual reproduction via budding
d. Obligate parasitism
e. None of the above

25

Plants

CHAPTER LEARNING OBJECTIVES—after studying this chapter you should be able to:

- ❑ List the basic taxonomic characteristics of the plant kingdom, and explain how plants differ from other organisms.
- ❑ Apply the concept of alternation of generations to plant life cycles and modes of reproduction.
- ❑ Correctly use the terms "gametophyte," "gamete," "antheridium," "archegonium," "sporophyte," "sporangia," "spore," "meiois," "mitosis," and "fertilization" when discussing plant life cycles.
- ❑ Distinguish between the nonvascular plants and the vascular plants and describe the anatomical differences between them.
- ❑ Explain current ideas regarding the evolutionary origins of the plant kingdom and the sequence in which nonvascular and vascular plants appeared on Earth.
- ❑ Understand the basic structure of plant vascular systems, and discuss the specific roles of xylem and phloem.
- ❑ Describe the basic characteristics of the nonvascular plants, and name the main phyla within this group.
- ❑ Explain the various stages in the moss life cycle and be able to define the terms "protonema," "rhizoid," "capsule," and "calyptra."
- ❑ Discuss the importance of tracheids in the evolution and function of vascular plants.
- ❑ Distinguish between the seedless and seed-bearing vascular plants, and name the major phyla within each group.
- ❑ Describe the evolution of true roots and leaves from the simple ancestor *Rhynia* to the modern vascular plants.
- ❑ Explain the difference between homospory and heterospory, and discuss the roles of megasporangia and microsporangia in these two processes.
- ❑ Discuss the fern life cycle, and explain how it differs from the moss life cycle.
- ❑ Differentiate between the gymnosperms and the angiosperms, and describe the importance of pollen in the reproduction of these plants.
- ❑ Name the four phyla of gymnosperms and describe in detail the life cycle of a typical conifer.
- ❑ Discuss the origin and development of gymnosperm seeds using the terms "integument," "ovule," and "micropyle."
- ❑ Cite the main anatomical and reproductive characteristics of the angiosperms, and discuss the importance of the flower in angiosperm reproduction.
- ❑ Describe the parts of a flower, and distinguish between perfect and imperfect flowers.
- ❑ Understand the difference between a monoecious and a dioecious angiosperm plant.
- ❑ Explain the importance of pollination and fruit development in angiosperm reproduction.
- ❑ Describe the basic anatomical differences between monocotyledonous and dicotyledonous angiosperms, and cite examples of each.
- ❑ List the main trends in plant evolution from the simple nonvascular plants through the angiosperms.

THE PLANT KINGDOM (pages 545–548)

Key Concepts

1. The ancestors of land plants left the water nearly 500 million years ago and underwent rapid evolutionary change. Modern plants became established during the last 100 million years.
2. Plants are defined as multicellular, photosynthetic eukaryotes whose embryos develop within protective parental tissues. Their cell walls are made of cellulose, they have chloroplasts that contain the pigments chlorophyll *a* and *b* as well as specific carotenoids, and their storage carbohydrate is starch.
3. Alternation of generations is a conspicuous feature of plant life cycles. A haploid spore develops by mitosis into the haploid multicellular plant, or *gametophyte*. This stage of the life cycle produces haploid gametes by mitosis.
4. Plant gametes fuse in the process of syngamy (fertilization) to form a diploid zygote, which then undergoes mitosis to form the multicellular diploid plant, or *sporophyte*. The sporophyte completes the plant life cycle by undergoing meiosis within specialized structures called *sporangia* to produce haploid spores.

5. The gametophyte and sporophyte stages of plants are always morphologically distinct. That is, plants exhibit heteromorphic alternation of generations.
6. The *nonvascular plants* include three phyla that lack internal tissues for water and nutrient transport. The remaining nine plant phyla comprise the *vascular plants;* all have well-developed transport tissues.
7. The early plant ancestors made the transition from a purely aquatic existence to life on land. Becoming terrestrial required that plants overcome problems of structural support and the transport of water and minerals throughout the body.
8. True plants likely arose from the protist phylum Chlorophyta (green algae). Supporting evidence includes similar pigment composition, starch as the storage carbohydrate, and cellulose cell walls. In addition, some green algae have plantlike oogamous reproduction, which features a large stationary egg.
9. The first nonvascular plants probably evolved from a green alga ancestor. One group then evolved into the modern nonvascular plants, the other group into the vascular plants.
10. Nonvascular plants (the mosses, liverworts, and hornworts) have little or no water-transporting tissue; they therefore do not attain a large body size but often occur in dense masses. Water moves through this mass by capillary action, and minerals are distributed by simple diffusion.
11. Ferns, conifers, and flowering plants are vascular plants. They have well-developed *vascular systems* with specialized conducting tissues: *xylem* (used for water and mineral transport) and *phloem* (used for transporting photosynthetic products).
12. Ironically, vascular plants actually appear first in the fossil record, predating the nonvascular plants by at least 60 million years. This does not mean, however, that vascular plants evolved before nonvascular plants. The discrepancy in fossil records results from the fact that vascular plants are more likely to form fossils because of their structure and chemical composition.
13. Plants are distinguished from the algae largely on the basis of adaptations for life on land. These adaptations include protective coverings to prevent drying, specialized structures such as roots and rhizoids to gather water, support mechanisms that utilize turgor pressure, woody stems, or thickened cell walls, and protection for the developing sporophyte embryo within parentally derived tissues.

Questions (for answers and explanations, see page 208)

1. In plants, the process of mitosis leads to the formation of
   a. gametes.
   b. the sporophyte plant.
   c. the gametophyte plant.
   d. spores.
   e. More than one answer is correct.
2. A plant specimen is brought to you for examination. You notice that it has eukaryotic cells whose chloroplasts contain both chlorophyll *a* and *b*. Careful analysis of its cell nuclei indicate that all cells are haploid. Based on this information, the most likely group to which this specimen belongs is the
   a. mosses.
   b. ferns.
   c. green algae.
   d. conifers.
   e. Cannot determine from information given
3. A moss possesses all but one of the following plant characteristics. Select the *exception*.
   a. Spores
   b. Xylem and phloem
   c. Gametophyte tissue
   d. Cellulose in cell walls
   e. Chlorophyll *a* in chloroplasts
4. While searching through a collection of 400-million-year-old fossils you find one that is clearly a vascular plant. However, no nonvascular plant fossils are present from this same time period. Which of the following is the best explanation for this observation?
   a. Nonvascular plants had not yet evolved 400 million years ago.
   b. Nonvascular plants do not coexist with vascular plants.
   c. Vascular plants fossilize better than nonvascular plants.
   d. Nonvascular plants were present at this time, but just not widely distributed.
   e. None of the above
5. The green alga *Ulva* has all but one of the following characteristics in common with a flowering plant. Select the *exception.*
   a. Similar photosynthetic pigments
   b. Cellulose cell walls
   c. Starch as a storage carbohydrate
   d. Heteromorphic alternation of generations
   e. Genetically different sporophytes and gametophytes
6. Which of the following plant groups lacks well-developed vascular tissue?
   a. Ferns
   b. Mosses
   c. Conifers
   d. Flowering plants
   e. None of the above

Activities (for answers and explanations, see page 208)

- List the similarities in structure and chemical composition that suggest an evolutionary link between the Chlorophyta and the true plants.

- On the following generalized diagram of a plant life cycle, identify the major morphological stages in the open boxes, and name the cell division process indicated by each arrow.

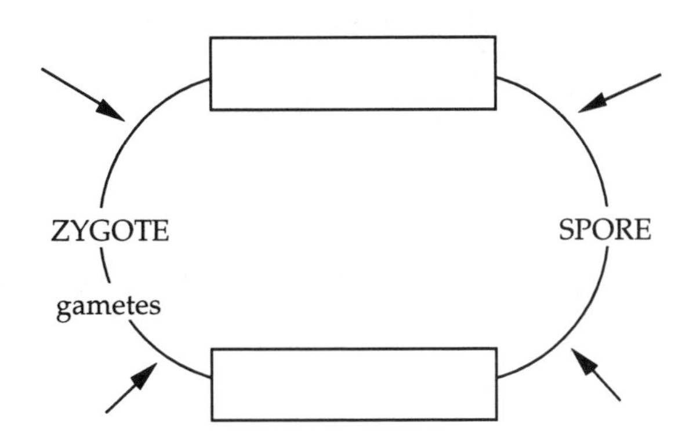

NONVASCULAR PLANTS (pages 548-552)

Key Concepts

1. Nonvascular plants are widely distributed and highly successful. They are typically found growing as dense mats in moist habitats. Some grow in the soil, some on rocks or buildings, and some even grow on other plants. They differ from vascular plants in that the conspicuous green plant that we recognize is the gametophyte generation.
2. Nonvascular plants are never large, mainly because they lack internal transport systems for water and minerals. Their size is also limited by reproductive needs; the sperm cells must swim through rainwater or dew in order to fertilize the stationary egg.
3. Like other plants, nonvascular plants possess a waxy cuticle that helps retard water loss. The developing embryo is also protected within layers of maternal tissue. Specific morphology of the nonvascular plants is dictated by habitat features, particularly water availability. The gametophyte stage is especially variable, while the sporophyte stage is less so and thus more diagnostic of each species.
4. Nonvascular plants have heteromorphic generations that are nutritionally distinct. The gametophyte stage is photosynthetic and therefore nutritionally independent. The sporophyte usually does not photosynthesize and is always dependent upon the gametophyte for its development and nutrition.
5. During reproduction, haploid spores divide by mitosis to give rise to the multicellular gametophyte, within which specialized sex organs develop. The *archegonium* produces a single female gamete (egg), while each *antheridium* produces many flagellated male gametes (sperm). Since swimming sperm are involved, water is essential for fertilization to occur.
6. Once a diploid zygote is formed by fertilization, mitotic divisions yield a multicellular sporophyte embryo, which remains protected within the parental gametophyte tissue. The embryo eventually elongates and becomes a mature sporophyte, which produces a *capsule* at its tip. Inside the capsule haploid spores develop by meiosis. These spores are released under appropriate conditions, germinate, and then mitotically divide to produce another gametophyte which completes the life cycle.
7. The nonvascular plants are subdivided into three phyla: *Hepaticophyta* (liverworts), *Anthocerophyta* (hornworts), and *Bryophyta* (mosses).
8. Liverworts often have flattened, lobed gametophytes, although this morphology is not universal. Sex organs are produced on top of these flat plates, and rhizoids below. The sporophyte stalk rises well above the gametophyte to elevate the capsule for spore dispersal. Specialized coil structures called *elaters* help to release the mature spores.
9. *Marchantia* is a well-known genus of liverwort that can reproduce vegetatively by forming *gemmae*, which are loose clumps of cells capable of developing into a whole new gametophyte plant.
10. The hornworts (so called because their sporophytes resemble horns) are structurally similar to liverworts. However, they differ from the liverworts and mosses in two ways: their archegonia are embedded in gametophyte tissue, and their sporophytes are capable of almost indefinite growth, making them the tallest (up to 20 cm) known nonvascular plants. Hornworts sometimes harbor cyanobacteria, which fix atmospheric nitrogen for use as a plant nutrient.
11. Mosses are the most diverse nonvascular plants. There are more species of mosses than liverworts and hornworts combined, and they can be found in almost every terrestrial habitat.
12. Moss spores germinate and mitotically develop into a filamentous structure called a *protonema,* which differentiates to produce nonphotosynthetic filaments (*rhizoids*) that attach to the substrate, and branched photosynthetic filaments that yield the "leafy" aerial portions of the gametophyte that we typically recognize as a moss.
13. Eggs and sperm develop in the archegonia and antheridia, respectively, located at the tips of the gametophyte leaves. The flagellated sperm require water in order to swim to the egg. After sperm and egg unite, the resulting zygote mitotically develops into the elongated sporophyte. A spore capsule develops at the tip of the sporophyte within which meiosis occurs to produce spores. Inside the pointed cap of the capsule are a set of humidity-sensitive teeth that facilitate spore dispersal during conditions favorable for germination.
14. Certain species of mosses depart from this method of spore dispersal. For example, members of the genus *Sphagnum*, which account for much of the biomass found in tundra and arctic environments, have pressure-sensitive spore chambers that explosively release their spores.
15. A structure that may represent the functional evolutionary precursor of vascular tissue exists in many mosses. When certain cell types, called hydroids, die in a moss they leave a hollow channel through which water can flow to other portions of the plant body, providing a limited method of water transport.

Activities (for answers and explanations, see page 209)

- Match the following nonvascular plant characteristics with the appropriate stage of the life cycle (i.e., sporophyte, gametophyte, both, or neither) in which they occur.

  ____________________Sperm and eggs
  ____________________Phloem
  ____________________Chloroplasts
  ____________________Waxy cuticle
  ____________________Capsule
  ____________________Archegonium
  ____________________Diploid tissue

- Refer to Figure 25.7(*a*) in your textbook, which depicts the moss life cycle. Approximately what percentage of this life cycle is dominated by haploid tissues? Explain your answer.

Questions (for answers and explanations, see page 209)

1. Which of the following factors either limit the size to which nonvascular plants can grow, or the type of habitat where they are found?
   a. Swimming sperm
   b. Lack of vascular tissue
   c. Possession of a diploid sporophyte stage
   d. Little specialized supporting tissue
   e. None of the above

2. Which of the following characteristics do liverworts and hornworts have in common?
   a. Well-developed vascular systems
   b. Possession of a waxy cuticle
   c. Nutritionally independent sporophyte stages
   d. Diploid spores
   e. Gametes that form through meiosis

3. In which of the following nonvascular plant structures would you expect to find a calyptra?
   a. Gametophyte of a moss
   b. Spore capsule of a liverwort
   c. Sporophyte of a moss
   d. Gametophyte of a hornwort
   e. Archegonium of a liverwort

4. Antheridia, archegonia, and haploid tissue cells all share which of the following?
   a. They do not occur at any stage in the life cycle of nonvascular plants.
   b. Each is directly involved in the production of liverwort spores.
   c. Hornworts have them, but mosses do not.
   d. All are found in the sporophyte stage of a hornwort.
   e. All are found in the gametophyte stage of a moss.

5. The easiest way to tell the difference between a hornwort and a moss is by examining
   a. the structure of the sporophyte.
   b. the structure of the gametophyte.
   c. the presence or absence of vascular tissue.
   d. Both *a* and *b*
   e. Both *b* and *c*

6. Indicate all of the following nonvascular plants in which you would expect to see a protonema at some stage in the life cycle.
   a. Marchantia
   b. Sphagnum
   c. All hornworts
   d. All mosses
   e. All liverworts

VASCULAR PLANTS • SURVIVING SEEDLESS VASCULAR PLANTS (pages 552-559)

Key Concepts

1. Vascular plants are a much larger and more diverse group than their nonvascular relatives. Nevertheless, all vascular plants are linked by a common anatomical feature, the presence of *tracheids*, which are the primary water-conducting elements in xylem.
2. Tracheids confer two significant advantages on terrestrial plants: a mechanism for internal transport and a means of rigid support against gravity. These advantages helped set the stage for the permanent invasion of land by plants.
3. Vascular plant phyla are divided into two groups; those that produce seeds and those that do not. The life cycles of seedless vascular plants are quite uniform, with the sporophyte constituting the large and obvious stage of the life cycle. The gametophyte stage is small but independent at maturity. The spores of seedless vascular plants represent a resting stage in the life cycle.
4. Seedless vascular plants still require an aqueous environment for reproduction because their flagellated sperm cells must swim to the egg.
5. Primitive vascular plants first appear in the fossil record over 400 million years ago (mya). Three surviving groups of seedless vascular plants, the club mosses, horsetails, and ferns are present as fossils between 345–400 mya.
6. Seed plants first appear in the fossil record 360 mya. As terrestrial plant size increased, various trees (both seedless and seed-bearing) became a dominant part of the landscape between 300–350 mya. This trend started with the large tree ferns, which subsequently gave way to the gymnosperms (conifers and their relatives). Angiosperm (flowering plant) fossils first appeared about 120 mya and then underwent an explosive evolutionary radiation to become the dominant plant life on Earth.
7. Although now extinct, members of the phylum Rhyniophyta were probably the first vascular plants.

They were clearly vascular because they possessed tracheids, but they had neither true leaves nor roots.

8. Rhyniophytes were anchored below the soil surface by horizontal protrusions of the stem called *rhizomes*. These stems had dichotomous aerial branching, and at the end of each branch the plant bore distinct sporangia, inside which clusters of four tightly packed spores developed. Fossil analysis of the genus *Rhynia* indicates that these structures represent the diploid sporophyte plant.
9. The seedless vascular plants that evolved from the Rhyniophytes show major changes in anatomy, including true leaves and roots, and two distinctly different spore types.
10. Roots are thought to have evolved from the dichotomous branching of early vascular plant stems, some of which likely bent downward and penetrated the soil. Adaptation of these underground branches to the unique subterranean environment eventually altered them into true roots.
11. True *leaves* are defined as flattened photosynthetic structures that attach laterally to the plant stem and possess vascular tissue. Two leaf types are found in vascular plants: *simple* and *complex*.
12. Today, simple leaves are found only in the club mosses and horsetails. They are usually small, scalelike outgrowths of the stem with rarely more than a single strand of vascular tissue. Their emergence from the vascular tissue of the stem causes little change in the conformation of the stem's vascular structure.
13. Complex leaves are found only in the ferns and seed plants. Flattened stems with dichotomous branching, inside which substantial vascular and photosynthetic tissue developed, are thought to be the evolutionary precursors of complex leaves. Unlike simple leaves, major modifications in the stem vascular structure occur at the points where complex leaves emerge from the stem.
14. Nonvascular plants and the ancient seedless vascular plants are *homosporous* (having spores of only one morphological type). All other plant groups either are exclusively *heterosporous* (having spores with two distinct morphologies), or include a mixture of species that employ one or the other of the two strategies. Heterospory evolved multiple times in several different plant lineages, underscoring its importance as a reproductive strategy.
15. Heterosporous plants produce *megaspores* and *microspores*. Megaspores are generated in small numbers and develop into the large female gametophyte (*megagametophyte*), which produces only eggs. Microspores are made in large numbers and develop into the smaller male *microgametophyte*, which produces only sperm.
16. Although extremely abundant in the Carboniferous period, today only a relatively few surviving species make up the phyla Lycophyta (club mosses) and Sphenophyta (horsetails or scouring rushes). They share a number of features, including having true roots and only simple leaves. Both have homosporous and heterosporous species, and exhibit heteromorphic alternation of generations. The sporophyte generation is the dominant stage in the life cycle.
17. Differences between the club mosses and horsetails include patterns of leaf arrangement, mechanisms of stem elongation, and the location and design of sporangia. Club moss sporangia assume a conelike appearance called a *strobilus*, whereas the sporangia of horsetails are arranged in whorls.
18. A somewhat problematic phylum Psilophyta (whisk ferns) is the taxonomic home for two genera of vascular plants that lack roots and bear spores. *Psilotum* has only minute scales instead of leaves, and *Tmesipteris* has well-developed vascular tissue in its flattened photosynthetic surfaces. Both were once thought to be living relics of the Rhyniophyta, but are now considered specialized offshoots of the more modern plants.
19. Ferns and seed plants typically have large leaves with more than one branching vascular strand. They also have stems and true roots. Their sporophytes are larger than, and independent of, the gametophyte. In ferns, the gametophyte is independent of the sporophyte, whereas in seed plants it is not.
20. The phylum *Pterophyta* (ferns) contains about 12,000 species of large-leaved, seedless vascular plants that require water for fertilization. The complex leaf of ferns, which is known as a frond, unfurls during development from a coiled "fiddlehead."
21. Fern sporangia are clustered on the underside of the frond in structures known as *sori*. Meiosis occurs in each sorus to produce haploid spores which may disperse considerable distances. Most ferns are homosporous. Their spores germinate to form small, independent gametophytes.
22. Antheridia and archegonia develop on the fern gametophyte. Male gametes must swim to find the female gamete (often on another gametophyte) and fertilize it. The resulting zygote then develops into a multicellular sporophyte embryo within the gametophyte tissue. Shortly after beginning its development, the sporophyte sprouts roots and becomes nutritionally independent of the gametophyte stage.
23. The dominant stage in the fern life cycle is the sporophyte, which can become quite large and persist for hundreds of years. The flattened gametophyte is usually small and rather delicate. Some ferns produce a fleshy, tuberous gametophyte that must form an obligatory mutualistic association with a fungus in order for the sporophyte to develop.

Activities (for answers and explanations, see page 209)

- Briefly describe how the Rhyniophyta, which have no true roots, could have given rise to the other seedless vascular plants, which do possess true roots.

- Compare and contrast the structure and function of simple leaves with that of complex leaves, and indicate the plant groups where each leaf type is found.

  Simple leaves:

  Complex leaves:

Questions (for answers and explanations, see page 209)

Refer to the following diagram of a plant life cycle when answering questions 1 and 2.

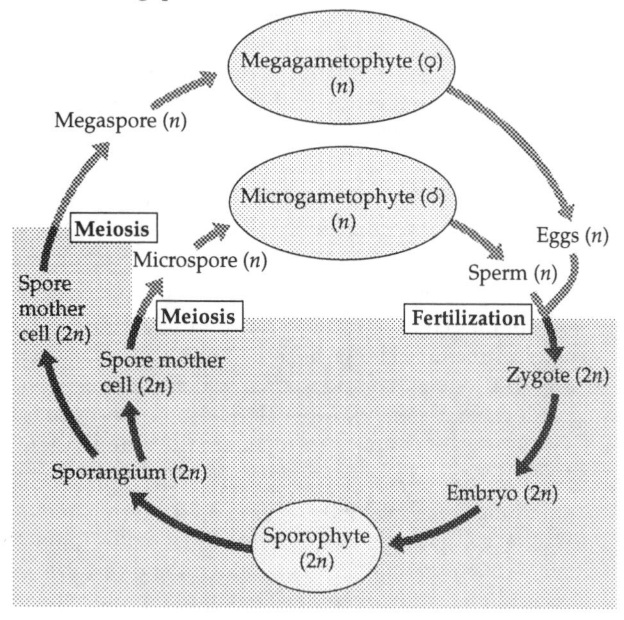

1. The type of reproduction depicted in this diagram is found in
   a. club mosses and hornworts only.
   b. ferns only.
   c. some ferns and all seed-bearing vascular plants.
   d. seed-bearing vascular plants only.
   e. seedless vascular plants only.

2. The specific events responsible for the switch between haploid and diploid stages in this life cycle are better known as
   a. meiosis and fertilization.
   b. mitosis and fertilization.
   c. meiosis and mitosis.
   d. meiosis and germination.
   e. fertilization and germination.

3. Match the characteristics on the left with the plant group(s) that possesses them on the right.

a. Tracheids	__________	Gymnosperms
b. Spores	__________	Psilopsids
c. True roots	__________	Club mosses
d. Megaphylls	__________	Horsetails
e. Sori	__________	Ferns

4. Megaspores are
   a. found in all plants.
   b. produced by the sporophyte generation of heterosporous plants.
   c. produced by the gametophyte generation of heterosporous plants.
   d. diploid structures that give rise to the megagametophyte.
   e. haploid structures that give rise to the sporophyte.

5. In all seedless vascular plants, meiosis takes place in
   a. sporangia.
   b. archegonia.
   c. antheridia.
   d. rhizomes.
   e. microspores.

6. Although they have significant anatomical advancements over the vascular plants, ferns are still limited to relatively moist habitats because
   a. their tracheids are rather poorly developed.
   b. their xylem tissue is seasonal.
   c. their reproduction requires water for fertilization.
   d. spore dispersal depends on heavy rains.
   e. frond tissue dries out easily.

THE SEED PLANTS • THE GYMNOSPERMS • THE ANGIOSPERMS: FLOWERING PLANTS (pages 559-571)

Key Concepts

1. The seed plants include the *gymnosperms* (conifers and relatives) and the *angiosperms* (flowering plants). The angiosperms are the most recently evolved of the vascular plants.
2. In both groups, the trend toward a reduced gametophyte stage has continued. The haploid gametophyte develops while still attached to the diploid sporophyte. Except for the most primitive seed plants, total reproductive independence from liquid water has been achieved by the elimination of swimming sperm.
3. Heterospory is the rule in seed plants. Megasporangia and microsporangia produce female and male spores, respectively, and are found on modified leaves grouped into strobili. Examples are cones and flowers.
4. The spores of seed plants are produced by meiosis within the sporangia, but the spores are not shed. Gametophyte development thus begins within the sporangium. Female gametophytes are retained within the megasporangium through maturity, while microgametophytes, commonly known as *pollen grains*, are released and distributed by wind or animal vectors.
5. Upon finding the appropriate portion of a sporophyte plant, the pollen grain forms a *pollen tube* that grows towards the female gametophyte. Pollen nuclei then travel down this tube to cause fertilization.
6. The diploid zygote which results from fertilization divides mitotically to form a sporophyte embryo while still inside the female gametophyte tissue. At some point this collection of sporophyte and gametophyte tissue goes dormant, resulting in the structure known as a *seed*.

7. Seeds often possess tissue from three plant generations. The protective seed coat is diploid sporophyte tissue from the original parent generation. Inside the seed coat is haploid gametophyte tissue from the second generation. Finally, the newly formed embryo is a third generation of diploid sporophyte tissue.
8. Unlike the embyros of seedless plants, which develop directly into sporophytes, the seeds of gymnosperms and angiosperms provide a tough, protective coating in which the embryo can "rest" for many months or years before developing into a mature sporophyte plant.
9. The ability of seeds to rest for long periods of time, the presence of stored food reserves that can be used to initiate development of the new sporophyte, and specialized structures to facilitate dispersal have combined to make the seed plants an incredibly successful group.
10. The gymnosperms yield only to the angiosperms in their dominance of the land. Approximately 750 species of gymnosperms are grouped into four living phyla. Cycadophyta (the cycads) are palmlike, Ginkgophyta (ginkgos) today exist as a single species of a once numerous group, Gnetophyta is a collection of diverse plants related by a unique type of vessel element cell, and Coniferophyta (the conifers), which is by far the most successful group of gymnosperms.
11. Gymnosperms exhibit secondary growth and include some of the tallest trees known. This despite the fact that tracheid cells are the only components in their xylem, which limits the extent of their water conduction and structural support.
12. A hint about gymnosperm evolutionary origins is provided by a transitional fossil called *Archaeopteris* from the Devonian period. Although it had distinct rhyniophyte and fernlike characteristics, the presence of tracheids in its woody tissue clearly allies it with the gymnosperms.
13. The Coniferophyta is the largest of the gymnosperm phyla. All conifer species are heterosporous, with spore formation occurring in specialized strobili borne as distinct male and female cones.
14. The life cycle of pine is representative of conifers generally. Pollen grains (male gametophytes) are dispersed in vast numbers by wind, with successful pollen eventually landing on the appropriate female sporophyte tissue. A pollen tube grows to provide the final route of travel for the sperm nucleus to the egg.
15. A special layer of sporophyte tissue called *integument* houses the megasporangium. This integument ultimately becomes the seed coat. An *ovule* consists of the megasporangium, the integument, and the sporophyte tissue connecting them to the maternal plant. The *micropyle* is a small opening at the apex of the ovule that provides access for the pollen tube.
16. Since most conifer seeds lack a protective covering of fleshy fruit (they are "naked" seeds), their only protection comes from the tight packing of scales within the female cone. Some scales are so tightly packed that only severe environmental disturbances such as fire cause them to open. The conifer seeds that do have fleshy fruits (e.g., juniper and yew) are dispersed mainly by animals who seek the fruit as food.
17. Angiosperms (phylum Anthophyta) represent the culmination of a trend toward increased size and independence of the sporophyte generation. In addition to the tracheids found in other vascular plants, their xylem also possesses specialized water-conducting cells called *vessel elements*, as well as *fibers* which improve support.
18. The 275,000 species of flowering plants derive their name because they all produce specialized reproductive structures called flowers. Seeds form within these flowers as part of a modified leaf known as the *carpel*.
19. The most unique reproductive characteristic of angiosperms is the process of *double fertilization*, in which the female gametophyte is penetrated by two male gamete nuclei (both from the same pollen grain). One sperm nucleus fertilizes the egg to form a diploid zygote which will become the next sporophyte generation. The other fuses with two haploid nuclei from the female gametophyte, resulting in a cell that is triploid ($3n$). Mitotic division of this triploid cell yields *endosperm* tissue, which provides nutrients for the developing embryo.
20. Flowers are a kind of strobilus derived from highly modified leaves. Flower parts bearing microsporangia are called *stamens* (male structures), and those bearing megasporangia are called *carpels* (female structures).
21. Each stamen contains two *anthers*, which house the many sporangia where pollen is produced. Anthers are attached to the rest of the flower by filaments.
22. One or more fused carpels is referred to as a *pistil*, at the base of which resides the *ovary*. An ovary contains one or several ovules, where the female gametophyte is located. The stalk of a pistil is called the *style*, and its sticky terminal surface where the pollen attaches is known as the *stigma*.
23. Flowers have nonreproductive parts that are also derived from modified leaves. The upper leaves of a flower are called *petals* (collectively = *corolla*), while the lower leaves are called *sepals* (collectively = *calyx*). Both play a role in attracting animal pollinators. Sometimes petals and sepals are indistinguishable, in which case they are called *tepals*, and at other times they are absent altogether. The stamens, carpels, corolla, and calyx are all attached to the main plant body by a central *receptacle*.
24. *Perfect* flowers contain both male and female reproductive parts as well as the characteristic nonreproductive structures. *Imperfect* flowers lack either carpels or stamens. *Monoecious* species have both male and female flowers on the same plant. *Dioecious* species have male and female flowers on separate plants.
25. Flowers often occur individually, but they may also be clustered together to form species-typical *inflorescences*.
26. Important evolutionary trends in flower structure include a reduction in the number and type of organs, specialization of petals and sepals, and a tendency toward bilateral symmetry accompanied by fusion of parts.

27. Selective pressures related to successful pollination have constrained flower morphology. Long pistils and filaments are adaptations to improve contact with wind-blown pollen or to help deposit pollen on animal bodies for transport to another flower. Competition among pollen tubes "racing" to grow down the style may also select for longer female structures.

28. Many flowering plants have coevolved with their animal pollinators, which include insects (especially bees, beetles, moths, and butterflies), birds, and bats. These relationships can be highly species-specific. Animals visit flowers for food (pollen and nectar), and in so doing efficiently distribute pollen from one flower to another. Flower shape, size, and color have all been influenced by the evolutionary selective pressures exerted by the plant's need to attract pollinators.

29. Fruit develops from the ovary of a flowering plant to surround its seeds. *Simple fruits* (cherry) are those that develop from a single ovary, *aggregate fruits* (raspberry) from multiple carpels in a single flower, *multiple fruits* (pineapple) from inflorescences, and *accessory fruits* (apples) from parts other than the ovary.

30. Like gymnosperms, angiosperms exhibit heterospory and extreme reduction of the female gametophyte. Male gametophytes are the pollen grains, and the ovules are located within fused carpels.

31. Angiosperm zygotes develop an embryonic axis with one or two *cotyledons*, or seed leaves, which may absorb endosperm nutrients or else be photosynthetically active after seed germination. Cotyledon structure is used to subdivide the angiosperms into classes. The *Monocotyledones* have a single embryonic leaf, and the *Dicotyledones* have paired embryonic leaves. Monocots and dicots also differ with respect to the number of flower parts, vascular tissue arrangement, patterns of leaf veins, and secondary growth characteristics. Monocots include the grasses, lilies, and palms. Dicots are far more diverse, and include most herbs, weeds, vines, shrubs, and trees.

32. Angiosperms, along with two groups of gymnosperms, evolved from a single ancestral species that produced no other groups. Ancient members of the magnolia and Chloranthaceae families are currently regarded as the most likely candidates for the first angiosperms. Evidence suggests that the first angiosperms were probably trees which evolved from gymnosperms. Later, small woody angiosperms evolved from these trees.

Questions (for answers and explanations, see page 209)

1. Which of the following statements about seed-bearing vascular plants is *not* true?
a. Heterospory is common in both gymnosperms and angiosperms.
b. The triploid endosperm is found only in angiosperm seeds.
c. Double fertilization occurs in most flowering plants.
d. Gymnosperm sporophytes are nutritionally dependent on the gametophyte.
e. The conifers are the largest and most diverse group of gymnosperms.

2. Raspberry is an example of
a. a plant with accessory fruit.
b. a plant with aggregate fruit.
c. a monoecious gymnosperm.
d. a seedless vascular plant.
e. a vascular plant that lacks vessel elements.

3. Which of the following events happens just prior to the fertilization of an egg cell in gymnosperms?
a. Pollen is released from male cones.
b. The female gametophyte withers and dies.
c. The pollen tube forms.
d. A sperm nucleus travels down the pollen tube.
e. Sporophylls on the female cone open.

4. Consider a honeybee carrying pollen from one flower to another. The only pollen grains which have a reasonable chance to fertilize the eggs contained in the ovules of the second flower are those that make contact with the
a. stigma.
b. anthers.
c. sepals.
d. receptacle.
e. filaments.

5. Which of the following would you *not* expect to find in the seed of a palm tree?
a. Haploid tissue
b. Integument
c. Two cotyledons
d. Triploid endosperm
e. An embryo

6. A certain flower has stamens, a corolla, and a calyx. It would therefore be classified as
a. a perfect flower.
b. an imperfect flower.
c. monoecious.
d. dioecious.
e. an inflorescence.

7. Ginkgos, cycads, and pines are all members of the
a. angiosperms.
b. gymnosperms.
c. Gnetophyta.
d. Anthophyta.
e. Coniferophyta.

Activities (for answers and explanations, see page 210)

- In the space provided below, draw a simple diagram of a gymnosperm seed and label the three generations of tissues it contains.

- On the following diagram of a generalized flower, label the structures indicated by blank lines.

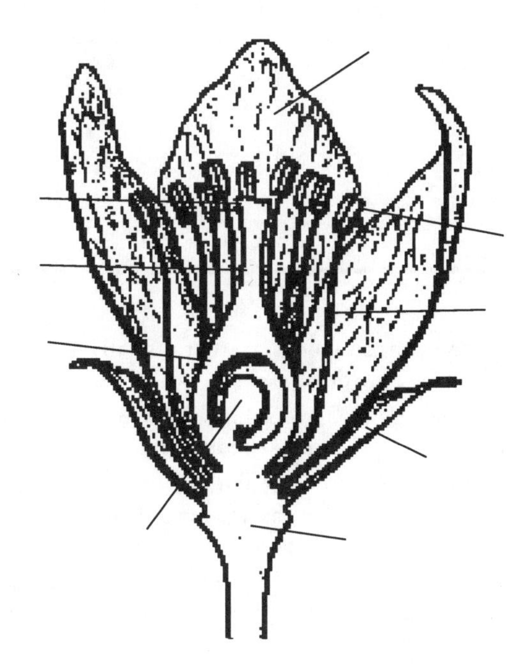

Integrative Questions (for answers and explanations, see page 210)

1. Which of the following plants lack flagellated sperm cells?
   a. Pine trees
   b. Roses
   c. Mosses
   d. Ferns
   e. More than one of the above

2. Compared to mosses, the anatomical structure in the angiosperm life cycle that has been most reduced in size is the
   a. sporophyte.
   b. gametophyte.
   c. root system.
   d. shoot system.
   e. sporangium.

3. Seeds were a significant evolutionary development in plants because they
   a. provided food for animal pollinators.
   b. increased dependence of the gametophyte on the sporophyte.
   c. improved the chances that a sporophyte embryo would survive and grow.
   d. provided a developmental link between megasporangia and microsporangia.
   e. None of the above

4. Which of the following statements most accurately describes a collection of organisms that contains a hornwort, a pine, and an apple tree?
   a. They are all nonvascular plants.
   b. They are all vascular plants.
   c. They are all gymnosperms.
   d. They are all angiosperms.
   e. They are all plants.

5. Which of the following is *not* an overall trend seen in the evolution of plants?
   a. Increasing size and dominance of the sporophyte generation
   b. Decreased dependence on liquid water for reproduction
   c. General trend towards increasing body size
   d. A shift from homospory to heterospory
   e. Decreasing complexity of vascular tissue

26

Sponges and Protostomate Animals

CHAPTER LEARNING OBJECTIVES—after studying this chapter you should be able to:

- ❑ Describe the general characteristics that define the animal kingdom and explain our current understanding of how animals evolved from protists.
- ❑ Differentiate between the protostomate and deuterostomate animals.
- ❑ Explain how characteristics of symmetry, embryology, morphology, and physiology are used to classify animals, and in what ways these factors have influenced the evolution of animal body plans.
- ❑ Differentiate between the different types of body symmetry in animals.
- ❑ Discuss the meaning of the terms "diploblastic" and "triploblastic," and know what types of animals fall into each category.
- ❑ Describe the different types of embryological development in animals using the terms "radial cleavage," "spiral cleavage," "determinate cleavage," and "indeterminate cleavage," and know the fate of the embryonic blastopore in protostomes and deuterostomes.
- ❑ Define the term "coelom," and describe the different types of animal body plans possible with respect to this structure.
- ❑ Explain how division of labor among colonial protists is thought to have influenced the evolution of animals.
- ❑ Discuss the evolutionary sequence of feeding adaptations in animals.
- ❑ List the basic characteristics of the sponges and explain why these animals are considered primitive.
- ❑ List the basic characteristics of the cnidarians and ctenophores and explain how these animals represent an evolutionary advancement over the sponges.
- ❑ Discuss alternation of generations as seen in the cnidarians, differentiate between polyp and medusa body forms, and explain the mechanisms of reproduction in each.
- ❑ Recognize common cnidarian representatives and list the basic characteristics of each.
- ❑ Describe the significance of bilateral symmetry in the evolution of complex animal body forms.
- ❑ Recognize common representatives of the flatworms, discuss their body plans, and characterize their reproductive and feeding lifestyles.
- ❑ Discuss the evolutionary significance of complete digestive systems and hydrostatic skeletons in the roundworms and rotifers.
- ❑ Identify common representatives of the annelid worms, and discuss how their body plans are an advancement over those of the simpler animal phyla.
- ❑ Explain the evolutionary significance of the exoskeleton, its advantages and disadvantages, and in what animal phyla it occurs.
- ❑ Provide reasons why arthropods, especially the insects, are such a diverse and successful assemblage of animals.
- ❑ Recognize and discuss the characteristics of representative arthropod phyla and classes.
- ❑ Discuss the phenomenon of metamorphosis, and describe the different kinds of metamorphosis that occur in arthropods.
- ❑ Describe the ways in which the molluscan body plan differs from that of arthropods.
- ❑ List the distinguishing characteristics of representative mollusk groups.
- ❑ Explain why the cephalopod mollusks are considered more advanced than other molluscan groups.
- ❑ Discuss the main trends in protostome evolution, and how these trends set the stage for even more advanced body plans in the deuterostomes.

HOW ARE ANIMALS CLASSIFIED? (pages 575–578)

Key Concepts

1. The animal kingdom consists entirely of multicellular, eukaryotic heterotrophs that ingest their food. The kingdom has its evolutionary roots in the animallike protists.
2. Two main animal groups, the *protostomates* and the *deuterostomates*, differ in the nature of their embryological development. They have been evolving separately since the Cambrian.
3. Because of their peculiar development, the simple animals known as sponges are classified separately from the protostomes and deuterostomes.
4. Animals are much more diverse organisms than plants. This high diversity is primarily linked to specializations for acquiring food, which have in turn lead to adaptations in body form, sensory systems, and behavior.
5. The relationship between an animal's overall *body plan* and the functioning of its component parts provides clues to the evolutionary relationships among animal groups.
6. Several important aspects of body plans are considered in animal classification. These include information from the fossil record, patterns of embryological development, and comparative morphology and physiology.
7. *Symmetry* is one aspect of an animal's overall body plan. Symmetric animals can be divided into similar halves along at least one plane.
8. *Asymmetric* animals, represented by the sponges, have no plane of symmetry. *Spherical symmetry* is seen mainly in protists, but is also present in some animals. *Radial symmetry* refers to bodies that yield similar halves when cut in any plane along the main body axis. If the body is modified so that only two planes will produce similar halves, the animal has *biradial symmetry*.
9. Another common body plan is *bilateral symmetry*, where mirror image halves result from a cut made along the main body axis. Bilateral symmetry is characteristic of animals that move through their environment. The development of sensory structures and central nervous system tissues in the anterior end of the animal (*cephalization*) is closely allied with bilateral symmetry.
10. Other important anatomical terms used to describe body plans include *anterior* (front), *posterior* (back), *dorsal* (top or side without a mouth), and *ventral* (bottom or side with a mouth).
11. The early developmental stages in animal embryos appear to have evolved more slowly, and are therefore said to be *evolutionarily conservative*. Embryology can thus reveal much about evolutionary relationships in animals. By contrast, traits showing great diversity among animals probably developed late in the evolutionary process.
12. The number of distinct cell layers present in animal embryos is diagnostic. *Diploblastic* animals have only an ectodermal layer and an endodermal layer, while *triploblastic* animals also have a middle layer of mesoderm. All deuterostomes and most protostomes are triploblastic.
13. The protostome lineage exhibits determinate cleavage of the fertilized egg, which means that early divisions irreversibly limit the fate of cells. If separated, individual protostome cells cannot produce an entire embryo. By comparison, cleavage in the deuterostomes is indeterminate. When separated, deuterostome embryonic cells can develop into complete embryos.
14. Cleavage pattern is another embryological difference between protostomes and deuterostomes, with the former showing spiral cleavage and the latter radial cleavage.
15. A third embryological difference between these lineages is the fate of the embryonic blastopore. In protostomes the mouth arises from the blastopore, whereas in deuterostomes the blastopore becomes the anus.
16. The presence or absence of a body cavity is yet another important feature of animal body plans. *Acoelomate* animals lack a body cavity altogether; their digestive systems are completely surrounded by mesenchyme tissue. The proboscis of ribbon worms is contained in a fluid-filled cavity called a *rhynchocoel*. *Pseudocoelomate* animals have a partial body cavity (a *pseudocoel*) derived from the embryonic blastocoel, and *coelomate* animals have a true body cavity (a *coelom*), which forms directly from embryonic mesoderm. The internal organs of coelomates hang within their body cavity inside pouches of mesodermal tissue called *peritoneum*.
17. Coelomate animals are found in both the protostomes and the deuterostomes. Protostome coeloms are formed when the embryonic mesoderm splits to form a cavity. The coelom arises as an outpocketing of the gut in deuterostome animals.

Questions (for answers and explanations, see page 210)

1. All but one of the following are factors that help determine animal body plans. Select the *exception*.
   a. Patterns of embryological development
   b. Type of body symmetry
   c. Whether or not the animal is heterotrophic
   d. Type of body cavity
   e. Specializations in feeding mechanism
2. An animal is known to be a deuterostome. Which of the following does *not* apply to this animal?
   a. It is diploblastic.
   b. If a coelom is present, it formed from the embryonic gut.
   c. Its embryos will exhibit indeterminate cleavage.
   d. Its early embryonic cleavage pattern was radial.
   e. Three distinct tissue layers were present while it was an embryo.
3. The primary reason why animals are more diverse than plants is because of considerable variation in animal
   a. body plans.
   b. embryology.
   c. symmetry.
   d. multicellularity.
   e. methods of food acquisition.

4. Cephalization is a characteristic mainly associated with which of the following types of body symmetry in animals?
   a. Asymmetric
   b. Radial
   c. Biradial
   d. Bilateral
   e. Spherical

5. If an animal is a protostome, then it also
   a. has spiral embryonic cleavage.
   b. has determinate embryonic cleavage.
   c. has a mouth that develops from the blastopore.
   d. is likely to be triploblastic.
   e. All of the above

6. An animal is divided along its main body axis to produce similar halves. Which of the following types of symmetry could apply?
   a. Spherical
   b. Radial
   c. Radial or biradial
   d. Bilateral
   e. Biradial or bilateral

Activities (for answers and explanations, see page 210)

- Complete the following table comparing characteristics of the protostomes and deuterostomes.

Characteristics	Protostome	Deuterostome
Cell type	deter.	indeter.
Mode of nutrition		
# of embryonic layers	3	3
Type of cleavage		
Cleavage pattern	spiral	radial
Fate of blastopore	mouth	anus

- Explain the basic differences between the acoelomate, pseudocoelomate, and coelomate body plans found in animals.

THE ORIGINS OF ANIMALS • SIMPLE AGGREGATIONS (pages 578-582)

Key Concepts

1. The advantages that accrued from the division of labor and specialization of cell types among colonial protists probably account for the evolutionary origins of the animals. Protistan ancestors of the animals may have resembled modern day colonies of *Volvox*, although multicellular animals probably emerged from the protists in at least three separate lineages leading to the sponges, flatworms, and cnidarians.
2. Animal origins date back nearly one billion years, with many millions of species having appeared during that time. Although incomplete, the animal fossil record yields much useful information about the evolution of this kingdom.
3. Physical and physiological issues related to *gas exchange* and the concentration of oxygen in the Earth's atmosphere have strongly influenced animal body form. Since surface-to-volume ratios decrease as individual cells or multicellular bodies get larger, specialized mechanisms have evolved in large animals to enhance gas exchange. Examples are structures such as gills, lungs, and circulatory systems.
4. About 700 million years ago (mya), when atmospheric oxygen concentrations became high enough to sustain the metabolic activity of large eukaryotic cells, simple animals appeared and began to diversify structurally. Although absent initially because low oxygen concentration limited body size and activity, the evolution of sexual reproduction in animals greatly increased the rate at which new forms appeared.
5. The primary food sources for early animals were *phytoplankton* (small floating algae) and the extensive *algal mats* that covered the shallow sea bottoms. Morphological adaptations for feeding on algae favored the evolution of larger, more mobile animals. As they evolved and became more abundant, these larger animals also began to feed on *zooplankton* (small floating animals).
6. The presence of large grazing animals opened evolutionary avenues for carnivores. This selective pressure in turn led to the development of protective shells, burrowing, and better locomotion among prey species.
7. One of the earliest animal groups to evolve from the colonial protists was the phylum *Porifera* (sponges). Sponges are exclusively aquatic aggregations of specialized cells that lead a *sessile* (attached to the substrate) lifestyle. Flagellated collar cells called *choanocytes* move water and food particles into the animal via numerous small pores on the body wall, and then out again through a larger opening at the top.
8. Wandering cells known as amoebocytes provide a primitive level of communication between sponge cells. Structural support is provided by mineralized or proteinaceous spicules.
9. Despite this cellular specialization, sponges lack true tissue-level organization or any regular form of body symmetry.
10. The most important evolutionary force shaping sponge morphology has been patterns of water movement. Shallow waters with strong wave action yield flattened forms, while larger, more upright forms are found in calm, deep waters.
11. Reproduction in sponges can be either sexual or asexual. A single individual produces both sperm and eggs,

which are released into the water column. Asexual reproduction is by budding.

12. The unusual phylum *Placozoa* consists of only two species of very small, asymmetric animals with no body cavity or distinct tissues and organs. They have a flattened body plan consisting of two layers of flagellated cells that cover a fluid-filled cavity. Placozoans feed on bacteria and protists by external digestion.

Activities (for answers and explanations, see page 210)

- Refer to Fig. 26.7 in your textbook and identify the function(s) of each of the following structural components of a sponge:

  Amoebocytes:

  Choanocytes:

  Incurrent pores:

  Osculum:

  Spicules:

- Explain why physical properties related to cell size and gas exchange limited the evolution of early organisms, and then list the ways in which multicellular animals solved this problem as atmospheric oxygen concentrations increased during Earth's history.

Questions (for answers and explanations, see page 210)

1. Which of the following is *not* a good explanation for why grazers, rather than carnivores, were the first animals to evolve?
   a. Animal biomass is generally not as good a form of nutrition as is plant biomass.
   b. The most abundant food at the start of animal evolutionary history was large populations of algae.
   c. Early animals were likely too small to have been very effective predators.
   d. Anatomical specializations for chasing down and catching prey had not yet evolved.
   e. All of the above

2. The increase in the diversity of simple grazing animals was followed by the rapid evolution of larger, more mobile predators. Which of the following is *not* a useful defense mechanism against predation?
   a. Improved locomotion
   b. Protective shells
   c. Burrowing in mud or sand
   d. Increased rate of new species formation

3. Sponges have, in general, a very simple body plan. Which of the following statements about sponge structure or function is *not* true?
   a. Choanocytes are flagellated cells that play a role in food acquisition.
   b. Large species are usually found in areas of heavy wave action, where more food can be obtained.
   c. Individual sponges are both male and female.
   d. As a whole, the phylum Porifera shows no regular pattern of body symmetry.
   e. Water enters a sponge through incurrent pores, and exits via one or more oscula.

4. Which of the following is *not* a trait shared by both sponges and placozoans?
   a. Asymmetric body plan
   b. No true tissues and organs
   c. Flagellated cells
   d. Mineralized structural support
   e. Limited numbers of specialized cell types

5. Which of the following is *not* a group of animals thought to have arisen from the ancestral colonial protists during the early stages of animal evolution?
   a. *Volvox*
   b. Placozoans
   c. Cnidarians
   d. Flatworms
   e. Sponges

THE EVOLUTION OF DIPLOBLASTIC ANIMALS (pages 582–588)

Key Concepts

1. The phyla *Cnidaria* and *Ctenophora* contain the only diploblastic animals, and they likely represent a distinct evolutionary lineage emerging from the ancestral protists. They have no body cavity and are radially symmetric.

2. The 10,000 species of cnidarians (sea anemones, jellyfish, corals, and their relatives) are primarily marine species named for the possession of a unique cell type called the cnidocyte, which contains a dischargable stinging structure known as a *nematocyst*. This stinging mechanism enables cnidarians to subdue large prey items, and provides better defense against predators.

3. The cnidarian body plan is more complex than that of sponges. Distinct tissue layers are present, including a modified epithelium, which forms tentacles and a nerve net to help integrate body activities. A single opening serves as both mouth and anus for a blind-ended sac called the *gastrovascular cavity*, which aids in digestion, gas exchange, and circulation.

4. Many cnidarians exhibit alternate body forms. The life cycle often begins with the asexual *polyp* stage, a sessile cylindrical stalk with mouth and tentacles directed upward. The other body form, or *medusa* stage, has a downward-directed mouth and tentacles and represents a free-living, sexual stage. Much of the medusa consists of a gelatinous material called *mesoglea* which forms a middle body layer almost devoid of cells and having very low metabolism.

5. The ciliated cnidarian larva that forms from the union of sperm and egg is called a *planula*. After a brief period spent swimming freely in the water column, it settles to the bottom and transforms into a polyp.
6. Hydrozoa is a class of cnidarians in which the polyp dominates the life cycle. This group contains the only freshwater members of the phylum. Some species are solitary, whereas others are colonial with a high degree of polyp specialization and a common gastrovascular cavity. A few hydrozoans, such as the siphonophores, combine medusae and polyps to form a complex, free-floating animal.
7. The medusa stage dominates the life cycle of the exclusively marine class Scyphozoa (jellyfish). These animals are active swimmers with long tentacles specialized for defense and prey capture. Individual animals are either male or female. Their planular larvae form through sexual reproduction and then undergo a brief polyp stage before once again becoming free-swimming medusae.
8. The class Anthozoa includes the sea anemones and corals, which are all marine. In this group the medusa stage is completely absent, with both asexual and sexual reproduction performed exclusively by the polyp.
9. Anemones are typically solitary and sedentary, although a few can crawl or swim, while corals are usually colonial and exclusively sessile. Corals produce an external skeleton of calcium carbonate that supports the colony. The buildup of these skeletons over geologic time has played an important role in formation of large oceanic reefs.
10. Many corals live in clear, nutrient-poor tropical waters, but survive by forming symbiotic relationships with photosynthetic dinoflagellates. These tiny protists significantly contribute to the nutritional base of the coral colony in return for the defense provided by the animal's stinging cells.
11. The long evolutionary history and success of cnidarians is largely attributed to their combination of simple body plans, low metabolic rates, and the ability to capture relatively large prey. In addition, associations with photosynthetic symbionts further enhance these other advantages for some cnidarians.
12. The diploblastic phylum *Ctenophora* (comb jellies) contains about 100 species of marine carnivores. Ctenophores are superficially similar to the cnidarians, but also have important structural differences. Ctenophore guts have one-way flow by virtue of two anal pores that open opposite the mouth. Body movement is by ciliary action, rather than by muscles as in the cnidarians.
13. The comb jellies lack stinging cells. They capture prey using sticky filaments attached to their tentacles. Ctenophores are primarily plankton feeders and do not pursue large food items as do the cnidarians. Sexual reproduction is simple: a miniature ctenophore develops directly from a fertilized egg and slowly grows to adult size.

Activities** **(for answers and explanations, see page 211)

- Compare and contrast the anatomy and foraging mechanisms of cnidarians with those of ctenophores.

- For each cnidarian class listed below, indicate with a checkmark whether the medusa or the polyp is the dominant stage in the life cycle.

Class	Medusa	Polyp
Hydrozoa		
Anthozoa		
Scyphozoa		

- On the diagram below showing the polyp stage in a generalized cnidarian life cycle, label the following structures: ectoderm, endoderm, mesoglea, gastrovascular cavity, mouth, and tentacles.

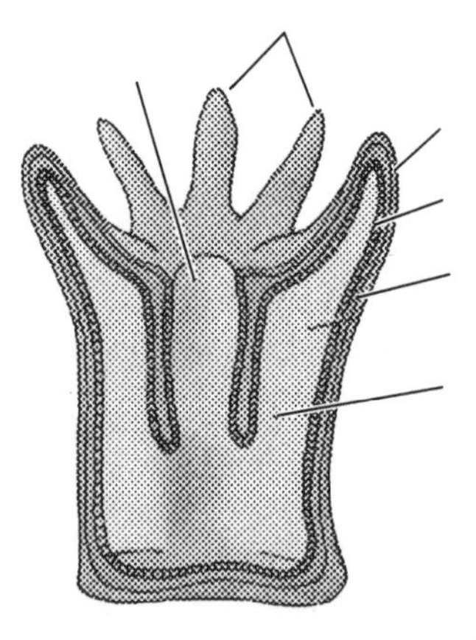

Questions** **(for answers and explanations, see page 211)

1. Compared to the sponges, which of the following does *not* represent an advancement in body plan found in the cnidarians?
   a. Presence of a distinct body cavity
   b. Radial symmetry
   c. Possession of nematocysts
   d. Alternation of body forms
   e. Gastrovascular cavity

2. Corals are common representatives of the class _________ within the phylum _________.
   a. Scyphozoa; Cnidaria
   b. Anthozoa; Ctenophora
   c. Hydrozoa; Cnidaria
   d. Scyphozoa; Ctenophora
   e. Anthozoa; Cnidaria

3. Which of the following traits is *not* shared by both sea anemones and jellyfish?
   a. Nematocysts located on tentacles
   b. Possession of a gastrovascular cavity
   c. Dominant stage in life cycle is a medusa
   d. Engage in sexual reproduction
   e. Are carnivorous

4. A major difference in feeding strategies between jellyfish and ctenophores is that
   a. jellyfish are predatory, while ctenophores are grazers.
   b. ctenophores are predatory, while jellyfish are grazers.
   c. ctenophores hunt as sessile polyps, jellyfish hunt as medusae.
   d. jellyfish have stinging cells on their tentacles, ctenophores catch prey with sticky filaments.
   e. ctenophores have stinging cells on their tentacles, jellyfish catch prey with sticky filaments.

5. Which of the following characteristics is unique to the phylum Cnidaria?
   a. Sexual reproduction
   b. Symbiotic associations with other organisms
   c. Sedentary body forms
   d. Nematocysts
   e. Nervous tissue

THE EVOLUTION OF BILATERAL SYMMETRY • THE DEVELOPMENT OF BODY CAVITIES • PSEUDOCOELOMATE ANIMALS (pages 588–592)

Key Concepts

1. Most animals exhibit bilateral symmetry, which probably first arose in simple, flattened, crawling organisms not unlike the modern placozoans.

2. All bilaterally symmetric animals are triploblastic. One early lineage is represented by the phylum *Platyhelminthes*, or flatworms. Although they have no body cavity, they do have more complex internal organs than either the cnidarians or ctenophores. Their flattened body plan permits efficient respiratory gas exchange because all cells are near the body surface.

3. The digestive system of flatworms contains a mouth that opens into a blind-ended sac, which is often highly branched to increase the surface area available for nutrient absorption.

4. The class Turbellaria is believed to most closely resemble ancestral flatworms. *Dugesia*, the common freshwater planarian, is representative of the group. Chemosensory organs are located on the anterior end of the animal, along with two light-sensitive eye spots. The presence of longitudinal nerve cords and a modest brain represent additional improvements for sensory control and locomotion in the flatworm body plan.

5. Some flatworms are carnivorous, others are scavengers. However, the bulk of the 25,000 species of flatworms are parasites. Tapeworms (class Cestoda) and certain flukes (class Trematoda) are common internal parasites of animals, especially vertebrates. Other flukes (class Monogenea) are important external parasites.

6. Complex life cycles with multiple hosts often characterize the existence of parasitic flatworms. These features are thought to help defeat the host's defense mechanisms and to increase the flatworm's likelihood of successful reproduction.

7. Fluid-filled body cavities first arose as adaptations to improve prey capture and avoid predators. Such *hydrostatic skeletons* work by creating forces through the action of opposing muscle groups on the constrained, incompressible fluid inside the body.

8. The 900 species of ribbon worms (phylum *Nemertea*) are mostly marine and resemble flatworms in many respects. A notable exception is that nemerteans possess a complete digestive tract with a mouth at one end and an anus at the other. This one-way flow of food permits specialization of the digestive tract, leading to increased digestive efficiency.

9. The fluid-filled body cavity, or *rhynchocoel*, of ribbon worms contains a proboscis armed with a sharp stylet which can be ejected with considerable force during prey capture. It also assists in burrowing.

10. An advancement on the rhynchocoel is found in other groups that evolved from the flatworms. The pseudocoelom is a fluid-filled cavity containing the internal organs which also serves as a hydrostatic skeleton. Such skeletons make locomotion more efficient by allowing forces generated by muscle contraction to be transferred from one portion of the body to another. Two examples of pseudocoelomate animals are the roundworms (phylum *Nematoda*) and the rotifers (phylum *Rotifera*).

11. Roundworms have a thick cuticle overlying their epidermis, beneath which are located well-developed longitudinal muscles. Circular muscles are lacking, resulting in somewhat inefficient thrashing movements. Roundworms are among the most abundant of all animals, with over 20,000 species known and many more yet to be identified. Tens of thousands or millions of individuals can be found inhabiting very small volumes of organic matter.

12. Some roundworms are predators on protists and very small animals, while others are significant parasites of plants and mammals, including humans. The roundworm *Trichinella spiralis* causes the sometimes fatal disease trichinosis. In the case of human infection, pigs usually serve as the intermediate host for the parasite.

13. Rotifers are tiny animals not much larger than ciliate protists. The 1,800 species possess well-developed internal organs including a complete digestive tract. Their pseudocoel is greatly reduced and does not function as a hydroskeleton. All rotifers are aquatic (mostly freshwater) and very active. Their locomotion is made possible by beating cilia.

14. Food is pulled into a rotifer's mouth on water currents established by the beating of cilia in the corona, a feeding organ located atop the animal's head. Organic particles are swept into the mastax, where grinding teeth help process the food.

Activities (for answers and explanations, see page 211)

- The following schematic diagram shows a longitudinal section through an unidentified bilaterally symmetric animal. Explain why, based solely on the information available in this diagram, you can conclude that this mystery animal is *not* a member of the phylum Platyhelminthes.

- Assuming that the above schematic animal is a roundworm, draw the orientation of the muscles that operate its hydroskeleton, and briefly explain the kind of body motions their contractions would produce.

Questions (for answers and explanations, see page 211)

1. Which of the following statements is true of all flatworms?
   a. Flatworms are biradially symmetric.
   b. Flatworms have a complete digestive system.
   c. Flatworms tend to have large, thickened bodies.
   d. Flatworms are triploblastic.
   e. All of the above

2. Although relatively simple, *Dugesia* shows significant advancements in body plan compared to the cnidarians and ctenophores. Which of the following is *not* one of those advancements?
   a. Longitudinal nerve cords
   b. Complete digestive tract
   c. Anterior chemosensory organ
   d. Light-sensitive eye spots
   e. Highly branched gastrovascular cavity

3. Ribbon worms are members of the phylum
   a. Platyhelminthes.
   b. Nemertea.
   c. Nematoda.
   d. Rotifera.
   e. Cestoda.

4. Hydrostatic skeletons are found in all but one of the following animal groups. Select the *exception.*
   a. Roundworms
   b. Nemerteans
   c. Rotifers
   d. Nematodes
   e. Ribbon worms

5. The pseudocoel is a body plan feature found in which of the following groups?
   a. Trematoda only
   b. Rotifera and Nematoda
   c. Rotifera only
   d. Nemertea and Cestoda
   e. Nemertea only

6. Consider the diagram of a rotifer shown below:

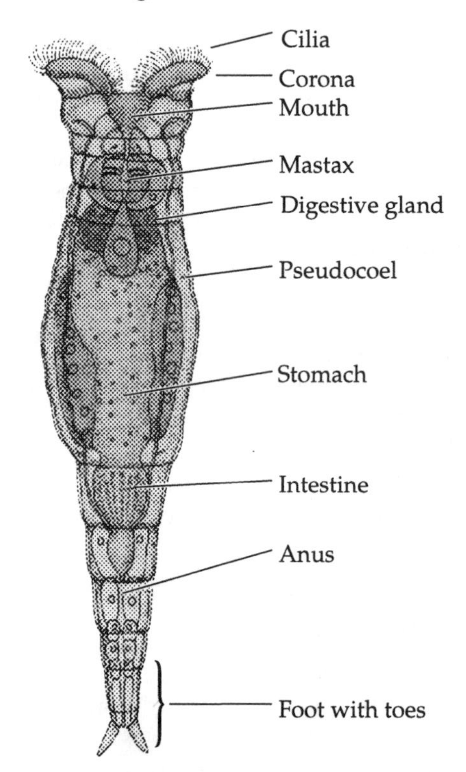

   Which aspects of this animal's body plan (apparent from this diagram) are also found in the phylum Nematoda?
   a. Complete digestive tract
   b. Bilateral symmetry
   c. Pseudocoelom body cavity
   d. Complex organ structure
   e. All of the above

COELOMATE ANIMALS • THE EVOLUTION OF EXTERNAL SKELETONS (pages 592–602)

Key Concepts

1. The complete encircling of the coelomic cavity by both longitudinal and circular muscles, combined with body segmentation for added precision of movement, gave animals a new and highly effective means of controlling body shape. This change in body plan occurred several times in widely divergent phyla.

2. Pogonophores (phylum *Pogonophora*) are deep-water, bottom-dwelling marine animals with tentacles that can be drawn down well into the coelom. There are about 145 species, all of which have been discovered only recently. The largest live near oceanic hydrothermal vents and use their large coelomic cavities to harbor colonies of specialized bacteria that fix carbon using energy obtained from hydrogen sulfide gas.
3. The 15,000 members of the phylum *Annelida* (annelid worms) are found in freshwater, marine, and terrestrial habitats. They exemplify the advantages gained by repeated segmentation of the coelom. Complex body shapes can be attained, the coordination of which is enhanced by separate nerve centers in each segment.
4. Annelids lack a rigid external covering, and their flexible body wall is key to the locomotory process. This thin wall is also essential for gas exchange, which limits annelids to moist habitats.
5. The majority of annelids belong to the class Polychaeta. The class is almost entirely marine, with well-developed anterior eyes and tentacles. Thin outgrowths known as *parapodia* emerge from each segment, serving respiratory as well as locomotory functions. Stiff bristles called setae also assist in locomotion.
6. Polychaetes usually have distinct male and female sexes. The larval stage, known as a *trochophore*, adds body segments as it grows, eventually metamorphosing into an adult worm.
7. The class Oligochaeta is composed mainly of terrestrial or freshwater annelids. The common earthworm is an example. They are virtually naked by comparison to polychaetes, lacking all external appendages except for a few setae. Most are scavengers.
8. All oligochaetes are *hermaphroditic* (an individual has both male and female sex organs). When two individuals copulate, sperm is exchanged between each member of the pair. Eggs are laid internally in a cocoon, which is later shed and hatches to release miniature worms.
9. The class Hirudinea (leeches) shows many similarities in body form and reproduction to the oligochaetes. However, leech body segments are not divided into distinct compartments as in the annelids. Their movement is thus quite different. Leeches move with the aid of suckers located at each end of the body.
10. Most leeches are external parasites of other animals. They have toothed jaws that can penetrate their host's outer surface. An anticoagulant is then injected into the wound to maintain blood flow.
11. The development of *exoskeletons* was a major evolutionary change in animal body form. Originating from a thickening of the outer body covering through incorporation of the strong, flexible polysaccharide chitin, exoskeletons later became hardened to assume supportive, protective, and locomotory functions.
12. The large grouping known collectively as the arthropods are among the most successful animals with exoskeletons. Their hardened external coverings are jointed, especially the appendages. This design, in conjunction with specialized internal musculature, permits rapid movement. Other jointed appendages acquired sensory, feeding, or reproductive functions.
13. A hardened exoskeleton makes hydrostatic movements virtually impossible. Arthropod coeloms have thus acquired a new function: they now act as body cavities in which blood bathes the animal's internal organs.
14. Exoskeletons are not without their limitations. Because they are impermeable to gases, specialized respiratory systems such as gills or trachea had to evolve in arthropods. Moreover, since they are of fixed size, exoskeletons constrain the animal's growth and must be molted periodically to accommodate increases in body size. This temporarily renders the animal unprotected and hinders its movement until a new skeleton forms.
15. Arthropods are a highly diverse lineage, due largely to the great adaptability of their jointed exoskeleton. The locomotory, supportive, protective, and antidesiccation benefits of this body plan allowed arthropods to successfully colonize terrestrial environments. The arthropod phylum accounts for the largest percentage of all animal species, numbering well over one million.
16. The 75 species of onychophorans (phylum *Onychophora*) may be modern examples of the ancestral arthropod body plan. Their bodies are not segmented, and they share anatomical traits with both the annelids (internal organ structure and flexible body covering) and the arthropods (circulatory system and simple appendages). Another modern group showing the primitive arthropod body plan includes the 550 living species of water bears (phylum *Tardigrada*). Their body plan is similar to oncychophorans, but since they are much smaller they lack circulatory and respiratory systems. Water bears live in marine sediments.
17. Trilobites (phylum *Trilobita*) are an extinct group of arthropods that flourished until the end of the Paleozoic; why they disappeared remains a mystery. Although heavily armored, they had a simple repetitive body plan reminiscent of the annelids.
18. All living arthropods are classified into one of three phyla: the *Chelicerata*, *Crustacea*, and *Uniramia*.
19. The classes Pycnogonida (sea spiders), Merostomata (horseshoe crabs), and Arachnida (spiders, scorpions, mites, and relatives) are all members of the Chelicerata. The bodies of all 63,000 known species are divided into two regions, an anterior portion with mouth parts and a posterior region with four pairs of walking legs. Sea spiders are a small group of exclusively marine chelicerates. Horseshoe crabs are locally common along certain seashores, and have remained essentially unchanged for many millions of years.
20. Arachnids are highly successful terrestrial relatives of the horseshoe crabs. Most have simple life cycles where miniature adult forms hatch directly from fertilized eggs. Others give birth to live young. Some are parasitic, but many are predators (e.g., spiders) and use their well-developed vision or elaborate silken webs to capture prey.

21. Crustaceans, with over 40,000 species, are the dominant marine arthropods, and include such familiar forms as copepods, crabs, shrimp, and lobsters. The body plans of most consist of a head, thorax, and abdomen, with appendages attached to each. A protective fold of the exoskeleton, the carapace, covers the head and other body segments. Separate sexes engage in copulation, and in some species the eggs hatch into a free-living larval form called a *nauplius*.

22. Barnacles (class Cirripedia) are specialized crustaceans in which the adult is housed inside a secreted calcium shell attached to the substrate. The animal stands on its head within this shell, and feeds by extending its modified appendages into the water currents to gather food particles.

23. The phylum Uniramia contains two important classes of primarily terrestrial arthropods, the Myriapoda and the Insecta. The most notable members of the myriapods are the millipedes and centipedes. The 13,000 species have long, flexible bodies with well-formed heads, segments, and appendages. Millipedes are scavengers, while centipedes are active predators.

24. The bodies of insects are divided into a head, thorax, and abdomen. A single pair of antennae is attached to the head and three pairs of legs are connected to the thorax. Specialized tubes called *tracheae* allow gas exchange by bringing air directly to the body tissues.

25. Insects are incredibly abundant and diverse. The 1.5 million described species probably represent only a fraction of those living today. They are found in virtually all habitats, although very few live in the oceans. Insects make their living through predation, scavenging, and parasitism.

26. Insects that hatch as miniature adults and then grow by simply increasing their size are said to have *simple development*. In other species, hatchlings are radically different from the adult form and must undergo a developmental change known as *metamorphosis*. In metamorphosis, subsequent molts produce new immature stages called *instars*. If changes between these instars are gradual, the metamorphosis is *incomplete*, but if the changes are dramatic, then the metamorphosis is said to be *complete*.

27. Associated with complete metamorphosis is an extreme specialization of each stage in the life cycle. Adults are often best adapted for reproduction and dispersal, whereas larvae are adapted for growth and feeding. A specialized inactive phase called the *pupa* represents a transitional form between the larva and the adult.

28. Entomologists recognize some 28 orders of insects, all of which can be classified as either wingless or winged. The winged insects can be further subdivided into those that do not fold their wings over the back, and those that do. The ability to fold the wings means that they are out of the way while the insect is not in flight. This adaptation has allowed the folded-wing groups to invade many more habitats than the fixed-wing groups, and they have thus become more abundant and diverse.

29. Insect diversity is in part explained by the ancient origins of this class, which permitted them to colonize relatively unexploited terrestrial habitats and thereby undergo extensive evolutionary adaptation in body form.

Activities (for answers and explanations, see page 211)

- On the following drawing of a crab larva, or nauplius, identify as many characteristics as you can that designate this animal as a member of the arthropod class Crustacea.

- List several ways in which the body plans of oligochaete and polychaete worms are similar, and several ways in which they differ.

  Similarities:

  Differences:

Questions (for answers and explanations, see page 211)

1. Lobsters, millipedes, and butterflies all share which of the following traits?
   a. Parapodia
   b. Setae
   c. Gas exchange across the skin
   d. Jointed appendages
   e. Complete metamorphosis

2. Suppose you have a box before you that contains an accurate representation of all annelid worm species. If you blindly reach into this box and pull out one individual, odds are that it will be a member of the class
   a. Oligochaeta.
   b. Polychaeta.
   c. Pogonophora.

d. Hirudinea.
e. Onychophora.

3. All but one of the following animals has an exoskeleton. Select the *exception*.
a. Millipede
b. Barnacle
c. Honeybee
d. Copepod
e. Earthworm

4. One day while vacationing at the beach, your friend brings you an unidentified animal that was found in a local tidepool. You note that it has a true coelom, numerous appendages, and only male sex organs. It also lacks any evidence of an exoskeleton. To which of the following groups does this animal most likely belong?
a. Polychaeta
b. Pogonophora
c. Oligochaeta
d. Crustacea
e. Insecta

5. When grasshoppers hatch from eggs, they look like miniature adults. Over time they pass through several molting stages until they eventually reach full maturity. This pattern of insect development is known as
a. complex development.
b. incomplete metamorphosis.
c. complete metamorphosis.
d. pupal development.
e. simple development.

6. Onychophorans and tardigrades are similar to earthworms in that they have
a. jointed appendages.
b. a hardened exoskeleton.
c. a flexible body covering.
d. a well-developed respiratory system.
e. All of the above

7. The bulk of arthropods belongs to the phylum Uniramia, and of these, the majority of species belong to the class
a. Myriapoda.
b. Insecta.
c. Cirripedia.
d. Tardigrada.
e. Hirudinae.

CALCIFIED PROTECTION • THEMES IN PROTOSTOME EVOLUTION (pages 602–607)

Key Concepts

1. The 100,000 species of modern mollusks (phylum *Mollusca*) possess heavily calcified shells into which most forms may retract their body to avoid predators. These shells may be large and rather ornate. Early mollusks were wormlike with only calcified spicules on their dorsal surface for protection.
2. An important diagnostic character of mollusks is the *radula*, a rasping feeding structure located in the anterior end of the animal. In some forms it has become modified for drilling, or else to act as a piercing dart which is attached to poison glands.
3. Although highly modified within specific groups, the basic molluscan body plan consists of a large muscular *foot* for locomotion, a covering of specialized tissue called the *mantle* which protects the internal organs and secretes the shell, and a *visceral mass* of internal organs proper.
4. Mollusks have a complete digestive tract and a simple nervous system. The fluid-filled sinuses of the open circulatory system are important elements in the animal's hydrostatic skeleton.
5. Chitons (class Polyplacophora) have conspicuous plates that connect along the back to form the shell. They are marine grazers that spend most of their lives firmly attached to algae-covered rocks.
6. The monoplacophorans (class Monoplacophora) have a single, caplike shell instead of a series of interconnected plates. Once thought extinct, they are believed to be the ancestors of all other mollusks. They have multiple gills, well-developed muscles, and a repeating excretory system.
7. During monoplacophoran evolution, one lineage developed a hinged shell that covered the entire body. The class Bivalvia (bivalves) includes the clams, oysters, scallops, and mussels. Many have large, heavy shells, which have so reduced locomotion that they are quite sedentary. Food is filtered from the water by the gills, which serve both respiratory and feeding functions.
8. A second lineage of monoplacophorans developed spiral shells. These gastropods (class Gastropoda) use their large muscular foot to move about slowly.
9. Gastropods undergo a characteristic torsion during larval development, which ends up rotating the visceral mass 180° relative to the foot. The result is that both the anus and mouth come to lie above the head, producing a U-shaped digestive tract and nervous system.
10. Species richness is high among gastropods. Familiar crawling forms include snails, slugs, whelks, limpets, abalones, and nudibranchs. Some are of substantial commercial value. Other, less familiar forms such as the sea butterflies and heteropods are active swimmers.
11. Cephalopod mollusks (class Cephalopoda) are the most advanced members of the phylum, and include such species as the squids and octopuses. They evolved relatively recently from yet a third lineage of monoplacophorans.
12. Cephalopods have sophisticated mechanisms for regulating the air/water mixture in their shell cavity, making them the first animals to move vertically within the water column. They also have an exit tube that can forcibly direct water out from the shell, allowing them to be rapid-swimming, active predators.
13. Advanced locomotory abilities and a pelagic (open ocean) lifestyle have favored the evolution of elaborate sensory organs among cephalopods. Their eyes are in many ways comparable to those of humans and other vertebrates.

14. The cephalopod foot is modified into a head with tentacles, a siphon for creating propulsive water jets, and a well-developed brain. Cephalopods have large gills inside the mantle cavity, and a beak for capturing and subduing prey.

15. Members of the genus *Nautilus* are living representatives of the earliest cephalopods. They possess an external shell and have distinct limitations for controlling their buoyancy. Evolution altered the body plans of subsequent cephalopod groups, substantially reducing the size of the shell and substituting other mechanisms of buoyancy control, such as fast swimming.

16. The fluid-filled cavities of protostomes were excellent precursors for hydrostatic skeletons which, with appropriate advancements in musculature, enabled sophisticated movements. Subsequent segmentation of the body further enhanced mobility, ultimately leading to the advanced swimming ability of cephalopods.

17. Hard, external protective coverings evolved mainly in response to predation pressure. Once in place, these coverings were incorporated as key elements of new locomotory systems.

18. Since many protostome lineages originated at times when only very small organisms or particulate food were available in the water, numerous sessile forms that actively extract food from moving water are still represented today. Despite the relative ease of feeding, sessile body forms require specializations for finding mates, the dispersal of gametes and larvae, and to deal with the intense competition for space. The emergence of coloniality was one evolutionary solution to these problems.

Activities (for answers and explanations, see page 212)

- Protostome evolution was intimately linked to the aquatic environment, and only much later did colonization of the land occur. List three lines of evidence that support this view of a watery origin for the protostomes.

  a.

  b.

  c.

- Next to the arthropods, mollusks are the most diverse group of invertebrate animals. List three reasons why the mollusks have been so evolutionarily successful.

  a.

  b.

  c.

Questions (for answers and explanations, see page 212)

1. Match each of the characteristics listed below with the molluscan group(s) that show those characteristics.

a. Advanced eyes	_____	Monoplacophorans
b. Larval torsion	_____	Cephalopods
c. Pelagic hunting	_____	Gastropods
d. Calcified shells	_____	Polyplacophorans
e. Multiple gills	_____	Bivalves

2. While dissecting a squid, you would expect to find all but one of the following. Select the *exception*.
   a. A beak
   b. Numerous interconnected plates forming a shell
   c. Well-developed gills inside the mantle cavity
   d. Large grasping tentacles
   e. A well-developed brain and visual system

3. The group of mollusks believed to have given rise to all other mollusks is the class
   a. Bivalvia.
   b. Polyplacophora.
   c. Gastropoda.
   d. Cephalopoda.
   e. Monoplacophora.

4. A meal containing snails, clams, and octopus represents which of the following classes of mollusks?
   a. Bivalvia only
   b. Polyplacophora only
   c. Gastropoda, Bivalvia, and Cephalopoda
   d. Cephalopoda and Gastropoda
   e. Monoplacophora only

5. Which of the following structures is *absent* from a typical gastropod mollusk such as a garden snail?
   a. Protective shell
   b. Head
   c. Radula
   d. Mantle
   e. None of the above

6. The cephalopods are highly advanced mollusks. Which of the following is *not* a characteristic of this group?
   a. Grasping tentacles
   b. Sophisticated vision
   c. Sessile lifestyle
   d. Propulsive water siphon
   e. Large brain

Integrative Questions (for answers and explanations, see page 212)

1. Which of the following statements is true of all bilaterally symmetric animals?
   a. They are ingestive heterotrophs.
   b. They have some form of digestive system.
   c. They are multicellular.
   d. They are triploblastic.
   e. All of the above

2. Ants, snails, and earthworms all share which of the following characteristics?
   a. They are protostomes.
   b. They are acoelomate animals.
   c. They are biradially symmetrical.
   d. They have excellent vision.
   e. None of the above

3. The body plans of rotifers and squid are alike in that they both
   a. have spherically symmetrical bodies.
   b. possess complete digestive tracts.
   c. are pseudocoelomate animals.
   d. have radial cleavage of the embryo.
   e. All of the above

4. A lobster eats a clam. A squid then eats the lobster. Which phyla of animals are involved in this feeding sequence?
   a. Crustacea only
   b. Mollusca only
   c. Mollusca and Uniramia
   d. Crustacea and Mollusca
   e. Uniramia and Crustacea

5. The extinct group of animals known as the trilobites are most closely related to which of the following?
   a. Crabs
   b. Earthworms
   c. Water bears
   d. Sponges
   e. Sea anemones

6. All but one of the following animals exhibits cephalization. Select the *exception.*
   a. Cockroach
   b. Nautilus
   c. Jellyfish
   d. Earthworm
   e. Lobster

27

Deuterostomate Animals

CHAPTER LEARNING OBJECTIVES—after studying this chapter you should be able to:

- ❑ List the basic characteristics used to classify deuterostome animals.
- ❑ Discuss the structure and function of a lophophore, along with the basic elements that make up the lophophorate body plan.
- ❑ Distinguish between representative lophophorate animals belonging to the phyla Phoronida, Bryozoa, and Brachiopoda.
- ❑ Describe how hemichordates differ from lophophorates.
- ❑ List the basic structural and lifestyle characteristics of echinoderms, and identify representative echinoderms to their appropriate subphylum and class.
- ❑ Explain the significance of the water vascular system in echinoderms, and describe how it works.
- ❑ Discuss how pharyngeal slits became modified for feeding in members of the phylum Chordata.
- ❑ List the main features of the chordate body plan, and explain how these represent improvements over the body plans of simpler deuterostomes.
- ❑ Describe the characteristics of the nonvertebrate chordates, and discuss how the vertebrates emerged from these animals.
- ❑ Present the main sequence of events that occurred in vertebrate evolution from the agnathans through the mammals.
- ❑ List the essential characteristics of each of the vertebrate classes and recognize representative members of each class.
- ❑ Discuss the importance of hinged jaws to the evolution of vertebrates.
- ❑ Describe the anatomical and physiological changes that occured as vertebrates made the transition from life in water to life on land.
- ❑ Explain the importance of the amniote egg to vertebrate evolution.
- ❑ Differentiate between the various reptilian lineages that gave rise to the dinosaurs, modern reptiles, birds, and mammals.
- ❑ List the characteristics that distinguish the mammals from all other vertebrates, and the primates from all other mammals.
- ❑ Discuss the evolutionary relationships of humans to the other primate groups.
- ❑ Describe the evolution of the genus *Homo* from the anthropoid primates and great apes.
- ❑ Explain in some detail the evolutionary history of the hominid lineage during the past 5 million years, including changes in species, tool use, and culture.
- ❑ List the main trends that have occurred in the evolution of the deuterostome animals.

TRIPARTITE DEUTEROSTOMES • INNOVATIONS IN FEEDING • ACTIVE FOOD SEEKERS (pages 610–614)

Key Concepts

1. Deuterostomes are characterized by having indeterminate cleavage, mesoderm that forms as an outpocketing of the embryonic gut, and an anus that forms directly from the blastopore. All are triploblastic and possess well-developed body cavities.
2. Although deuterostome species diversity is low compared to protostomes, the deuterostomes are much more variable in their body plans.
3. The lophophorate animals are sessile, filter-feeding, mainly marine deuterostomes that possess a specialized feeding and respiratory structure called a *lophophore.* It typically consists of one or two rows of ciliated hollow tentacles, which are located near the mouth and are under muscular control.
4. The lophophorate body plan consists of an anterior prosome, middle mesosome, and posterior metasome. Each of these body compartments has its own coelomic cavity, known respectively as the protocoel, mesocoel, and metacoel. The gut is U-shaped, with the anus emerging near the mouth but outside the tentacles.
5. The phylum *Phoronida* is a small group (15 species) of bottom-dwelling, marine worms that live in secreted chitinous tubes. Particulate food is filtered through the external lophophore and transferred to a stomach located at the animal's posterior end.

6. The moss animals (phylum *Bryozoa*, or *Ectoprocta*) are colonial lophophorates. They secrete "houses" from which individuals emerge to feed. Food is passed internally between individual animals by connecting tissues. The colony forms through the asexual reproduction of a single founding member. Sexual reproduction yields free-ranging larvae that ultimately settle to form new colonies.
7. The brachiopods (phylum *Brachiopoda*) superficially resemble bivalve mollusks, but they are actually sedentary lophophorates whose feeding structure is housed inside a hinged shell. Once very numerous, today only about 350 species survive in marine habitats.
8. One lineage of deuterostomes modified the lophophore and its coelomic cavity to create a new tripartite body plan consisting of a proboscis, collar, and trunk. These hemichordates (phylum *Hemichordata*) include two groups of animals with widely divergent feeding modes.
9. One small group of hemichordates, the pterobranchs (class *Pterobranchia*), is made up of 10 species of mostly colonial, bottom-dwelling worms that live inside their secreted tubes. The proboscis is surrounded by a collar of tentacles which serves both respiratory and feeding functions.
10. The other group of hemichordates comprises the 70 species of acorn worms (class *Enteropneusta*), whose lophophore has been replaced by a proboscis modified for digging. A sticky mucus coating on the proboscis facilitates prey capture. Food is funneled to the mouth, behind which are located numerous pharyngeal slits that once filtered sand but now function exclusively in respiration.
11. Although the basic lophophorate body plan is mainly associated with sedentary or only slightly mobile lifestyles, other tripartite deuterostomes such as the arrow worms (phylum *Chaetognatha*) have become quite mobile.
12. The 100 species of arrow worms have a streamlined, bilaterally symmetric body with two pairs of lateral fins and a taillike caudal fin. These small, marine carnivores can thus swim in rapid, short bursts which allows them to feed among masses of plankton and larval fish. Some arrow worms also engage in daily vertical migrations of several hundred meters to avoid predators.

Activity (for answers and explanations, see page 212)

- For each of the deuterostome groups listed below, indicate with a "yes" or "no" which groups have a lophophore.

________Pterobranchs

________Bryozoans

________Arrow worms

________Phoronids

________Brachiopods

________Acorn worms

Questions (for answers and explanations, see page 213)

1. The general evolutionary trend among the deuterostomes can best be stated as
   a. a decrease in the number of body segments from three to one.
   b. an increase in the size and importance of the lophophore.
   c. a reduction in the size and importance of the lophophore combined with a more mobile lifestyle.
   d. a strong trend toward a more sedentary lifestyle.
   e. a loss of mobility due to the enlargement of the lophophore.

2. All but one of the following are valid characteristics of the acorn worms. Select the *exception*.
   a. Determinate cleavage of the early embryo
   b. Tripartite body plan
   c. Segmented coelomic cavity
   d. Mucus-coated proboscis used for feeding
   e. Absence of a lophophore

3. All but one of the following are valid characteristics of the moss animals. Select the *exception*.
   a. Indeterminate cleavage of the early embryo
   b. Tripartite body plan with presence of a well-developed lophophore
   c. Segmented coelomic cavity
   d. Asexual formation of larvae, which disperse to form new colonies
   e. Common tissue connections between colonial individuals

4. The phoronids differ from the arrow worms mainly in that
   a. the phoronids are completely sessile, whereas arrow worms are active swimmers.
   b. arrow worms lack the obvious lophophore present in the phoronids.
   c. phoronids are filter feeders and arrow worms are predators.
   d. phoronids secrete chitinous tube shells and arrow worms do not.
   e. All of the above

5. The brachiopods bear a superficial resemblance to which of the following protostome groups?
   a. Annelid worms
   b. Round worms
   c. Bivalve mollusks
   d. Cephalopod mollusks
   e. Crustaceans

CALCIFYING THE SKELETON (pages 614–617)

Key Concepts

1. Although radial symmetry is most often associated with sedentary animals, the 7,000 living species of echinoderms (phylum *Echinodermata*) have successfully combined a radially symmetric body plan with an active lifestyle.
2. A significant echinoderm advancement is calcification of an internal skeleton, which confers substantial protection from predators.

3. The *water vascular system* is another important trait of echinoderms. Seawater fills a system of modified coelomic cavity channels, which are under muscular control. This system coordinates the action of hundreds of *tube feet*, which cause movement. A connection to the outside environment is provided by way of a pore called the sieve plate.
4. Nearly all living echinoderms have bilaterally symmetric, ciliated larvae that feed in the plankton before settling to the substrate, where they metamorphose into radial adult forms.
5. *Pelmatozoans* are a subphylum of the echinoderms that were once very abundant, but today are represented by about 80 species in two groups known as the sea lilies and feather stars (class Crinoidea).
6. The sea lilies attach themselves to the substrate and stand upright by means of a flexible stalk of calcareous discs from which extend many jointed arms. Feeding is accomplished by the muscular contraction of the arms and tube feet, which funnel food into the animal's mouth. The tube feet are also used for gas exchange and excretion.
7. The 600 species of feather stars are much like sea lilies, except that they have flexible appendages that can grasp the substrate during feeding or resting, and later let go to allow walking or swimming movements.
8. *Eleutherozoans* represent the other subphylum of modern echinoderms. The 2,000 species of brittle stars (class Ophiuroidea) are like crinoids in outward appearance, but usually have many fewer arms (often only five). They are deposit feeders, and have only a single opening to the digestive system that acts as both mouth and anus, a characteristic not shared with other eleutherozoans.
9. Sea stars (class Asteroidea) are probably the most familiar echinoderms. Although they somewhat resemble the brittle stars, the arms of sea stars are much less flexible. They are important predators in many marine communities.
10. The tube feet of sea stars use both adhesive substances and hydraulic suction to facilitate locomotion and feeding. Suction is generated by the interplay of internal musculature and the water vascular system. The combined force of many hundreds of tube feet working in unison can eventually exhaust the powerful muscles of bivalves, gastropods, fish, and worms.
11. Some sea stars have an additional adaptation for feeding on bivalves: they evert their stomach into an opening in the shell of their prey. They then secrete digestive enzymes into the soft tissues of the bivalve, ingest the food, and retract their stomach.
12. All other echinoderms lack arms. Sea daisies (class Concentricycloidea) are a little-known group that apparently feeds on bacteria found near deposits of rotting wood in deep ocean waters.
13. Sea urchins (class Echinoidea) are round-bodied, herbivorous echinoderms that possess a fused internal skeleton, which is covered with defensive spines.
14. Sea cucumbers (class Holothuroidea) have a more elongated, soft body. They lack spines and possess large feathery feeding tentacles that protrude from the mouth. The tentacles are covered with a sticky mucus to which food adheres.

Activities (for answers and explanations, see page 212)

- What is the major functional difference between the feeding mechanism of a sea star and that of a brachiopod?

- Which group tends to be the more mobile, lophophorates or echinoderms? What anatomical feature is responsible for this marked difference in locomotory ability?

Questions (for answers and explanations, see page 213)

1. During the dissection of a sea lily, you would expect to find all of the following, *except* for
   a. a water vascular system.
   b. diploid cells.
   c. some form of coelomic cavity.
   d. a lophophore.
   e. tube feet.
2. Anatomically speaking, the echinoderms that are most similar to feather stars are the
   a. sea lilies.
   b. sea stars.
   c. sea urchins.
   d. sea cucumbers.
   e. brittle stars.
3. All but one of the following echinoderm classes belong to the Eleutherozoan lineage. Select the *exception*.
   a. Asteroidea
   b. Concentricycloidea
   c. Crinoidea
   d. Ophiuroidea
   e. Holothuroidea
4. A friend returns from the beach one day and shows you the skeleton of what he claims is a sea star. Upon close examination you inform him that what he really has is the internal skeleton of a sea urchin. Which of the following observations allows you to make this discrimination?
   a. There are numerous tube feet still attached to the skeleton.
   b. The skeleton shows evidence of once having had many large arms attached.

c. There is a long flexible stalk made of calcareous discs.
d. There is a large hole at one end through which feeding tentacles once protruded.
e. The skeleton is rounded, completely enclosed, and has five rows of holes for the tube feet.

5. The fossil record of echinderms is substantial. Which of the following are reasons for this large number of echinoderm fossils?
   a. Their internal skeletons fossilize well.
   b. They are abundant in the world's oceans.
   c. They have been successful deuterostomes for over 400 million years.
   d. They are bottom-dwellers and live in areas that promote fossilization.
   e. All of the above

EVOLUTION OF THE CHORDATE PHARYNX • SUCKING MUD: THE RISE OF THE VERTEBRATES (pages 617–621)

Key Concepts

1. A strategy used by some deuterostomes to exploit planktonic food sources was the modification of *pharyngeal slits,* formerly used only for respiration, into large filtering devices.
2. The chordates (phylum Chordata) were the main beneficiaries of this new trend in deuterostome evolution. All chordates are bilaterally symmetric, have gill slits at some developmental stage, possess an internal skeleton, a dorsal hollow nervous system, and a ventral heart.
3. Chordates also possess a dorsal supportive rod called the *notochord* at some point during their life cycle.
4. Tunicates (subphylum *Urochordata*) are a group of 2,500 sessile marine species that most likely represent the ancestors of all other chordate groups. As adults they bear little resemblance to other chordates, but as free-swimming larvae they possess pharyngeal gill slits, a hollow nerve cord, and a notochord.
5. Sea squirts (class Ascidiacea) are the most common tunicates. Two large siphons direct water in and out of the large pharyngeal basket, where respiration and feeding take place. Some species are solitary, others colonial.
6. The larvaceans (class Appendicularia) are another small group of tunicates that spend their entire lives as free-roaming individuals in the plankton.
7. Lancelets (subphylum Cephalochordata) are a group of small, fishlike, sediment-dwelling deuterostomes that have pharyngeal gill baskets used to filter food from the mud. They retain a notochord throughout life.
8. Filtering food from mud is difficult because so much nonfood must be ingested. However, pharyngeal baskets with their many openings and large surface area are ideal for collecting small food particles and eliminating the indigestible sediments.
9. About 500 mya, several additional chordate changes in body plan improved mud feeding. The *vertebral column,* a jointed dorsal supporting structure, replaced the more primitive notochord in the animals called *vertebrates.* The advent of external armor also allowed vertebrates to feed on the surface mud with less concern for predators.
10. The typical chordate body plan is a centralized vertebral column with two pairs of appendages attached. Chordates possess a large coelom in which the internal organs, including well-developed circulatory and respiratory systems, are grouped.
11. The first vertebrates were of the bottom-dwelling, mud-sucking variety, which ultimately gave rise to the jawless fishes (class Agnatha). Early agnathans such as the ostracoderms were small and heavily armored, and probably made their living feeding on carrion.
12. Lampreys and hagfishes are the only modern representatives of the agnathans, a once abundant group. One reason for the evolutionary demise of these fishes was the development of jaws in other groups of fish.
13. Jaws evolved from the cartilaginous, or bony, gill arches of ancestral agnathans. Jaws represent a major change in the vertebrate body plan because they permit the capture and handling of living prey. The development of teeth further enhanced the usefulness of jaws by allowing hard body parts, as well as soft tissues, to be consumed.
14. Placoderms (class Placodermi) were the most important group of early jawed fishes. Although now extinct, these heavily armored, often large, and highly maneuverable animals were important predators of the Devonian seas.
15. The sharks, skates, rays, and their relatives comprise the cartilaginous fishes (class Chondrichthyes), a largely marine group characterized by a skeleton made entirely of pliable *cartilage*. Their skin is tough and leathery, and is covered with hardened projections that increase its durability.
16. The loss of heavy armor seen in the placoderms coincided with an increase in rapid swimming ability in the Chondrichthyes. Also contributing to increased speed and maneuverability was the evolution of paired pelvic and pectoral fins, which improved balance and steering, along with a large caudal fin to provide propulsion.
17. Although most of today's sharks, skates, and rays are predatory, the largest living species are actually filter feeders that specialize in collecting enormous quantities of plankton.
18. The other major lineage of fishes evolved in fresh water. The bony fishes (class Osteichthyes) have skeletons made of calcified bone rather than cartilage. There is a single gill chamber covered by a hardened flap, the movement of which increases water flow over the gill surfaces.
19. Early bony fishes had lunglike sacs that supplemented gill respiration. Subsequent evolution altered these sacs to become *swim bladders,* which now serve as adjustable gas-filled organs for buoyancy.
20. In an effort to improve swimming speed and maneuverability, bony fishes have lost much of the heavy armor characteristic of ancient fishes. Their external surface is now covered by smooth, flat, thin scales, which provide adequate protection with a minimum of weight.
21. Remarkable diversity characterizes the 20,000 species of bony fishes. Although a majority still inhabit fresh water, many have reinvaded the ocean, and virtually every feeding style including predation, grazing, and scavenging is represented.

22. Because their eggs are heavier than water and therefore tend to sink, fishes often return to shallow waters for breeding purposes.

Activities (for answers and explanations, see page 213)

- Describe the major elements of the skeleton of a modern bony fish. What one skeletal factor differs most notably between these animals and the agnathan fishes?

- Discuss the main advantages of hinged jaws, and why this evolutionary advancement prompted an extensive radiation of the ancient agnathan fishes into more modern jawed forms.

Questions (for answers and explanations, see page 213)

1. The primary anatomical characteristic that pre-adapted pharyngeal slits for use as feeding structures was their
   a. large surface area.
   b. location anterior to the mouth.
   c. ability to provide support for the internal skeleton.
   d. bilateral symmetry.
   e. None of the above

2. Lancelets are like larvaceans in that
   a. the adults of both are primarily sedentary.
   b. neither retains a notochord throughout life.
   c. both are vertebrates.
   d. both have jaws.
   e. each evolved from some ancestral tunicate.

3. Which of the following statements correctly describes the placoderm fishes?
   a. Placoderm fishes arose directly from the tunicates.
   b. Placoderm fishes are members of the class Chondrichthyes.
   c. Placoderm fishes had large, formidable teeth.
   d. Placoderm fishes were heavily armored and jawless.
   e. A few placoderm species still live in modern oceans.

4. Which of the following groups of chordates evolved prior to the appearance of cartilaginous fishes?
   a. Bony fishes and sea squirts
   b. Tunicates, lancelets, and agnathans
   c. Tunicates, lancelets, and bony fishes
   d. Agnathans and bony fishes
   e. None of the above

5. The most probable origin of fish jaws was from the
   a. paired pelvic fins of bony fishes.
   b. pharyngeal slits of tunicates.
   c. notochord of larvaceans.
   d. gill arches of agnathans.
   e. supportive cartilage of the caudal fin in sharks.

6. The lungs of early bony fishes evolved into swim bladders. This change had the primary effect of
   a. improving feeding efficiency.
   b. supplementing gill respiration.
   c. aiding reproduction.
   d. providing buoyancy.
   e. making it easier to capture large prey.

7. Which of the following is *not* a key element in the basic vertebrate body plan?
   a. Internal skeleton
   b. Pharyngeal basket
   c. Dorsal spinal cord
   d. Segmented muscles
   e. Organs suspended in the coelom

BREATHING AIR AND EXPLORING THE LAND • THE ORIGINS OF MAMMALS (pages 621–632)

Key Concepts

1. The crossopterygians, or lobe-finned fishes (subclass Crossopterygii), were the first land vertebrates. The structure of their limbs, as well as their possession of lunglike sacs, preadapted them for this transition. The only known living crossopterygian is *Latimeria chalumnae*.
2. Amphibians (class Amphibia) evolved from some ancient crossopterygian-like ancestor during the Devonian. Early amphibians were probably quite clumsy on land, but rapidly evolved limb modifications that improved their locomotion.
3. Three orders of amphibians, comprising about 4,500 species, still exist today: caecilians (order Gymnophiona), frogs and toads (order Anura), and salamanders (order Urodela). They are widely distributed in tropical and temperate regions.
4. Amphibian reproduction and respiration remain tightly linked to the aquatic environment. The larvae of virtually all species are exclusively aquatic, eventually metamorphosing into adults that spend some portion of their life on land. However, even this land-based portion of the life cycle is limited to relatively moist habitats because amphibian skin cannot withstand prolonged drying.
5. Although the causes are unknown, amphibian populations are currently declining on a worldwide basis.
6. The transition from an amphibious existence to a fully terrestrial lifestyle was made possible by two events: the development of tough, water-impermeable skin, and the advent of a shelled egg through which gases, but not water, could freely pass. The reptiles (class Reptilia) were the first vertebrates to develop both traits.

7. The reptilian amniote egg houses an embryo surrounded by three membranes: the *amnion*, the *chorion*, and the *allantois*. This fluid-filled package is encased inside a leathery, water-retarding shell. Stored food supplies in the form of yolk provide the developing embryo with nutrients. Usually, little parental care is provided once the eggs are laid.
8. The heavily keratinized skin of reptiles is relatively water-impermeable, so gas exchange takes place exclusively via the lungs. Advancements in respiratory and circulatory musculature, as well as additional heart chambers, make gas exchange more efficient than in amphibians. Reptiles can thus sustain more activity than amphibians.
9. The roughly 6,000 species of modern reptiles belong to three subclasses. The *Chelonia*, which includes the turtles and tortoises, have a heavy shell into which the head and limbs can be withdrawn for protection. Most are slow-moving vegetarians that are at least partially aquatic, but a few are exclusively terrestrial.
10. The subclass *Sphenodontida* contains only two lizardlike species known as the *tuataras*. The subclass *Squamata* contains all true lizards, as well as the snakes. Most lizards have four legs, which show improvements for terrestrial motion compared with amphibian limbs. Lizard feeding modes include insectivory, herbivory, and carnivory. Snakes are completely legless, and have therefore evolved entirely new methods for locomotion; they are mainly carnivorous.
11. Another reptilian lineage, the thecodonts (class Archosauria), diverged from the other reptiles during Mesozoic times and gave rise to the subclass Crocodylia, a collection of semi-aquatic predators that includes the crocodiles, alligators, gavials, and caimans. A second group of thecodonts produced the dinosaurs (subclasses Ornithischia and Saurischia). These diverse and sometimes enormous animals ruled the earth for many millions of years inhabiting land, sea, and air.
12. One branch of the dinosaur lineage lives on today by virtue of a highly modified assemblage of dinosaur descendants called the birds (subclass Aves).
13. Birds are characterized by having feathers: highly modified reptilian scales that serve as insulators and increase the available surface area for flight. The transition between reptiles and birds is represented by the fossil *Archaeopteryx*, which had distinct feathers but still retained obvious reptilian characteristics in its skull, limbs, breastbone, and tail.
14. Many anatomical features of modern birds are adaptations for weight reduction associated with flight. Radical changes in bone structure, breast musculature, metabolism, brain structure, respiration, and circulation evolved among birds. Many of these changes helped to make flight possible or more efficient.
15. Like that of reptiles, the avian amniote egg is freed from a requirement for external water. It has a rigid calcified shell and the same three internal membranes found in reptilian eggs. Most bird eggs are laid in nests which the parents tend. Some species hatch *altricial* (helpless) young, while others produce *precocial* (independent) offspring. Low reproductive rates with high levels of offspring survival are characteristic of birds.
16. As a group, birds have a highly varied diet. Herbivores and insectivores are common, and among the carnivorous species both active predators and scavengers exist. Sizes range from tiny hummingbirds to several extinct forms that were considerably larger than humans.
17. The mammals (class Mammalia) emerged from yet another reptilian lineage during the Mesozoic. A major evolutionary adjustment involved skeletal simplification through loss or fusion of bones, particularly those of the jaw. Limbs came to be positioned under the body more than in reptiles, allowing improved locomotion.
18. Other key mammalian features include hair or blubber for insulation, mammary glands, sweat glands, and a four-chambered heart. The number of teeth is reduced in mammals compared to reptiles, but mammalian dentition is more diverse.
19. Milk secreted by the female's mammary glands nourishes the young, a trait unique to mammals. Fertilization is internal, and most mammals undergo at least some development within the female's uterus prior to birth.
20. There are about 4,000 living species of mammals. Collectively, they show the largest size range of any animal group, with representatives as small as shrews and as large as the blue whale, the largest animal ever to live on Earth.
21. The mammalian subclass Prototheria is represented by a single order Monotremata, which contains the duck-billed platypus and spiny anteaters. These mammals lay an external, reptilelike egg. After hatching, the young nurse from their mother as do other mammals.
22. The subclass Theria contains two other mammal groups, the marsupials and the eutherians. Marsupials (e.g., koalas and kangaroos) give birth to their young very early in development. The young then crawl into the mother's ventral pouch, or marsupium, to complete their development. Marsupials are well adapted for prolonged droughts and variable habitats. Although once widely distributed, they are now restricted mainly to Australia and South America.
23. Eutherians represent by far the largest number of mammal species, and are arranged into 16 orders. Unlike marsupials, they undergo extensive development within the female's uterus prior to birth.
24. The largest eutherians are marine mammals (whales), but the most abundant are small terrestrial species such as rodents, bats, and insectivores. These latter groups alone account for more than 50% of all eutherian species.

Activities (for answers and explanations, see page 213)

- Use the following abbreviations to indicate which vertebrate groups possess(ed) the listed characteristics.
(**A** = amphibians, **B** = birds, **D** = dinosaurs, **L** = lobe-finned fishes, **M** = mammals, **R** = modern reptiles)

_____ Lungs

_____ Aquatic larvae

_____ Wings

_____ Feathers

_____ Uterine development

_____ Terrestrial egg

_____ Water-impermeable skin

- *Archaeopteryx*, although generally accepted as the earliest fossil bird, still exhibits many reptilian features. List four of these features and tell how each has been modified in modern birds.

a.

b.

c.

d.

Questions (for answers and explanations, see page 213)

1. The transition from aquatic to terrestrial lifestyles required many adaptations in the vertebrate lineage. Which of the following is *not* one of those adaptations?
a. Switch from gill respiration to air-breathing lungs
b. Improvements in water resistance of skin
c. Development of feathers for insulation
d. Alteration in mode of locomotion
e. Adjustments to reproductive physiology

2. The eggs of reptiles and birds have many anatomical similarities. Which of the following is *not* a common feature of both?
a. Developing embryo resides in an air-filled chamber
b. Tough exterior shell for protection of embryo
c. Ability to pass gases between egg and environment
d. Complex system of internal membranes
e. Abundant supply of yolk to nourish developing embryo

3. Dinosaurs dominated the Earth for nearly 150 million years before becoming extinct. Which of the following vertebrate groups is arguably a living representative of the dinosaur lineage?
a. Amphibians
b. Birds
c. Lobe-finned fishes
d. Snakes
e. Mammals

4. All mammals nourish their young with milk manufactured by the mother, but not all are born after an extended period of development inside the female's uterus. Which of the following mammals emerges from the uterus in the least developed condition?
a. Mouse
b. Dolphin
c. Human
d. Kangaroo
e. Bat

5. Modern amphibians include all of the following groups except for the
a. caecilians.
b. frogs.
c. salamanders.
d. toads.
e. caimans.

6. Refer to the following diagram of an amniote egg:

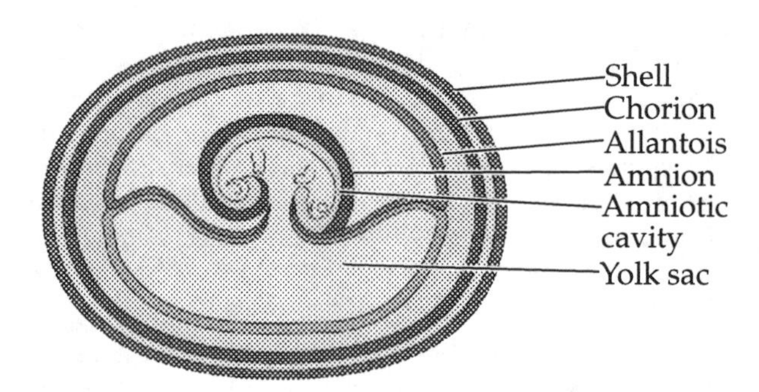

Of the structures depicted in this diagram, which is responsible for providing the developing embryo with food?
a. Amnion
b. Allantois
c. Yolk sac
d. Shell
e. Chorion

7. Consider the evolutionary sequence of change in vertebrate limb structure shown by Fig. 27.25 in your textbook. Of what significance is the trend toward limbs being located more directly under the body?
a. Improved thermoregulation
b. More efficient feeding
c. Better locomotion
d. Increased egg laying ability
e. None of the above

8. Amphibians remain tightly linked to moist habitats and bodies of water because of their
   a. food requirements.
   b. respiratory requirements.
   c. reproductive requirements.
   d. Both *a* and *b*
   e. Both *b* and *c*

HUMAN EVOLUTION • THEMES IN DEUTEROSTOME EVOLUTION (pages 632–641)

Key Concepts

1. The primates are the order of mammals to which the human species belongs. Primates have undergone significant radiation and evolutionary change since their appearance in the late Cretaceous.
2. Primate characteristics stem largely from adaptations for an arboreal lifestyle. Flexible limbs, dexterous hands with opposable thumbs, nails instead of claws, forward-facing eyes for improved depth perception, and small litters all facilitated life in the trees.
3. Prosimians are rather similar to the earliest primates, and are represented today by the lemurs, tarsiers, and their relatives. Anthropoid primates, which include the monkeys, apes, and humans, diverged from the prosimians some 55 mya. New World and Old World lineages emerged within the anthropoid lineage, and it is from the Old World group that apes evolved.
4. The first apes were largely arboreal, but later came down to the ground as climates cooled and forested habits began to fragment 15–20 mya. Some of these apes underwent morphological changes. For instance, their muzzle size was reduced and their canines shortened.
5. Humans and their relatives (family Hominidae) began diverging from an apelike ancestral primate approximately 5 mya. The oldest known hominid fossil is "Lucy," a member of the species *Australopithecus afarensis*, which dates back 3.5 million years. Her species is thought to be the progenitor of all subsequent hominid groups.
6. Other australopithecine species also existed, apparently occurring simultaneously with *A. afarensis* but in different parts of the African continent. These include a robust form, and a more slender form called *A. africanus*. Both disappeared suddenly about 1.5 mya.
7. The australopithecines evolved *bipedalism*, in which only the hindlimbs are used for locomotion. Ground mobility was improved by modification of the pelvis and hindlimb musculature. Forelimbs became free to carry large objects, and better vision was facilitated by being able to see over the surrounding tall grasses.
8. Coincident with the evolution of bipedalism were changes in hand structure, which allowed for precise grasping motions. Foot modifications also improved walking and running. Both advancements increased the ability of hominids to exploit their ground habitat.
9. Members of the genus *Homo* evolved from the australopithecines about 2.5 mya. *Homo* was a contemporary of *A. africanus* (which was perhaps the ancestral species of *Homo*) for at least half a million years.
10. *Homo habilis* was the first member of the genus to appear. Its body and brain size had nearly doubled over that of the australopithecines, and it was the first hominid to use tools for obtaining food.
11. From a basic diet of plant matter, *H. habilis* expanded its feeding habits to include animals. An increased division of labor between the sexes, with males hunting cooperatively and females harvesting plant foods while simultaneously tending infants, laid the foundation for more modern hominid societies.
12. The next hominid species, *Homo erectus*, appeared 1.6 mya and spread rapidly throughout Africa and Asia. *H. erectus* was as large as modern humans, cooked its food (which included large mammals), and made specialized stone tools.
13. Our own species, *Homo sapiens*, replaced *H. erectus* in Africa 500,000 years ago. Remnant populations of *H. erectus* survived for another quarter million years before disappearing completely. *H. sapiens* continued earlier hominid evolutionary trends of increasing brain size and social complexity.
14. With increased sociality came improved communication, leading to better coordination of group hunting and exploitation of variable or patchy resources. More effective communication skills may have also enhanced social status and thus reproductive success.
15. A new trait that appeared with *H. sapiens* was religion and a concept of life after death. Grave sites were provisioned with food, clothing, and tools, presumably intended for use in the afterworld.
16. Several groups of *H. sapiens* have existed. Neanderthal people lived between 75,000 and 30,000 years ago. They were short, stocky, and created excellent hunting weapons.
17. The more recent Cro-Magnon people had bodies much like our own. They appeared between 50,000 to 100,000 years ago and briefly overlapped with Neanderthals, whom they may have eliminated through competition. About 20,000 years ago their range expanded across Asia and into North America.
18. Cro-Magnons used sophisticated tools and were responsible for elaborate cave paintings depicting the hunting of animals. Their efficient cooperative hunting probably caused the extinction of numerous large mammals of that time.
19. Increasing brain size and the development of formal language ultimately lead to culture, which then became an integral component of human evolution. Culture allowed traditions and knowledge to be passed rapidly from one generation to the next through teaching and observation, bypassing the slower process of evolutionary change through genetic mechanisms.

20. Changes in the pattern of human tool use reflect the role of culture in moving us from our cooperative hunting origins to our current lifestyle of *pastoralism* (herding and farming). *Agriculture* benefited greatly from cultural transmission, originating independently 10,000 years ago in the Middle East and Asia and then spreading across Europe to the Americas. Steady increases in human population size are directly related to more reliable and abundant food resources.

21. Techniques of molecular biology involving mitochondrial DNA are now being used to trace the path of human evolution. Mitochondrial DNA is particularly useful because, coming only from the maternal parent, it is not subject to changes brought on by sexual recombination. Changes in mitochondrial DNA thus reflect the progression of a molecular clock, which can be used to directly measure evolutionary time.

22. Molecular work indicates a common ancestral population for all modern humans that existed only 200,000 years ago, with a subsequent radiation in Africa within the last 100,000 years.

23. Deuterostome evolution parallels that of protostomes in several ways: bottom feeding in marine sediments dominated the early evolution of each group; compartmentalization of the coelomic cavity and improvements in feeding structures occurred; and advancements in locomotion and buoyancy opened up new habitats for exploitation.

24. Both groups invaded the terrestrial world, but with different results. Exoskeletons provided protection and support for protostomes, but limited growth. The internal skeletons of deuterostomes also provided support against gravity, but did not hinder growth. The consequence is that vertebrates have become the dominant, large animals on land.

25. Some lineages of terrestrial vertebrates have returned to an aquatic lifestyle. Many of these aquatic reinvasions produced specialized filter-feeding forms.

26. Terrestrial herbivores must be relatively large in order to handle and process tough-bodied land plants in sufficient quantities to sustain growth and metabolism. The appearance of large herbivores in turn selected for large carnivores. This evolutionary cycle was at least temporarily halted by the appearance of humans and their use of tools for cooperative hunting and defense.

Activities (for answers and explanations, see page 213)

- Rank the following hominid species in terms of their appearance in the fossil record from most ancient (**1**) to most recent (**6**).

  ____ *Homo erectus*
  ____ *Homo sapiens* - Neanderthal
  ____ *Homo sapiens* - Cro-Magnon
  ____ *Australopithecus africanus*
  ____ *Australopithecus afarensis*
  ____ *Homo habilis*

- Which characteristics that predisposed the early primates for life in the trees are still present in modern humans?

Questions (for answers and explanations, see page 213)

1. The species to which the hominid fossil called "Lucy" belongs possessed all but one of the following traits. Select the *exception*.
   a. Bipedal locomotion
   b. Well-developed language and culture
   c. Hands capable of precision grasping
   d. Feet modified for walking and running
   e. Ability to carry objects with forelimbs

2. The first hominid species to regularly eat animal foods collected through cooperative hunting was
   a. A. afarensis
   b. A. africanus
   c. A. robustus
   d. H. habilis
   e. H. sapiens

3. The development of language and culture in hominids was beneficial because
   a. it provided each species with a foundation for more efficient social interactions.
   b. it helped increase the status of individuals with the group.
   c. hunting and farming methods could be passed easily from one generation to the next.
   d. it complemented the division of labor between the sexes.
   e. All of the above

4. The now-extinct hominid most like ourselves is
   a. H. habilis.
   b. H. erectus.
   c. Cro-Magnon.
   d. Neanderthal.
   e. A. robustus.

5. Which of the following statements about human evolution is *not* true?
   a. Bipedalism was a hominid adaptation for life on the ground.
   b. Increases in the size of hominid brains preceded the appearance of language and culture.
   c. The spread of pastoralism is a relatively recent human event that is largely the result of culture.
   d. The extinction of the Cro-Magnon people was caused by the emergence of the Neanderthals.
   e. Humans are not the direct descendants of modern-day chimpanzees.

6. Mitochondrial DNA techniques are good for measuring the rate of evolution in humans because
   a. mitochondrial DNA is maternally inherited and not subject to changes from recombination.
   b. mitochondrial DNA mutates very rapidly and shows changes in a short time period.
   c. only humans have mitochondrial DNA.
   d. the mitochondrial DNA of humans and other primates is very different.
   e. None of the above

7. All but one of the following is a trend in deuterostome evolution. Select the *exception.*
   a. Advancements in locomotion
   b. Development of a rigid exoskeleton
   c. Compartmentalization of the coelomic cavity
   d. Improvements of feeding structures
   e. Invasion of the terrestrial world

Integrative Questions (for answers and explanations, see page 214)

1. Match the relevant method of feeding with each deuterostome group listed.

a. Filters with lophophore	_____	Squamata
b. Bites with toothed jaws	_____	Agnatha
c. Sucks mud or blood	_____	Asteroidea
d. Digging proboscis	_____	Brachiopoda
e. Extrudes stomach into prey	_____	Enteropneusta

2. Which of the following statements about deuterostome evolution is true?
   a. Lophophores are common in the phylum Echinodermata.
   b. Amphibians are the immediate evolutionary ancestors of the fishes.
   c. The evolution of jaws is but one of several important changes in feeding specializations.
   d. Larvaceans and lancelets gave rise to the echinoderms.
   e. A major advance in deuterostome evolution was the loss of the coelomic cavity.

3. A bony fish eats a frog. The fish is eaten by an eagle. The eagle is eaten by a snake. How many different classes of vertebrates are involved in this feeding sequence?
   a. One
   b. Two
   c. Three
   d. Four
   e. Five

4. Which of the following characteristics is present at some stage in the life of modern humans?
   a. Bilateral symmetry
   b. Pharyngeal slits
   c. Internal skeleton
   d. Notocord
   e. All of the above

5. Choose the one set of characteristics that applies to both humans and sea stars.
   a. Eukaryotic, ingestive heterotroph, tube feet, radial symmetry, internal skeleton
   b. Chordate, radial symmetry, ingestive heterotroph, triploblastic
   c. Bilaterally symmetrical, lophophorate, vertebrate, produces terrestrial egg
   d. Deuterostomate, indeterminate cleavage of embryo, true coelomic cavity
   e. Water vascular system, internal skeleton, jaws, triploblastic

28

Patterns in the Evolution of Life

CHAPTER LEARNING OBJECTIVES—after studying this chapter you should be able to:

- ❑ Describe how radioactive isotopes can be used to age fossils and rocks, and give details of this technique as it relates to the use of ^{14}C.
- ❑ Differentiate among the geologic terms "eons," "eras," "periods," and "epochs."
- ❑ Explain the concept of continental drift, and describe how it has affected life on Earth.
- ❑ Discuss how and when the supercontinent Pangaea formed, and how it split to form Laurasia and Gondwanaland.
- ❑ Explain how Laurasia and Gondwanaland eventually split to form the modern continents.
- ❑ Describe how climate change can influence the patterns of evolution.
- ❑ Differentiate between background extinction and mass extinction, and give the geologic times of several important mass extinction events.
- ❑ Discuss how and where fossils are formed, and why such a small percentage of the organisms that ever lived are known from the fossil record.
- ❑ Explain the various problems associated with interpreting the fossil record.
- ❑ Differentiate between the terms "flora," "fauna," and "biota," and explain what is meant by "provincialization of biota."
- ❑ Name the major eras of life on Earth, their historical times in millions of years, and the main evolutionary events that occurred in each.
- ❑ Discuss the evidence supporting the hypothesis that the great Cretaceous mass extinction was caused by Earth's collision with a large meteorite.
- ❑ Explain our current understanding of the rate at which evolutionary change occurs, and describe the ways in which such change can be measured.
- ❑ Provide possible reasons for why mass extinctions, and the subsequent radiative explosions of new organisms that followed these extinctions, took place.

HOW EARTH HAS CHANGED • THE FOSSIL RECORD (pages 644–649)

Key Concepts

1. The age of fossils is determined by dating the rocks where those fossils occur. One of the most remarkable achievements of modern science has been to develop technologies that accurately date rock formations. Radioactive decay of isotopes is used for this purpose.
2. Isotopes decay at regular rates, so that during any given time a fixed fraction of the material will change from one form to another. Isotopes are commonly categorized by their half-lives, the period of time it takes for an amount of radioactive material to be reduced by one-half. Many kinds of radioactive isotopes are used to date rocks, including phosphorus, tritium, potassium, uranium, and carbon.
3. The isotope ^{14}C can be used to determine the age of organisms that died within the past 15,000 years. The age of these fossils is then used to infer the age of the surrounding sedimentary rock.
4. In living organisms, ^{14}C converts to ^{12}C at the same rate as ^{14}N changes to ^{14}C in the atmosphere. Because carbon is continuously exchanged between the atmosphere and organisms, new ^{14}C is constantly acquired and a steady state is reached. But as soon as an organism dies, this exchange ceases and the total amount of ^{14}C in the organism's remains starts to decline. By measuring the ratio of ^{14}C to ^{12}C, the time since the organism died can be calculated.
5. Dates obtained from radioactive decay measurements such as ^{14}C can sometimes be confirmed by independent forms of evidence, such as known-age cultural artifacts.
6. Earth's history can be divided into four *eons*: the Hadean, the Archean, the Proterozoic, and the Phanerozoic. These eons are further subdivided into *eras* and *periods*, whose names are typically taken from the first fossils discovered in rocks formed during those time periods.
7. Earth's crust is made of a number of thick, solid plates that float atop a liquid mantle. Movement of these plates is caused by sea floor spreading. Fault lines and mountain ranges are produced where moving plates intersect.

8. Movement of these crustal plates has had profound influences on climate, sea level, and the distribution of organisms.

9. During the late Cambrian period (550 mya), six continents were spread across Earth's surface at equatorial latitudes. Some 300 million years later, during the Permian period, the continents united to form a single large continent called Pangaea. Shallow seas formed and receded several times during the Paleozoic era to alternately cover and then expose substantial portions of Pangaea.

10. A process known as *continental drift* both formed and initiated the breakup of Pangaea during the Permian. As Pangaea broke up, two smaller continents were formed. Laurasia included all of the northern land masses, and Gondwanaland the southern ones.

11. The mid-Cretaceous period witnessed a widening of oceanic channels that forced the breakup of Gondwanaland into South America, Africa, and a third landmass that would ultimately divide during the late Cretaceous to yield Australia, Antarctica, and India. The Himalayas were formed when India eventually collided with the Asian continent.

12. Over the next 70–100 million years the continents continued to separate, reaching roughly their current positions by the mid-Tertiary period. At several times, land bridges of various sizes and durations formed that connected formerly isolated land masses. Some, such as the Isthmus of Panama, still exist today.

13. Climate also changed as the continents shifted positions. Much of Earth's history reveals that climates were considerably warmer than today. Other periods show pronounced cold, which spawned extensive glacial ice sheets. Most climate changes were gradual, but a number were quite rapid, occurring during only a few hundred years. These appear as "instantaneous" events in the geologic record, and have often been responsible for the sudden extinctions of organisms.

14. *Mass extinctions* occur periodically in Earth's history when relatively more species disappear from the fossil record than during intervening periods. One such event occurred 65 mya at the end of the Cretaceous, ushering in the demise of the dinosaurs. This event may have been caused by Earth's collision with a large meteorite. Although still controversial, considerable chemical and physical evidence has recently accumulated in support of this hypothesis, showing that evolutionary change can perhaps be dramatically influenced by extraterrestrial events.

15. Fossilization is the critical process that leaves a biological record of ancient times. The process occurs only under favorable conditions, which include low oxygen concentrations and rapid sedimentation due to the action of wind or water.

16. Although we have no direct way of knowing the total number of species that have ever lived on Earth, the number of species that have so far been identified in the fossil record (~300,000) certainly represents only a tiny fraction of those that ever existed.

17. Estimates of the number of species that have ever lived, based on the current *biota* (present species in all kingdoms) and the rate of turnover in the fossil record, lead to the conclusion that less than 2 percent of all species are known from fossils.

18. Marine animals with hard skeletons, such as mollusks, often fossilize well. Nearly two-thirds of all fossil species are from nine phyla of hard-shelled organisms. By contrast, soft-bodied organisms, many of which also do not live in the best habitats for fossil formation, are not particularly well represented in the fossil record.

19. Another potential problem with interpreting the fossil record is that most fossils come from limited geographic areas. Often information from widely separated geographic locations is necessary in order to decide when and where organisms first appeared during evolution.

20. For example, we know that horses evolved slowly over millions of years in North America, but periodically crossed the Bering Land Bridge into Asia. If only the Asian fossils were available, one could erroneously conclude that horses evolved very rapidly on several different occasions in Asia.

Questions (for answers and explanations, see page 214)

1. The half-life of an isotope is best defined as the
   a. time a fixed fraction of isotope material will take to change from one form to another.
   b. amount of time it takes for an herbivore to eat half of this material in its food.
   c. age over which the isotope is useful for dating rocks.
   d. ratio of one isotope species to another in a sample of organic matter.
   e. None of the above

2. Earthquakes and mountain ranges are ultimately the result of
   a. plates in Earth's crust that move against one another on top of a liquid mantle.
   b. climate changes and glacial ice sheets movement.
   c. the rising and falling of shallow oceans.
   d. leftover vibrations and debris from ancient collisions with an asteroid or meteor.
   e. the breakup of Laurasia and Gondwanaland.

3. Which of the following would likely be the *best* setting for fossilization to occur?
   a. The surf zone along a sand beach
   b. A shallow, cool swamp with good deposition rates of mud sediments
   c. The bottom of a hot, dry cave with no running water
   d. A fast running mountain stream
   e. Dry desert areas with rapidly shifting sand dunes

4. Although the continents drifted slowly apart, creating large oceans in the process, terrestrial animals could still sometimes get from one major land mass to another by means of
   a. glacial ice sheets.
   b. deep ocean trenches.
   c. high mountain ranges.
   d. fault lines.
   e. narrow land bridges.

5. The Himalayas formed when
   a. the Indian subcontinent collided with Asia.
   b. the Bering Land Bridge sank below the ocean's surface.
   c. Gondwanaland separated from Laurasia.
   d. sea levels fell in the Cambrian period.
   e. None of the above

6. Despite being incomplete as a whole, the fossil record is rather detailed for
   a. soft-bodied insects.
   b. cnidarians and sponges.
   c. most terrestrial animals.
   d. hard-shelled mollusks.
   e. All of the above

7. The best estimate for the total number of species that have ever lived on Earth is approximately
   a. two hundred thousand.
   b. half a million.
   c. one million.
   d. five million.
   e. more than fifteen million.

LIFE IN THE REMOTE PAST (pages 649–655)

Key Concepts

1. The fossil record is sketchy prior to the Cambrian period. Nevertheless, we know that some forms of protists, fungi, and animals, but not plants, already existed by that time.
2. Biologists use the word *fauna* to refer to the animals living in a given area or time, and *flora* to describe the plant life. Often the names of places where fossil species were first discovered are associated with a particular fauna or flora.
3. During the early Cambrian (600 mya), a rich fauna had already developed, including early forms of annelids, arthropods, echinoderms, and cnidarians.
4. By 530 mya (mid-Cambrian) all current animal phyla existed, along with species from ten now-extinct phyla. Many of these animals had hard external skeletons, suggesting a strong evolutionary influence of predation at that time.
5. The Ordovician period (500–440 mya) witnessed a great expansion of classes and orders. Many filter feeding marine animals flourished, and on land primitive ancestors of the club mosses and horsetails appeared. This period ended when sea levels dropped, driving many groups to extinction.
6. The Silurian period (440–400 mya) saw the continuation of abundant marine communities and more progress towards colonization of the land. Simple marine vertebrates first appeared, as did the initial groups of terrestrial arthropods, the millipedes and scorpions.
7. Another major radiation occurred during the Devonian period (400–345 mya). Cephalopods, corals, and land plants in particular proliferated. The evolution of jaws in fishes and the loss of heavy armor plating were major milestones in the evolution of vertebrates. Gymnosperms appeared toward the end of the Devonian, along with the first insects.
8. The Carboniferous period (345–290 mya) is named for the extensive terrestrial forests of the time, which gave rise to modern coal deposits. The rocks associated with these deposits are rich in fossils, which document a great increase in the diversity of terrestrial animals. Amphibians, some of them quite large, were the dominant land vertebrates of the time.
9. Giant insects dominated invertebrate life on land in the Permian period (245–220 mya), and reptiles replaced amphibians as the ruling vertebrates. Aquatic habitats experienced an explosion in the number of bony fishes.
10. The largest of all mass extinctions took place at the end of the Permian when 95 percent of all invertebrate species disappeared. Abundant forms such as the trilobites vanished, and substantial reductions in species diversity for many other groups also occurred. On land, ginkgos, cycads, and conifers replaced the great carboniferous trees. The most probable cause for the Permian mass extinction was a radical shift in climate as Pangaea coalesced.
11. Earth's biota, which had previously been rather homogeneous, became increasingly *provincialized* during the Mesozoic era as the continents spread further apart and acquired their own unique geographic and climatic characteristics. As a result, distinctive terrestrial floras and faunas arose around the world.
12. The Triassic period (245–195 mya) saw the corals, mollusks, and other aquatic invertebrate groups increase dramatically in species number. On land, the main event was the rise in prominence of reptiles. The first mammals also appeared late during this period.
13. Another mass extinction occurred at the end of the Triassic and the beginning of the Jurassic period (194–138 mya). Many groups underwent major radiations in the wake of this extinction event, including large numbers of shelled aquatic invertebrates. Bony fishes came to dominate the seas, and the modern amphibians appeared. The terrestrial dinosaurs became abundant, as well as aquatic and aerial forms. One lineage of dinosaurs gave rise to the first birds in the mid-Jurassic.
14. The Cretaceous period (138–66 mya) witnessed the continued diversification of dinosaurs and increased provincialization of terrestrial biotas. Most modern mammal groups appeared at this time, as did the first flowering plants. Coincident with the emergence of angiosperms was a major radiation of their insect pollinators. The Cretaceous ended with another mass extinction event, probably caused by Earth's collision with a large meteorite.
15. The most recent 65 million years in Earth's history is called the Cenozoic era, and it is here that mammals came to dominate the terrestrial vertebrate fauna. Because it is the most recent time, the Cenozoic is rich in fossils. This allows paleontologists to more finely divide the Cenozoic into epochs as well as periods.

16. During the Tertiary period (65–2 mya) angiosperm diversification continued. Herbaceous forms of flowering plants appeared in response to cooler climates. Most of the modern groups of invertebrates were present by the beginning of the Tertiary, and the remaining modern vertebrate groups appeared as well.
17. Birds in particular radiated dramatically during the Tertiary. Mammalian lineages generally increased in body size, and great herds of grazing animals appeared in response to the herbaceous plants that evolved in the drier climates. In addition to hoofed mammals, bats, rodents, and primates increased dramatically in numbers.
18. The Quaternary period represents the latest geologic time slice in evolutionary history. The Pleistocene epoch commenced about 2 mya and was highlighted by many glacial episodes, the most recent occurring less than 10,000 years ago. The climatic changes associated with these events produced few extinctions, but did dramatically alter the range of many organisms.
19. The majority of human evolution has taken place during the Quaternary. From their beginnings it appears that humans have been, and continue to be, an important factor in determining the evolutionary fate of other organisms.

Questions (for answers and explanations, see page 214)

1. When biologists say that Earth's biota became provincialized, they mean that
   a. mass extinctions took place.
   b. continental drift did not occur.
   c. distinctive floras and faunas arose in isolated areas.
   d. reductions in species diversity took place.
   e. All of the above
2. All but one of the following occurred during the Ordovician period. Select the *exception*.
   a. Marine filter feeders flourished
   b. Club mosses appeared on land
   c. The number of classes and orders increased
   d. Modern mammals appeared
   e. Many groups became extinct at the end of the period
3. Which of the following events are coincident with the rapid diversification of the angiosperms during the Cretaceous?
   a. A sudden decrease in species diversity of mollusks
   b. A sudden increase in species diversity of corals
   c. A sudden increase in species diversity of amphibians
   d. A sudden decrease in species diversity of humans
   e. A sudden increase in species diversity of insects
4. Most of human evolution has occured during the
   a. Paleozoic era.
   b. Devonian period.
   c. Quaternary period.
   d. Carboniferous period.
   e. Mesozoic era.
5. The most recent mass extinction event that occurred *prior* to the evolution of humans took place approximately
   a. 10 mya.
   b. 65 mya.
   c. 220 mya.
   d. 400 mya.
   e. None of the above
6. The main reason(s) why the Cenozoic era can be more finely divided into geologic epochs and periods is because
   a. it is more recent.
   b. there are relatively more Cenozoic fossils present than in other eras.
   c. geologic processes have not had time to destroy these fossils.
   d. Cenozoic rock layers tend to be closer to the surface.
   e. All of the above

THE TIMING OF EVOLUTIONARY CHANGE (pages 655-660)

Key Concepts

1. Rates of evolutionary change are often difficult to measure due to the incompleteness of the fossil record and the fact that only the hard parts of an organism's body are preserved. Many characteristics are therefore unavailable for analysis.
2. Nevertheless, the fossil record does indicate that frequent long periods of evolutionary *stasis* have occurred, during which rates of morphological change are slow. These periods are interrupted, or punctuated, by occasional episodes of more rapid evolution.
3. Evolutionary rates have varied dramatically among species. Some have undergone rapid changes in form while others have remained unchanged over equivalent timespans.
4. Organisms that have changed very little through evolutionary time are often found in environments that have also remained unchanged for millenia. Modern examples of such "living fossils" are horseshoe crabs and chambered nautiluses.
5. Alternatively, if the physical or biological environment changes suddenly, then morphological evolution can also be rapid. An example comes from work on a small, typically marine fish called the three-spined stickleback, which has undergone morphological changes in pelvic girdle structure and spine size upon invading freshwater habitats. Current population information and known predator pressures can be combined with fossil evidence to help understand the rapid and repeated evolution of differences in stickleback morphology.
6. Evolutionary changes in morphology may be most rapid in association with speciation events. However, speciation data needed to test this hypothesis are few. Moreover, if environmental change is too rapid, the result may be extinction rather than speciation.
7. Rapid rates of evolutionary change can be accounted for by microevolutionary mechanisms. Using a unit called the *darwin*, defined as a change of magnitude *e* (base of natural logarithms) per million years, evolutionary rates can be measured. Artificial selection can yield short-

term rates of change greater than 60,000 darwins. But natural rates of evolution as measured from fossils are typically less than 20 darwins, showing that known rates of natural evolution are compatible with, but not necessarily proof of, microevolutionary mechanisms.

8. Although extinction is the norm in evolution (>99% of all species that ever lived are now extinct), our understanding of the causes for extinction is still limited. It is known, however, that rates of extinction have not been constant throughout evolutionary history.
9. Background extinction is the rate at which species disappear under "normal" conditions, whereas mass extinction is characterized by a much higher rate of disappearance than normal.
10. There have been numerous mass extinctions on Earth. The earliest, at the end of the Cambrian, eliminated nearly half of the known animal families. The most recent, at the end of the Cretaceous, claimed the dinosaurs as well as many marine groups.
11. The primary effect of mass extinctions is to radically alter the flora and fauna of the next period. Those traits that enhance survival during background extinction periods may not insure survival during a mass extinction event.
12. During some extinctions, terrestrial organisms are particularly hard hit, while in others, aquatic organisms suffer most. Still other extinctions have strong effects on both environments.
13. Mass extinctions do not appear to have a single cause. Some were probably caused by natural events such as climate change, mountain building, volcanic activity, and changes in sea level. Others may have been caused by events of an external origin, such as collision with a large meteorite.
14. Large numbers of new evolutionary lineages emerged on three separate occasions in Earth's history; during the Cambrian, the Permian, and the Triassic. Why these explosions of new lineages occurred when they did remains a paleontological puzzle, but probably had to do with two factors. One was the availability of newly vacated ecological niches resulting from the demise of older lineages during preceding mass extinction events. The other was the presence of a wide array of body forms in organisms, which eliminated the need for major evolutionary innovations.

Questions (for answers and explanations, see page 214)

1. One of the main factors that distinguishes the Cambrian explosion from all others is that
   a. evolutionarily, it was the most recent explosion.
   b. many novel lineages (particularly phyla) appeared at this time, but not during other explosions.
   c. it was the time when the dinosaurs became extinct.
   d. there was a dramatic drop in species diversity, especially among marine organisms.
   e. it was the time when the mammals first appeared.
2. During which one of the following geologic times would the most new kinds of body plans have appeared?
   a. Paleozoic
   b. Triassic
   c. Jurassic
   d. Cambrian
   e. Cretaceous
3. The event that precipitated the sudden disappearance of the dinosaurs some 55 mya may have been the result of
   a. Earth's collision with a large meteorite.
   b. slow climate changes due to planetary cooling.
   c. competition from better adapted organisms.
   d. the rise of birds and mammals.
   e. All of the above
4. Which of the following statements about evolution is *not* true?
   a. Rates of evolutionary change are known to have varied throughout Earth's history.
   b. Background extinction has been occurring ever since the appearance of the first species.
   c. Marine mollusks provide a good example of the increase in species diversity through time.
   d. There is no evidence in the fossil record for periods of rapid evolutionary change.
   e. Humans appear to have been influencing the evolution of other animals for at least 2 million years.
5. The biological unit known as a "darwin" is used to
   a. measure mutation rates.
   b. measure extinction rates.
   c. measure rates of evolutionary change.
   d. radioactively date fossils and rocks.
   e. None of the above

Integrative Questions (for answers and explanations, see page 214)

1. Which of the following groups of organisms has undergone an overall decrease in body size during evolution?
   a. Mammals
   b. Fish
   c. Mollusks
   d. Plants
   e. None of the above
2. Which of the following statements about patterns or processes in the evolution of life is *not* true?
   a. ^{14}C can be used to date the ages of dinosaur bones.
   b. Mass extinctions have occurred numerous times in Earth's history.
   c. The supercontinent Pangaea formed during the Permian.
   d. Simple marine vertebrates first appeared in the Silurian.
   e. Rates of evolutionary change can be rapid during times of dramatic change in physical environments.
3. Which of the following pairs of organisms were *not* present on Earth in their living forms at the same time?
   a. Humans and angiosperms
   b. Dinosaurs and bony fish
   c. Amphibians and birds
   d. Gymnosperms and insects
   e. Humans and trilobites

Answers and Explanations

Chapters 22–28

CHAPTER 22 – BACTERIA AND VIRUSES

Note : The page numbers listed with the section titles refer to the Study Guide; numbers in parentheses for each question refer to the relevant key concepts.

GENERAL BIOLOGY OF THE BACTERIA (pages 148–150)

- (2, 4, 7, 15–17, 21)
  1. Invasiveness is a measure of how easily bacteria can multiply within a host, while toxigenicity is a measure of the level of injurious chemicals the bacterium produces.
  2. Facultative anaerobes are those bacteria that can switch between aerobic and anaerobic metabolism, depending on environmental conditions.
  3. Chemoautotrophic bacteria use energy obtained from the oxidation of inorganic molecules to fix CO_2.
  4. Gram staining is a technique used to identify different categories of bacteria based on the chemical nature of their cell walls, which differentially bind to the staining molecules.
  5. These are the possible shapes of bacterial cells. "Coccus" = spherical, "bacillus" = rod-shaped, and "spiral" or "curved" = curvilinear.

1. Choice *c*. Koch's postulates test for the pathogenic nature of microorganisms. In bacteria, both gram-positive and gram-negative bacteria may be pathogenic. (15–16, 20)
2. Choice *a*. The purple color indicates a positive reaction to the Gram stain. The rod shape indicates a bacillus type of bacteria. (15, 17)
3. Choice *d*. All three items represent important differences between eukaryotic and prokaryotic flagella. (14)
4. Choice *a*. The cyanobacteria use chlorophyll *a* as their primary photosynthetic pigment. (5)
5. (4–8)
   b: photoautotrophy
   d: photoheterotrophy
   c: chemoautotrophy
   a: chemoheterotrophy
6. Choice *c*. Although prokaryotes lack membranous internal organelles, many nevertheless have extensive internal membrane systems to carry out metabolic functions. (13)
7. Choice *a*. The phyla of Archaebacteria are subdivided by habitat type into thermoacidophiles, methanogens, and strict halophiles. (12)

PHYLA OF THE ARCHAEBACTERIA • PHYLA OF THE EUBACTERIA (pages 150–152)

- (3–4, 13–14, 16–17, 19)
  a. Eubacteria, mycoplasmas: lack cell walls, smallest cells known, parasites
  b. Eubacteria, gram-positive: rods, some make endospores, some do not, some toxic
  c. Eubacteria, gram-negative: photosynthetic, fermentation, nitrogen fixers
  d. Archaebacteria, methanogens: obligate anaerobes, methane production
  e. Eubacteria, gram-negative: round, nitrifiers, gonorrhea
  f. Archaebacteria, thermoacidophiles: tolerate low pH and high temperatures
- (7–8, 14, 18)
  a. A unique protein structure composed of flagella that assists the movement of curved or spiral bacteria.
  b. Chemicals manufactured by some Eubacteria that inhibit cell wall formation in other prokaryotic cells. Humans have harnessed this method of chemical defense for use in medical treatment against bacterial infections.
  c. Fruiting bodies form from aggregations of gliding bacteria and produce thick-walled spores or cysts, which are highly resistant to drying out.
  d. Heterocysts are found in filamentous forms of bacteria. They serve as break points during reproduction and as important storage sites for oxygen-sensitive enzymes needed in nitrogen fixation.

1. Choice *e*. All of these groups have at least some toxic or disease-causing members. (8–9, 12, 17)
2. Choice *a*. The methanogens are members of the Archaebacteria. (2, 6)
3. Choice *d*. The absence of any kind of cell wall is unique to this group of bacteria. (19)
4. Choice *e*. All of the statements listed apply to methanogens. (1, 4, 15–17)
5. Choice *d*. (1, 6, 10)
6. Choice *e*. The antibiotic streptomycin was first isolated from these bacteria. (18)

VIRUSES (pages 152–154)

- Viruses are acellular because they lack the typical cellular characteristics of either prokaryotes or eukaryotes. The term obligate intracellular parasite means that viruses are unable to replicate by themselves, so must commandeer the metabolic machinery of a host cell in order to make new viruses. (2–3)
- A virus is defined as an acellular, obligate intracellular parasite of prokaryotes and eukaryotes. The term "virion" is used to describe the virus when it is outside the host cell. Virions/viruses consist of a central core of DNA or RNA surrounded by a protein coat, or capsid, which has a characteristic shape for that virus. The capsid's role is to interact with the host cell surface and allow infection. (1, 4)

1. Choice *a*. The ability to self-replicate is an essential feature of living entities. Viruses lack this property and must therefore invade living cells in order to commandeer their metabolic machinery for replication purposes. (3, 6)
2. Choice *b*. Viroids consist of small segments of RNA that lack any trace of a protein capsid or tail. (16–17)
3. Choice *b*. Since all viruses are intracellular parasites, this is not a very useful criterion for classifying them. (3)
4. Choice *c*. Unlike the arboviruses, viroids are believed to be transmitted by mechanical contact between plants or through plant reproductive processes. (16–17)
5. Choice *e*. Arboviruses are transmitted to their mammalian hosts through the bite of an insect vector. (15)
6. Choice *d*. Scrapie-associated fibrils, or prions, have been implicated as a possible causative agent in all of these diseases. (18)

Integrative Questions

1. Bacteria are primitive in the sense that they are prokaryotic cells, which were probably the first types of cells to arise on Earth. They lack the sophisticated membrane-bounded organelles found in eukaryotic cells, which somewhat limits their cellular capacities. Referring to bacteria as primitive is also misleading because these organisms exhibit a wide array of metabolic adaptations and are capable of living in a number of rather harsh environments. The bacteria are extremely abundant and evolutionarily very successful organisms.
2. Even the largest viruses are VERY much smaller than the smallest cells!
3. *One*. Since viruses are acellular, they do not have kingdom status. The gram-positive bacterium is a member of the kingdom Eubacteria.
4. *a*. Hot springs: ***thermoacidophiles***
   b. Intestines of a cow: ***methanogens, gram-negative rods***
   c. Intracellular parasite: ***many forms, excluding the archaebacteria***
   d. Skin surface: ***gram-positive cocci***
   e. Alkaline lake: ***strict halophiles***

CHAPTER 23 – PROTISTS

Note : The page numbers listed with the section titles refer to the Study Guide; numbers in parentheses for each question refer to the relevant key concepts.

PROTISTA AND THE OTHER EUKARYOTIC KINGDOMS • GENERAL BIOLOGY OF THE PROTISTS (pages 155–157)

- (3)

ANIMALIA

PLANTAE | FUNGI

PROTISTA

EUBACTERIA | ARCHAEBACTERIA

- The contractile vacuole is used to regulate water balance in some protists. In fresh water, the environment is hypotonic to the cell and water thus enters the protist's body. The contractile vacuole then pumps this excess water out. In an isotonic solution, there should be no net accumulation of water, so the vacuole should contract only rarely. When moved to a hypertonic environment the cell will lose water through osmosis, so no pumping activity should be seen. (10)

1. Choice *a*. All protists are eukaryotic. (1, 2)
2. Choice *c*. Multicellularity is the main difference between animals and animallike protists. (6)
3. Choice *b*. (13)
4. Choice *c*. Increased pumping of the contractile vacuole indicates that the cell is in a solution that is even more hypotonic than pond water; in this case, distilled water is the best answer. (10)
5. Choice *e*. (14)
6. Choice *b*. The sporophyte stage does indeed produce spores, but through meiosis and not mitosis. (17)

PROTOZOANS (pages 157–159)

- (16–18)

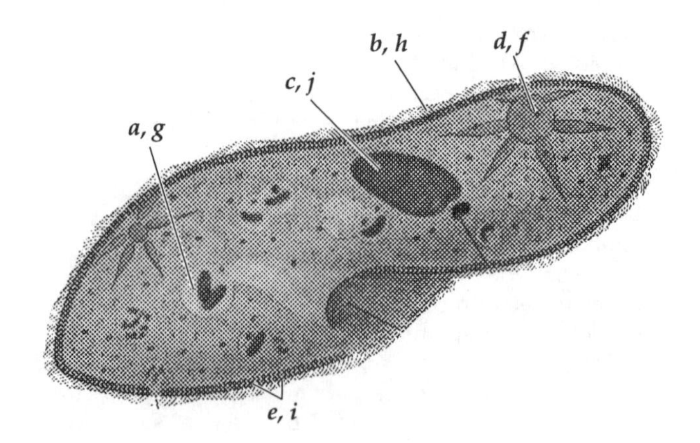

- Cilia are used for locomotion in the ciliates, as are cirri and their associated neurofibrils and myonemes. (17–18, 20)

1. Choice *b*. The Apicomplexa contains only parasitic forms of protists. (12)
2. Choice *e*. *Plasmodium* is an apicomplexan. (4–5, 14)
3. Choice *b*. Pseudopods are a prominent feature of actinopod anatomy. (9–11)
4. Choice *d*. Trichocysts are the defensive structures of ciliates such as *Paramecium*. (17)
5. Choice *a*. Pseudopods are found in actinopod protists, not ciliates. (9, 17, 20)
6. Choice *d*. Both phyla have an amoeboid body plan. (6, 12)

FUNGUSLIKE PROTISTS (pages 159–160)

- A protist is coenocytic when it takes the form of a wall-less mass of cytoplasm that contains numerous diploid nuclei. This aptly describes the plasmodium, or vegetative body form, of an acellular slime mold. The pseudoplasmodium of a cellular slime mold is not coenocytic because, although individual myxamoebas do aggregate to form a large mass, or fruiting body, they retain their individual cellular identities by virtue of the walls that remain between individual cells. (2–3, 8–9)
- Alternation of generations is the process of switching between haploid (n) and diploid ($2n$) stages in the life cycle. In some funguslike protists, both stages are multicellular with distinctly different morphologies (heteromorphic alternation of generations). In some, such as *Allomyces*, the two generations may be morphologically indistinguishable (isomorphic alternation of generations). (12)

1. Choice *d*. (3, 8–9)
2. Choice *a*. Isomorphic means "same body form." These protists have indistinguishable haploid and diploid morphologies. (12)
3. Choice *b*. The hypochytrids belong to the phylum Protomycota. (11, 15–16)
4. Choice *e*. Conjugation is common among certain animallike protists, bacteria, and fungi, but not the funguslike protists. (1–14)
5. Choice *c*. Gametangia are found in the Protomycota. (13)
6. Choice *a*. (14)

ALGAE (pages 160–163)

- They are all eukaryotic and photosynthetic, exhibit alternation of generations (often complex), and have a zygotic stage that is independent of parental protection. (1–4, 18)
- When diatoms reproduce asexually, each half of the shell is used as a template for a new complementary half. If the smaller half were used indefinitely, the cell would eventually become too small to be functional. Periodic sexual reproduction avoids this problem by forming a zygote, which grows substantially in size before resuming the process of asexual reproduction. (10, 11)
- The life cycle is isomorphic because the mature gametophyte and sporophyte stages are morphologically indistinguishable. It is anisogamous because the male and female gametes are of different sizes. (26)

1. Choices *a* and *c*. Although most chlorophytes are heteromorphic, *Ulva* is a genus with isomorphic forms. (18, 26–27)
2. Choice *b*. Chromatic adaptation alters the amount of phycoerythrin, thereby maximizing photosynthetic efficiency at different water depths (i.e., different light conditions). (21)
3. (6, 10, 16, 19, 20, 24)
   a: Rhodophyta
   f: Chlorophyta
   d, *e*: Phaeophyta
   b: Chrysophyta
   c: Pyrrophyta
4. Choice *e*. Dinoflagellates, a kind of pyrrophyte algae, are the cause of red tides. (8)
5. Choice *a*. Chlorophyll *a*, while not exclusive to the green algae, is the only characteristic listed that is found in all members of Chlorophyta. (24)
6. Choice *c*. (15, 16)

Integrative Questions

1. Choice *c*. As nebulous as it sounds, this is the only description given that fits all protists! The other choices either contain characteristics not found in protists (e.g., prokaryotic), or characteristics found in only some protists (e.g., flagellated).
2. Choice *b*. Only certain members of the kingdom Eubacteria are capable of nitrogen fixation.
3. Choice *b*. In haplontic life cycles, the only diploid stage is the zygote.
4. Choice *e*. The green alga is clearly Chlorophyta. Since the amoeba is said to be nonparasitic, it is probably Rhizopoda instead of Apicomplexa.

CHAPTER 24 – FUNGI

Note : The page numbers listed with the section titles refer to the Study Guide; numbers in parentheses for each question refer to the relevant key concepts.

GENERAL BIOLOGY OF THE FUNGI (pages 164–166)

1. Choice *d*. The cells of fungi have no flagella. (5)
2. Choice *b*. (8)
3. Choice *e*. A large surface-to-volume ratio maximizes contact between fungal cells and their food source. (10)
4. Choice *d*. Since the original cells are of different mating types, and because the new cell has two distinct nuclei, the new cell is both heterokaryotic and dikaryotic. (19)
5. Choice *b*. Conidia are also known as naked spores because they form at the tips of hyphal cells and not in specialized sporangia. (18)
6. Choice *a*. An ascus is the site where haploid nuclei fuse to produce a diploid zygote that undergoes meiosis to form spores. (20)

- (19–20)

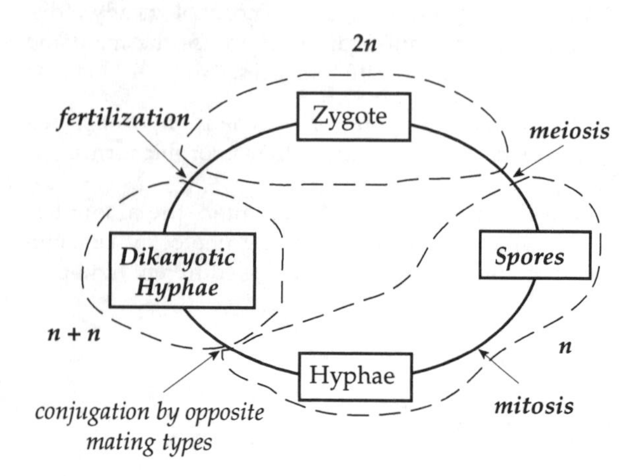

- Hyphae consist of haploid fungal cells that result from the germination and growth of spores. As they grow and multiply, the hyphae form a mass of threadlike filaments called a mycelium, which may have a cottony texture or form a more solid structure like a fruiting body. Rhizoids are extensions of the mycelium that anchor some fungi to the substrate, while haustoria are specialized hyphae found in parasitic fungi that invade the living cells of a host to help procure nutrients. (8)

PHYLUM EUMYCOTA • LICHENS (pages 166–168)

1. Choices *a, c, d*. The fungal groups that might appear on your dinner plate include the edible mushrooms in the class Basidiomycetes, fungi involved in producing certain cheeses, such as Roquefort and Camembert, soy sauce in the class Deuteromycetes, and truffles and morels in the class Ascomycetes. (9, 11, 15)
2. Choice *d*. All three are caused by, or come from, deuteromycete fungi. (14–15)
3. Choice *a*. The pileus, or cap, is a common characteristic of the club fungi (Basidiomycetes). The gills are where the spores form, and basidiomycete hyphal cells are septate. (11–13)
4. Choice *c*. As the name suggests, karyogamy is the fusion of two nuclei. It occurs prior to ascus formation in euascomycetes, as well as in conjunction with zygote formation in other groups of fungi. (4, 8, 12)
5. Choice *d*. Lichens reproduce vegetatively by breaking off a piece of their existing structure (the thallus), or by producing a special dispersal structure called the soredium. Fungal fruiting bodies are not observed in the reproduction of lichens. (19–20)
6. Choice *a*. Since lichens are an association between a fungus and some photosynthetic alga or bacterium, they are therefore excluded from true organismal status. (16)
7. Choice *a*. (8)
8. Choice *e*. All of the traits listed are common to the class Ascomycetes, of which truffles and bread molds are members.(4, 8–9)

- The yeasts are generally very small and unicellular. They reproduce primarily by asexual means (either by fission or budding), and there is never a dikaryotic stage. They also engage in alcohol fermentation. (7)
- The bracket is the fruiting body, or basidium, of this fungus. It forms as a specialized outgrowth of the mycelium. Within certain hyphae of the basidium, dikaryotic cells undergo nuclear fusion to form zygotes. The zygotes then undergo meiosis to form basidiospores, which are shed from the underside of the bracket. (11–13)
- Since lichens have a photosynthetic component to their mutualistic relationship (the alga or bacterium), it makes sense that they should grow best on the side of trees that get the most sun. In the Northern Hemisphere this is the southern side, in the Southern Hemisphere the northern side. (16–17)

Integrative Questions

1. Choice *e*. Some fungi, such as the yeasts, are unicellular.
2. Choice *c*. The description provided best describes a foliose (leafy) species of lichen.
3. Choice *e*. Saprobes feed on all sorts of dead organic matter.
4. Choice *a*. Most fungi, including the two groups mentioned here, have haploid hyphal cells.

CHAPTER 25 – PLANTS

Note : The page numbers listed with the section titles refer to the Study Guide; numbers in parentheses for each question refer to the relevant key concepts.

THE PLANT KINGDOM (pages 169–171)

1. Choice *e*. Mitosis occurs in several phases of the plant life cycle (see above diagram). Of the choices mentioned, the only place it does not occur is during the formation of spores, which is achieved through meiosis. (3–4)
2. Choice *e*. The information given is insufficient to determine a specific plant group. All plants are eukaryotic, have both chlorophyll *a* and *b*, and have both single-celled as well as multicelled haploid forms. (2–4)
3. Choice *b*. In the vascular plants xylem tissue conducts water and minerals, while phloem carries the products of photosynthesis. Since mosses are nonvascular, they lack these tissues. (6, 10–11)
4. Choice *c*. Nonvascular plants did evolve before vascular plants. However, because vascular plants fossilize more easily due to their structure, in older rock formations they are more likely to be preserved than are nonvascular plants. (12)
5. Choice *d*. Unlike plants, *Ulva* has isomorphic alternation of generations. (2–5, 8)
6. Choice *b*. Mosses are nonvascular plants. (6, 10)

- Similar photosynthetic and accessory pigments; starch as the storage carbohydrate; cellulose cell walls; oogamous reproduction. (8)

- (3–4)

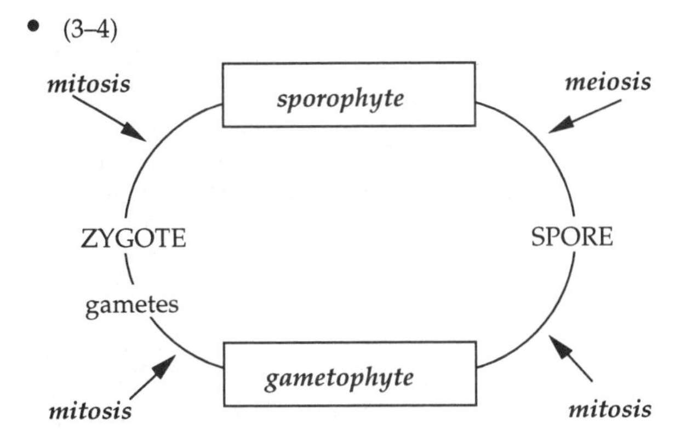

NONVASCULAR PLANTS (pages 171–172)

- *Gametophyte:* sperm and eggs (5)
  Neither: phloem (2)
  Gametophyte: chloroplasts (4)
  Both: waxy cuticle (3)
  Sporophyte: capsule (6)
  Gametophyte: archegonium (5)
  Sporophyte: diploid tissue (6)
- More than 75% of the life cycle is dominated by the haploid gametophyte, which forms the "leafy" part typically recognized as a moss. The sporophyte is smaller and shorter-lived. (1, 5–6)

1. Choice *c*. Water-related reproductive factors and those concerning physical support are what limit the absolute size of nonvascular plants; having a diploid sporophyte stage is not one of these factors. (1–2)
2. Choice *b*. All plants, including the hornworts and liverworts, have a water-resistant waxy cuticle. (3)
3. Choice *c*. The calyptra is a protective cap that forms atop the spore capsule of the sporophyte stage in mosses. It opens at maturity to release the spores. (13)
4. Choice *e*. Antheridia and archegonia are the gamete-producing structures that develop on the haploid gametophyte stage of mosses and other nonvascular plants. (5–6, 8–13)
5. Choice *d*. Both sporophytes and gametophytes are distinctly different in these two groups of nonvascular plants. Neither possess vascular tissue. (10, 12)
6. Choice *d*. A protonema is the filamentous structure that forms from the mitotic divisions of a germinating moss spore. (12)

VASCULAR PLANTS • SURVIVING SEEDLESS VASCULAR PLANTS (pages 172–174)

- The branching of Rhyniophyte rhizoids was dichotomous. Given the unique selective pressures to which these subterranean rhizoids were subjected, any adaptations for increased surface area and uptake of water, or support of the aerial portions of the plant, would have been greatly favored by evolution. In time, these adaptations would have accumulated to produce the features shown by the true roots of higher plants. (10)
- Simple leaves and complex leaves are flattened, vascularized photosynthetic structures attached to the plant stem. Simple leaves are small and scalelike, and hardly distort the stem vascular tissue at their point of separation. They are found only in the club mosses and horsetails. Complex leaves are present only in the ferns and seed plants. Their large photosynthetic surfaces are dichotomously branched and highly vascularized. The vascular tissue radically alters the stem vascular system at its point of departure. (11–13)

1. Choice *c*. This is alternation of generations with heterospory, which is found in some ferns and all seed-bearing vascular plants. (14–15)
2. Choice *a*. Meiosis is how the plant goes from diploid to haploid, and fertilization is how it goes from haploid to diploid. (3–4, 14–15)
3. (1–3, 9, 13, 19, 21).
   Gymnosperms: ***a, b, c, d***
   Psilopsids: ***a, b***
   Club mosses: ***a, b, c***
   Horsetails: ***a, b, c***
   Ferns: ***a, b, c, d, e***
4. Choice *b*. Megaspores are the large spores produced by heterosporous plants that subsequently develop into the haploid female gametophyte (megagametophyte). They are produced through meiosis by the diploid sporophyte plant. (14–15)
5. Choice *a*. Meiosis is the process that produces spores in plants. Spores are made in specialized structures called sporangia. (14, 19–21)
6. Choice *c*. Ferns have a vascular system, so water transport is not the issue here. The fact that ferns have swimming sperm cells is the reproductive factor that still limits them to moist habitats. (22)

THE SEED PLANTS • THE GYMNOSPERMS • THE ANGIOSPERMS: FLOWERING PLANTS (pages 174–177)

1. Choice *d*. The sporophyte plant is the dominant stage in gymnosperm life cycles. It is the gametophyte that is dependent on the sporophyte. (2, 4)
2. Choice *b*. Raspberry plants have fruits formed from the multiple carpels of a single flower. (29)
3. Choice *d*. The pollen (sperm) nucleus must travel down the pollen tube before entering the female gametophyte tissue. (4–6)
4. Choice *a*. The receptive female part of a flower is the sticky stigma. Pollen grains that alight here can form a pollen tube and compete for access to the ovules contained in the ovary. (22, 27)
5. Choice *c*. Palms belong to the monocotyledones subdivision of the angiosperms. They possess everything in the list except for a second cotyledon. (31)
6. Choice *b*. Since it lacks carpels, this flower is imperfect. (20–24)
7. Choice *b*. (10, 17)

- (7)

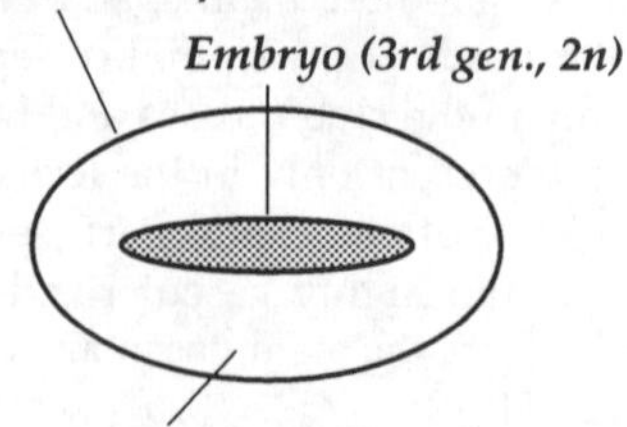

- (20–24)

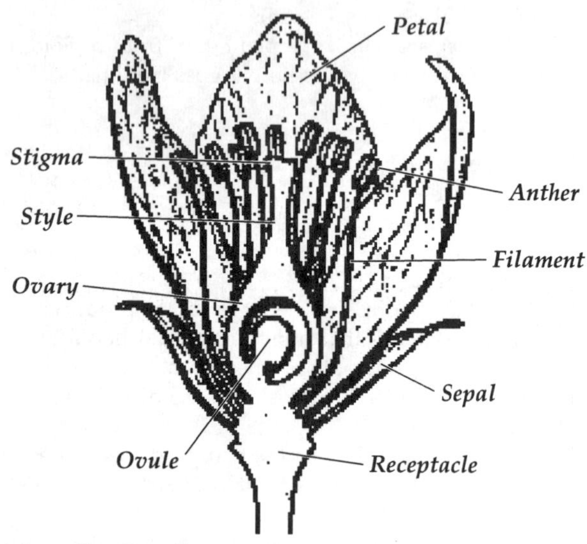

Integrative Questions

1. Choice *e*. Both the angiosperms and gymnosperms lack swimming sperm cells, and thus have been freed of the requirement for liquid water to reproduce. Pine trees and roses are examples.
2. Choice *b*. There has been a strong tendency in plant evolution to reduce the size of the gametophyte generation, a pattern that reaches its culmination in the angiosperms.
3. Choice *c*. Seeds are a way of protecting and providing initial nourishment to the sporophyte embryo.
4. Choice *e*. Hornworts are nonvascular, pines are gymnosperms, and apple trees are angiosperms. The best answer is therefore that they are all plants.
5. Choice *e*. The first four answers are all valid trends in plant evolution. The last choice is just the opposite of the actual trend observed in plant vascular tissue.

CHAPTER 26 – SPONGES AND PROTOSTOMATE ANIMALS

Note : The page numbers listed with the section titles refer to the Study Guide; numbers in parentheses for each question refer to the relevant key concepts.

HOW ARE ANIMALS CLASSIFIED? (pages 179–180)

1. Choice *c*. All animals are heterotrophic, so this criterion does not really contribute to body plan, while all of the others do. (1, 4–6)
2. Choice *a*. All deuterostomes are triploblastic. (12)
3. Choice *e*. Diversity in the methods for how food is acquired is thought to have influenced much of animal evolution. (4)
4. Choice *d*. Bilateral animals tend to move through the environment. Cephalization is important in the control and coordination of this locomotion. (9)
5. Choice *e*. (12–15)
6. Choice *c*. Similar body halves could be obtained with either radial or biradial symmetry. Spherical symmetry has no main body axis along which to cut, and bilateral symmetry produces mirror-image halves. (8–9)

- (1, 2, 4, 12, 14–15)

Characteristics	Protostome	Deuterostome
Cell type	-- *both eukaryotic* --	
Mode of nutrition	-- *both ingestive heterotrophs* --	
# of embryonic layers	*two or three*	*three*
Type of cleavage	*determinate*	*indeterminate*
Cleavage pattern	*spiral*	*radial*
Fate of blastopore	*mouth*	*anus*

- (16) The body cavity, or coelom, is a diagnostic feature of animal bodies. Acoelomate animals have no cavity at all. Pseudocoelomate animals have a partial fluid-filled cavity in which internal organs reside. Coelomate animals have a complete body cavity, with internal organs slung in sacs of mesentery tissue.

THE ORIGINS OF ANIMALS • SIMPLE AGGREGATIONS (pages 180–181)

- (7–8)

  Amoebocytes are wandering cells that provide simple communication within the sponge body.
  Choanocytes are flagellated collar cells that move water and food into the body cavity of a sponge for feeding.
  Incurrent pores are the numerous small openings in a sponge's body through which water enters.
  Osculum is the larger opening in a sponge's body through which water, wastes, and gametes exit.
  Spicules are mineral- or protein-based structural elements that help support a sponge's body.
- Surface-to-volume relationships limit the size of cells or bodies that can get by on simple diffusion, which was the only option available in the low oxygen concentrations that prevailed during the early history of animals. Thus, early animals could not attain large body sizes. When atmospheric oxygen levels rose, animal bodies grew also, but eventually alternatives to simple diffusion became necessary as multicellular bodies became large. At that point, specialized respiratory structures such as gills, lungs, and other gas exchange mechanisms evolved. (3–4)

1. Choice *a*. In fact, the opposite is often true. The main point here is that algae were about the only food available to early animals, and so grazers were the first to evolve. Predators on these grazers came later. (5–6)
2. Choice *d*. While the rate of new species formation may lead to diversity, it does not protect the individual animal from being attacked or eaten. (5–6)
3. Choice *b*. Because they are not structurally robust, heavy wave action would destroy large, upright sponges. Thus we find smaller, flattened forms in near-shore or intertidal areas subjected to waves. (10)
4. Choice *d*. Placozoans do not have support structures like the spicules of sponges. (8, 12)
5. Choice *a*. *Volvox* is a modern colonial protist. (1)

THE EVOLUTION OF DIPLOBLASTIC ANIMALS (pages 181–183)

- Cnidarians have tentacles complete with stinging cells to subdue and capture prey. Ctenophores use sticky filaments to capture much smaller prey items. (2, 13)
- (6–8)

Class	Medusa	Polyp
Hydrozoa		✔
Anthozoa		✔
Scyphozoa	✔	

- (3–4)

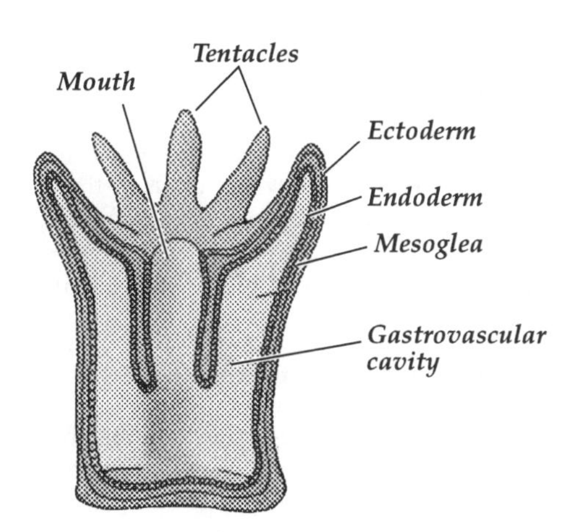

1. Choice *a*. Cnidarians have made several body plan improvements, including radial symmetry, nematocysts, alternating body forms, and a gastrovascular cavity. They do not, however, possess a true coelom, or body cavity. (1–4)
2. Choice *e*. Corals and sea anemones are members of the phylum Cnidaria, class Anthozoa. (2, 8)
3. Choice *c*. While the dominant stage in the life cycle of jellyfish is the medusa, anemones have lost this stage altogether. (7–8)
4. Choice *d*. Jellyfish use the stinging cells of their tentacles for prey capture and defense. Ctenophores lack these nematocysts, and thus capture prey by extruding sticky filaments which they reel in when full. (7, 13)
5. Choice *d*. Nematocysts, the stinging cells found in cnidocytes, are the only item in this list that the cnidarians possess exclusively. (2)

THE EVOLUTION OF BILATERAL SYMMETRY • THE DEVELOPMENT OF BODY CAVITIES • PSEUDOCOELOMATE ANIMALS (pages 183–184)

- The main clue to why this animal cannot be a flatworm is the presence of a continuous digestive tract running from the anterior end of the animal to its posterior end. The phylum Platyhelminthes lacks a complete digestive tract. (3)
- Roundworms have only longitudinal muscles, which run the length of the animal's body. They contract in alternating waves down each side, resulting in a rather ungainly, but nevertheless effective type of side–to–side writhing motion. (11)

1. Choice *d*. Choices *a–c* are all false; exactly the opposite is true in each case. Flatworms are, however, triploblastic animals. (1–3)
2. Choice *b*. Although having changed from the cnidarian body plan, flatworms have yet to develop a complete digestive system. (2–4)
3. Choice *b*. (8–9)
4. Choice *c*. Hydrostatic skeletons are fluid-filled sacs which, when combined with appropriate musculature, can produce motion. Rotifers, however, rely on the beating of cilia for most of their body movements. (7–10, 13)
5. Choice *b*. (10)
6. Choice *e*. All of these characteristics are found in the rotifers and roundworms. (11–14)

COELOMATE ANIMALS • THE EVOLUTION OF EXTERNAL SKELETONS (pages 184–187)

- The fact that we are told this is a crab larva indicates that it is also a crustacean arthropod. From the diagram itself we see evidence of a jointed exoskeleton; notice that there are distinct segments to the body, in particular, an obvious head and abdomen (there is also a thorax, but it is less evident). Moreover, there are paired appendages which attach to specific body segments. Other obvious features are the well-developed sensory organs (eyes, antennae) located on the anterior end of the animal. (12–15, 21)
- Polychaetes are mostly marine, segmented, have parapodia and well-developed sensory systems. Their sexes are distinct and they have a trochophore larval stage which undergoes metamorphosis to become an adult. Oligochaetes are aquatic and terrestrial, lack parapodia, and have limited sensory abilities. They are hermaphrodites, and their young hatch as fully developed, miniature worms. Both groups have hydrostatic skeletons with circular and longitudinal muscles. (5–8)

1. Choice *d*. Lobsters, millipedes, and butterflies are all arthropods, and therefore have exoskeletons with jointed appendages. (11–12)
2. Choice *b*. The majority of annelids are polychaetes, so if the box does contain a fair representation of all annelid species, odds are the one selected will be a polychaete. (5)
3. Choice *e*. Earthworms have a very thin cuticular layer that functions in gas exchange and is also part of their hydroskeleton. (3–4, 7)
4. Choice *a*. All of the animals considered in this section have true coeloms, so that piece of information is not useful. The observation that this animal lacks an exoskeleton eliminates all of the crustaceans and insects from consideration. The presence of appendages and the fact that the specimen came from a shallow tidepool eliminates the pogonophores as well as the oligochaetes. Only the polychaetes have single-sexed individuals that also meet the other criteria. (2, 5–7, 12)
5. Choice *e*. (26)
6. Choice *c*. (16)

CALCIFIED PROTECTION • THEMES IN PROTOSTOME EVOLUTION (pages 187–189)

- *a*. The presence of fluid-filled cavities, which make for effective hydroskeletons. (16)

  b. Hard external protective coverings for defense against large, rapidly swimming aquatic predators. (17)

  c. Numerous sessile (and still mainly aquatic) representatives. (10)
- *a*. Heavily calcified shells offering protection against all but the toughest predators. (1)

  b. Advanced body plans that make for evolutionarily flexible lifestyles. (2–3)

  c. Complete digestive tract, improved circulatory system, and better sensory organs. (4)

1. (5–16)
   Monoplacophorans: ***d,e***
   Cephalopods: ***a,c***
   Gastropods: ***b,d***
   Polyplacophorans: ***d***
   Bivalves: ***d***
2. Choice *b*. Interconnected plates that form a shell are characteristic of the polyplacophorans. All the other traits are found in cephalopods such as the squid. (5, 12–14)
3. Choice *e*. (6)
4. Choice *c*. Snails are gastropods, clams are bivalves, and octopuses are cephalopods. (7–8, 11)
5. Choice *b*. Snails have no heads. (3, 8–9)
6. Choice *c*. Cephalopods are predatory, active swimmers. (14–15)

Integrative Questions

1. Choice *e*. All of these statements correctly apply to every bilaterally symmetrical animal.
2. Choice *a*. None are acoelomates, all are bilaterally symmetrical, and neither snails nor earthworms have very good vision. However, all three are indeed protostome animals.
3. Choice *b*. Neither are spherically symmetrical, only rotifers are pseudocoelomates, and since neither is a deuterostome, so they can't have radial cleavage of the embryo. However, both do possess complete digestive tracts.
4. Choice *d*. Lobsters are crustaceans, while clams and squids are mollusks.
5. Choice *a*. Of those listed, crabs are the most closely related to trilobites by virtue of the fact that they are both arthropods.
6. Choice *c*. Cephalization is the accumulation of neural tissue in the anterior end of the animal. The only animal listed which lacks some obvious form of this characteristic is the jellyfish, a cnidarian.

CHAPTER 27 – DEUTEROSTOMATE ANIMALS

Note : The page numbers listed with the section titles refer to the Study Guide; numbers in parentheses for each question refer to the relevant key concepts.

TRIPARTITE DEUTEROSTOMES • INNOVATIONS IN FEEDING • ACTIVE FOOD SEEKERS (pages 190–191)

- ***no***: Pterobranchs (9)
  yes: Bryozoans (6)
  no: Arrow worms (11–12)
  yes: Phoronids (5)
  yes: Brachiopods (7)
  no: Acorn worms (10)

1. Choice *c*. Most simple deuterostomes are filter-feeding lophophorates. As other feeding mechanisms evolved, animals have tended to become larger and more mobile. (2–12)
2. Choice *a*. Determinate cleavage is characteristic of the protostomes. (1)
3. Choice *d*. While new bryozoan colonies do in fact form from a single larva that settles out from the plankton, the larva itself is generated by *sexual* reproduction. (6)
4. Choice *e*. (5, 11–12)
5. Choice *c*. The hinged shells of the two groups appear somewhat similar. (7)

CALCIFYING THE SKELETON (pages 191–193)

- Brachiopods are filter feeders; they use the cilia on their internalized lophophore to strain food from the water. Sea stars are active predators and scavengers that use the action of their tube feet to overpower prey such as bivalve mollusks; they may also evert their stomachs to digest this food externally. (9–11)
- The echinoderms are much more mobile than the mostly sedentary lophophorates. The echinoderm water vascular system and specialized internal musculature coordinates the action of the tube feet to achieve motion. (3)

1. Choice *d*. Lophophores are characteristic of several other phyla, but not the echinoderms. (6–14)
2. Choice *a*. The sea lilies and feather stars together make up the class Crinoidea. (5–7)
3. Choice *c*. Crinoids belong to the Pelmatozoans. All other echinoderms are Eleutherozoans. (5, 8)

4. Choice *e*. Sea stars do not have a rounded, fused internal skeleton. They also have distinct arms, which are lacking in the sea urchins. (2, 9, 13)
5. Choice *e*. (1–14)

EVOLUTION OF THE CHORDATE PHARYNX • SUCKING MUD: THE RISE OF THE VERTEBRATES (pages 193–194)

- The skeleton of a modern bony fish would have a well-formed skull with hinged jaws, a vertebral column, two sets of paired pectoral and pelvic fins, a caudal (tail) fin, and perhaps a dorsal fin. It is the presence of hinged jaws that makes this skeleton so obviously different from that of agnathan (jawless) fishes. (11–13, 18–20)
- Hinged jaws are advantageous because they allow the bearer to catch and manipulate larger, more mobile prey. They also might be used as weapons for self-defense. Hinged jaws were much more adaptable than the sucking aparatus of the agnathans, and so the early jawed fishes underwent rapid evolutionary radiation into previously unexploited feeding habitats. (13–15)

1. Choice *a*. Filter feeding requires a large surface area in order to gain enough food to sustain larger animals. (8)
2. Choice *e*. Larvaceans are a type of tunicate, and the lancelets (like all other chordates) evolved from tunicate ancestors. (4, 6–7)
3. Choice *c*. Placoderms were among the first large predatory fishes with hinged jaws; formidable teeth were an important part of their feeding apparatus. (14)
4. Choice *b*. Cartilaginous fishes include the sharks and their relatives. Bony fishes either evolved simultaneously with these fishes, or perhaps slightly after. Lancelets, tunicates, and agnathans all appeared before the sharks. (4, 7, 15)
5. Choice *d*. The hyoid gill arches of agnathan fishes are believed to be the anatomical precursors of jaws in vertebrates. (13)
6. Choice *d*. Some of the other answers are perhaps justifiable as indirect effects, but improved buoyancy is the primary role of the swim bladder. (19)
7. Choice *b*. Pharyngeal baskets are present in the prevertebrate animals. (8–11)

BREATHING AIR AND EXPLORING THE LAND • THE ORIGINS OF MAMMALS (pages 194–196)

- (Refer to all key concepts from this section)
  Lungs: ***A, B, D, L, M, R***
  Aquatic larvae: ***A, L***
  Wings: ***B, D, M***
  Feathers: ***B***
  Uterine development: ***M***
  Terrestrial egg: ***B, D, M, R***
  Water-impermeable skin: ***B, D, M, R***
- (13)
  a. Teeth have been completely lost.
  b. Bones of the tail have been lost.
  c. Breast bone (sternum) has been enlarged to accommodate flight muscles.
  d. Digits of hand have been reduced and fused to form base for attachment of flight feathers.

1. Choice *c*. Feathers are certainly useful for insulation, but had nothing to do with the transition to life on land, since birds evolved from terrestrial vertebrates. (13)
2. Choice *a*. The embryo resides inside a *fluid*-filled chamber created by the amnion. (7, 15)
3. Choice *b*. Many vertebrate systematists now believe that the avian lineage is a direct descendant of the dinosaurs. (12)
4. Choice *d*. Kangaroos are marsupials whose young are born after only a minimum of uterine development. They then crawl into the mother's ventral pouch to complete their growth. (22)
5. Choice *e*. Caimans are a type of semi-aquatic reptile. (3, 12)
6. Choice *c*. (7)
7. Choice c. The limb position of mammals raises the body off of the ground for better locomotion than is possible in the reptiles. (17)
8. Choice *e*. Amphibian eggs and larvae must develop in water, and adult amphibians must stay moist because they lack desiccation-resistant skin. (4)

HUMAN EVOLUTION • THEMES IN DEUTEROSTOME EVOLUTION (pages 197–199)

- (5–17)
  4: *Homo erectus*
  5: *Homo sapiens* – Neanderthal
  6: *Homo sapiens* – Cro-Magnon
  2: *Australopithecus africanus*
  1: *Australopithecus afarensis*
  3: *Homo habilis*
- (2)
  Flexible limbs, opposable thumbs, depth perception, flattened nails, and small numbers of offspring in humans are all holdovers from the early primate arboreal lifestyle.

1. Choice *b*. Lucy was a very early hominid who lived long before formal language and culture appeared. (5, 14–15, 19)
2. Choice *d*. One of the important expansions in lifestyle of *Homo habilis* was the incorporation of animal protein in the diet, which provided better nutrition and spurred cooperative hunting. (10–11)
3. Choice *e*. (14, 19)
4. Choice *c*. Both Cro-Magnon and Neanderthal were of our own species, *Homo sapiens*, but Cro-Magnon was anatomically more like us and definitely survived into the more recent past. (16–18)
5. Choice *d*. Actually, just the reverse is true, although exactly how big an influence Cro-Magnon had on the disappearance of Neanderthal is not entirely clear. (17)
6. Choice *a*. (21)
7. Choice *b*. Exoskeletons are a characteristic of certain protostome phyla. (23–24)

Integrative Questions

1. *b*: Squamata
   c: Agnatha
   e: Asteroidea
   a: Brachiopoda
   d: Enteropneusta

2. Choice *c*. Although jaws were a very important event in vertebrate evolution, significant feeding modifications also occurred in other deuterostome lineages (e.g., loss of lophophores in hemichordates, mobile mud-sucking in agnathans, etc.).
3. Choice *d*. Fish, amphibian, bird, reptile.
4. Choice *e*. They are all chordate characteristics, and humans are chordates.
5. Choice *d*. This is the only set of characteristics listed which accurately fits *both* humans and sea stars.

CHAPTER 28 – PATTERNS IN THE EVOLUTION OF LIFE

Note : The page numbers listed with the section titles refer to the Study Guide; numbers in parentheses for each question refer to the relevant key concepts.

HOW EARTH HAS CHANGED • THE FOSSIL RECORD (pages 200–202)

1. Choice *a*. Radioactive decay is measured as the time it takes for one-half the amount of a substance to spontaneously convert into another substance. (2)
2. Choice *a*. As the crustal plates move past each other, the friction they cause generates earthquakes and pushes rocks upward to make mountains. (7)
3. Choice *b*. Fossilization occurs best in areas of low oxygen concentration and rapid sedimentation, where scavengers cannot destroy the body. (15)
4. Choice *e*. Although continental drift caused land masses to separate by forming large oceans, periodic land bridges allowed animals to overcome this separation. (12)
5. Choice *a*. The forces produced when India "rammed" into Asia caused uplifting that resulted in the tallest mountain range on Earth. (11)
6. Choice *d*. Hard-shelled animals such as the mollusks are good candidates for fossilization because their shells can withstand decay long enough to become buried. Many also tend to live in quiet, shallow waters. (18).
7. Choice *e*. If about 300,000 species are currently known as fossils, and only 2 percent of all species become fossilized, then at least 15 million species must have existed throughout evolutionary history. (16–17)

LIFE IN THE REMOTE PAST (pages 202–203)

1. Choice *c*. Coming from the same root word as "province," which is a specific area within a country or state possessing unique characteristics, provincialized biotas are geographically unique assemblages of organisms produced by a variety of physical and biological isolating factors. (11)
2. Choice *d*. The modern mammals did not appear until the Tertiary, over 350 million years after the Ordovician. (5, 12–14)
3. Choice *e*. An integral part of the success of angiosperms has been their coevolution with insect pollinators. (14)
4. Choice *c*. The genus *Homo* first appeared about 2 mya, at the start of the Quaternary. (19)
5. Choice *b*. The mass extinction at the end of the Cretaceous, 65 mya, was the most recent before the evolution of humans. (14, 19)
6. Choice *e*. (15–19)

THE TIMING OF EVOLUTIONARY CHANGE (pages 203–204)

1. Choice *b*. The Cambrian explosion occurred when many open habitats were available for exploitation. Thus, the largest numbers of new phyla were produced then, with later explosions causing increases in diversity only within already existing phyla. (14)
2. Choice *d*. For the reasons outlined in the previous answer, the Cambrian is the time of choice. (14)
3. Choice *a*. Although there is still much debate, many paleontologists believe that an asteroid's collision with Earth so altered conditions that the dinosaurs rapidly succumbed during what we know as the great Cretaceous mass extinction. (13)
4. Choice *d*. Evolutionary rates apparently have changed through time. Long periods of stasis may be punctuated by periods of rapid change. (2–3, 6–7)
5. Choice *c*. Darwins measure the magnitude of evolutionary change per million years, and can be used in both living and fossilized organisms. (7)

Integrative Questions

1. Choice *e*. Looking at evolution from an overall viewpoint, all groups of organisms have increased in size.
2. Choice *a*. Carbon dating is generally only reliable for fossils less than 15,000 years old.
3. Choice *e*. The trilobites disappeared in the Permian, over 200 million years before humans evolved.

29

The Flowering Plant Body

CHAPTER LEARNING OBJECTIVES—after studying this chapter you should be able to:

- ❑ Describe the characteristic features of the two classes of angiosperms, the Dicotyledones and the Monocotyledones.
- ❑ List the three principle plant organs and describe their functions.
- ❑ Describe the characteristic features of stems, including nodes, internodes, apical buds, lateral buds, and leaf primordia.
- ❑ Describe the general structure of roots and the specific adaptations seen in taproot and fibrous root systems.
- ❑ Characterize the structure of a typical leaf and differentiate between simple and compound leaves
- ❑ Name and characterize the general functions of the three tissue systems in vascular plants.
- ❑ Name and discuss the various modifications of roots, stems, and leaves.
- ❑ Explain the relationships between primary cell walls, secondary cell walls, and the middle lamellae.
- ❑ Describe the features and functions of pit pairs in plants.
- ❑ Describe the characteristic features and functions of the following plant cell types and be able to recognize these cell types: parenchyma, sclerenchyma, collenchyma, tracheids, vessel elements, sieve tube elements, and companion cells.
- ❑ Name and characterize the common simple and compound tissue types seen in the vascular plants.
- ❑ Explain what a meristem is, characterize where the major meristems are located in the plant body, and differentiate between primary and secondary growth.
- ❑ Describe the formation of the procambium, the ground meristem, and the protoderm, and explain the activities of these meristems in producing the mature tissues of the plant root and stem.
- ❑ Characterize and be able to identify the major structural features of the root system and discuss typical differences seen in monocot and dicot roots.
- ❑ Characterize and be able to identify the major structural features of the stem and discuss typical differences seen in monocot and dicot roots.
- ❑ Characterize and be able to identify the major structural features associated with secondary growth in woody plants.
- ❑ Characterize and be able to identify the major structural features associated with leaves.
- ❑ Explain the significance of the differences in the leaf anatomy of C_3 and C_4 plants.
- ❑ Differentiate between the two types of reaction wood, tension wood and compression wood, and discuss the occurrence and use of each.

FLOWERING PLANTS • FOUR EXAMPLES • CLASSES OF FLOWERING PLANTS • AN OVERVIEW OF THE PLANT BODY • ORGANS OF THE PLANT BODY • LEVELS OF ORGANIZATION IN THE PLANT BODY (pages 665–673)

Key Concepts

1. The *flowering plants* or *angiosperms* (phylum *Anthophyta*) have vascular tissue, including xylem composed of vessel elements and fibers, sexual reproduction involving double fertilization, and seeds with endosperm. Monocotyledones (*monocots*) and Dicotyledones (*dicots*) are the two classes of the Anthophyta.
2. The monocots (grasses, lilies, orchids, and palms) are the narrow-leaved flowering plants with one *cotyledon* (embryonic leaf).
3. Monocots usually have leaves with parallel venation, vascular bundles scattered throughout the stem, and floral parts in multiples of three.
4. *Dicots* have two cotyledons, and include the remaining broad-leaved flowering plants.
5. Dicots usually have leaves with netlike venation, vascular bundles arranged in a ring within the stem, and floral parts in fours or fives.
6. Plant parts such as branches or leaves act as semi-independent modules. As a result of this modular design, control systems in plants are decentralized.
7. Plant growth is restricted to specific regions of active cell division called *meristems*. Meristems at the tips of shoots and roots result in elongation of the plant body.

8. The vascular plant body consists of three organs, the *root system, leaves,* and *stem.* Leaves and stem combined are called the *shoot system.*
9. Main functions of these organs are; leaves: sites of photosynthesis; roots: anchorage and water/nutrient absorption; stem: display of leaves and two-way transport of materials between the roots and leaves.
10. *Nodes* are the attachment sites of leaves to stems; stem regions between nodes are called *internodes.*
11. *Taproot* and *fibrous* are the two main root system types found in flowering plants. The taproot system, common in dicots, has a single main root that often serves as a food storage organ. Fibrous root systems, seen in monocots and some dicots, create very large surface areas for absorption of water and minerals.
12. The end of each root is covered by a *root cap* that protects it from abrasion. Behind the root cap is the *root apical meristem* where dividing cells cause the root to grow in length. Some new cells also replace lost root cap cells.
13. *Adventitious roots* arise from stems or leaves and may help to prop up the plant (corn, for example) or to establish a detached plant part in the soil.
14. *Lateral buds* are found at nodes on stems and can develop into branches.
15. Stems are highly modified in many species. Potato *tubers* are portions of stems; strawberry *runners* are horizontal stems that can form new plants asexually.
16. The *shoot apical meristem* within the terminal or *apical bud* at the tip of each stem causes the stem to grow in length.
17. *Leaf primordia* within each bud develop into mature leaves.
18. In addition to photosynthesis, leaves also carry out other important metabolic activities and, in succulent plants, store water. Tendrils in peas and squash and spines in cacti are modified leaves.
19. A typical leaf consists of a flattened *blade* and a *petiole,* or stalk, that attaches the blade to the stem.
20. *Simple leaves* have a single blade; *compound leaves* have multiple blades that can be pinnately or palmately arranged. The venation of a leaf blade can be parallel or netlike in arrangement.
21. A *tissue* is an organized collection of functionally and structurally similar cells. *Simple tissues* consist of only one cell type; *compound tissues* are composed of several cell types. *Tissue systems* include two or more different tissues.
22. The vascular plant body consists of three tissue systems, the *vascular tissue system,* the *dermal tissue system,* and the *ground tissue system.*
23. The vascular tissue system consists of *xylem* and *phloem.* Its function is to conduct materials throughout the plant body. The dermal tissue system protects the body surface. The ground tissue system has a diversity of roles.

Questions (for answers and explanations, see page 248)

1. Match the plant organs on the left with the functions they perform on the right.

______	Root system	*a.* Water absorption
______	Leaves	*b.* Photosynthesis
______	Stem	*c.* Nutrient absorption
		d. Transport
		e. Anchorage

2. A plant living in an environment with low soil moisture and abrasive soil particles would likely have
   a. a taproot system with well-developed root caps.
   b. a taproot system with poorly developed root caps.
   c. a fibrous root system with well-developed root caps.
   d. a fibrous root system with poorly developed root caps.
   e. either a taproot or a fibrous root system, but with no root apical meristems.
3. Which of the following is *not* typical of the plant group to which rice and palms belong?
   a. More than one embryonic leaf
   b. Leaves with parallel veins
   c. Fibrous root systems
   d. Scattered vascular bundles
   e. Floral parts in multiples of three
4. Characterize each of the following plant structures as modified roots, stems, or leaves.
   a. Potato tubers ________________
   b. Runners in strawberry plants ________________
   c. Cactus spines ________________
   d. Tendrils in peas ________________
   e. Prop roots in corn ________________

Activities (for answers and explanations, see page 248)

- On the following diagram of a simple plant, identify a node, an internode, a lateral bud, an apical bud, a leaf petiole, and a leaf blade, and circle the location of a root apical meristem.

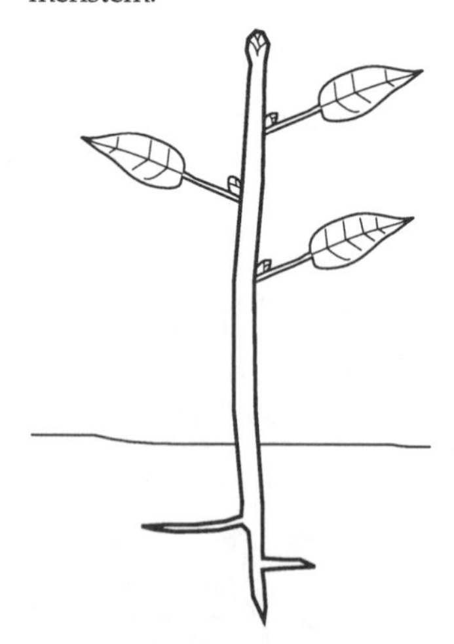

- Contrast control systems and growth in plants and animals and explain evident differences in terms of the structural organization of plants.

PLANT CELLS • PLANT TISSUES AND TISSUE SYSTEMS • GROWTH AND MERISTEMS (pages 673–677)

Key Concepts

1. The polysaccharides (primarily cellulose) secreted by the cell prior to cell expansion become the *primary cell wall*. Additional layers of polysaccharides (including other materials like lignin and suberin) laid down by some cells after elongation become the *secondary cell wall*.
2. The *middle lamella* cements plant cell walls together.
3. Cell walls lie outside the plasma membrane of the cell. Openings in the wall are called *pits*.
4. Pits usually occur in pairs (called *pit pairs*) with a middle lamella, a thinned primary wall, and no secondary wall between the two cells. Pit pairs are traversed by *plasmodesmata*, facilitating the movement of materials between neighboring cells.
5. *Parenchyma cells* are the most numerous cells in the plant body and are alive when their primary functions are performed. Parenchyma cells are not elongated, have a large central vacuole, and lack secondary cell walls.
6. A major function performed by parenchyma cells is storage of lipids as droplets within the cytoplasm or of starch in specialized plastids called *leucoplasts*. Parenchyma cells containing chloroplasts occur in the leaves and are the primary sites for photosynthesis.
7. In some regions, called *meristems*, parenchyma retains an ability to divide mitotically.
8. *Sclerenchyma* cells are dead at functional maturity. They provide structural support by means of their thick secondary cell walls. Sclerenchyma cells are of two types: elongated *fibers*, usually occurring in bundles, and variously shaped *sclereids*, particularly common in bark.
9. *Collenchyma* cells are elongated living cells with thickened primary cell walls and no secondary walls. The flexibility of petioles and stems is due in part to tissues rich in collenchyma cells.
10. *Tracheids* are spindle-shaped xylem cells found mostly in gymnosperms. Dead at maturity, tracheids, interconnected by numerous pits, form the hollow conduits through which water and minerals travel.
11. *Vessel elements* are a type of xylem cell found mostly in angiosperms. Vessel elements are dead at functional maturity, and larger than tracheids. Vessel elements lose most of their end walls and form hollow tubes called *vessels* for water transport.
12. Unlike xylem, phloem cells are alive at functional maturity. The *sieve tube element* is the main phloem cell type. *Sieve tubes* are stacks of sieve tube elements arranged end-to-end and interconnected through their perforated end walls, called *sieve plates*.
13. As sieve tube elements mature, the central vacuole membranes break down, nuclei and some other organelles disappear, and the sieve tube elements become specialized for transport of organic nutrients. In some flowering plants small, nucleated *companion cells* may regulate the activities of adjacent sieve tube elements.
14. Simple tissues consist of a single cell type. Simple tissues include parenchyma, sclerenchyma, and collenchyma.
15. Xylem and phloem are compound tissues. In addition to tracheids and vessel elements, xylem may include parenchyma cells and sclerenchyma cells, and may perform the functions of support, transport, and storage.
16. Phloem includes sieve tube elements, companion cells, fibers, sclereids, and parenchyma cells.
17. The vascular tissue system has two main functions: the transport of food from sites of production *(sources)* in the leaves to sites of utilization or storage *(sinks)*, and movement of water and minerals upward from the roots to the stem and leaves.
18. The dermal tissue system includes the cell layers covering the plant. *Epidermis* is a layer one or more cells thick covering the entire plant body. Epidermis is covered by a waxy *cuticle* in the stems and leaves, but not the roots. *Periderm* covers the roots and stems of older woody plants.
19. The ground tissue system includes mainly parenchyma cells, with some collenchyma and sclerenchyma. Its functions include storage, support, photosynthesis, and other activities.
20. *Meristems* are regions where cells divide to form new tissues. *Primary growth*, due to the activity of the root and shoot apical meristems, produces the primary tissues of the plant.
21. A *vascular cambium*, a cylinder of tissue located between the xylem and phloem, produces new xylem cells to the inside and new phloem cells to the outside and results in growth in circumference — *girth* — of roots and stems.
22. A *cork cambium* located in the bark of woody plants produces cells impregnated with the waxy substance *suberin* to augment the dermal tissue system as girth increases.
23. *Secondary growth*, due to the activity of lateral cambia, such as the vascular and cork cambia, produces the secondary tissues of the plant.
24. Unlike animals, plants show *indeterminate growth*, because they continue to grow throughout their lives. Based on the growth strategies of plants, we can categorize plants as *annuals, biennials*, or *perennials*.

Questions (for answers and explanations, see page 248)

1. What would be the correct order of structures you would encounter in moving from inside a parenchyma cell into a living, immature sclerenchyma cell?
   a. Primary cell wall, secondary cell wall, middle lamella, secondary cell wall, primary cell wall
   b. Primary cell wall, middle lamella, secondary cell wall, primary cell wall
   c. Primary cell wall, middle lamella, primary cell wall, secondary cell wall
   d. Secondary cell wall, primary cell wall, middle lamella, primary cell wall, secondary cell wall
   e. Primary cell wall, middle lamella, primary cell wall

2. To which of the following cell types would this description apply: cell alive at maturity, elongated shape, and a thick primary cell wall, but lacking a secondary cell wall?
   a. Collenchyma cell
   b. Sclerenchyma cell
   c. Parenchyma cell
   d. Tracheid
   e. Vessel element

3. Which of the following characteristics do tracheids, sclerenchyma, and vessel elements share in common?
   a. Dead at functional maturity
   b. No primary cell walls
   c. No secondary cell walls
   d. Mainly transport water
   e. Mainly transport organic nutrients

4. Which of the following characteristics do parenchyma, collenchyma, and sieve tube elements share in common?
   a. Contain leucoplasts
   b. No primary cell walls
   c. Lack nuclei
   d. All transport water
   e. Alive at functional maturity

5. Choose *all* of the following that are primary functions of the dermal tissue system.
   a. Protection
   b. Transport
   c. Food storage
   d. Reduction of water loss
   e. Absorption of water and minerals

6. Categorize the life cycle of each of the following plants as either annual, perennial, or biennial.

   _____________ Carrot
   _____________ Red maple
   _____________ Rice
   _____________ Coconut palm

Activities (for answers and explanations, see page 248)

- In the following figure showing several plant cells, label an example of a primary cell wall, a secondary cell wall, and the middle lamella.

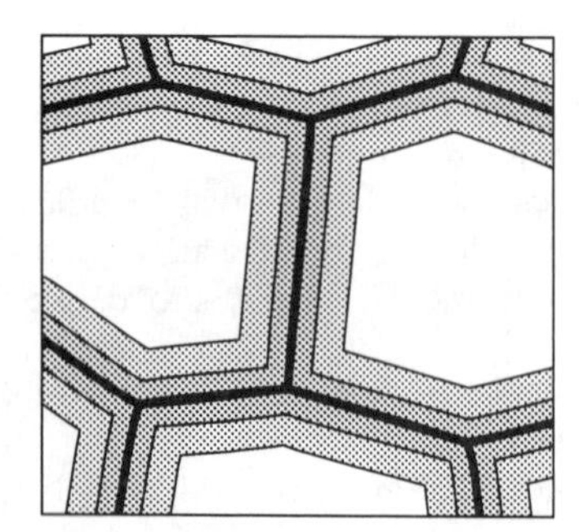

- In the following figure showing cells of the phloem, label a companion cell, a sieve tube element, and a sieve plate.

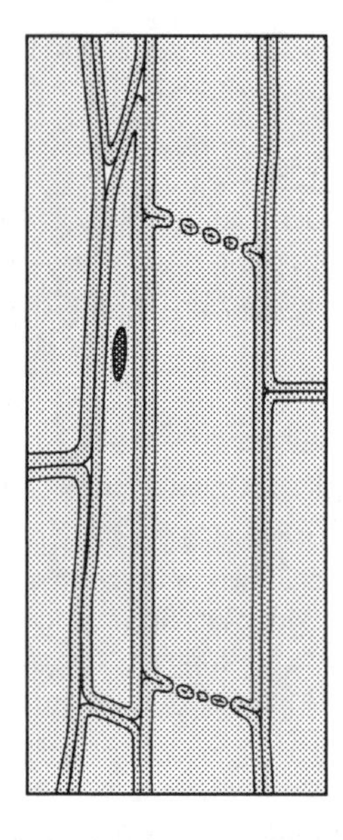

THE MERISTEMS AND THEIR PRODUCTS • SUPPORT IN A TERRESTRIAL ENVIRONMENT (pages 677–686)

Key Concepts

1. The root apical meristem gives rise to three primary cylindrical meristems, the inner *procambium*, the *ground meristem*, and the outermost *protoderm*. The procambium produces the vascular tissues, the ground tissues arise from the ground meristem, and the protoderm develops into the dermal tissue system.
2. The apical and primary meristems constitute the root's *zone of cell division*. Just above this zone is the *zone of elongation* where cells are increasing in length. The older cells are found in the *zone of differentiation*, where they become structurally and functionally specialized.
3. The epidermis of the root develops from the protoderm. Epidermal cells are flattened and may possess *root hairs* that increase surface area and aid in absorption of water and minerals.
4. The *root cortex* contains parenchyma cells that frequently serve as food storage sites and may form an association with fungal species to produce a *mycorrhiza*. Mycorrhizae improve absorption of water and minerals by the root.
5. The innermost cell layer of the cortex is the *endodermis*, a single layer of cells each ringed by a waterproof seal called a *Casparian strip* that prevents the inward passage of water. To enter the *vascular cylinder*, water must pass through the plasma membrane of an endodermal cell.
6. The vascular cylinder, or *stele*, consists of three tissues: the *pericycle*, the *xylem*, and the *phloem*.
7. Undifferentiated cells making up the pericycle give rise to lateral roots and provide additional cells to increase the girth of the root.
8. Xylem occupies the center of the root in dicots; in monocots a layer of parenchyma cells called *pith* lies internal to the xylem.
9. Phloem is found in cylinders between the arms of xylem.

10. Like the root apical meristem, the shoot apical meristem also gives rise to a procambium, ground meristem, and protoderm.
11. Growth of dicot stems occurs in a zone of elongation below the shoot apical meristem.
12. Grasses and some other monocots have *basal meristems*, an adaptation that helps the plant deal with grazing by herbivores.
13. Leaves develop from *leaf primordia* laid down by the shoot apical meristem.
14. Stem vascular tissue occurs in bundles, generally forming a cylinder in dicots, and uniformly distributed throughout the stem in monocots.
15. In dicots, a storage tissue called *pith* is located internal to the vascular tissue. Another cylinder of storage tissue called *cortex* is located external to the vascular tissue. Pith, cortex, and tissue between the vascular bundles make up the ground tissue system of the stem.
16. In stems, the vascular cambium is a single layer of undifferentiated cells located between primary xylem and phloem, but also extending between the vascular bundles to form a continuous cylinder of meristematic tissue. The vascular cambium produces secondary xylem (*wood*) to the inside, and secondary phloem to the outside.
17. *Vascular rays*, formed by cells derived from the vascular cambium, are rows of radially arranged parenchyma cells that connect the storage tissues of the stem (cortex and pith) with the vascular tissues.
18. As the stem increases in diameter, some cells within the vascular cambium divide tangentially to give rise to additional cambium cells, thus increasing the cylinder's circumference.
19. Trees from temperate regions have wood with *annual rings* that result from differences in xylem cell size due to seasonal variation in the availability of water.
20. The xylem tracheids and vessel elements of *spring wood* are larger and have thinner walls than the xylem cells of *summer wood*.
21. *Heartwood* results from deposition of resins in the older, more central xylem. The younger, more peripheral xylem called *sapwood* is mainly responsible for water and mineral transport in the stem.
22. Secondary growth in both the root and stem displaces the cortex and epidermis, both of which are eventually lost.
23. A *cork cambium* produces layers of cork cells containing suberin to replace the lost epidermis and provide waterproofing.
24. Cells produced to the inside by the cork cambium are called the *phelloderm*.
25. Cork cells, cork cambium, and phelloderm together constitute the *periderm*.
26. *Lenticels* are openings in the epidermis/periderm of the stem that allow gas exchange.
27. Chloroplast-containing parenchyma cells within the leaf form two layers of photosynthetic cells, an upper layer of elongated cells called the *palisade mesophyll* and a lower layer of irregularly shaped, loosely packed cells called the *spongy mesophyll*.
28. Vascular bundles form an extensive network of veins that bring the plant's transport system to within a short distance of each cell.
29. A nonphotosynthetic epidermal layer, covered with a waxy cuticle, prevents water loss from the leaf.
30. Openings called *stomata* (sing. *stoma*), are located in the epidermis. They permit the needed exchange of CO_2 and O_2 between the leaf interior and the external atmosphere.
31. Specialized pairs of *guard cells* change shape to regulate the size of the stomatal opening.
32. C_4 plants have a modified leaf anatomy in which photosynthetic cells are arranged concentrically around each vein. The outer mesophyll layer and the inner *bundle sheath* layer have different types of chloroplasts that carry out different biochemical reactions.
33. Terrestrial plants depend on two types of support systems: turgor within cells due to a relatively high pressure potential, or cells containing walls reinforced with lignin, such as secondary xylem (wood).
34. *Reaction wood* is specialized secondary xylem that maintains trees in an upright position. It occurs in two different forms, *compression wood* and *tension wood*.
35. Compression wood, found in gymnosperms, occurs in compressive areas (like the lower sides of limbs). It contains thicker and shorter tracheids, with more lignin and less cellulose, and expands as it develops to push the branch upward.
36. Tension wood, found in angiosperms, occurs in expansive areas (like the upper side of limbs). It contains fibers with thicker walls, containing less lignin and more cellulose, and contracts as it develops, pulling the branch upward.
37. Rapidly growing tree species (like balsa) tend to consist of wood with relatively large, thin-walled cells; slow-growing species (like mahogany) consist of wood with small, thick-walled cells.

Questions (for answers and explanations, see page 249)

1. Choose *all* of the following structures that did *not* result from the activity of the procambium.
   a. Epidermal cell
   b. Sieve tube element
   c. Cortex cell
   d. Vessel element
   e. Collenchyma

2. Which of the following does *not* show a correct derivation of cellular types?
   a. Root apical meristem → procambium → tracheid
   b. Root apical meristem → ground meristem → parenchyma
   c. Root apical meristem → ground meristem → companion cell
   d. Root apical meristem → protoderm → periderm
   e. Root apical meristem → procambium → sclerenchyma

3. Which one of the following structures found in roots performs a function analogous to the the lateral buds of stems?
   a. Epidermis
   b. Cortex
   c. Endodermis
   d. Pericycle
   e. Vascular tissue

4. Which of the following is *not* a complex tissue?
   a. Epidermis
   b. Phloem
   c. Endodermis
   d. Xylem
   e. Pericycle

5. In which of the following regions does most water transport occur?
   a. Sapwood
   b. Spring wood
   c. Summer wood
   d. Heartwood
   e. Bark

6. Based on your knowledge of the location of xylem and phloem within the vascular cylinder of the stem, which of the following would best describe the locations of these tissues in vascular bundles of veins within leaves?
   a. Phloem should be above xylem
   b. Xylem should be above phloem
   c. Veins only contain xylem
   d. Veins only contain phloem
   e. Xylem and phloem are in separate veins

7. Lenticels are functionally similar to which of the following structures?
   a. Guard cells
   b. Root hairs
   c. Stomata
   d. Vascular rays
   e. Periderm

8. In proceeding from the vascular cambium toward the periphery of a mature woody stem, which of the following could be the correct sequence of tissue types encountered?
   a. Primary phloem → secondary phloem → cortex → phelloderm → cork cambium → cork
   b. Secondary phloem → primary phloem → phelloderm → cork cambium → cork
   c. Secondary phloem → primary phloem → cortex → phelloderm → cork cambium → cork
   d. Secondary phloem → primary phloem → pith → phelloderm → cork cambium → cork
   e. Primary phloem → secondary phloem → cortex → epidermis

9. Plants with the C_4 photosynthetic pathway (C_4 plants) differ in leaf anatomy from plants using the C_3 photosynthetic pathway in the following way:
   a. C_4 plants have specialized guard cells.
   b. C_4 plants have two concentric layers of photosynthetic cells around each vein.
   c. C_4 plants only have stomata on the lower leaf surface.
   d. C_4 plants only have a palisade mesophyll layer.
   e. C_4 plants have a much thicker waxy cuticle covering the epidermis of the leaf.

10. Which of the following statements about compression wood is *not* true?
   a. Compression wood has thicker and shorter tracheids than tension wood.
   b. Compression wood is found on the underside of limbs.
   c. Compression wood tracheids have more lignin and less cellulose.
   d. Compression wood is found in gymnosperms.
   e. Compression wood contracts as it develops.

Activities ***(for answers and explanations, see page 249)***

- On this photomicrograph of a portion of a typical dicot root, label the location of the cortex, endodermis, pericycle, xylem, and phloem.

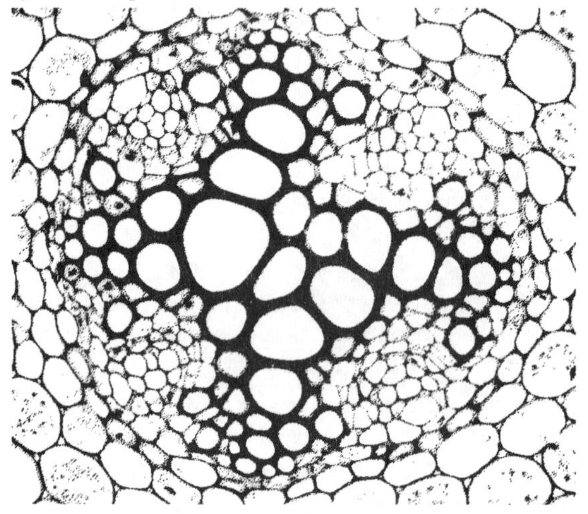

- On the following figure of a cross-sectioned leaf, label the location of (A) a palisade mesophyll cell, (B) a spongy mesophyll cell, (C) a guard cell, (D) a stoma, (E) an epidermal cell, (F) the cuticle, (G) a vein, and (H) a bundle sheath cell.

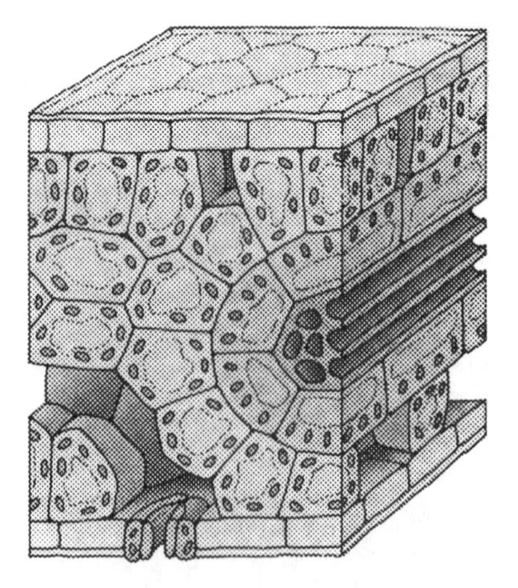

Is this leaf from a C_3 or a C_4 plant? ________________

- On the following photomicrograph of a vascular bundle of a monocotyledonous stem, label the location of (A) a parenchyma cell, (B) a vessel element, (C) a companion cell, (D) a sieve tube cell, and (E) a sclerenchyma cell.

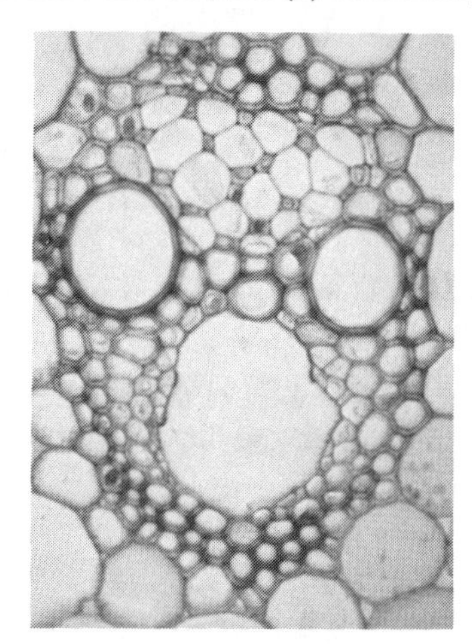

- On the following figure showing the tissues and regions of the root tip, label the three developmental zones, the epidermis, a root hair, the stele, the cortex, the protoderm, the ground meristem, the procambium, the apical meristem, and the root cap.

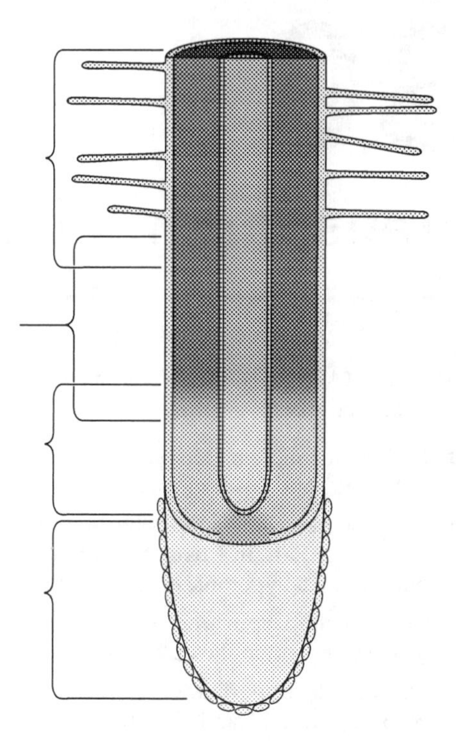

30

Transport in Plants

CHAPTER LEARNING OBJECTIVES—after studying this chapter you should be able to:

- ❑ Define the following terms: "water potential," "pressure potential," and "osmotic potential."
- ❑ Describe the relationship between water potential, pressure potential, and osmotic potential in a plant cell placed into a hypotonic environment.
- ❑ Explain the importance of facilitated diffusion and active transport in the uptake of minerals by plant cells.
- ❑ Describe the importance of proton pumps in movement of ions such as K^+ and Cl^- into plant cells.
- ❑ Describe the patch clamping technique and explain how it has been used to better understand how ions move through membranes.
- ❑ Explain what the apoplast and the symplast are and how water moves through these two routes in the plant.
- ❑ Describe the importance of the endodermis to water and mineral uptake by the plant.
- ❑ Characterize the structural features of transfer cells and their role in movement of water into the xylem conduits.
- ❑ Describe the movement of water from outside a root hair to inside a xylem conduit.
- ❑ Explain the role of capillary action in the movement of water up the xylem.
- ❑ Explain what root pressure is, its relationship to guttation, and its importance in the movement of water up the xylem.
- ❑ Describe the evaporation–cohesion–tension hypothesis for movement of water up the xylem.
- ❑ Explain what a pressure bomb is and how it was used to measure the tension of xylem sap.
- ❑ Define transpiration and list some adaptations that leaves show for limiting transpiration.
- ❑ Describe the sequence of events that causes the opening and closing of stomata and the control mechanisms involved.
- ❑ Explain the unique stomatal cycles seen in CAM plants, the biochemical features underlying these differences, and how their stomatal cycles adapt CAM plants for life in hot, dry habitats.
- ❑ Contrast phloem transport with xylem transport.
- ❑ Differentiate between sources and sinks relative to phloem transport.
- ❑ Explain how the pressure flow model accounts for the characteristic features of phloem transport.
- ❑ Describe movement of a sucrose molecule from a mesophyll cell to a root parenchyma cell.

UPTAKE AND TRANSPORT OF WATER AND MINERALS (pages 689–696)

Key Concepts

1. Autotrophs (plants) preceded heterotrophs (animals) in the colonization of the terrestrial environment.
2. The buoying action of water helps support aquatic organisms. In a terrestrial environment, support is provided by a framework of rigid material, such as bone or wood, or by the hydrostatic pressure of water in tissues.
3. The *osmotic potential* of any solution is negative and depends on the solute concentration of the solution.
4. Water moves from a solution with a lesser solute concentration (less negative osmotic potential) to a solution with a greater solute concentration (more negative osmotic potential) if no other factors are operating.
5. As water enters a plant cell, a positive *pressure potential* (sometimes called *turgor pressure*) develops because of the expansion of the cytoplasmic portion of the cell against the rigid cell wall.
6. Equilibrium is reached when the tendency of water to enter the plant cell due to osmotic potential is balanced by the repulsion of water from the cell due to the pressure potential.
7. The overall tendency of a solution to take up water is called its *water potential*. Water potential is the sum of the osmotic potential (usually negative) and the pressure potential (usually positive).
8. The water potential, osmotic potential, and pressure potential of pure water under atmospheric pressure are all equal to zero. Water always moves toward the region of most negative water potential.

9. Facilitated diffusion by *carrier proteins* moves some minerals across cell membranes and down their concentration gradient. Carrier proteins are also used in active transport to move minerals against a concentration gradient.

10. In plants, active transport involves *proton pumps* that move hydrogen ions (H^+) from the inside to the outside of the cell. Cations, such as K^+, move through membrane channels into the cell due to the net negative charge of the inside of the membrane. Other ions, such as Cl^-, are cotransported into the cell with H^+ by a symport system.

11. In *patch clamping,* mild suction is used to attach a small section of membrane to the end of a micropipette. Ion channels in the isolated membrane can be studied by measuring changes in electric potential across the membrane.

12. Water and minerals can enter the plant from the soil by diffusing through the meshwork of cell walls and intercellular spaces called the *apoplast*.

13. The meshwork of living cells and their interconnections (*plasmodesmata*) is called the *symplast*. Entrance to the symplast is controlled by selectively permeable cell membranes.

14. As water enters the plant from the soil, it follows a water potential gradient. Water has a tendency to flow from the cortex to the stele because the water potential of the cortex is less negative than the water potential of the stele.

15. Because the suberin-containing *Casparian strips* of the endodermis seal off the apoplast of the cortex from the apoplast of the vascular cylinder, water and its dissolved minerals must enter the symplast by passing through the cytoplasm of endodermal cells.

16. Endodermal membrane proteins regulate the kinds and amounts of minerals entering the stele.

17. Xylem cells (tracheids and vessel elements) are dead at maturity and, thus, a part of the apoplast.

18. Minerals are actively pumped out of living parenchyma and pericycle cells within the stele and enter the xylem. Parenchyma cells with numerous mitochondria and specializations to increase the surface area of the plasma membrane are called *transfer cells.*

19. As transfer cells pump minerals into the xylem conduits, the water potential within the xylem becomes more negative and water passively reenters the apoplast by osmosis.

20. *Eduard Strasburger* conducted research demonstrating that the rise of sap in the stem can occur without the presence of roots. The rise is not caused by the pumping of living cells within the stem but rather seems to involve the leaves.

21. *Capillary action* by itself can cause water to rise only to about 40 cm within the xylem conduits of the plant.

22. *Root pressure* results because of the passive movement of water into the roots due to a water potential gradient.

23. *Guttation,* the loss of liquid water through openings in the leaves, occurs under conditions when water is in abundance and is evidence for the existence of root pressure. Root pressure also causes sap to ooze from cut stumps.

24. *Transpiration* is evaporative water loss from leaves.

25. As water evaporates from the surface of a mesophyll cell and is lost from the leaf, the water potential of the mesophyll cell becomes more negative. As water diffuses into the mesophyll cell from an adjacent xylem conduit, a *tension* or "pull" results on the water column within the xylem.

26. Tension develops within water columns because of the great *cohesion* of water molecules created by hydrogen bonding between them (refer to Chapter 2).

27. As water moves throughout the plant, from the soil into the root cortex, into the stele, up the xylem conduits, into mesophyll cells, and out into the dry air surrounding the leaf, it is following a water potential gradient, always going from regions of less negative water potential to regions of more negative water potential.

28. Minerals are moved passively with water in the xylem.

29. *Per Scholander* used a pressure bomb to show that the xylem sap was under tension in all cases where sap was ascending in a plant. The sap tension measurements were great enough to account for the movement of sap in the tallest plants.

Questions (for answers and explanations, see page 249)

1. In a plant root, a cell type in which water *cannot* move via the apoplast is the
a. epidermis.
b. endodermis.
c. pericycle.
d. vessel element.
e. parenchyma.

2. In a plant root, a cell type in which water *cannot* move via the symplast is the
a. epidermis.
b. endodermis.
c. pericycle.
d. vessel element.
e. parenchyma.

3. If the pressure potential is +1.6 atmosphere and the osmotic potential is –2.4 atmosphere, then the water potential would be
a. 4.0 atmosphere.
b. 0.8 atmosphere.
c. –0.8 atmosphere.
d. +1.6 atmosphere.
e. –2.4 atmosphere.

4. If you were to order the water potential of the following root cells/regions from least to most negative, which cell/region would be third?
a. Xylem conduit
b. Soil next to root
c. Cortex apoplast
d. Stele apoplast
e. Transfer cell

5. If solutions A and B are separated by a membrane permeable only to water, and if the following are the osmotic and pressure potentials of the two solutions (in atmospheres), in which direction will there be a net movement of water?

Solution	Osmotic Potential	Pressure Potential
A	–1.0	0.0
B	–2.0	1.0

a. From A to B
b. From B to A
c. No net movement
d. First from A to B, then from B to A
e. First from B to A, then from A to B

6. What is the *minimum* number of cell membranes that a water molecule must move through in getting from the soil into a xylem vessel element?
a. 0
b. 1
c. 2
d. 6
e. Many more than 6

7. Root pressure is caused by
a. the active transport of water into xylem by living cells next to the xylem.
b. the passive movement of water into the xylem because of the less negative water potential there.
c. the passive movement of minerals into the xylem from living cells next to the xylem.
d. the active transport of minerals into the xylem by living cells next to the xylem and the passive movement of water into the xylem.
e. the passive movement of water into the xylem because of the greater pressure potential within living cells next to the xylem.

8. Examine the following diagram of a leaf cross-section. Which letter (a-e) indicates the area where the water potential would be *least* negative?

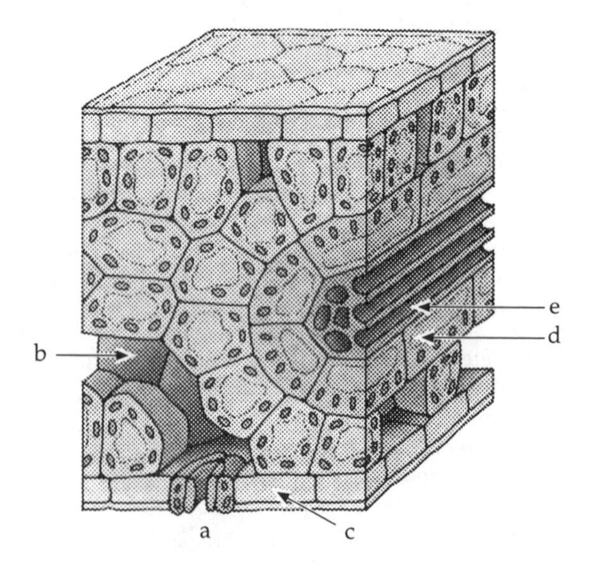

9. The movement of water up the stems of tall plants is *least* dependent on which of the following factors?
a. Guttation
b. Transpiration
c. Cohesiveness of water molecules
d. Tension within columns of water molecules
e. Evaporative water loss from leaves

Activity (for answers and explanations, see page 250)

- In the following figure showing a portion of a cross-section through a root, label regions corresponding to the epidermis, cortex, endodermis, pericycle, and xylem. Also, draw the paths that a water molecule could take from a root hair to the xylem by remaining entirely within the symplast or by traveling through both the symplast and apoplast.

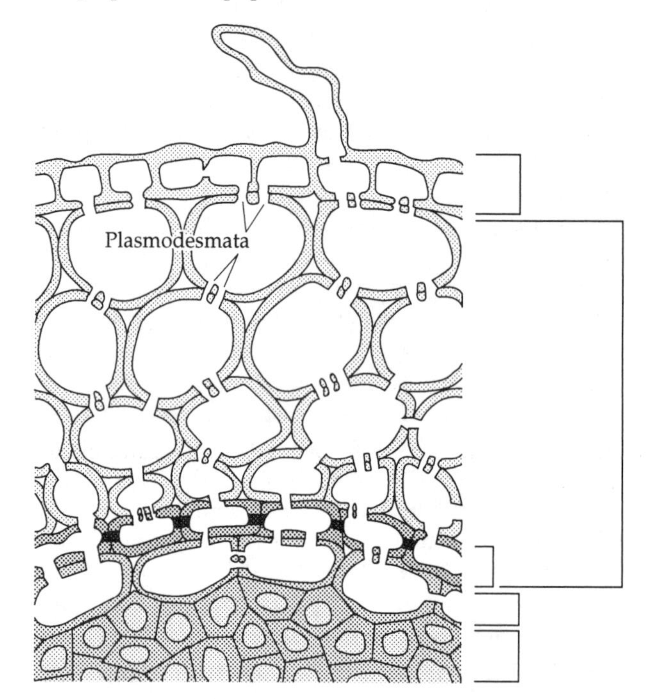

TRANSPIRATION AND THE STOMATA • CRASSULACEAN ACID METABOLISM AND THE STOMATAL CYCLE • TRANSLOCATION OF SUBSTANCES IN THE PHLOEM (pages 696–701)

Key Concepts

1. The waxy *cuticle* covering the leaf epidermis minimizes transpiration, but also prevents exchange of carbon dioxide and oxygen between leaf tissues and the atmosphere.
2. *Stomata* (singular: *stoma*) are openings in the leaf created by a pair of specialized epidermal cells called *guard cells.* Unlike other epidermal cells, guard cells have chloroplasts. To conserve water, stomata are closed at night or when water is limited.
3. Potassium ions move into guard cells and make their water potential more negative. Water then passively diffuses into the guard cells, which swell, become turgid, and change shape so as to create the stomatal opening.

As potassium ions move out of the guard cells, water passively follows, the cells become less turgid, and the stoma is closed or reduced in size.

4. Shape change of guard cells is regulated by special radially oriented cellulose microfibrils so that stomatal opening occurs when the cells take up water and stomatal closing occurs when they lose water.
5. The carbon dioxide level within the intercellular spaces of the leaf influences potassium ion movement into guard cells. Low carbon dioxide levels lead to stomatal opening; high carbon dioxide levels lead to stomatal closing.
6. *Abscisic acid* is released by certain cells within the leaf in response to water loss and an increasingly negative water potential. Abscisic acid may cause guard cells to release potassium ions, thus closing the stomata and preventing further water loss.
7. Light may also influence the opening and closing of stomata due to light activation of proton pumps recently discovered in guard cells. The activity of the pumps would influence the movement of K^+ and Cl^- ions into guard cells.
8. Unlike most plants, the *succulent plants* in the family Crassulaceae have their stomata open at night and closed during the day.
9. At night when stomata of crassulaceans are open, carbon dioxide combines with *PEP* (phosphoenolpyruvic acid) to become chemically fixed in the form of certain organic acids (e.g., malic acid and aspartic acid). This causes the plant tissues to become acidic at night.
10. During the day, when stomata of crassulaceans are closed, the organic acids are broken down to release the carbon dioxide and make it available for photosynthesis. The set of chemical reactions for fixing and releasing carbon dioxide is referred to as *crassulacean acid metabolism (CAM)*.
11. Because CAM plants can keep their stomata closed during the day they lose less water through transpiration. This adapts them for life in hot, dry habitats.
12. Unlike transport in the xylem, phloem transport involves living cells, can occur in both directions simultaneously, and is an energy-requiring process.
13. Sap-feeding aphids have been useful for obtaining samples of phloem sap. After they insert their stylet into a sieve tube, the aphid can be detached from its stylet, which continues to exude sap.
14. The *pressure flow model* is the most widely accepted explanation for transport of phloem sap. It depends on loading of sugars into the phloem sieve tubes in *source regions* (usually the leaves) and unloading of sugar from the phloem in *sink regions* (usually the roots). Phloem loading and unloading is an ATP-requiring process.
15. Sugars and other solutes produced in leaf mesophyll cells leave the symplast and enter the apoplast near a vein.
16. Specific organic molecules are selectively transported from the apoplast into the symplast of phloem cells, frequently companion cells.
17. Loading of phloem with organic materials makes the osmotic potential and the water potential of these cells more negative. Water moves osmotically into the phloem sieve tubes and increases their pressure potential.
18. In sink regions, solutes are actively unloaded from the sieve tubes and this makes the water potential of these cells more negative. Water passively moves from the sieve tubes into the surrounding tissue and reduces the pressure potential in the sieve tubes.
19. The gradient of pressure potential from source regions to sink regions causes the bulk flow of water and dissolved solutes.
20. In some plants, like sugar maples, excess photosynthate stored in the trunk and twigs over the winter is moved into the xylem the following spring.

Questions (for answers and explanations, see page 250)

1. As a leaf loses water, a series of events typically occurs leading to the closing of stomata. Which of the following shows the correct sequence of these events?
   a. Potassium ions enter guard cells, water passively enters guard cells, abscisic acid released by leaf cells, pressure potential in guard cells decreases
   b. Potassium ions enter guard cells, water passively leaves guard cells, pressure potential in guard cells decreases
   c. Water passively leaves guard cells, pressure potential in guard cells decreases, abscisic acid released by leaf cells, potassium ions leave guard cells
   d. Potassium ions leave guard cells, water passively leaves guard cells, pressure potential in guard cells decreases, abscisic acid released by leaf cells
   e. Abscisic acid released by leaf cells, potassium ions leave guard cells, water passively leaves guard cells, pressure potential in guard cells decreases

2. Which of the following explains why the leaf cells of plants within the family Crassulaceae have a lower pH at night than during the day?
   a. These plants keep their stomata open at night, but closed during the day.
   b. Carbon dioxide in these plants is chemically fixed in organic acids during the night, but these acids are metabolized during the day.
   c. These plants have a buildup of abscisic acid at night.
   d. Because their stomata are closed during the day, excess carbon dioxide within the leaf forms carbonic acid.
   e. The premise statement is incorrect, the leaves of CAM plants have a lower pH during the day than at night.

3. Which of the following characteristics applies to both xylem and phloem transport?
   a. Follows a water potential gradient
   b. Involves only living cells
   c. Can occur in both directions
   d. Involves active transport of solute with passive movement of water
   e. Is a passive, non-energy-requiring process

4. Which of the following cell types would be the *third* cell type a sugar molecule is likely to encounter on its route from its site of production in a chloroplast to its site of storage in the root?
   a. Mesophyll cell
   b. Companion cell
   c. Sieve tube element
   d. Parenchyma
   e. Endodermis

5. According to the pressure flow model, select *all* of the following choices where active transport is involved in movement of sugar and other organic solutes.
   a. From mesophyll cell to apoplast within the leaf
   b. From apoplast to companion cell within the leaf
   c. From companion cell to sieve tube element within the leaf
   d. Between sieve tube elements within the stem
   e From sieve tube element to surrounding root cells

Activities (for answers and explanations, see page 250)

- In the following figure showing two stomatal cycles, label the cycle typical of a CAM plant and the cycle typical of most other plants. Circle on the appropriate cycle the times when you would expect the cytoplasm of leaf tissue to have its highest pH.

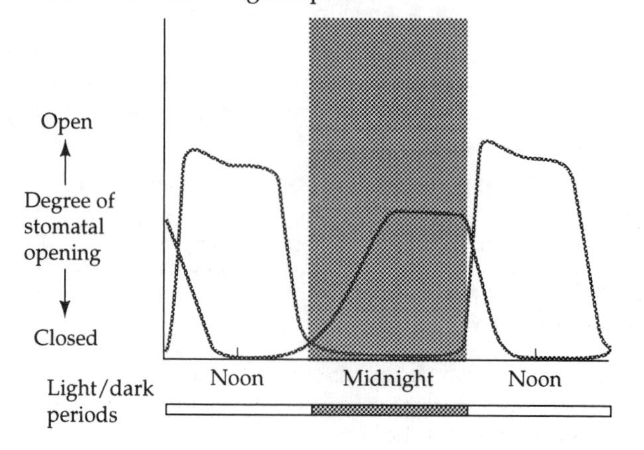

- In the figure below showing movement of sucrose through the symplast and apoplast of the leaf, label a phloem parenchyma cell, a companion cell, a mesophyll cell, a bundle sheath cell, and a sieve-tube element. Indicate on the drawing where you would expect to find cell membranes with an abundance of sucrose–proton symports.

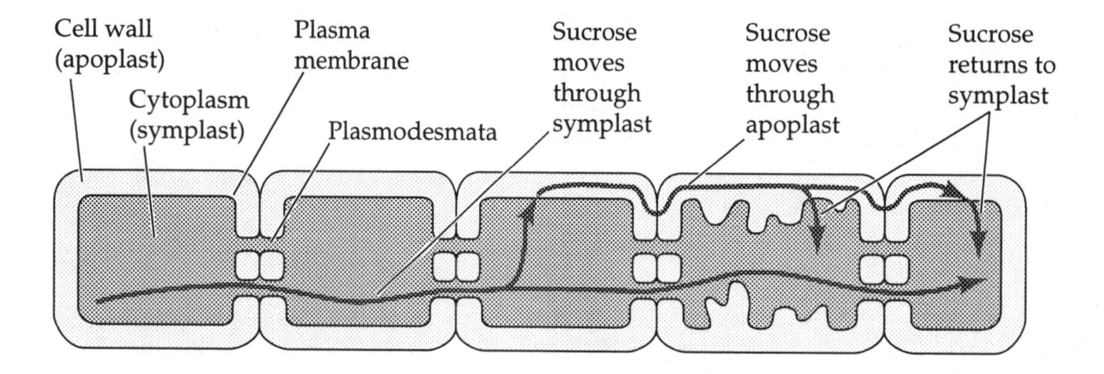

31

Environmental Challenges to Plants

CHAPTER LEARNING OBJECTIVES—after studying this chapter you should be able to:

- ❑ Define what a xerophyte is and compare the rate of growth and photosynthetic efficiency of xerophytes relative to other plants.
- ❑ Describe important adaptations seen in the leaves of xerophytes.
- ❑ Describe important adaptations seen in the roots of xerophytes.
- ❑ Discuss the problems faced by plants living in waterlogged soils and some of their adaptations such as pneumatophores.
- ❑ Describe important adaptations, such as aerenchyma, that are seen in submerged plants.
- ❑ Define what a halophyte is and discuss important adaptations seen in halophytes.
- ❑ Explain why halophytes and xerophytes have many similar adaptations and describe these adaptations.
- ❑ Explain what the major processes are that result in soils laden with heavy metals.
- ❑ Characterize the responses that plant species have shown to life in soils contaminated with heavy metals.
- ❑ Characterize important adaptations that plants show to serpentine soils.
- ❑ Explain how grazing by herbivores affects the photosynthetic production of plants.
- ❑ Describe some common adaptations of plants to grazing by herbivores.
- ❑ Define what secondary products are and list the major types of secondary products produced by plants.
- ❑ Explain how the secondary product canavanine protects plants from herbivores.
- ❑ Contrast how animals and plants react to infection.
- ❑ Explain what phytoalexins are and their role in the hypersensitive reaction in plants.
- ❑ Describe three major ways that plants protect themselves from their own toxic secondary products.

THREATS TO PLANT LIFE • DRY ENVIRONMENTS • WHERE WATER IS PLENTIFUL AND OXYGEN IS SCARCE • SALINE ENVIRONMENTS • HABITATS LADEN WITH HEAVY METALS (pages 704–710)

Key Concepts

1. *Xerophytes* are plants with structural or biochemical adaptations for life in dry environments. Some desert plants do not show any xerophytic adaptations and complete their life cycles during periods of the year when water is available.
2. Xerophytic leaf adaptations include a thickened waxy cuticle, stomata in sunken, hair-bordered pits, fleshy leaves in which water is stored, or loss of leaves during dry periods. Xerophytic adaptations reduce water loss, but also restrict photosynthesis, so xerophytes tend to grow slowly.
3. Some xerophytes also show root adaptations, including those with a very long taproot that is able to reach underground water supplies, species in which the root system dies back during dry periods, and species, such as the cacti, with very shallow but extensive fibrous root systems.
4. In general, xerophytes fix more grams of carbon by photosynthesis per gram of water lost through transpiration than nonxerophytes do.
5. The osmotic potential in root cells of water-stressed plants tends to be very negative due to the accumulation of the amino acid *proline* within these cells. The increased water potential gradient allows these plants to extract more water from the soil.
6. Floating plants tend to have their stomata restricted to the upper epidermis of the leaf, unlike the more normal situation where stomata are generally found on the lower surface.
7. The leaves of many submerged or floating plants have mesophyll layers containing large air spaces. Such layers are called *aerenchyma* and provide buoyancy as well as oxygen storage.

8. Plant species living in waterlogged soils may have slow-growing root systems that do not penetrate deeply. Alcoholic fermentation may supply some ATP in these root cells when insufficient oxygen is available for aerobic respiration.

9. *Pneumatophores* are extensions of the root systems of swamp-dwelling plants that emerge from the water and help to absorb oxygen for the submerged root system.

10. *Halophytes* are plants that have evolved adaptations for growing in saline (salty) environments, such as land bordering the sea or agricultural areas where *salinization* has occurred.

11. The large negative water potential of saline soils makes it more difficult for plant root systems to absorb water. Also, increased sodium ion (and sometimes chloride ion) concentration is toxic to many plants.

12. Many halophytes have evolved a resistance to salt toxicity so they can maintain relatively high levels of sodium ions in their cells. The accumulation of sodium in the cells of halophytes makes their own water potential more negative and facilitates water absorption from the soil. Some halophytes have salt glands in their leaves that secrete excess salt.

13. Some halophytes, like xerophytes, accumulate proline within their cells. Other adaptations seen in both halophytes and xerophytes are succulence, high root-to-shoot ratios, sunken stomata, reduced leaf area, and thick cuticles.

14. *Heavy metals* such as copper, nickel, lead, cadmium, chromium, and zinc usually make soil toxic to most plants.

15. Heavy metal contamination can be due to natural geological processes, acid rain, which releases toxic aluminum ions in soil and water, or the deposition of tailings from mining operations.

16. In some plant species, individuals become genetically adapted to tolerate contamination by a specific metal, usually the most common one. Even though they take up the contaminant and accumulate it in their tissues, they are tolerant to it.

17. Plant species that evolve tolerance to heavy metals are able to thrive in those habitats because of reduced competition with other, nontolerant species.

Questions (for answers and explanations, see page 250)

1. Which of the following is *not* an adaptation seen in a typical xerophyte?
   a. Leaves with aerenchyma
   b. Stomata in pits
   c. Leaves that can be curled up
   d. Ability to accumulate proline in root cells
   e. Thickened cuticle

2. Mangrove trees grow submerged in salt water. Which of the following adaptations would *not* be useful given their environment?
   a. Salt glands
   b. Ability to develop large negative water potential within root cells
   c. Stomata on upper leaf surface
   d. Pneumatophores
   e. Thickened cuticle

3. Xerophytes have many features in common with halophytes. Which one of the following is *not* a feature that these two plant types have in common?
   a. Accumulation of proline
   b. Succulence
   c. Thick cuticle
   d. Tolerance of high sodium ion concentration
   e. Large root-to-shoot ratio

4. Match the term on the left with the appropriate description on the right.

a. Proline	______	Seen in mangrove roots
b. Aerenchyma	______	Can result from mining activities
c. Pneumatophore	______	Can increase aluminum content of soils
d. Tailings	______	Many cacti show this
e. Succulence	______	Seen in floating plants
f. Acid rain	______	Affects the osmotic potential of the cell

5. Describe what salinization is, its principal causes, and the problems that it creates for plants.

6. List three examples of heavy metals:

   ____

   ____

   ____

SERPENTINE SOILS • PLANTS AND HERBIVORES • PROTECTION AGAINST FUNGI, BACTERIA, AND VIRUSES (pages 710–715)

Key Concepts

1. *Serpentine soils* are deficient in calcium, have higher levels of magnesium than more fertile soil, and may also contain relatively high levels of heavy metals. Some plant species are adapted to living on serpentine soils by being highly efficient at absorbing calcium and excluding or tolerating magnesium.

2. Grazing tends to increase the photosynthetic production of certain plant species. In a grazed plant, minerals

absorbed by the root system need not be divided among as many leaves, changes in the source-to-sink ratio favor transport of photosynthetic products away from the leaves, and removal of the older, outer leaves makes more light available to the inner, younger leaves.

3. Many plants have developed adaptations to grazing by herbivores. Grasses, for example, grow from the bases of the shoots, not the tips. Removal of apical buds by grazing usually leads to the development of lateral buds and results in a bushier plant.
4. *Secondary products* are special chemicals produced by plants that influence the behavior of other organisms. Some secondary products serve as attractants of animal pollinators, repellents of herbivore grazers or fungi, or growth inhibitors of competing plant species.
5. Most secondary plant products are of low molecular weight and help plants compensate for their inability to move.
6. Alkaloids affect herbivore nervous systems.
7. Some nitrogen- and sulfur-containing secondary plant products are carcinogenic or cause nerve damage and pain in herbivores.
8. Phenolics are distasteful to herbivores, interfere with the functioning of the nervous system, or act as fungicides.
9. Quinones reduce competition by inhibiting growth of other plants.
10. Terpenes act as fungicides and insecticides, but some also attract pollinators.
11. Steroids mimic animal hormones, thereby preventing normal development of insect herbivores.
12. Flavonoids attract pollinators and seed dispersers.
13. Canavanine is an amino acid chemically very similar to arginine. Some plants use it as a nitrogen storage product. Herbivores that consume it are poisoned when the canavanine is incorporated into their proteins instead of arginine.
14. The cutin, suberin, and waxes covering the outer, epidermal surface of plants physically isolate them from potential pathogens.
15. Whereas animals generally attempt to repair infected tissue, plants prevent the spread of pathogens by isolating the diseased area with nonliving tissue, such as wood in trees. Plants also produce substances that resist growth of microorganisms.
16. *Phytoalexins* are a class of chemicals providing plants with some nonspecific protection against fungi and some bacteria.
17. A hypersensitive reaction in plants occurs when cells near the site of infection release phytoalexins and then die. The remaining "dead spot," or necrotic lesion, isolates the microbial infection, while the rest of the plant remains infection-free.
18. Export of salicylic acid from an infected plant part may serve as a signal for the production of pathogenesis-related proteins (PR proteins). PR proteins may limit the spread of the infection.
19. In order to isolate toxic secondary plant products, they are usually stored in vacuoles within cells if hydrophilic, or within special tubes called *laticifers* if hydrophobic.
20. Some plants store precursors of toxic secondary plant products in one compartment and the enzymes that will activate them in a different compartment. Mixing of enzyme and substrate occurs only when herbivores damage tissue.
21. Some plants have modified enzymes or receptor molecules that do not recognize the toxic secondary plant product. For example, many canavanine-producing plants have tRNA-charging enzymes that can distinguish between canavanine and arginine and only bind arginine.

Questions (for answers and explanations, see page 251)

1. Which of the following statements about serpentine soils or the plants that are adapted to live on them is *not* true?
   a. Serpentine soils are deficient in magnesium.
   b. Plant growth on serpentine soils is sparse.
   c. Plants adapted to grow on serpentine soils are efficient at absorbing calcium.
   d. Heavy metals may be abundant in serpentine soils.
   e. Some plants will grow on serpentine soils if the mineral content is adjusted.

2. Which of the following is *not* a plant adaptation to grazing by herbivores?
   a. Production of spines and thorns
   b. Production of canavanine
   c. Resistance to salt toxicity
   d. Stem growth from basal meristems
   e. A bushy growth form.

3. Which of the following is *not* a typical use of secondary plant products?
   a. Reduce competition from neighboring plants
   b. Promote pollination
   c. Attract animals
   d. Energy storage within the plant
   e. Repel animals
   f. Inhibit fungal growth
   g. Affect the normal development of insects

4. Order the following events that are hypothesized to occur in the hypersensitive reaction following microbial infection in plants.

   ____ Production of a necrotic lesion

   ____ Synthesis of pathogenesis-related proteins

   ____ Release of salicylic acid

   ____ Secretion of phytoalexins

   ____ Death of infected tissue

5. Which of the following is *not* a typical mechanism used by plants to protect themselves from their own toxic secondary products?
   a. Storage of hydrophobic secondary products in laticifers
   b. Production of the secondary product in an inactive form
   c. Activation of the secondary product only in damaged tissue
   d. Modified enzymes or receptors that are able to recognize the secondary product
   e. Isolation of the secondary product in a special compartment

6. Discuss some of the ways in which grazing may actually increase photosynthetic production in plants.

7. Discuss how the prevalent production of secondary products by plants may be an adaptation to their sedentary lifestyle.

8. Explain how the secondary product canavanine protects plants from herbivores.

32

Plant Nutrition

CHAPTER LEARNING OBJECTIVES—after studying this chapter you should be able to:

- ❑ Define what a nutrient is and differentiate between a macronutrient and a micronutrient.
- ❑ Describe the method plant physiologists use to determine which nutrients are essential.
- ❑ Enumerate the mineral nutrients required by plants and explain their principal functions.
- ❑ Explain how carbon, hydrogen, oxygen, nitrogen, and the mineral elements enter living systems.
- ❑ Define and relate the following terms: "autotroph," "photosynthetic," "chemosynthetic," "heterotroph," "herbivore," and "carnivore."
- ❑ Discuss the features of nutrient acquisition that relate to a plant's status as a sessile organism.
- ❑ Explain what deficiency symptoms are and how they are useful for diagnosis of nutrient-related problems in plants.
- ❑ Describe the biologically important characteristics of soil, how soil composition varies with depth, and the features of the three soil horizons.
- ❑ Describe the phenomenon called ion exchange and explain its biological importance.
- ❑ Describe various agricultural methods for delivering essential nutrients to plants, including application of organic and inorganic fertilizers, liming, and foliar sprays.
- ❑ Characterize the major processes of soil formation and discuss the type of soil formation called laterization, seen in wet tropical areas, as well as soil formation in semiarid regions.
- ❑ Discuss humus formation in soils and differentiate between mull and mor humus.
- ❑ Describe changes in the clay, soluble mineral, biomass, and humus content of soils as soils age.
- ❑ Explain what is meant by nitrogen fixation and list the types of organisms that are able to fix nitrogen.
- ❑ Describe the nitrogen fixation reactions involving the enzyme nitrogenase and the important features of the enzyme.
- ❑ Describe the mutualistic symbiosis that exists between the nitrogen-fixing bacteria *Rhizobium* and certain plant species.
- ❑ Describe *Rhizobium*–legume nodule formation.
- ❑ Discuss problems inherent in the use of biotechnology to "equip" plants with the the ability to fix their own nitrogen.
- ❑ Explain the roles of nitrogen fixation, nitrification, nitrate reduction, and denitrification in the nitrogen cycle, and characterize the types of bacteria that facilitate the cycling of nitrogen.
- ❑ Describe sulfur metabolism and the types of bacteria involved in the modification of sulfur-containing ions.
- ❑ Discuss the two major types of heterotrophic seed plants and provide several examples of each.

ACQUIRING NUTRIENTS • WHICH NUTRIENTS ARE ESSENTIAL? • SOILS (pages 717–724)

Key Concepts

1. *Nutrients* are chemical raw materials obtained by organisms from their environment.
2. *Deficiency symptoms* result when an organism lacks enough of a needed nutrient. For example, nitrogen deficiency in plants causes a yellowing of leaves called *chlorosis* because nitrogen is essential for chlorophyll synthesis.
3. Carbon and oxygen enter living systems through photosynthesis by *autotrophs* (mostly plants, but also some protists and bacteria). Hydrogen is supplied by water, and also fixed into biological form by photosynthesis.
4. Nitrogen from the inert dinitrogen gas N_2 is fixed into organic form by specialized soil bacteria.
5. *Mineral nutrients,* including sulfur, phosphorus, magnesium, and iron, are obtained by plants in ionic form from the soil.
6. *Autotrophs* make their own organic food from inorganic nutrients. A *heterotroph* is any organism that requires preformed organic molecules to live. *Herbivores* eat plants and *carnivores* eat herbivores or other carnivores.
7. Most autotrophs are photosynthetic and use light energy to produce organic compounds. Some autotrophs are *chemosynthetic* and use the energy stored in reduced inorganic molecules, like H_2S, to synthesize organic compounds. All chemosynthesizers are bacteria.

8. Plants are *sessile* (stationary) organisms that must grow in order to move and exploit new resources.
9. Plants are more selective than heterotrophs in their procurement of nutrients, eliminating much of the need for waste removal.
10. An *essential element* is an element required directly for normal growth and reproduction, and for which there is no substitute. Lack of an essential element results in characteristic *deficiency symptoms.*
11. In addition to hydrogen, oxygen, carbon, and nitrogen, essential elements include magnesium (chlorophyll constituent; cofactor for many enzymes), iron (constituent of oxidation–reduction molecules), phosphorus (ATP constituent; involved in enzyme regulation), potassium and chlorine (maintenance of electric neutrality and proper osmotic balance) and calcium (processing of hormonal and environmental cues; affects membranes and the cytoskeleton; constituent of the middle lamella).
12. *Macronutrients* are essential elements needed in concentrations of at least 1 mg/g dry matter; *micronutrients* are needed in concentrations of less than 100 μg/g dry matter.
13. Plant physiologists identify essential elements by transplanting seedlings into solutions lacking a single ingredient and looking for deficiency symptoms.
14. Soil includes both living (roots, bacteria, and fungi) and nonliving components and can be divided into horizontal layers called *horizons,* which vary with depth.
15. The *A horizon* is the uppermost soil zone containing most of the living and dead organisms. The A horizon is depleted of minerals by leaching.
16. The *B horizon* is the next deepest zone, where minerals leached from the A horizon accumulate.
17. The *C horizon* is the the deepest zone, containing the original parent material from which the soil develops.
18. Most plant roots reside in the A horizon. Some deep-rooted plants extend into the B horizon, but rarely into the C horizon.
19. *Ion exchange* refers to the the exchange of protons, either released by roots or derived from the ionization of carbonic acid ($H_2CO_3 \rightarrow HCO_3^- + H^+$), with K^+, Mg^{2+}, and Ca^{2+} ions that are bound to negatively charged clay particles, making these minerals available to plants.
20. Inorganic fertilizers are usually described by their N–P–K values, where, for example, a 5–10–5 fertilizer contains 5% nitrogen, 10% phosphorus, and 5% potassium.
21. Application of calcium-containing material, usually calcium carbonate, or liming can reverse acidification of soil caused by decomposition of organic material. Liming releases protons from soil particles by ion exchange and the excess protons can then be leached from the soil. Liming also increases availability of Ca^{2+}.
22. *Mechanical* and *chemical weathering* of rock produces soil. Hydrolysis is a key first step in chemical weathering of soil minerals.
23. The physical and chemical properties of soils depend on the amounts and types of clay particles they contain.
24. *Laterization,* common in tropical soils, results in infertile soils because silica and soluble nutrients are leached, while insoluble aluminum and iron compounds remain in the A horizon.
25. In semiarid areas, a hard layer of calcium carbonate may form in the B horizon because water is removed rapidly from the soil by evaporation or plant roots leaving mineral nutrients behind. Because of the high nutrient content of these soils, irrigation can support very successful agriculture.
26. *Humus* is the carbon-rich, dark material resulting from the death and decay of organisms, especially plants. *Mull* is a fertile, alkaline humus; *mor* is an infertile, acidic humus.
27. As soils age, clay content tends to increase, soluble minerals decrease, and biomass and humus reach a peak and then decrease.

Questions (for answers and explanations, see page 251)

1. Match the appropriate letters for the roles in plant metabolism listed on the right to the plant nutrients listed on the left.

______Magnesium	*a.* Oxidation–reduction molecules
______Iron	*b.* Chlorophyll constituent
______Phosphorus	*c.* Important enzyme cofactor
______Potassium	*d.* Helps maintain osmotic balance
______Nitrogen	*e.* Constituent of all amino acids
	f. Needed for chlorophyll synthesis
	g. ATP constituent

2. In the beginning of the growing season, deep plowing is a common agricultural practice in many parts of the world. Which of the following statements best explains the need for this practice?
   a. Deep plowing mixes mull and mor soils.
   b. Deep plowing reduces ion exchange in soils.
   c. Deep plowing promotes laterization.
   d. Deep plowing mixes the A and B horizons.
   e. Deep plowing brings the C horizon to the surface.

3. Which of the following statements about the cycling of nutrients is *not* true?
   a. Nitrogen is fixed into organic form by bacteria.
   b. Some chemosynthetic autotrophs use H_2S to synthesize organic compounds.
   c. Heterotrophs must obtain their organic compounds directly from autotrophs.
   d. Carbon, hydrogen, and oxygen first become part of living material via an autotroph.
   e. Both photosynthetic and chemosynthetic autotrophs can fix carbon.

4. Which of the following statements correctly characterizes changes in $CaCO_3$, clay, and biomass during the long-term aging of soil?

a. $CaCO_3$ gradually decreases, clay gradually increases, and biomass peaks and then declines.
b. $CaCO_3$ gradually decreases, clay and biomass gradually increase.
c. Clay gradually increases, $CaCO_3$ and biomass peak and then decline.
d. All three gradually decrease.
e. All three gradually increase.

5. Which of the following is *not* involved in the phenomenon called ion exchange?
a. Negatively charged clay particles
b. pH
c. Cations like K^+, Mg^{2+}, and Ca^{2+}
d. H_2S
e. Ionization of carbonic acid

6. Match the appropriate letters for characteristics listed on the right with the soil horizons listed on the left.

______A horizon	*a.* Leaching occurs here
______B horizon	*b.* Most plant roots are here
______C horizon	*c.* Most plant nutrients are here
	d Soil formation occurs here
	e. Most humus is here
	f. Fewest roots are here

7. In the following figure showing aging of soils, label the curves representing changes in biomass, humus, $CaCO_3$, and clay.

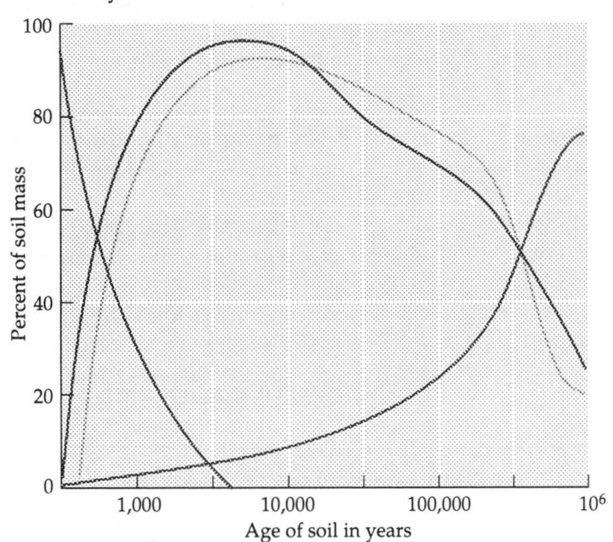

8. Discuss the features of plant nutrient acquisition that relate to their status as sessile organisms.

9. In terms of their potential for supporting agriculture, compare the characteristics of tropical soils subject to laterization with soils more typical of semiarid regions.

NITROGEN FIXATION • DENITRIFICATION • NITRIFICATION • NITRATE REDUCTION • SULFUR METABOLISM • HETEROTROPHIC SEED PLANTS (PAGES 724–731)

Key Concepts

1. Dinitrogen gas (N_2) is highly unreactive and only a few species of bacteria, called the *nitrogen fixers*, are able to convert N_2 to ammonia (NH_3).
2. All organisms depend on the nitrogen fixers for the world's supply of chemically available nitrogen. Some nitrogen is fixed through industrial means, and some is fixed in the atmosphere by lightning, volcanic eruptions, and forest fires.
3. *Rhizobium* and several other species of soil bacteria can fix nitrogen when they form *root nodules* in association with certain plants, like the legumes.
4. Species of cyanobacteria are important nitrogen fixers in the oceans (along with some other photosynthetic bacteria), in fresh water, and in association with fungi (lichens), or with ferns, cycads, or bryophytes.
5. The filamentous bacteria called *actinomycetes* fix nitrogen in association with the roots of many shrub species, such as alder.
6. Shrubs with nodules containing nitrogen-fixing bacteria are important pioneering species in many plant communities.
7. *Nitrogenase* is the enzyme responsible for catalyzing the nitrogen-fixing reaction. Nitrogenase contains molybdenum and iron atoms.
8. While dinitrogen is bound to nitrogenase, three successive pairs of hydrogens are added to it. With each pair added, one covalent bond between the two nitrogen atoms breaks until two ammonia molecules are formed and released from the enzyme.
9. Nitrogen fixation requires much ATP and strong reducing agents to transfer the hydrogens to dinitrogen.
10. Nitrogenase is sensitive to oxygen and is only catalytically active in an anaerobic environment.
11. *Mutualistic symbiosis* occurs when two different species associate closely and interact in some way so that both species benefit from the interaction. Nodule formation and nitrogen fixation depend on mutualistic symbiosis between *Rhizobium* and an appropriate plant species, usually a legume.

12. *Rhizobium*–legume nodule formation begins when root hairs stimulated by *Rhizobium* cells invaginate to form an *infection thread*. At the same time, the bacteria produce substances that cause formation of a special region in the cortex called a *primary nodule meristem*.
13. The infection thread with enclosed bacteria grows into the root cortex where it enters a cell in the primary nodule meristem. The thread now bursts and releases the bacteria into the cell's cytoplasm.
14. Bacteria increase in size and develop an elaborately folded internal membrane system to become *bacteroids*.
15. The cortex cell produces a type of hemoglobin, called leghemoglobin, creating a low oxygen area around the bacteroids.
16. Most industrial nitrogen fixation now occurs by an energy-intensive method called the *Haber process*, in which dinitrogen and hydrogen gases are combined to form ammonia. Genetic engineering of plants may someday change this situation.
17. However, providing adequate ATP and reducing power and creating an oxygen-free environment within plant cells will be much more difficult than simply inserting the nitrogenase genes into a higher plant.
18. *Denitrification* carried out by aerobic bacteria (genera *Bacillus* and *Pseudomonas*) living anaerobically converts fixed nitrogen, usually in the form of nitrate (NO_3^-), to dinitrogen gas in the atmosphere.
19. Plants can use the ammonium ion (NH_4^+) resulting from ammonia produced by the nitrogen fixers, but often prefer nitrogen in the form of nitrate (NO_3^-).
20. *Nitrification* is the oxidation of NH_4^+ to nitrate by certain chemosynthetic soil bacteria called the *nitrifying bacteria*.
21. Other nitrifying bacteria include *Nitrosomonas* and *Nitrosococcus*, which convert NH_4^+ into nitrite, NO_2^-, and *Nitrobacter*, which oxidizes nitrite to nitrate (NO_3^-).
22. Plants do not incorporate the nitrogen in nitrate directly into amino acids; instead they reduce nitrate to ammonia, a process called *nitrate reduction*.
23. The steps in which nitrite is reduced to ammonia take place within chloroplasts and result in the synthesis of various amino acids.
24. All organically fixed nitrogen needed by animals must be obtained from plants.
25. Nitrogen fixation, nitrification, nitrate reduction, denitrification, and other processes constitute the *nitrogen cycle*.
26. Sulfur is a constituent of the amino acids cysteine and methionine, as well as several other important molecules, like coenzyme A.
27. Plants take up sulfur in the form of sulfate ions (SO_4^{2-}). Sulfate is reduced and incorporated into cysteine.
28. *Chemosynthetic sulfur bacteria* oxidize hydrogen sulfide (H_2S) to sulfate, deriving ATP to fix carbon dioxide, just as the nitrifying bacteria use NH_3 to make their own food.
29. Some *heterotrophic plants* are parasites, such as the mistletoes, dodders, and Indian pipe.
30. Some heterotrophic plants are carnivores: they obtain additional nitrogen and energy by capturing and digesting insects. Examples include the Venus's-flytrap, sundews, and pitcher plants.

Questions (for answers and explanations, see page 252)

1. Which of the following groups contain *no* species that are able to fix nitrogen?
   a. Cyanobacteria in the ocean and fresh water
   b. Soil bacteria including *Rhizobium*
   c. Cyanobacteria in lichens
   d. Aerobic bacteria in the genera *Bacillus* and *Pseudomonas*
   e. Filamentous bacteria called actinomycetes

2. Nitrogenase
   a. is insensitive to oxygen.
   b. contains magnesium.
   c. releases two NH_3 molecules as products.
   d. requires an aerobic environment.
   e. catalyzes an oxidation reaction that yields ATP.

3. Which of the following statements about *Rhizobium*–legume nodule formation is *not* true?
   a. Rhizobium can only fix nitrogen after it becomes a bacteroid within a root cortex cell.
   b. Rhizobium induces invagination of root hairs.
   c. Within an infection thread, *Rhizobium* is still extracellular to the plant.
   d. The infection thread can fuse with any root cortex cell of an appropriate legume species.
   e. Bacteroids are surrounded by plant-produced hemoglobin.

4. The least challenging problem facing biotechnologists trying to develop a recombinant DNA alternative to the Haber process is
   a. inserting the nitrogenase gene into a plant cell.
   b. creating an anaerobic environment within the plant cell.
   c. providing adequate reducing power for the reaction.
   d. providing an adequate ATP supply for the reaction.

5. Which of the following statements about nitrification is *not* true?
   a. Nitrobacter oxidize nitrite to nitrate.
   b. Nitrosomonas and *Nitrosococcus* convert ammonium ions to nitrite.
   c. Nitrification reactions are energy producing (exergonic) reactions.
   d. The nitrifying bacteria are chemosynthetic autotrophs.
   e. Heterotrophic plants are more dependent on the nitrifying bacteria for usable nitrogen then autotrophic plants.

6. In the next set of figures showing nodule formation, label the following structures: root hair, cortex cell, *Rhizobium* bacteria, infection thread, bacteroids, and nodule.

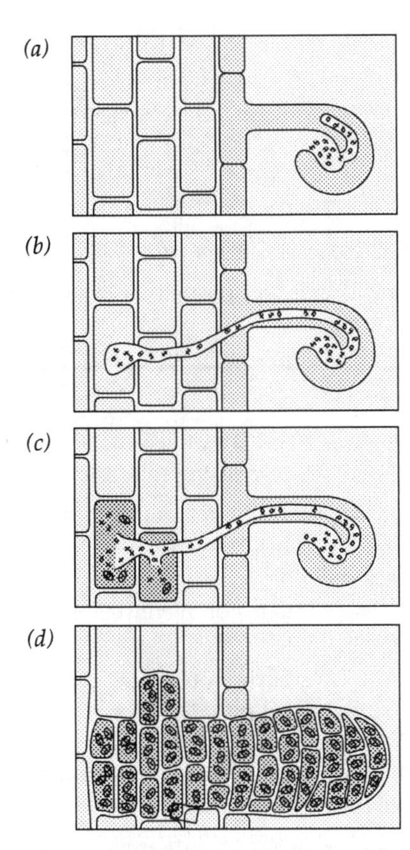

In plant material shown in which of these figures would you expect leghemoglobin to be most abundant? _____

7. In what way are the chemosynthetic sulfur bacteria and the nitrifying bacteria most alike?
   a. Both groups are symbiotic with the roots of higher plants.
   b. Both groups produce a reduced inorganic nutrient, which plants then proceed to oxidize.
   c. Both groups are autotrophs.
   d. Both groups are nodule-forming organisms.

8. The nitrogen-containing molecule that is most typically incorporated into amino acids by plants is ________. The organelle in the plant where this process occurs is the ____________________.

9. Discuss the problems inherent in current attempts to use biotechnology to "equip" plants with the the ability to fix their own nitrogen.

10. In the figure below, label the boxes to show the location of each of the following steps in the nitrogen cycle: nitrification, nitrate reduction, denitrification, and nitrogen fixation.

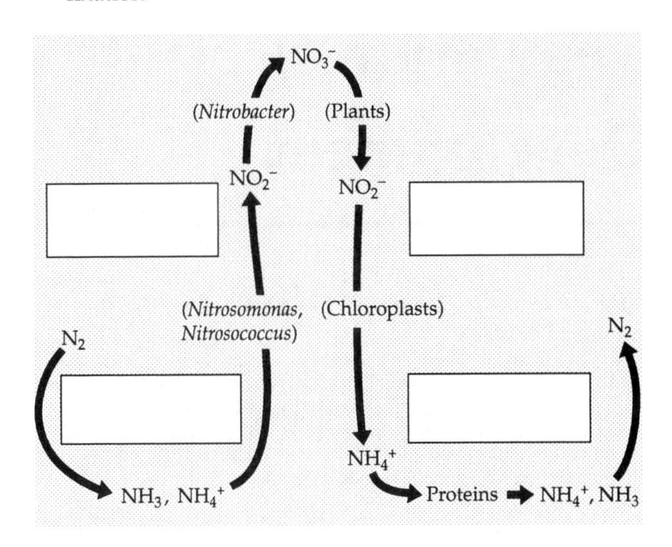

Which of the compounds shown in this cycle is most reduced? _____________

Integrative Question (for answers and explanations, see page 252)

Match the appropriate letters for the generalized chemical reactions listed below with the organisms that can accomplish them.

Denitrifying bacteria ____________

Nitrifying bacteria ____________

Nitrogen fixers ____________

Chemosynthetic sulfur bacteria ____________

Plants ____________

Heterotrophs ____________

a. Dinitrogen to ammonia
b. Ammonium to nitrite
c. Nitrite to nitrate
d. Nitrate to nitrite
e. Nitrite to ammonium
f. Ammonia to amino acids
g. Hydrogen sulfide to sulfate
h. Ammonia to dinitrogen
i. Sulfate to hydrogen sulfide

33

Regulation of Plant Development

CHAPTER LEARNING OBJECTIVES—after studying this chapter you should be able to:

- ❑ Describe the interactions between the four factors involved in the regulation of development.
- ❑ Describe the cellular processes that are affected by the factors involved in the regulation of development.
- ❑ Define what a hormone is and list the major plant hormones.
- ❑ Explain why phytochrome is not classified as a hormone.
- ❑ List major events in the life cycle of a flowering plant that are subject to developmental regulation.
- ❑ Describe the process of germination in flowering plants.
- ❑ Characterize some of the factors that regulate the breaking of dormancy in seeds, such as abscisic acid, scarification, stratification, and afterripening, and describe three advantages of dormancy.
- ❑ Contrast some important characteristics of light-requiring and dark-requiring seeds.
- ❑ Characterize the events that occur after imbibition and explain the role of gibberellins in mediating this phase of germination.
- ❑ Describe some of the major roles that the plant hormones called gibberellins have in the regulation of development.
- ❑ Explain what phototropism is and describe some of the early studies of phototropism using coleoptiles.
- ❑ Characterize our current understanding of the role of auxins in phototropism, gravitropism, apical dominance, and fruit development.
- ❑ Describe the effects of auxins on plant cell walls in the process of cell elongation.
- ❑ Characterize the interactions between auxins and cytokinins in regulating differentiation and organ formation in plants.
- ❑ Describe some of the major roles that cytokinins have in the regulation of plant development.
- ❑ Describe some of the major roles that ethylene has in the regulation of plant development.
- ❑ Describe some of the major roles that abscisic acid has in the regulation of plant development.
- ❑ Explain some of the important changes associated with the onset of winter dormancy in plants.
- ❑ Describe the effects of light on phytochrome.
- ❑ Discuss the involvement of etiolation in seedling development and the role of phytochrome in moderating the process.
- ❑ Characterize the features of the small plant *Arabidopsis thaliana* that make it a key organism in current attempts to better understand plant development.

WHAT REGULATES PLANT DEVELOPMENT? • AN OVERVIEW OF DEVELOPMENT • SEED DORMANCY AND GERMINATION • GIBBERELLINS (pages 733–742)

Key Concepts

1. Plant development—the progressive changes in the plant through time—is regulated by the environment, hormones, the pigment phytochrome, and the plant's genome.
2. Environmental cues, such as changes in light or temperature, are mediated by plant hormones and phytochrome.
3. *Plant hormones* are produced in specific parts of the plant and are transported to different parts of the plant where they have their effects. Plant hormones usually exert multiple influences on plant development. Major plant hormones include auxin, gibberellins, cytokinins, abscisic acid, and ethylene.
4. Phytochrome is acted on directly by light and, unlike hormones, is not transported beyond the cell where it resides, but has its action locally.
5. Development is ultimately determined by the plant's genome. The genome encodes phytochrome and the enzymes that catalyze reactions producing the plant's hormones and mediating the plant's response to its hormones.
6. Many events in the life cycle of flowering plants are subject to developmental regulation, including seed dormancy, germination, utilization of endosperm, cell division and elongation, flowering, pollen tube development, fruit development, fruit ripening, plant dormancy, and senescence.

7. Plant development results from the effects of the environment, hormones, and phytochrome on three cellular processes: division, expansion, and differentiation.
8. In some species, seed *germination* can only occur after a period of *dormancy*.
9. Modification of the seed coat, called *scarification*, may break dormancy by allowing water to pass through the impermeable seed coat, or by removing the restraining effect of the coat on embryo growth.
10. Agents of scarification include abrasion by the environment, digestive enzymes, and fire.
11. Chemical *inhibitors* may also maintain dormancy. The concentration of inhibitors is reduced by leaching or, as with the common inhibitor *abscisic acid*, by competition with growth *promoters*.
12. Some seeds need to be dried or may require a period of cold temperatures before they will germinate. Refrigerating seeds to hasten germination is called *stratification*.
13. Some seeds will not germinate until a specific amount of time passes (called *afterripening*); others simply require water, presence or absence of light, or some other environmental factor.
14. Light-requiring seeds usually germinate near the soil surface. They tend to be small and have little stored food.
15. Dark-requiring seeds must be buried in order to germinate. They tend to be large and contain much stored food.
16. Dormancy seems to be advantageous in three different ways: it helps plants survive adverse conditions, increases chances of finding suitable new habitats, and aids in dispersal.
17. Seeds initiate germination by absorbing water in a process called *imbibition*.
18. After imbibition, metabolic changes occurring in the seed include increases in the rates of respiration, RNA synthesis, and protein synthesis. After the radicle emerges from the seed, DNA synthesis begins.
19. Lipids (fats and oils) are the main energy storage molecules in the majority of seeds; starch is also a common energy storage molecule in seeds.
20. Amino acids are usually stored in seeds in polymerized protein form. When the embryo needs amino acids, it hydrolyzes protein.
21. In some species, developing embryos secrete *gibberellins*, which diffuse through the endosperm to the *aleurone layer*, located beneath the seed coat.
22. Under stimulation by gibberellin, proteins in bodies called *aleurone grains* break down into constituent amino acids. These amino acids are used to synthesize digestive enzymes.
23. Digestive enzymes are released from the aleurone layer into the endosperm, where they catalyze the hydrolysis of energy storage polymers. Building block molecules can now be absorbed by the embryo.
24. *Gibberellins* are a family of chemically similar plant hormones produced by higher plants and some fungi, such as the "foolish seedling" disease fungus, *Gibberella fujikuroi*.
25. In 1925 *Eiichi Kurosawa* showed that cell-free extracts of culture media in which *Gibberella fujikuroi* had been growing caused stem elongation in rice seedlings.
26. The gibberellin known as *gibberellin* A_1 controls stem elongation in many species; many other gibberellins are intermediates in the synthesis of gibberellin A_1. *Gibberellin* A_3 is a commercially available gibberellin.
27. Dwarf plant strains have a recessive mutant allele that prevents them from making their own gibberellin A_1. They grow normally in response to external gibberellin application.
28. In some species, gibberellin secreted by seeds promotes development of the fruit containing the seeds.
29. During their second year, many biennial plants secrete gibberellins that induce rapid shoot elongation and development of flowers, a process called *bolting*.
30. Other effects of gibberellins in different species include fruit development from unfertilized flowers, seed germination, and budding in spring.
31. Gibberellin secreted by the embryo leads to metabolism of the seed endosperm in some species (as considered in Chapter 31).
32. Stem elongation in most species results from interaction between gibberellin and *auxin*.

Questions (for answers and explanations, see page 252)

1. Four players in regulation of plant development are the environment, hormones, phytochrome, and the plant's genome. Discuss their interaction.

2. Which of the following is *not* an agent of scarification?
   a. Digestive enzymes
   b. Fire
   c. Afterripening
   d. Abrasion by soil particles
   e. Stream bed abrasion

3. In the spaces provided below, arrange the letters for the following events in their correct sequence as they would occur during normal seed germination.

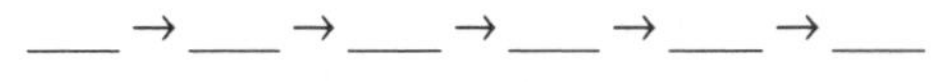

a. Emergence of radicle	*b.* Breaking of dormancy
c. DNA synthesis	*d.* Imbibition
e. RNA synthesis	*f.* Protein synthesis

4. In the spaces provided below, arrange the letters for the following events in their correct sequence as they would occur during the phase of seed germination leading up to the metabolism of endosperm nutrients in a species with an aleurone layer.

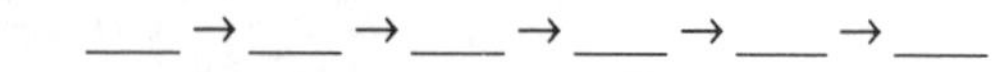

a. Aleurone grains release amino acids

b. Gibberellin secretion by embryo

c. Building-block molecules absorbed by embryo

d. Synthesis of digestive enzymes

e. Hydrolysis of endosperm molecules

f. Diffusion of gibberellins to aleurone

5. Which of the following is *not* an effect of gibberellins on plant development in at least some species?
   a. Stem elongation
   b. Fruit development
   c. Induction of bolting
   d. Phototropism
   e. Metabolism of endosperm in seeds

6. Which of the following were *not* observed in studies done on certain dwarf strains of plants?
   a. Applications of extracts of normal strains promoted growth of dwarf strains.
   b. Dwarf strains grew normally if additional fertilizer was applied.
   c. Application of gibberellin A_1 promoted growth of dwarf strains.
   d. Gibberellin caused little additional growth of normal strains.
   e. Dwarf strains are homozygous recessive for genes involved in gibberellin synthesis.

7. The following figure shows a seed in the process of germinating shortly after it has imbibed water. In this figure, label the aleurone layer, the endosperm, and the embryo.

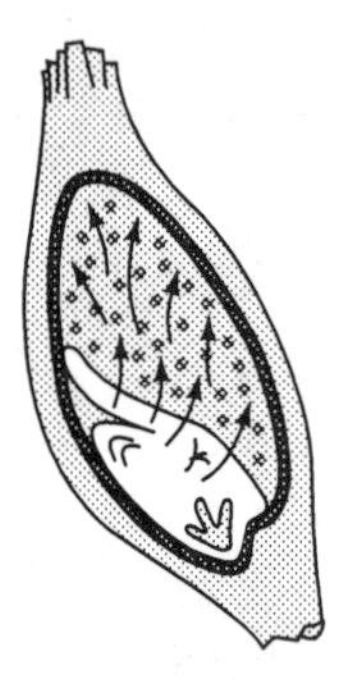

What do the small arrows shown in the figure represent?

What will be the next step in this process, due to the movement shown by these arrows?

AUXIN • CYTOKININS • ETHYLENE (pages 742–750)

Key Concepts

1. Auxin, or *indoleacetic acid*, mediates a variety of plant responses as diverse as *phototropism* and *apical dominance.*
2. Phototropism is a light-directed growth response in plants, either toward light, as in shoots, or away, as in roots.
3. Charles Darwin and his son studied phototropism using the leaf sheaths of grass seedlings, called *coleoptiles*, and determined that the coleoptile tip senses light direction.
4. In 1926, *Frits Went* showed that a chemical diffuses from the tip of the coleoptile and that its distribution in the coleoptile below the tip results in phototropism.
5. The tip produces auxin and the cells in the growth region below the tip respond to the chemical by elongating.
6. Illumination from the side causes auxin produced in the tip to move laterally toward the shaded side. The unequal auxin distribution reaching the growth region causes more elongation on the shaded side, making the tip bend toward the light.
7. *Gravitropism*, a plant growth response directed by gravity, results because auxin produced in the shoot tip moves toward the lower side of a tilted shoot. Greater elongation on the lower side causes the tip to curve up and away from gravity.
8. The effects of auxin on vegetative plant development include the induction of root formation in cut stems, the delay of petiole *abscission* by auxin produced in the leaf, and the maintenance of dormancy in lateral buds by auxin produced in the apical bud, a process called *apical dominance.*
9. The chemical *2, 4-D* is a selective herbicide that kills dicots at concentrations that do not affect monocots.
10. In strawberries, the fruits, called *achenes*, produce auxin, which induces growth of the receptacle. Auxin treatment can bring on fruit development without fertilization, a process called *parthenocarpy*.
11. Plant cell walls are composed of cellulose *microfibrils* embedded in a matrix of other polysaccharides. Each microfibril consists of about 250 cellulose molecules.
12. In growing plant cells, uptake of water and expansion of the vacuole increase hydrostatic pressure within the cell and stretch the cell wall. Walls must be loosened before they can be stretched.
13. Auxin loosens cell walls, perhaps by increasing the hydrogen ion concentration within the cell wall. Decreased pH may activate an enzyme in the cell wall that breaks linkages of microfibrils to one another or with components of the polysaccharide matrix.
14. Recent evidence suggests that auxin, and perhaps other plant hormones, may not act directly in causing their effects, but first may need to bind to specific receptor proteins.

15. Work with tissue cultures and intact plants clearly shows that auxin can cause undifferentiated cells like pith to develop into specific cell types.
16. The ratio of auxin and *cytokinins* determines the nature of organ formation in many plants; a high proportion of auxin leads to root formation, a high proportion of cytokinins leads to bud and shoot formation.
17. Whereas auxin stimulates cell elongation, cytokinins stimulate cell division; an appropriate combination of the two produces rapid growth of tissues.
18. Cytokinins are produced in the roots and moved to other parts of the plant.
19. *Kinetin*, a synthetic cytokinin extracted from aged herring sperm DNA, increases the rate of cell division in cultures of carrot root cells.
20. *Zeatin* and *isopentenyl adenine* are two cytokinins naturally found in plants.
21. Some of the following are effects of cytokinins: germination of light-requiring seeds kept in the dark; inhibition of stem elongation; stimulation of lateral bud growth; expansion of leaves; delayed senescence of leaves; alteration of the distribution of nutrients within the plant.
22. The gas *ethylene* promotes senescence in plants, including abscission of leaves and ripening of fruit.
23. In response to ethylene, chlorophyll degrades and cell walls break down in ripening fruit.
24. All parts of the plant produce ethylene, especially ripening fruit.
25. Carbon dioxide has effects antagonistic to those of ethylene; ethylene promotes fruit ripening, carbon dioxide delays it.
26. The apical hook of many dicot seedlings kept in the dark (i.e., underground) is maintained by production of ethylene in the hook area.
27. Under stimulation from ethylene, stems are inhibited from elongating and lose their normal gravitropism.
28. Leaf senescence and abscission are helpful in areas with cold winters where photosynthesis would be minimal and leaves could be damaged by freezing.
29. During leaf senescence, resources in the leaves (like amino acids) are reclaimed by the plant.
30. During reproduction in many species, adult plants die because all available resources have been allocated to seed production.

Questions (for answers and explanations, see page 253)

1. Experiments done by Charles Darwin and his son on plant phototropism showed that
   a. auxin is produced in the tip of the coleoptile.
   b. the tip of the coleoptile is the light receptor of the plant.
   c. within coleoptiles, auxin moves laterally away from the source of the light.
   d. more cell elongation takes place on the shaded side of the plant.
   e. phototropism would occur even if the coleoptile tip was covered.

2. A coleoptile is suspended in a horizontal position and illuminated as shown below. Which of the following arrows shows the expected growth response of the coleoptile under these conditions?

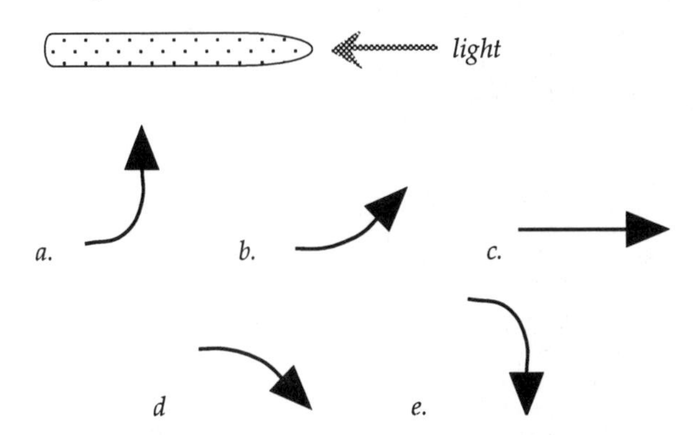

3. A coleoptile tip is severed from its base and an agar block is placed between the tip and base. After a period of time with illumination from the right side, the block is removed and cut into two pieces, *x* and *y*. Pieces *x* and *y* are positioned on the cut bases of two other *dark-grown* coleoptiles. From the following diagrams, select the letter showing the expected response of the coleoptile bases.

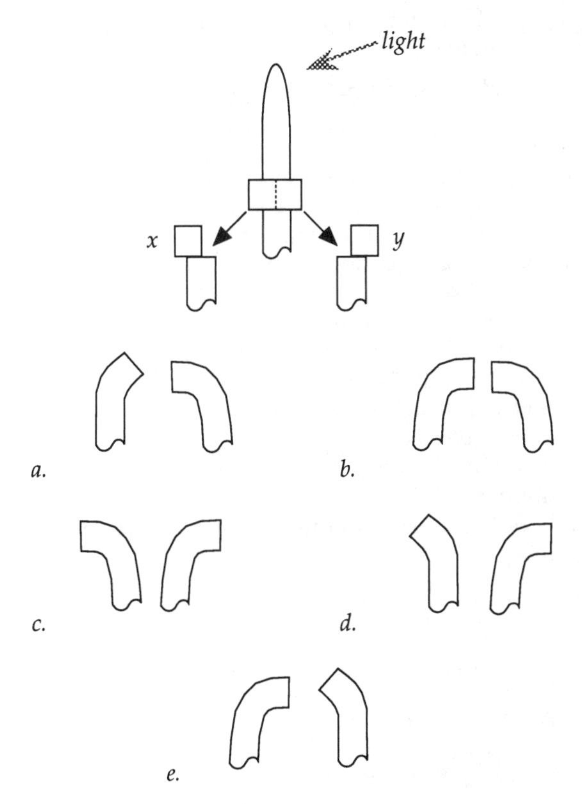

4. You mix auxin of different concentrations with agar and make three types of blocks (*x*, *y*, *z*), but you get the blocks mixed up. To determine the relative concentrations of auxin in the blocks, you place the blocks on the cut bases of coleoptiles as shown below (note carefully the positioning of the blocks). Auxin concentration in the blocks, from highest to lowest, is:

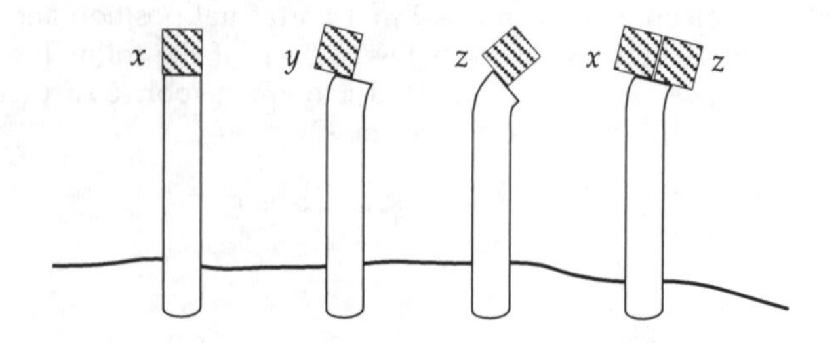

a. x, y, z
b. x, z, y
c. y, x, z
d. y, z, x
e. z, y, x

5. Number the following steps associated with auxin-induced cell elongation in the sequence in which they are hypothesized to occur.

_____ New microfibrils deposited on the inner cell wall surface
_____ Activation of an enzyme within the cell wall
_____ Decrease in pH within cell walls
_____ Hydrostatic pressure in vacuole stretches the cell wall.
_____ Binding of auxin to receptor protein
_____ Linkages between microfibrils are broken.

6. Which of the following events would likely occur if a plant tissue culture is treated with a solution containing a relatively high concentration of cytokinin and a relatively low concentration of auxin?
a. Bud and shoot formation; rapid cell expansion; decreased cell division
b. Bud and shoot formation; slow cell expansion; increased cell division
c. Bud and shoot formation; rapid cell expansion; increased cell division
d. Root formation; rapid cell expansion; decreased cell division
e. Root formation; slow cell expansion; increased cell division

7. Which of the following is *not* a normal plant response to cytokinins?
a. Seed germination
b. Lateral bud growth
c. Abscission of leaves
d. Leaf expansion
e. Redistribution of nutrients

8. You wish to delay the ripening of stored fruit. Which of the following would *not* help accomplish this objective?
a. Increasing the ventilation within the storage bins
b. Removal of ripe fruit from the bins
c. Increasing the atmospheric CO_2 concentration where the fruit is stored
d. Controlling the photoperiod where the fruit is stored
e. Decreasing the atmospheric ethylene concentration where the fruit is stored

9. Explain the observation that in the past, trees near gas street lights would lose their leaves sooner than trees farther from the lamps.

10. Compare the example presented in the textbook of using biotechnology to make more effective use of herbicides with older approaches to using herbicides.

ABSCISIC ACID • ENVIRONMENTAL CUES • LIGHT AND PHYTOCHROME • STUDYING THE PLANT GENOME (pages 750–754)

Key Concepts

1. *Abscisic acid* is a plant growth substance that is in high concentration in winter-dormant buds and also in water-stressed leaves.
2. In preparation for winter, plants undergo many changes leading to winter dormancy.
3. Formation of dormant buds with wax-covered bud scales protects the growth regions of stems.
4. Other features of winter dormancy are cessation of stem growth, leaf abscission, and an increase in the solute concentration of transport systems, lowering the freezing point.
5. Night length is monitored by plants and used to initiate the process of winter dormancy when nights become longer.
6. Studies done on plant species whose seeds require exposure to light in order to germinate show that blue light and red light induces germination, but green light does not.
7. Far-red light reverses the effect of previous red light exposure and prevents germination.
8. Seeds given successive exposures to red and far-red light respond to the last treatment received: if red, they germinate; if far-red, they remain dormant.
9. *Phytochrome*, a bluish pigment, is a protein that exists in two interconvertible forms, P_r and P_{fr}.
10. P_r absorbs best at a wavelength of 660 nm (red light) and is converted into P_{fr}; P_{fr} absorbs best at a wavelength of 730 nm (far-red light) and is converted into P_r.

11. P_r is stable in the dark, but P_{fr} is either converted into P_r or is destroyed.
12. P_{fr}–P_r may exert its effects on plant development by functioning as a ion channel within membranes.
13. Flowering plant seedlings kept in the dark do not synthesize chlorophyll and are called *etiolated*.
14. Etiolated shoots elongate more rapidly than shoots of seedlings kept in the light. This response helps bring the shoot to the soil surface rapidly.
15. The hook on the shoot of a etiolated dicot seedling helps protect the apical bud from abrasion by pulling instead of pushing the apical bud through the soil.
16. In some dicots, the cotyledons remain in the soil and the hook forms on the epicotyl. In other dicots, the hook forms in the hypocotyl, and the cotyledons and enclosed apical bud are pulled upward.
17. In some monocots, the growing shoot is protected by the *coleoptile*, the leaf sheath that grows until it reaches light; the shoot then emerges from the coleoptile.
18. Phytochrome in an etiolated seedling is in the P_r form. Light exposure converts phytochrome P_r to P_{fr} and causes stem elongation to slow, the hook to straighten, chlorophyll synthesis to begin, and leaves to expand.
19. The small flowering plant *Arabidopsis thaliana* is being studied extensively to identify developmental mutants. It produces very many small seeds, has a short generation time, and a genome about the size of *Drosophila*.

Questions (for answers and explanations, see page 253)

1. Which of the following is *not* characteristic of changes associated with winter dormancy in many species of temperate plants?
   a. Development of wax-covered bud scales
   b. High concentration of abscisic acid in buds
   c. Decrease in solute concentration of sap
   d. Leaf abscission
   e. Decrease in the rate of stem growth

2. Seedling A was placed in light of wavelength 660 nm for several minutes; seedling B was placed in light of wavelength 730 nm for several minutes. Both seedlings were then placed into the dark for 12 hours. Which of the following statements comparing the amount of P_{fr} and P_r in the seedlings is *not* true?
   a. Before being placed in the dark, A has more P_{fr} than B.
   b. Before being placed in the dark, B has more P_r than A.
   c. A had more P_{fr} before being placed in the dark than after 12 hours in the dark.
   d. A had more P_r before being placed in the dark than after 12 hours in the dark.
   e. A had more P_{fr} before being placed in the dark than did B after 12 hours in the dark .

3. Which of the following does *not* normally occur when an etiolated seedling is exposed to light?
   a. Chlorophyll synthesis begins.
   b. The rate of shoot elongation increases.
   c. The hypocotyl hook straightens (in species with a hypocotyl hook).
   d. Phytochrome P_r is converted to P_{fr}.
   e. Leaf expansion begins.

4. Dormant lettuce seeds were exposed to the following succession of brief periods of red light (R) and far-red light (FR) light: FR-R-FR-R-FR-R-FR. Which of the following is most likely to occur?
   a. Most of the seeds will germinate.
   b. Most of the seeds will not germinate.
   c. About one-half will germinate.
   d. The seeds will germinate but development will cease unless the epicotyl is exposed to red light.
   e. The seeds will germinate but development will cease unless the epicotyl is exposed to far-red light.

5. What are some of the features of *Arabidopsis thaliana* that are making it the plant biologist's *E. coli*?

Integrative Questions (for answers and explanations, see page 253)

1. Match the site of production on the right with the plant growth substance produced at that site on the left.

Abscisic acid	______	*a.* Embryo
Auxin	______	*b.* Young leaves
Cytokinin	______	*c.* Old leaves
Ethylene	______	*d.* Ripening fruit
Gibberellin	______	*e.* Root tips
		f. Shoot tips

2. Match the effect on the right with the plant hormones on the left.

Abscisic acid	______	*a.* Promotes germination
Auxin	______	*b.* Promotes seed dormancy
Cytokinin	______	*c.* Stimulates cell division
Ethylene	______	*d.* Stimulates cell elongation
Gibberellin	______	*e.* Inhibits lateral buds
		f. Stimulates lateral buds
		g. Breaks winter dormancy
		h. Imposes winter dormancy
		i. Promotes fruit development
		j. Promotes fruit ripening
		k. Inhibits leaf abscission
		l. Promotes leaf abscission

34

Plant Reproduction

CHAPTER LEARNING OBJECTIVES—after studying this chapter you should be able to:

- ❑ Compare and contrast the advantages and disadvantages of sexual and asexual reproduction in plants.
- ❑ Identify and describe the following parts and regions of a complete flower: carpels, stamens, petals, sepals, pistil, ovary, style, stigma, stamens, filament, anthers, corolla, calyx, and receptacle.
- ❑ Identify and describe the structure of the ovule.
- ❑ Identify and describe the structure of the male and female gametophytes in flowering plants.
- ❑ Describe the process of pollination and some of the methods that plants use for transport of pollen to female flower parts.
- ❑ Describe the events and consequences of double fertilization in flowering plants, including formation of the sporophyte embryo and endosperm.
- ❑ Identify and describe the structures and stages associated with early development in a typical dicot.
- ❑ Identify and describe the structures associated with a dicot seed.
- ❑ Describe some of the variation associated with fruit structure and dispersal in flowering plants and give an example of each.
- ❑ Differentiate between indeterminate and determinant growth relative to the activities of vegetative apical meristems, floral meristems, and inflorescence meristems.
- ❑ Explain how photoperiodism is involved in controlling the onset of flowering in many plants.
- ❑ Define the terms "day-neutral plant," "short-day plant,"and "long-day plant," and interpret the results of data from flowering studies to classify plants relative to these terms.
- ❑ Explain how it was determined that photoperiodic plants measure night length rather than day length.
- ❑ Describe the involvement of phytochrome in the photoperiodic timing mechanism.
- ❑ Describe the characteristic features of circadian rhythms.
- ❑ Characterize the evidence that argues for the existence of a flowering hormone.
- ❑ Define the term "vernalization" and give an example of it.
- ❑ Explain what is meant by the term "vegetative reproduction" and give some common examples of it.
- ❑ Define the term "apomixis" and give an example of it.
- ❑ Describe important agriculture practices involving asexual reproduction, such as the production of slips and grafts, and the prospects for the development of artificial seeds.

MANY WAYS TO REPRODUCE • SEXUAL REPRODUCTION (pages 757–764)

Key Concepts

1. Meiosis and recombination of gametes associated with sexual reproduction leads to increased genetic diversity. Increased genetic diversity may help to better adapt a population to new environments.
2. *Carpels*, *stamens*, *petals*, and *sepals* are the parts (modified leaves) of a complete flower.
3. Female flower parts are called *pistils*. Each pistil consists of one or more carpels and can be divided into an *ovary* (containing one or more *ovules*), an elongated *style*, and a pollen-receptive *stigma*.
4. Ovules consist of a *megasporangium*, a *stalk* suspending the ovule within the ovary, and protective layers called *integuments*. The opening through the integuments into the megasporangium is called the *micropyle*.
5. Male flower parts are called *stamens* and consist of a *filament* and two *anthers*, each with two *microsporangia*.
6. Whorls of petals, often colored, are called the *corolla*. Whorls of sepals, usually green, are called the *calyx*.
7. The stem tip bearing the flower parts is called the *receptacle*.
8. Within the megasporangium, the diploid megasporocyte undergoes meiosis, producing four haploid cells, three of which die. The remaining cell divides mitotically three times to produce eight nuclei.
9. The female gametophyte, called an *embryo sac*, consists of eight nuclei organized into seven cells. The *egg* and

two *synergids* are located at the micropyle end of the embryo sac. Two *polar nuclei* are located within a large central cell. Three *antipodal cells* are located at the other end of the sac.

10. Within anthers, *microsporocytes* divide meiotically, each producing four haploid *microspores*. Each microspore divides once mitotically to form a pollen grain. Thus, the male gametophyte consists of one cell with two nuclei.

11. One nucleus within the pollen grain, called the *generative nucleus*, divides to produce two *sperm nuclei*; the remaining nucleus is the *tube nucleus*.

12. Transfer of pollen to the stigma of different flowers can involve transport by wind (grasses), water (some aquatic angiosperms), or animal pollinators.

13. *Self-pollinating* species transfer pollen directly from anther to stigma within the same flower.

14. The pollen tube grows from the surface of the stigma down through the style and into the female gametophyte, usually following a chemical gradient. The tube nucleus remains close to the growing end of the pollen tube.

15. The two sperm nuclei are released into the embryo sac. One sperm nucleus fuses with the egg to form the diploid zygote; the other sperm nucleus fuses with the two polar nuclei to form a triploid nucleus. This process, called *double fertilization*, is only found in angiosperms.

16. The zygote divides mitotically to form the new *sporophyte embryo*; the triploid nucleus divides to form the nutritive tissue called *endosperm*. The remaining cells in the embryo sac degenerate.

17. After fertilization, the endosperm accumulates starch, lipids, and proteins, the integuments develop into a seed coat, and the carpel becomes the wall of the fruit surrounding the seed.

18. The first division of the zygote produces one cell that will form a filamentlike supporting structure called the *suspensor*, and another cell that becomes the *embryo*.

19. The single cell layer covering the embryo becomes the *protoderm*; the protoderm develops into the *epidermis*.

20. In dicots, two embryonic leaves called *cotyledons* develop and give the embryo a heart-shaped appearance.

21. During the *torpedo stage*, the embryo elongates and internal tissue differentiation begins.

22. The elongated region below the attachment of the cotyledons is called the *hypocotyl*.

23. The *shoot apex* is located between the cotyledons. The *root apex* is on the opposite end of the hypocotyl. Apical meristems on the shoot and root apex result in *primary growth* of the plant.

24. During seed development in some species (like peas and peanuts), the cotyledons absorb most of the endosperm; in other species (like castor beans), the endosperm remains intact.

25. Minimally, *fruit* includes the ovary wall and its seeds, but also may consist of other parts of the flower.

26. Some fruit is structurally modified to facilitate dispersal by wind (dandelions), animals (burdocks), or water (coconuts).

27. Nutritive material associated with the fleshy fruit may be used to attract animal species, which then act as agents of dispersal for the seeds.

Activities (for answers and explanations, see page 254)

- Label all cells and nuclei of the embryo sac shown below with the correct numbers of the following terms: 1) synergids, 2) egg cell, 3) polar nuclei, and 4) antipodal cells.

- Label the drawing of a developing plant embryo shown below with the correct letters for the following parts: a) suspensor, b) root apex, c) seed coat, d) hypocotyl, e) shoot apex, f) cotyledon, g) endosperm.

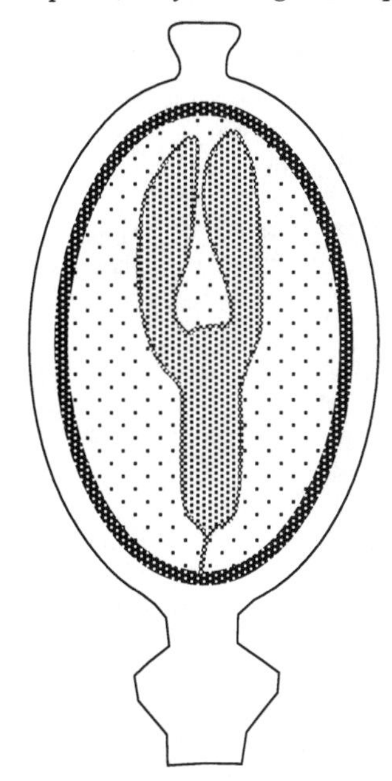

Questions (for answers and explanations, see page 254)

1. Pistils are formed from one or more
a. ovaries.
b. ovules.
c. carpels.
d. anthers.
e. megagametophytes.

2. A flower without a calyx would lack
a. carpels.
b. stamens.
c. petals.
d. sepals.
e. a receptacle.

3. A species has flowers with 12 petals, six sepals, six stamens, three carpels, 18 ovules, and one ovary. How many seeds can this species produce per flower?
   a. 1
   b. 3
   c. 6
   d. 12
   e. 18

4. Crested wheat grass has a diploid number ($2n$) of 14. Which of the following is the correct sequence of numbers for the chromosomes you would find in an egg cell, an endosperm cell, and a megagametocyte from this species?
   a. 7, 14, 28
   b. 7, 21, 14
   c. 7, 28, 14
   d. 14, 28, 21
   e. 7, 21, 7

5. Which of the following is *not* a mechanism used for pollination of cross-pollinating species?
   a. Wind
   b. Insects
   c. Direct transfer from anther to sigma
   d. Water
   e. Birds

6. The male gametophyte in flowering plants consists of a maximum of ____ cell(s) with ____ nuclei. The number of chromosomes in the nuclei is (n or $2n$) ____ .

7. Compare and contrast the relative ease with which in vitro fertilization has been accomplished in animals and plants.

8. Comment on the statement that in many flowering plant species, paternal genes disperse a greater distance from the parental plant than do maternal genes.

TRANSITION TO THE FLOWERING STATE • PHOTOPERIODIC CONTROL OF FLOWERING (pages 764–769)

1. During vegetative growth the plant shows indeterminate growth—the apical meristems continue to produce leaves, lateral meristems, and internodes.
2. Under certain conditions, an apical meristem becomes an inflorescence meristem. Inflorescence meristems produce leafy structures called bracts, internodes, and new lateral meristems that are either additional inflorescence meristems or floral meristems.
3. Floral meristems produce whorls of sepals, petals, stamens, and carpels. Floral meristems and some inflorescence meristems show determinant growth.
4. Homeotic genes work together to determine the types of floral whorls (sepals, petals, stamens, or carpels) produced by floral meristems. Mutations of these genes result in plants with incorrectly constructed flowers.
5. *Photoperiodism* is the regulation of a biological process by the daily light–dark period.
6. Many plants time the occurrence of their flowering through photoperiodism and will only flower when the day length is greater or less than some *critical day length.*
7. Plants whose flowering is unaffected by day length are called *day-neutral* plants.
8. *Short-day plants (SDP)* flower when the day length is less than the critical day length. They usually flower in the late summer or early spring.
9. *Long-day plants (LDP)* flower when the day length is greater than the critical day length. They usually flower in the midsummer.
10. Some plants, called *short-long-day plants,* only flower when long days follow short days. They usually flower before midsummer.
11. Some plants, called *long-short-day plants,* only flower when short days follow long days. They usually flower in the fall.
12. Experiments with *cocklebur,* a SDP, showed that photoperiodic plants actually measure the length of darkness, not light, in determining the photoperiod.
13. In these experiments, either the day or night period was varied while the other period was kept constant. Cocklebur only flowers if the night length is less than 9 hours. A single short night will cause flowering in this species.
14. Other experiments showed that brief exposure to light during the dark period caused a LDP to flower and prevented a SDP from flowering, but a brief period of darkness during the light period had no effect.
15. That red light is the most effective wavelength in night-interruption experiments, and that far-red light nullifies a previous exposure to red light, suggests the involvement of phytochrome in night length determination.
16. *Circadian rhythms* are cycles of biological events that occur approximately every 24 hours. *Sleep movements* are examples of such rhythms in plants.

17. The period of a circadian rhythm (time from one cycle to the next) is insensitive to temperature, although the amplitude (difference between high and low measurements) is reduced at lower temperatures.
18. Circadian rhythms can persist in constant light or darkness, but they can also be *entrained* by light–dark periods different than 24 hours.
19. Circadian rhythms can be permanently phase-shifted by exposure to light during the dark period.
20. The molecular and cellular basis of the *biological clock* remains unknown, as does how light can reset the clock.
21. Night length determination occurs in the leaves. A single leaf can cause flowering if exposed to an appropriate night length.
22. Grafting experiments have suggested that leaves produce a hypothetical chemical messenger called *florigen,* which moves to other parts of the plant and induces flowering there. Grafting of a single induced leaf can cause flowering in the recipient plant in some species.

Questions (for answers and explanations, see page 254)

1. In the following figure showing shoots resulting from three types of meristems (*a*, *b*, *c*), label the following structures: vegetative meristem, inflorescence meristem, floral meristem, leaf, internode, bract, carpel, petal, sepal, and stamen.

(*a*)

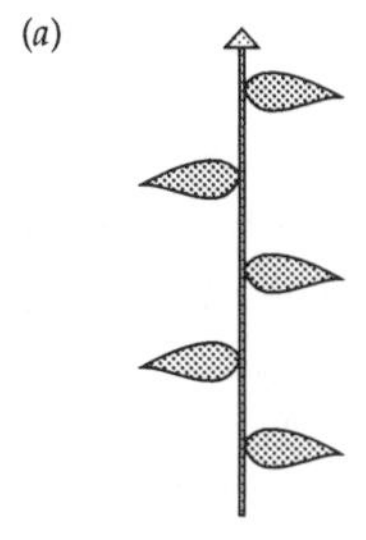

(*b*)

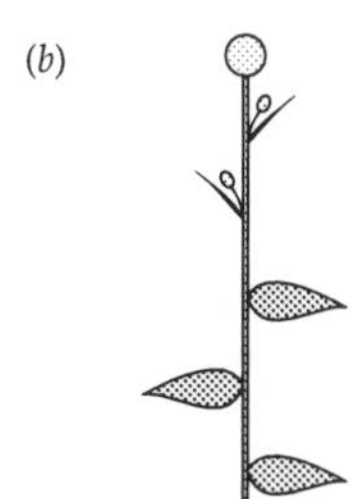

(*c*)

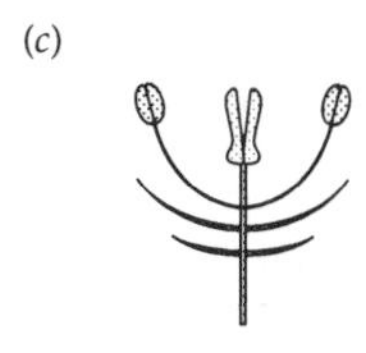

Which shoot system (*a*, *b*, or *c*) would be abnormal in a plant with a homeotic mutation? _____

Explain which shoot systems can show indeterminate growth, determinant growth, or both.

2. The next figure shows the results of exposing four plants to different light–dark periods to discover if photoperiod affects flowering in this species. Day and night length are shown as the white–dark portions of each 24-hour bar. Plant 3 was exposed to a 15-hour night interrupted by a brief flash of white light shown as a small white rectangle.

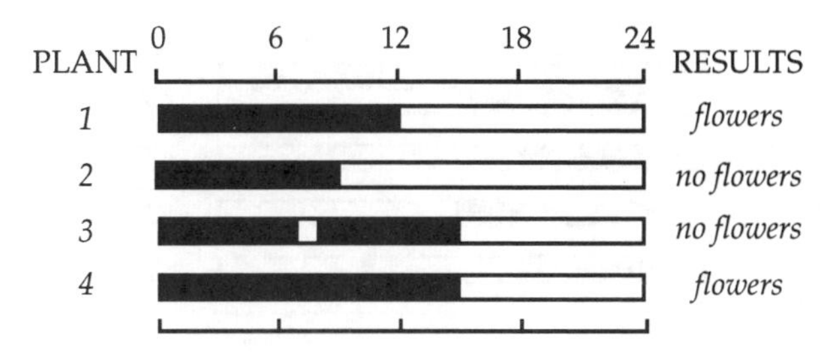

Which of the following statements is true? This species is a

a. short-day species with critical day length of 12 hours.
b. short-day species with a critical day length between 12 and 15 hours.
c. short-day species with a critical day length between 9 and 12 hours.
d. long-day species with a critical day length between 12 and 15 hours.
e. long-day species with a critical day length between 9 and 12 hours.

3. A long-day species has a critical day length of 14 hours. Select letters of all of the following treatments likely to induce flowering in this species. Note: Day and night length are shown as the white–dark portions of each 24-hour bar; *fr* indicates a flash of far-red light, *r* indicates a flash of red light.

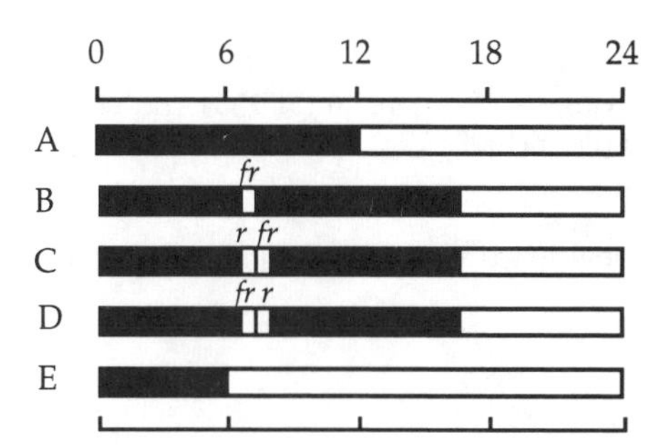

4. If a species is a long-short-day plant, it is likely to flower in

a. early spring.
b. before midsummer.
c. midsummer.
d. late summer.
e. fall.

5. The following figure shows the results from a series of experiments on henbane. Day and night length are shown as the white–dark portions of each 24-hour bar. Plant 3 was exposed to a short flash of white light during its dark period; the lights were briefly turned off during the light period for plant 4.

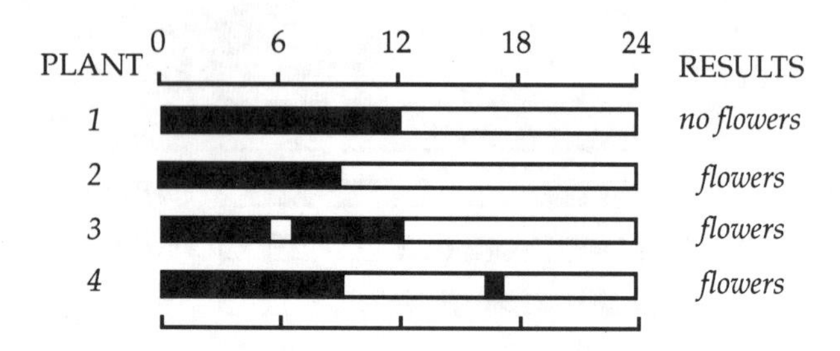

Based on these results, which of the following conclusions is *not* valid?

a. Henbane is a long-day plant (LDP).
b. The critical period for henbane is greater than 12 hours.
c. To determine when to flower, henbane measures day length.
d. Henbane probably flowers in midsummer.
e. Henbane is not day-neutral.

6. Which of the following conclusions is *not* verified by studies done on the regulation of flowering?
a. Phytochrome may be involved in the determination of night length by plants.
b. Night length is determined within the leaves.
c. Plants exposed to the correct photoperiod produce a flowering hormone, florigen.
d. If a flowering long-day plant (LDP) is grafted to a short-day plant (SDP) kept under long days, the SDP will flower.
e. If a flowering short-day plant (SDP) is grafted to a long-day plant (LDP) kept under short days, the LDP will flower.

7. Discuss the evidence for a flowering hormone in plants.

VERNALIZATION AND FLOWERING • ASEXUAL REPRODUCTION (pages 769–772)

Key Concepts

1. *Vernalization* represents the events that occur in some plants during periods of cold temperature that eventually lead to normal flowering.
2. A self-fertilizing individual heterozygous at one locus can produce offspring with all three genotypes, but it cannot produce individuals with new alleles.
3. A *clone* is a group of individuals with identical genotypes, usually all produced by the same parent.
4. Asexually reproducing individuals produce a clone of offspring. Asexual reproduction is a good strategy for a well adapted species in a stable environment.
5. *Vegetative reproduction* involves propagation by the vegetative parts of a plant: the stems, leaves, or roots.
6. Horizontal parts of the stem that have become modified for vegetative reproduction are *stolons* (strawberries), *tubers* (potatoes), and *rhizomes* (bamboo).
7. Vertical parts of the stem that have become modified for vegetative reproduction are *bulbs* (lilies and onions) and *corms* (crocuses, gladioli, etc.).
8. Leaves can also produce new plantlets (*Kalanchoe*), as can roots, as in *root suckers* (grasses, aspens, poplars).
9. *Apomixis* is asexual production of seeds. The ovule develops into a seed; the ovary wall develops into a fruit. Pollination is sometimes needed for endosperm production or to synchronize the timing of development.
10. *Slips* are cut pieces of stem that can be placed into soil and allowed to develop roots and grow.
11. In *grafting*, a piece of plant, called the *scion*, is attached to a root-bearing host plant, called the *stock*. The cambium of the scion eventually becomes joined to the cambium of the stock via fused wound tissue.
12. Grafting is very important in agriculture because it allows the combination of a shoot system that produces desirable fruit with a hardy root system.
13. Artificial seeds may be important in the future as a way to deliver plant embryos resulting from recombinant DNA and tissue culture techniques.

Questions *(for answers and explanations, see page 255)*

1. A self-pollinating plant is heterozygous for a single gene. What percentage of the offspring of this plant will have genotypes different from that of this plant?
a. 0%
b. 25%
c. 50%
d. 75%
e. 100%

2. Match the plants on the right to the plant parts on the left that they use in vegetative reproduction.

______Stolons	*a.* Potatoes
______Rhizomes	*b.* Onions, lilies
______Bulbs	*c.* Crocuses
______Tubers	*d.* Strawberries
______Corms	*e.* Bamboo
______Root suckers	*f.* Aspens, poplars

3. Which of the following statements about apomictic reproduction (apomixis) is *not* true?
   a. Apomixis is a form of asexual reproduction.
   b. Apomixis is likely to occur in a species that is well adapted to a stable environment.
   c. Apomixis produces offspring that are genetically identical to the parents.
   d. Pollination is never required in apomixis.
   e. Apomixis results in a clone of offspring.

4. Which of the following is *not* characteristic of the application of grafting?
   a. The cambium of the scion and stock must join.
   b. The stock usually has a hardy root system.
   c. Both scion and stock must produce wound tissue.
   d. The scion usually has desirable fruit.
   e. The slip is usually treated with root-inducing chemicals.

5. Discuss the construction of an artificial seed as presented in the textbook.

Answers and Explanations

Chapters 29–34

CHAPTER 29 – THE FLOWERING PLANT BODY

Note : The page numbers listed with the section titles refer to the Study Guide; numbers in parentheses for each question refer to the relevant key concepts.

FLOWERING PLANTS • FOUR EXAMPLES • CLASSES OF FLOWERING PLANTS • AN OVERVIEW OF THE PLANT BODY • ORGANS OF THE PLANT BODY • LEVELS OF ORGANIZATION IN THE PLANT BODY (pages 215–216)

1. (8, 9)
   Root system: ***a, c, e***
   Leaves: ***b***
   Stem: ***d***
2. Choice ***c***. A fibrous root system provides the plant with a large absorptive surface area and well-developed root caps would protect the root tip meristems from abrasion. (11, 12)
3. Choice ***a***. Rice and palms are monocots. As the name implies, monocots have a single embryonic leaf or cotyledon. They also usually have leaves with parallel venation, vascular bundles scattered throughout the stem, and floral parts in multiples of three. (1–5)
4. (13, 15, 18)
   Potato tubers: ***stem***
   Runners in strawberry plants: ***stem***
   Cactus spines: ***leaves***
   Tendrils in peas: ***leaves***
   Prop roots in corn: ***roots***

- (10, 12, 14–17, 20)

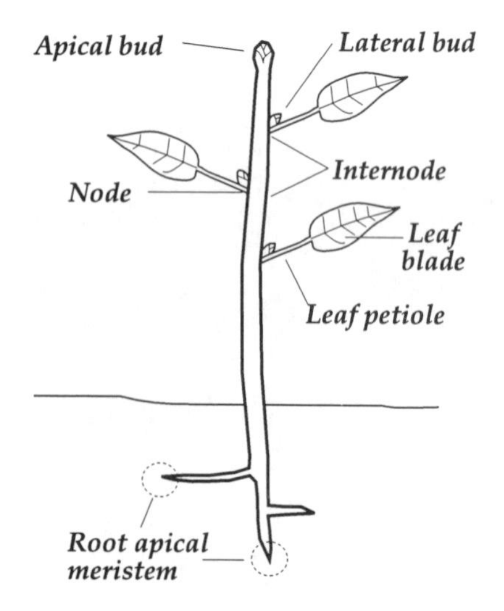

- (6, 7)
  Because plant parts such as branches and leaves act as somewhat autonomous modules, the control mechanisms in plants are more decentralized than in animals. Also, plant growth is restricted to specific meristematic areas; growth in animals is not as regionalized.

PLANT CELLS • PLANT TISSUES AND TISSUE SYSTEMS • GROWTH AND MERISTEMS (pages 217–218)

1. Choice ***c***. Parenchyma cells have only primary cell walls; sclerenchyma cells have both, with the secondary wall laid down to the inside of the primary cell wall. (1, 2, 5, 8)
2. Choice ***a***. (5–11)
3. Choice ***a***. (8, 10, 11)
4. Choice ***e***. (5, 6, 9, 13)
5. Choices ***a, d, e***. Epidermis protects the plant body, retards evaporative water loss in the shoot system, and absorbs water and minerals in the root system. (18)
6. (24)
   Carrot: ***biennial***
   Red maple: ***perennial***
   Rice: ***annual***
   Coconut palm: ***perennial***

- (1, 2)

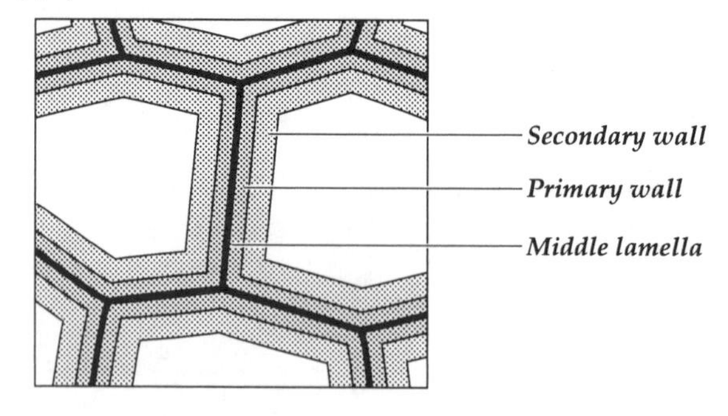

• (12, 13)

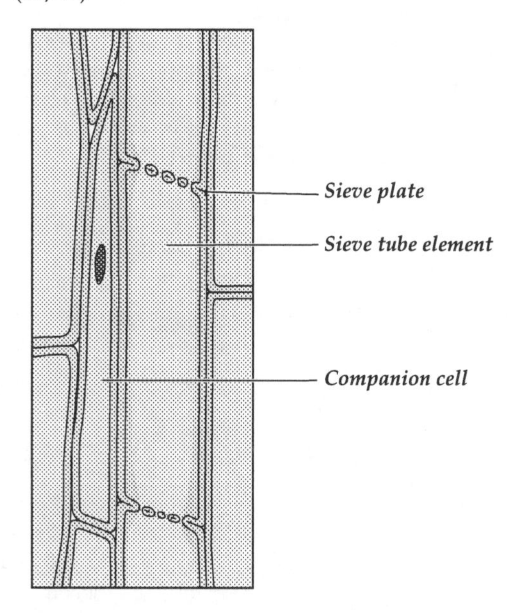

THE MERISTEMS AND THEIR PRODUCTS • SUPPORT IN A TERRESTRIAL ENVIRONMENT (pages 218–221)

1. Choices *a, c, e*. Epidermal cells are derived from the protoderm; cortex and collenchyma cells come from the ground meristem. (1–3, 10, 11)
2. Choice *c*. Companion cells are vascular tissue and derive from the procambium. (1–3, 10, 11)
3. Choice *d*. Both lateral buds and pericycle are regions that give rise to appendages of the plant body. (7)
4. Choices *a, c,* and *e*. Complex tissues are composed of more than one cell type. This includes xylem and phloem. (3–6)
5. Choices *a* and *b*. Water transport does not occur in bark and very little occurs in heartwood. Spring wood and summer wood are types of sapwood; more water transport can occur in spring wood, because its tracheids and vessel elements are larger that those of summer wood. (19–21)
6. Choice *b*. Within stems, xylem is located to the inside of phloem, so within the vascular bundle of a vein, xylem will be above phloem. (6–9, 16)
7. Choice *c*. Like stomata on leaves, lenticels are openings on the surface of stems. (26, 30)
8. Choice *b*. Secondary phloem is interior to primary phloem, and usually the cortex has been lost and replaced by tissues derived from the cork cambium, with phelloderm to the inside and cork to the outside. (14–25)
9. Choice *b*. C_4 plants have an outer layer of mesophyll cells and an inner bundle sheath layer around each vein. (32; also refer to Chapter 8)
10. Choice *e*. Compression wood expands as it develops and helps hold the branch in position. (34–37)

• See Figure 29.18 (*c*) on page 679 in the textbook. (4–9)

• (27–32)

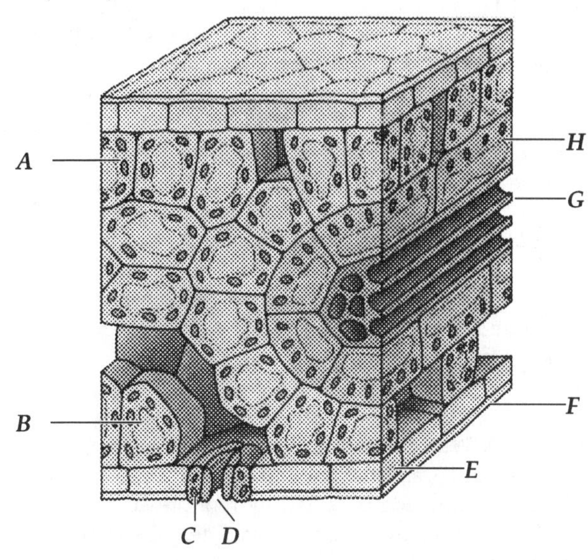

This is a C_4 plant—it has the outer layer of mesophyll and inner bundle sheath layer around each vein seen in C_4 plants.

• See Figure 29.20 on page 680 in the textbook. (14–16)
• (1–4)

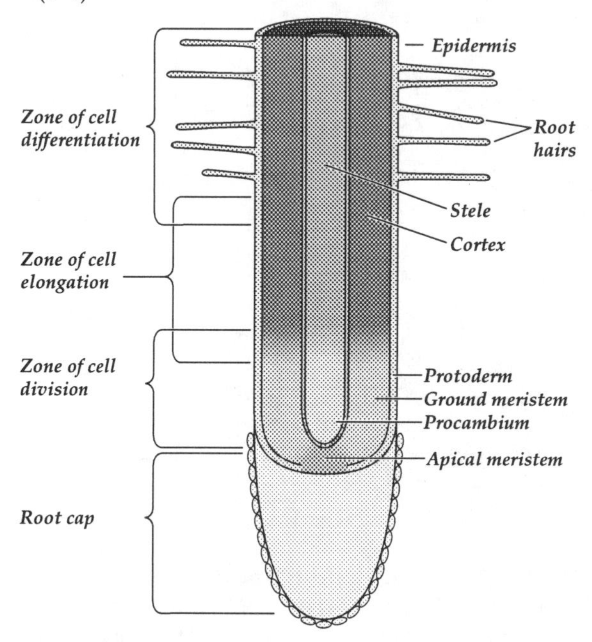

CHAPTER 30 – TRANSPORT IN PLANTS

Note : The page numbers listed with the section titles refer to the Study Guide; numbers in parentheses for each question refer to the relevant key concepts.

UPTAKE AND TRANSPORT OF WATER AND MINERALS (pages 222–224)

1. Choice *b*. Endodermal cells are surrounded by a watertight endodermal "gasket" that isolates the apoplast of the stele from the apoplast of the root cortex. (12–19)
2. Choice *d*. Since vessel elements are dead at maturity, they are not part of the root's symplast. (12–19)

3. Choice *c*. The water potential is the sum of the osmotic potential (usually negative) and the pressure potential (usually positive), so WP = –2.4 + 1.6 = –0.8 atmosphere. (3–8)
4. Choice *e*. There is a water potential gradient from outside to inside the root. The correct ranking is (from less to more negative): soil next to root, cortex apoplast, transfer cell, stele apoplast, xylem conduit. (3–8, 14–19)
5. Choice *c*. Since the water potential is the same on both sides of the membrane, the net movement of water will be zero. (3–8)
6. Choice *c*. Water could travel to the endodermis layer via the apoplast, pass through the outer endodermis cell membrane, and enter the stele apoplast by diffusing through the inner endodermis membrane. (14–19)
7. Choice *d*. Minerals are actively pumped into the xylem conduits; water follows osmotically. (3)
8. Choice *e*. Water potential will be least negative in the xylem conduits and most negative outside the leaf. (6, 8)
9. Choice *a*. Guttation is caused by root pressure. It is only important to the movement of water in quite small plants. (3–7)

- (12–19)

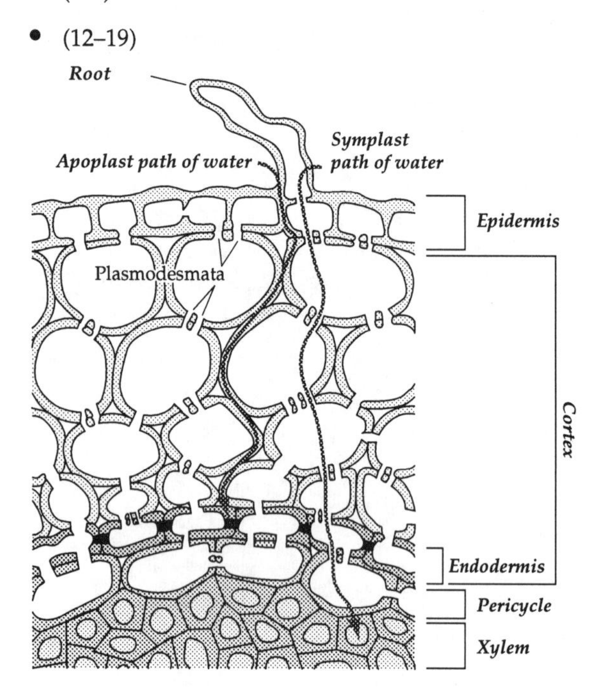

TRANSPIRATION AND THE STOMATA • CRASSULACEAN ACID METABOLISM AND THE STOMATAL CYCLE • TRANSLOCATION OF SUBSTANCES IN THE PHLOEM (pages 224–226)

1. Choice *e*. As leaf cells lose water, they release abscisic acid, which causes potassium ions to leave guard cells, decreasing their pressure potential and closing stomata. (2–7)
2. Choice *b*. As carbon dioxide is fixed into organic acids at night the pH decreases; metabolism of these acids during the day (with the carbon dioxide going into the citric acid cycle) causes the pH to increase. (8–11)
3. Choice *a*. Choices *b*, *c*, and *d* are true for phloem transport only; choice *e* is true for xylem transport only. (12)
4. Choice *c*. Sugar molecules diffuse through the symplast of mesophyll cells, are pumped into the apoplast near a vein, reenter the symplast of a companion cell, and then move into a sieve tube element. (14–19)
5. Choices *b* and *e*. Active transport is used to selectively move sugars and other organic compounds from the leaf apoplast into the phloem symplast. Unloading of sugars from the phloem sieve tube elements in sink areas such as the roots also occurs by active transport. (14–19)

- (8–11)

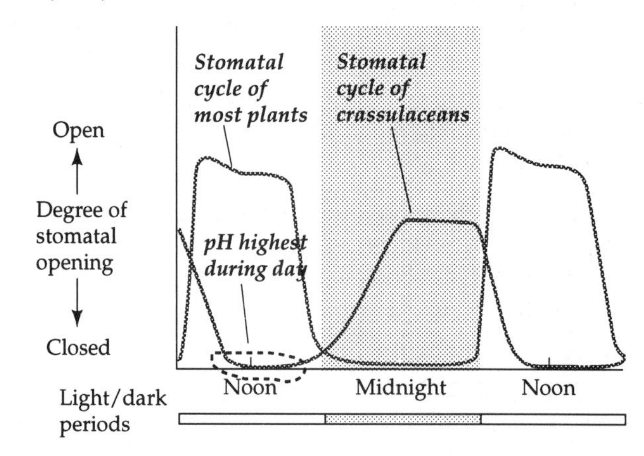

- (14–19)

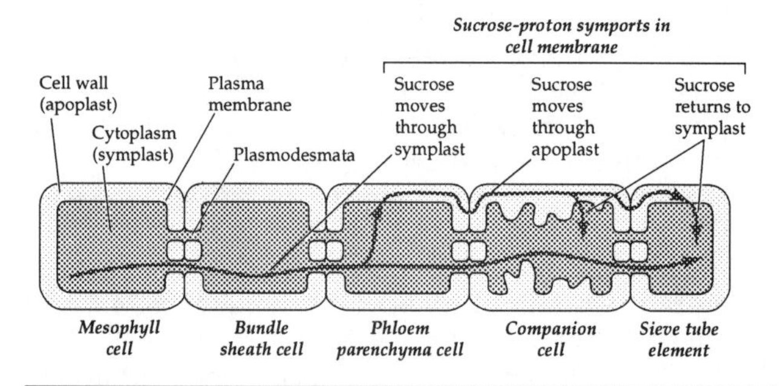

CHAPTER 31 – ENVIRONMENTAL CHALLENGES TO PLANTS

Note : The page numbers listed with the section titles refer to the Study Guide; numbers in parentheses for each question refer to the relevant key concepts.

THREATS TO PLANT LIFE • DRY ENVIRONMENTS • WHERE WATER IS PLENTIFUL AND OXYGEN IS SCARCE • SALINE ENVIRONMENTS • HABITATS LADEN WITH HEAVY METALS (pages 227–228)

1. Choice *a*. Mesophyll with large air spaces, called aerenchyma, is characteristic of submerged plants with floating leaves, not xerophytes. (1–7)
2. Choice *c*. Any halophytic adaptation would be useful; stomata in the upper epidermis is characteristic of submerged plants with floating leaves. (6–13)
3. Choice *d*. Tolerance to high sodium ion concentration is useful to halophytes, but not xerophytes. All other adaptations would serve both plant types. (10–13)
4. (5, 9, 13, 15)
   Seen in mangrove root: ***c. Pneumatophore***
   Can result from mining activities: ***d. Tailings***
   Can increase aluminum content of soils: ***f. Acid rain***
   Many cactus show this: ***e. Succulence***

Seen in floating plants: ***b. Aerenchyma***
Affects the osmotic potential of the cell: ***a. Proline***

5. Salinization is any process that increases the salt content of soil. Agricultural salinization results from irrigation in hot, dry regions with high evaporation rates. In these regions, salt buildup produces soils with large negative osmotic potentials which create osmotic problems for plants. Sodium and chloride is also toxic to plants in high concentration. (10–13)
6. Heavy metals include such elements as copper, nickel, lead, cadmium, chromium, and zinc. (14)

SERPENTINE SOILS • PLANTS AND HERBIVORES • PROTECTION AGAINST FUNGI, BACTERIA, AND VIRUSES (pages 228–230)

1. Choice *a*. Serpentine soils are deficient in calcium and have too much magnesium. (1)
2. Choice *c*. Resistance to salt toxicity is an adaptation seen in halophytes and is not related to grazing by herbivores. (2, 3)
3. Choice *d*. Secondary plant products are chemicals produced by the plant, but used to influence another organism. They are not used by the plant directly, as in energy storage. (4–13)
4. (15–18)
   3: Production of a necrotic lesion
   5: Synthesis of pathogenesis-related proteins
   4: Release of salicylic acid
   1: Secretion of phytoalexins
   2: Death of infected tissue
5. Choice *d*. Plants with this adaptation usually have modified enzymes and receptor molecules that do not recognize the toxic secondary product. That is, they respond to the normal product and can distinguish it from the toxic secondary product, as in canavanine-producing plants. (19–21)
6. After a plant has been grazed, minerals absorbed by the roots (especially nitrogen) need not be shared by as many leaves, so the remaining leaves are better supplied. Grazing often removes the older, slower growing, outer leaves so that more light is available to the younger, faster growing, inner leaves. Also, many grazed plants respond by producing multiple shoots and becoming bushier. (2, 3)
7. The major functions of secondary products in plants are to attract, resist, and inhibit other organisms. Since plants cannot move, they cannot flee from herbivores or disease-causing organisms and are dependent on animals for pollination and seed dispersal services. They need these secondary products to influence the behavior of these animals with whom they are coevolving. (4, 5)
8. Canavanine is very similar to arginine. In many herbivores, the enzyme responsible for charging the tRNA molecule specific for arginine cannot distinguish between arginine and canavanine. Consequently, canavanine is occasionally substituted for arginine during protein synthesis. This produces proteins with modified tertiary structures and reduced biological activities. (13, 21)

CHAPTER 32 – PLANT NUTRITION

Note : The page numbers listed with the section titles refer to the Study Guide; numbers in parentheses for each question refer to the relevant key concepts.

ACQUIRING NUTRIENTS • WHICH NUTRIENTS ARE ESSENTIAL? • SOILS (pages 231–233)

1. (11)
   Magnesium: *b, c*
   Iron: *a, f*
   Phosphorus: *g*
   Potassium: *c, d*
   Nitrogen: *a, b, e, g*
2. Choice *d*. There is a constant leaching of nutrients from the A horizon downward into the B horizon. Deep plowing moves some of these nutrients back up into the A horizon. (14–17, 19, 22, 24)
3. Choice *c*. Some heterotrophs, like the carnivores, can obtain organic compounds by consuming other heterotrophs. (3, 4, 6, 7)
4. Choice *a*. During the long-term aging of soil, $CaCO_3$ gradually decreases, clay gradually increases, and biomass peaks and then declines. (27)
5. Choice *d*. Hydrogen sulfide has little to do with ion exchange. (19)
6. (14–18, 26)
   A horizon: ***a, b, e***
   B horizon: ***c***
   C horizon: ***d, f***
7. (27)

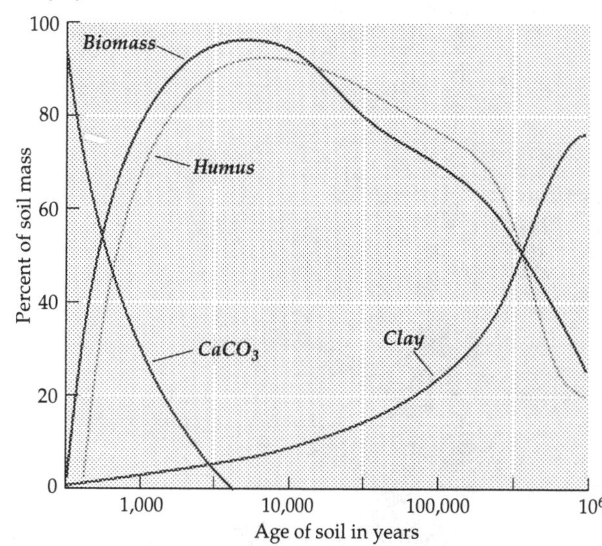

8. Most sessile organisms are aquatic and depend on the movement of water to bring them nutrients. Plants avoid depleting their supply of nutrients by growth of roots, leaves, and stems, and by dispersal of their seeds. (8)

9. Laterization involves the removal of mineral nutrients by leaching so that only oxidized, insoluble iron and aluminum compounds remain. Consequently, these soils are very poor in nutrients and are not suitable for agriculture. In contrast, soils in semiarid regions have retained their soluble nutrients because of the limited rainfall and high rates of evaporation. These soils are suitable for agriculture provided water is available for irrigation. (24, 25)

NITROGEN FIXATION • DENITRIFICATION • NITRIFICATION • NITRATE REDUCTION • SULFUR METABOLISM • HETEROTROPHIC SEED PLANTS (pages 233–235)

1. Choice *d*. These organisms are denitrifying bacteria. (3–8, 19)
2. Choice *c*. All other statements are false. (9–12)
3. Choice *d*. Infective threads fuse preferentially with tetraploid (4n) root cortex cells. (13–17)
4. Choice *a*. The technology for inserting genes into plasmids and transforming plant cells is well established; changing the genetics of higher plants to create a suitable environment for nitrogenase (no oxygen, lots of ATP and reducing agents) will be far more difficult. (13–18)
5. Choice *c*. Both chemosynthetic sulfur bacteria and the nitrification bacteria are (chemosynthetic) autotrophs; all other statements are false. (1–3, 9)
6. (12–15)

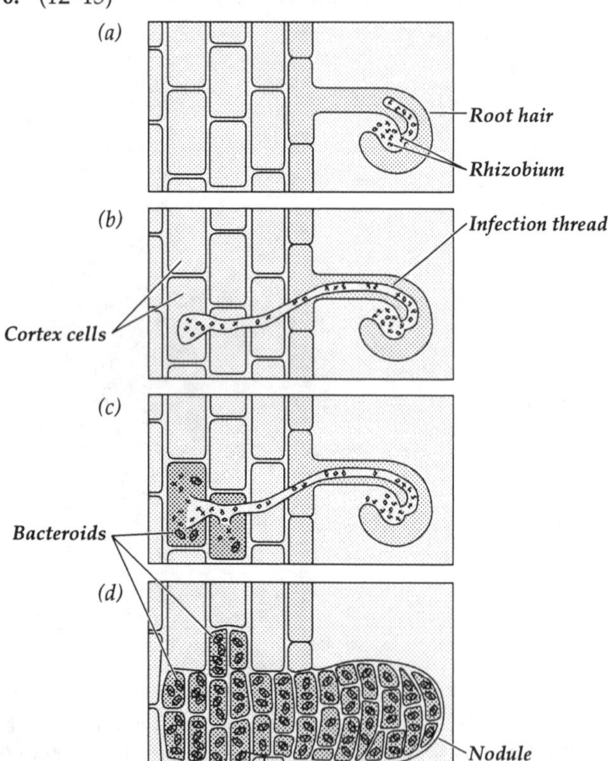

Nodules shown in figure *d*.

7. Choice *b*. Recall that the light compensation point is the light intensity at which the photosynthesis rate equals the respiration rate (net photosynthesis = 0). Because shade leaves have a lower respiration rate than sun leaves, they have a lower light compensation point; all other statements are false. (10–15)
8. The nitrogen-containing molecule that is most typically incorporated into amino acids by plants is *ammonia*. The organelle in the plant where this process occurs is the *chloroplast*.
9. Insertion of the genes to produce nitrogenase into crop plants will be much less difficult than providing the necessary anaerobic conditions for nitrogenase to function properly. Strong reducing agents and an energy source are also required. (16, 17)
10. (18–25)

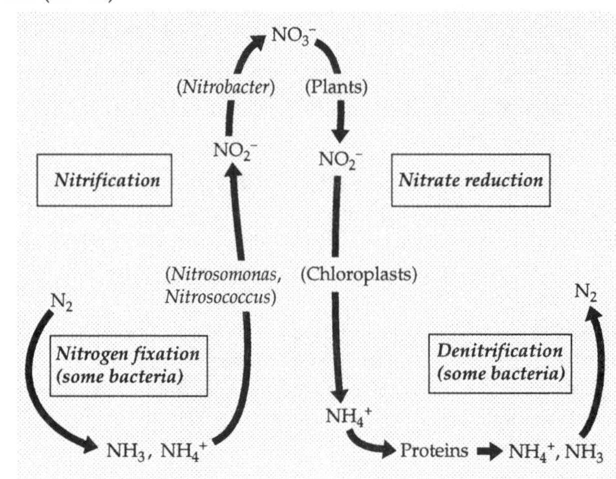

NH_3 and NH_4^-

Integrative Question

Denitrifying bacteria:	*h*
Nitrifying bacteria:	*b, c*
Nitrogen fixers:	*a*
Chemosynthetic sulfur bacteria:	*g*
Plants:	*d, e*
Heterotrophs:	*none*

CHAPTER 33 – REGULATION OF PLANT DEVELOPMENT

Note : The page numbers listed with the section titles refer to the Study Guide; numbers in parentheses for each question refer to the relevant key concepts.

WHAT REGULATES PLANT DEVELOPMENT? • AN OVERVIEW OF DEVELOPMENT • SEED DORMANCY AND GERMINATION • GIBBERELLINS (pages 236–238)

1. Environmental cues such as temperature or light changes cause changes in the plant's production of specific hormones; in the case of phytochrome, this may cause the conversion of one form of phytochrome to another. Hormones and phytochrome affect rates of cell division, elongation, and differentiation. The plant genome codes for phytochrome and for the enzymes that moderate the plant's developmental responses to environmental cues. (1–5)
2. Choice *c*. Scarification is the chemical or mechanical modification of the seed coat. Afterripening refers to the amount of time that must pass before seeds can germinate in some species, regardless of whether scarification is required. (9, 10)
3. *b*. Breaking of dormancy → *d*. imbibition → *e*. RNA synthesis → *f*. protein synthesis → *a*. emergence of radicle → *c*. DNA synthesis (8, 17, 18)

4. *b.* Gibberellin secretion by embryo → *f.* Diffusion of gibberellins to aleurone layer → *a.* Aleurone layer grains release amino acids → *d.* Synthesis of digestive enzymes → *e.* Hydrolysis of endosperm molecules → *c.* Building-block molecules absorbed by embryo (19–23)
5. Choice *d.* Gibberellins are involved in all listed phenomena except phototropism. (21, 24–32)
6. Choice *b.* Dwarf strains were genetically unable to produce gibberellin, so additional fertilizer would have no effect. (24–27)
7. (21–23)

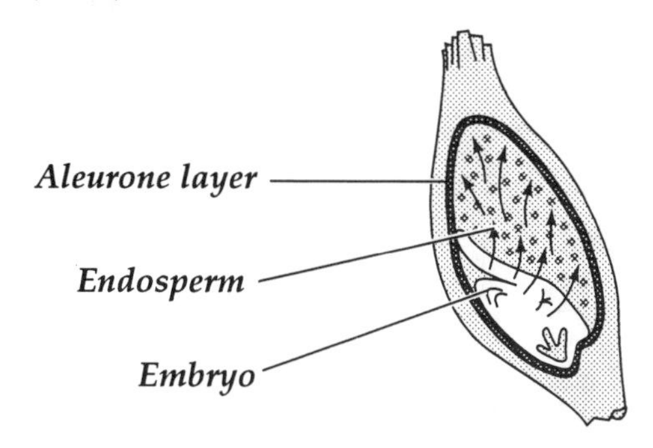

The small arrows represent gibberellins released from the embryo diffusing through the endosperm to the aleurone layer.

Gibberellins cause protein-containing bodies called aleurone grains to be hydrolyzed into amino acids. These amino acids are then assembled into digestive enzymes in the aleurone layer. The digestive enzymes are secreted into the endosperm and the reserve polymers located there are broken down.

AUXIN • CYTOKININS • ETHYLENE (pages 238–240)

1. Choice *b.* Statements *a*, *c*, and *d* are true, but were not demonstrated by the Darwin's' studies. Statement *e* is false. (2, 3)
2. Choice *b.* Phototropism and gravitropism will both be at work in determining the response of the coleoptile. As the coleoptile grows upward in response to gravity, its lower surface will become more illuminated than the upper surface, so it will assume an intermediate position relative to gravity and light gradients. (5–7)
3. Choice *e.* After illumination from the side, more auxin will be on the left side of the tip than the right side. Therefore, more auxin will diffuse into block *x* than block *y*. With the blocks positioned as shown, the left base should bend to the right more than the right base should bend to the left. (5–7)
4. Choice *b.* Clearly, the auxin concentration of block *z* is greater than *y*, since the base with *z* bends more than the base with *y*. The rightmost base shows that more bending occurs under *x* than *z*, so the correct order is *x*, *z*, *y*. (5–7)
5. *1:* Binding of auxin to receptor protein
   2: Decrease in pH within cell walls
   3: Activation of an enzyme within the cell wall
   4: Linkages between microfibrils are broken
   5: Hydrostatic pressure in vacuole stretches the cell wall
   6: New microfibrils deposited on the inner cell wall surface (11–14)
6. Choice *b.* Cytokinin promotes bud and shoot formation and increased cell division; auxin promotes root formation and increased cell elongation. (15–18)
7. Choice *c.* Cytokinin *delays* senescence (and abscission) of leaves. (16–21)
8. Choice *d.* Anything to reduce atmospheric ethylene or increase atmospheric carbon dioxide concentration would help. Photoperiod does not normally effect ripening of fruit. (22–25)
9. Trees near old gas lamps were exposed to some ethylene from the lamps. Ethylene promotes leaf senescence and loss (abscission). (22)
10. The old strategy was to search for herbicides that only kill certain types of undesirable plants while sparing desirable plants. A new approach is to employ a broad-spectrum herbicide that kills all plants, but to find a gene that confers resistance to the herbicide and to use biotechnology to insert that gene into the plants we want to cultivate.

ABSCISIC ACID • ENVIRONMENTAL CUES • LIGHT AND PHYTOCHROME • STUDYING THE PLANT GENOME (pages 240–241)

1. Choice *c.* Solute concentration of the sap is *increased* to prevent freezing during the winter. (1–4)
2. Choice *d.* Since light of wavelength 660 nm converts P_r to P_{fr} and light of 730 nm converts P_{fr} to P_r, initially A has mostly P_{fr} and B, mostly P_r. Since P_{fr} is converted to P_r or degrades in the dark, but Pr is stable, only statement d is false. (7–12)
3. Choice *b.* The rate of shoot elongation is greatest when the seedling is in the dark, and *decreases* upon exposure to light. (13–18)
4. Choice *b.* Far-red light prevents seed germination. Plants respond to the last light treatment they received (far-red in this case). (6–8)
5. *Arabidopsis thaliana* is small and produces many seeds that can be grown in petri dishes. It has a short generation time, a small genome (about the size of *Drosophila*), and many developmental mutants have thus far been identified. (19)

Integrative Questions

1. Abscisic acid: *b, c, j*
   Auxin: *a, e, f, g, k, m*
   Cytokinin: *a, d, e, h*
   Ethylene: *k, l*
   Gibberellin: *a, d, e, f, i, k*
2. Abscisic acid: *c, g, h*
   Auxin: *a, b, f*
   Cytokinin: *g*
   Ethylene: *d*
   Gibberellin: *a, b, e, f*

CHAPTER 34 – PLANT REPRODUCTION

Note : The page numbers listed with the section titles refer to the Study Guide; numbers in parentheses for each question refer to the relevant key concepts.

MANY WAYS TO REPRODUCE • SEXUAL REPRODUCTION (pages 242–244)

Activities

- (9–11, 15, 16)

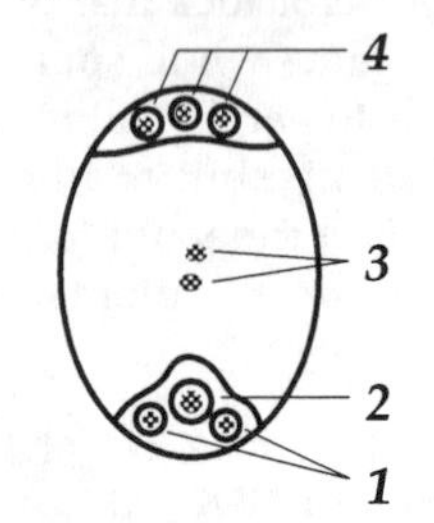

- (15–23)

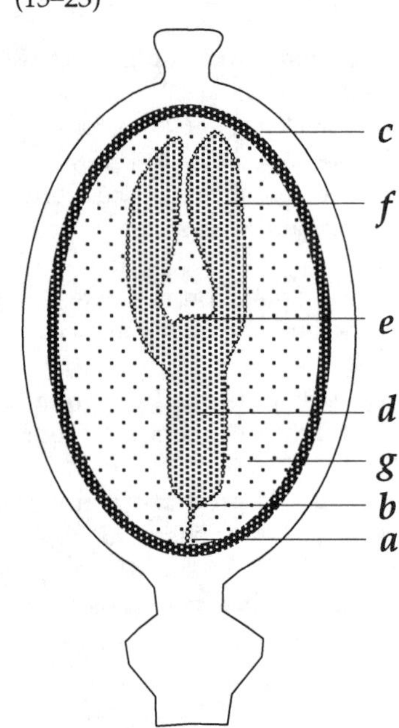

1. Choice *c*. A pistil is a single fused carpel or a group of them. Carpels can include one or more ovules. (2–7)
2. Choice *d*. The calyx consists of the sepals. (2–7)
3. Choice *e*. Each ovule contains one megagametophyte and can develop into one seed. (2–9)
4. Choice *b*. An egg cell is haploid (n), endosperm is triploid ($3n$), and a megagametocyte, being part of the sporophyte generation, is diploid ($2n$). (8, 9, 15, 16)
5. Choice *c*. Direct transfer of pollen from anther to stigma would most likely result in self-pollination, not cross-pollination. (12, 13)
6. The male gametophyte in flowering plants consists of a maximum of *one* cell with *three* (one tube and two sperm) nuclei. The number of chromosomes in the nuclei is n. (10, 11)
7. *In vitro* fertilization is relatively easy to accomplish in animals because their gametes are released from the organism and easy to collect and combine. Plant gametes are integral parts of other structures (pollen tubes and the embryo sacs) and are difficult to separate and may depend on interaction with these structures to develop normally.(9, 11, 14)
8. Paternal genes are transported in the pollen grain, but also in the seed. Maternal genes are transported only in the seed. In general, pollen is dispersed greater distances than are seeds. (11–13, 26)

TRANSITION TO THE FLOWERING STATE • PHOTOPERIODIC CONTROL OF FLOWERING (pages 244–246)

1. (1–4)

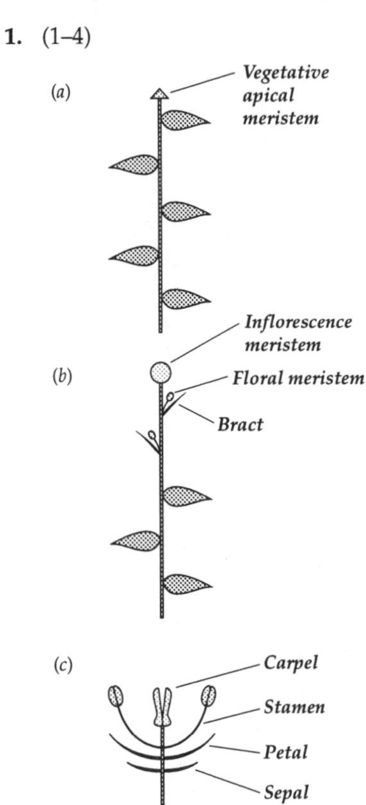

Shoot c, the shoot (flower) produced by a floral meristem would be abnormal in a plant mutant for a homeotic gene.

Vegetative meristems show indeterminate growth; they can continue indefinitely to give rise to leaves, internodes, and additional vegetative meristems. Floral meristems produce only flowers and no additional meristems so they show only determinant growth. Inflorescence meristems can show either indeterminate or determinant growth, depending on whether they produce additional inflorescence meristems or only floral meristems.

2. Choice *b*. Since the day length has to be less than a minimum value, this is a short-day plant. Since it flowers with days of 12 but not 15 hours, the critical day length is greater than or equal to 12 but less than 15 hours. (5–9)
3. Choices **D** and **E**. A long-day species with a critical day length of 14 hours requires more than 14 hours of light (less than 10 hours of dark) to flower. E provides those conditions, but so does D, where a brief red flash creates two short nights. The far-red flash in B has no effect and negates the red flash in C. (5–9, 12–15)

4. Choice *e*. Long-short-day plants flower in the fall when short days follow long days. (10, 11)
5. Choice *c*. Interrupting a long day with a brief period of dark, as in plant 4, did not prevent flowering, but interrupting a long night with a flash of light did, as in plant 3. These observations suggest that night length, not day length, is being measured. (12–15)
6. Choice *c*. The existence of florigen is only a hypothesis. (24, 30, 31)
7. If a photoperiodically induced leaf is immediately removed from the plant, it will not flower. If the photoperiodically induced leaf is grafted to another plant, the second plant will flower. In studies of plants grafted together, induction of flowering in one plant causes the other plants to flower as well. All of these studies suggest that a hormone is produced in photoperiodically induced leaves, and then transported to the rest of the plant. (22)

VERNALIZATION AND FLOWERING • ASEXUAL REPRODUCTION (pages 246–247)

1. Choice *c*. Since the plant is self-pollinating, the genetic cross is a heterozygote mated with a heterozygote. This will produce about 50% homozygous individuals. (2)
2. Stolons: ***d. Strawberries***
   Rhizomes: ***e. Bamboo***
   Bulbs: ***b. Onions, lilies***
   Tubers: ***a. Potatoes***
   Corms: ***c. Crocuses***
   Root suckers: ***f. Aspens, poplars*** (6–8)
3. Choice *d*. Pollination is required in some species to initiate the development of endosperm or to time developmental events. (9)
4. Choice *e*. Slips are cut stems that are used to form new plants, without grafting, by inducing root formation directly, typically using auxin. (10–12)
5. Artificial seeds would contain a multicellular somatic embryo, derived asexually in tissue culture, embedded in a water-soluble gel, and encapsulated by a protective covering. The gel and covering would dissolve after planting; materials such as nutrients, fungicides, and pesticides could be added to the gel. (13)

35

Physiology, Homeostasis, and Temperature Regulation

CHAPTER LEARNING OBJECTIVES—after studying this chapter you should be able to:

- ❑ Describe the four basic tissue types found in vertebrate animals and give examples of each.
- ❑ List the main organ systems of the human body and provide the important characteristics and functions of each.
- ❑ Define "homeostasis" and explain its importance to animal physiology.
- ❑ Differentiate between the concepts of control and regulation in homeostasis.
- ❑ Explain how set points and feedback control influence animal thermoregulation.
- ❑ Discuss the difference between negative and positive feedback control mechanisms, and compare them to the process of feedforward control.
- ❑ Explain the physiological reasons why animals need to control their body temperature.
- ❑ Describe the concept of Q_{10} and relate this concept to animal thermoregulation.
- ❑ Differentiate between acclimitization and metabolic compensation in animals.
- ❑ Define and correctly apply the terms "homeotherm," "poikilotherm," "heterotherm," "ectotherm," and "endotherm" to animals.
- ❑ Interpret and explain graphs depicting the results of laboratory or field experiments on the metabolic response of endotherms and ectotherms to changes in environmental temperature.
- ❑ Describe the differences between, and provide specific examples of, physiological and behavioral mechanisms for thermoregulation in animals.
- ❑ Explain the basic structure and function of the thermoregulatory mechanism known as the countercurrent heat exchanger.
- ❑ For a representative endotherm, define and explain the terms "thermoneutral zone," "basal metabolic rate," "upper critical temperature" and "lower critical temperature," and "summit metabolism."
- ❑ Discuss the various physiological and morphological adaptations shown by animals for life in hot and cold climates.
- ❑ Explain the central role of the hypothalamus in controlling animal thermoregulatory activity.
- ❑ Describe the role of hypothalamic thermal set points in normal thermoregulatory behavior and the fever response.
- ❑ Define "hypothermia" and explain how shallow torpor and hibneration are used as thermoregulatory mechanisms by some animals in cold climates.
- ❑ Explain the role of circannual rhythms in regulating animal hibernation.

HOMEOSTASIS • ORGANS AND ORGAN SYSTEMS (pages 777-782)

Key Concepts

1. Physiology is the study of how animals work. Physiological systems try to maintain stable internal body conditions optimal for cell function, a concept called *homeostasis.*
2. Homeostasis is under precise regulatory control through a complex network of organs and organ systems that gather information about relevant environmental stimuli and then coordinate appropriate physiological and behavioral responses.
3. Cells associate with one another according to common functions. Masses of specialized cells with a common purpose are called tissues, of which four general types are recognized in vertebrate animals. Most organs (e.g., the human stomach) contain examples of all four tissue types.
4. *Epithelial tissues* form dense sheets of tightly compacted cells that cover the outer body surface and the lining of

hollow internal organs. Epithelial cells are modified for use in movement, secretion, absorption, and protection.

5. *Connective tissues* provide support and reinforcement for other tissue types. They consist of loose arrays of cells embedded in some kind of extracellular matrix. Strong, elastic fibers provide matrix flexibility in some connective tissues (e.g., tendons, ligaments), while others are mineralized and rigid (e.g., bone).
6. *Muscle tissue* is made of cells that contract, which permits the movement of organs, limbs, and other body parts. Muscles are one type of *effector*, a structure that enables the body to do things.
7. *Nervous tissue* consists mainly of *neurons*, highly diverse cells whose primary job is the processing and integration of electrochemical information. Neurons are specialized for many different information-handling tasks, including detection, relay, analysis, and reaction.
8. The nervous system includes all of the nervous tissue in a body. In vertebrates, the brain and spinal cord make up the central nervous system. The peripheral nervous system conveys information between sensory organs, the central nervous system, and effectors.
9. The endocrine system transfers information through specialized chemical signals. Molecules called hormones are manufactured at one location and then transported (usually via the blood) to target cells found elsewhere in the body. Because of their roles in communication, many important connections exist between the nervous and endocrine systems.
10. The skin system is the largest organ of the body. It acts as a first line of defense against disease-causing organisms, and serves as a barrier to water loss. A rich complement of nervous tissue also gives skin a sensory function.
11. The skeletal system supports and protects soft body parts. It interacts with the voluntary muscle system to produce body movements. Involuntary muscle movements are not associated directly with skeletal elements; they generate the movements of the gut and other internal organs, as well as the beating of the heart.
12. Organs of the reproductive system produce both the sex cells and the sex hormones of animals. Additional organs for storage and delivery of sex cells exist in many species, as do specialized internal and external organs used for offspring development.
13. The digestive system, or gut, is a continuous tube divided into specialized segments. Each segment performs different nutrient-processing functions such as secretion, digestion, movement, absorption, and elimination of solid wastes.
14. The respiratory, or gas exchange, system provides a way for oxygen to enter the body and for carbon dioxide to exit. Complex air passages and large exchange surfaces are common features of respiratory systems.
15. The circulatory system connects all cells and organ systems in the body via a continuous system of vessels. Blood, the fluid transport medium, consists of both cellular and plasma components. Nutrients, oxygen, hormones, heat, and wastes are pumped through this system by the action of a muscular heart.
16. The lymphatic system represents a second set of circulatory organs, although it lacks a muscular pump and does not form a complete circuit of vessels. Its job includes mobilizing defense mechanisms during immune responses and returning lost circulatory fluids to the blood.
17. The excretory system removes nitrogenous wastes from the body, and also helps to maintain a proper water and salt balance. The kidney is the primary excretory organ of vertebrates.

Activities (for answers and explanations, see page 355)

- Your skin, like most other organs in the human body, contains representatives of all four vertebrate tissue types. Suggest functions for each of the following tissues found in human skin:

  Epithelium:

  Connective:

  Muscle:

  Nervous:

- For each of the organ systems listed on the left, choose its primary function from the list on the right.

____	Excretory	*a.* Electrical communication
____	Circulatory	*b.* Remove nitrogenous waste
____	Skeletal	*c.* Process and absorb food
____	Muscle	*d.* Chemical communication
____	Nervous	*e.* Produce and deliver gametes
____	Endocrine	*f.* Protection and support
____	Digestive	*g.* Contractile movements
____	Reproductive	*h.* Deliver nutrients to body cells

Questions (for answers and explanations, see page 355)

1. Homeostatic control involves the coordinated response of many physiological processes. Which of the following contributes to the proper maintenance of homestasis in animal bodies?
   a. pH buffering of the blood
   b. Control of body temperature
   c. Elimination of nitrogenous wastes
   d. Production of sex hormones
   e. All of the above

2. Your eyes are ________ elements of the ________ system.
   a. regulatory; skeletal
   b. sensory; nervous
   c. control; excretory
   d. regulatory; respiratory
   e. sensory; lymphatic

3. Endocrine organs release hormones that influence many of the physiological systems involved in homeostasis. The tissue type that produces these hormones is
   a. epithelium.
   b. connective.
   c. muscle.
   d. nervous.
   e. All of the above

4. Which of the following is a characteristic unique to connective tissue?
   a. Highly modified cells that show the special property of contractility
   b. Diverse anatomy with distinct specializations for information transfer
   c. Lines the inner surfaces of the intestines and lungs in mammals
   d. Loose array of cells embedded in some kind of extracellular matrix
   e. None of the above

5. The two human organ systems that communicate via electrical signals and transfer essential materials between all other organ systems are the
   a. respiratory and digestive systems.
   b. lymphatic and circulatory systems.
   c. endocrine and lymphatic systems.
   d. muscular and endocrine systems.
   e. nervous and circulatory systems.

6. Which one of the following animal organs or organ systems is *not* directly involved in the functions of information control, protection, support, or movement?
   a. Skeletal
   b. Excretory
   c. Muscle
   d. Skin
   e. Nervous

GENERAL PRINCIPLES OF HOMEOSTASIS • THE EFFECTS OF TEMPERATURE ON LIVING SYSTEMS • THERMOREGULATORY ADAPTATIONS (pages 782–790)

Key Concepts

1. Maintaining homeostasis requires both *control* and *regulation*. Control allows for a change in system status, while regulation maintains a system within certain limits. Regulation requires control mechanisms.

2. A crucial factor in the regulatory process is the need for information concerning the current status of both external and internal environments.

3. A *set point* is the desired condition of a system. *Feedback* mechanisms provide information about current conditions relative to the set point. The difference between the set point and the current condition of a system is the *error signal*, the nature of which indicates what corrective measures are needed to restore homeostasis.

4. *Controlled systems* include those molecules, cells, tissues, organs, and organ systems that change their characteristics in response to physiological regulation. Feedback mechanisms are essential for regulation since they detect, process, integrate, and act upon relevant information.

5. *Negative feedback* (the most common form of regulation) occurs when the end result of a process feeds back to reduce or reverse the process, leading to a stable condition approximating the set point. The regulatory principles involved in negative feedback are exemplified by a household thermostat.

6. Thermostats have an adjustable set point that is compared to current room temperature through information conveyed by a sensor. Depending on the type of error signal produced, either the furnace or the air conditioner is turned on. When room temperature returns to the set point, negative feedback shuts down the system.

7. *Positive feedback* does not regulate a system within stable limits. Rather, it produces an amplification of the initial perturbation which is reinforced until the process achieves its end. Examples include sexual arousal and the action potentials produced by nerve cells.

8. *Feedforward* information is used to change the set points of regulatory mechanisms by anticipating physiological change and causing responses to counteract those changes.

9. Maintenance of proper body temperature is of great physiological importance to animals. They must avoid having body temperatures drop below 0°C (where ice crystals form) or rise above 45°C (where most proteins denature). Even within this 45-degree window, small changes in temperature can significantly affect physiological function.

10. The temperature dependence of a system is represented by its Q_{10} value. Q_{10} values result from dividing the rate of a process at temperature T (°C) by the rate of the same process at temperature $T - 10$°C.

11. Most physiological processes have Q_{10} values between 2 and 3, meaning that for every 10°C rise in temperature, the rate of the process doubles or triples. $Q_{10} = 1$ indicates that a process is not temperature-sensitive.

12. Q_{10} values can differ for the individual reactions in complex metabolic pathways. Thus, even minor changes in body temperature can lead to severe disruptions of homeostasis. This places a premium on animals being able to compensate for even the slightest change in body temperature.

13. Animals can adjust for seasonal fluctuations in environmental temperature by making specific biochemical or physiological changes, collectively referred to as *acclimatization*. This is often done using *metabolic compensation*, in which cellular biochemistry is adjusted by the use of different enzymes with appropriate temperature optima.

14. Physiologists do not use the terms "cold-blooded" or "warm-blooded" when discussing animal thermoregu-

lation. Instead, animals that regulate their body temperature at constant levels are called *homeotherms*. Animals whose body temperature changes with fluctuations in environmental temperature are called *poikilotherms*. *Heterotherms* are animals that regulate their body temperature at certain times, but not at others.

15. Animals are also classified by thermoregulatory mechanisms. *Ectotherms* depend on external sources of heat to maintain an elevated body temperature. *Endotherms* produce and retain enough metabolic heat to effectively regulate their body temperature without external heating.
16. Under strict laboratory conditions, an ectotherm's body temperature and metabolic rate will rise and fall in direct proportion to increases or decreases in environmental temperature. By contrast, an endotherm of similar mass will exhibit a constant body temperature over the same range of environmental temperatures because it is able to adjust its metabolic rate to compensate for heat loss.
17. Under real-world conditions, ectotherms can maintain relatively constant body temperatures through behavioral thermoregulatory mechanisms such as shuttling between sunlight and shade. Endotherms also engage in various types of behavioral thermoregulation, including sophisticated activities such as nest construction and huddling.
18. Body temperature in both endotherms and ectotherms is strongly influenced by regulation of blood flow to the skin. Marine iguanas retain heat while foraging in cold water by constricting peripheral blood flow. They obtain additional heat by shunting blood to the skin while basking in the sun. Humans can likewise lose or retain metabolic heat by regulating peripheral blood flow.
19. Metabolic heat production is not unique to endotherms. Some insects rely on muscle contraction to generate heat for flight on cold nights. Honeybees engage in clustering behavior to metabolically heat the sensitive brood portions of their hive in winter. Some snakes can increase the incubation temperature for their eggs through isometric muscle contractions.
20. Sustained muscle activity produces large amounts of heat. If retained within the body core, this heat helps maintain body temperature in excess of ambient temperature.
21. Countercurrent heat exchangers are found in some animals, including several species of fast-swimming tuna and sharks that live in cold waters. Vessels carrying warm blood from the swimming muscles run parallel to vessels carrying cold blood into the muscles at the periphery. Heat is transferred from the outgoing vessels to the incoming ones, thereby concentrating heat in the body core. Increased core temperture allows the muscles to provide much higher levels of sustained power output.

Activities (for answers and explanations, see page 355)

- Refer to Figure 35.9 in your textbook, which shows a schematic diagram of a home heating and cooling system. Explain what would constitute an error signal for this system, and describe how the thermostat would respond to this signal to appropriately adjust the internal temperature of the house.
- Discuss the difference between the concepts of acclimitization and metabolic compensation as they relate to temperature regulation in animals.

Questions (for answers and explanations, see page 355)

1. Resetting the thermoregulatory set points in an animal about to undergo hibernation is an example of
   a. homeostatic control.
   b. positive feedback.
   c. countercurrent exchange.
   d. feedforward regulation.
   e. More than one of the above

2. During childbirth, the pressure exerted on the mother's cervix by the emerging infant leads to increased contraction of the uterus. This interaction could be explained by
   a. negative feedback.
   b. feedforward information.
   c. positive feedback.
   d. controlled homeothermy.
   e. metabolic compensation.

3. The reaction rate for an important metabolic process in human digestion is measured as 200 units at a temperature of 30°C. If the Q_{10} for this reaction is 1.5, what would the rate be at 20°C?
   a. 300
   b. 200
   c. 150
   d. 133
   e. Cannot determine from information given

4. Lizards cannot metabolically produce or retain enough heat to maintain continuous elevated body temperatures, but they are able to hold their body temperatures somewhat constant by appropriately positioning themselves in the environment. Select the terms that *best* apply to these thermoregulatory capabilities in lizards.
   a. Endothermic, physiological heterotherm
   b. Endothermic, behavioral poikilotherm
   c. Endothermic, physiological homeotherm
   d. Ectothermic, behavioral poikilotherm
   e. Ectothermic, behavioral homeotherm

5. In an environment with an ambient temperature less than an animal's core body temperature, which of the following would be an *inappropriate* physiological or behavioral response if the animal needed to eliminate excess body heat?
   a. Wallowing in a pool of water
   b. Preventing blood flow to peripheral vessels
   c. Reducing the metabolic activity of voluntary skeletal muscles
   d. Seeking shelter in a shaded area
   e. All of the above

6. In fast-swimming, cold-water fish such as sharks and tuna, which of the following contribute to the generation and maintenance of body core temperatures in excess of ambient water temperatures?
   a. High rates of activity in their powerful swimming muscles
   b. An ability to rapidly acclimatize to the surrounding water
   c. Specializations in the size and location of their blood vessels
   d. Metabolic rates that are insensitive to temperature change
   e. Both *a* and *c*

7. In a laboratory experiment, metabolic rate as a function of environmental temperature is measured for a mouse and a lizard of comparable size. The results are shown in the graph below. Which of the following statements about these results is *not* true?

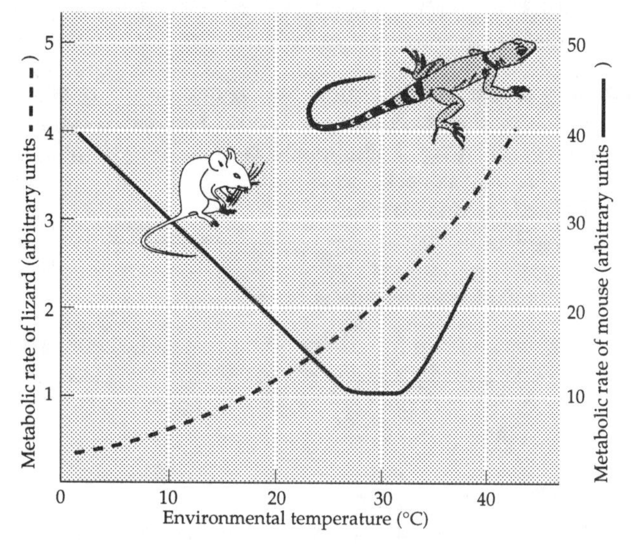

   a. Below 25 °C, the lizard's metabolic rate is directly porportional to environmental temperature.
   b. Below 25 °C, the mouse's metabolic rate is inversely proportional to environmental temperature.
   c. The overall pattern of change in metabolic rate for the mouse is characteristic of ectotherms.
   d. The overall pattern of change in metabolic rate for the lizard is characteristic of ectotherms.
   e. The mouse and the lizard have different physiological mechanisms for thermoregulation.

THERMOREGULATION IN ENDOTHERMS • THE VERTEBRATE THERMOSTAT • TURNING DOWN THE THERMOSTAT (pages 790–797)

Key Concepts

1. Endotherms adjust their metabolic rate to maintain a constant body temperature.
2. The *thermoneutral zone* is a range of ambient temperatures within which an endotherm's metabolic rate is low and independent of temperature. The *basal metabolic rate*, which represents the rate of metabolism needed to carry out essential physiological functions when the animal is at rest, does not change within the thermoneutral zone.
3. Endotherms are capable of greater heat production than ectotherms. Endotherms routinely maintain large temperature differentials between their bodies and the environment.
4. Lower and upper critical temperatures bound the thermoneutral zone. Metabolism increases as environmental temperature drops below the lower critical limit until a maximum, or summit, metabolism, is reached. Continued decreases in ambient temperature result in net heat loss and a reduction in body temperature. Metabolic rate increases above the upper critical temperature are due to increased energy expenditures associated with active mechanisms of heat loss such as panting and sweating.
5. Both mammals and birds produce metabolic heat through shivering. Heat forms during the chemical breakdown of ATP as opposing muscle groups are simultaneously contracted.
6. Mammals also engage in nonshivering heat production during which *brown fat* is metabolized. In brown fat cells, the protein thermogenin uncouples oxidative phosphorylation to release heat without the production of ATP.
7. Special morphological adaptations for thermoregulation characterize animals that live in cold climates. Most are adaptations for reducing heat loss, either by improving characteristics of insulation or by reducing surface-to-volume ratios through rounder bodies and shorter appendages.
8. Endothermic animals can regulate heat loss by controlling peripheral blood flow. Wolves conserve heat by regulating blood flow to their feet. In this way just enough heat is transferred to keep the foot tissue from freezing. Excess heat can be lost by exposing patches of bare skin to air, ice, or cool water.
9. Evaporative cooling is another mechanism for heat loss. Because liquid water requires a substantial heat input to become gas, sweating or panting can remove large amounts of body heat. Drawbacks to this thermoregulatory strategy include excessive water loss and significant energy requirements. The latter factor explains the rise in metabolic rate observed at temperatures above the upper critical limit.
10. Thermoregulatory mechanisms are ultimately under the control of a small region of the brain called the *hypothalamus*. In addition to its role as a thermostat, the vertebrate hypothalamus also regulates many other homeostatic systems.
11. The hypothalamus generates thermal set points against which its own temperature is compared. Temperature information is also obtained from peripheral nervous connections.
12. Different thermoregulatory processes often have different hypothalamic set points. For example, the temperature at which sweating occurs differs from the one that stimulates constriction of peripheral blood vessels.
13. Unlike a simple household thermostat, the hypothalamus can respond proportionally to the difference between set points and the hypothalamus temperature. The thermoregulatory responses of the animal are thus

proportional to how much the hypothalamus temperature differs from a specific set point.

14. Information about peripheral skin temperature is used to anticipate possible changes in body temperature, an example of feedforward regulation.
15. Fevers represent upward shifts in the thermoregulatory set point that cause an increase in body temperature. They result from exposure to *pyrogens*, substances from viruses or bacteria that stimulate a change in the set point. Although dramatic increases in body temperature are potentially life-threatening, slight fevers probably aid the body's efforts to ward off infection.
16. Temperature settings can also be lowered on the biological thermostat. Hypothermia (below normal body temperature) results either from natural downward adjustments in set points or from physiological insult. Extreme hypothermia can be lethal to most endotherms.
17. Because metabolism is temperature-dependent, hypothermia conserves metabolic energy. Many endothermic species engage in regulated hypothermia to survive prolonged periods of cold and food shortages.
18. Hummingbirds, pocket mice, and other small endotherms show an adaptive form of hypothermia called *shallow* (or daily) *torpor*. They lower their body temperature as much as 10–15°C below normal during their daily period of inactivity in order to avoid exhausting their energy reserves.
19. *Hibernation* is a seasonal form of regulated hypothermia which can last for a few days or many weeks. It is used by animals whose habitat gets too cold or food-depleted for them to remain active. Only one bird hibernator is known; all others are mammals. During hibernation body temperatures as low as 2–4°C are common.
20. Hibernation is controlled by inherent *circannual rhythms*, which divide the animal's year into the active season (where hibernation is not possible) and the hibernation season. Hibernating animals periodically revive from their state of reduced metabolism, at which time they may ingest stored foods to replenish fat reserves.
21. Mammalian hibernation likely evolved as a natural extension of reductions in hypothalamic set points that occur in all animals during sleep.

Activities (for answers and explanations, see page 356)

- On the axes provided in the next figure, draw a graph of metabolic rate versus environmental temperature for an endothermic animal and indicate each of the following:

  ✔ upper and lower critical temperatures

  ✔ thermoneutral zone

  ✔ basal metabolic rate

  ✔ summit metabolism

- The graph you drew should show an increase in metabolic rate both above and below the thermoneutral zone. It is easy to understand why an endothermic animal's metabolic rate should rise when environmental temperature falls below the lower critical temperature (i.e., increased metabolism generates more body heat). What physiological phenomenon explains the apparently counterintuitive observation that metabolic rate also rises when environmental temperature exceeds the upper critical limit?

Questions (for answers and explanations, see page 356)

1. One cold winter day your little brother urges you to come see a small animal shivering outside on the front porch. Before arriving, you suspect that the animal you are about to see is
   a. a bird or mammal approaching the upper end of its thermoneutral zone.
   b. a lizard that has exhausted its supply of brown fat.
   c. a bird or mammal whose temperature is below the lower critical limit.
   d. a bird that has exceeded its summit metabolism.
   e. a mammal with a severely damaged hypothalamus.
2. The summit metabolism of an endothermic animal is
   a. its maximum rate of thermoregulatory heat production.
   b. the maximum metabolic rate produced above the upper critical temperature.
   c. equivalent to its basal metabolic rate during hibernation.
   d. the metabolic rate at which its cells function optimally.
   e. a measure of its nonshivering heat production.

3. Evaporative cooling is an effective way to increase heat loss, but carries with it the important physiological drawback(s) of
   a. increasing water uptake, which can lead to bloating.
   b. accelerating the rate of brown fat deposition.
   c. dramatically lowering the hypothalamic thermal set point.
   d. stimulating the release of pyrogens that lead to fever.
   e. increased use of ATP and substantial loss of water.

4. Which of the following statements about hypothalamic function is *not* true?
   a. Circulating blood temperature is monitored by the hypothalamus.
   b. Hypothalamic thermal set points never change.
   c. The hypothalamus is a part of the central nervous system.
   d. Hibernation and torpor involve downward shifts in hypothalamic thermal set points.
   e. Different thermoregulatory responses have different hypothalamic set points.

5. The difference between the concepts of daily torpor and hibernation is that
   a. daily torpor is a rhythmic phenomenon and hibernation is not.
   b. daily torpor involves an upward shift in thermal set points and hibernation a downward shift.
   c. daily torpor occurs only in mammals and hibernation only in birds.
   d. daily torpor lasts for shorter periods than does hibernation.
   e. All of the above

6. Substances called ________ are released when an animal is exposed to certain bacteria or viruses, resulting in an upward shift in the hypothalamic thermal set point that leads to ________ .
   a. pyrogens; fever
   b. pyrogens; summit metabolism
   c. brown fat; fever
   d. thermogenins; basal metabolism
   e. ATP; summit metabolism

7. Which of the following is *not* a morphological adaptation for life in the cold?
   a. Decreased surface-to-volume ratio of body
   b. Shortened appendages
   c. Thicker insulation
   d. Metabolism of brown fat
   e. Waterproof feathers and fur

Integrative Questions (for answers and explanations, see page 356)

1. In which of the following animals might you expect to see some type of behavioral thermoregulation?
   a. Humans
   b. Elephants
   c. Marine iguanas
   d. Birds
   e. All of the above

2. Some animals rely heavily on the metabolism of brown fat to assist in their thermoregulation. Brown fat is a type of ____________ tissue.
   a. epithelial
   b. connective
   c. nervous
   d. muscle
   e. None of the above

3. A small rodent is at rest on a day when the environmental temperature is 30°C. Its body temperature is 37°C, and it is expending no energy for digestion or reproduction. Metabolically speaking, this individual is most likely
   a. in its thermoneutral zone.
   b. exhibiting a fever.
   c. at its summit metabolism.
   d. hibernating.
   e. in shallow torpor.

36

Animal Hormones

CHAPTER LEARNING OBJECTIVES—after studying this chapter you should be able to:

- ❑ Define the term "hormone" and explain the relationship between hormones, target cells, and receptors.
- ❑ Explain the difference between a local and a circulating hormone.
- ❑ Discuss the action of the local hormone histamine and relate this action to the phenomenon of anaphylactic shock.
- ❑ Describe the difference between endocrine and exocrine glands.
- ❑ Discuss the sites of production and the functional importance of neurohormones.
- ❑ Cite the experimental evidence that explains the role of ecdysone and juvenile hormone in the control of insect growth and development.
- ❑ Describe the hormonal differences in growth and development between insects that show simple development and those that exhibit complete metamorphosis.
- ❑ List the major endocrine glands of humans, their primary hormone products, and the main function of each hormone.
- ❑ Discriminate between the anterior and posterior lobes of the pituitary and describe their anatomical connections with the hypothalamus.
- ❑ Explain the role of the portal blood vessels in mediating the interaction of the hypothalamus and pituitary glands.
- ❑ Differentiate between releasing hormones and release-inhibiting hormones.
- ❑ Identify abnormalities in human growth and metabolism resulting from the action of hormones.
- ❑ Explain the negative feedback control mechanism regulating the action of thyroxine and discuss the causes of hypo- and hyperthyroidism.
- ❑ Describe the hormonal feedback control mechanisms for glucose metabolism and blood calcium availability in humans.
- ❑ Identify the major sex steroids for both males and females, where they are produced in the body, and their effects on growth and development.
- ❑ Describe the sequence of hormonal events that triggers the onset of puberty and sexual development.
- ❑ Explain the hormonal and genetic conditions that produce hermaphrodites and pseudohermaphrodites in humans.
- ❑ Discuss important biochemical differences between water-soluble and lipid-soluble hormones.
- ❑ Compare and contrast the cellular modes of action of water-soluble and lipid-soluble hormones.
- ❑ Describe the general features of hormone second messenger systems with special emphasis on the roles of receptor proteins, G-proteins, cAMP, cGMP, adenylate cyclase, protein kinases, and ATP.
- ❑ Explain the phenomenon of signal amplification that results from the cascade of reactions triggered by a hormone.
- ❑ Provide a functional explanation for why lipid-soluble hormones act more slowly and produce a longer-lasting cellular response than water-soluble hormones.

CHEMICAL MESSAGES • HORMONAL CONTROL IN INVERTEBRATES (pages 800–805)

Key Concepts

1. *Endocrinology* is the study of hormones and their actions. *Hormones* are chemical signals that mediate communication between cells in multicellular animals.
2. Hormones are released by cells in one part of the body and have their effects on *target cells* located elsewhere. Only those target cells that possess appropriate chemical *receptors* respond to hormones. This response may be developmental, physiological, or behavioral.
3. Local hormones are released in small quantities and stimulate effects in the immediate vicinity of their release point. They rarely enter the blood supply in significant quantity because of rapid uptake or enzymatic degradation by nearby target cells.
4. *Histamine* is a local hormone that promotes the inflammation associated with immune responses. Histamine makes blood vessels more permeable, allowing protective blood proteins and white blood cells to enter damaged tissues.

5. Local response to histamine is protective, but side effects may occur if the response becomes widespread. Allergy symptoms, such as those of hay fever, are due to excessive histamine activity. In some cases, histamine response is so dramatic that the circulatory and respiratory systems cease functioning, causing life-threatening *anaphylactic shock.*
6. Circulating hormones are transported by the blood and can therefore travel substantial distances from their site of production. The same circulating hormone can have different effects on different target cells depending on the type of receptors present. For example, epinephrine causes heart muscle to beat faster but stimulates metabolic changes in liver tissue.
7. Many local and circulating hormones are released from discrete clusters of secretory cells called *glands*. *Endocrine* glands release their products inside the body. *Exocrine* glands produce nonhormone substances for release outside of the body.
8. Hormones such as epinephrine and those of the digestive tract are not produced by glands; they are made by isolated cells scattered throughout various tissues. Specialized *neurohormones*, secreted by cells in the nervous system, can either act locally or be taken up by the blood to produce effects at a distance.
9. The episodic molting of insect exoskeletons (growth), and the transition from juvenile instars to adult forms (development) are controlled by three hormones, as demonstrated by Wigglesworth's experiments with the bug *Rhodnius*. His experiments showed that insect brain cells produce a *brain hormone*, which is stored in paired head structures called *corpora cardiaca*. After the animal has been stimulated by a blood meal, the corpora cardiaca release their store of brain hormone.
10. Brain hormone triggers the *prothoracic gland* to produce and release a second hormone called *ecdysone*, which is directly responsible for initiating the physiological events associated with molting. Wigglesworth demonstrated the chemical basis of this interaction by noting the effects of connecting decapitated instars of various ages and feeding status (see textbook Figures 36.3 and 36.4).
11. Wigglesworth also showed that other brain structures called *corpora allata* are responsible for determining the development of a *Rhodnius* instar. These structures produce *juvenile hormone*; when present, juvenile hormone leads to the production of another juvenile instar.
12. Normally juvenile hormone production ceases between the fourth and fifth instars in *Rhodnius*, which causes the transition from juvenile to adult forms. However, if early instars have their corpora allata removed, they will immediately molt into adults. By contrast, if older instars are attached to younger ones still possessing active corpora allata, the older animals molt into juvenile instars and not the adult form.
13. Juvenile hormone has a more complex interaction with development in insects that undergo complete metamorphosis, where distinct larval, pupal, and adult forms exist. Larval endocrine cells produce high concentrations of juvenile hormone; thus successive molts produce more larval forms. When larvae reach a certain age, juvenile hormone concentrations fall and the pupal molt follows. Pupae produce no juvenile hormone, so the product of the next molt is the adult form.
14. Juvenile hormone provides a natural mechanism for controlling insect pest species. Exposing larval forms to artificially high concentrations of synthetic juvenile hormone prevents them from maturing into adults, and thus no reproduction occurs.

Activities (for answers and explanations, see page 356)

- Match the following terms with the appropriate description

_____	Histamine	*a.* Produces secretions for use inside the body
_____	Endocrine gland	*b.* Contains receptors specific for only certain hormones
_____	Target cell	*c.* Young insects remain subadults in its presence
_____	Ecdysone	*d.* A local hormone mediating inflammatory responses
_____	Juvenile hormone	*e.* Triggers physiological responses leading to molting

- Based on the results of Wigglesworth's experiments, indicate on the following diagram of *Rhodnius* the relative anatomical positions of the neuroendocrine cells that produce brain hormone, the corpora allata, and the prothoracic gland.

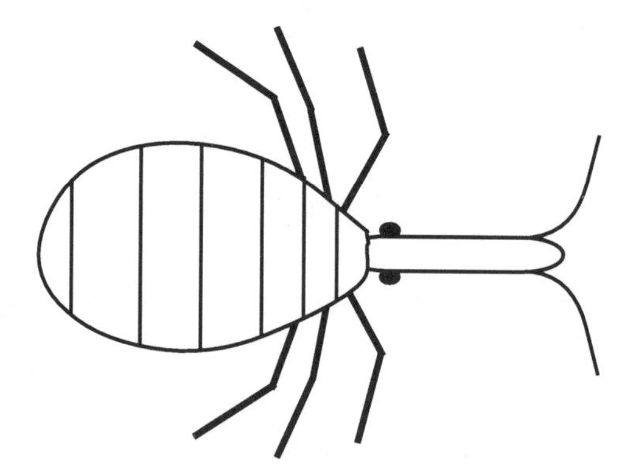

- Examine Figure 36.3 on page 803 in your textbook. Part *a* of this figure shows the results of an experiment in which fourth instar *Rhodnius* larvae are subjected to different types of decapitation one week after receiving a blood meal. Explain the endocrinological basis for these results.

Questions (for answers and explanations, see page 356)

1. The hormone vasopressin is produced by a gland located at the base of the human brain. Its primary effect is to influence water reabsorption by the kidneys. Which of the following accurately describes vasopressin and its action?
   a. It is a local hormone whose concentration is rarely detectable in the blood.
   b. It circulates through the blood until encountering receptors in the brain.
   c. It is a local hormone that acts in much the same way as histamine does on blood vessels.
   d. It is a circulating, endocrine hormone whose receptors are located on target cells in the kidneys.
   e. It is an exocrine hormone produced by target cells in the brain.

2. Histamine is correctly described as a
   a. circulating hormone.
   b. local hormone.
   c. receptor protein found on traget cells in blood vessels.
   d. hormone responsible for the molting of insect exoskeletons.
   e. chemical released by the corpora allata of *Rhodnius*.

3. The same circulating hormone can have different effects on adjacent target cells in an animal's body because
   a. different cells can have different receptor proteins that initiate specific biochemical responses to the hormone.
   b. the hormone is modified by the blood and can therefore influence many cell types.
   c. the extracellular matrix is modified by the hormone, which influences the response of target cells.
   d. histamine is released in response to the hormone, which changes the permeability of blood vessels.
   e. None of the above

4. If the corpora cardiaca of a fourth instar of *Rhodnius* are surgically removed and the bug is then provided with a blood meal, it will
   a. develop into an adult.
   b. molt but remain a juvenile.
   c. not molt and remain a fourth instar.
   d. molt and return to a third instar.
   e. None of the above

5. If a fourth instar of *Rhodnius* is partially decapitated one week after receiving a blood meal, it will develop into a fifth instar instead of an adult. This occurs because
   a. the bug makes no ecdysone.
   b. the source of brain hormone has been removed.
   c. the corpora allata are no longer present.
   d. juvenile hormone concentrations remain high.
   e. None of the above

6. Synthetic forms of juvenile hormone can be used as an effective weapon against insect pests because
   a. juvenile hormone poisons the adult insects.
   b. adults will not mate when juvenile hormone is present.
   c. larvae refuse to feed in the presence of juvenile hormone and die before molting into adults.
   d. juvenile hormone prevents the molting of successive larval instars.
   e. high concentrations of juvenile hormone prevent insect larvae from ever molting into adults.

VERTEBRATE HORMONES (pages 805–817)

Key Concepts

1. Many vertebrate hormones have different effects in different species. For example, in frogs thyroxine is essential for metamorphosis from tadpole to adult, while in mammals it helps regulate metabolism. Prolactin stimulates milk production in female mammals, but in salmon it controls osmotic balance as the fish migrate between salt- and freshwater environments.

2. At least nine distinct endocrine glands are recognized in vertebrates, many of which produce more than one hormone. Nonglandular tissues, such as the lining of the intestinal tract or cells of the nervous system, also secrete hormones.

3. The *pituitary* is the "master gland" of the vertebrate endocrine system, producing or storing nearly a dozen different hormones. It is a blueberry-sized organ located at the base of the skull near the back of the roof of the mouth.

4. The pituitary is divided into two lobes. The *posterior pituitary* derives from an outpocketing of the developing brain, and retains neural connections via the pituitary stalk with nerve cell bodies in the hypothalamus. The *anterior pituitary* is made primarily of epithelial tissue and has endocrine functions.

5. *Vasopressin* (antidiuretic hormone) is a neurohormone released by the posterior pituitary. It is a small peptide manufactured by nerve cells in the hypothalamus and released from the posterior pituitary in response to low blood pressure or high salt content in the blood. Its primary action is to facilitate water reabsorption by the kidneys.

6. *Oxytocin,* the other neurohormone of the posterior pituitary, is also a small peptide. It plays a crucial role in human reproduction by stimulating uterine contractions during childbirth and by promoting the release of milk from the mother's mammary glands.

7. The anterior pituitary produces two classes of protein-based hormones. *Tropic hormones* control the activity of other endocrine glands in the body; they include thyrotropin, adrenocorticotropin, luteinizing hormone, and follicle-stimulating hormone.

8. The second class of anterior pituitary hormones exerts influence on nonendocrine tissues. Chief among these is *growth hormone,* which indirectly stimulates growth by causing the liver to produce growth-regulating factors called *somatomedins* that influence the development of bone and cartilage.

9. Overproduction of growth hormone in children leads to gigantism, while underproduction causes dwarfism. Dwarfism is now routinely treated with human growth hormone produced by genetically engineered bacteria. Recent evidence also suggests that growth hormone may help to reverse the aging process in older adults.

10. *Prolactin* is another anterior pituitary hormone. Depending on the species, it has effects on the production and secretion of milk, the maintenance of pregnancy in females, and the endocrine function of testes in males.
11. *Melanocyte-stimulating hormone* is yet another anterior pituitary product. Its function is obscure in humans, but in fishes, amphibians, and reptiles that are able to change color, it stimulates the redistribution of melanin in pigmented skin cells.
12. Hormones called *endorphins* and *enkephalins* are also produced by the anterior pituitary, and serve as the body's natural painkillers. They are derived by precise enzymatic cleavage of a large parental protein called proopiomelanocortin, which also gives rise to several other anterior pituitary hormones.
13. The anterior pituitary really acts more like an intermediary during hormonal control, rather than a master gland. Neurohormonal *releasing hormones* and *release-inhibiting hormones* are manufactured by the hypothalamus, secreted, and then travel through *portal blood vessels* that connect the hypothalamus with the anterior pituitary. These neurohormones then stimulate or inhibit the release of pituitary hormones into the general circulatory system.
14. The *thyroid gland* is a bilobed structure surrounding the trachea. It produces two important peptide hormones. *Thyroxine* regulates cell metabolism in mammals. It also influences growth and development by enhancing amino acid uptake and protein synthesis.
15. The release of thyroxine is controlled by both the pituitary and hypothalamus. Hypothalamic cells release thyrotropin-releasing hormone, which stimulates the pituitary to release the circulating hormone *thyrotropin*. Thyrotropin causes target cells in the thyroid gland to release thyroxine. Thyroxine feeds back negatively onto the pituitary, making it less sensitive to thyrotropin-releasing hormone and slowing the release of thyrotropin.
16. Malfunction of the thyroid can lead to a condition called goiter in which the gland increases to enormous proportions. This happens either because of *hyperthyroidism* (excess thyroxine production) or *hypothyroidism* (depressed thyroxine production). Both conditions involve abnormalities in the negative feedback control of thyroid activity.
17. Hyperthyroid goiter occurs when the pituitary does not respond to the negative influence of thyroxine. As a result, thyrotropin levels remain high and the thyroid gland becomes overactive. Hypothyroid goiter occurs when circulating thyroxine levels are too low, usually because iodide (an important component of the thyroxine molecule) is lacking in the diet. The gland grows larger through overstimulation because there is little functional thyroxine to inhibit thyrotropin production by the pituitary.
18. The second thyroid hormone is *calcitonin*, which regulates blood calcium levels in mammals. Calcitonin stimulates osteoblast cells, which lay down new bone, and decreases the activity of osteoclast cells, which normally stimulate the breakdown of existing bone. Calcitonin thus acts to reduce the level of circulating calcium in the blood.
19. The *parathyroid glands* are located on the surface of the thyroid. They secrete a single hormone called *parathormone*, which functions in the control of blood calcium by stimulating osteoclasts to dissolve bone and release more calcium. Parathormone also promotes absorption of calcium by the kidneys and digestive tract.
20. The *pancreas* plays an important role in glucose metabolism through the action of two hormones, *insulin* and *glucagon*. Insulin, produced by cells in the pancreatic *islets of Langerhans*, promotes both the use of blood glucose by cells and the storage of glucose as glycogen or fat. Glucagon, made by other cells in the islets, acts to mobilize stored carbohydrates for use when blood glucose levels fall.
21. The disease diabetes mellitus results from a failure of the pancreas to manufacture sufficient quantities of functional insulin. This causes abnormally high blood glucose levels and deprives cells of the fuel they need for metabolism. The condition is treated with insulin replacement therapy.
22. *Somatostatin* is a pancreatic hormone that inhibits the release of insulin and glucagon and slows digestive activity within the gut. This functionally extends the time over which nutrients are absorbed and used by the body. It also serves as a hypothalamic neurohormone that inhibits the release of growth hormone from the anterior pituitary.
23. An adrenal gland rests atop each kidney, and is divided into two regions. The *adrenal medulla* forms the core of this gland. Its primary hormone product is *epinephrine*, which is associated with the fight-or-flight reactions to stress. Epinephrine promotes increased heart and breathing rates, elevates blood pressure, and redirects blood flow to skeletal muscles. The adrenal medulla also manufactures *norepinephrine*, which counteracts the physiological effects of epinephrine.
24. The *adrenal cortex* makes up the outer layer of each adrenal gland. It produces many steroid hormones that fall into three classes: *glucocorticoids*, *mineralocorticoids*, and *sex steroids*. Steroid hormones are modified forms of cholesterol.
25. Glucocorticoids such as cortisol influence blood sugar levels and general metabolism. Cortisol contributes to short-term management of stress by reducing glucose demand in nonessential cells, increasing blood pressure, promoting fat mobilization, and blocking immune system reactions.
26. Cortisol release is regulated by the pituitary hormone *adrenocorticotropin*, which is in turn regulated by its own hypothalamic releasing hormone. Chronic stress, which causes prolonged elevated cortisol levels, does permanent damage to the brain regions involved in the negative feedback regulation of cortisol release.
27. Mineralocorticoids such as aldosterone help regulate salt and ion balance, thus ensuring the proper osmotic conditions necessary for homeostasis.

28. Sex steroids are also manufactured by the adrenal cortex, but their roles in promoting sexual development and reproductive activity are relatively minor compared to the gonadal sex hormones.

29. *Androgens* are the male sex steroids, of which the dominant one is *testosterone*. *Estrogens* and *progesterone* constitute the female steroids, with *estradiol* being the main form of estrogen.

30. The presence or absence of sex hormones determines whether a developing fetus will become male or female. After birth, sex hormones regulate sexual maturation and the development of secondary sexual characteristics.

31. In humans, an XY combination of sex chromosomes causes the undifferentiated gonads of a seven-week old fetus to begin producing androgens, which leads to a male reproductive system. If the fetus is genetically XX, no androgens are made and a female system forms. Thus, in humans the default developmental pattern is female in the absence of androgens.

32. In other vertebrates, where the presence of estrogens initiates female development, the default sex in the absence of appropriate female steroids is male.

33. Abnormalities in sexual development can occur due to errors in hormonal control. *Hermaphrodites* are individuals who possess both male and female gonads, while *pseudohermaphrodites* are those with the external genitalia of one sex and the gonads of another.

34. Sex hormones have their most dramatic developmental effects at the time of puberty. Hypothalamic-releasing hormones cause the pituitary to secrete the gonadotropins *luteinizing hormone* and *follicle-stimulating hormone*. In males these hormones cause the testes to produce androgens, and in females they stimulate the ovaries to begin secreting estrogens and progesterone.

35. These hormonal events lead not only to gonadal maturation, but also produce dramatic changes in secondary sexual characteristics. In males, the voice deepens, skeletal muscle mass increases, the penis enlarges, and facial, pubic, and body hair develops. In females, the breasts, vagina, and uterus all enlarge, the pelvis broadens, there is increased subcutaneous fat deposition, pubic hair develops, and the menstrual cycle commences.

36. There has been a dramatic increase in the use of synthetic androgens (*anabolic steroids*) by young men and women seeking to increase muscle mass for athletic or aesthetic purposes. The presence of these hormones interferes with normal sexual development and also poses serious health risks in the forms of increased rates of cancer, heart disease, kidney damage, sterility, and mental disorders.

Questions (for answers and explanations, see page 356)

1. Which of the following sets of vertebrate hormones are all produced by the anterior pituitary?
a. Somatostatin, vasopressin, insulin
b. Oxytocin, prolactin, adrenocorticotropin
c. Prolactin, growth hormone, enkephalins
d. Thyroxine, calcitonin, thyrotropin
e. Estrogen, glucagon, progesterone

2. A drop in blood temperature would be detected by the ____________, which would chemically stimulate the ____________ to release the hormone ____________, leading eventually to a rise in metabolic rate.
a. hypothalamus; anterior pituitary; thyrotropin
b. posterior pituitary; adrenal glands; adrenaline
c. anterior pituitary; hypothalamus; thyroxine
d. hypothalamus; pancreas; insulin
e. anterior pituitary; posterior pituitary; prolactin

3. If you wanted to incapacitate a male dog's ability to regulate water reabsorption in its kidneys, and at the same time minimize the impairment of other physiological functions, which of the following structures would you need to selectively destroy?
a. Parathyroid glands
b. Anterior pituitary
c. Hypothalamus
d. Posterior pituitary
e. Adrenal glands

4. A blood sample taken from a normal woman 30 minutes after eating a large meal would likely reveal elevated quantities of
a. glucagon.
b. cortisol.
c. insulin.
d. parathormone.
e. aldosterone.

5. A rat is kept in a small cage with many other rats and subjected to mild electric shocks every half-hour for several days. Which of the following hormones would you expect to find in abnormally high concentrations in this animal?
a. Cortisol
b. Testosterone
c. Vasopressin
d. Calcitonin
e. Progesterone

6. A woman conceives a child who is genetically XY and whose body cells are insensitive to female sex hormones. Which of the following best describes this child on its eighteenth birthday?
a. It will be a hermaphrodite.
b. It will be a pseudohermaphrodite.
c. It will be a normal female.
d. It will be a normal male.
e. It will be look like a normal male, but be sterile.

7. At the onset of puberty in both males and females, large amounts of ________________ are released by the ________________.
a. testosterone and estrogen; anterior pituitary
b. aldosterone and cortisol; adrenal glands
c. gonadotropin-releasing hormones; hypothalamus
d. somatostatin and insulin; pancreas
e. oxytocin and vasopressin; posterior pituitary

8. All but one of the following hormones are related by a common chemical characteristic. Select the *exception*.
   a. Epinephrine
   b. Estrogen
   c. Progesterone
   d. Testosterone
   e. Cortisol

Activities (for answers and explanations, see page 357)

- In the following space, diagram the negative feedback control system that regulates the release of thyroxine in humans. Be sure your drawing includes all of the relevant hormones, releasing factors, neural control centers, and endocrine glands involved in this process.

- Based on your understanding of the anatomical relationship between the hypothalamus and the pituitary gland, suggest a mechanism for how a baby's cry can rapidly stimulate a nursing mother to release milk from her breasts.

RECEIVING AND RESPONDING TO HORMONES (pages 817-821)

Key Concepts

1. For hormones to cause a response, they must alter the internal biochemistry of their target cells. This is accomplished in two fundamentally different ways, depending on whether the chemical messenger is a *water-soluble hormone* (peptides and proteins) or a *lipid-soluble hormone* (steroids and thyroxine).
2. Because water-soluble hormones do not readily pass through the membranes of target cells, their receptors are on the cell surface. Each receptor has a binding domain projecting outside the cell membrane, a transmembrane domain anchoring it to the membrane, and a catalytic domain extending into the cytoplasm. When the hormone attaches to the binding domain it either directly or indirectly causes the catalytic domain to activate a *second messenger* molecule. Cyclic adenosine monophosphate, or *cAMP*, is a common second messenger molecule.
3. When the hormone binds to its receptor protein on the cell membrane, the receptor undergoes a conformational change that causes a second membrane protein to become active. This activated G-protein binds guanosine triphosphate (GTP) inside the cell. One subunit of the G-protein then dissociates and activates the enzyme *adenylate cyclase*, which converts ATP into cAMP. Active G-proteins are inactivated when GTP is hydrolyzed to form GDP, ending the hormone's effect on its target cell.
4. G-proteins are diverse. Some can even inactivate adenylate cyclase, producing an inhibitory effect for the hormone. G-proteins also represent important control points in cell metabolism. If the activated G-protein fails to hydrolyze GTP, it will continue to stimulate cAMP production which can lead to excessive metabolic activity that may severely disrupt homeostasis.
5. The cAMP second messenger stimulates intracellular enzymes called protein kinases which are responsible for catalyzing the transfer of phosphates from ATP to specific proteins. cAMP also interacts with phosphoprotein phosphatases, another class of enzymes which remove the phosphates from phosphorylated proteins.
6. The influence of epinephrine on liver cells demonstrates the dramatic effects of second messenger systems. cAMP initiates a chain of events that ultimately results in the inactivation of glycogen synthetase (which converts glucose into glycogen) and the activation of phosphorylase kinase, which in turn activates glycogen phosphorylase (the enzyme that converts glycogen into glucose). Thus, cAMP stimulates events that both shut down glycogen production and stimulate glycogen breakdown when the body's need for glucose is high.
7. Extreme amplification of hormone signals is possible with second messengers. At each step in the biochemical cascade, the number of molecules activated is increased by 10–100 fold. Thus, for each molecule of epinephrine that binds to the cell membrane, a billion glucose molecules are made available for use by the cell.
8. As the hormone concentration is reduced near the cell surface, the chain of events that produced cAMP is disrupted, leading to a decline in the intracellular concentration of cAMP through enzymatic degradation by phosphodiesterases. This removes the inhibition on glycogen formation, which in turn lowers the concentration of free glucose. Moreover, as cAMP levels fall, phosphoprotein phosphatase inhibition is removed and the enzymes controlling glycogen metabolism are dephosphorylated.
9. Other second messenger molecules are used with hormones such as norepinephrine and vasopressin. *Inositol*

trisphosphate (IP3) and *diacylglycerol* (DAG) are second messengers derived from membrane phospholipids, called phosphoinositols, through the action of G-proteins. DAG activates additional protein kinases, while IP3 activates protein kinases and changes membrane permeability to calcium ions.

10. Cyclic guanosine monophosphate, or *cGMP*, is a chemical analog of cAMP which is used in still other second messenger systems. For example, the effects of insulin on certain target cells are mediated by cGMP. In many cases the effects of cGMP are opposite to those of cAMP.
11. Calcium ions also interact with specific binding proteins such as *calmodulin* to mediate the biochemical responses of many cells, including activation of certain protein kinases. They are probably not, however, true second messenger molecules.
12. The lipid-soluble hormones, which include thyroxine and the steroid hormones produced by the gonads and adrenal cortex, do not bind to receptors on the plasma membrane. Instead, they pass through the membrane and enter the cell cytoplasm.
13. Lipid-soluble hormones bind with specific receptor proteins inside the cytoplasm. These proteins then undergo conformational changes which allow the protein–hormone complexes to associate with acidic chromosomal proteins. Once associated, the hormone can influence DNA transcription and protein synthesis.
14. Because they penetrate the cell and directly influence gene expression, the actions of lipid-soluble hormones tend to be slower and longer-lasting than those of water-soluble hormones, which bind to their receptors on the cell surface and initiate short-lived changes in membrane ion permeability or enzyme activity.

Questions (for answers and explanations, see page 357)

1. Which of following statements about hormones and second messenger systems is *not* true?
   a. cAMP is one of the most common second messenger molecules.
   b. Activated G-proteins help to activate the enzyme adenylate cyclase.
   c. Second messenger systems are not employed by most lipid-soluble hormones.
   d. Calcium ions probably do not function as true second messenger molecules.
   e. Thousands of hormone molecules are required to activate a few intracellular enzymes.

2. The cascade of regulatory steps associated with second messenger systems results in
   a. the conversion of cAMP to ATP, which is then used by the cell for energy.
   b. conformational changes in numerous protein kinases within the cell.
   c. a remarkable amplification of the effect of a single hormone molecule.
   d. Both *a* and *b*
   e. Both *b* and *c*

3. The role of phosphodiesterases in the second messenger system of hormone action is to
   a. dramatically increase the yield of cAMP.
   b. facilitate the enzymatic degradation of cAMP.
   c. move cAMP molecules from one location to another within the cell.
   d. convert cAMP into cGMP.
   e. None of the above

4. Lipid-soluble hormones differ from water-soluble hormones in all but one of the following ways. Select the *exception*.
   a. Mechanism of cellular action
   b. Chemical structure
   c. Location of receptor binding site on the target cell
   d. Basic role as a communication signal
   e. Ability to penetrate the cell membrane

5. Which of the following is *not* considered a type of second messenger molecule in the biochemical responses of cells to water-soluble hormones?
   a. DAG
   b. cAMP
   c. IP3
   d. cGMP
   e. ATP

6. Target cells in animals tend to respond faster and for a shorter time to the presence of water-soluble hormones than to lipid-soluble hormones. This is mainly due to differences in the
   a. type of target cells involved.
   b. amount of hormone released.
   c. cellular mechanism of action of the two hormone types.
   d. availability of calcium ions in the cell.
   e. None of the above

Activities (for answers and explanations, see page 357)

- Using the list of molecules given below, generate the proper sequence of events that occurs when a water-soluble hormone stimulates the production of a cAMP-based second messenger signal inside a cell.

____ Inactive G-protein

____ Active adenylate cyclase

____ Hormone

____ cAMP

____ ATP

____ Active G-protein

____ Receptor molecule

____ Protein kinase

____ GTP

____ Phosphoprotein phosphatase

____ Inactive adenylate cyclase

- The following two diagrams (*a* and *b*) show the second messenger mechanism initiated by a water-soluble hormone. Identify the specific substances that represent the first and second messengers in this system, describe the biochemical role of the second messenger, and explain why a second messenger substance is necessary for this particular class of hormones.

a.

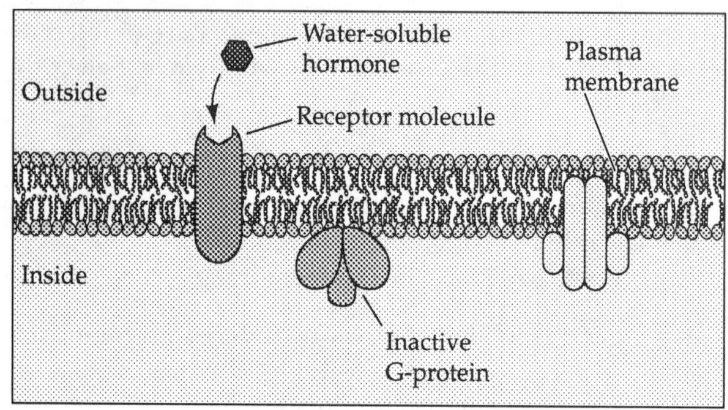

b.

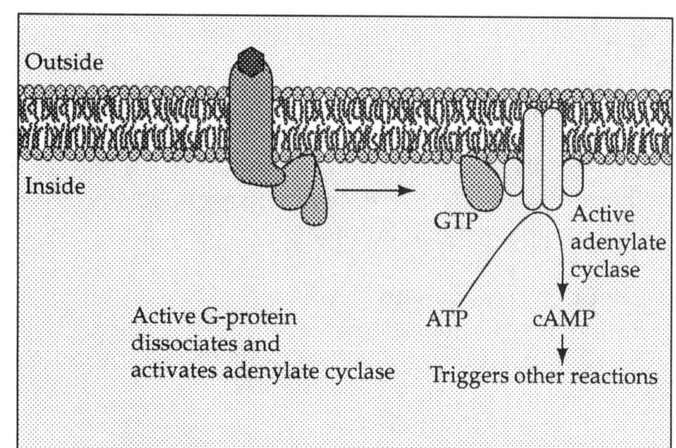

Integrative Questions (for answers and explanations, see page 357)

1. Consider the following vertebrate hormones: oxytocin, adrenocorticotropin, prolactin, insulin, and somatostatin. Which of the following do all five have in common?
   a. They are all released by the posterior pituitary.
   b. They are all released by the anterior pituitary.
   c. They are all neurohormonal releasing factors.
   d. They are all made of amino acids.
   e. They are all sex hormones.

2. A sample of radioactively labeled prolactin is injected into an adult female mouse, who is then analyzed for the location of this hormone in her tissues. This marked sample would most likely be found
   a. attached to the chromosomal proteins of kidney cells.
   b. inside the nucleus of hypothalamic cells.
   c. bound to protein receptors on the plasma membrane of mammary gland cells.
   d. attached to molecules of cAMP within the pancreas.
   e. None of the above

3. Pregnant women require the proper functioning of many important hormones in order to maintain their own health and that of their child. In humans, the hormone __________ is largely responsible for sustaining pregnancy, while ___________ helps protect the calcium content of the mother's bones.
   a. progesterone; calcitonin
   b. testosterone; parathormone
   c. aldosterone; thyroxine
   d. estrogen; cortisol
   e. oxytocin; insulin

4. If you wanted to interfere with the cellular mode of action of the hormone thyroxine, which of the following would be most effective?
   a. Block action of G-proteins
   b. Inhibit hormone receptors in cytoplasm
   c. Inactivate calmodulin
   d. Inactivate adenylate cyclase
   e. Remove all GTP

37

Animal Reproduction

CHAPTER LEARNING OBJECTIVES—after studying this chapter you should be able to:

- ❑ Differentiate between the various forms of asexual and sexual reproduction seen in animals, and explain the important advantages and disadvantages of each.
- ❑ Explain the difference between a germ cell line, gametes, and the other cells of an animal's body.
- ❑ Distinguish between the asexual forms of reproduction known as budding, regeneration, and parthenogenesis.
- ❑ Explain the various cellular stages in the processes of spermatogenesis and oogenesis and determine whether each stage is diploid or haploid.
- ❑ Differentiate among the following terms: "gonads," "testes," "ovaries," "sperm," "eggs," "ova," "zygotes," and "fertilization."
- ❑ Discuss the difference between dioecious and monoecious animals, and explain how these terms apply to simulanteous and sequential hermaphrodites.
- ❑ Describe the advantages and disadvantages of internal and external fertilization, and discuss the various mating strategies animals use to effect internal fertilization.
- ❑ Describe the anatomy and functions of all the major structures in the reproductive systems of human males and females.
- ❑ Differentiate between emission and ejaculation in male mammals.
- ❑ Relate the hormonal changes that take place during the human menstrual cycle to the ovarian and uterine changes that occur in this process.
- ❑ Characterize the differences between a menstrual cycle and an estrous cycle.
- ❑ Describe the four phases of human sexual response and the physiological traits associated with each phase for both men and women.
- ❑ Discuss the cellular events that occur in both sperm and egg during fertilization.
- ❑ Explain the difference between a fast and a slow block to polyspermy.
- ❑ Describe the major forms of human birth control and their relative effectiveness at preventing pregnancy.
- ❑ Distinguish between oviparity, viviparity, and ovoviviparity.
- ❑ Identify the extraembryonic membranes present in the amniotic egg and discuss the function of each.
- ❑ Discuss the importance of the medical techniques known as amniocentesis and chorionic villus sampling.
- ❑ Describe the major events that occur in both mother and baby during each trimester of a human pregnancy.
- ❑ Explain the sequence of events that occur during human birth and describe how specific hormones help to regulate this process.
- ❑ Describe the roles of prolactin and oxytocin in the development and release of milk from a mammalian mother's breasts.

ASEXUAL REPRODUCTION • SEXUAL REPRODUCTIVE SYSTEMS OF ANIMALS (pages 824–829)

Key Concepts

1. Sexual reproduction is the norm for animals. Gametes are produced that contain a precise subset of the parental genome. When two gametes combine, a new individual develops whose genetic makeup is unlike that of either parent. This genetic diversity is then operated upon by natural selection during periods of environmental change.
2. Sperm and eggs are part of an almost continuous line of germ cells whose existence is interrupted only by periodic episodes of meiosis. In this sense, animals are simply devices for the reproduction of their germ cell lines.
3. Many animals also reproduce asexually. Asexual reproduction generates new individuals directly from preexisting ones. These offspring are genetically identical to their parents.
4. Efficiency is a hallmark of asexual reproduction since little energy is required, mates need not be found, and all members of a population are potentially able to have offspring. The only limiting factor is the animal's ability to convert available resources into new offspring.
5. Sponges and cnidarians use a form of asexual reproduction called *budding,* where new individuals arise as mitotic outgrowths of older ones. Buds are usually produced externally, but some sponges produce internal buds called *gemmules* which are released as small, undifferentiated masses of cells and later become free-living individuals.

6. Animal cells capable of asexually producing a complete, new individual are said to be *totipotent*. Totipotency is rare in animals, but some clusters of mature cells can make a whole new organism through *regeneration*. For example, sea stars and some annelid worms can lose a portion of their body and, over time, regenerate the lost part or have the lost part regenerate a new body.
7. A third kind of asexual reproduction is *parthenogenesis*, the development of offspring from unfertilized eggs. Many arthropods and some species of fish, amphibians, and reptiles utlilize this process.
8. Most parthenogenetic species also reproduce sexually. Which mechanism they use depends on environmental conditions. In aphids, parthenogenesis occurs when the habitat is stable and favorable. During times of environmental instability, aphids employ sexual reproduction to promote genetic variability among their offspring.
9. Paradoxically, some parthenogenetic species still require mating. The mating process itself serves only to stimulate egg development, and no fertilization actually takes place. A few parthenogenetic species have no males in their populations; the females mate with males of closely-related species.
10. Sex is the primary mode of animal reproduction. It produces enormous genetic variability among offspring due to the random assortment of alleles into gametes which occurs during meiosis. These genetically unique gametes combine during *fertilization* to form a *zygote*, the genotype of which is also unique. By selecting for offspring whose genetic makeup best adapts them for a particular environment, natural selection operates on the genotypic diversity produced by sexual reproduction.
11. The tiny haploid gametes of males are called *sperm*. They are motile cells powered by a beating flagellum. *Eggs*, or *ova*, are produced by females. They are nonmotile cells with relatively large stores of cytoplasm, organelles, and energy reserves.
12. Sperm and eggs are produced by the *gonads*, or primary sex organs. Male gonads are called *testes*, female gonads are *ovaries*. Accessory sex organs are usually present in the form of various ducts, canals, glands, and structures for the delivery and receipt of gametes. The primary and accessory sex organs combined represent the animal's reproductive system.
13. The diploid *germ cell line* proliferates mitotically from early embryonic tissues, and only later comes to reside within the gonads. *Oogonia* are the germ cells of females and *spermatogonia* those of males.
14. The process of *spermatogenesis* yields new sperm cells. Within the seminiferous tubules of mammalian testes, diploid *primary spermatocytes* undergo the first division of meiosis to produce haploid *secondary spermatocytes*. The second meiotic division results in four haploid *spermatids*. Complete spermatogenesis takes about ten weeks in humans, during which spermatocytes and spermatids are nourished by Sertoli cells lining the seminiferous tubules.
15. *Oogenesis* is the process of gamete formation in females. *Primary oocytes* develop from a subset of oogonia and enter meiosis I, where they arrest during prophase. Each oocyte then grows in size, acquiring all the nutrients and cytoplasmic structures needed for development should fertilization occur.
16. Once a month throughout the reproductive lifespan of human females, at least one primary oocyte resumes meiosis. After meiosis I, one daughter cell retains a majority of the cytoplasmic material and develops into a large *secondary oocyte*; the other becomes the *first polar body*. Meiosis II also divides the secondary oocyte unequally, producing a large, haploid *ootid* as well as a *second polar body*. The polar bodies then degenerate, leaving a well-provisioned ovum ready for the cleavage stages of early development.
17. *Dioecious* animals are those with distinct males and females. *Monoecious*, or *hermaphroditic*, species are those where the same individuals possess both male and female reproductive structures. Hermaphrodites can also result from developmental abnormalities in characteristically dioecious species.
18. *Simultaneous hermaphrodites* are both male and female at the same time. They usually cannot fertilize themselves, and thus require another individual for mating. However, hermaphroditic species that have a low probability of encountering potential mates, such as tapeworms, are able to self-fertilize.
19. *Sequential hermaphrodites* are first one sex, then later in life change to the other sex. This strategy helps prevent inbreeding in situations where siblings live together, or where a dominant male controls all the matings within a group of females, as with some tropical fish.
20. Fertilization can either be external or internal. External fertilization only occurs in aquatic habitats. Adaptations for external fertilization include the production of large numbers of gametes, mechanisms to synchronize mating attempts, and highly coordinated sexual behaviors such as the long migrations of salmon.
21. Internal fertilization is essential for terrestrial animals in order to avoid drying out the sperm, but it is also used by many aquatic species. A major advantage of internal fertilization is protection of the offspring during early development.
22. Special accessory sex organs facilitate *copulation*; the process by which sperm move from the male's reproductive tract into the female's. Most males of internally fertilizing species have a tubular *penis* which is inserted into the female's *vagina* (specialized sperm deposition canal) or *cloaca* (combined urinary, digestive, and reproductive chamber) prior to sperm release.
23. A few animals that use internal fertilization forego direct copulation. In species such as salamanders and scorpions, males deposit a protected *spermatophore* (sperm packet) somewhere in the environment. The female then collects and inserts this packet into her reproductive tract. Male squids and spiders also use spermatophores, but have specially modified tentacles or legs which they use to insert the sperm packet into a female.

24. Even though they deliver sperm in spermatophores, most male insects also possess a tubular penis that fits precisely into the female's reproductive opening. This system may exist because of the prolonged copulatory times required to achieve successful fertilization, or to prevent cross-species fertilization attempts. Insect penises are often equipped with devices that scoop out the spermatophores left by other males, thus increasing the chances that the new male will fertilize the female's eggs.

Activities (for answers and explanations, see page 358)

- For each of the stages in spermatogenesis shown below, indicate whether the nucleus of the cell is diploid ($2n$) or haploid (n).

  _____ Primary spermatocytes

  _____ Spermatids

  _____ Spermatogonia

  _____ Secondary spermatocytes

  _____ Functional sperm

- How is it possible for a sea star to reproduce both regeneratively and by sexual means? Which of these two processes results in "offspring" that are genetically identical to the parent?

Questions (for answers and explanations, see page 358)

1. Asexual reproduction is a good strategy to use in stable environments because
   a. gametogenesis is most efficient under these conditions.
   b. offspring are genetically pre-adapted for this type of environment in that they are identical to their parents.
   c. parthenogenesis can produce a large amount of genetic diversity.
   d. animal cells tend to be more totipotent under stable conditions.
   e. All of the above

2. If you compared the genetic makeup of an animal produced by parthenogenesis with that of its father, which of the following results would you expect?
   a. About a 75% degree of genetic similarity.
   b. 100% genetic similarity, except in cases of highly variable environments.
   c. No genetic similarity at all, except for that due to mutations.
   d. Genetic similarity would vary depending on the species.
   e. Not possible to compare, since parthenogenetic animals have no father.

3. An important difference between a sperm and an egg is
   a. their relative sizes.
   b. the amount of cytoplasm they contain.
   c. whether or not they are motile.
   d. which sex produces them.
   e. All of the above

4. If you were examining the gonads of a squirrel and found Sertoli cells, what would this *unambiguously* tell you about the animal?
   a. Its age
   b. Its sex
   c. How many offspring it has already produced
   d. Whether or not it is hermaphroditic
   e. None of the above

5. Select the set of terms that best describes a male squid.
   a. Monoecious species, asexual, parthenogenetic
   b. Totipotent, regenerative, sequential hermaphrodite
   c. Dioecious species, makes spermatogonia, utilizes a spermatophore
   d. Internal fertilizer, simultaneous hermaphrodite, makes oogonia
   e. Produces diploid gametes, external fertilizer, copulates

6. Which of the following statements about animals that utilize external fertilization is *not* true?
   a. They are divided about equally between terrestrial and aquatic species.
   b. Many produce large numbers of gametes to ensure successful reproduction.
   c. The behaviors associated with mating are often highly synchronized.
   d. The probability of any one egg being fertilized and developing into an adult is low.
   e. None of the above

7. Which of the following statements about animal reproduction is *not* true?
   a. Species that reproduce sexually cannot also reproduce asexually.
   b. Males of a species tend to produce many more gametes than do females.
   c. Male insects can remove spermatophores deposited in a female by other males.
   d. Oogenesis and spermatogenesis both occur in simultaneous hermaphrodites.
   e. The cells that initiate budding in sponges and cnidarians are totipotent.

8. The biggest *disadvantage* of asexual reproduction is
   a. having to find a mate.
   b. the large amount of energy it requires.
   c. the complexity of the developmental process involved.
   d. the lack of genetic diversity produced in the offspring.
   e. None of the above

REPRODUCTIVE SYSTEMS IN HUMANS AND OTHER MAMMALS • FERTILIZATION (pages 830-837)

Key Concepts

1. In most mammals, the male's testes are located in an external pouch of skin called the *scrotum*. Involuntary muscle contractions adjust scrotum size to maintain proper thermal conditions for spermatogenesis.

2. Each testis contains a large volume of tightly coiled *seminiferous tubules* where spermatogenesis takes place. Surrounding the tubules are clusters of cells where male sex hormones are produced. *Sertoli cells* nurture the developing spermatids near the lumen of the tubules. Fully formed sperm move into a storage organ called the *epididymis*, where they mature and become motile.
3. Newly formed spermatids undergo substantial modification before becoming functional sperm. They acquire a flagellum for movement, a midpiece packed with mitochondria to generate the ATP needed for swimming, and a head region containing the genetic material. At the tip of the sperm head is the *acrosome*, which carries enzymes needed to penetrate the egg cell.
4. The *vas deferens* is a tube connecting the epididymis with the *urethra*. The urethra runs the length of the penis and also serves as the exit for urine from the bladder. The penis is largely made of spongy connective tissue. When engorged with blood it becomes erect to facilitate insertion into the female's vagina. Some mammalian species (not humans) also have a bone in the penis.
5. The *glans penis* is an area of thin, highly sensitive skin at the tip of the penis. In humans it is covered by the foreskin, which is often removed at birth by circumcision. The glans penis is responsive to sexual stimulation.
6. The male sex act occurs in two phases. During *emission*, sperm mixes with fluids from the *seminal vesicles* and *prostate gland* to form *semen*, which collects in the urethra at the base of the penis. Seminal fluid contains mucus, proteins, and fructose to fuel the sperms' mitochondria, as well as prostaglandins which stimulate contractions of the female's reproductive tract. Prostate fluid is alkaline to neutralize the acidity of the female's reproductive tract. It also contains clotting enzymes that make the semen a gelatinous mass.
7. *Ejaculation* follows emission and occurs through wave-like muscle contractions that begin at the base of the rigid penis. As they move up the penis, the contractions force semen through the urethra and out of the body.
8. The female gonads (ovaries) are paired structures that release eggs (ova) directly into the lower body cavity. The egg is then collected into one of the paired *oviducts* (fallopian tubes) whose ciliated lining propels the egg into the *uterus*, or womb. This thick-walled, muscular cavity is where the infant develops. An opening at the bottom of the uterus called the *cervix* leads to the *vagina*, which serves as the birth canal as well as a receptacle for the penis and sperm during copulation.
9. The vagina is surrounded by two sets of folded skin, the delicate *labia minora* and the thicker *labia majora*. The anterior ends of the labia minora enclose a small bulb of erectile tissue called the *clitoris* which is the sensory homolog of the glans penis. The clitoris plays an important role in female sexual arousal.
10. Sperm deposited in the vagina swim through the cervix, uterus, and most of the oviduct, where they fertilize the egg. Early embryonic development produces a *blastocyst*, which moves down the oviduct and implants itself in the internal lining of the uterus, or *endometrium*. This initiates development of the *placenta*, which exchanges nutrients and wastes between the mother and her developing baby.
11. Human ovaries contain nearly a million primary oocytes at birth. This number is reduced to about 200,000 by the onset of puberty. During a woman's reproductive lifespan, only about 450 oocytes actually mature into functional eggs. *Menopause*, which occurs near age 50, marks the end of female fertility, as the maturation of oocytes ceases.
12. A developing egg and the layer of ovarian cells surrounding it form a functional unit called the *follicle*. Up to a dozen follicles develop in a woman's ovaries each month during her reproductive lifespan, but usually only one ruptures to release its egg in the process called *ovulation*. After ovulation, the remaining follicle cells form the *corpus luteum*, which remains inside the ovary and functions as an endocrine gland producing estrogen and progesterone.
13. Ovulation is only one of several events in the human female's menstrual cycle, so-named for the conspicuous sloughing off of the endometrial lining that leads to *menstruation*. The menstrual cycle really consists of two coordinated cycles: one in the ovary which prepares an egg for release, the other in the uterus which prepares it for pregnancy.
14. The reproductive cycles of female mammals vary by species; the human cycle averages 28 days from one menstruation event to the next. Unlike humans, most female mammals reabsorb their endometrial lining rather than sloughing it off through menstruation.
15. *Estrous cycles* occur in mammals that do not menstruate. A striking behavior in these species is the increased sexual receptivity of females around the time of ovulation, referred to as estrus, or "heat." In humans, sexual receptivity is not physiologically correlated with stages of the reproductive cycle or season (i.e., there is no estrus).
16. Several hormones released by the hypothalamus and anterior pituitary play critical roles in coordinating the timing of the human menstrual cycle. The hypothalamus increases its output of gonadotropin-releasing hormone at the time of puberty in females, which leads to increased outputs of follicle-stimulating hormone (FSH) and luteinizing hormone (LH) by the anterior pituitary. Ovarian development follows, and the rapid rise in estrogen and progesterone secretion which results initiates the cycles of follicular growth and uterine development that continue monthly until the onset of menopause.
17. The ovarian and uterine cycles commence at the onset of menstruation. Just prior to this event, the anterior pituitary secretes increased levels of FSH and LH, which causes the development of new follicles within the ovaries and the production of estrogen. Eventually one of these follicles matures and the others wither away.
18. The mature follicle begins producing larger amounts of estrogen, which causes the endometrium to grow. Increased estrogen levels negatively feed back on the anterior pituitary to reduce gonadotropin release. About

day 12 of the cycle this negative feedback is reversed and a surge of LH is released by the pituitary, triggering ovulation.

19. The corpus luteum, which develops from the remaining follicle cells in the ovary, continues to secrete estrogen and rapidly increases its production of progesterone to help maintain the endometrial lining. Both hormones have negative effects on pituitary gonadotropin release, effectively shutting down the production of new follicles.

20. If the egg is not fertilized, the corpus luteum degenerates by about day 26 of the cycle, dramatically reducing estrogen and progesterone concentrations. This initiates the breakdown of the endometrial lining, leading to the onset of menstruation. It also removes the inhibition on pituitary gonadotropins, which allows new follicles to start developing.

21. If fertilization does occur, the developing blastocyst begins secreting a new hormone called *human chorionic gonadotropin*. This hormone maintains the corpus luteum in an active state of steroid production so that the newly implanted embryo can survive until a functional placenta has formed. The placenta then takes over the production of progesterone, which maintains the uterine lining and prevents the pituitary from secreting additional gonadotropins, thus shutting down the ovarian cycle for the duration of pregnancy.

22. Human sexual responses are basically similar in men and women. They consist of four distinct phases: excitement, plateau, orgasm, and resolution.

23. Excitement is marked by elevated heart and breathing rates, increased blood pressure, and enhanced muscle tension. In women the clitoris swells and becomes more sensitive, and the vaginal walls secrete lubricating fluids. In men the penis fills with blood and becomes erect.

24. The plateau phase in women shows further increases in heart and breathing rates. The clitoris now retracts, and the sensitivity once focused there spreads across the external genitals. Men also show increased breathing and heart rates, the diameter of the glans increases, testes swell as the scrotum tightens, and small amounts of lubricating fluids are released from the penis.

25. Orgasm in women begins with several long contractions (2-4 seconds each) in the outer third of the vagina, followed by a rapid series of shorter ones. Unlike men, women may experience multiple orgasms in rapid succession which can last for several minutes. A man's orgasm is stimulated by frictional pressure on the shaft and glans of the penis. This stimulates massive contractions of muscles in the genital area that expel the semen.

26. During resolution, blood flow and breathing rates return to normal in both sexes. This takes from 5–10 minutes if a woman has had an orgasm, but may take 30 minutes or more if she has not. Resolution in men is more immediate, marked by rapid shrinking in penis size within minutes after ejaculation. Unlike women, men experience a refractory period of 20 minutes or more during which they are physiologically incapable of achieving a full erection or ejaculation regardless of the intensity of sexual stimulation.

27. Fertilization involves a complex series of events that produce a diploid zygote and initiate embryonic development. The sequence begins with the juxtaposition of sperm and egg. Like all other cells, the mammalian egg is bounded by a plasma membrane. Immediately surrounding this membrane is a glycoprotein layer called the *zona pellucida*, which is covered by follicle cells within a jellylike matrix collectively known as the *cumulus*.

28. When sperm initially enter the female's reproductive tract they are incapable of penetrating the protective layers surrounding an egg. Sperm activation occurs inside the uterus through a process called *capacitation*. The response of capacitated sperm when they encounter an egg in the oviduct is mediated by the *acrosomal reaction*.

29. The details of acrosomal reactions differ among species, but all begin with the breaking down of the membranes surrounding the acrosome, thereby freeing digestive enzymes. *Hyaluronidase* and other acrosomal enzymes help disperse the cells of the cumulus, improving the chances of sperm penetration.

30. Once inside the cumulus, linkage reactions occur between enzymes on the sperm head and substrate molecules in the zona pellucida. Additional enzymes from the acrosome digest their way through this layer, and the sperm eventually fuse with the egg plasma membrane. Several important events occur at this point. First, *blocks to polyspermy* are initiated which prevent more than one sperm nucleus from entering the egg.

31. The fast block to polyspermy happens within a few seconds of egg penetration and involves the rapid uptake of sodium ions. This changes the electrical potential across the egg's plasma membrane and prevents the entry of additional sperm.

32. A slow block to polyspermy also occurs over a period of 20–30 seconds. In sea urchins, where a *vitelline envelope* replaces the zona pellucida of mammals, cortical vesicles residing just below the plasma membrane are stimulated to release their enzymes when sequestered calcium ions flood the egg cytoplasm following sperm penetration. This causes the bond between the plasma membrane and the vitelline envelope to break. Water flows by osmosis into this newly created space, raising the vitelline envelope and forming the *fertilization membrane*, further inhibiting sperm entry.

33. Postfertilization release of calcium within the egg also triggers increased metabolic activity. Cytoplasmic pH changes, oxygen consumption, and protein synthesis increase. The egg nucleus then completes its second meiotic division before fusing with the sperm nucleus.

34. The simplest human birth control techniques are *coitus interruptus* (where the penis is withdrawn before ejaculation occurs) and the *rhythm method* (avoiding sex around the time of ovulation), but failure rates are high for both. Most other forms of birth control interfere with the events leading to fertilization and have higher success rates. Forms of *sterilization* such as vasectomy or tubal ligation are highly effective, but are usually irreversible. Slightly less effective and reversible ways to prevent the union of sperm and egg include mechanical barriers such as *condoms*, *diaphragms*, and *cervical caps*, along with the use of chemicals such as contraceptive foams and jellies.

35. Methods for preventing the implantation of a fertilized egg are also used in birth control. The intrauterine device (*IUD*) interferes with implantation. The drug *Ru486* achieves the same contragestational effect through chemical means.

36. Another highly effective method of contraception is the *birth control pill*, which uses hormonal manipulations of the ovarian cycle to prevent maturation and release of eggs. The pill does, however, carry a risk of significant medical side effects for some women.

Questions (for answers and explanations, see page 358)

1. Mature mammalian sperm are stored in the _________ prior to their release during ejaculation.
   a. seminiferous tubules
   b. scrotum
   c. vas deferens
   d. epididymis
   e. testes

2. The function of the prostate gland in human males is to:
   a. initiate the muscular contractions leading to emission.
   b. initiate the muscular contractions leading to ejaculation.
   c. produce a fructose solution used to fuel the mitochondria in the sperm's midpiece.
   d. swell with blood and cause the penis to become erect.
   e. secrete alkaline fluids that neutralize the acidity of the female's reproductive tract.

3. Which of the following best represents the normal path of a sperm cell as it makes its way from the point of entry into the female's reproductive tract to the place where fertilization typically occurs?
   a. Cervix, vagina, ovary, oviduct
   b. Vagina, cervix, uterus, oviduct
   c. Uterus, oviduct, vagina, cervix
   d. Ovary, oviduct, cervix, vagina
   e. Vagina, uterus, cervix, oviduct

4. For approximately how long during the human female's menstrual cycle are progesterone concentrations high enough to maintain the uterus in a proper condition for pregnancy?
   a. All of the cycle
   b. None of the cycle
   c. Less than one-half of the cycle
   d. More than one-half of the cycle
   e. Two days either side of midcycle

5. If you needed to take a blood sample during the time in a woman's menstrual cycle when the concentrations of her gonadotrophic hormones would be at their *lowest* levels, which of the following days would be the best choice for sampling?
   a. 1–5
   b. 5–10
   c. 10–15
   d. 15–20
   e. 20–25

6. A couple have been trying to conceive a child for over a year without success. Recently, the man's doctor informed him that his seminal vesicles are defective. This means that
   a. the man makes functional sperm, but they are not delivered because he is unable to ejaculate.
   b. the man's sperm lack functional flagella.
   c. the fluid needed to lubricate the man's urethra for easier sperm travel is defective.
   d. the man's semen lacks the fructose fuel needed for the sperm to swim.
   e. the acidity in the woman's reproductive tract kills the sperm before they can reach the egg.

7. Which of the following contraceptive methods prevents the implantation of a fertilized egg?
   a. IUD
   b. Diaphragm
   c. Birth control pill
   d. Condom
   e. Vaginal jelly

8. Which of the following statements about human sexual response is *not* true?
   a. Increased breathing rate is common during the excitement phase of both men and women.
   b. Like women, men often experience multiple orgasms in rapid succession.
   c. The plateau phase in women is marked by retraction of the clitoris.
   d. A man's penis engorges with blood and becomes rigid during the excitement phase.
   e. The resolution phase is typically shorter in women who have achieved orgasm.

Activities (for answers and explanations, see page 358)

- Where applicable, indicate the structure in the reproductive systems of both male and female mammals that corresponds with each of the following.

  1. Site of gametogenesis:

     Male _______________ Female _______________

  2. Primary source of sex steroids:

     Male _______________ Female _______________

  3. Main organ of sexual sensory stimulation:

     Male _______________ Female _______________

  4. Storage organ for mature gametes:

     Male _______________ Female _______________

  5. Organ where most fetal development occurs:

     Male _______________ Female _______________

- The following graph depicts the average blood concentrations of three circulating hormones collected from 100 healthy adult women. Each hormone plays a role in regulating some aspect of the menstrual cycle. Identify the most likely hormone that each curve represents, and explain the criteria used in making your choices.

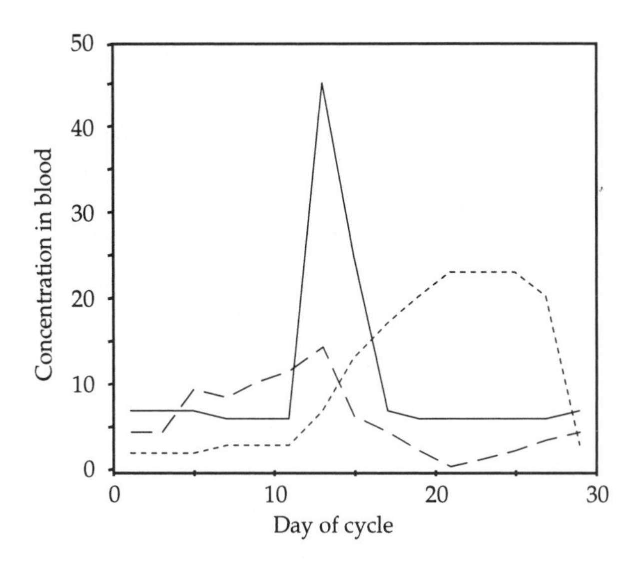

CARE AND NURTURE OF THE EMBRYO • BIRTH (pages 837–846)

Key Concepts

1. Developing embryos require oxygen and other essential nutrients, metabolic waste removal, and a suitable physical environment. Animals use two basic strategies to care for their developing embryos, oviparity and viviparity.
2. *Oviparous* animals lay external eggs that develop independently of the female parent. Terrestrial oviparous animals such as insects, reptiles, birds, and a few mammals coat their eggs with tough external shells for protection from desiccation and predators. They may also provide parental care in the form of nests or guarding behavior. After hatching, the offspring may receive additional care until independence is achieved.
3. *Viviparous* animals retain the embryo within the mother's body for some portion of its development. Most viviparous animals are mammals. Marsupial mammals give birth to very underdeveloped offspring, which then complete their development in an external pouch located on the mother's abdomen.
4. *Eutherian mammals* complete their embryonic growth inside the mother's uterus and are born in a relatively advanced state. A distinguishing feature in the uterine development of these animals is the presence of a placenta, which connects the fetal and maternal blood supplies.
5. Some species of fish, amphibians, and reptiles are *ovoviviparous*. They produce shelled eggs, but instead of laying them externally, the eggs are retained inside the female's body until the young hatch.
6. A set of extraembryonic membranes surrounds the developing embryos of reptiles, birds, and mammals. These membranes help regulate the exchange of nutrients and wastes between the embryo and its environment.
7. In bird eggs, the *yolk sac* is the first extraembryonic membrane to form. It surrounds the large yolk mass and develops from embryonic cells that eventually become the digestive system.
8. The *amnion* is an extraembryonic membrane that surrounds the entire embryo, encasing it in a fluid-filled chamber that physically cushions the developing animal and buffers it from environmental changes.
9. The *allantois* forms a cavity for the deposition of metabolic wastes. The *chorion*, which is the outermost membrane, forms the inner lining of the egg shell.
10. All four membranes are integral components of the *amniotic egg*, a major adaptation for terrestrial life in the vertebrate lineage. The development of the amniotic egg was one of the most significant events in the evolutionary transition from amphibians to reptiles because it freed terrestrial animals from the need for aquatic reproduction. Birds have improved upon the basic design of this egg by calcifying the external shell.
11. Although viviparous, mammals also possess extraembryonic membranes. The first one to develop is the chorion, which facilitates implantation of the blastocyst into the endometrial lining and stimulates development of the placenta.
12. The amnion completely surrounds the mammalian embryo, providing a fluid-filled chamber inside which the developing animal floats. A portion of the allantois forms an *umbilical cord* connecting the embryo to the placenta, its source of nutrients and waste removal. The yolk sac is a rudimentary structure in mammals which degenerates early in development.
13. A useful medical technique for assessing the genetic makeup of a developing embryo is to sample sloughed-off cells floating free in the amniotic fluid. *Amniocentesis* allows doctors to examine the chromosomes of an unborn child and advise its parents of any genetic defects that may be present.
14. A newer technique called *chorionic villus sampling*, which takes embryonic cells from the outer surface of the chorionic membrane, is a safer and much faster way of obtaining the same information as amniocentesis. If parents should choose to terminate the pregnancy, CVS information allows the abortion to be performed earlier, which presents a much lower health risk to the mother.
15. *Gestation*, or pregnancy, represents the time from fertilization to birth. Its length generally correlates positively with body size in mammals. Human gestation is 266 days, or about nine months, and is divided into three, roughly 90-day trimesters.
16. The first trimester begins with fertilization and is initially characterized by a series of rapid embryonic cell divisions called cleavage, all of which occur prior to implantation. *Organogenesis* follows implantation and is the main event of the first trimester. At this time the limbs and all major organ systems develop.

17. By the eighth week of pregnancy the embryo resembles a miniature human and is called a *fetus*. Because of the high rate of cell division at this stage, the fetus is highly susceptible to drug use or nutritional deficiencies in the mother's diet. At this time a multitude of hormonal changes occur in the mother's body which cause major physiological shifts, producing the classic "symptoms" of pregnancy such as morning sickness, mood swings, and unusual food cravings.
18. The second trimester is a period of rapid fetal growth, along with the first signs of fetal movement. As estrogen and progesterone secretions increase in the placenta, chorionic gonadotropin levels drop and the corpus luteum degenerates, taking with it many of the unpleasant symptoms of pregnancy.
19. Fetal growth continues throughout the third trimester. Final maturation of the fetus' internal organs occurs in preparation for birth. Important among these are the complete functioning of its digestive system, glycogen storage in the liver, urine production by the kidneys, and regular sleep/wake cycles in the brain.
20. Unlike the slow, rhythmic Braxton–Hicks contractions (or false labor) that occur throughout pregnancy, *parturition*, or childbirth, begins with the onset of major contractions in the uterine muscles, otherwise known as true labor. The mucous plug which has been covering the cervix is shed, and the amnion ruptures.
21. Labor is influenced by numerous physiological factors, including mechanical stimulation of the uterine musculature, hormonal shifts leading to an abundance of estrogen, and the release of oxytocin from the posterior pituitary. Positive feedback loops stimulate and intensify labor contractions, which often last up to 15 hours in first-time pregnancies.
22. Delivery is the second stage of labor. It commences when the cervix is fully dilated and the baby's head passes through the vaginal opening. As soon as it clears the birth canal, the infant begins breathing and its umbilical cord is clamped and cut. Expulsion of the placenta and remaining portions of the umbilical cord follows.
23. *Caesarian sections*, in which the baby is removed from the uterus by abdominal surgery, are performed when any of a number of complications arise during childbirth. Situations that may warrant a caesarian section include an excessively large child, small pelvis size in the mother, prolonged labor, failure of the cervix to dilate, or threats to the health of either mother or child.
24. All mammalian mothers nurse their babies with milk from mammary glands. Breast development is stimulated throughout pregnancy by estrogen and progesterone, but milk production is inhibited. A few days prior to birth the female begins secreting *colostrum*, a low-fat fluid precursor of milk.
25. The levels of estrogen and progesterone decrease dramatically after birth, allowing prolactin (which had previously been inhibited by the presence of sex steroids) to stimulate milk production. Oxytocin stimulates the release of milk from the mammary glands through a neural reflex initiated by the infant's suckling.

***Activities** (for answers and explanations, see page 359)*

- Label the following diagram of a generalized amniotic egg. Why was the development of this structure so important to the evolution of terrestrial vertebrates?

- Describe the difference between the two main forms of embryo development that occur in animals, oviparity and viviparity. How does ovoviviparity differ from these two patterns?

***Questions** (for answers and explanations, see page 359)*

1. During its development, the human fetus is contained within a fluid-filled chamber bounded by the
   a. yolk sac.
   b. endometrium.
   c. amnion.
   d. chorion.
   e. allantois.

2. Which of the following statements about human pregnancy and fetal development is *not* true?
   a. The blastocyst is a stage in fetal development that appears early during the third trimester.
   b. A vital connection between the fetal and maternal blood supplies occurs at the placenta.
   c. Most of the growth in fetal size occurs during the second and third trimesters.
   d. All major organ systems have formed by the end of the fetus' first three months of life.
   e. The corpus luteum degenerates sometime during the second trimester.

3. Which of the following is an example of a eutherian, viviparous mammal?
   a. Kangaroo
   b. Duck-billed platypus
   c. Spiny anteater
   d. Cow
   e. None of the above

4. Human organogenesis occurs
   a. entirely during the last trimester of pregnancy.
   b. after implantation and during the first trimester of pregnancy.
   c. shortly before the onset of true labor.
   d. due to the stimulation generated during Braxton–Hicks contractions.
   e. under the influence of oxytocin and prolactin.

5. An important factor contributing to the onset and maintenance of labor contractions is
   a. increased contractility of the uterine muscles from increased oxytocin levels.
   b. mechanical pressure on the cervix generated by the emerging baby's head.
   c. stretching of the uterus during the final stages of fetal growth.
   d. increased contractility of the uterine muscles due to elevated estrogen levels.
   e. All of the above

6. The extraembryonic membranes known as the chorion and allantois are found in all but one of the following animals. Select the *exception*.
   a. Trout
   b. Bear
   c. Chicken
   d. Alligator
   e. Human

Integrative Questions (for answers and explanations, ***see page 359)***

1. Which of the following animals qualifies as a sexually reproducing, oviparous species?
   a. Human
   b. Chicken
   c. Kangaroo
   d. Sea star
   e. None of the above

2. If you examine the reproductive system of a normal human female before she undergoes puberty, which of the following would you *not* expect to find?
   a. Corpus luteum
   b. Ovaries
   c. Primary oocytes
   d. Oviducts
   e. Estrogen

3. In sea urchins, all of the following are somehow involved in the events associated with fertilization, *except* for
   a. acrosomal enzymes.
   b. calcium ions.
   c. the vitelline envelope.
   d. the zona pellucida.
   e. sperm-binding receptors.

4. Keeping a normal, sexually-mature woman in her mid-twenties on continuous high doses of progesterone will
   a. cause her to ovulate repeatedly within each menstrual cycle.
   b. stimulate the production of mature follicles.
   c. prevent her from producing mature follicles and ovulating.
   d. cause the endometrial lining of her uterus to slough off.
   e. stimulate milk production in her mammary glands.

38

Neurons and the Nervous System

CHAPTER LEARNING OBJECTIVES—after studying this chapter you should be able to:

- ❑ Distinguish between the terms "sensor," "effector," "central nervous system," and "peripheral nervous system."
- ❑ Explain in a general way how the complexity of nervous systems varies among the different phylogenetic groups of animals.
- ❑ Describe the structure of a typical neuron, and explain the roles of synapses and neurotransmitters in neuron function.
- ❑ Discuss the function of glial cells in nervous systems and explain the role of astrocytes in creating the blood–brain barrier.
- ❑ Explain the difference between a nerve and a neuron, and describe the function of nerves.
- ❑ Differentiate between afferent and efferent divisions of the peripheral nervous system, and explain their actions with respect to the control of voluntary and involuntary (autonomic) functions.
- ❑ Identify the structural and functional subdivisions of the human brain and describe the relative importance of these subdivisions in nonhuman vertebrates.
- ❑ Describe the structure and function of the human spinal column and its associations with afferent neurons, efferent neurons, and interneurons.
- ❑ Discuss the roles of the reticular and limbic systems in vertebrate behavior.
- ❑ Explain the anatomy of the human cerebral cortex and discuss the functional subdivisions (lobes) within this structure.
- ❑ Distinguish between a resting membrane potential and an action potential in neurons and explain the ionic basis for how each is produced.
- ❑ Identify and describe the roles of the various ion channels and pumps involved in neuron function and discuss the ionic basis for a neuron's refractory period.
- ❑ Explain the difference between depolarization and hyperpolarization in neurons and how these ideas relate to EPSPs and IPSPs.
- ❑ Discuss the mechanism for how action potentials are propagated in both myelinated and unmyelinated axons.
- ❑ Describe some of the basic techniques used to study neuron function.
- ❑ Discuss the structure and function of both a neuromuscular junction and a neuron-to-neuron chemical synapse.
- ❑ Identify common vertebrate neurotransmitters that operate in the peripheral and central nervous systems and explain their functions.
- ❑ Describe the mechanisms known for how neurotransmitters are removed from chemical synapses and explain the problems that can occur if these mechanisms fail.
- ❑ Distinguish between the sympathetic and parasympathetic nervous systems and explain their functions.
- ❑ Distinguish between monosynaptic and polysynaptic reflex loops and explain how they operate.
- ❑ Discuss the various stages of sleeping and dreaming, the neuromuscular events associated with them, and how they are studied experimentally.
- ❑ Explain the difference between simple forms of learning such as habituation and sensitization, and the more complex forms associated with higher brain functions.
- ❑ Differentiate between immediate, short-term, and long-term memory and explain the role of the hippocampus in mediating these phenomena.
- ❑ Discuss the influence of lateralization in the human brain and the role of the corpus callosum in regulating the flow of information between the cerebral hemispheres.
- ❑ Explain the neural disability known as aphasia and indicate how this condition provides information about higher brain functions.

COMMUNICATION AND COMPLEXITY • CELLS OF THE NERVOUS SYSTEM • FROM STIMULUS TO RESPONSE: INFORMATION FLOW IN THE NERVOUS SYSTEM (pages 849-858)

Key Concepts

1. In animals that have one, the brain is the coordinating center for all organ and muscle functions. Some of these functions are under conscious, or voluntary, control while others are controlled by involuntary mechanisms.
2. *Sensors* (e.g., eyes and ears) convert stimuli into messages that the nervous system can use to cause behavioral or physiological responses by way of *effectors* (e.g., muscles and glands). In humans and other vertebrates, the brain and spinal cord make up the *central nervous system*. It connects with the *peripheral nervous system*, which transfers neural information to and from the central system.
3. Information transfer in nervous systems is both electrical and chemical. Electrical signals are used to transfer information within nerve cells, chemical signals are used to communicate between cells.
4. The structure and complexity of nervous systems varies with species. In general, larger and more complex nervous systems permit more sophisticated functional control. However, small animals with "simple" nervous systems are often capable of quite remarkable control and behavior. Much of our knowledge about nerve cell function comes from studies of these simple animals.
5. *Neurons* are the functional units of nervous systems. These cells have electrically excitable plasma membranes that generate nerve impulses called *action potentials*. Neurons typically consist of four regions: a cell body, dendrites, an axon, and axon terminals.
6. The *cell body* of a neuron contains the nucleus and most of the cell's organelles. A number of long extensions, or processes, emerge from the cell body.
7. *Dendrites* are the most abundant neuronal processes. They are usually fairly short and collect incoming signals from other neurons.
8. *Axons* are typically elongated processes that connect the cell body with its target cell(s). There is usually only one axon per neuron, and its length may range from less than a millimeter to many meters, depending on function and species. Axons can conduct their electrical signals at speeds of many meters per second, making the nervous system a much more rapid communication system than the hormonal system.
9. *Axon terminals* are swellings found at the end of tiny processes that branch off the ends of each axon. They are storage sites for special molecules called *neurotransmitters* which are used to transfer information to the target cell.
10. Chemical communication occurs at *synapses* between the membranes of two cells. A 25 nanometer intercellular gap, or synaptic cleft, exists between the axon terminal and the adjacent cell membrane. Neurotransmitter is released from the presynaptic membrane. After diffusing across the cleft, the neurotransmitter binds to receptors on the target cell and initiates new electrical events. By summing these synaptic events, which occur thousands of times per second between neurons throughout the body, nervous systems are able to rapidly integrate large quantities of information.
11. *Glial cells* differ from neurons by lacking the ability to generate and conduct electrical impulses (although some do communicate electrically through specialized contacts called *gap junctions*). Glia serve a number of accessory roles for neurons, including physical support and spatial orientation, electrical insulation, nutrient supply, maintenance of proper ionic balance, and defense against foreign particles or cell debris.
12. *Astrocytes* are specialized glial cells that create the *blood–brain barrier*. The smallest blood vessels supplying the brain are covered by astrocytes, which provide protection against the leakage of toxic substances from the blood that might cause cellular damage to sensitive brain cells.
13. Since the number of brain cells is fixed at birth and damaged ones cannot be regenerated, the blood–brain barrier helps reduce the natural death rate of these cells. The system is not infallible, however, because lipid-soluble toxic substances such as anesthetics and alcohol can penetrate the blood–brain barrier and produce a number of abnormal effects.
14. Even the simplest animal nervous systems have many tens of neurons and hundreds of synapses. In the complex nervous systems of vertebrates, there are billions of neurons and millions of billions of synapses. These multitudes of cells are grouped into well-defined regions that control specific body functions.
15. *Nerves* carry information to and from various regions within the body. A nerve is a bundle of axons which can simultaneously convey many kinds of information to and from different locations.
16. The peripheral nervous system is functionally divided into two components. The *efferent* side carries information from the central nervous system to effector organs in the periphery. The *afferent* side relays information from sensors located in the periphery to the central nervous system.
17. Each of the peripheral subsystems is further subdivided into two components. The afferent system has one division that carries conscious signals from the eyes, ears, and other peripheral sensors, while the unconscious division relays information from internal organs and metabolic processes. The efferent system likewise has a voluntary component under conscious control, and an autonomic component that controls involuntary functions.
18. The first evidence of a developing nervous system in vertebrates is the embryonic neural tube. The anterior end of this tube subdivides into three regions: the forebrain, midbrain, and hindbrain. The remaining part of the neural tube forms the spinal cord, from which the cranial and spinal nerves emerge.

19. The hindbrain gives rise to structures that maintain basic physiological functions. The *medulla* lies just below the *pons*; both control essential functions such as breathing and circulation. The *cerebellum* attaches dorsally to the medulla and pons; it coordinates and refines behavior patterns.
20. The midbrain develops above the hindbrain along the neural tube and becomes a processing center for auditory and visual information.
21. The forebrain is the superiormost structure of the brain, resting atop the midbrain region. The lower portion of the forebrain is called the *diencephalon*. It contains an important sensory relay station called the *thalamus* and one of the body's main regulatory centers of physiological function, the *hypothalamus*.
22. The diencephalon is surrounded by two *cerebral hemispheres* (collectively the *cerebrum*) that make up the *telencephalon*. The size of the telencephalon increases as one follows the evolutionary path from simple vertebrates to birds and mammals. It is the largest of all the brain structures in higher vertebrates, and is responsible for controlling sensory perception, learning, memory, and conscious behavior.
23. Primitive autonomic functions are localized either in the *brain stem* (combination of the medulla, pons, midbrain, and diencephalon) or even lower down the spinal cord. More complex control functions reside in the telencephalon. As a result, animals high on the phylogenetic scale (e.g., mammals) rely heavily on telencephalic functions for their behavior, while those on the lower end (e.g., fishes) function with little or no telencephalic control.
24. The spinal cord is a conduit for information between the periphery and the brain, integrating much of this information along the way. In cross section, the spinal cord shows a butterfly-shaped gray matter, which contains neuron cell bodies, surrounded by white matter made of axons that run up and down the cord.
25. The gray matter is divided into *dorsal* and *ventral horns* which are connected to the spinal nerves by way of the dorsal and ventral roots, respectively. Each spinal nerve carries both incoming and outgoing information, with afferent signals entering through the dorsal roots and efferent signals exiting via the ventral roots.
26. *Interneurons* connect afferent axons with the cell bodies of efferent neurons. They also relay information up and down the spinal cord between the brain and those cell bodies whose axons exit through spinal nerves. A large amount of information processing and integration goes on in the cord due to the activity of interneurons.
27. The *reticular system* (also called the reticular activating system) runs through the center of the medulla, pons, and midbrain. It mainly controls arousal state, with high levels of activity during wakefulness and low activity during sleep or coma.
28. The *limbic system* is the primitive portion of the forebrain that controls basic physiological drives, instincts, and emotions such as pain, pleasure, or rage. It constitutes the entire telencephalon of fishes, amphibians, and reptiles. In birds and mammals it still plays an important regulatory role despite the elaboration of other telencephalic regions. In humans, the *hippocampus* is the portion of the limbic system that mediates the transfer of short-term memory to long-term memory.
29. The cerebral hemispheres dominate the brains of higher vertebrates, and in humans cover all other brain regions except for the cerebellum. The thin, highly folded outer *cortex* forms a sheet of gray matter containing 80 percent of all the neuronal cell bodies found in the human nervous system.
30. Regions, or lobes, within the cerebral cortex are dedicated to specific control functions. The *temporal* lobe processes auditory information and controls language. The *occipital* lobe processes visual information.
31. The *parietal* and *frontal* lobes are separated by a deep cleft called the *central sulcus*. A portion of the frontal cortex sends motor commands to the voluntary muscles. Each body region is mapped onto a portion of this area called the *primary motor cortex*. The amount of motor cortical space devoted to a specific body region is positively correlated with the degree of fine motor control in that region.
32. An area of the parietal lobe called the *primary somatosensory cortex* receives information about touch and pressure from the whole body surface by way of the thalamus. Like the motor cortex, the somatosensory cortex accurately maps the relative sensory importance of various body regions.
33. Substantial portions of the cerebral hemispheres are devoted to the *association cortex*. As the name suggests, these areas receive input from many sensory channels and associate this information to produce output for different motor regions.
34. Cortical size has increased not only within the vertebrate lineage as a whole, but also within subsets of the mammalian lineage. For example, an increase in overall cortical size can be observed going from rodents to primates. This increase also parallels advancements in behavioral repertoire between the groups.
35. The most dramatic increase in cortical size and function has occurred in humans over the last several million years, and correlates with advances in the complexity of our behavior. While humans do not have the largest brains in absolute size, we do have the largest ratio of brain size to body size found in vertebrates. We also have the most highly developed cerebral cortices, including the largest relative amount of association cortex.

Questions (for answers and explanations, see page 359)

1. The secretion of hormones from an endocrine gland is most likely under the control of which of the following components of your nervous system?
   a. Autonomic
   b. Voluntary
   c. Dendritic
   d. Limbic
   e. Reticular

2. The portion of a neuron that stores and releases chemicals used to communicate with other cells is called the
a. dendrite.
b. cell body.
c. axon.
d. axon terminal.
e. synapse.

3. The portion of a neuron that would be most commonly encountered in any random cross section through a sensory nerve would be the
a. dendrite.
b. cell body.
c. axon.
d. axon terminal.
e. synapse.

4. Most toxic substances are prevented from entering the brain by the blood–brain barrier. The cells responsible for creating this barrier belong to a type of neural tissue called
a. neurons.
b. nerves.
c. glia.
d. thalamus.
e. midbrain.

5. A nerve impulse traveling up the spinal cord to the brain would first pass through the _________ before reaching the telencephalon.
a. pons
b. medulla
c. midbrain
d. diencephalon
e. All of the above

6. You stub your toe and immediately perceive the sensation of pain, which is a function of neural activity in your brain. This sensation would not be possible without the action of
a. afferent neurons.
b. efferent neurons.
c. interneurons.
d. afferent neurons and interneurons.
e. efferent neurons and interneurons.

7. A blow to the back of the head would most likely impair which of the following functions of the human cerebral cortex?
a. Hearing
b. Vision
c. Speech
d. Hand movements
e. Pressure sensation in the feet

8. Which of the following animals would likely have the *smallest* ratio of telencephalon size to body size?
a. Fish
b. Amphibian
c. Reptile
d. Bird
e. Mammal

Activities (for answers and explanations, see page 359)

- On the neuron diagram that follows, label the four basic anatomical regions and briefly describe the function of each.

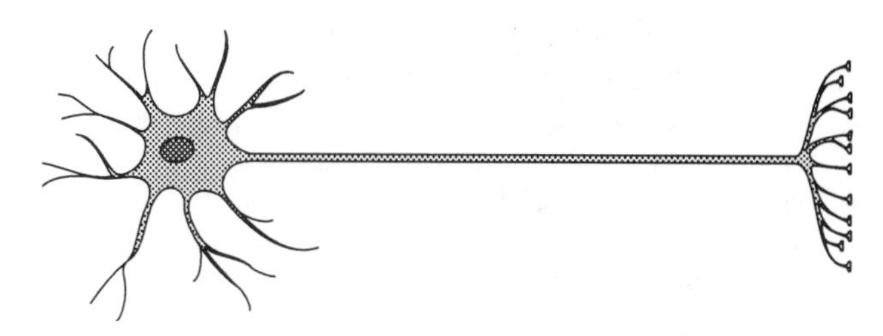

- On the lateral view of an adult human brain shown below, label the indicated structures and briefly describe their functions.

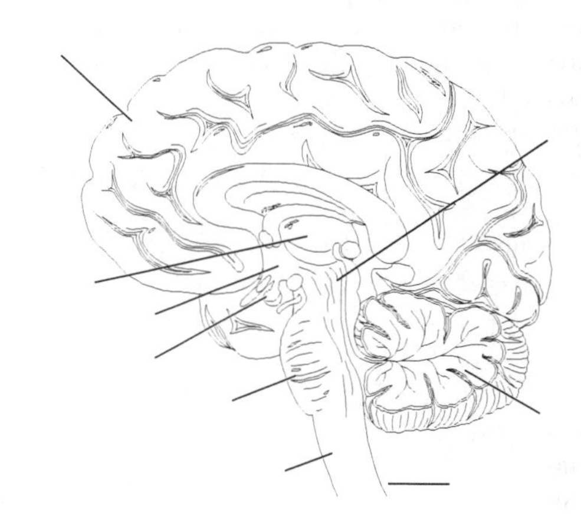

THE ELECTRICAL PROPERTIES OF NEURONS (pages 858-866)

Key Concepts

1. The *resting potential* of a neuron is defined as the electric charge across the plasma membrane when the cell is inactive. Most neurons are about 70 millivolts (mV) more negative on the inside of the cell when at rest.

2. *Action potentials* represent a rapid reversal of the resting membrane potential in response to an electrical, chemical, or physical stimulus. During this event the inside of the cell becomes temporarily positive with respect to the outside. Action potentials last for only 1–2 milliseconds and sweep down the length of a neuron, providing the basis for electrical communication.
3. Voltage is the tendency for electrical charges to move between two points, such as the poles of a battery or opposite sides of a membrane. In cells, the electrical charges that flow across membranes are caused by the movement of charged ions such as Na^+, K^+, and Ca^{2+}.
4. The hydrophobic portions of a phospholipid bilayer will not allow charged ions to freely pass through the membrane. Ions must therefore pass through specialized proteins called pumps, or channels, in order to cross the neuron membrane.
5. Pumps move charged ions across membranes against the ion's concentration gradient. This is an energy-requiring process. In neurons, the sodium–potassium pump is the primary pumping mechanism.
6. The action of the sodium–potassium pump expels three Na^+ ions from the cell for every two K^+ ions that enter. This exchange works to maintain high K^+ concentrations and low Na^+ concentrations inside the cell. Outside the cell these relative concentrations are reversed. Although this difference in ion concentration influences the electrical charge across the membrane, it does not create the resting potential.
7. The resting potential results from the interaction of charged ions with protein pores called *ion channels*. These channels selectively permit specific charged ions to pass in either direction across the membrane.
8. Many ion channel proteins change their shape, which inhibits or permits ion flow; they are said to be gated. *Voltage-gated* channels alter their shape in response to changes in the electrical state of the membrane. Other channels are *chemically gated*, changing shape when they bind to appropriate molecules.
9. Open potassium channels are more abundant in the neuron membrane than are open sodium channels, and K^+ ions therefore diffuse out of the cell more rapidly than Na^+ diffuses in. This produces an excess negative charge on the inside of the cell, which eventually becomes a sufficiently large voltage difference to restrict further K^+ diffusion. This phenomenon is largely responsible for the membrane resting potential.
10. Resting membrane potentials can be approximated using the *Nernst equation* and the *potassium equilibrium potential*. This calculation requires a knowledge of potassium concentrations both inside and outside the cell, and assumes that the cell membrane is permeable only to K^+ ions (See Box 38.A for details). The *Goldman equation* considers all possible ions and their permeabilities, and thus produces a more precise estimate of the true resting membrane potential.
11. The presence of a resting membrane potential in neurons means that they are polarized. This unequal separation of electrical charge provides a way for neurons to respond to chemical, electrical, or other types of stimuli. If the inside of a neuron becomes more negative than normal, the membrane is said to be *hyperpolarized*. If the inside becomes less negative (i.e., more positive), the membrane has *depolarized*. Whether hyperpolarization or depolarization takes place depends on the action of specific gated ion channels.
12. Hyperpolarization typically results from the opening of gated potassium channels. This allows more K^+ ions to diffuse out of the cell (down their concentration gradient), thus increasing the negative charge on the inside of the membrane. Gated chloride channels also exist in neuron membranes. When they open, Cl^- ions flow into the cell, causing hyperpolarization. Depolarization occurs when gated sodium channels open, allowing Na^+ ions to enter the cell where their concentration is lower. This reduces the magnitude of the negative charge on the inside of the membrane, eventually turning it into a positive charge.
13. The coordinated action of gated ion channels producing an alternation between depolarization and hyperpolarization that forms the basis of action potentials, the language of the nervous system. However, because membranes are inherently poor conductors of electricity, the action potential must be continuously regenerated in successive patches of membrane in order to successfully pass down the entire length of an axon.
14. Action potentials are brief phenomena that can be measured by inserting a microelectrode into a neuron and comparing its readings with one placed outside the cell. If the readings are referenced to the external electrode, we see that an action potential first involves a rapid depolarization from the resting potential of about –70 mV to a peak reversal state of about +40 mV. This is followed by a somewhat slower hyperpolarization that eventually restores the original resting membrane potential.
15. If measured at successive locations along the length of the axon, the action potential follows exactly the same series of events at each location. Thus, action potentials are an all-or-nothing, self-regenerating phenomena.
16. The depolarization phase of an action potential is due to the rapid opening of gated sodium channels. When a threshold voltage more positive than the resting potential is reached, all the sodium gates in a patch of membrane open. Because the action of the sodium–potassium pump has produced a high concentration of Na^+ ions on the outside of the cell, sodium rushes into the cell. In less than a millisecond, the membrane potential is reversed to the peak positive voltage of the action potential.
17. At the peak of the action potential spike, gated sodium channels close and remain closed for at least a few milliseconds, which is long enough for the normal leakage of K^+ ions out of the cell to restore the resting potential. In many neurons, voltage-gated potassium channels also open, allowing K^+ ions to rush out of the cell along the concentration gradient established by the sodium–potassium pump. This flow hyperpolarizes (or repolarizes)

the membrane. As the membrane nears its resting potential, increasing numbers of the potassium channels close until the normal membrane voltage is restored.

18. A *refractory period,* during which no new action potentials can be produced, occurs immediately after an action potential occurs. This effect is explained by assuming the presence of two types of voltage-sensitive gated sodium channels, activation and inactivation gates. Activation gates open and inactivation gates close at the onset of an action potential, but because the activation gates respond faster, the membrane depolarizes. Eventually the activation gates switch back to their closed condition, halting the influx of Na^+ ions. However, the inactivation gates remain closed for several milliseconds before spontaneously reopening. This renders the membrane temporarily incapable of generating a new action potential.

19. The concentration difference of Na^+ ions across the neuron membrane represents the "battery" that powers action potentials. Because few ions must move relative to the available ion pools inside and outside the cell for an action potential to occur, the battery is not easily run down. The action of the sodium–potassium pump is more than sufficient to restore ionic balance even in the face of hundreds of action potentials per second.

20. Action potentials propagate along an axon because the current flow produced by charged ion movements stimulates the depolarization of adjacent sections of membrane. The current produced by action potentials always brings adjacent membrane patches to the threshold voltage, and so the action potential is self-regenerating.

21. Action potentials propagate in only one direction along the axon because of the refractory period experienced by the previously stimulated section of membrane. The temporary inactivation of gated sodium channels in these areas prevents the action potential from spreading in both directions.

22. Axon size is one factor that determines the conduction velocity of action potentials. For invertebrates such as squids or insects, the diameter of the axon is positively correlated with the speed of action potentials. Neurons involved in quick-response behaviors such as escape tend to have large diameter axons.

23. However, there are anatomical limits on the absolute diameter that axons can achieve. In large animals such as vertebrates, increased conduction velocity is achieved not by increasing axon diameter, but rather by surrounding axons with an insulating membranous wrapping called *myelin*. This material, which is manufactured by specialized glial cells known as *Schwann cells*, gives the shiny appearance to white matter in the central nervous system.

24. The myelin cover along an axon is not continuous. It is absent at regular intervals, exposing the bare axon membrane at sites known as *nodes of Ranvier*. Action potentials are generated only at these nodes, where ions can move across the membrane due to the lack of myelin insulation. Current flows within the axon between successive nodes to stimulate a new action potential, giving the appearance that the regenerating action potential is jumping from one node to the next. Such *saltatory conduction* can increase action potential conduction velocities to speeds in excess of 100 m/sec.

25. In humans and other vertebrates, high-speed myelinated axons are part of the sensory systems that detect sharp pain. Sensory axons that transmit information about temperature or aching pain tend to be slow-speed and unmyelinated.

26. Two important techniques for studying neuron function are *voltage clamping* and *patch clamping*. Voltage clamps permit the experimenter to hold (clamp) the voltage of a cell at some constant value and measure the ionic currents that flow across a section of cell membrane. Patch clamps make it possible to study ion movements through single ion channels, providing insights about the molecular basis of electrical properties in membranes. (See Box 38.B for details).

Questions (for answers and explanations, see page 360)

1. The rapid depolarization of a neuron during the first half of an action potential is due to the
   a. exit of K^+ ions from the cell through gated potassium channels.
   b. rapid reversal of ion concentration caused by the action of the sodium–potassium pump.
   c. movement of both Na^+ and K^+ ions through appropriate open channels.
   d. entry of Na^+ ions into the cell through gated sodium channels.
   e. None of the above

2. One microelectrode is inserted into a healthy neuron and another is placed outside the cell. If the neuron is then hyperpolarized by a chemical stimulus, what will happen to the membrane potential recorded by the electrodes?
   a. It will become more positive.
   b. It will become more negative.
   c. It will first become more positive, then return to normal.
   d. It will not change.
   e. Cannot determine from the information given

3. If a chemical is applied to a neuron that inactivates all voltage-gated potassium channels, and then immediately thereafter the resting membrane potential is depolarized beyond threshold, which of the following will occur?
   a. A normal action potential will be generated.
   b. The neuron will hyperpolarize.
   c. Open potassium channels will compensate for the activity of the voltage-gated channels.
   d. The cell will depolarize, but not repolarize to the resting membrane potential.
   e. K^+ ions will diffuse into the cell in large numbers.

4. The refractory period of a neuron results from
   a. the shutting down of the sodium–potassium pump.
   b. activation of voltage-gated potassium channels.
   c. closing of inactivated voltage-sensitive gated sodium channels.
   d. activation of membrane receptors on the neuron surface.
   e. the inability of charged ions to pass through the myelin covering the axon.

5. Which of the following statements about the behavior of action potentials or neurons is *not* true?
   a. In ants, small diameter axons have higher conduction velocities than large diameter axons.
   b. In humans, myelinated axons have higher conduction velocities than unmyelinated axons.
   c. Current flow inside the axon is an important mechanism in saltatory conduction.
   d. In the myelinated axons of fish, action potentials are only generated at the nodes of Ranvier.
   e. Refractory periods occur in both myelinated and unmyelinated neurons.

6. The unmyelinated axon of a vertebrate neuron
   a. could possibly be associated with a slow pain sensor.
   b. conducts action potentials more slowly than a myelinated axon.
   c. has gated ion membrane channels all along its length.
   d. lacks nodes of Ranvier.
   e. All of the above

7. If you wanted to study the effects of a snake neurotoxin on ion current flow across individual voltage-sensitive gated sodium channels in a mammalian neuron, which of the following techniques would be most suitable?
   a. Standard microelectrodes
   b. Voltage clamping
   c. Patch clamping
   d. Nernst equation calculations
   e. Goldman equation calculations

Activities (for answers and explanations, see page 360)

- On the diagram of an action potential shown below, properly label the axes of the graph and indicate the parts of the trace where the gated sodium channels and gated potassium channels are mostly open. Also indicate the resting membrane potential and the apparent threshold voltage for this neuron.

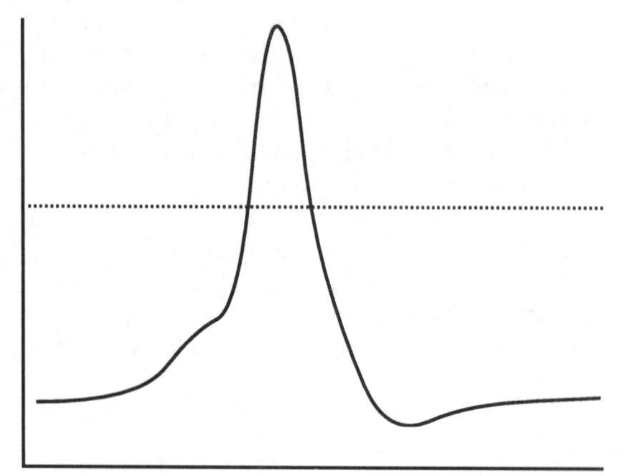

- The diagram below depicts a portion of membrane from a neuron. Explain the role of the various ions, channels, and pumps shown in creating the resting membrane potential of a neuron.

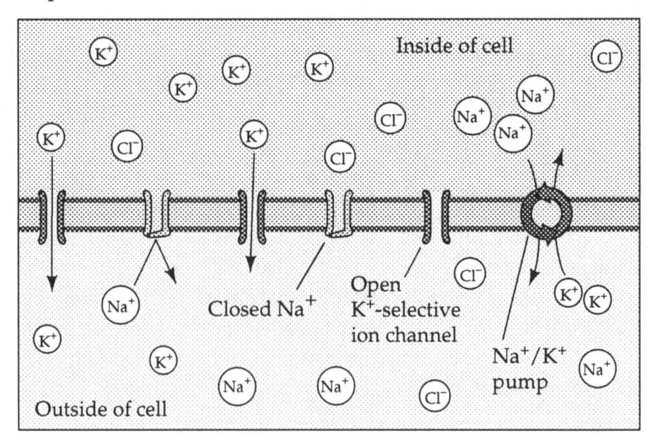

SYNAPTIC TRANSMISSION (pages 866-870)

Key Concepts

1. Neurons interact with neurons and other target cells at *synapses.* A one-way flow of information occurs at chemical synapses from *presynaptic cell* to *postsynaptic cell.*
2. *Neurotransmitters* are the chemical messengers used at synapses. They are stored within vesicles located in the axon terminal of the presynaptic cell, and released into the synaptic cleft upon arrival of an action potential. Neurotransmitters bind to receptors on the postsynaptic membrane. Some cause specific chemically gated ion channels to open. The opening of these channels in turn affects the cell's polarization.
3. Acetylcholine is a common vertebrate neurotransmitter used at modified synapses known as *motor end plates,* which form at the junctions of axons and skeletal muscle (a neuromuscular junction). Motor end plates have highly convoluted surfaces to maximize the area for acetylcholine receptor sites. Relatively few action potentials can stimulate the contraction of a muscle fiber.
4. The release of neurotransmitter is mediated by voltage-gated calcium channels, which occur at no other location in neurons except the synapse. Action potentials cause calcium channels to open, flooding the axon terminal with Ca^{2+} ions. This causes the synaptic vesicles to fuse with the presynaptic membrane and release their contents into the synaptic cleft. The neurotransmitter then moves to the postsynaptic cell by simple diffusion.
5. Postsynaptic cell membranes, such as those found at motor end plates, dendrites, and most regions of nerve

cell bodies, have few voltage-gated sodium channels. They are thus unable to produce action potentials. Instead, the chemically gated channels at these sites perturb the postsynaptic cell's resting potential just enough so that the stimulation spreads across a portion of the postsynaptic cell. If the stimulus is sufficiently large, this perturbation reaches an area containing more voltage-gated sodium channels where an action potential can be generated.

6. Each action potential arriving at a neuromuscular junction stimulates the release of about 100 synaptic vesicles, each containing some 10,000 neurotransmitter molecules. In the case of acetylcholine, this quantity is sufficient to generate an action potential and produce a muscle fiber twitch.
7. Although some synapses are always excitatory (e.g., the motor end plate), others cause an inhibition of the postsynaptic cell. *Excitatory postsynaptic potentials* (EPSPs) occur at synapses that cause a depolarization of the postsynaptic membrane, making it easier for an action potential to occur. *Inhibitory postsynaptic potentials* (IPSPs) are generated at synapses that produce hyperpolarizations of the postsynaptic cell, which make action potentials less likely to occur.
8. Whether a synapse is excitatory or inhibitory depends on the chemical nature of the postsynaptic receptors and not on the neurotransmitter itself. Thus, the same neurotransmitter can be both inhibitory (stimulates gated potassium channels) and excitatory (stimulates gated sodium channels), depending on its location within the nervous system.
9. The decision-making process of an individual neuron is accomplished by summing all of its excitatory and inhibitory inputs, and using this information to vary the rate at which it produces action potentials. This summation and integration of incoming information occurs in both space and time. It takes place at the *axon hillock*, where the axon emerges from the cell body.
10. High concentrations of voltage-gated sodium channels occur at the axon hillock, making it the first place on the neuron cell body where the spreading EPSPs and IPSPs encounter membrane that is capable of generating action potentials. Because of the decrease in intensity of postsynaptic potentials that occurs over distance, synapses that make contact with the postsynaptic cell nearer the axon hillock have more influence on whether the cell fires an action potential.
11. Many chemical synapses are fast-acting, responding within a few milliseconds of neurotransmitter contact. Others act more slowly, taking hundreds of milliseconds to respond. This occurs because the binding of neurotransmitter to receptor causes the activation of second messenger systems involving cAMP. Chemically mediated slow synapses are involved in the opening of ion channels, the control of membrane pumps, enzyme activation, and the induction of gene expression.
12. Some synapses form between axon terminals of one neuron and the axon of another. These synapses influence the amount of neurotransmitter released from the presynaptic cell. Depending on their effect, such influences are either termed *presynaptic excitation* or *presynaptic inhibition*.
13. Electrical synapses, or gap junctions, also occur between some neurons. Here the neurons have an intercellular gap of only 2–3 nanometers and are physically linked by connexons. Electrical currents or chemical ions actually pass through the connexons, permitting extremely rapid transmission in either direction.
14. Although very fast, gap junctions are not common in vertebrate nervous systems because they do not allow for summation, and they limit the number of synaptic contacts that can be made with any given neuron. Both of these features run counter to the high degree of integration characteristic of complex nervous systems.
15. More than 25 vertebrate neurotransmitter substances have been identified to date, but none are better understood than *acetylcholine*, which functions at all the voluntary (skeletal) neuromuscular synapses in the the peripheral nervous system. Both acetylcholine and epinephrine control the autonomic branch of the peripheral nervous system, and each plays a role in the central nervous system as well. However, neither is the dominant neurotransmitter molecule of the brain.
16. The primary neurotransmitters of the vertebrate brain are simple amino acid substances such as glutamic acid and aspartic acid, which are excitatory, and glycine and gamma-aminobutyric acid (GABA), which are inhibitory. Amino acid derivatives such as dopamine, norepinephrine, and serotonin are also important to brain function, along with a number of peptide molecules. Recent evidence indicates that the gases carbon monoxide and nitric oxide are also used as chemical messengers between some neurons.
17. Neuropharmacology is the study and development of drugs that affect the nervous system. Acetylcholine is a good example of how neurotransmitters can have powerful and divergent effects depending on the type of receptors present in their target tissues.
18. Within the central nervous system *nicotinic* acetylcholine receptors are excitatory, while *muscarinic* receptors are inhibitory. Chemical analogs of acetylcholine (e.g., nicotine from cigarettes or the psychoactive drugs in certain mushrooms) can have dramatic influences on nervous system activity when they bind to these receptors, possibly leading to addictive physiological or behavioral responses.
19. *Curare* is another example of a natural substance that influences acetylcholine activity. Curare binds to acetylcholine receptors in skeletal muscle, inactivating them and rendering the muscle incapable of contraction. It is used to tip the poisoned hunting arrows used by some native peoples in South America. A medicinal compound called *atropine* blocks the muscarinic receptors in heart and other muscle tissue. It can be used to increase heart rate, dilate pupils in the eye, or decrease gut spasms.

20. An important part of synaptic control is the need to inactivate used neurotransmitters, and to then clear the synapse of these molecules before the arrival of subsequent chemical signals. Three mechanisms are commonly used to inactivate neurotransmitters.

21. The first involves the actual chemical destruction of the neurotransmitter while it is still in the synaptic cleft. For example, acetylcholine is inactivated by the enzyme *acetylcholinesterase*. The absence of acetylcholinesterase leads to uncontrolled muscle spasms, and eventually death, due to continuous stimulation of the neuromuscular junction. Many potent nerve toxins and insecticides have acetylcholinesterase inhibitors as their active ingredient.

22. A second approach is to actively reabsorb the neurotransmitter back into the presynaptic cell, where it is metabolically degraded or repackaged into new synaptic vesicles. A third method is to simply allow the neurotransmitter to diffuse away from the synaptic cleft.

Activities (for answers and explanations, see page 360)

- Match the following chemicals with their appropriate functions in the vertebrate nervous system.

_____Curare	*a.* Inhibitory neurotransmitter in the brain
_____Acetylcholine	*b.* Stimulates fusion of synaptic vesicles with presynaptic membrane
_____Glycine	*c.* Blocks inhibitory effects of muscarinic receptors in heart muscle
_____Calcium	*d.* Inactivates acetylcholine receptors in skeletal muscle
_____Atropine	*e.* Has both excitatory and inhibitory influences

- The active ingredients in many nerve gases belong to a class of chemicals called anticholinesterases. Suggest a possible synaptic mechanism to explain how these chemicals can do such dreadful damage to an animal's nervous system.

Questions (for answers and explanations, see page 360)

1. Some neurotransmitters affect the behavior of postsynaptic neurons by
a. altering their membrane potentials through the action of gated ion channels.
b. stimulating the release of calcium into the postsynaptic cell.
c. initiating the bonding of membrane proteins that form connexons.
d. causing the active uptake of neurotransmitter molecules across the postsynaptic membrane.
e. None of the above

2. Which of the following statements about the process of summation in a neuron is *not* true?
a. Slight perturbations of the membrane potential spread across the postsynaptic cell body.
b. Axons that terminate closer to the axon hillock have more influence on the summation process.
c. Summation consists mostly of comparing the total of excitatory and inhibitory postsynaptic inputs.
d. The concentration of gated sodium channels is highest in the dendrites of the postsynaptic cell.
e. All of the above

3. All but one of the following chemicals are neurotransmitters that function in the human brain. Select the *exception*.
a. Dopamine
b. Glycine
c. Atropine
d. Glutamic acid
e. Serotonin

Refer to the following diagram when answering questions 4 and 5.

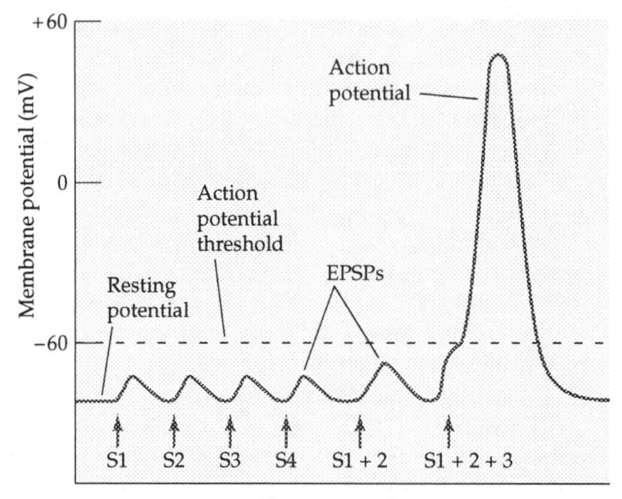

4. The neural phenomenon depicted by this diagram is best described as
a. spatial summation.
b. temporal summation.
c. presynaptic excitation.
d. presynaptic inhibition.
e. synaptic transmission.

5. The electrical events labeled as EPSPs are the result of
a. hyperpolarizations of the postsynaptic membrane.
b. depolarizations of the postsynaptic membrane.
c. hyperpolarizations of the presynaptic membrane.
d. depolarizations of the presynaptic membrane.
e. the action of anticholinesterases.

6. If you could selectively neutralize the active transport mechanisms at the axon terminals of a neuron, which of the following processes would you most significantly impact?

a. Fusion of synaptic vesicles with the presynaptic membrane
b. Uptake of released neurotransmitter by the presynaptic cell
c. Operation of voltage-gated potassium channels
d. Release of calcium within the postsynaptic cell
e. Diffusion rate of neurotransmitter across the synaptic cleft

7. The one-way flow of information between presynaptic and postsynaptic cells is a characteristic of most kinds of neuron interactions. A notable exception to this rule is the
a. neuromuscular junction.
b. integration of postsynaptic potentials by the axon hillock.
c. nicotinic receptor system of autonomic neurons.
d. motor end plate.
e. gap junction.

NEURONS IN CIRCUITS • HIGHER BRAIN FUNCTIONS (pages 870-878)

Key Concepts

1. The autonomic nervous system, which controls gland and involuntary muscle activity, includes the *sympathetic* and *parasympathetic* divisions. These divisions work opposite one another to cause increases or decreases in effector activity.

2. One common example is the "fight-or-flight" response. Here the sympathetic pathways stimulate increases in blood pressure, heart rate, and cardiac output in preparation for defense or escape. The parasympathetic pathways cause these same physiological parameters to decrease, returning them to normal levels after the threat has passed.

3. It is not always the sympathetic pathway that increases activity, nor the parasympathetic that decreases it. For example, digestion is speeded up by the parasympathetic and slowed by the sympathetic. Nevertheless, it is always true that these pathways work in opposition to one another, regardless of which division has what effect.

4. The neural circuits of parasympathetic and sympathetic pathways consist of two cell types. The first cell in the circuit is located in either the brain stem or the spinal cord and is called a *preganglionic neuron*. Its axon uses acetylcholine as a neurotransmitter.

5. Preganglionic cells synapse onto *postganglionic neurons*, the second cell in an autonomic circuit. These cells reside in a cluster (ganglion) of functionally related neurons located somewhere outside the spinal cord. The axons of postganglionic cells terminate on effector organs. If they are part of the sympathetic division they use norepinephrine as their synaptic neurotransmitter, if part of the parasympathetic, they use acetylcholine.

6. In effectors that receive input from both parasympathetic and sympathetic pathways, the target cells respond in opposite ways to acetylcholine and norepinephrine. One neurotransmitter causes depolarization and an increase in activity, and the other leads to hyperpolarization and a decrease in activity. The control of cardiac pacemakers and digestive muscles are examples of this interplay.

7. Anatomical differences also exist between the two autonomic divisions. Preganglionic neurons of the sympathetic system are located in the thoracic and lumbar regions of the spine, and those of the parasympathetic in the brain stem and lower spinal segments. Postganglionic sympathetic neurons are mostly situated in chains adjacent to the spinal cord, while those of the parasympathetic are found closer to the target organ.

8. *Monosynaptic reflex loops* are the simplest forms of neural circuits. They have a single synapse that separates the afferent and efferent sides of a pathway. The involuntary knee-jerk response is an example.

9. When the doctor's hammer taps the tendon covering the knee, stretch receptors made of *muscle spindles* stimulate action potentials in sensory neurons. Their axons enter the spinal cord via the dorsal root to synapse on individual motor neurons in the ventral horn of the spinal cord. These motor neurons send action potentials through their axons, which exit the spinal cord via the ventral root. This stimulates contraction of the leg extensor muscle and extends the leg.

10. Although this reflex loop is involuntary, signals reaching the brain also provide conscious information that the knee has been struck. This occurs through spinal interneurons, which are stimulated by synaptic input from branches of the entering sensory axons. The axons of interneurons travel up the spinal cord to appropriate brain centers, which make decisions about this input and then relay information back to the motor neurons. The system provides voluntary control of leg position.

11. *Polysynaptic reflex loops* are also common in involuntary responses. In the knee-jerk example, at the same time that excitatory signals are sent to the extensor muscle, inhibitory signals are sent to the leg flexor by way of additional sensory input onto other motor neurons. This prevents the flexor from contracting, a motion that would oppose the action of the extensor.

12. More complex reflexes such as escape responses result from the layering of additional polysynaptic circuits. The initial control of these behaviors is, however, still orchestrated primarily within the spinal cord, and usually does not require conscious intervention through higher brain functions.

13. Only the most basic of neural circuits have been studied thoroughly. The neural mechanisms of complex activities such as reasoning, perception, and memory remain largely unknown. Nevertheless, patterns of neural activity associated with certain nervous system functions have been documented.

14. Sleep occurs in all birds and mammals, and probably all other vertebrates as well. Yet the neural reasons for why animals sleep, or how sleep cycles are controlled, are not clear. The *electroencephalogram* (EEG), which monitors regional brain activity, and the *electromyogram* (EMG), which records the electrical activity associated with muscle movements, provide tools for studying these questions.

15. EEG and EMG traces from sleeping mammals reveal several different stages of sleep. *Slow-wave sleep* accounts for about 80 percent of sleep time. It is characterized by low-frequency, high-amplitude EEG waves. In humans it is divided into four stages of progressively deeper sleep.
16. The other 20 percent of sleep time is spent in *rapid-eye-movement* (REM) sleep, during which the ocular muscles cause rapid eye twitches. REM is the sleep period during which humans experience vivid dreams and nightmares. During this time descending neural input from the brain inhibits the movement of voluntary muscles, presumably to prevent the acting out of dreams.
17. The function of sleep is not entirely clear, but most researchers believe it is some kind of restorative process, or a way of consolidating memory patterns so we can forget incorrect or unnecessary mental associations. The neural control of sleep resides in the brain stem, centered on the pons and the medulla.
18. Learning is a higher brain function that allows behavioral modification, but not all learning is complex. *Habituation* is a simple form of learning where a repeated stimulus is ignored over time if it provides little useful information (e.g., crowd noises at a party). *Sensitization* is also simple learning in which the animal becomes increasingly aware of a stimulus if it provides particularly useful information (e.g., an infant's cry at the same noisy party).
19. The synaptic basis of habituation and sensitization has been studied in the sea hare *Aplysia*. This mollusk has a monosynaptic gill siphon withdrawal reflex which protects the delicate gill tissue from damage.
20. Habituation of the withdrawal reflex to repeated touching of the siphon is achieved by decreasing the amount of neurotransmitter released at the axon terminals of presynaptic cells, thus reducing the level of response. The reflex can also be sensitized by mild electric shocks to the animal's tail. In this case a third neuron releases serotonin onto the presynaptic cells, which stimulates a second messenger system that causes calcium channels to remain open longer for any one action potential. This means that more neurotransmitter is released and a larger postsynaptic response occurs.
21. *Associative learning* occurs when unrelated stimuli are linked to the same response. A simple example is the *conditioned reflex*, where a novel stimulus is associated with another stimulus that normally triggers the response. Eventually, the novel stimulus will by itself stimulate the response, as in the case of Pavlov's experiments on the neural control of salivation in dogs.
22. Like other complex neural phenomena, human memory is not particularly well understood. However, it is clear that stimulation of certain brain regions will evoke vivid memories in humans, although memories do not appear to reside in any one particular brain location. It seems that the neural connections responsible for stimulating memory have a multiplicity of pathways within the brain.
23. There are several categories of memory. *Immediate memory* recalls events that are happening now; it is nearly perfect but lasts only a few seconds. *Short-term memory* retains less information, but can last for many minutes.
24. Repetitive use or reinforcement facilitates the transfer of short-term memory into *long-term memory*, which can last for years. The hippocampus is an integral part of the neural circuitry that allows the formation of long-term memory in humans. Without it the individual cannot remember learning tasks or facts for more than 10–15 minutes, although the motor skills associated with these actions is retained.
25. The ability of humans to acquire and use language resides in the cerebral hemispheres. More precisely, in 97 percent of people these skills reside in the left hemisphere, a neural phenomenon known as *lateralization*.
26. The human cerebral hemispheres are connected by a mass of white matter called the *corpus callosum*, which allows the transfer of neural information from one hemisphere to the other. Surgical or accidental severing of the corpus callosum produces "split-brain" individuals.
27. Cases of split-brain patients whose corpus callosum is cut to control severe epilepsy show that mechanical knowledge contained in the right hemisphere cannot be expressed in language, which is controlled by the left hemisphere, but the mechanical act itself may still be carried out since this motor function is governed by the right hemisphere. In some instances language cannot even be used as a mechanism to communicate with the right hemisphere in these patients.
28. Further experiments with patients suffering from *aphasias* (deficits in the ability to understand or use words) indicate that normal language skills depend on the proper flow of information between different language areas in the left cerebral cortex. Broca's area is a part of the motor cortex that is essential for speech. Wernicke's area is part of the temporal lobe related to sensory aspects of language perception and production.
29. If damage occurs to the left cerebral hemisphere while a person is still a child and the process of language lateralization has not been completed, the right hemisphere will take over the motor and sensory functions related to speaking and writing with little or no detectable loss of language skills.

Activities (for answers and explanations, see page 361)

- The following diagram represents a monosynaptic reflex loop. Label the sensory neuron, motor neuron, and effector in this simple neural circuit. Also indicate which part of the circuit is the afferent side, and which the efferent.

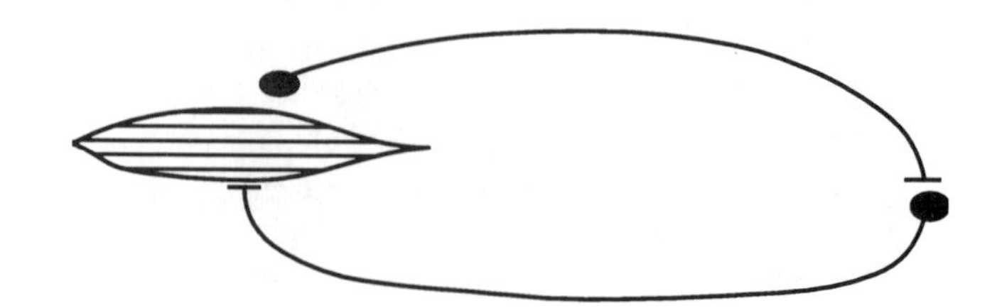

- Explain the neural reason why it is impossible for a person to sleepwalk while dreaming. What unique facial behavior characterizes the portion of the human sleep cycle during which dreams occur?

Questions (for answers and explanations, see page 361)

1. In the fight-or-flight response, input from the ________ pathway stimulates an increase in heart rate and blood pressure, whereas input from the ________ pathway returns these parameters to normal levels.
   a. sympathetic; parasympathetic
   b. parasympathetic; sympathetic
   c. autonomic; central
   d. central; autonomic
   e. postganglionic; preganglionic

2. In which of the following structures would you expect to find the axon terminals of a postganglionic cell from the sympathetic division of the autonomic nervous system?
   a. Lungs
   b. Heart
   c. Pupil of the eye
   d. Stomach
   e. All of the above

3. Which of the following statements about the neural basis of the knee-jerk reflex in humans is *not* true?
   a. The leg extensor is stimulated to contract by the motor neuron in this reflex loop.
   b. The only synapses between neurons involved in this loop occur in the spinal cord.
   c. The basic response is a polysynaptic circuit containing several interneurons.
   d. Conscious action can modify the basic reflex because of the action of interneurons.
   e. It is very simple by comparison to other neural circuits such as those involved in learning or memory.

4. A friend wakes you from your sleep, and you immediately have the sensation of having just experienced a vivid dream. If you had been making EMG and EEG recordings of muscle and neural activity at the time you were awakened, which of the following would be true?
   a. Your hands and feet might have been twitching slightly.
   b. Your eyes would have been twitching.
   c. Most of your voluntary body muscles would have been inactive.
   d. Your cerebral cortex would have been active as though you were awake.
   e. All of the above

5. Which of the following represents an example of what is called *habituation*?
   a. You learn to pick out a particular color pattern from among a large collection of cloth styles.
   b. A friend tells you his phone number, and within minutes you have memorized it.
   c. The smell of apple pie causes your stomach to growl in anticipation of eating.
   d. You enter a room with a loud ticking clock, and after a while you do not notice the ticking.
   e. None of the above

6. Memories associated with the skills that allow a person to sight-read and perform a piece of music which they have never seen before reside in
   a. Wernicke's area.
   b. Broca's area.
   c. the corpus callosum.
   d. the right cerebral hemisphere only.
   e. many areas of the brain.

7. Observations of patients with aphasias indicate that
   a. only Broca's area is essential for normal language skills.
   b. language skills depend on proper neural flow between the temporal lobes and motor cortex.
   c. the right hemisphere dominates the production and use of language in humans.
   d. the corpus callosum is not necessary for interhemispheric transfer of information in the brain.
   e. most people can control language functions from either cerebral hemisphere.

Integrative Questions (for answers and explanations, see page 361)

1. You see a piece of chocolate and reach down to pick it up. As you begin chewing the candy, glands in your mouth start secreting saliva. Which of the following elements of your nervous system is involved in this sequence of events?
   a. Central
   b. Peripheral
   c. Autonomic
   d. Parasympathetic
   e. All of the above

2. If you electrically stimulate a resting neuron in the middle of its axon and cause it to fire an action potential, which of the following will happen?
   a. The action potential will propagate towards the axon terminal.
   b. The action potential will propagate towards the cell body.
   c. The action potential will propagate towards both the axon terminal and the cell body.
   d. The action potential will not propagate at all.
   e. The neuron will hyperpolarize.

3. The neural process of speaking a written word begins in ____________ of the human brain.
   a. the primary visual cortex
   b. Wernicke's area
   c. Broca's area
   d. the motor cortex
   e. the corpus callosum

4. The resting membrane potential for neuron A is –70 mV, while the resting potential for neuron B is –50 mV. The threshold voltage for the production of an action potential is –35 mV for both neurons. Which of the following statements is *not* true?
   a. Neuron A must depolarize by 35 mV to reach the threshold voltage.
   b. Neuron B must hyperpolarize by 15 mV to reach the threshold voltage.
   c. The inside of both neurons is negatively charged with respect to the outside.
   d. A single EPSP received by neuron A would cause it to depolarize slightly.
   e. A single IPSP received by neuron B would cause it to hyperpolarize slightly.

39

Sensory Systems

CHAPTER LEARNING OBJECTIVES—after studying this chapter you should be able to:

- ❑ Appreciate the biological basis for why the sensory worlds of different animal species are unique.
- ❑ Define "sensory transduction" and explain how it leads to receptor, generator, and action potentials.
- ❑ Discuss the concept of sensory adaptation and explain why different types of sensors show different levels of adaptation.
- ❑ Describe the function of chemosensors and explain the molecular mechanism for how they transduce chemical stimuli.
- ❑ Differentiate between olfaction and gustation, and discuss the role of pheromones in arthropod chemosensation.
- ❑ Provide the sensory explanation for human smell and taste.
- ❑ Describe the function of mechanosensors and explain the various ways that mechanoreception is used in sensory systems.
- ❑ Discuss the types of mechanoreceptors present in human skin.
- ❑ Identify the various sensory systems that employ hair cells for mechanoreception, and explain the role of hair cells in these sensors.
- ❑ Describe the parts of the mammalian ear and explain their functions.
- ❑ Discuss the structure and function of the mammalian cochlea, and explain how different pitches of sound are transduced by this organ.
- ❑ Differentiate between conduction deafness and nerve deafness in humans and provide possible causes for each.
- ❑ Describe the function of photosensors and explain the various ways that photoreception is used by animals.
- ❑ Explain the molecular mechanism of photodetection by rhodopsin and the second messenger system by which receptor potentials are generated in photoreceptors.
- ❑ Discuss the structure and function of the arthropod compound eye, and compare it to the structure and function of the vertebrate eye.
- ❑ Describe the process of visual accommodation in the vertebrate eye.
- ❑ Differentiate between the fovea and blind spot of the vertebrate eye and discuss the function of each.
- ❑ Explain the difference between a rod and a cone, discuss their functions, and describe their relative importance in nocturnal versus diurnal animals.
- ❑ Name the five layers of neural tissue in the vertebrate eye and describe the anatomical and functional relationships among these layers.
- ❑ Compare and contrast the phenomena of ganglionic and cortical receptive fields, and distinguish between the roles of simple and complex cells in the visual cortex.
- ❑ Explain the anatomical and neural basis for binocular vision.
- ❑ Briefly describe the roles of ultraviolet, infrared, electric, and magnetic sensitivities in animals.

WHAT IS A SENSOR? • CHEMOSENSORS (pages 881-888)

Key Concepts

1. The sensory world is different for each animal species. How the world "appears" to a given animal depends on the specific environmental stimuli its sensors detect.
2. Sensors are cells that convert, or transduce, physical energy into action potentials which can be understood by the nervous system. Sensors are keyed to specific forms of energy such as chemicals, mechanical forces, or light. They detect stimuli from both the external and internal environments.
3. The perceived sensation caused by a stimulus depends on where in the nervous system the action potentials from a sensor are sent. Sensory nerves travel through afferent pathways to various regions in the central nervous system, where they are interpreted as pain, sound, light, etc. The phantom limb phenomenon, where a person perceives sensations as coming from an amputated limb, exemplifies this concept.
4. Specialized sensory organs which gather and focus stimuli may be associated with sensory cells. These organs facilitate the transduction process and help to amplify or filter the environmental stimulus.

5. During transduction, a receptor protein is activated by the stimulus and changes the ion permeability of the sensor cell membrane. This happens either as a direct result of the receptor protein's opening or closing an ion channel, or because the protein sets into motion a series of intracellular events that eventually influence ion channels. Changes in ion movement alter the membrane potential of the sensor cell to produce a *receptor potential,* the magnitude of which is directly proportional to stimulus intensity.
6. Receptor potentials are converted into action potentials in two ways. The first is for the sensor itself to generate an action potential. Receptor potentials in these sensors are called *generator potentials,* because they spread to a region of cell membrane containing voltage-gated sodium channels and generate an action potential. Crayfish stretch sensors provide an example of this mechanism.
7. The second way receptor potentials lead to an action potential is for the sensor to release neurotransmitter onto an adjacent neuron, which causes the production of an action potential in the postsynaptic cell. Photosensors are good examples of this transduction mechanism. In both transduction mechanisms, the magnitude of receptor or generator potentials determines the frequency of action potentials fired by the sensor or its associated neuron.
8. *Adaptation* occurs when sensors reduce their response to a continuous source of stimulation. By ignoring constant background conditions, the sensor is able to respond to changes in the stimulus which represent new information. Different categories of sensors have different degrees of adaptation. Sensors that monitor critical parameters such as pain and balance do not adapt at all, or adapt more slowly than those that monitor less critical stimuli such as light and sound.
9. *Chemosensors* are responsible for detecting stimuli that produce the senses of taste and smell. They are the oldest and most ubiquitous of all sensors, and can have powerful effects on behavior and certain autonomic functions.
10. Arthropods use chemical signals called *pheromones* (external hormones) to attract mates. When a female silkworm moth secretes the pheromone bombykol, males detect the stimulus using sensors on their antennae and fly towards its source. The male's response is so acute that one molecule of bombykol can generate several action potentials in a single chemosensor. Arthropods may also have chemosensory hairs located on their bodies, especially the feet. These hairs contain sensor cells that detect foods such as salts, sugars, and amino acids, as well as other important chemical features of the environment.
11. *Olfaction* is the sense of smell. Vertebrate olfactory receptors are located on sensor cells embedded in epithelium that lines the nasal cavity. The dendrites (olfactory hairs) of these cells project from this lining and are covered by a thin layer of mucus, into which chemicals diffuse to make contact with the hairs. Nasal congestion causes a loss of smell due to the proliferation of the mucus coating, making it difficult for chemicals to contact the hairs.
12. A second messenger signalling system mediates the transduction of olfactory stimuli. Odor molecules bind to specific receptors on the sensor hairs. This activates G-proteins, which in turn activate adenylate cyclase, causing an increase in the concentration of the second messenger, cAMP. Gated sodium channels open in response to cAMP, which depolarizes the cell and generates an action potential.
13. Odor intensity is coded for by the number of action potentials generated by sensor cells. Increasing concentrations of odorant molecules yield more action potentials, and the perception of a more intense smell results. The subjective impression of different kinds of smells comes from the activation of various combinations of receptors and sensors. Recent evidence indicates that a large family of genes codes for the wide variety of olfactory receptor types present in vertebrates.
14. *Taste buds* are the chemosensors responsible for *gustation,* the sense of taste. They are found primarily in the mouths of terrestrial vertebrates, although they also occur on the skin of aquatic animals that forage in murky waters where vision is of little use.
15. Clusters of taste sensors are embedded in the tongues of humans and other mammals. Many are associated with the raised papillae that give the tongue its bumpy texture. Tiny hairlike projections called microvilli are connected to the taste sensors and emerge through a pore in the top of each taste bud.
16. Unlike olfactory sensors, taste sensors are not neurons. The cell bodies of the sensory neurons that actually relay taste information to the central nervous system are located at the base of each cluster of taste sensor cells.
17. As chemicals make contact with receptors on the microvilli of taste sensor cells, receptor potentials are generated. The sensor cells release neurotransmitter onto the dendrites of the sensory neurons below each taste bud. These neurons then generate the action potentials that are sent to the central nervous system.
18. The epithelial surface of the tongue is subjected to much physical abuse because of the abrasion that occurs during eating. Taste buds and their sensor cells are therefore replaced every few days, but the sensory neurons remain to form synapses with new sensor cells.
19. In general, specific regions of the human tongue are responsible for recognizing the four basic categories of taste (bitter, sour, salty, and sweet). In reality, however, these areas overlap to a large extent and many varieties of molecules can be detected by our taste buds. Subtle discriminations within each basic flavor category are possible due to the interplay of gustatory and olfactory senses, which is why your sense of taste suffers during a cold.
20. Snakes use their tongues to smell. The flicking tongue gathers airborne molecules and delivers them directly onto olfactory receptors located in the roof of the mouth. In this way the animal makes repeated, rapid samplings of the air which are not otherwise possible due to the slow respiratory rate of reptiles.

Activities (for answers and explanations, see page 361)

- Describe the difference between a receptor potential and a generator potential in a sensory cell. What are the main functions of these potentials?

- Using your understanding of the anatomy and transduction mechanisms of human olfaction and gustation, explain why your senses of smell and taste suffer when you have a cold.

Questions (for answers and explanations, see page 361)

1. Which of the following would qualify as a form of sensory transduction?
   a. Sex pheromone from a female moth produces action potentials in the antennal nerve of a male.
   b. Receptor potentials in taste sensors stimulate action potentials in associated sensory neurons.
   c. An active G-protein in olfactory neurons causes production of cAMP.
   d. Sugars in a food source bind to receptors on the chemosensory hairs of a fly's foot.
   e. All of the above

2. An electrode is inserted into a chemosensory nerve leading away from a taste bud in the mouth of a dog. A mild acid solution is then continuously flushed over the sensors associated with this nerve. Initially the nerve responds to this stimulation, but over time ceases to carry action potentials. This observation would best be explained by
   a. transduction.
   b. adaptation of the sensor cell.
   c. depletion of neurotransmitter in the sensory nerve.
   d. second messenger influences that increase cell membrane potentials.
   e. death of the sensor cell.

3. Which of the following is *not* a neuron?
   a. Human olfactory sensor
   b. Moth antennal chemosensor
   c. Human taste sensor
   d. Grasshopper foot chemosensor
   e. Snake olfactory receptor

4. If the concentration of cAMP suddenly increases in a human olfactory sensor cell, which of the following will happen?
   a. The membrane potential will become more negative.
   b. The membrane potential will become more positive.
   c. The membrane potential will remain unchanged.
   d. The sensor will show immediate signs of adaptation.
   e. The person will perceive a complex mixture of smells.

5. The reason that food loses much of its flavor when you have a cold is because
   a. adaptation of chemosensors is more rapid.
   b. the transduction process in your taste buds has been inhibited.
   c. you cannot make sufficient quantities of cAMP.
   d. sensitivity of olfactory chemosensation is decreased by nasal congestion.
   e. mucus from the olfactory epithelium coats your taste buds.

6. Which of the following statements about human gustation is *not* true?
   a. Taste sensors are relatively short-lived cells due to the high degree of abrasion they encounter.
   b. A single sensory neuron is associated with each taste bud in the vertebrate tongue.
   c. Lemon juice is tasted most clearly on the sides of the tongue.
   d. Receptors on the microvilli of taste sensors bind to food molecules.
   e. Humans perceive only three categories of tastes: sweet, sour, and bitter.

7. The chemosensory molecules used for mate attraction in arthropods are otherwise known as
   a. hormones.
   b. pheromones.
   c. protein receptors.
   d. G-proteins.
   e. second messengers.

8. Which of the following sensory structures is capable of transduction?
   a. Snake olfactory sensor
   b. Dog chemosensor
   c. Insect chemosensory hair
   d. Human taste bud
   e. All of the above

MECHANOSENSORS (pages 888-894)

Key Concepts

1. *Mechanosensors* respond to mechanical forces that distort cell membranes. Some mechanosensors are associated with the senses of touch, pain, pressure, or position. Others act as stretch receptors in blood vessels, muscles, or joints. Highly modified mechanosensors called hair cells are involved in balance and hearing.

2. When mechanical forces destort the sensor membrane, changes in ion permeability occur that alter the cell's membrane potential. The number of action potentials fired in response to this deformation is directly proportional to the stimulus intensity.

3. Several types of mechanoreceptors are found in human skin. Near the surface are *Meissner's corpuscles*, sensitive mechanoreceptors that detect light touch and adapt rapidly to a stimulus. *Expanded-tip tactile sensors* adapt more slowly than Meissner's corpuscles; they provide steady-state information about touch. Different body areas have different densities of these receptors, explaining why the fingertips are more touch-sensitive than the skin on a person's back.

4. Other mechanosensors are found deep within the skin. Physical displacement of epidermal hairs is detected by nerve endings that wrap their dendrites around the bases of hair follicles. Rapid skin vibrations are detected by sensory neurons called *Pacinian corpuscles*, which are encased in concentric layers of connective tissue. They adapt rapidly to stimuli because this layering efficiently redistributes vibrations away from the nerve endings.

5. *Stretch sensors* respond to the forces associated with the stretching of vessel walls, bone joints, and muscles. Muscle spindles are examples of stretch sensors that contribute to the maintenance of body position through their connection with neural reflex loops. *Golgi tendon organs* are stretch sensors found in the tendons and ligaments that monitor the forces exerted by a contracting muscle.

6. *Hair cells* are a non-neuronal class of mechanosensors present in many animal sensory systems. All hair cell varieties possess *stereocilia* that project from the cell surface. When bent in one direction they hyperpolarize the cell membrane. If bent the other way they depolarize it. Depolarization causes the release of neurotransmitter onto neighboring sensory neurons, which fire action potentials that are carried to the brain.

7. The lateral line system of fishes uses hair cells to detect pressure changes associated with the movement of water or nearby objects. Lateral line hair cells are located in a canal just beneath the skin surface. Their stereocilia protrude into the canal and are encased by a gelatinous structure called the cupula. Water movement through the canal causes the cupulae to move and the stereocilia to bend. This induces electrical activity in associated sensory neurons.

8. *Statocysts* are the position-sensing organs of many invertebrates. Hair cells line the closed chambers of these organs, inside which are dense structures called *statoliths*. As the animal changes position, gravity causes the statoliths to move and the stereocilia to bend. Statolith position, as measured by the response of hair cells, indicates the animal's body position in space.

9. Vertebrate equilibrium is monitored by two other types of hair cell organs. The inner ear has three fluid-filled *semicircular canals* with terminal swellings called ampullae. Each *ampulla* contains many hair cells encased in a cupula. As the head moves, so does the fluid in the canals. This causes movement of the cupula and bending of the hair cells. A chain of events is initiated that results in the production of action potentials by associated sensory neurons.

10. The *vestibular apparatus* possesses two statocystlike chambers used to detect head position. The floors of these chambers are covered with hair cells, on top of which lie many calcium carbonate *otoliths*. Gravity pulls on the otoliths to bend the stereocilia, generating action potentials that inform the nervous system about head position.

11. The sense of hearing (audition) also uses hair cells as transducers. Airborne pressure waves are converted into action potentials within the ear. These are then interpreted by the brain as sound.

12. In humans and other vertebrates, several accessory structures are used to collect, direct, and amplify sound waves. *Ear pinnae* are the flaps of skin that collect sound, funneling it into the auditory canal where it causes the *tympanic membrane* to vibrate. The tympanic membrane separates the auditory canal from the air-filled middle ear. Air pressure in the middle ear is equalized through a connection with the oral cavity called the *eustachian tube*, which keeps the tympanic membrane from inappropriately bulging in or out.

13. Connected to the tympanic membrane in the middle ear is a series of three small bones called *ear ossicles*. Membrane vibrations are first transferred to the *malleus* (hammer) which is physically attached to the tympanic membrane, then to a fulcrum bone called the *incus* (anvil), and finally to the *stapes* (stirrup). This system of bony levers acts to amplify the sound 20-fold.

14. The stapes is attached to the *oval window*, a second membrane which connects the middle ear with the fluid-filled inner ear. Vibration of the stapes causes the oval window to vibrate. Because fluids are relatively incompressible, this creates movement within the inner ear. Thus, airborne pressure waves are converted into the oscillatory movements of inner ear fluids.

15. The inner ear is known as the *cochlea* because of its coiled, snaillike shape. It is within the cochlea that the transduction of pressure waves into action potentials takes place.

16. The cochlea consists of three parallel chambers separated by several membranes. Two of these chambers are interconnected at the distal end of the cochlea, making them one large chamber. Atop the basilar membrane rests the *organ of Corti*, the structure that contains the hair cells. Reissner's membrane serves a nutritional function for the organ of Corti but is not involved in sound transduction.

17. Sounds of different frequencies, or pitches, cause the oval window to produce unique fluid oscillations inside the cochlea, which are absorbed by displacement of the *round window*. These oscillations cause specific regions of the basilar membrane to flex and stimulate the organ of Corti. The hair cells in these areas are attached to another, fixed membrane called the tectoral membrane. This produces a shearing force that bends the sterocilia. The deformation that results in these hair cells leads to the production of action potentials that travel to the brain via the auditory nerve.

18. The vibration of the basilar membrane is "tuned" to specific frequencies of sound vibration. Low frequency sounds cause the distal end of the membrane to vibrate, while high frequency sounds stimulate vibrations in the

proximal end. Only those hair cells located in the organ of Corti near regions of the basilar membrane that are stimulated by particular frequencies will respond. The action potentials produced in these areas are perceived as different sounds in the brain.

19. The auditory nerve makes synaptic connections with many areas in the brain, including the reticular activating system in the brain stem which is responsible for arousal state. This explains why loud or unusual sounds get our attention.

20. Neural information from the cochlea eventually reaches the temporal lobes of the cerebral cortex. Just as with the somatosensory and motor cortex regions of the brain, the primary auditory cortex contains detailed tone maps, with high frequency sounds represented in one location and low frequency sounds in another.

21. The loss of auditory ability, or deafness, can result from two general causes. *Conduction deafness* is a mechanical problem caused by the failure of the tympanic membrane or ear ossicles to properly transfer sound vibrations from the middle ear to the inner ear. It is usually the result of repeated ear infections which damage the middle ear tissues, or age-related stiffening of the movable ear ossicles.

22. *Nerve deafness* is caused by physical damage to the delicate transduction mechanism within the cochlea, usually as a result of prolonged exposure to excessively loud sounds such as those from jet engines or highly amplified music. It can occur at any age and is irreversible. Nerve deafness may also result from the failure of the auditory nerve to transmit the action potentials produced by the organ of Corti to the auditory cortex.

Questions (for answers and explanations, see page 362)

1. All but one of the following mechanosensor systems employs hair cells as its transducer. Select the *exception*.
a. Organ of Corti
b. Statocyst
c. Vestibular apparatus
d. Meissner's corpuscle
e. Lateral line

2. In which of the following would you find the malleus?
a. Knee joint of a human
b. Middle ear of a dog
c. Surface skin layer of a fish
d. Inner ear of a bat
e. Skeletal muscle of a bird

3. Mechanosensors have which of the following features in common with other types of sensory transducers?
a. Mechanical deformation of the membrane is the transducing stimulus.
b. Hair cells are utilized to convert a physical stimulus into action potentials.
c. The stimulus leads to a change in ion permeability of the sensor cell membrane.
d. They are integral sensory components of monosynaptic reflex loops.
e. None of the above

4. Statocysts are to the vestibular system as expanded-tip tactile receptors are to
a. Pacinian corpuscles.
b. stretch receptors.
c. the organ of Corti.
d. the tympanic membrane.
e. hair cells.

5. Which of the following statements about mechanosensors is *not* true?
a. Hair cells are found in the semicircular canals of the human ear.
b. Cupulae are present both in the vestibular apparatus of mammals and the lateral line system of fishes.
c. Different body areas have different proportions of the various skin touch receptors.
d. Hearing and balance in humans are both mediated by the same kind of mechanosensor.
e. All mechanosensors are neurons that fire action potentials in response to a physical force.

6. A man acquires a viral infection of the inner ear that permanently damages only the proximal end of the basilar membrane. Which of the following conditions would result?
a. He would suffer conduction deafness.
b. He would permanently lose some of his high-frequency hearing.
c. He would permanently lose some of his low-frequency hearing.
d. He would have difficulty maintaining proper balance.
e. He would become painfully sensitive to even very low-intensity sounds.

7. Action potentials produced by sensory neurons in the cochlea could end up influencing other neurons in which of the following brain areas?
a. Reticular activating system
b. Temporal lobes of the cerebral cortex
c. Primary auditory cortex
d. Brain stem
e. All of the above

8. A constant light touch on the skin of your finger is at first sensed by your brain, but then rapidly fades away. This effect is the result of adaptation by ___________.
a. Reissner's membrane
b. Pacinian corpuscles
c. Meissner's corpuscles
d. expanded-tip tactile sensors
e. the vestibular system

Activities (for answers and explanations, see page 362)

- Match each of the following mechanosensors with the most appropriate description.

_____Organ of Corti	*a.* Pressure sensor in fishes
_____Pacinian corpuscle	*b.* Monitors forces muscles exert on bones
_____Golgi tendon organ	*c.* Senses balance in humans
_____Statocyst	*d.* Structure of the vertebrate cochlea
_____Vestibular apparatus	*e.* Skin vibration sensor
_____Lateral line	*f.* Position sensor in lobster

- The diagram below shows some of the structures associated with the mammalian inner ear. Indicate which of these structures help detect head position, acceleration, or deceleration, and explain specifically how this is accomplished.

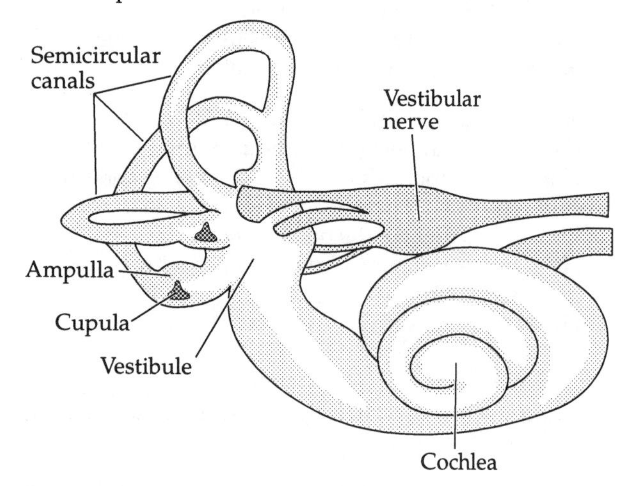

- The heavy line below represents the basilar membrane of a human ear encased within the cochlea. (Note that it has been stretched out flat, similar to Figure 39.15 in your textbook). The position of other relevant inner ear structures is also shown.

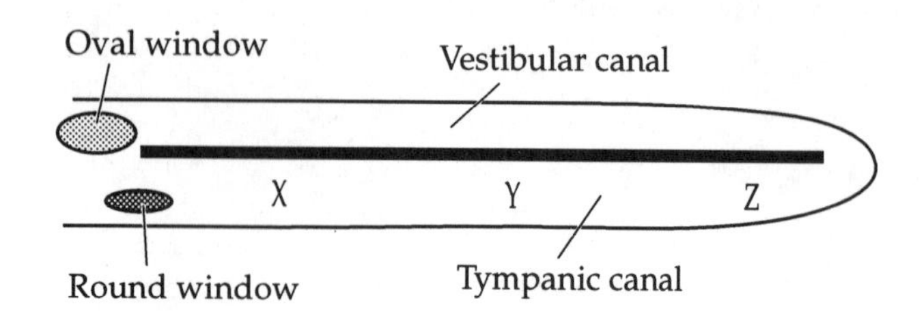

Predict the relative pitch of the three sounds that a person would hear if the basilar membrane were to vibrate first at point X, then Y, and finally Z.

Location	Perceived pitch of sound
X	______________
Y	______________
Z	______________

PHOTOSENSORS AND VISUAL SYSTEMS • OTHER SENSORY WORLDS (pages 894-906)

Key Concepts

1. Photosensitivity is the ability to detect light. Although the sophistication of photosensors ranges from those capable only of simple light–dark discriminations to those that perceive complex visual images, all animals detect light using a single photoreactive molecule called *rhodopsin.*
2. Rhodopsin consists of a nonabsorbing protein called *opsin* coupled with a light-absorbing side group called *11-cis-retinal.* The whole molecule is situated in the plasma membrane of a photosensor cell. When rhodopsin absorbs a photon of light, the 11-*cis* retinal undergoes a conformational change to become the isomer all-*trans* retinal. This strains the bond between opsin and retinal, causing the molecule to split apart in a process known as bleaching. The bleached molecule is no longer photosensitive. After bleaching, retinal spontaneously reconverts to the 11-*cis* isomer. This allows it to combine with the isolated opsin and regenerate a functional rhodopsin molecule.
3. Each vertebrate photoreceptor has thousands of rhodopsin molecules embedded in the membranes of its outer segments. The membrane potential of an inactive photosensor is considerably less negative than most neurons because the membrane is more permeable to Na^+ ions. Sodium channels are kept open in the resting state by the presence of a high concentration of cyclic GMP (cGMP) molecules.
4. Photoreceptor cells do not produce action potentials. Instead, they hyperpolarize when stimulated by light. This occurs because the photoactivation of rhodopsin causes a cascade of chemical events that leads to the closing of sodium channels.
5. Photoexcited rhodopsin activates a G-protein called transducin, which in turn activates a phosphodiesterase enzyme. This enzyme converts cGMP to 5′-GMP, removing the molecule needed to keep the sodium channels open. With Na^+ ion flow into the cell interrupted, the membrane hyperpolarizes.
6. Although this mechanism seems a roundabout way to change membrane potential, its power lies in the enormous signal amplification possible through the repeated action of each transducin and phosphodiesterase molecule. Thus, a single photon of light can result in the closing of over a million sodium channels.
7. Simple invertebrates such as flatworms have photosensitive cells organized into primitive eye cups. The system does not provide image vision, but does allow the

animal to discriminate between areas of light and dark in order to seek shelter and avoid predators.

8. Arthropods have more complex visual systems capable of detecting patterns, images, and movement. These animals have *compound eyes* that consist of many individual optical units known as *ommatidia*. Depending on the species, the number of ommatidia in each eye varies from only a few to over 10,000.

9. Each ommatidium has a crystalline lens that focuses light on photosensitive retinula cells. Microvilli on each retinula cell contain rhodopsin molecules that capture photons. In some insects, the microvilli overlap to form a central structure called the rhabdom. The receptor potentials generated by these combined retinular cells are fed onto a single nerve fiber that connects each ommatidium with the brain. Because a single ommatidium is fixed within the compound eye and views only a limited portion of space, the animal probably sees a fragmented image of the world.

10. The best image-forming eyes are found in mollusks and vertebrates. Their eyes operate like a camera, and are strikingly similar in both structure and function considering the independent evolution of each group.

11. Vertebrate eyes are fluid-filled spheres bounded by a sheet of tough, opaque connective tissue called the sclera. An area in the front of the sclera is transparent, forming the *cornea* which allows light to enter the eye.

12. Immediately inside the cornea is the *iris*, a pigmented tissue that is under autonomic nervous control. Changes in the size of the iris regulate the diameter of its central aperture called the *pupil*. Pupillary size determines the amount of light that passes through the eye to reach the photosensors.

13. A crystalline *lens* lies behind the iris, helping to focus light rays on the photosensors. An important function of the lens is accommodation, which is how the eye adjusts its focus for objects at different distances from the observer. Strong ciliary muscles attached to the lens either move it back and forth within the eye (fishes, amphibians, and reptiles) or cause it to change shape (birds and mammals). This adjusts the optical properties of the lens and refocuses the light rays. In humans, age contributes to a reduction in lens elasticity, making accommodation difficult and producing the need for corrective lenses.

14. The retina is comprised largely of neural tissue which is connected directly to the brain via the optic nerve. It contains the photoreceptor cells, the density of which varies across the retinal surface. A *fovea* is the place on the retina where photoreceptors are in particularly high concentration, thus providing good visual acuity. The number, location, and shape of the fovea varies with species.

15. The neurons within the retinal tissue overlie the photoreceptor cells. No photoreceptors occur at the spot on the retina where the axons from these neurons join to exit the eye through the optic nerve. This area is known as the *blind spot*. You do not normally detect the visual "hole" created by your blind spot because the brain fills in this space based on information from photoreceptors in the surrounding area.

16. Two major types of photoreceptors are found in vertebrate retinas, each named because of its shape. *Rod cells* are the more abundant, numbering nearly 100 million in humans. They are very sensitive to low levels of light, but make no contribution to color vision. *Cone cells* are much less numerous (about 3 million in humans). They are only activated by relatively high light levels and produce color vision.

17. The distribution and number of rods and cones in the retina varies with the animal's lifestyle. Diurnal animals have cone-dominated foveas, with rod cells pushed mostly to the peripheral areas of the retina. This provides excellent visual acuity during daylight hours, but not at night. The entire retina of some diurnal animals contains only cone cells. Nocturnal animals tend to have rod-dominated foveas, and so their visual acuity is best at night. Some nocturnal animals have only rods in their retinas, and thus possess no color vision.

18. There are at least three, and possibly five or six, types of vertebrate cones. Each detects a different color because of the slightly different wavelength absorption properties of their opsin molecules. Retinal actually absorbs the photon, but it is the opsin that determines where in the color spectrum retinal absorbs best. Human cones absorb maximally in the blue (455 nm), green (530 nm), and red (625 nm) wavelengths.

19. Five layers of neurons (four in addition to the rods and cones) are found in the vertebrate retina. These neurons overlie the photoreceptors, so light must pass through them before encountering the rods and cones. Behind the rods and cones of diurnal animals is a dark layer of epithelium that captures stray light and prevents backscattering, which would blur the image.

20. Many nocturnal animals, for which efficient light gathering is a must, have a highly reflective layer behind the photoreceptors called the *tapetum*. This structure returns uncaptured photons to the rods for possible detection. While it enhances visual sensitivity, acuity is decreased. Light reflected from the tapetum back through the eye is what produces the "eye shine" of animals illuminated by a car's headlights at night.

21. Photoreceptors are the outermost layer of retinal neurons. The next inner layer consists of bipolar cells, which make direct synaptic contact with more than one rod or cone. As the membrane potential of the photoreceptors changes, neurotransmitter is released onto the bipolar cells. Like rods and cones, the bipolar cells do not fire action potentials.

22. Several bipolar cells make synaptic contact with a single ganglion cell, whose axon exits the eye through the optic nerve. When a bipolar cell is stimulated, it releases neurotransmitter onto the ganglion cell, which may then fire an action potential that travels directly to the brain. The rate at which action potentials are produced by ganglion cells varies directly with the pattern and quantity of light striking the photoreceptors it monitors.

23. Sandwiched between the bipolar and ganglion cell layers are amacrine and horizontal cells, which communicate laterally across the retina to increase sharpness, contrast, and overall light sensitivity. Horizontal cells link neighboring photoreceptor–bipolar cell pairs, while amacrine cells connect adjacent bipolar–ganglion cell combinations.
24. The information processing achieved by these complex retinal connections helps reduce the total amount of neural information sent to the brain. Each ganglion cell receives input from hundreds of adjacent photoreceptors and tens of bipolar cells, which together form the *receptive field* of that ganglion cell.
25. Ganglionic receptive fields are mapped by measuring the cell's response to a spot of light passing over the section of retina it monitors. These fields are circular and consist of two concentric rings called the center and the surround. The cell's response to a stimulus varies depending on whether the spot hits the center or the surround.
26. Some ganglion cells are excited by light in the center of their receptive fields but not the surround; these cells have on-center receptive fields. If a stimulus in the surround but not the center is excitatory, the receptive field is called off-center. For large spots of light that cover portions of both the center and surround, the response of the center dominates. This pattern allows the ganglion cells to inform the brain about light–dark contrasts.
27. The interconnection of bipolar and ganglion cells made possible by the horizontal cells provides the underlying circuitry for ganglionic receptive fields. Since any one photoreceptor or bipolar cell can be connected to several ganglion cells, the receptive fields of adjacent ganglion cells often overlap.
28. The brain analyzes visual images in the occipital cortex, where it combines information from many ganglion cells, forming cortical receptive fields. Several types of receptive fields are found in the visual cortex. The fields of *simple cells* respond best to a bar of light with fixed orientation that stimulates one particular location on the retina. *Complex cells* also respond to bars of fixed orientation, but they do so at many locations across the retina. Some complex cells also prefer bars that move in a particular direction.
29. The ganglionic receptive fields that supply information to simple cortical cells are arranged close together on the retina, while those of complex cortical cells are more widely scattered. It appears that the brain creates its image of the visual world by analyzing the patterns of light edges that stimulate various regions on the retina.
30. Binocular vision is a product of the brain's integrating visual information from both eyes, which are placed on the head so as to view the same object from slightly different angles. The *optic chiasm* is the point just below the hypothalamus where the optic nerves join and then separate again. The axons of cells that view the right visual field in both eyes go to the left cerebral hemisphere, and those that view the left visual field go to the right hemisphere.
31. Both simple and complex cells in the primary visual cortex are organized into alternating columns that gather information from the left eye, then the right, and so on. Those cells on the border between columns receive input from both eyes, and are thus *binocular cells*. They measure distance by determining the disparity between where an object falls on the retinas of the two eyes.
32. Specific visual images are likely the product of simultaneous interactions among many retinal and cortical cells. This process is enhanced by neural input from memory and other sensory channels.
33. Beyond the visual spectrum to which our photoreceptors respond, other forms of electromagnetic radiation can be sensed by animals. Ultraviolet radiation contains shorter wavelengths than our visual system can respond to, but many insects see quite well in this region of the spectrum. Flowering plants take advantage of this sensitivity and use ultraviolet markers to attract insect pollinators.
34. Wavelengths longer than those we can detect are also sensed by some animals. Infrared radiation, which we sense as heat, is detected by pit vipers using special organs located near their eyes. This information can be used to hunt endothermic prey even when there is not enough light to see.
35. Echolocation is a mechanism used by bats, dolphins, and a few other vertebrates to create acoustic images of the world. These animals emit high-intensity (and often high-frequency) sounds and then listen for the echoes that return as those sounds reflect off of objects in the environment.
36. Electroreceptors are modified mechanoreceptors found in certain fish that live in murky waters. Some are associated with the lateral line system and others are separate organs. They produce weak electric fields that are used to detect and capture prey, as a means of defense, or to communicate with other individuals.
37. Magnetic sensitivity is also a viable sensory system in some animals, although the mechanism for magnetoreception remains unclear. Its primary function appears to be the gathering of directional information from the Earth's magnetic field for use in orientation.

Questions ***(for answers and explanations, see page 362)***

1. In which of the following animals would you expect to find the protein opsin?
   a. Human
   b. Bird
   c. Salamander
   d. Butterfly
   e. All of the above
2. Photopigments from the eye of a cow are extracted and labeled with a substance that turns bright blue in the presence of 11-*cis* retinal. A sample of these labeled pigments is then exposed to white light for 20 minutes. Which of the following statements about the results of this treatment is *not* true?
   a. The retinal side group will have dissociated from the opsin protein.
   b. The sample will be bright blue.
   c. The sample will show little or no blue color.
   d. No light-absorbing rhodopsin will be present.
   e. Most of the retinal will be in the all-*trans* isomeric form.

3. From which of the following cells in the human visual system could you *not* record an action potential?
   a. Ganglion cells
   b. Simple cells in the visual cortex
   c. Complex cells in the visual cortex
   d. Axons in the optic nerve
   e. Bipolar and photoreceptor cells

4. A photon of light striking a human rod cell causes that cell to hyperpolarize because
   a. the rate of conversion of 5′-GMP to cGMP is increased.
   b. the flow of Na^+ ions into the cell rises dramatically.
   c. transducin causes inactivation of the enzyme phosphodiesterase.
   d. cGMP production is halted, causing sodium channels to close.
   e. None of the above

5. Which of the following is *not* an element of the vertebrate eye?
   a. Lens
   b. Horizontal cell
   c. Retinula cell
   d. Iris
   e. Sclera

6. Through which of the following cell layers must a photon of light pass before striking a cone cell in the eye of a hawk?
   a. Amacrine
   b. Bipolar
   c. Horizontal
   d. Ganglion
   e. All of the above

7. Changing the focus of a vertebrate eye by making adjustments to the shape or position of the lens is called
   a. bleaching.
   b. isomerization.
   c. hyperpolarization.
   d. accommodation.
   e. echolocation.

8. Which of the following statements about animal vision is *not* true?
   a. The tapetum aids in the rapid regeneration of bleached rhodopsin molecules.
   b. There are several types of cones in vertebrate retinas, each of which detects different wavelengths.
   c. An ommatidium is only one visual element within the compound eye of an arthropod.
   d. Photosensors in simple invertebrates detect light–dark contrasts, but do not produce image vision.
   e. Although retinal actually absorbs the photon, the wavelength it absorbs is determined by opsin.

9. Simple cells in the mammalian visual cortex
   a. respond to stationary or moving light bars of fixed orientation at several locations across the retina.
   b. respond only to stationary light bars of fixed orientation at one specific location on the retina.
   c. analyze input from a single ganglionic receptive field.
   d. utilize the photosensor information from a very limited number of rods and cones.
   e. probably do not contribute to binocular vision.

***Activities** (for answers and explanations, see page 363)*

- Label each of the following structures on the diagram of a human eye shown below: sclera, cornea, iris, pupil, lens, retina, fovea, blind spot, optic nerve.

- Below is the receptive field of a retinal ganglion cell.

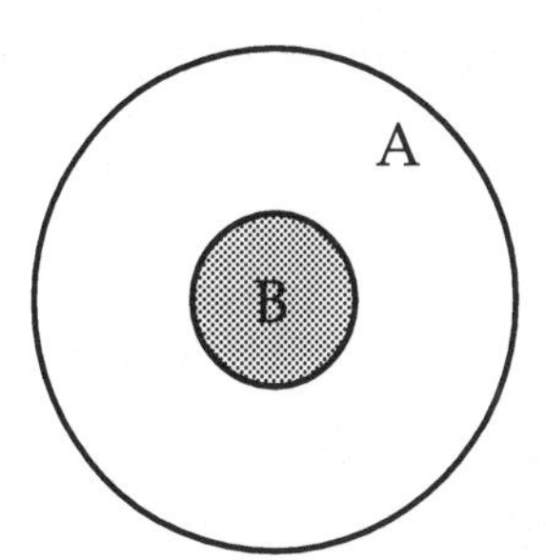

Previous studies have revealed that this is an off-center receptive field of a ganglion cell. Suppose you stimulate the portion of the retina associated with this receptive field using a small spot of light, and then record the action potentials the ganglion cell produces. Predict the cell's response to each of the following experimental conditions:

1. The light is flashed briefly at point A.

2. The light is flashed briefly at point B.

3. The light is moved slowly from point A to point B.

Integrative Questions (for answers and explanations, see page 363)

1. Which of the following sensory systems would you *not* expect to find in a pit viper?
   a. Olfaction
   b. Infrared detection
   c. Echolocation
   d. Gustation
   e. Vision

2. Hearing, touch, balance, and temperature sensitivity are all related by a common transduction mechanism known as
   a. photoreception.
   b. mechanoreception.
   c. chemoreception.
   d. electroreception.
   e. magnetoreception.

3. After a day of vigorous sailing you come ashore and find that your walk is somewhat unsteady. When you sit down to rest, the world seems to spin a little even though you are perfectly still. The sensory basis underlying this uneasy feeling is related to overstimulation of
   a. the visual system.
   b. the vestibular system.
   c. the olfactory system.
   d. the auditory system.
   e. None of the above

4. You discover a new species of animal which appears to be a very primitive type of invertebrate. Which of the following kinds of sensory systems would you almost certainly expect it to possess?
   a. Chemical
   b. Visual
   c. Auditory
   d. Magnetosensory
   e. All of the above

40

Effectors

CHAPTER LEARNING OBJECTIVES—after studying this chapter you should be able to:

- ❑ Define the term "effector" and give examples of the kinds of things that effectors do in animals.
- ❑ Compare and contrast the molecular structure of microfilaments with that of microtubules, and discuss the roles of each in controlling animal effectors.
- ❑ Describe the structure and function of cilia and flagella in animals.
- ❑ Explain the molecular basis for the functioning of axonemes.
- ❑ Discuss the roles of microfilaments, cytoplasm, plasmagel, and plasmasol in controlling the amoeboid movements of cells.
- ❑ Describe the structural and functional differences between the three types of vertebrate muscle tissue.
- ❑ Differentiate between muscles, muscle fibers, and myofibrils.
- ❑ Explain the structure and function of a sarcomere.
- ❑ Discuss the roles of actin, myosin, troponin, and tropomyosin in the sliding filament theory of muscle contraction.
- ❑ Describe the molecular events that occur when a sarcomere contracts and the importance of ATP to this process.
- ❑ Explain the role of a motor unit in muscle contraction.
- ❑ Describe the events that lead to a muscle twitch and explain how twitches are related to the phenomena of tetanus and tonus.
- ❑ Discuss the anatomical and physiological differences between fast twitch and slow twitch fibers.
- ❑ Explain how the special anatomy of cardiac muscle leads to its myogenic properties.
- ❑ Compare the structural and functional properties of hydroskeletons, exoskeletons, and endoskeletons.
- ❑ Discuss the advantages and disadvantages associated with hydroskeletons, exoskeletons, and endoskeletons.
- ❑ In humans, distinguish between the bones of the axial skeleton and the appendicular skeleton.
- ❑ Describe the roles of osteoblasts, osteoclasts, and osteocytes in regulating the calcium content of bone.
- ❑ Identify the various types of bones and joints in a human skeleton and explain the structural and functional importance of each.
- ❑ Explain the concept of antagonistic muscle action and distinguish between a flexor and an extensor muscle.
- ❑ Discuss the roles of ligaments and tendons in the action of endoskeletons.
- ❑ Describe how endoskeletal action can be analyzed using the concept of a lever, and what general principles of skeletal design emerge from this analysis.
- ❑ List several of the nonmuscle effectors found in animals and give their essential functions.

HOW ANIMALS RESPOND TO INFORMATION FROM SENSORS • CILIA, FLAGELLA, AND MICROTUBULES • MICROFILAMENTS AND CELL MOVEMENT (pages 909-914)

Key Concepts

1. *Effectors* allow animals to respond to sensory stimuli. Both internal and external responses are possible, and can take the form of locomotion, chemical secretion, or some sort of energy emission.
2. Cell movements range from the actions of cilia and flagella to the contractions of muscle. *Microtubules* and *microfilaments* are the cellular structures responsible for creating movement. Microtubules make cilia and flagella move, while microfilaments generate larger-scale movements of muscle. Both microtubules and microfilaments consists of long protein molecules that slide past one another. When linked with another cell structure or a skeletal system against which they can pull, coordinated locomotion becomes possible.
3. *Cilia* are tiny hairlike appendages used to move entire cells, push mucus or fluids past the linings of epithelial tissue in animals, or filter food from the environment.
4. Cilia push against a fluid like a swimmer's arms push against water during the breaststroke. The *power stroke* occurs when the outstretched cilium moves backwards through the water. The cilium folds during the *recovery stroke*, creating minimal drag as it is pulled forward through the water.

5. If cilia are fixed in place, as is the case for cells lining the respiratory tract and portions of the female reproductive system, the fluids that overlie them are moved in the same direction as the power stroke.
6. In eukaryotes, *flagella* are essentially long cilia that occur singly or in small clusters. Flagella move in a whiplike fashion that creates more power than can be produced by an individual cilium. Flagella can be used to move entire cells such as sperm.
7. Microtubules are responsible for the movement of cilia and flagella. They contain long chains of the globular polypeptide *tubulin* and are always found in a characteristic ring of nine pairs. Sometimes one or two extra microtubules are found in the center of this ring, connected to it by radial spokes. The entire complex is called an *axoneme*.
8. Axoneme movement occurs when adjacent microtubules slide past one another. Chemical crossbridges between microtubules are made of the protein *dynein*, a mechanoenzyme that generates force when it hydrolyzes ATP in the presence of Ca^{2+} ions and undergoes a conformational change. Because microtubules are fixed to the cell at their bases, the sliding movement produces a bend in the cilium or flagellum rather than an elongation. This motile mechanism is an inherent property of the axoneme itself.
9. Microtubules also cause changes in cell shape through mechanical effects on the cytoskeleton. By changing the length of tubulin chains, intracellular events such as spindle formation during mitosis and directed cell growth in neurons can be coordinated.
10. Microfilaments are made largely of the protein *actin*. They are used to support cilia and to direct cell movements. Acting in combination with another microfilament made of the protein *myosin*, actin generates the contractile forces used to cleave dividing cells, facilitate phagocytosis and pinocytosis, and cause muscle contraction.
11. *Amoeboid movement* is another microfilament-controlled activity characteristic of certain protists and cells that migrate within animal bodies. Actin and myosin microfilaments contract in precise patterns just beneath the plasma membrane of these cells to regulate the distribution of thick *plasmagel* and thin *plasmasol*. The sol is pushed out to form a pseudopod which eventually stops growing as the cytoplasm turns to gel.
12. Controlled amoeboid motion is based on the cell cytoplasm's ability to alternate between sol and gel forms, combined with the coordinated contraction of microfilaments, which directs cytoplasmic streaming to form a pseudopod.

Activities (for answers and explanations, see page 363)

- In mammals, portions of the respiratory tract are lined with cilia. Suggest a physiological function for these cilia and briefly explain the molecular basis for how that function is accomplished.

- Match each term with its best description.

____	Axoneme	*a.* Mechanoenzyme bridge between microtubules in cilia
____	Recovery stroke	*b.* Central structure of a flagellum
____	Myosin	*c.* Cell shape change caused by microfilaments
____	Dynein	*d.* Cilium bends to reduce drag forces in water
____	Pinocytosis	*e.* Microfilament protein often associated with actin

Questions (for answers and explanations, see page 363)

1. ATP is needed for the movements of cilia and flagella because
   a. the microtubules are made of ATP.
   b. ATP is a precursor of tubulin, from which the microtubules are then made.
   c. its hydrolysis provides the energy needed to change the shape of dynein.
   d. ATP is used to connect the base of the cilium or flagellum with the cell.
   e. None of the above
2. Detailed analysis of a cilium from the human respiratory tract would reveal all but one of the following. Select the *exception*.
   a. Tubulin
   b. Plasmagel
   c. Microtubules
   d. Axoneme
   e. Dynein
3. Which of the following statements is *not* true?
   a. Effectors are involved only in the production of cell movement.
   b. Microtubules play more of a role in flagellum movement than do microfilaments.
   c. Cilia are often found associated with epithelial tissue in vertebrates.
   d. Flagella are essentially large cilia found singly or in small clusters.
   e. Axonemes are present both in cilia and flagella.
4. Microfilaments play a major role in all but one of the following cell activities. Select the *exception*.
   a. Phagocytosis
   b. Muscle contraction
   c. Amoeboid movement
   d. Flagellum movement
   e. Cell cleavage
5. Which of the following always occurs when an axoneme moves?
   a. Microfilaments form crosslinkages with microtubules.
   b. Tubulin dissociates under the regulation of mechanoenzymes.
   c. A flagellum completes its recovery stroke.
   d. Phagocytosis or pinocytosis occurs.
   e. Adjacent microtubules slide past one another.

6. A scientist gently treats a flagellar axoneme with digestive proteins that separate the cross-links and spokes but leave the microtubules and dynein arms intact. If she then exposes the isolated microtubules to ATP and calcium ions the result should be
   a. the complete breakdown of dynein but not tubulin.
   b. the complete breakdown of tubulin but not dynein.
   c. the breakdown of both tubulin and dynein.
   d. bending of the whole structure.
   e. elongation of the whole structure.

7. If the connection that anchors the microtubules of a cilium to a cell is severed, then the
   a. cilium will continue to function normally.
   b. cilium will beat like a flagellum.
   c. cilium will no longer bend.
   d. microtubules will no longer slide past one another.
   e. power stroke will occur but not the recovery stroke.

MUSCLES (pages 914–920)

Key Concepts

1. Muscle cells are specialized for contraction. Their diversity and effectiveness makes them the most important effectors in animals.
2. Muscle cells have high densities of the microfilaments actin and myosin. All three vertebrate muscle types share a common mechanism of contraction that depends on the sliding action of these microfilaments.
3. *Smooth muscle* is under autonomic control and provides the contractile activity of internal organs. It helps regulate food digestion, blood flow, and excretion.
4. Smooth muscle cells are simple compared to other types of muscle cells. They are long and spindle-shaped, contain a single nucleus, and possess a loose network of microfilaments.
5. Smooth muscle occurs in sheets with gap junctions between adjacent cells. Action potentials can therefore travel rapidly through the entire sheet. Smooth muscle cells also respond to stretching by depolarizing their membranes. This makes it easier for an action potential to occur and leads to contraction. Smooth muscle thus has the capacity to contract in proportion to the magnitude of a stretching stimulus.
6. *Skeletal* (or *striated*) *muscle* controls all voluntary movements. Skeletal muscle cells, also known as *muscle fibers*, are large and contain many nuclei. Bundles of muscle fibers constitute a whole muscle (e.g., biceps).
7. Just like a whole muscle is made of many muscle fibers, a fiber is made of many *myofibrils*. Each myofibril is in turn a bundle of microfilaments arranged in a highly regular fashion.
8. The myofibril consists of an array of thick myosin filaments surrounded by several thin filaments of actin. The two filament types overlap in places, creating a banded pattern along the myofibril. Repeating units of banding, called *sarcomeres*, are evident.
9. The sarcomere is the basic unit of muscle contraction. The widths of certain characteristic bands making up each sarcomere shorten as the microfilaments slide past one another.
10. Z lines on either end of the sarcomere represent the outer boundary anchor points for the actin filaments. A band of myosin filaments (the A band) is centered in the middle of the sarcomere.
11. As the actin and myosin move past each other, the entire sarcomere shortens. The light spaces where actin and myosin filaments do not overlap (H band and I band) shorten as the A band approaches the Z lines. The M band, a dark stripe within the H band, contain proteins that hold the myosin filaments in a hexagonal arrangement.
12. The fine structure of actin and myosin suggests a mechanism for how sarcomeres contract. Myosin molecules are made of two long polypeptides coiled together, each with a globular head at one end that contains a binding site capable of forming a crossbridge with actin. A myosin filament contains many myosin molecules. Actin filaments consist of two chains of monomers wound together in a helix. Two strands of *tropomyosin*, a third protein, are coiled around the actin chains and block access to the myosin binding sites.
13. The *sliding-filament theory* explains the molecular events that occur when actin and myosin interact and a sarcomere shortens. The head regions on each myosin subunit bind to actin molecules on an adjacent microfilament, forming a chemical crossbridge between the two chains. When this bond forms, the myosin heads change orientation and the two filaments move past each other by some 5–10 nm.
14. ATP binds to myosin at these crossbridge sites, causing the myosin to release the actin. Energy gained from the subsequent hydrolysis of ATP reorients the myosin head so that it can form another bond with actin. ATP is thus required to break the bonds between microfilaments, not to make them. This explains the phenomenon of rigor mortis, where the body stiffens following death because ATP is no longer being produced to break the actin and myosin bonds.
15. A *motor unit* includes all of the muscle fibers innervated by a single motor neuron. The timing and coordination of actin–myosin interactions within a motor unit begins at the nerve–muscle synapse. Neurotransmitter released at the neuromuscular junction stimulates changes in the membrane potential of muscle fibers in the motor unit. This leads to depolarization and the firing of an action potential, which spreads rapidly across the entire cell surface.
16. The plasma membrane of muscle fibers plunges deep into the cytoplasm to form a system of *T-tubules*, which conducts action potentials throughout the muscle cell. The T-tubules communicate with a membranous network called the *sarcoplasmic reticulum*, which surrounds every myofibril and serves as a storage site for Ca^{2+} ions.
17. Action potentials travel quickly through the T-tubules and cause the sarcoplasmic reticulum to release Ca^{2+} ions. This stimulates another chemical reaction that directly mediates the interaction of actin and myosin.

18. At rest, tropomyosin covers the myosin binding sites on each actin molecule within the filament. Another protein called *troponin* is found attached at regular intervals along the tropomyosin strand. Calcium entering the muscle cell upon depolarization binds to troponin and causes it to change shape. This causes the attached tropomyosin to move away from the actin binding sites it was covering, thus preparing the actin for interaction with myosin.
19. When the muscle cell repolarizes, the calcium ion channels close. Molecular pumps then recollect Ca^{2+} ions into the sarcoplasmic reticulum. This returns troponin to its original conformation, allowing tropomyosin to once again cover the myosin binding sites. The muscle fiber can no longer sustain contraction and returns to its resting condition (i.e., it relaxes).
20. *Twitches* are the minimal units of contraction caused by a single action potential within the muscle fiber. Their strength is measured in terms of the tension they generate.
21. If the action potentials that generate twitches are separated by enough time, each twitch appears as an all-or-none event (i.e., a complete contraction and relaxation cycle). However, if the action potentials arrive in rapid succession, the cell does not have enough time to complete its relaxation before the next contraction phase begins. The muscle tension created increases in a graded response proportional to the level of neural stimulation.
22. If the stimulation rate is very high, the muscle cell never has time to relax and remains in a completely contracted state called *tetanus*. Tetanus can be maintained only as long as the supply of ATP persists. If the energy supply fails, the muscle fatigues and tension declines.
23. The graded response of whole muscle depends on the number of motor units activated. Low levels of activation are continuously maintained in many muscles, a condition known as *tonus*. Muscle tone is possible without fatigue because precise neural control means that only a small number of fibers are involved at any one moment.
24. Muscle fibers are categorized according to the speed of their twitch. *Fast twitch fibers* are found in white muscle, which lacks oxygen-binding myoglobin and has few mitochondria or blood vessels. The white breast muscle of a domestic chicken can rapidly produce high maximum tension, but it fatigues quickly and thus cannot sustain this tension. White muscle is useful during short bursts of powerful anaerobic muscle activity, which explains why chickens are not strong flyers. Fast-twitch fibers are also abundant in human sprinters and weight-lifters.
25. *Slow twitch fibers* are found in dark muscle, which has high concentrations of myoglobin, mitochondria, and blood vessels. They are slow to reach maximum tension, but capable of sustaining that tension for long periods of time without fatigue. They are also well supplied with fat and glycogen reserves for ATP production. Dark muscle is the endurance muscle associated with the chicken's legs. It is used in sustained aerobic activity, such as when the bird runs or when a human swims long distances.
26. Although there can be some adjustment through physical training, the proportion of fast- and slow-twitch fibers in human skeletal muscle is determined primarily by a person's genetic background.
27. *Cardiac muscle* is the third type of vertebrate muscle; it is found only in the heart. Cardiac muscle is striated like skeletal muscle, but does not occur in linear chains of cells. Instead, it forms a meshwork of fibers that is not easily torn by the high pressures developed within the heart.
28. Cardiac muscle fibers are uninucleate cells joined together by *intercalated disks*, which provide good adhesion. Gap junctions are also present, allowing direct electrical contact for rapid transmission of action potentials.
29. The contractions of certain cardiac muscle fibers are *myogenic*; they can act as a pacemaker for the heart. This is possible because of a unique class of potassium channels found only in these cardiac muscle cells. These channels remain somewhat open after an action potential, but then slowly close. In response the cell slowly depolarizes, eventually reaching threshold and producing another action potential which leads to contraction.
30. Although influenced by the autonomic nervous system, the beating of the heart does not depend on direct neural influence. A completely isolated vertebrate heart can beat for hours due to its own inherent pacemaking properties. This physiological trait is what makes heart transplant surgery possible.

Questions (for answers and explanations, see page 363)

1. Smooth muscle controls the action of all but one of the following effectors. Select the *exception*.
   a. Human thumb
   b. Rat intestine
   c. Chicken blood vessel
   d. Frog urinary bladder
   e. Cow stomach

2. Bundles of _________ make up a _________, which in turn are bundled to create a _________.
   a. muscle cells; myofibril; whole muscle
   b. microtubules; muscle cell; myofibril
   c. microtubules; whole muscle; microfilament
   d. microfilaments; myofibril; whole muscle
   e. None of the above

3. Microfilaments can be found in
   a. skeletal muscle.
   b. cardiac muscle.
   c. smooth muscle.
   d. all three vertebrate muscle types.
   e. every muscle type except cardiac.

4. When a sarcomere contracts,
   a. the H band and I band both widen.
   b. the A band stays the same, but the Z lines get closer together.
   c. the Z lines move farther apart.
   d. the A band widens.
   e. the H band shortens and the I band widens.

5. Which of the following statements about the sliding-filament theory is *not* true?
   a. Ca^{2+} ions play a crucial role in facilitating the binding of actin and myosin.
   b. The separation of myosin from actin requires ATP.
   c. Troponin is a double-stranded protein that covers the myosin binding sites on actin microfilaments.
   d. The headpiece of a myosin molecule binds to special receptor sites on the actin filament.
   e. Rigor mortis results from a breakdown in the mechanism that releases bound microfilaments.

6. Before stimulating the release of calcium ions that causes muscle contraction, an action potential must pass through the
   a. sarcoplasmic reticulum.
   b. T-tubules and sarcoplasmic reticulum.
   c. plasma membrane and T-tubules.
   d. plasma membrane, T-tubules, and sarcoplasmic reticulum.
   e. None of the above

7. The release of calcium ions is essential for initiating the sliding action of microfilaments because
   a. calcium is needed to complete the structural organization of the myofibril.
   b. the enzyme that produces ATP is calcium-dependent.
   c. it is the calcium that depolarizes the muscle cell.
   d. tropomyosin cannot bind to actin in the absence of Ca^{2+} ions.
   e. calcium binds to troponin, causing it to change shape and expose myosin binding sites on the actin.

8. When a person accidentally touches a live electric wire and passes current through his body, many muscles become locked in a state of continuous contraction known as
   a. tonus.
   b. tetanus.
   c. twitching.
   d. depolarization.
   e. myogenesis.

9. The leg muscles of champion marathon runners have a large proportion of _______ fibers, as well as ________.
   a. slow twitch; little or no glycogen and fat reserves
   b. fast twitch; lots of glycogen and fat reserves
   c. slow twitch; very few mitochondria
   d. fast twitch; abundant mitochondria
   e. slow twitch; abundant mitochondria

10. The intercalated discs of ________ muscle ________ .
   a. smooth; make possible the movements of intestines and glands
   b. smooth; provide strong mechanical adhesion and rapid electrical communication
   c. skeletal; are the basis for all voluntary muscle action
   d. skeletal; make possible both fast twitches and slow twitches.
   e. cardiac; provide strong mechanical adhesion and rapid electrical communication

Activities (for answers and explanations, see page 364)

- In the following diagram of a relaxed sarcomere, label the actin and myosin filaments, as well as all the major bands and lines visible.

- Now sketch this same sarcomere as it would appear in a contracted state. What basic molecular event has happened to cause this change in the shape of the sarcomere?

SKELETAL SYSTEMS • OTHER EFFECTORS (pages 920-929)

Key Concepts

1. For muscles to generate useful motion they must have supporting structures against which to direct their contractile forces. Such support is provided by *skeletal systems,* of which there are three basic types.
2. *Hydrostatic skeletons* result from the containment of incompressible fluids inside a muscular body cavity. When the muscles contract, the fluid moves and changes the body shape. Opposing sets of circular and longitudinal muscles produce alternating elongation and shortening of the body, allowing controlled movements such as those of a sea anemone.
3. Earthworms have a compartmentalized hydrostatic skeleton made of many repeating segments. Locomotion is achieved by passing alternating waves of circular and longitudinal muscle contraction down the length of this segmented body.
4. Another useful adaptation of hydroskeletons is the ability to create jet propulsion. By contracting muscles surrounding an open-ended cavity, squids and octopuses are able to force water out of their bodies under pressure. The force generated by this moving water pushes the animal in the opposite direction.
5. *Exoskeletons* are hardened external shells to which muscles attach. Muscle contraction causes the exoskeleton to bend at hinged areas called joints.
6. The shell of a mollusk is the simplest kind of exoskeleton. It is made of one or only a few heavily mineralized plates. These plates cover most of the animal's body and

can be quite massive. Shells offer good protection for the soft hydroskeletal components and delicate internal body organs.

7. The complex exoskeletons of arthropods are made of a layered cuticle that forms tough plates covering all body areas and appendages. Hardening material reinforces the plates everywhere except the joints, which remain flexible. The thin outer epicuticle is covered with a waxy coating to prevent desiccation. The thicker, tougher endocuticle contains the polysaccharide chitin for strength and may be further reinforced with calcium salts. Muscles attached to the inner surfaces of the exoskeleton contract to create movement around the hinged joints.
8. Although it provides excellent protection against drying and physical damage, once in place the arthropod exoskeleton cannot increase its size. To grow, an animal must *molt* its exoskeleton and secrete a larger one. Immediately after molting the animal passes through a brief period of vulnerability until its newly formed exoskeleton hardens.
9. The vertebrate *endoskeleton* is an internal framework of hinged bones that attach to the skeletal muscles. Platelike, rodlike, and tubelike bones articulate at joints to permit a wide range of motion.
10. The human skeleton contains over 200 bones and is subdivided into the trunk region, or *axial skeleton*, and the outer region, or *appendicular skeleton*. The appendicular skeleton attaches to the axial skeleton at two locations, the pectoral and pelvic girdles.
11. Endoskeletons are not as protective as exoskeletons, but this disadvantage is more than offset by the endoskeleton's ability to grow continuously throughout life.
12. Cartilage and bone are the connective tissues that make up the endoskeleton. They are good structural tissues because of their tough extracellular matrices, which are reinforced with collagen fibers.
13. Cartilage is a strong yet flexible material that confers stiffness, resiliency, and padding to endoskeletons. It is used to shape certain features such as the nose, larynx, and ears, or to join bone surfaces together in regions that experience high impact, strain, and repeated movement. Cartilage is the principle component of the embryonic skeleton in most vertebrates, but is replaced with bone as development progresses. An exception is found in the cartilaginous fishes (sharks and rays) which retain a cartilage-based skeleton throughout life.
14. The extracellular matrix of bone is heavily mineralized with calcium phosphate, making it a hard and rigid structural material. Endoskeletons also represent the body's major reserve of calcium ions, which are used in muscle and nerve function as well as many other physiological activities. Soluble calcium in body fluids and skeletal calcium are in a dynamic equilibrium under hormonal control.
15. Young bone-producing cells called *osteoblasts* lay down new matrix on bone surfaces. Eventually they become encased in bone, but remain connected to other living cells via canals that pass through the bone matrix. At this older stage the cells are called *osteocytes*, and they help to regulate the dynamic equilibrium of calcium balance in the extracellular fluid.
16. *Osteoclasts* are the cells that erode or reabsorb bone. By tunneling into the bone, osteoclasts create room for osteoblasts to remodel old bone with new matrix. How this activity is controlled is not understood, but it may be related to the pattern of stress forces experienced by the bone.
17. Membranous bone develops as plates of connective tissues directly under the skin (e.g., skull), whereas cartilage bone begins as a cartilage mold of the skeletal structure and then eventually ossifies into bone (e.g., long bones of limbs).
18. Bone may be *compact* (solid and hard) or *cancellous* (spongy). The location and function of the bone usually determines whether it will be compact or cancellous, and bones with mixtures of both designs are common.
19. Cancellous bone is lightweight yet rigid. It is well suited for the large heads of long bones, which must provide surface area for articulation and muscle attachment without adding weight. The internal meshwork of cancellous bone also provides high resistance to the compression forces encountered at skeletal joints.
20. Compact bone makes up the outer shaft of the hollow long bones. Maximum strength against bending, coupled with the minimum weight that results from having a hollow center, provides excellent biomechanical advantage to these tubelike bones.
21. Compact bone is also called Haversian bone because of the concentric rings of bone matrix that form around central blood vessels. Osteocytes within each *Haversian system* are connected to one another, but isolated from other systems. Because of these independent areas of fused bone matrix, Haversian bone is highly resistant to fracturing.
22. Since muscles can only exert force while contracting, endoskeletal joints must be controlled by *antagonistic pairs* of muscles. A *flexor* muscle bends the *joint*, where two bones come together, and the *extensor* straightens it.
23. Bones are held together at joints by *ligaments*, and muscles are attached to bone by *tendons*. Both are flexible, strong straps of connective tissue. Ligaments may also hold tendons in place to redirect the force being exerted by a muscle.
24. Two basic joint types are found in the human endoskeleton. Hinged joints such as the knee or elbow flex well but do not rotate, whereas ball-and-socket joints like the shoulder and hip permit movements in almost all directions. A compromise between these two extremes is found in the bones of the hand, which have a great range of movements.
25. Important mechanical features of endoskeletons are understood by viewing them as levers, which are simple machines that distribute forces over a distance. All levers have a pivot point (fulcrum), a load-bearing arm, and a power arm.

26. Levers with a short load arm and a long power arm produce strength, which is ideal for activities such as digging and chewing. If the load arm is long and the power arm short, strength gives way to speed. The limb bones of running and flying animals tend to be of this second design. Comparing the mechanical design of the jaw and the lower leg clearly demonstrates these concepts.
27. Effectors not related to movement are also used under special circumstances. Some are specialized for food capture, others for defense, and still others for communication.
28. *Nematocysts* are a type of effector used by cnidarians to capture prey or repel enemies. These cells contain coiled harpoons, often tipped with toxin, that are forcibly ejected from the animal's body.
29. Changes in body color are accomplished by pigment-containing effector cells called *chromatophores*. The response of these cells is under neural and hormonal control. Most chromatophore cell types are fixed in place, but others are capable of amoeboid movement. Still others change shape via the action of attached muscles.
30. Glands also qualify as effectors because they release chemicals that produce responses. Chemicals such as pheromones are used in communication, others regulate physiological functions, and still others play a role in prey capture or defense. Some defensive chemicals are deadly nerve toxins, such as snake venom or pufferfish poison, but many others are only offensive or mildly irritating substances.
31. A number of communication effectors produce nonchemical signals. Some make sounds by passing air over a vibratory surface, others by rubbing objects together. Light is made by a few animals through the use of enzyme-catalyzed reactions between ATP and certain bioluminescent proteins.
32. Some effectors produce electric pulses used to sense the environment, communicate, capture prey, and repel predators. Most electrical probing of the environment is done using low levels of voltage and power. However, the modified muscle cells in the electric organs of certain eels can generate pulses of high-voltage, high-power electricity strong enough to stun a person.

Activities (for answers and explanations, see page 364)

- Select from the right column all those terms that apply to each item in the left column.

_____	Osteoblast	*a.* Cell that lays down new bone
_____	Nematocyst	*b.* Muscle that causes the leg to straighten
_____	Extensor	*c.* Component part of an endoskeletal system
_____	Exoskeleton	*d.* Found in crabs and insects
_____	Chromatophore	*e.* Found only in cnidarians

- The following is a drawing of a greyhound dog, a breed prized for its impressive running speed. Even though you cannot see its bones, what obvious endoskeletal features related to speed are apparent in the drawing?

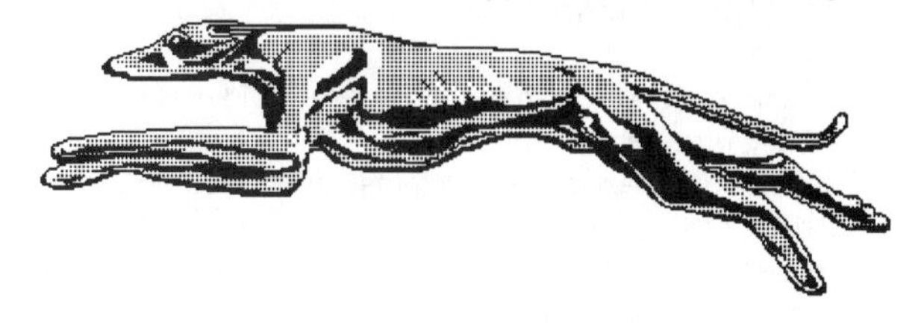

- On the diagram of a human endoskeleton shown below, circle the names of the labeled bones belonging to the axial skeleton.

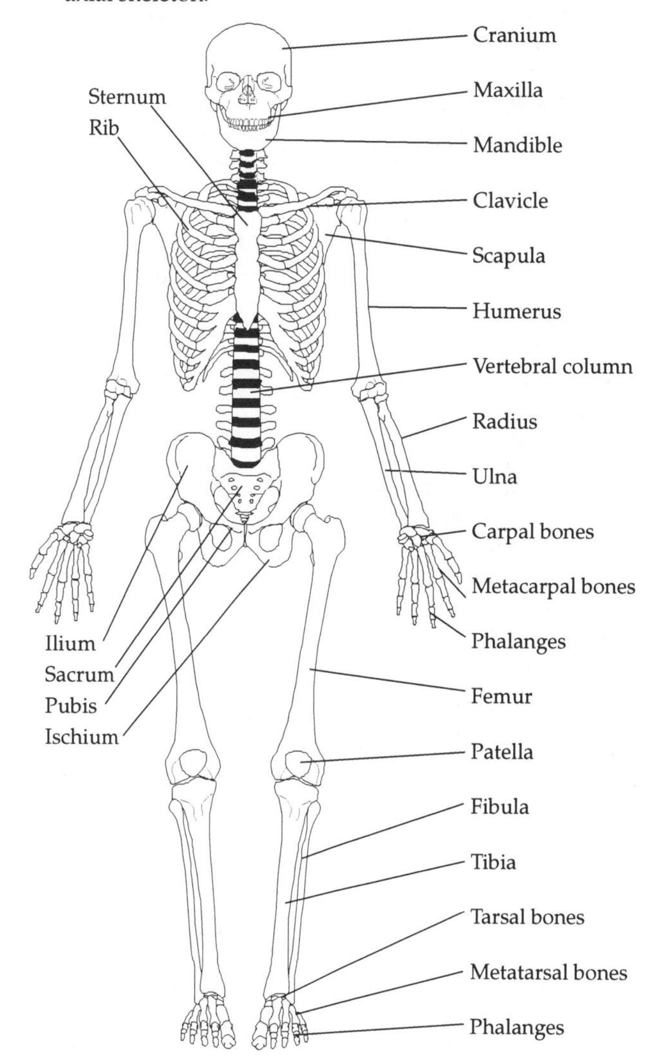

Questions (for answers and explanations, see page 364)

1. Which of the following statements about animal skeletons is *not* true?
   a. Skeletal motions are mainly the result of smooth and cardiac muscle activity.
   b. Muscle forces are exerted against some sort of supporting structure.
   c. Articulated joints allow motion to occur in endoskeletons and exoskeletons.
   d. Only endoskeletons contain bone.
   e. Exoskeletons must be periodically molted for the animal to grow in size.

2. The extension and retraction of a sea anemone's body is mechanically analogous to ____________, and demonstrates the workings of a(n) ____________.
   a. tightening a screw; endoskeleton
   b. moving a hinge; exoskeleton
   c. using a pulley and rope; endoskeleton
   d. squeezing a water balloon; hydroskeleton
   e. None of the above

3. The primary difference between an endoskeleton and an exoskeleton is
   a. whether both circular and longitudinal muscles are present.
   b. the kind of bones each skeleton has.
   c. the presence or absence of joints.
   d. the specific locations of ligaments and tendons.
   e. whether or not the skeleton is on the outside of the body.

4. All but one of the following bones are found in the human appendicular skeleton. Select the *exception.*
   a. Finger bones
   b. Upper arm bone
   c. Jawbone
   d. Ankle bones
   e. Kneecap

5. __________ are responsible for regulating extracellular calcium balance by releasing or sequestering Ca^{2+} ions in the bone matrix.
   a. Osteoblasts
   b. Osteoclasts
   c. Osteocytes
   d. Osteoclasts and osteocytes
   e. Osteoblasts, osteoclasts, and osteocytes

6. A football player suffers a knee injury that damages the tissue holding his upper and lower leg bones together. The damaged tissue is probably a kind of
   a. muscle.
   b. tendon.
   c. ligament.
   d. cuticle.
   e. cartilage.

7. Which of the following statements about skeletons and muscles is *not* true?
   a. Bones with long power arms and short load arms are found in the skeletons of powerful animals.
   b. The limb bones of fast-flying birds tend to have long load arms.
   c. When a flexor muscle contracts, its opposing extensor muscle relaxes and is stretched.
   d. A common category of appendicular skeletal injury in humans is damage to the vertebral disks.
   e. Both extensor and flexor muscles are needed to operate regions of the skeleton adapted for power.

8. Which of the following is probably the most common type of animal effector organ not related to movement?
   a. Light organ
   b. Chemical organ
   c. Auditory organ
   d. Chromatophore
   e. Electric organ

9. All but one of the following animals possesses a type of hydroskeleton. Select the *exception.*
   a. Butterfly
   b. Squid
   c. Earthworm
   d. Sea anemone
   e. Snail

Integrative Questions (for answers and explanations, see page 364)

1. In a hydrostatic skeleton, antagonistic control of movement is achieved by
   a. the use of ligaments as straps over tendons to redirect forces from a skeletal muscle.
   b. flexible hinge joints where the skeleton has not hardened.
   c. repeated segmentation of the body.
   d. waves of action potentials sweeping through the smooth muscle of the body cavity.
   e. the opposing action of circular and longitudinal muscles.

2. Which of the following effectors is *not* found in humans?
   a. Smooth muscle
   b. Electric organ
   c. Glands
   d. Jointed endoskeleton
   e. Cilia

3. Cilia bend in a rowing motion, while flagella create whiplike movement. This difference is due mainly to
   a. the differences in length between the two types of effectors.
   b. differences in the arrangement of their microtubules.
   c. the kind of muscles that cause the motion to occur.
   d. differences in the rate at which each effector takes up Ca^{2+} ions.
   e. None of the above

4. Which of the following would you *not* expect to find in both a skeletal muscle cell and a cardiac muscle cell?
   a. Actin filaments
   b. Myosin binding sites
   c. Potassium channels that close slowly after an action potential
   d. Large numbers of mitochondria
   e. Troponin

41

Gas Exchange in Animals

CHAPTER LEARNING OBJECTIVES—after studying this chapter you should be able to:

- ❑ Explain the physiological function of respiration and the importance of diffusion in this process.
- ❑ Describe what is meant by the term "respiratory gas-exchange surface" and give examples from different kinds of animals.
- ❑ Discuss the physical limitations on diffusion rate that influence animal respiration and explain how gas-exchange systems have evolved to minimize these limitations.
- ❑ Differentiate between barometric pressure and partial pressure, and explain the importance of these two concepts to animal respiration.
- ❑ Discuss the relevance of Fick's law of diffusion to animal respiration.
- ❑ Describe the basic components of any gas-exchange system and state the roles of ventilation and perfusion in these systems.
- ❑ Compare the structure and function of external and internal gills in aquatic animals.
- ❑ Describe the anatomy of a fish gill and explain the basic principle of countercurrent flow upon which it is based.
- ❑ Differentiate between positive-pressure pumps and negative-pressure pumps in animal respiration.
- ❑ Describe the anatomy and function of a bird's respiratory system and explain how it allows for the high metabolic activities characteristic of these animals.
- ❑ Explain how tidal respiration differs from unidirectional respiratory flow.
- ❑ Discuss the terms "tidal volume," "inspiratory reserve volume," "expiratory reserve volume," "residual volume," and "total volume" as they apply to the mammalian respiratory system.
- ❑ Explain the meaning of the term "dead space" in a respiratory system.
- ❑ Describe the anatomy of the human respiratory system and explain the mechanisms by which inhalation and exhalation occur.
- ❑ Discuss the roles of mucus and surfactant in the human respiratory system, and describe the problems that result when these substances do not function properly.
- ❑ Distinguish between the functional roles of blood plasma and red blood cells in respiration.
- ❑ Describe the molecular structure of hemoglobin and explain how it binds to oxygen.
- ❑ Explain the concept of positive cooperativity with respect to hemoglobin and describe how it affects the shape of the oxygen-dissociation curve for hemoglobin.
- ❑ Discuss the functional differences in oxygen affinity between hemoglobin and myoglobin.
- ❑ Explain how altitude, pH, and developmental stage can affect the oxygen-carrying capacity of hemoglobin.
- ❑ Discuss the molecular mechanisms for CO_2 transport in the blood and explain the roles of carbonic anhydrase and carboxyhemoglobin in these processes.
- ❑ Describe the neural control of breathing in humans and indicate which brain centers regulate respiration.
- ❑ Explain the ways in which O_2 and CO_2 concentrations are monitored by the human body, and compare the relative accuracy of each monitoring system.

BREATHING AND LIFE • LIMITS TO GAS EXCHANGE (pages 933-936)

Key Concepts

1. Metabolic production of ATP is accomplished through the oxidation of nutrient molecules by cells. This process requires oxygen gas (O_2) and produces the gaseous waste product carbon dioxide (CO_2).
2. Specialized cells such as muscle can incur an oxygen debt for short periods of time and still function properly. However, most cells cease metabolic activity and die shortly after their oxygen or ATP supplies are eliminated.
3. Whole-animal gas exchange (not to be confused with the biochemical process of cellular respiration) is how gases are exchanged between animals and their environments.
4. Many animals use breathing to move gases in and out of their bodies, but the actual exchange of gases takes place across specialized internal respiratory membranes. Other animals accomplish gas exchange without breathing, absorbing and releasing gases across external respiratory membranes.

5. The rate of gas exchange necessary to support metabolism varies greatly with species and physiological conditions. For example, different species subjected to identical physiological conditions can have very different respiratory rates, as can the same species subjected to different conditions.

6. Regardless of respiratory anatomy, simple *diffusion* is the only mechanism for gas exchange in animals; active transport is never involved. Gas exchange is therefore subject to a variety of important physical limitations.

7. Whether an animal respires in air or water is one physical limitation that influences the structure of gas-exchange mechanisms. It is easier to respire in air because air contains more oxygen by volume than does water. In addition, oxygen diffuses faster in air than in water. It also takes more energy to move water across respiratory exchange surfaces than it does to move an equal volume of air.

8. Physical limits on the rate of O_2 diffusion in water also affect the size, shape, and location within the body of animal cells. All respiring eukaryotic cells must be located within a millimeter or less of the nearest oxygen source, otherwise the slow diffusion of O_2 through an aqueous medium severely limits the rate of cellular respiration.

9. Many of the smaller invertebrates have solved this diffusion problem by having a thin body plan. Others have a centralized hollow body cavity through which water circulates. Larger, more complex animals such as vertebrates have specialized gas-exchange tissues with a very large surface area, combined with an internal circulatory system for transporting gases throughout the body.

10. Temperature is another factor limiting respiration in aquatic breathers. As temperature rises, the O_2 content of water drops. However, because of Q_{10} effects, the metabolic rate of aquatic ectotherms increases as water temperature rises, creating a need for more O_2. Water-breathing animals are thus forced to extract greater amounts of O_2 from the water as temperature rises, or else decrease their activity.

11. Air-breathing animals face similar O_2 constraints with respect to altitude. *Barometric pressure* is the pressure caused by the weight of all atmospheric gases combined. Since the quantity of atmosphere overhead decreases with increasing altitude, both barometric pressure and O_2 concentration decrease with height.

12. The *partial pressure* of a gas indicates what proportion of total barometric pressure is due to that gas. Since barometric pressure decreases with altitude, partial pressure does also. Because gas exchange occurs by diffusion, which is driven by concentration or pressure differences across respiratory membranes, the reduction in P_{O_2} that occurs at high altitude makes gas exchange more difficult.

13. Diffusion rates for O_2 and CO_2 are similar given eqivalent gas concentrations. However, CO_2 typically leaves the body more readily than O_2 enters. This occurs for air breathers because CO_2 concentrations are very low in the atmosphere, and for water breathers because CO_2 solubility in water is high.

Activities (for answers and explanations, see page 364

- Explain why the partial pressure of O_2 is lower on top of Mount Everest than it is on top of Mount Kilimanjaro, and what effect this would have on an air-breathing animal.

- What physical process is the primary rate-limiting step for gas exchange in animals? Explain your answer.

Questions (for answers and explanations, see page 364)

1. Which of the following statements about gas exchange or respiration is *not* true?
   a. Whole-body respiration refers to gas exchange between animals' bodies and their environments.
   b. Temperature is an important limiting influence on gas exchange in air-breathing animals.
   c. The chemical breakdown of nutrients to form ATP is an oxygen-requiring metabolic process.
   d. The diffusion of gases in air is faster than their diffusion in water.
   e. The concentration of O_2 in air is greater than in a similar-sized volume of water.

2. As water temperature increases,
   a. O_2 content of the water decreases.
   b. barometric pressure decreases.
   c. CO_2 content of the water increases sharply.
   d. the diffusion rate of O_2 decreases.
   e. All of the above

3. All but one of the following are valid reasons why breathing air is generally easier than breathing water. Select the *exception*.
   a. A given volume of air is easier to move through the respiratory system than the same volume of water.
   b. Air holds more oxygen per unit of volume than does water.
   c. O_2 diffuses faster in air than it does in water.
   d. Temperature increases affect the O_2 content of water more than they do that of air.
   e. Water breathers have more difficulty getting rid of CO_2 because it does not dissolve well in water.

4. The typical barometric pressure at the top of Mt. Kilimanjaro (about 5800 meters) is 400 mm Hg. What is the P_{O_2} at this altitude?
   a. 420.9 mm Hg
   b. 379.1 mm Hg
   c. 205. 3 mm Hg
   d. 83.6 mm Hg
   e. 19.1 mm Hg

5. An aquatic animal has no specialized respiratory surfaces, and must rely on simple diffusion alone to carry out gas exchange. Which of the following would *not* be characteristic of this animal?
   a. Its body would be very thin or possibly have a hollow, water-filled central cavity.
   b. Its metabolic rate would be rather low.
   c. It would be a fast-swimming, active predator.
   d. None of its cells would be more than one millimeter from the outside world.
   e. Water temperature would have a large influence on its rate of gas exchange.

RESPIRATORY ADAPTATIONS (pages 936–946)

Key Concepts

1. *Fick's law of diffusion* describes the rate at which substances diffuse between two locations (see page 936 of the textbook for details of the equation). Since substances diffuse faster in air than in water, animals can increase the diffusion rate by breathing air whenever possible. The only other parameter which animals can directly influence to increase diffusion rate is the size of the respiratory surface area over which gas exchange occurs.
2. *External gills* are specialized elaborations of the body surface which greatly enlarge the membrane area available for gas exchange in water. They are efficient but delicate, and their external position exposes them to many hazards. *Internal gills* are protected by body coverings and water is pumped over them.
3. Air-breathing animals must prevent their respiratory membranes from drying out. The actual gas-exchange surfaces of vertebrate air breathers are elaborate, highly divided air sacs contained within expandable bags called *lungs*. The gas-exchange surfaces of air-breathing insects are finely branched tubes called *tracheae* which penetrate the animal's entire body.
4. Increased respiratory membrane surface area is not the only way to improve gas exchange. Fick's law suggests that animals can also maximize the concentration gradient of gases across these membranes by 1) continuously exposing the external side to fresh respiratory medium, and 2) maintaining the lowest possible O_2 and the highest possible CO_2 concentrations on the internal side of the membrane. Finally, membranes can also be kept very thin to minimize the diffusion distance necessary for gas exchange.
5. An animal's *gas-exchange system* includes the respiratory membranes where diffusion occurs, the mechanisms used to *ventilate* the external surfaces of those membranes with respiratory medium, and the mechanisms used to *perfuse* the internal side of the respiratory surfaces with blood.
6. In the tracheal system of insects, air enters the body through openings called *spiracles*. It then travels through the tracheae until it reaches finely branched tubes known as *air capillaries*, where diffusion actually takes place. Insect tracheal systems are so extensive that all body cells are well supplied with O_2. The high rates of cellular respiration this system permits allow insects to maintain very active lifestyles. Some air-breathing insects can even survive underwater by carrying a bubble of oxygen with them.
7. Fish have paired internal gills that are supported by bony *gill arches* and covered with protective *opercular flaps*. Lining the arches are hundreds of interdigitated rows of *gill filaments* pointing in the direction of water flow. Each filament has many tiny, delicate surface folds called *lamellae*, which greatly increase the surface area for gas exchange.
8. Gas exchange in fishes is maximized by ventilating both sides of the respiratory membranes. The external surface is ventilated with water using a two-pump mechanism. The mouth cavity acts as a *positive-pressure pump* to push water over the gills. Slightly out of phase with the mouth, the opercular flaps act as a *negative-pressure pump* to pull water over the gills. The result is a virtually continuous, unidirectional flow of water over the gill filaments.
9. On the internal side of the gill respiratory membranes, *afferent blood vessels* carry blood to the lamellae and *efferent blood vessels* take it away. Since blood perfusion is in the opposite direction of water movement, a *countercurrent flow* is established. This antiparallel flow creates more efficient gas exchange between the water and blood than would be possible with parallel flow.
10. *Countercurrent exchangers* are the physiological mechanisms and anatomical structures associated with antiparallel flow in animal bodies. They are used not only for gas exchange, but also for concentrating urine in the kidneys and for retaining body heat. (See Box 41.A, page 941 of your textbook for details.)
11. Most air-breathing vertebrates have lungs with branching airways that terminate in clusters of blind-ended microscopic sacs where gas exchange takes place. Air movement must therefore be tidal (moving in and out of the lungs along the same path) rather than unidirectional. Tidal respiration is relatively inefficient compared to one-way air flow with countercurrent exchange.
12. Birds are unique among vertebrate air breathers in having unidirectional air flow through their lungs. This is possible using tiny airways known as *parabronchi*, from which emerge countless air capillaries that greatly increase the surface area for gas exchange
13. Bird lungs expand and contract very little during breathing. To help move air through the lungs, birds have large *air sacs* at the front and back of the body which act as bellows to create unidirectional flow. A single breath of air remains in the avian respiratory system for two complete inhalation–exhalation cycles.

14. Upon first inhalation, air passes through the tubular *trachea* (windpipe) and into smaller airways called *bronchi* before entering the posterior air sacs. During first exhalation, this air moves from the posterior air sacs into the lungs via the parabronchi. The second inhalation forces the air from the lungs into the anterior air sacs while new air refills the posterior sacs. During the second exhalation the old air leaves the anterior air sacs and exits the body.

15. Birds enhance gas exchange by having crosscurrent blood flow on the internal side of the respiratory membranes. With blood flow and air flow operating at right angles, more O_2 can be extracted from a given volume of air than is possible in the tidal breathing of mammals, reptiles, or amphibians. This increased respiratory efficiency permits the sustained high metabolic rates and energy-intensive behaviors (e.g., flight at high altitudes) characteristic of birds.

16. Vertebrate lungs have their evolutionary origins as blind-ended outpocketings of the digestive tract. In bony fishes these outpocketings serve as air bladders. In terrestrial vertebrates they have evolved into lungs.

17. Humans (like all other terrestrial vertebrates except birds) utilize tidal breathing. The *tidal volume*, or amount of air moved per breath during a normal breathing cycle, averages about 500 ml in adult humans. The *inspiratory reserve volume* represents the air gathered in excess of the tidal volume during deep inhalation. More than the tidal volume can also be exhaled; this constitutes the *expiratory reserve volume*.

18. Some *residual volume* remains in the lungs even after extreme exhalation because the lungs do not collapse completely. *Total lung capacity* is therefore the sum of tidal volume, residual volume, and inspiratory and expiratory reserve volumes.

19. Tidal breathing constrains respiratory efficiency. Fresh air entry is limited to one-half of the respiratory cycle, and once inside the lungs, the fresh air mixes with stale air from the residual volume. *Dead space* refers to parts of the lung volume such as airways where gas exchange does not occur. Mixing of fresh and stale air greatly reduces P_{O_2}, and therefore limits the diffusion rate compared to the continuous airflow systems of fish and birds.

20. Tidal breathing eliminates the possibility of countercurrent exchange. This means the P_{O_2} in blood leaving the lungs of tidal breathers can never exceed that of the exhaled air, whereas in fishes and birds the blood leaving the respiratory surfaces can have a P_{O_2} almost as high as the environmental air or water. Despite these limitations, tidal breathing mechanisms are adequate to serve the respiratory needs of mammals.

21. Air enters the mammalian respiratory system through the oral or nasal cavities, which join in the pharynx. One opening from the pharynx leads to the esophagus, which passes food to the stomach. The other opening leads to the trachea, a cartilage-reinforced tube that directs air toward the lungs. At the head of the trachea is the *larynx*, or voice box, which contains the vocal cords.

22. The trachea branches into two smaller airways called bronchi, one of which leads to each lung. The bronchi branch repeatedly and decrease in size, eventually becoming small-diameter *bronchioles*. Located at the end of the bronchioles are clusters of blind-sac respiratory surfaces called *alveoli*.

23. The airways connecting mammalian lungs with the outside world act as conduits for gases (i.e., they are dead spaces). It is only in the alveoli where actual diffusion of gases occurs between the lungs and the blood.

24. Although individually tiny, the over 300 million alveoli in human lungs collectively possess an enormous surface area (~70 m^2). Each alveolus is made of extremely thin cells and is surrounded by small, thin-walled blood vessels. This arrangement produces diffusion distances of less than 2 µm, which greatly enhances the rate of gas exchange.

25. Surfactant is a chemical substance produced by certain cells in the alveoli. Its function is to reduce the surface tension of liquids covering the alveolar walls. Surface tension results from the cohesive forces between molecules of a liquid, and it can make fluids act like an elastic membrane. Surfactant interferes with the cohesive forces that produce surface tension and thus improves gas diffusion. Surface tension creates problems when the individual trys to inflate its alveoli during inhalation. The condition known as *respiratory distress syndrome* in premature infants is an example.

26. Mucus is also produced by the epithelial cells lining mammalian airways. Beating cilia in these airways create a mucus escalator that moves the mucus, along with any trapped particulate matter, away from the delicate alveoli. This material eventually reaches the pharynx where it is swallowed. Inhaled pollutants such as cigarette smoke immobilize the cilia and cause the mucus escalator to cease operation. Hacking and coughing result as the body attempts to clear the airways of accumulated debris.

27. Mammalian lungs are suspended inside the *thoracic cavity*, and are virtually surrounded by the ribs. The thoracic cavity is lined by the *pleural membranes*, which enclose each lung within its own *pleural cavity*. Below the lungs is located the dome-shaped muscular *diaphragm*, which separates the thoracic cavity from the abdominal cavity.

28. Inhalation is an active process in mammals. As the diaphragm contracts, it flattens and moves downward into the abdominal region. At the same time, the rib cage expands because of the contraction of *intercostal muscles* between the ribs. Both actions increase the volume of the pleural cavities, which creates a negative pressure (suction) that pulls air into the lungs.

29. If the seal formed by the pleural cavity is penetrated, as can occur with a puncture wound to the thoracic region, air leaks into the cavity as it tries to expand. This prevents the lung from inflating properly, and may lead to its collapse.

30. Exhalation in mammals is usually a passive process. It begins when the diaphragm and intercostals relax, there-

by reducing the volume of the thoracic cavity; this pushes air out of the lungs. However, there are muscle groups located between the ribs that can contract to forcibly expel air from the lungs during periods of strenuous activity.

Activities (for answers and explanations, see page 365)

- If you were to try to design an apparatus that would enable divers to rebreathe the same air by allowing that volume of air to exchange gases with the surrounding water, what three essential design features would you need to build into your apparatus?

a.

b.

c.

- Match each of the following respiratory adaptations with its best description.

____ Internal gills	*a.*	Where diffusion occurs in birds and insects
____ Air capillaries	*b.*	Reduces surface tension in mammalian lungs
____ Air sacs	*c.*	Respiratory structures of water breathers
____ Surfactant	*d.*	Blind sacs found in clusters within the human lung
____ Alveoli	*e.*	Act as bellows for the one-way flow of air in bird lungs

- Below is a diagram of a countercurrent exchanger from a fish gill. The values shown indicate percent saturation of O_2. Based on this information alone, indicate which tube represents the flow of blood and which represents the flow of water. Indicate the direction in which each of these fluids is flowing, and show the direction of net O_2 diffusion. Explain the reasoning behind your choices.

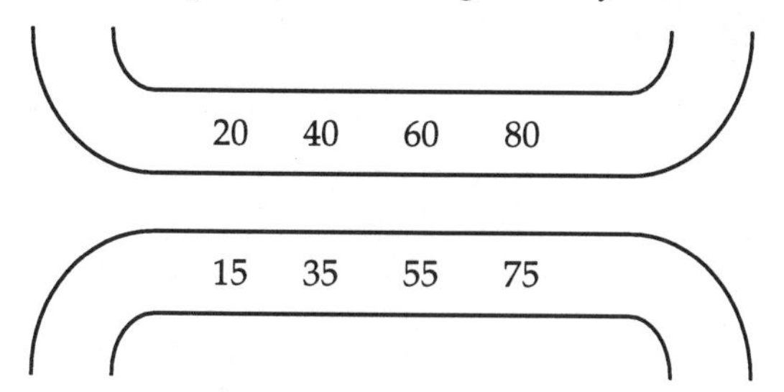

Questions (for answers and explanations, see page 365)

1. External gills, tracheae, and lungs all share which of the following sets of characteristics?
   a. Part of gas exchange system; exchange both CO_2 and O_2; increase surface area for diffusion
   b. Used by water breathers; based on countercurrent exchange; use negative pressure breathing
   c. Exchange only O_2; are associated with a circulatory system; found in vertebrates
   d. Are associated with spiracles; used only by air breathers; relatively inefficient respiratory rates
   e. Found in insects; employ positive-pressure pumping; based on crosscurrent flow

2. Which of the following would you expect to find in an insect's respiratory system?
   a. Lamellae
   b. Gill arches
   c. Spiracles
   d. Bronchioles
   e. Pleural cavity

3. A water-borne molecule of oxygen arriving at the gills of a fish would encounter which of the following structures *prior* to entering the fish's blood?
   a. Opercular flap, gill arch, gill filament, pleural cavity
   b. Gill filament, air capillary, parabronchus
   c. Tracheae, bronchioles, alveoli
   d. Gill filaments, lamellae, blood capillary wall
   e. None of the above

4. In terms of the design of their gas-exchange systems, fish are to birds as
   a. insects are to cats.
   b. elephants are to people.
   c. insects are to frogs.
   d. llamas are to birds.
   e. All of the above

5. During the second inhalation in a sequence of breaths taken by a bird, air is
   a. exiting the body directly from the posterior air sacs.
   b. exiting the body directly from the lungs.
   c. entering the mouth and going straight to the lungs.
   d. filling both the anterior and posterior air sacs.
   e. entering the mouth and going straight to the anterior air sacs.

6. All but one of the following animals utilizes tidal breathing. Select the *exception*.
   a. Bear
   b. Salmon
   c. Frog
   d. Lizard
   e. Whale

7. Which of the following represents a *larger* volume of air than is normally found in the resting tidal volume of a human lung?
   a. Residual volume
   b. Inspiratory reserve volume
   c. Expiratory reserve volume
   d. Total lung capacity
   e. All of the above

8. Which of the following statements about gas exchange in vertebrates is *not* true?
   a. Blood leaving the gills of fish normally has a P_{O_2} about equal to the environmental P_{O_2}.
   b. Surfactant reduces the surface tension in the alveoli.
   c. Inhalation is a passive process in mammalian lungs, while exhalation requires muscle contraction.
   d. The mucus escalator is a mechanism for removing foreign particles from the respiratory system.
   e. The diaphragm and the intercostal muscles help produce negative pressure for human breathing.

9. Which of the following structures is *not* directly involved with gas exchange in the vertebrate respiratory system?
   a. Bronchus, bronchiole, pulmonary venule and arteriole
   b. Bronchus, alveoli, and pulmonary arteriole
   c. Bronchus and bronchiole
   d. Alveoli and blood capillaries
   e. Alveoli only

TRANSPORT OF RESPIRATORY GASES BY THE BLOOD • REGULATION OF VENTILATION (PAGES 946–953)

Key Concepts

1. The blood circulatory system perfuses the internal surface of respiratory membranes. By removing O_2 as fast as it diffuses across the membrane, the circulatory system helps keep the O_2 concentration low on the internal surface, which encourages further diffusion. O_2 is then circulated throughout the body for use by cells.

2. The liquid fraction of blood, called *blood plasma*, carries only a small fraction of the O_2 picked up at the respiratory membranes. The O_2 carrying capacity of blood is greatly enhanced by the oxygen-binding pigment molecule hemoglobin, large numbers of which are contained within specialized *red blood cells.*

3. Four polypeptide subunits make up a molecule of hemoglobin, each of which surrounds an iron-containing ring structure called a heme group. The heme group reversibly binds O_2 and holds it within the red blood cell, thereby maximizing the concentration difference of free O_2 between the two sides of the respiratory membrane.

4. The P_{O_2} of the local environment determines the ability of hemoglobin to pick up or release oxygen. An S-shaped oxygen-dissociation curve describes this relationship. It shows that when P_{O_2} is high, hemoglobin molecules carry their maximum load of O_2, and that at low P_{O_2} values hemoglobin releases its O_2 cargo.

5. This relationship exists because of shape changes that occur when hemoglobin subunits bind to oxygen. At low P_{O_2} levels, only one subunit of hemoglobin will bind to O_2. Once a single molecule of O_2 is bound, the hemoglobin changes shape, which makes it easier for the next O_2 to bind. Further binding requires only slight increases in P_{O_2}, and each addition of O_2 causes further shape changes in the hemoglobin molecule. This interaction between the binding of O_2 to one subunit of hemoglobin and the change in binding affinity of another subunit is known as *positive cooperativity.*

6. The mechanism is actually very adaptive because of the steep slope produced in the oxygen-dissociation curve. When fully loaded hemoglobin molecules reach normal body tissues, they usually give up about 25 percent of their O_2 load. However, it only takes a slight drop in the P_{O_2} of those tissues to stimulate a massive unloading of O_2 from the hemoglobin. Such drops in P_{O_2} often occur because of increased metabolic activity by cells, exactly the situation where more O_2 is required.

7. *Myoglobin* is a special oxygen-carrying pigment found in muscle. It has only one polypeptide chain, and its affinity for O_2 is much higher than hemoglobin's, meaning that it will pick up and retain O_2 at much lower P_{O_2} values (i.e., it plots to the left of the hemoglobin oxygen-dissociation curve). Myoglobin thus provides an oxygen reserve when metabolic demands are high and blood flow may be restricted because of muscle activity. Dark muscle, which is used for extended bouts of flying, running, or diving, contains much myoglobin.

8. Hemoglobin function is regulated by structural aspects of the constituent polypeptides that make up the molecule, as well as by chemical characteristics of the cellular environment in which hemoglobin is found.

9. In adult humans, hemoglobin is made up of two pairs of different polypeptides. Two α chains and two β chains are found in each molecule. They combine to produce the characteristic oxygen-binding properties of human hemoglobin described above.

10. Fetal hemoglobin differs from adult hemoglobin in that the β chains are replaced by two γ chains. This alteration allows fetal hemoglobin to pick up O_2 from the mother's blood even though both are at the same P_{O_2}. Thus, like myoglobin, the fetal hemoglobin oxygen-dissociation curve plots to the left of the adult curve.

11. Mammals adapted for life at high altitudes, such as llamas and vicuñas, also have left-shifted oxygen-dissociation curves. This allows their hemoglobin to saturate at the very low P_{O_2} levels characteristic of the high mountains where they live.

12. The *Bohr effect* describes the influence of pH on hemoglobin function. In tissues undergoing high metabolic activity, acidic metabolites are produced that lower pH. Under these conditions, the oxygen-dissociation curve shifts to the right, which means that hemoglobin will more easily give up its O_2 in areas of high cell metabolism.

13. *Diphosphoglyceric acid*, a normal cell metabolite, is found in especially high concentrations within human red blood cells. The concentration of diphosphoglyceric acid increases in response to sustained exercise and during acclimation to high altitudes. It reversibly binds to deoxygenated hemoglobin and causes it to change shape, making the hemoglobin more likely to release O_2. This produces a shift to the right in the oxygen-dissociation curve.

14. Humans therefore employ a fundamentally different strategy from llamas and vicuñas for altering hemoglobin function at high altitude. However, the end result (delivering more O_2 to the body cells) is the same.

15. The blood circulatory system also has responsibility for removing waste CO_2 from the tissues and transporting it to the gas-exchange surfaces. Although CO_2 readily

dissolves in water, very little of it is actually transferred in the blood as dissolved CO_2.

16. Most CO_2 is transported in the blood in the form of *bicarbonate ions* (HCO_3^-). CO_2 reacts with water to form carbonic acid (H_2CO_3), which then dissociates into H^+ and HCO_3^-. This reaction proceeds rather slowly in the blood plasma, but is greatly accelerated inside red blood cells by the enzyme *carbonic anhydrase*.

17. As CO_2 enters the blood from metabolizing tissues, it diffuses into red blood cells and is quickly converted into bicarbonate ions, which then return to the plasma. This keeps the CO_2 concentration low in the blood, promoting further diffusion from the tissues into the blood. A small amount of CO_2 is also carried directly by deoxygenated hemoglobin in the form of *carboxyhemoglobin*.

18. Because of breathing, CO_2 concentration in the airspace of the lungs is kept low relative to its concentration in the blood. Thus, upon arriving at the alveoli, CO_2 in the blood readily diffuses into the alveolar air space. This stimulates the movement of HCO_3^- from the plasma into the red blood cells, where H^+ and HCO_3^- reassociate to form H_2CO_3, which converts to CO_2 and H_2O, catalyzed by carbonic anhydrase. The newly reformed CO_2 leaves the red blood cells and enters the aleveoli from the plasma.

19. Although breathing results from many complex, coordinated movements involving several different muscle groups, it is nevertheless an autonomic function. Ventilatory rhythms are automatically maintained by the central nervous system to provide the proper depth and frequency of breathing for current metabolic conditions.

20. Neurons in the *medulla* of the brain stem are primarily responsible for the control of breathing. These cells fire as the active phase of inhalation begins. At the peak of inhalation the neurons cease firing, and the passive recoil of respiratory musculature produces exhalation. In situations that demand high respiratory rates, expiratory neurons in the medulla also fire to drive active exhalation.

21. The *Hering–Breuer reflex* is an override mechanism that is activated during intense respiration to prevent the ventilatory muscles from damaging the lung tissue. Stretch receptors in the lungs detect overextension and send action potentials to the brain along the vagus nerve, inhibiting the activity of respiratory neurons in the medulla.

22. Neurons located in higher brain centers also influence breathing rhythm. The control provided by these neural elements allows for proper breathing during speech, ingestion of food, and coughing.

23. Humans and most other mammals are not well-equipped to detect low levels of blood P_{O_2}. Instead, mammals are very sensitive to increases in the blood P_{CO_2}, because it is this gas that varies most dramatically in response to changing metabolic or environmental conditions.

24. Blood P_{CO_2} is sensed primarily by detectors located in the vicinity of the medullary neurons controlling ventilatory rhythm. What P_{O_2} sensitivity there is resides in nodes of tissue found on the aorta and carotid arteries of the heart. These vessels continuously receive large volumes of blood and are equipped with chemosensors to detect drops in P_{O_2}. If activated, the chemosensors send signals directly to the respiratory centers of the brain.

Activities (for answers and explanations, see page 365)

- Match each of the following features of respiratory control with its best description.

____Carbonic anhydrase	*a.* Dark muscle pigment with high oxygen-binding affinity
____Myoglobin	*b.* Tendency for hemoglobin to release O_2 as P_{O_2} decreases
____Oxygen dissociation	*c.* Brain stem structure responsible for regulating ventilatory rhythm
____Carboxyhemoglobin	*d.* Deoxygenated blood pigment bound to CO_2
____Medulla	*e.* Important gas exchange enzyme found in red blood cells

- Examine the family of oxygen-dissociation curves shown below.

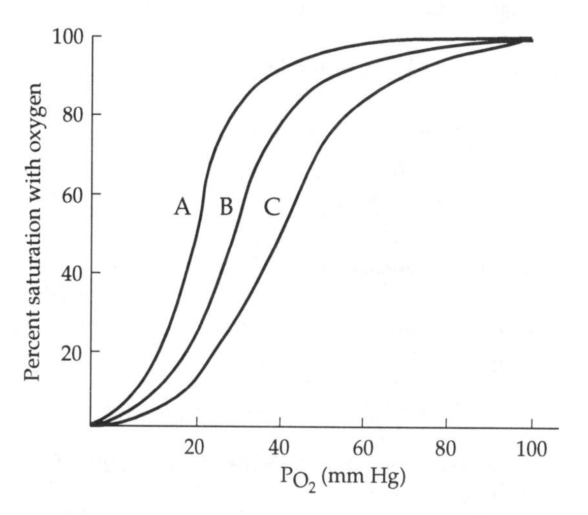

Which of these curves probably belongs to a normal adult human? _____

Which of these curves probably belongs to an adult human acclimated to high altitude? _____

Which of these curves probably belongs to a llama? _____

Explain the reasoning behind your choices.

Questions (for answers and explanations, see page 365)

1. Which of the following statements about hemoglobin is *not* true?
   a. Hemoglobin is what allows the blood to carry a large volume of O_2.
   b. Its capacity to hold O_2 is influenced by pH and chemical features of the environment.
   c. It contains a single polypeptide chain with very high affinity for O_2.
   d. In humans, fetal hemoglobin is structurally different from adult hemoglobin.
   e. It is packaged inside of red blood cells.

2. Which of the following would normally be present in the blood plasma of a mammal?
   a. Red blood cells
   b. Bicarbonate ions
   c. Water
   d. Oxygen
   e. All of the above

3. The oxygen-dissociation curve for myoglobin is steep and shifted to the left of the same curve for hemoglobin. This indicates that
   a. people with myoglobin are acclimated for life at high altitudes.
   b. myoglobin occurs in high concentration in the human fetus.
   c. myoglobin does not function well at high percent saturation of oxygen.
   d. myoglobin picks up and releases O_2 at lower P_{O_2} values than does hemoglobin.
   e. None of the above

4. A shift in the O_2 binding capacity of hemoglobin as a function of pH is known as
   a. oxygen dissociation.
   b. the Hering–Breuer reflex.
   c. the Bohr effect.
   d. respiratory ventilation.
   e. None of the above

5. The primary respiratory effect of diphosphoglyceric acid is to
   a. shift the oxygen-dissociation curve for myoglobin to the right of normal.
   b. shift the oxygen-dissociation curve for hemoglobin to the left of normal.
   c. convert CO_2 into HCO_3^- so that more carbon dioxide can be removed from body tissues.
   d. alter the pH of the blood plasma during times of reduced metabolic activity.
   e. change the shape of hemoglobin and make it more likely to release O_2 at the P_{O_2} of the tissues.

6. The main reason why CO_2 concentrations in the alveoli of mammalian lungs are usually rather low is because
   a. CO_2 is rapidly removed from the lungs through breathing.
   b. carbonic anhydrase converts the CO_2 into carbonic acid.
   c. CO_2 rapidly diffuses from the alveolar air into the blood.
   d. CO_2 spontaneously converts to bicarbonate ions once in the alveolar air.
   e. carboxyhemoglobin binds to CO_2 as soon as it enters the alveolar air.

7. All but one of the following are involved in the neural control of ventilation. Select the *exception*.
   a. Neurons in the medulla
   b. The vagus nerve
   c. Chemosensors on the surface of the medulla
   d. Contraction state of the diaphragm
   e. Stretch receptors in the lungs

8. A person receives a serious cut on the back of the neck that is deep enough to severely damage the brain stem just below the medulla. The respiratory effect of this type of wound would likely be
   a. catastrophic, and it might result in a complete cessation of breathing.
   b. to reduce respiration to an irregular rhythm.
   c. to shift the oxygen-dissociation curve of hemoglobin to the left.
   d. to stimulate the production of additional diphosphoglyceric acid.
   e. almost undetectable.

Integrative Questions (for answers and explanations, see page 365)

1. An animal has external gills. Which of the following *must* also be true of this animal?
   a. It is a fish.
   b. It is aquatic.
   c. It is an invertebrate.
   d. It lives at high altitude.
   e. It is an insect.

2. All but one of the following respiratory structures or functions is present in birds. Select the *exception*.
   a. Lungs
   b. Air sacs
   c. Tidal respiration
   d. Positive cooperativity in hemoglobin function
   e. Parabronchi

3. Refer to textbook Fig. 41.3, page 936. A mammal species that has evolved to live on top of which of the following mountains would have the left-most shifted oxygen-dissociation curve for its hemoglobin?
   a. Mt. Kilimanjaro
   b. Mt. Blanc
   c. Mt. Whitney
   d. Mt. Fuji
   e. There should be no difference in hemoglobin function among these mammals.

4. Which of the following is *smaller* than an alveolus from a human lung?
   a. Red blood cell
   b. Molecule of hemoglobin
   c. Molecule of oxygen
   d. Diameter of smallest blood capillary in lung
   e. All of the above

42

Internal Transport and Circulatory Systems

CHAPTER LEARNING OBJECTIVES—after studying this chapter you should be able to:

- ❑ Compare the circulatory needs of simple animals with those of more complex animals.
- ❑ Describe the basic anatomical differences between gastrovascular cavities, open circulatory systems, and closed circulatory systems.
- ❑ Explain the structure and function of a representative insect circulatory system.
- ❑ Describe the role of the heart, vessels, and blood in a closed circulatory system, and discuss what circulatory advantages closed systems have over open systems.
- ❑ Distinguish between the pulmonary and systemic circuits of a closed circulatory system.
- ❑ Compare the structure and function of arteries, veins, arterioles, venules, and capillaries.
- ❑ Describe the evolution of vertebrate hearts and circulatory systems from fish through mammals, and state the advantages that accrued with each advancement.
- ❑ Identify the major structures of the human heart, give their functions, and trace the flow of blood through the heart.
- ❑ Describe the various phases of the cardiac cycle and explain the importance of systole and diastole in this process.
- ❑ Discuss the use of an electrocardiogram to measure heart function and explain several abnormal heart conditions that can be identified using this technique.
- ❑ Describe the anatomical structures and neural connections that produce the myogenic properties of the human heart.
- ❑ Explain how heartbeat rate is controlled by the autonomic nervous system.
- ❑ Discuss the cause and prevention of stroke as well as several forms of heart disease.
- ❑ Describe the control mechanisms for regulating the exchange of materials between capillaries and body tissues.
- ❑ Explain how the architecture of capillaries influences their function and regulation.
- ❑ Compare the structure and function of the lymphatic system with that of the blood circulatory system.
- ❑ Explain how blood contained within the venous system is returned to the heart for recirculation.
- ❑ Name the cellular and acellular parts of blood and explain their functions.
- ❑ Describe the cellular and biochemical events that lead to blood clotting and vessel repair.
- ❑ Explain what is meant by "autoregulation" of the circulatory system.
- ❑ Discuss the hormonal and neural feedback control mechanisms that influence cardiovascular function in humans.
- ❑ Explain the physiological changes that occur during the diving response of marine mammals, and discuss why these responses are advantageous.

TRANSPORT SYSTEMS (pages 956–960)

Key Concepts

1. Multicellular organisms depend heavily on a division of labor among their body cells. In all but the simplest animals, supplying the needs of distantly separated cells is accomplished by internal transport systems.
2. Most internal transport systems consist of a muscular pump called the heart, a network of interconnected vessels, and a transport fluid called blood. Collectively, these components are referred to as a circulatory or cardiovascular system.
3. Simple transport systems do not have vessels; some do not even have pumping mechanisms. If the animal is small, unassisted environmental fluid movements are sufficient to transport substances past all body cells.
4. Many cnidarians utilize nonvascular transport. For example, every cell in a hydra's body is in direct contact with the environment either through its external surface or through its *gastrovascular cavity*, a blind-ended sac that serves as both a digestive and transport organ.
5. The gastrovascular cavities of some invertebrates are highly branched. For very thin animals, this puts every cell within diffusion distance of the transport system. Flatworms are examples of this body design.

6. Larger animals, whose bodies cannot be adequately supplied by simple diffusion from a gastrovascular cavity, must have some type of circulatory system that transports fluids to specific regions within the body.
7. Increased body size and the inability of simple diffusion to handle transport needs is a particularly acute problem for terrestrial animals, because the external medium (air) can no longer serve a transport function without causing the animal to dry out. Terrestrial animals must have an internal transport system to maintain the proper composition of their *interstitial fluids*.
8. In some internal transport mechanisms, body movements cause the interstitial fluid to percolate through the tissues outside of the circulatory vessels. This is known as an *open circulatory system*, and is characteristic of animal groups such as mollusks and arthropods.
9. The movement of fluids through an open circulatory system is usually aided by the action of a muscular pumping heart, which propels the fluid through short vessels that open directly into major regions of the body. After trickling through the tissues, the fluid collects in the body cavity and reenters the heart through valved holes called *ostia*.
10. The relative inefficiency of open circulatory systems limits the metabolic activity and body size of animals. However, groups such as insects have gotten around this problem to a degree by decoupling their respiratory system (tracheal tubes) from the circulatory system. This anatomical separation permits insects to lead more active lives than would otherwise be possible with an open circulatory system that served both respiratory and circulatory functions.
11. *Closed circulatory systems* keep the transport fluid separate from the interstitial fluid in a network of closed vessels. One or more pumping hearts create pressure to move blood within the vascular network.
12. Closed circulatory systems have important design advantages that lead to an overall metabolic superiority over open circulatory systems. Closed systems provide more rapid transport, they direct blood flow to specific areas of the body, and they provide a boundary that constrains the movement of specialized cellular elements and large blood molecules.
13. Different levels of sophistication are seen in closed circulatory systems. Earthworms have a relatively simple closed system that consists of large dorsal and ventral vessels which direct the blood along the animal's body. These vessels are connected by five contractile vessels that act as pumping hearts. Within each body segment, the large vessels branch into smaller vessels where exchange occurs.
14. Vertebrates have sophisticated closed circulatory systems. The multichambered vertebrate heart has valves that prevent the backflow of blood when the heart chambers contract. There has been an increase in the number of heart chambers during evolution so that modern fishes have two chambers, amphibians and most reptiles have three, and crocodilian reptiles, birds, and mammals have four.
15. Increased heart complexity allows for the separation of blood flow into two distinct pathways. All vertebrates except fishes have a *pulmonary circuit* that sends blood to the respiratory surfaces and then returns it to the heart. The *systemic circuit* then sends this oxygenated blood to the body cells. Waste-laden, deoxygenated blood is collected from the body tissues and returned to the heart before once again entering the pulmonary circuit for reoxygenation.
16. Vessels of the vascular system are classified in terms of their size and whether they direct blood to or from the heart. *Arteries* are large-diameter vessels that take blood away from the heart. *Veins* are the large vessels that return blood to the heart. Arteries and veins branch repeatedly to become smaller *arterioles* and *venules*, respectively.
17. *Capillaries* are the smallest diameter vessels; they connect the arterioles to the venules in both the pulmonary and systemic circuits. The exchange of materials between the interstitial fluids and the vascular system takes place only across the capillary walls.
18. The fish heart has two chambers, consisting of a heavily muscularized pump called the *ventricle* attached to a less muscular pump known as the *atrium*. Deoxygenated blood from the body returns to the heart through veins and collects in the atrium, which then pumps it into the ventricle. The blood is then pumped from the ventricle to the gills, where it flows through extensive capillary beds to accomplish gas exchange. Once oxygenated, the blood travels through a large dorsal artery called the *aorta*, which distributes blood to other arteries and arterioles that branch throughout the body.
19. Most of the pressure generated by ventricular contraction in fishes is dissipated by the high resistance of the tiny gill capillaries. Blood traveling through a fish's body is therefore under relatively low pressure, putting distinct upper limits on the rate of metabolic exchange possible.
20. Lungfish provide transitional anatomical evidence for how other vertebrate groups evolutionarily solved the problem of low systemic blood pressure. Lungfish have two atria and an air bladder that is modified into a primitive lung. The lung forms as an outpocketing of the gut. Oxygenated blood leaving the gills enters the lung and is then directed back to the heart, forming a rudimentary pulmonary circuit. The two atria serve to mostly separate oxygenated blood bound for the body from deoxygenated blood heading to the gills; also keep the systemic blood pressure higher than in other fish.
21. Adult amphibians have three-chambered hearts. Separate atria receive deoxygenated blood from the body and oxygenated blood from the lungs. The two types of blood are then partially mixed as they flow through the single ventricle. However, the anatomy of the ventricle keeps the blood types mostly separated.
22. The main advantage of the amphibian system is that oxygenated blood returning to the heart from the lungs can be repressurized before traveling to the body tissues. This dramatically increases the rate at which nutrients and oxygen can be delivered to body cells, thereby increasing metabolic activity.

23. Most reptiles also have three-chambered hearts, but a septum partially divides their ventricle to reduce the mixing of pulmonary and systemic blood supplies. This further increases the pressure and efficiency of the entire circulatory system. One group of reptiles, the crocodilians, has a complete partition in the ventricle which yields a four-chambered heart.

24. Because their metabolic rates are typically lower than that of birds or mammals, reptiles do not have to breath continuously. All reptiles therefore have the ability to shunt blood away from the pulmonary system when they are not breathing. Special anatomical features of the heart and arteries permit this switching of blood flow between pulmonary and systemic circuits.

25. Four-chambered hearts are characteristic of all birds and mammals. The complete separation of pulmonary and systemic circuits creates independent pumping mechanisms for the lungs and the body tissues, which greatly increase transport efficiency.

26. Other important advantages of four-chambered hearts include: providing the highest possible oxygen concentration for the systemic blood supply, maximizing gas exchange by sending to the lungs blood that has the lowest possible O_2 concentration and the highest possible CO_2 concentration, and permitting differential pressurization of the pulmonary and systemic circuits.

Questions (for answers and explanations, see page 366)

1. All but one of the following structures can be found in the transport system of a butterfly. Select the *exception*.
   a. Heart
   b. Artery
   c. Ostium
   d. Capillary
   e. Transport fluid

2. Like all other animals, hydras and chickens need to obtain oxygen from their environments and eliminate the waste products of their cellular metabolism. Which of the following transport structures is found in *both* of these animals?
   a. Complex network of transport vessels which supply all regions of the body
   b. Gastrovascular cavity to facilitate transport to internal body tissues
   c. Powerful muscular pump to move transport fluids
   d. Atria, ventricles, and capillaries
   e. None of the above

3. For a variety of physiological reasons, closed circulatory systems are generally considered to be more efficient than open circulatory systems. Which of the following is *not* one of those reasons?
   a. Closed systems rely exclusively on simple diffusion for transport, whereas open systems rely on pumping mechanisms.
   b. Transport within closed systems is more rapid than in open systems.
   c. Blood is directed to specific areas of need in open systems, but not in closed systems.
   d. Closed systems create a boundary between transport fluid components and the body tissues.
   e. The interstitial fluid is separated from the transport fluid in closed systems.

4. Through which of the following vessels would the largest volume of blood flow during a one-minute period?
   a. A capillary in the lung
   b. A venule leaving a capillary bed in the right foot
   c. An artery connected to the left ventricle
   d. An arteriole feeding a capillary bed in the liver
   e. Cannot determine from the information given

5. An important difference between a fish heart and a lizard heart is that the
   a. fish heart is part of an open circulatory system, and the lizard heart is not.
   b. fish heart has two chambers, and the lizard heart has three.
   c. fish heart has many ostia, whereas the lizard heart has none.
   d. fish heart pumps blood to the lungs, but the lizard heart pumps blood to the gills.
   e. fish heart pumps oxygenated blood, and the lizard heart pumps deoxygenated blood.

6. In which of the following would you likely record the highest blood pressure?
   a. The ventricle supplying blood to the gills of a fish
   b. The anterior dorsal artery of an ant
   c. The pulmonary vein of a frog
   d. The ventricle supplying blood to the systemic circuit of a bird
   e. All of these locations would have the same pressure.

7. Between the time it leaves the heart and the next time it reenters the heart, blood traveling through the systemic circuit of a cow will pass through
   a. arteries and veins.
   b. arteries, veins, and capillaries.
   c. arteries, arterioles, and veins.
   d. arteries, arterioles, venules, and capillaries.
   e. arteries, arterioles, capillaries, venules, and veins.

8. Which of the following statements about the transport systems of birds and mammals is *not* true?
   a. Blood pressure on the arterial side of the systemic system is higher than that of fishes.
   b. Only oxygenated blood travels through the systemic circuit.
   c. The pulmonary circuit carries blood to and from the lungs.
   d. There are functionally independent pumping systems for the pulmonary and systemic circuits.
   e. Gas exchange in bird and mammal lungs is more efficient than gas exchange in a frog lung.

9. During the period of time when a reptile is not breathing,
   a. little blood is pumped from the heart to the lungs.
   b. large amounts of blood are shunted from the systemic to the pulmonary circuit.
   c. the metabolic rate of the animal increases dramatically.
   d. the resistance in its pulmonary circuit is lower than in its systemic circuit.
   e. All of the above

Activities (for answers and explanations, see page 366)

- Match each of the following cardiovascular structures with its best description.

_____	Ventricle	*a.* Muscular pump that sends blood to the lungs or gills
_____	Ostium	*b.* Smallest diameter vessels in the pulmonary circuit
_____	Atrium	*c.* Vessel through which blood passes just prior to entering a capillary
_____	Venule	*d.* Collecting hole in the pumping structure of insect cardiovascular systems
_____	Capillaries	*e.* Receives blood returning to the heart from the lungs of a frog
_____	Arterioles	*f.* Blood passes through this vessel just prior to entering a vein

- Explain how the open circulatory system of a grasshopper functions, and how it differs from the closed circulatory system of an earthworm.

- Below is a schematic diagram of a mammalian circulatory system. Indicate with arrows the correct flow of blood through this system. Identify the atria and the ventricles of the heart, the pulmonary and systemic circuits, and show one example of an artery and a vein. Also indicate the location of a capillary bed.

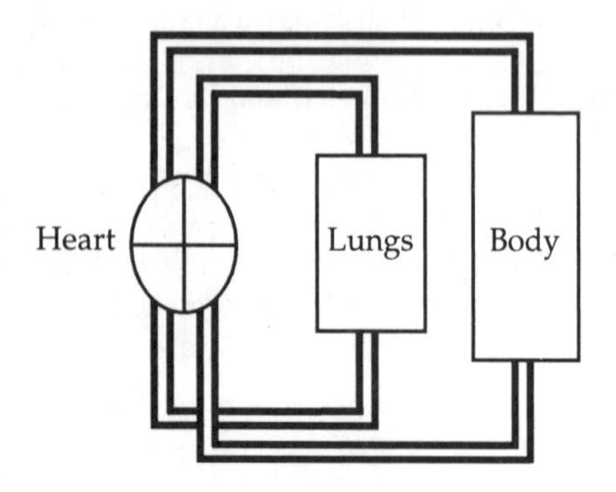

THE HUMAN HEART • THE VASCULAR SYSTEM (pages 961–971)

Key Concepts

1. The human heart has four chambers: a right and left atrium, and a right and left ventricle. Each atrium is paired with a ventricle to form two independent pumping systems. Atria collect blood returning to the heart by veins, and ventricles pump blood away from the heart through arteries.
2. Because it is a pressurized system, valves are needed to prevent backflow of blood when the atria and ventricles relax. *Atrioventricular valves* prevent backflow between the atria and ventricles. The *pulmonary* and *aortic valves* prevent backflow between the ventricles and their respective arteries.
3. The right atrium receives systemic blood from veins called the *superior* and *inferior vena cavae,* which drain the upper and lower body, respectively. This blood is pumped into the right ventricle, which sends it to the lungs via the *pulmonary artery.* Blood returns from the lungs through the *pulmonary vein* and is collected by the left atrium, which pumps it into the left ventricle. The blood is then pumped back through the systemic circuit via the aorta.
4. High resistance caused by the large number of arterioles and capillaries in the systemic circuit requires the left ventricle to pump harder than the right ventricle, which services the smaller quantity of capillary beds in the lungs. Higher pumping pressures are therefore produced in the systemic circuit than in the pulmonary circuit.
5. The *cardiac cycle* involves alternating contractions of the atria and ventricles. The period of ventricular contraction is called *systole,* while ventricular relaxation is called *diastole.* Blood pressure measured in large arteries located near the heart is reported as a ratio of systolic to diastolic pressure, which for a young adult is typically around 120/80 mm Hg.
6. The beating sound made by a human heart is caused by the closing of heart valves. Atrioventricular valves close with a "lub" sound as the ventricles contract. When the ventricles relax, the aortic and pulmonary valves close with a "dub" sound. Defective valves produce abnormal sounds called murmurs.
7. An *electrocardiogram* (EKG or ECG) measures the collective electrical activity of cardiac muscle during contraction and relaxation. Heart cells are joined by gap junctions, which allow simultaneous contraction of entire heart regions to facilitate coordinated pumping of the blood.
8. EKG waves are divided into regions that correspond to various stages in the depolarization and repolarization of muscle cells within the heart. By analyzing the pattern of these waves, abnormal heart conditions such as tachycardia (heart rate over 100 beats/min), ventricular fibrillation (uncoordinated contraction of ventricles), and heart block (failure of ventricular stimulation) can be detected. See Box 42.A on page 964 in your textbook for details.

9. Some cardiac muscle cells are capable of spontaneous, or myogenic, contraction without direct stimulation by the nervous system. The resting membrane potential of these pacemaker cells slowly depolarizes until an action potential is initiated, which then rapidly spreads throughout the entire heart to produce coordinated, synchronous contractions.
10. The *sinoatrial node* is the collection of pacemaker cells that initiates contraction of the atria. It is located at the junction between the superior vena cava and the right atrium.
11. Since there are no gap junctions between the atria and the ventricles, a second pacemaker region called the *atrioventricular node* located at the base of the atria regenerates action potentials in the ventricles. The *bundle of His* channels these action potentials from the AV node down the length of the ventricles. *Purkinje fibers* then disperse the signal throughout the ventricular muscle mass.
12. The slight time delay caused by the regeneration of action potentials in the AV node separates the contractions of the atria and ventricles. The atria thus contract first, filling the ventricles with blood. The ventricles then contract to expel blood from the heart.
13. Heartbeat rate is controlled in nodal tissue by the release of neurotransmitters from the autonomic nervous system. Acetylcholine released from parasympathetic nerves slows the depolarization rate of pacemaker cells, and leads to a reduction in heartbeat rate. If overactive, this control system can produce a *vagal reaction*, where heartbeat rate slows to the point that fainting results.
14. Sympathetic nerves release norepinephrine at the pacemaker nodes. This causes the depolarization rate and contraction strength of muscle cells to increase, leading to increased heartbeat rate and ventricular pressure. The release of epinephrine (adrenaline) from the adrenal glands is also stimulated by these changes, which further contributes to the excitatory effect on the heart.
15. The walls of arteries and arterioles contain elastic fibers which permit vessel expansion under the increased pressure of ventricular contraction. These fibers recoil when the ventricles relax, helping to smooth the flow of blood. Arteries and arterioles are known as *resistance vessels* because their diameters, and thus their resistances, can be regulated through the action of smooth muscles located in the vessel walls.
16. Arterial disease is now responsible for one-half of all human deaths in developed countries. *Atherosclerosis*, or hardening of the arteries, occurs over many years through the accumulation of fatty deposits called *plaque* at sites of endothelial damage within the arterial system. As the vessel diameter narrows, calcium deposits form and reduce the elasticity of the vessel wall.
17. A blood clot, or *thrombus*, forms if clotting agents in the blood come in contact with exposed plaque. The result can be complete blockage of the vessel. If the blockage develops in a *coronary artery* supplying blood to the heart, then the condition is called a *coronary thrombosis*, which often leads to a *coronary infarction*, or heart attack.
18. A dislodged piece of a thrombus, called an *embolus*, can travel through the vascular system and block other vessels, producing an *embolism*. Small-diameter arteries of the brain are particularly susceptible. When these vessels are blocked, brain cells normally nourished by their blood supply will die; this is the physiological basis of a stroke.
19. Both genetic and lifestyle factors influence the likelihood of developing arterial disease. High-fat diets, smoking, lack of exercise, obesity, certain medical conditions, and high blood pressure (*hypertension*) are all risk factors.
20. Cells exchange nutrients, oxygen, and wastes with the blood across capillary walls. Arterioles are connected with venules by capillary beds. These beds are so extensive that no body cell is more than one or two cell diameters away from a capillary.
21. Thin, permeable walls characterize capillaries. Because the total cross-sectional area of a capillary bed is large compared to the arterioles feeding it, flow rate is greatly reduced and pressure lowered in the capilllaries. This facilitates diffusion across the capillary wall.
22. Capillaries are permeable to small molecules and water, but not to large molecules such as blood proteins. Pressure tends to force water and small molecules out of the capillary in a process called filtration. The large molecules that remain inside the vessel create an osmotic potential that pulls water back into the capillary.
23. Since blood pressure is highest on the arterial side of a capillary, water loss is also greatest there. By contrast, osmotic forces dominate on the low-pressure venous side of the bed. The net movement of water within a capillary bed is determined by the interaction between these opposing forces.
24. During the inflammatory response caused by *histamine*, arterioles expand and blood flow increases to the injured area. This increases blood pressure in capillaries. At the same time the permeability of capillaries and venules increases, allowing water to move out of these vessels and into the surrounding tissues. The result is swelling, or *edema*, which can be countered by the use of *antihistamine* drugs.
25. Decreases in the osmotic potential of blood can also lead to capillary water loss. The disease kwashiorkor is characterized by extreme edema and results from severe protein deficiency. In response to this deficiency, blood proteins are degraded, removing the molecules responsible for creating the osmotic potential that draws water back into the blood.
26. Capillary architecture, concentration gradients, and the chemical characteristics of substances transported in the blood all determine whether small molecules will cross the capillary wall. The capillaries of some tissues, such as those associated with the blood–brain barrier, are particularly selective about what will pass. Capillaries found in the digestive tract and kidneys allow large molecules to pass through pores in the capillary membrane. Still others utilize endocytosis.
27. The *lymphatic system* returns interstitial fluid lost by seepage from capillaries back to the blood. *Lymph* (as this fluid is known once it enters the lymphatic system) is chemically similar to blood but has less protein and con-

tains no red blood cells. Lymph is returned to the blood by a separate system of lymphatic vessels which begin as fine capillaries in the tissues and eventually collect into the *thoracic duct*, which empties into the superior vena cava. The propelling force for lymph movement is the contraction of skeletal muscle. Backflow is limited by valves within the lymphatic vessels.

28. Lymph nodes are associated with the major lymphatic vessels of mammals and birds, and play an important role in immune defenses. In addition to being sites for lymphocyte production and phagocytosis, they also act as mechanical filters to trap foreign particles and metastasized cancer cells.

29. Blood returns to the heart through the venous system, which is characteristically under low pressure. The low pressure allows blood to accumulate in the veins, which are referred to as *capacitance vessels* because of their capacity to store blood. In humans, up to 80 percent of total blood volume is in the veins at any one time.

30. Gravity helps move blood through veins located above the heart. However, venous blood below the heart must be pushed or pulled against gravity. If this does not occur, blood accumulates in the lower body, the brain becomes deprived of oxygen, and fainting results. Fainting is a self-correcting response because it changes body position and thus redistributes the blood.

31. Like the lymphatic system, the action of skeletal muscle is important for moving blood through the venous system. Valves keep blood flow unidirectional as the milking action of the muscles gradually pushes the blood back to the heart.

32. During vigorous exercise, muscular pumping action on the veins is enhanced. This pushes an increased volume of blood back to the heart, causing the cardiac muscle to be stretched. The heart responds by contracting more forcefully, a relationship known as the *Frank–Starling law*.

33. A ventilatory pump helps return venous blood to the heart as the negative pressure created by contraction of the diaphragm and intercostal muscles pulls blood and lymph toward the chest. A few veins, especially large ones located near the heart, also contain smooth muscle which can assist in the return of venous blood.

Questions (for answers and explanations, see page 366)

1. The inferior and superior vena cavae of humans have which of the following in common?
a. Carry venous blood
b. Empty into the right atrium
c. Are capacitance vessles
d. Drain the systemic blood circuit
e. All of the above

2. A man visits his doctor and finds that his blood pressure is 125/50 mmHg. This indicates that his blood pressure is
a. slightly higher than normal when the ventricles are relaxed.
b. slightly higher than normal when the atria are contracted.
c. lower than normal when the ventricles are contracted.
d. lower than normal when the ventricles are relaxed.
e. All of the above

3. During tachycardia,
a. heartbeat rate exceeds 100 beats/min.
b. the ventricles contract in an uncoordinated fashion.
c. blood pressure drops precipitously.
d. the ventricles fail to contract at all.
e. coronary arteries are blocked.

4. A artificial electric stimulus is simultaneously provided to both the sinoatrial node and the atrioventricular node. This would cause
a. the atria to beat first, then the ventricles.
b. the ventricles to beat first, then the atria.
c. all four chambers of the heart to contract simultaneously.
d. the left side of the heart to beat first, then the right side.
e. heart murmurs.

5. The coronary arteries of humans are susceptible to all but one of the following conditions. Select the *exception*.
a. Atherosclerosis
b. Stroke
c. Plaque deposits
d. Embolism
e. Thrombosis

6. Which of the following statements about capillaries is *not* true?
a. They directly link the arteries of the systemic blood supply with the veins in the same circuit.
b. Diffusion of gases and nutrients occurs between the blood and tissues across capillary walls.
c. Capillary walls are highly permeable to small molecules, but not to large blood proteins.
d. Histamines increase the permeability of capillary walls.
e. Filtration is the process where blood pressure forces water and small molecules out of capillaries.

7. All of the following are associated with the lymphatic system, *except*
a. the thoracic duct.
b. specialized transport vessels.
c. red blood cells.
d. large proteins.
e. elements of the immune system.

8. Which of the following is *not* a mechanism that helps move venous blood back to the heart in humans?
a. The pull of gravity
b. Activity of parasympathetic nerves
c. Pumping action of skeletal muscle
d. Specialized valves which prevent backflow of blood
e. Ventilatory pumping

9. A patient is brought into a hospital suffering from severe histamine-induced edema. This means that
a. the patient has suffered a stroke.
b. the patient has suffered a coronary infarction.
c. too much blood has collected in the venous system.
d. interstitial fluids have accumulated in the body tissues due to leaky capillaries and venules.
e. None of the above

Activities (for answers and explanations, see page 367)

- On the following diagram, label the indicated structures of the human heart.

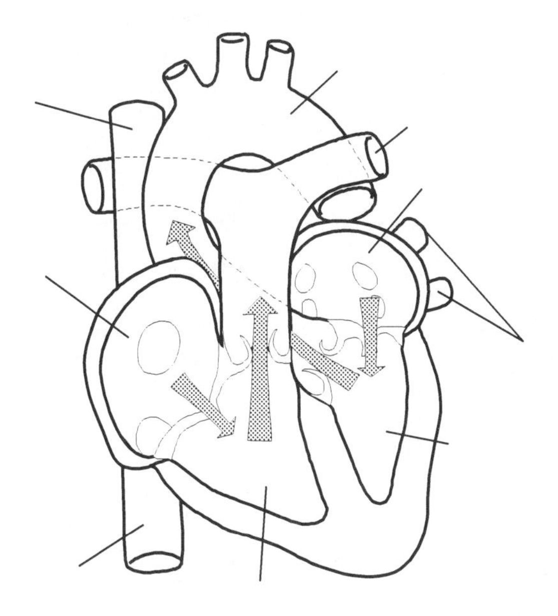

- Number the following cardiovascular elements in the sequence that they would be encountered while traveling through the bloodstream of the human body. Begin your journey at the left ventricle.

___ Left ventricle ___ Right ventricle

___ Atrioventricular valve ___ Aortic valve

___ Pulmonary valve ___ Pulmonary arterioles

___ Pulmonary venules ___ Aorta

___ Vena cavae ___ Left atrium

___ Right atrium ___ Systemic capillaries

___ Pulmonary capillaries ___ Pulmonary artery

___ Pulmonary vein ___ Systemic arterioles

___ Systemic venules

- On the following graph of human blood pressure (measured in the left ventricle), indicate which sections of the curve correspond to systole and which to diastole.

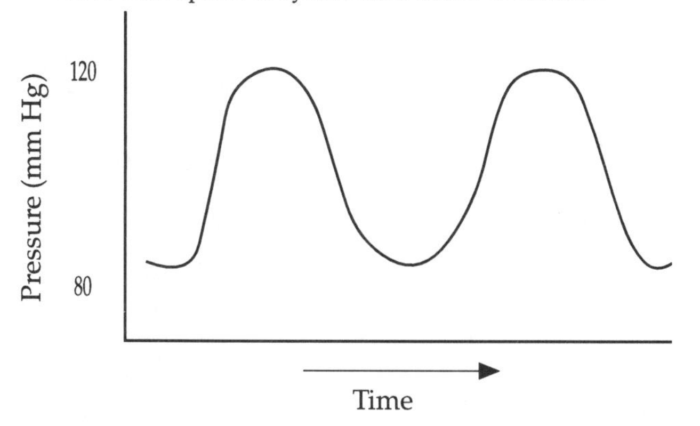

THE BLOOD • CONTROL AND REGULATION OF CIRCULATION (pages 971–976)

Key Concepts

1. Blood is a type of connective tissue. It consists mainly of *plasma,* an aqueous fraction of complex chemical composition, within which are suspended a variety of cellular elements.
2. When a sample of whole blood is centrifuged, the cellular fraction moves to the bottom of the tube and the plasma remains at the top. The *packed cell volume,* or *hematocrit,* is normally about 40 percent in humans (less in women than in men) It can vary widely with health, physical conditioning, and acclimation to high altitude. The cellular fraction of blood contains three important cell types.
3. *Erythrocytes,* or red blood cells, comprise the largest percentage of the hematocrit. Packed with hemoglobin, their function is to transport gases between the respiratory surfaces and the body cells.
4. Red blood cells are produced at a rate of over two million every second by *stem cells* located in the bone marrow. Production rate is controlled by the hormone *erythropoietin,* which is released by the kidney in response to low oxygen levels.
5. Just before entering a capillary from the stem cells, red blood cells lose many of their internal organelles as an adaptation for increased hemoglobin content. Mature cells are shaped like biconcave discs. This design increases the surface area available for gas exchange, and makes the cell flexible so it can more easily squeeze through capillaries.
6. *Leukocytes,* or white blood cells, are less abundant than red blood cells. They are important components of the body's defense mechanisms, functioning in the search for foreign particles and the phagocytosis and destruction of those particles. They also play a role in antibody formation.
7. White blood cells are capable of amoeboid movement. They spend much of their time outside the vascular system following chemical cues to sites of infection or cell damage.
8. *Platelets* are also produced by cells of the bone marrow. Their concentration in the blood is intermediate to that of red and white blood cells. Platelets are tiny fragments of cells that are packed with the enzymes and chemicals needed to seal leaks in the vascular tissue.
9. Platelets are activated by contact with the collagen fibers that are exposed when blood vessels are damaged. Upon activation the platelet swells, becomes irregularly shaped, and sticks to the damaged site. Activated platelets working together create a plug at the damaged site, which temporarily patches the leak.
10. Many chemical *clotting factors* are involved in the complex cascade of reactions that ultimately leads to the complete repair of damaged vascular tissue. The process begins with platelet activation, which stimulates the circulating inactive enzyme *prothrombin* to be converted into its active form *thrombin.*

11. Thrombin causes the blood protein *fibrinogen* to polymerize and form threads of *fibrin*, which settle at the injury site to form a framework upon which a stable clot can develop. This promotes the growth of scar tissue.
12. Plasma has a large concentration of dissolved ions, gases, proteins, and nutrient molecules. Important ions include Na^+ and Cl^-, and common nutrients are glucose, amino acids, lipids, and lactic acid. Plasma is very similar in composition to interstitial fluid, except that the plasma has a higher concentration of protein.
13. In addition to the proteins involved in clotting, the plasma also includes albumin, the protein responsible for maintaining proper osmotic potential in the capillaries. Antibodies, hormones, and various carrier molecules such as *transferrin*, which moves iron from the gut to its site of usage, are also present in the plasma.
14. Autoregulatory mechanisms in each capillary bed control the constriction and dilation of arterioles supplying the bed. Changes in oxygen content, waste production, pH, and blood flow in the capillaries influence the arterial pressure of blood leaving the heart.
15. Autoregulation is intrinsic to the tissue served by a capillary bed. However, blood flow through capillaries is also influenced by neural and hormonal input. These control mechanisms affect the contraction of smooth muscle in the arteries and the arterioles that supply the capillary bed.
16. Activities that increase cell metabolism in tissues also increase blood flow to those tissues. Precapillary sphincter muscles form cuffs around the arterioles feeding capillary beds. As O_2 concentration decreases, or CO_2 and metabolite concentrations increase, the smooth muscle sphincters relax to allow more blood into the bed.
17. Most arteries and arterioles receive direct neural input from the sympathetic division of the autonomic nervous system. In the vessels of skeletal muscle, sympathetic nerves release acetylcholine and cause the smooth muscle to relax, leading to dilation and increased blood flow. Elsewhere, sympathetic nerves release norepinephrine, which causes smooth muscle to contract and thus restrict blood flow.
18. Hormones such as epinephrine, angiotensin, and vasopressin also cause arterioles to constrict. In particular, they influence the arterioles feeding peripheral capillary beds, reducing blood flow to the extremities. During times of metabolic stress this response channels more blood to essential internal organs such as the brain, heart, and kidneys.
19. Cardiovascular control centers in the medulla are the source of autonomic nerve signals that regulate blood flow. These centers receive input about changes in blood pressure from stretch receptors located in the aorta and the carotid arteries, which supply blood to the brain. This information helps determine whether blood flow to the periphery will be restricted.
20. Chemosensors in the carotid arteries and aorta detect low O_2 levels and stimulate the cardiovascular control centers to increase blood pressure. Neural input from higher brain centers, which regulate emotions and anticipate physiological needs, can also affect cardiovascular control.
21. Heartbeat rate is slowed by neural inputs during the diving reflex of marine mammals. These animals can dramatically reduce heartbeat rate, which, when coupled with a restriction of peripheral blood flow, helps to conserve O_2 for the operation of heart and brain.
22. Overall cardiovascular control is thus maintained by two mechanisms: autoregulatory systems that provide for the needs of local tissues, and the control of autonomic nerves and hormones that respond to whole-body metabolic demands.

Questions (for answers and explanations, see page 367)

1. The rate of new red blood cell production in the __________ is controlled by the __________.
   a. stem cells of bone marrow; blood protein fibrinogen
   b. blood plasma; nutrients glucose and lactic acid
   c. packed cell volume; hormone erythropoietin
   d. capillaries; protein transferrin
   e. stem cells of bone marrow; hormone erythropoietin
2. Which of the following statements about leukocytes is *not* true?
   a. They are capable of amoeboid movement.
   b. They are tiny cell fragments that initiate the chain of events leading to clot formation.
   c. They often leave the vascular system following chemical cues toward damaged cells.
   d. They engulf foreign particles in the blood through the process of phagocytosis.
   e. There are many fewer of them in the packed cell volume compared to erythrocytes.
3. Before fibrin threads are produced to lay the groundwork for clot formation,
   a. fibrinogen causes platelets to become activated.
   b. red blood cells lose their hemoglobin and stimulate the activation of erythropoietin.
   c. platelets are activated and prothrombin is converted to thrombin.
   d. transferrin is converted into prothrombin, which activates the platelets.
   e. None of the above
4. The autoregulatory mechanisms that control the dilation and constriction of capillaries are influenced by the
   a. pH of the blood.
   b. O_2 content of the blood.
   c. production of metabolic wastes in neighboring tissues.
   d. rate of blood flow through the capillaries.
   e. All of the above
5. Which of the following statements about the control of circulation in humans is *not* true?
   a. Sympathetic nerve input to skeletal muscle causes the blood vessels in the muscle to dilate.
   b. Adrenaline produces arteriole constriction, especially in peripheral capillary beds.
   c. Blood flow can be regulated by autonomic nerve signals emanating from cardiovascular control centers in the medulla of the brain.
   d. Hormones such as angiotensin and vasopressin cause venules to constrict.

e. Carotid artery chemosensors detect low O_2 levels in the blood and promote increased blood pressure.

6. A whale dives for food while carrying a radio transmitter that allows you to measure a number of parameters related to its cardiovascular function. Which of the following is likely to be true about this animal during its dive?
   a. Heartbeat rate will decrease, and peripheral blood flow will decrease
   b. Heartbeat rate will increase, and peripheral blood flow will decrease
   c. Heartbeat rate will decrease, and the number of new red blood cells produced will increase
   d. Heartbeat rate will increase, and activity of autoregulatory mechanisms in the brain will decrease
   e. Heartbeat rate will decrease, and peripheral blood flow will increase

7. You perform an experiment in which a drug is supplied to a normal human that causes selective dilation of *only* the precapillary sphincter muscles. Which of the following effects is most likely to occur?
   a. Blood flow through the capillary beds will decrease.
   b. The permeability of the capillary walls will decrease.
   c. A chemical cascade leading to clotting will be initiated.
   d. Blood flow through the capillary beds will increase.
   e. Heartbeat rate will slow dramatically and capillary blood pressure will rise.

8. The medullary cardiovascular control center receives neural input from
   a. higher brain centers involved with emotions and stress.
   b. chemosensors on the aorta and carotid arteries.
   c. stretch sensors in the arteries.
   d. chemosensors in the medulla.
   e. All of the above

Activities (for answers and explanations, see page 367)

- Match each of the following blood components with its best description.

_____	Plasma	*a.* Protein used to regulate osmotic potential in capillaries
_____	Erythrocyte	*b.* Causes polymerization of molecules needed to stabilize a blood clot
_____	Leukocyte	*c.* Contains large concentration of hemoglobin
_____	Platelet	*d.* Aqueous component of blood containing many dissolved chemicals
_____	Thrombin	*e.* Cell fragments that initiate the events associated with blood clotting
_____	Albumin	*f.* Important blood element used in the body's defense mechanisms
_____	Hematocrit	*g.* Contains red and white blood cells after whole blood is centrifuged

- Diving is a strenuous activity, yet when marine mammals dive, their heart rate actually decreases and blood flow to most of their internal organs is reduced. Explain the cardiovascular reasons why this type of diving response is adaptive for marine mammals.

Integrative Questions (for answers and explanations, see page 368)

1. If you compare blood from the superior vena cava of a human with blood from the aorta of a dog, the main difference between the two samples of blood would be
   a. the types of cells in the cellular fraction.
   b. the presence or absence of platelets.
   c. the concentration of O_2.
   d. the presence or absence of plasma.
   e. the presence or absence of prothrombin.

2. Which of the following characteristics of animal circulatory systems can be found in a clam?
   a. Closed circulatory system
   b. Pumping heart
   c. Red blood cells
   d. Pulmonary valves
   e. None of the above

3. Which of the following statements about animal circulatory systems is *not* true?
   a. Leukocytes typically contain large amounts of hemoglobin.
   b. The hormone erythropoietin controls red blood cell production.
   c. Circulation is not necessary in cnidarians because they can utilize direct diffusional exchange with their environment.
   d. In arthropods, circulatory fluids return to the heart through valved holes called ostia.
   e. The right and left atria in a human heart receive blood from the venous supply.

4. If too much blood collects in a person's venous supply because of prolonged inactivity, this may
   a. stimulate the chemical cascade that leads to blood clotting.
   b. cause the person to suffer a brain embolism and stroke.
   c. lead to the development of atherosclerosis or hypertension.
   d. deprive the brain of oxygen and cause the person to faint.
   e. block neural activity in the Purkinje fibers of the heart.

43

Animal Nutrition

CHAPTER LEARNING OBJECTIVES—after studying this chapter you should be able to:

- ❑ Explain the differences between the terms "autotrophy," "heterotrophy," "saprophyte," "detritivore," "predator," "carnivore," "herbivore," "omnivore," "filter feeder," and "fluid feeder."
- ❑ Distinguish between the processes of intracellular and extracellular digestion.
- ❑ Describe how the term "calorie" is used to understand nutrition, and distinguish it from a "kilocalorie" and a "Calorie."
- ❑ Discuss the relative importance of carbohydrates, fats, and proteins as energy sources for animals.
- ❑ Explain the difference between the terms "undernourished," "overnourished," and "malnourished," and cite examples of the problems caused by these conditions in humans.
- ❑ Discuss the metabolic reasons why certain amino acids and fats are essential in the diets of animals.
- ❑ Distinguish between macro- and micronutrients in animals, and explain their roles in nutrition.
- ❑ Describe the role of vitamins in animal nutrition, explain how they function, and give examples of some problems that result from vitamin deficiencies in humans.
- ❑ Provide examples of common feeding adaptations in carnivores and herbivores.
- ❑ Describe the feeding mechanisms of both small and large filter-feeding animals.
- ❑ Explain how the structure and function of teeth provide useful information about the diet of vertebrates.
- ❑ Describe the basic structure and function of a complete digestive tract in animals.
- ❑ Discuss the role of hydrolytic enzymes in animal digestion, and explain how these enzymes are stored and activated.
- ❑ Describe the structure and function of the vertebrate gut, with special emphasis on the human digestive system.
- ❑ Explain the role of the liver and pancreas in aiding vertebrate digestion.
- ❑ Discuss the ways that carbohydrates, proteins, and fats are digested and absorbed in the vertebrate gut.
- ❑ Describe the modifications to the basic vertebrate gut design that are seen in ruminant herbivores and explain the function of these modifications.
- ❑ Explain how an intrinsic neural reflex differs from a reflex mediated by the central nervous system, and give examples of each from the human digestive system.
- ❑ Discuss the roles of the hormones gastrin, secretin, cholecystokinin, glucagon, and insulin in the control of digestion and nutrient availability in humans.
- ❑ Explain the digestive purpose of lipoproteins and differentiate among the various kinds of lipoproteins found in humans.
- ❑ Describe the differences between the absorptive and postabsorptive states in human digestion with regard to hormonal control and nutrient availability .

DEFINING ORGANISMS BY WHAT THEY EAT • NUTRIENT REQUIREMENTS (pages 979–988)

Key Concepts

1. *Heterotrophs* obtain energy and structural molecules from preformed organic nutrients. *Autotrophs* harness solar or inorganic chemical energy to manufacture their own organic nutrients. All animals are heterotrophs, but not all heterotrophs are animals.
2. Autotrophs form the nutrient base of ecosystems, funneling energy from nonliving sources into the living food web. Heterotrophs gain access to this energy by consuming the autotrophs, either directly or indirectly. Heterotrophic organisms show diverse adaptations for gaining nutrient energy.
3. The fungi are *saprophytes;* they obtain nutrients through absorption by feeding on dead organisms. Some animals are *detritivores,* gathering energy by ingestion from the remains of dead organisms. Animals that ingest other living organisms are called *predators.*
4. *Carnivores* prey on other animals, while *herbivores* prey on plants. *Omnivores* prey on both animals and plants. Within these categories other feeding distinctions are made. For example, *filter feeders* strain small organisms or particles from the surrounding medium, while *fluid feeders* draw blood or other body fluids from their prey.

5. Most of the food ingested by heterotrophs is too large to be absorbed directly by body cells. Food must therefore undergo *digestion*, a process which chemically breaks down nutrients into their molecular subunits, before it can be absorbed.

6. A few animals, such as sponges, engulf food particles and carry out intracellular digestion. Most animals, however, first perform *extracellular digestion*. This occurs inside a specialized digestive compartment, or gut, within the body.

7. Various levels of gut complexity are found in animals. Some guts are saclike structures with only a single opening. Others are hollow tubes with an entrance at one end and an exit at the other end. This design allows for specializations that accommodate the handling of food at one end, storage and digestion in the middle, and excretion at the other end.

8. Even though they are contained within the gut, nutrients undergoing digestion are still technically outside of the body because they have not yet crossed cell membranes. Once digested, these materials are absorbed by cells.

9. Energy in the chemical bonds of food is ultimately transferred to ATP, which is used in all aspects of cell metabolism. These energy conversions are never 100 percent efficient, some energy is lost in the form of heat.

10. A useful physiological unit of heat energy is the *calorie*, defined as the energy needed to raise the temperature of one gram of water one degree Celsius. A calorie is equivalent to 4.184 joules. Most physiological processes produce or use large amounts of energy. It is therefore convenient to think in terms of a *kilocalorie* (one thousand calories). The dietician's *Calorie*, indicated with a capital "C," equals one kilocalorie.

11. An animal's overall metabolic rate is the amount of energy it must acquire through feeding and digestion. Basal metabolic rate is the amount of energy needed to sustain basic cell metabolism. Any activity beyond this basic maintenance requires an additional energy input through the metabolism of fats, carbohydrates, or proteins.

12. Fats provide roughly twice as much energy per gram (9.5 kcal) as do carbohydrates (4.2 kcal) or proteins (4.1 kcal). Nutrients are often metabolized immediately after uptake by body cells. However, since most animals do not eat continuously, yet require energy at a steady rate, nutrients must also be stored for later use.

13. Liver and muscle cells store carbohydrates in the form of glycogen. However, glycogen's structural properties do not allow it to be conveniently stored in large quantities. Fats, on the other hand, not only provide more energy than do carbohydrates but can also be stored more compactly. Fats therefore represent a better long-term source of energy reserves. Protein is not used for long-term energy storage, although it is broken down to yield energy under conditions of metabolic stress.

14. An animal is said to be *undernourished* if its food intake does not provide sufficient energy to meet its metabolic needs. This shortfall must be made up by self-consumption, which begins by using stored glycogen and fat, and if necessary proceeds to the breakdown of body proteins. Since proteins are integral components of enzymes, muscles, neurons, and blood, this level of undernourishment rapidly leads to impaired body function, and eventually death.

15. Human undernourishment occurs most often in underdeveloped countries, and now affects well over one billion people. However, developed nations are not immune to the problem. For example, the self-imposed starvation known as *anorexia nervosa*, which results from a psychological aversion to fat, disables thousands of people in Western nations.

16. *Overnourishment* occurs when an animal consistently acquires more energy than it needs for metabolism. The excess is stored as glycogen and fat, which may be of great importance to seasonal hibernators and migratory species that rely on these reserves to survive extended periods of food deprivation.

17. Human overnourishment presents serious health risks. High blood pressure, heart attack, and diabetes are a few of the problems associated with abnormal weight gain. Excess weight puts a heavy burden on the heart, which must supply more blood at higher pressures to sustain the extra body mass.

18. Unlike plants, animals cannot directly synthesize required organic molecules. Carbon skeletons, such as the acetyl group ($COCH_3$), must first be obtained through the metabolism of fats, carbohydrates, and proteins. These skeletons are then rearranged to form the complex energy and structural molecules needed by animals.

19. *Malnutrition* occurs when an animal cannot obtain sufficient quantities of a required carbon skeleton. Some skeletons, such as the acetyl group common to all carbohydrates and fats, are rarely in short supply. Others, such as the amino acids used in protein synthesis, are often limited in available food resources.

20. Certain amino acids are synthesized by animals through the transfer of an amino group to the acetyl skeleton. Most animals, however, cannot directly synthesize all needed amino acids. These must therefore be obtained preformed through the diet.

21. Required dietary amino acids are called *essential amino acids*. If one or more essential amino acids are absent from the diet, protein formation is impaired. Different species have different essential amino acid requirements. Herbivores generally require fewer essential amino acids than do carnivores.

22. Eight essential amino acids are required in the human diet: isoleucine, leucine, lysine, methionine, phenylalanine, threonine, tryptophan, and valine. All are made available to the body when meat, eggs, and milk are eaten regularly. However, one or more is usually lacking in any particular type of plant food. Vegetarians must therefore carefully select food combinations to insure a proper balance of essential amino acids.

23. The human body does not store large quantities of free amino acids, yet it is continuously synthesizing new proteins. Thus, all of the amino acids needed for protein synthesis must be made available in each meal.

24. Several factors prevent animals from using dietary protein to make their own protein directly. These include the fact that 1) macromolecules do not readily cross cell membranes, 2) protein structure is highly species-specific, and 3) foreign proteins are attacked by the animal's immune system. These problems are avoided by digesting dietary proteins into their component amino acids, and then reassembling them inside the cells.
25. Virtually all the lipids needed for the construction of cell membranes and sex hormones can be assembled by combining acetyl groups with a dietary source of essential fatty acids, notably linoleic acid.
26. Certain mineral elements are also required by animals, and these needs are highly species-specific. Elements needed in large quantities (e.g., calcium, phosphorus, and sodium) are called *macronutrients.* Those needed only in trace amounts (e.g., iron, copper, and zinc) are termed *micronutrients.*
27. Some minerals are used directly as ions to stimulate the contraction of muscle or to generate action potentials in the nervous system. Others function as enzyme cofactors, structural components of important electron carrier molecules and hemoglobin, or help to maintain proper osmotic balance in tissues.
28. A special group of essential nutrients are known as *vitamins.* They are generally small, organic molecules that function mainly as coenzymes; all are required for normal physiological function. Unlike essential amino acids and fatty acids, vitamins are only needed in small amounts. They are divided into two broad categories based on solubility properties.
29. *Water-soluble vitamins* include the B vitamins: thiamin (a coenzyme necessary for cellular respiration whose absence leads to the disease *beriberi*), niacin (a component of the electron transfer molecules NAD and NADP), and riboflavin (which forms the oxidation–reduction site of FAD and FMN). Vitamin C (ascorbic acid) plays an important role in the formation of collagen; a lack of vitamin C in the diet leads to the disease *scurvy,* characterized by massive tissue breakdown.
30. *Fat-soluble vitamins* include vitamin A (retinol) which is essential for visual pigment formation, vitamin D (calciferol) which regulates calcium metabolism, vitamin E which may protect cellular membranes from oxidation, and vitamin K which plays a crucial role in blood clotting.
31. If water-soluble vitamins are acquired in excess of metabolic needs, the excess is simply excreted through the urine. By contrast, excess fat-soluble vitamins tend to accumulate in body tissues, and can eventually build to toxic levels.
32. Deficiency diseases develop when animals experience chronic nutrient shortages. The human disease kwashiorkor results from severe protein deficiencies which cause extreme edema, breakdown of the immune system, and the degeneration of metabolic and neural functions.
33. Vitamin deficiencies produce the characteristic disease symptoms of beriberi and scurvy. *Pellagra* is another vitamin-deficiency disease caused by a lack of nicotinamide; it is often associated with alcoholism. *Rickets* results from a vitamin D deficiency that causes skeletal deformation as bones soften from decreased calcium absorption. *Pernicious anemia,* a blood disease caused by a lack of the animal vitamin B_{12}, occurs in strict vegetarians who do not take vitamin supplements.
34. Deficiencies in mineral nutrition can also result in disease. For example, iodine is a component of the thyroid hormone thyroxine. If there is a dietary deficiency of iodine, a *goiter* develops as the thyroid gland enlarges to try and compensate for the lack of functional thyroxine.

Activities (for answers and explanations, see page 368)

- Match each of the following nutrients, nutritional processes, or nutritional diseases with its best description.

____	Rickets	*a.* Breakdown of nutrients into chemical building blocks
____	Malnutrition	*b.* Results from a dietary deficiency of thiamin
____	Vitamin C	*c.* Found in citrus fruits and also known as ascorbic acid
____	Heterotrophy	*d.* Condition resulting from an insufficient intake of calories
____	Digestion	*e.* Results from an insufficient intake of essential nutrients
____	Undernourished	*f.* Bone disease caused by a lack of vitamin D
____	Beriberi	*g.* Process of obtaining energy from preformed organic food

- Is it possible for an animal to be simultaneously overnourished and undernourished? Explain your answer.

- Is it possible for an animal to be overnourished and at the same time malnourished? Explain your answer.

Questions (for answers and explanations, see page 368)

1. For which of the following organisms do the terms "heterotroph," "carnivore," and "filter feeder" all apply?
a. Mushroom
b. Blue whale

c. Lion
d. Bison
e. Termite

2. Consider the digestive characteristics of a sheep. Which of the following sets of descriptions applies to this animal?
   a. Saprophyte, engulfs food, performs intracellular digestion
   b. Autotroph, synthesizes organic nutrients, performs extracellular digestion
   c. Heterotroph, absorbs organic nutrients, carnivore
   d. Detritivore, ingests food, performs intracellular digestion
   e. Herbivore, ingests food, performs extracellular digestion

3. A chipmunk takes in 1,000 kcal of food per day, but this food is completely lacking in the B vitamin thiamin. It also has a total metabolic rate of 1,050 kcal per day. Which of the following appropriately describes the energetic and nutrient status of this animal?
   a. Slightly undernourished and also malnourished
   b. Greatly undernourished but not malnourished
   c. Slightly overnourished but not malnourished
   d. Greatly overnourished and also malnourished
   e. Perfectly balanced nutritional and energetic condition

4. Certain amino acids are essential in the diet of animals because
   a. they prevent overnourishment.
   b. they are cofactors and coenzymes that are required for normal physiological function.
   c. the animal cannot directly synthesize them through the transfer of an amino group to an appropriate carbon skeleton.
   d. animals need these substances in order to make the stored fats used during hibernation and migration.
   e. None of the above

5. All but one of the following diseases is caused by a vitamin deficiency. Select the *exception.*
   a. Rickets
   b. Pernicious anemia
   c. Scurvy
   d. Kwashiorkor
   e. Beriberi

6. Sodium, a substance important for the proper functioning of muscle and nerves, is nutritionally classified as
   a. a mineral micronutrient.
   b. an essential amino acid.
   c. an essential fatty acid.
   d. a water-soluble vitamin.
   e. a mineral macronutrient.

7. Which of the following statements about animal nutrition is *not* true?
   a. Essential fatty acids from the diet are combined with acetyl groups to form lipids.
   b. Excess fat-soluble vitamins (such as A, D, and K) are largely excreted through the urine.
   c. A person on a 1,000-Calorie per-day diet is actually taking in 1 million calories of energy.
   d. Glycogen is a common energy storage molecule found in animal liver and muscle cells.
   e. All animals are heterotrophs, but not all heterotrophs are animals.

8. A person has been starving for five days. Which of the following energy reserves would probably be present in the *lowest* concentration in this person's body?
   a. Carbohydrates
   b. Proteins
   c. Fats
   d. Both proteins and fats
   e. Both carbohydrates and proteins

ADAPTATIONS FOR FEEDING • DIGESTION (pages 988–993)

Key Concepts

1. An animal's ecological niche is largely determined by its morphological, behavioral, and biochemical adaptations for feeding.
2. Carnivores have evolved remarkable powers of prey detection, and have coupled this sensory ability with stealth, speed, and power. Some carnivores have specialized adaptations for immobilizing, killing, and ingesting their prey through the use of sticky surfaces, poisons, and external digestive enzymes.
3. Because their prey is less likely to resist being eaten, herbivores have less dramatic feeding adaptations than carnivores. Simple grazing is a common form of herbivory, but specialized feeding and locomotory structures such as an elephant's trunk or a hummingbird's wings are also used. Some species of ants and termites actively tend fungal gardens that they harvest for food.
4. Most small, sessile aquatic animals are filter feeders, using specialized structures to strain small organisms or food particles from large volumes of water that pass through the front end of their gut. A number of larger, more mobile animals are also filter feeders, including several kinds of aquatic insect larvae and many species of fish.
5. Baleen whales are the most dramatic example of how big animals can be sustained by filter feeding. Large *baleen plates* hang down from either side of the upper jaw to form a netlike filtering device. Enormous quantities of water containing small crustaceans and fish are filtered daily to satisfy the huge metabolic demand of these animals.
6. A hallmark of vertebrate predators is the possession of specialized teeth which aid in food acquisition and the initiation of digestion. Although vertebrate teeth are sufficiently distinct to allow identification of individual taxa, they nevertheless all share a common structural plan.
7. Each tooth has a crown, or biting surface, and a root which anchors the tooth into the jaw. A hard calcium phosphate layer called *enamel* covers the crown. A layer of bony material known as the *dentine* underlies the enamel and extends into the root. Beneath the dentine is the pulp cavity, which contains nerves, blood vessels, and the cells that produce dentine.

8. The specific function of a tooth is reflected in its shape. Incisors are the sharp, flattened teeth used for cutting, chopping, and gnawing. Canines are daggerlike and are used for stabbing, ripping, and shredding. Cheek teeth (also known as molars and premolars), are used for shearing, crushing, and grinding.
9. The relative number and size of each tooth type varies with the animal's age and its diet. Humans have a set of 20 milk teeth during the first few years of life. Later, as our omnivorous adult diet takes shape, a generalized set of 32 teeth containing representatives of each tooth type replaces the juvenile set.
10. Although animals take their food into a digestive cavity, or gut, they must still utilize extracellular digestion. Degradative enzymes are secreted into the gut and catalyze the breakdown of nutrients into their building block molecules, which are then absorbed across cell walls.
11. The simplest digestive system is the gastrovascular cavity of cnidarians. This gut type consists of a blind-ended sac with a single opening that serves as both an entrance for food and as an exit for wastes.
12. Flatworm gastrovascular cavities are slightly more complex than those of cnidarians. They are highly branched to increase the surface area for nutrient absorption. There is still only a single opening that attaches to the mouth via an extendable *pharynx*.
13. Many internal parasites, such as tapeworms, have no digestive system at all. Instead, they live inside the gut of their hosts and absorb the predigested food that surrounds them directly across their body cells.
14. All animals more advanced than the sponges, cnidarians, and flatworms have a tubular gut with two distinct openings. Undigested food enters through the mouth, and digestive wastes called feces exit through the anus.
15. Tubular guts show regional specializations for particular functions. Food enters at the anterior end, or *foregut*, through the mouth and *buccal cavity*. The initial breakdown of food begins through the action of teeth or mandibles, although some animals simply ingest large chunks of food which are then swallowed whole. Additional mechanical breakdown can occur in the gizzard (if present), where muscular contractions, aided by small stones, grind food into even smaller pieces.
16. Food then passes into the *stomach* and/or *crop*, which function mainly as storage chambers. Depending on the species, some chemical digestion and absorption may also occur in these chambers.
17. Most digestion and absorption occurs in the *midgut* or *intestine*, which receives the well-mixed and partially digested food from the storage chambers. The intestinal wall is lined with cells that secrete digestive enzymes; it is also associated with accessory glands that produce other digestive chemicals.
18. The last stop for food in the digestive system is the *hindgut*, which reclaims most of the water and ions released during digestion. The hindgut also acts as a temporary storage site for feces, which are eventually passed out of the body through the anus. This occurs by contractions of the muscular *rectum* in a process called *defecation*.
19. The midgut and hindgut regions of many animal digestive systems harbor colonies of symbiotic bacteria or protists which cooperate in the digestion of food. These colonies are especially important to herbivores, which feed exclusively on plant materials containing high concentrations of hard-to-digest cellulose.
20. Absorptive surfaces in the midgut of many animals contain elaborate structures to increase surface area. Earthworms have a simple longitudinal infolding of the intestinal wall called a typhlosole. Sharks have a spiral valve which forces food to take a longer path through the intestine. Other vertebrates have highly folded intestinal walls lined with *villi*, which are in turn covered with microvilli that greatly increase the available area for nutrient absorption.
21. Proteins, carbohydrates, and fats are all digested by specific *hydrolytic enzymes* which catalyze the breakdown of macromolecules into their constituent amino acids, monosaccharides, monoglycerides, and fatty acids.
22. Digestive enzyme names all end with the suffix "-ase," and each is named according to its function. Thus, a carbohydrase degrades carbohydrates, a protease hydrolyzes proteins, etc. Whether the enzyme cleaves the macromolecule on the outside or the inside is indicated by the prefixes "exo-" and "endo-," respectively.
23. Many digestive enzymes are produced in an inactive form called a *zymogen*, and then stored within secretory cells of the gut lining or accessory organs. This protects the animal's body cells from self-digestion. The inactive enzyme is later released into the gut, where it is activated by changes in pH or, more commonly, by the action of other enzymes.
24. The gut also avoids digesting itself by virtue of a lubricating, protective mucus layer which isolates the gut cells from the action of digestive enzymes. *Ulcers* develop in the gut wall whenever this protective layer fails. Instead of mucus, insects secrete a thin layer of tough chitin which lines the gut and prevents damage from digestion or abrasion.

Questions (for answers and explanations, see page 368)

1. Which of the following characteristics is *not* common to all filter-feeding organisms?
   a. Specialized structures that strain small food items from the external medium
   b. Ability to move large volumes of water through the anterior portion of the gut
   c. Heterotrophic nutrient procurement
   d. Small body size and sessile lifestyle
   e. Predatory feeding strategy

2. The toothed whales (e.g., dolphins) usually have a large number of conically shaped teeth, but no flattened molars like those found in humans. This suggests that the toothed whales
   a. carefully chew their food into a fine pulp before swallowing.
   b. rip or tear their food and probably swallow it whole.
   c. are herbivorous and feed mostly on easily digested algae.

d. strain large volumes of water through their teeth and filter out the small fish.
e. None of the above.

3. Which of the following statements about the feeding and digestive adaptations of animals is *not* true?
a. Feeding morphology and the biochemistry of digestion largely determine an animal's ecological role.
b. Some herbivores engage in a sort of agriculture by actively tending renewable sources of non-animal food.
c. The simplest animal digestive systems are the gastrovascular cavities of cnidarians and flatworms.
d. The digestive system is entirely absent from some species of internal parasites that live in the host's digestive tract.
e. Herbivores possess more diverse feeding adaptations than carnivores because their prey is so elusive and difficult to capture.

4. The ________ is primarily a storage chamber within the digestive system, while the _________ reabsorbs water, ions, and generates the feces.
a. buccal cavity; midgut
b. crop; midgut
c. stomach; hindgut
d. foregut; hindgut
e. hindgut; rectum

5. A drug is given to an animal that completely blocks the absorption of nutrients from the digestive system. Which of the following structures is most severely impacted by this treatment?
a. Intestine
b. Buccal cavity
c. Crop
d. Stomach
e. Esophagus

6. Although their feeding strategies are rather divergent, the digestive systems of elephants and hummingbirds both contain all but one of the following materials. Select the *exception.*
a. Zymogens
b. Feces
c. Proteases
d. Dentine
e. Nucleases

7. After a lunch that includes a cheeseburger with "the works" and fries, which of the following can be found actively participating in the digestive process within your gut?
a. Peptidase
b. Lipase
c. Protease
d. Carbohydrase
e. All of the above

8. The typhlosole of an earthworm and the spiral valve of a shark are both adaptations for
a. capturing and ingesting prey items.
b. improving absorption of nutrients by the intestine.
c. protecting the gut from its own digestive enzymes.
d. filtering food items from the environment.
e. preventing ulcers.

Activities (for answers and explanations, see page 368)

- In the following table, list the important structures associated with each section of the tubular gut, and indicate the basic function(s) of these structures.

Section	Structures	Functions
Foregut		
Midgut		
Hindgut		

- Examine Figure 43.10 on page 991 of your textbook. Diagrams *a* and *c* show the skulls and teeth of a human and a saber-tooth cat. What structural differences in dentition exist between these two species, and what is the functional significance of these differences?

STRUCTURE AND FUNCTION OF THE VERTEBRATE GUT (pages 993–1001)

Key Concepts

1. Although the digestive tracts of different vertebrates are highly specialized, all of the segments have the same general tissue organization. Beginning with the *lumen,* or central cavity of the gut, and working outward, four distinct tissue layers can be identified.
2. Immediately adjacent to the lumen is the *mucosa.* In certain portions of the gut the mucosa is highly folded to increase the surface area for absorption. Many mucosal cells also secrete lubricating mucus or digestive enzymes.
3. The *submucosa* lies just outside the mucosal layer. It contains the blood and lymphatic vessels that transport absorbed nutrients to other parts of the body. Sensory and regulatory nerves are also found in the submucosa.
4. Surrounding the submucosa are two layers of smooth muscle oriented at right angles to one another. *Circular muscle* surrounds the circumference of the gut tube, while *longitudinal muscle* runs down its length. Alternating contractions of these muscles produce wavelike churning movements in the stomach and intestines, which mixes food and moves it through the digestive tract.

5. A final layer of fibrous tissue called the *serosa* surrounds and stabilizes the gut tube. The digestive system is also covered and supported by *peritoneum* tissue.

6. Food enters the vertebrate digestive tracts via the mouth, where it is usually chewed and then swallowed. Swallowed food passes through the *pharynx* and into the *esophagus*, a tube that connects the oral cavity with the remainder of the gut. The pharynx is located at the junction between the oral cavity, the trachea (windpipe), and the esophagus. The pharynx directs air and food into their proper pathways.

7. Smooth muscle contractions begin in the esophagus and normally progress toward the anus, creating a wavelike *peristalsis* which pushes food through the digestive tract. The presence of food causes the gut to stretch, which stimulates muscle contraction. Sometimes the direction of peristalsis is reversed, as when the stomach is too full (e.g., heartburn) or during the violent contractions of the vomiting reflex.

8. Reversed movement of food within the digestive tract is normally prevented by thick rings of smooth muscle called *sphincters*. They are found at the junction of the esophagus and the stomach (esophageal sphincter) and where the stomach attaches to the intestine (pyloric sphincter). The release of feces through the anus is also regulated by sphincter muscles.

9. Chemical digestion of food is initiated in the mouth. The most notable enzyme in this process is amylase, which initiates the breakdown of carbohydrates. The tongue then pushes a bolus of food to the back of the mouth, where a complex set of neural reflexes causes swallowing.

10. Although animals typically consume their food rapidly, digestion is a relatively slow process. Food must therefore be stored in the stomach and metered into the digestive tract at a slow, steady pace. While awaiting release into the intestine, the churning motions of the stomach continue the mechanical breakdown of food.

11. The stomach secretes *hydrochloric acid*, which creates a very low pH that kills ingested microorganisms and is essential for the proper digestion of proteins. Stomach cells located in deep folds of the stomach lining called *gastric pits* secrete the zymogen *pepsinogen*, which is converted into the active endopeptidase *pepsin* by the acid environment of the stomach. This conversion process is known as *autocatalysis*.

12. The combined action of acid, muscular action, and pepsin activity causes the breakdown of ingested foods (especially proteins), exposing more nutrient surface area to enzymatic degradation. The resultant mixture, called *chyme*, is pushed to the bottom of the stomach where it is slowly released into the small intestine through the pyloric sphincter.

13. Little absorption occurs across the cell walls of the stomach, due mostly to the protective mucus coating. However, a few substances such as alcohol, aspirin, and caffeine can enter the stomach cells, which explains the rapid physiological effects these drugs have on the body.

14. Attached to the stomach is the *small intestine*. Named for its diameter, it is actually a very long, coiled tube with many minute foldings of the mucosal lining which create an enormous absorptive surface area.

15. The small intestine is divided into three sections. The *duodenum* is the initial portion responsible for most digestion. The *jejunum* and the *ileum* make up the bulk of the intestine; they perform 90 percent of the nutrient absorption.

16. The liver and pancreas are two accessory organs that produce digestive enzymes and other secretions used in the small intestine. The liver produces *bile*, a cholesterol derivative used to emulsify fats. Bile is secreted from the liver via the *hepatic duct* and stored in the gall bladder. In response to the presence of fat in the small intestine, bile travels from the gall bladder through the *common bile duct* and is released into the duodenum.

17. Emulsification is necessary because large, hydrophobic clusters of fat molecules are not easily digested. They must first be broken up into very small fat clusters called *micelles* to facilitate the action of *lipases* (fat-digesting enzymes). Bile molecules emulsify fats by inserting their lipophilic ends into the hydrophobic portions of the large fat clusters, leaving their lipophobic (hydrophilic) ends sticking out into the aqueous solution. This prevents the fat molecules from coalescing into large clusters, and provides more surface area for the action of lipases.

18. A wide variety of exocrine products of digestive importance are made in the *pancreas*. They are released as zymogens in order to prevent self-digestion of the pancreas, and are activated inside the small intestine. One especially important zymogen is *trypsinogen*, which is activated by an *enterokinase* to become the functional protease *trypsin*. Trypsin then cleaves more trypsinogen molecules through autocatalysis to produce additional trypsin, and also activates other kinds of zymogens.

19. In addition to digestive enzymes, the pancreas releases bicarbonate ions which neutralize the chyme entering the small intestine from the stomach. This is necessary because pancreatic enzymes, unlike the stomach protease, do not function well at low pH.

20. The small molecules produced during the digestion of nutrient macromolecules eventually find their way to the cell walls of the microvilli lining the small intestine. These cells release dipeptidases and the enzymes maltase, lactase, and sucrase. The individual amino acids and monosaccharides produced during this final digestion are then absorbed across the mucosal layer.

21. Although the mechanism is not yet completely understood, amino acids and monosaccharides are probably absorbed across the mucosal layer by active transport. A *sodium cotransport* system plays a role in this process, most likely by providing energy in the form of a sodium concentration difference between the mucosa and the submucosa to actively transport nutrients through the mucosal membranes.

22. The fatty acids and monoglycerides produced by fat digestion can dissolve through the lipid membranes of mucosal cells. Inside the cells they are reassembled and combined with other molecules to form water-soluble *chylomicrons*, which are then transported by the lymph and released into the blood.

23. Bile molecules are reused by shuttling back and forth between the microvilli and the gut contents. Eventually they reach the ileum, where many are reabsorbed and transported back to the liver. Some bile loss is considered healthy, however, since it helps remove cholesterol from the body.

24. The large intestine, or *colon*, receives undigested food by peristalsis from the ileum. Water and inorganic ions are removed in the colon, producing the semisolid waste called feces. These are stored near the end of the colon and periodically excreted. If the colon removes too much water from the feces, constipation results; too little water removal produces diarrhea.

25. Large numbers of beneficial bacteria such as *E. coli* live within the colon, feeding on undigested food and producing beneficial products such as vitamin K and biotin for the host. Some animals enhance these nutritional benefits by eating their feces, a behavior called *coprophagy*.

26. The *appendix* of humans is a small, blind-ended sac that emerges from the colon. It is a *vestigial* (nonfunctional remnant) of the much larger caecum found in herbivores, which functions in cellulose digestion. Although it still plays a role in immune activity, the human appendix is not essential and is often surgically removed if it becomes infected.

27. Ironically, most herbivores are incapable of producing *cellulases*, the enzymes needed to digest their primary food source, cellulose. These animals therefore harbor vast numbers of microorganisms inside their digestive tracts which can manufacture cellulases.

28. *Ruminant* animals such as cows and sheep have modified digestive anatomies that enhance the activity of these microorganisms. The typical mammalian stomach has been replaced in these animals by a four-chambered structure, each part of which is specialized for a particular function in cellulose digestion.

29. The first two chambers are the *rumen* and the *reticulum*. They are packed with anaerobic organisms that carry out fermentation of cellulose to produce fatty acids. Periodically this fermenting mixture (the cud) is regurgitated into the mouth and rechewed before being sent back to the chambers for additional digestion. The gaseous byproducts of the fermentation process (CO_2 and CH_4) are expelled by belching.

30. Partially digested materials from the rumen and reticulum are passed on to a third chamber called the *omasum*, which is the site of water reabsorption. The food then enters the *abomasum*, or true stomach, where high acidity and proteases kill the microorganisms. The fatty acids, proteins, and carbohydrates obtained from both the digested plant materials and microorganisms are then absorbed into the body by the intestine.

31. Nonruminant herbivores such as rabbits maintain their own microbial fermentation vats inside the caecum. There is, however, no small intestine after this structure, so the nutritionally beneficial materials produced by microbial activity can only be made available for absorption through coprophagy.

Activities (for answers and explanations, see page 369)

- Label the tissue layers and gut space indicated in the following schematic cross section of a vertebrate small intestine.

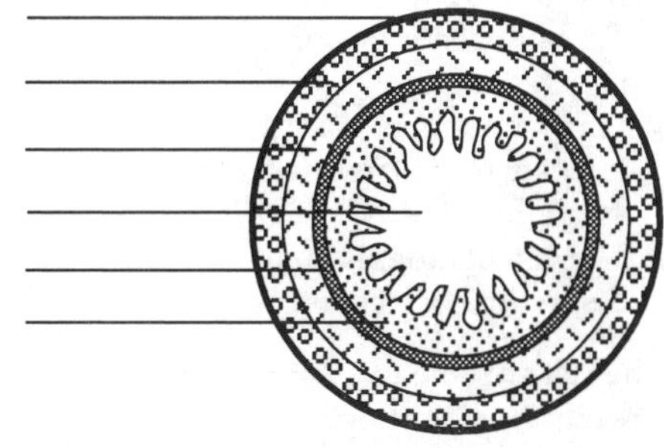

- Explain the functional significance of the large size of the rumen and the reticulum, compared with the omasum and the abomasum, in the digestive tracts of ruminant mammals.

Questions (for answers and explanations, see page 369)

1. A blood vessel entering the small intestine from the peritoneum will pass through the _________ before reaching the submucosa.
   a. lumen and serosa
   b. serosa and mucosa
   c. serosa and two smooth muscle layers
   d. two smooth muscle layers and mucosa
   e. two smooth muscle layers only

2. The contraction of smooth muscle is important in which of the following digestive processes?
   a. Swallowing
   b. Peristalsis and sphincter control
   c. Secretion of digestive enzymes
   d. Zymogen formation
   e. All of the above

3. Carbohydrate digestion first begins in the human ________, but occurs mostly in the _________ .
   a. mouth; stomach
   b. stomach; small intestine
   c.. small intestine; colon
   d. mouth; small intestine
   e. stomach; colon

4. The reabsorption of water within the vertebrate digestive system
   a. takes place mainly in the large intestine.
   b. occurs only in the duodenum.
   c. occurs both in the jejunum and the ileum.
   d. takes place through the gastric pits of the small intestine.
   e. is regulated by the sodium cotransport system.

5. Which of the following statements about the digestive roles of the gall bladder and pancreas is *not* true?
   a. The exocrine products of the pancreas empty into the duodenum.
   b. The gall bladder stores carbohydrases for subsequent release into the small intestine.
   c. Trypsinogen is one of the zymogens released from the pancreas.
   d. Large amounts of bicarbonate ions are released from the pancreas to neutralize the chyme entering the small intestine.
   e. The gall bladder is an important element in the process of lipid digestion.

6. The last enzyme to act upon the remnants of an ingested protein before its component amino acids cross the membranes of cells lining the intestinal mucosa is
   a. a carbohydrase.
   b. an enterokinase.
   c. a dipeptidase.
   d. pepsinogen.
   e. amylase.

7. A biochemical test for the presence of chylomicrons within the vertebrate body would find them located mainly in the
   a. stomach and colon.
   b. mucosal cells of the small intestine.
   c. mucosal cells of the small intestine, the lymph fluid, and the blood.
   d. duodenum, jejunum, and the ileum.
   e. abomasum and reticulum of ruminant mammals.

8. Which of the following statements about the digestion of cellulose is *not* true?
   a. Cellulose can only be digested by highly specialized mammalian herbivores.
   b. Cellulases are the enzymes that hydrolyze cellulose.
   c. Cellulose fermentation occurs in the rumen and reticulum of a bison.
   d. Two byproducts of cellulose digestion in mammals are carbon dioxide and methane gas.
   e. The caecum is used to carry out cellulose digestion in nonruminant, herbivorous mammals.

CONTROL AND REGULATION OF DIGESTION • CONTROL AND REGULATION OF FUEL METABOLISM (pages 1001–1005)

Key Concepts

1. The digestive tract is unique in possessing an intrinsic nervous system which sends signals from one part of the gut to another without the signal having to pass through the central nervous system. An example is the *gastrocolic reflex,* which initiates bowel movements through direct stimulation of the colon when food load increases in the stomach.

2. Neural connections mediated by the central nervous system also help control digestive activity. As demonstrated by Pavlov's classic experiments with dogs, learning to associate an unrelated stimulus such as a bell with the presentation of food can produce a *conditioned reflex* appropriate to feeding. Over time, the bell alone is sufficient to elicit a physiologically relevant digestive response.

3. In addition to neural control, hormones also influence the activity of digestive systems. Indeed, the first-ever demonstration of hormonal influences within the body came from experiments by Bayliss and Starling showing that the chemical *secretin* travels from the mucosal lining, through the bloodstream, and causes the release of pancreatic secretions.

4. Other important digestive hormones include *cholecystokinin,* which is secreted by the intestinal mucosa to stimulate the release of bile from the gall bladder. *Gastrin,* which is released by the stomach in response to the presence of food, causes secretions from and muscular contractions of the upper stomach walls.

5. The regulation of nutrients for use in energy metabolism is mediated primarily by the liver, which stores large quantities of glycogen and fat and synthesizes important blood proteins. The liver also interconverts carbohydrates with fats, and regulates the conversion of amino acids, pyruvate, and lactate into glucose through a process known as *gluconeogenesis.*

6. The liver produces *lipoproteins* that help control fat metabolism and transport in the blood. Different forms of lipoproteins, which differ mainly in their densities, are believed to have an influence on the rate of cholesterol accumulation in the human body. See Box 43.C on page 1003 in your textbook for details.

7. During the *absorptive period,* when food is present in the gut and nutrients are being absorbed into the circulatory system, the liver converts glucose into glycogen and fat. At this time, body cells store fatty acids in the form of fat, and preferentially use glucose as their metabolic fuel.

8. These processes are reversed during the *postabsorptive period,* when free nutrients are no longer available in the gut or circulatory system. The lone exception to this reversal is nervous tissue, which requires glucose for energy metabolism at all times.

9. The switch between absorptive and postabsorptive energy metabolism is largely controlled by the hormones *glucagon* and *insulin,* which are released from the pancreas. Insulin is secreted in response to elevated blood glucose levels. It facilitates the entry of glucose into cells, stimulating the use of glucose as metabolic fuel and its conversion into fat or glycogen.

10. Conversely, glucagon is released when blood glucose falls below normal postabsorptive levels. Its effects are opposite those of insulin, stimulating the conversion of glycogen and fat into glucose, which is then released into the blood.

11. The primary control mechanism in this system is a reduced concentration of insulin in the blood; the presence of glucagon is a secondary stimulus regulating fuel metabolism.

12. Under stressful conditions, postabsorptive blood glucose levels are also influenced by the hormones cortisol and epinephrine, which are released by the adrenal cortex and the adrenal medulla, respectively. Epinephrine has similar effects on liver cells such as glucagon, stimu-

lating glycogen breakdown and increasing gluconeogenesis. Cortisol inhibits glucose metabolism and promotes the use of fats as metabolic fuel.

13. Thus, the absorptive state is characterized by the channeling of glucose and other fuel molecules toward a variety of storage forms. The postabsorptive state is marked by the conversion of stored fuels into glucose for use by the nervous system, with other tissues switching over to the use of fatty fuels.

Activity (for answers and explanations, see page 369)

- Match each of the following hormones involved in the control and regulation of energy metabolism with its best description.

____	Secretin	*a.* Secreted by mucosal lining and causes release of pancreatic secretion
____	Glucagon	*b.* Causes muscular contraction of the stomach wall in response to food
____	Gastrin	*c.* Adrenal gland hormone that inhibits glucose metabolism
____	Epinephrine	*d.* Causes the gall bladder to release bile into the small intestine
____	Insulin	*e.* Has same effects as glucagon under conditions of physiological stress
____	Cholecystokinin	*f.* Pancreatic hormone that stimulates the conversion of glycogen to glucose
____	Cortisol	*g.* Released from the pancreas in response to elevated blood glucose levels

Questions (for answers and explanations, see page 369)

1. A fundamental difference between the gastrocolic reflex of humans and Pavlov's conditioned salivation reflex in dogs is that
   a. one reflex involves only the peripheral nervous system, the other only the central nervous system.
   b. one reflex involves only the intrinsic nervous system of the gut, while the other is mediated by the central nervous system.
   c. one reflex is influenced by hormonal regulation, the other is not.
   d. one reflex produces an appropriate digestive response, the other does not.
   e. None of the above

2. Damage to the mucosal lining of the gut would directly impair the control function of which of the following hormones?
   a. Gastrin and secretin
   b. Secretin and insulin
   c. Gastrin, secretin, and cholecystokinin
   d. Cholecystokinin and glucagon
   e. Cholecystokinin only

3. The rate of gluconeogenesis is affected by which of the following hormones?
   a. Glucagon
   b. Insulin
   c. Epinephrine
   d. Cortisol
   e. All of the above

4. During the absorptive period, when food is present in the gut, ________ is released by the pancreas and stimulates the ________ in the liver.
   a. insulin; conversion of glucose to glycogen and fat
   b. glucagon; conversion of glucose to glycogen and fat
   c. cortisol; conversion of glucose to glycogen and fat
   d. insulin; conversion of glycogen and fat to glucose
   e. glucagon; conversion of glycogen and fat to glucose

5. The postabsorptive period of vertebrate digestion is characterized mainly by
   a. elevated blood glucose levels and the rapid secretion of insulin.
   b. accelerated rates of glycogen storage in liver and muscle cells.
   c. decreased levels of gluconeogenesis.
   d. conversion of stored fuels into glucose, and a switch to most cells using fatty fuels.
   e. an overall state of metabolic quiet.

6. Chylomicrons, VLDL, LDL, and HDL are related to one another in that they are all
   a. proteins.
   b. lipids.
   c. lipoproteins.
   d. digestive hormones.
   e. None of the above

Integrative Questions (for answers and explanations, see page 369)

1. A cow, an eagle, a snake, and a human have which of the following digestive characteristics in common?
   a. They are ingestive heterotrophs.
   b. They are predators.
   c. They have a stomach and a small intestine.
   d. They utilize autocatalysis.
   e. All of the above

2. Which of the a following is *not* a type of nutritional problem in animals?
   a. Anorexia nervosa
   b. Malnutrition
   c. Undernourishment
   d. Beriberi
   e. Defecation

3. The human liver carries out all but one of the following functions. Select the *exception*.
   a. Insulin production
   b. Synthesis of plasma proteins
   c. Gluconeogenesis
   d. Glycogen storage
   e. Interconversion of fuel molecules

4. In which of the following animals would you find both a mouth and an anus?
   a. Nematode
   b. Snail
   c. Cockroach
   d. Baleen whale
   e. All of the above

44

Salt and Water Balance and Nitrogen Excretion

CHAPTER LEARNING OBJECTIVES—after studying this chapter you should be able to:

- ❑ Describe the primary roles of excretory systems and the three mechanisms employed in their operation.
- ❑ Contrast the osmoregulatory problems associated with life in fresh water, life in salt water, and a terrestrial existence.
- ❑ Differentiate between osmoconformers and the two types of osmoregulators, hypertonic osmoregulators and hypotonic osmoregulators.
- ❑ Contrast ionic conformers with ionic regulators.
- ❑ Describe the concepts of osmoregulation illustrated by the brine shrimp *Artemia* living in different environments.
- ❑ Describe the three major types of nitrogenous waste products that result from the metabolism of the major classes of macromolecules.
- ❑ Define the terms "ammonotelic," "ureotelic," and "uricotelic" and classify animals relative to these terms.
- ❑ Describe the structure and functioning of the protonephridia excretory system based on the flame cell seen in flatworms.
- ❑ Describe the structure and functioning of the metanephridia excretory system seen in most of the annelid worms.
- ❑ Describe the structure and functioning of Malpighian tubule-based excretory systems as seen in the insects.
- ❑ Describe the structure and functioning of the green gland-based excretory system seen in many crustaceans.
- ❑ Describe the structure of a typical vertebrate nephron and associated circulatory elements and explain how they participate in filtration, secretion and reabsorption, and urine processing.
- ❑ Explain what can be inferred about the evolution of the nephron based on a study of the kidneys of living vertebrates.
- ❑ Discuss osmoregulatory adaptations seen in the following groups: marine bony fish, freshwater fish, cartilaginous fish, amphibians, reptiles, and birds.
- ❑ Describe the components of the mammalian excretory system and the major circulatory elements serving it.
- ❑ Describe the components of the mammalian nephron and the capillary beds associated with it.
- ❑ Describe the regions of the mammalian kidney and explain what nephron components are found in these regions.
- ❑ Explain the quantitative relationship between the amount of blood entering the kidney, the amount of filtrate formed, and the amount of urine produced.
- ❑ Relate the characteristics of the cells lining the different regions of the nephron to the activities occurring in those regions.
- ❑ Explain how the countercurrent multiplier system in the vertebrate kidney establishes a NaCl concentration gradient in the medulla.
- ❑ Describe the roles of each of the following nephron components in the production of concentrated urine: proximal convoluted tubule, descending limb of the loop of Henle, ascending limb of the loop of Henle, distal convoluted tubule, collecting duct.
- ❑ Describe the relationship between the following variables: concentration of urine formed, magnitude of the NaCl concentration gradient in the medulla, length of the loop of Henle.
- ❑ Explain the osmoregulatory principles involved in the operation of the artificial kidney.
- ❑ Discuss the major mechanisms involved in the autoregulation of the glomerular filtration rate, especially the role of renin, angiotensin, and aldosterone.
- ❑ Discuss the central control mechanisms involving antidiuretic hormone (ADH) that regulate the concentration of the urine and the osmolarity and volume of the blood.

INTERNAL ENVIRONMENT • WATER, SALTS, AND THE ENVIRONMENT • THE EXCRETION OF NITROGENOUS WASTES (pages 1008–1012)

Key Concepts

1. *Excretory systems* regulate the volume, composition, and osmotic potential of body fluids. Excretion of excess and toxic molecules and retention of needed molecules is part of the responsibility of excretory systems.
2. The functioning of excretory organs is closely tied to the environment in which the animal lives, be it marine, freshwater, or terrestrial.
3. Since water cannot be moved by active transport, all excretory mechanisms employ pressure or osmotic gradients to move water.
4. Most excretory organs include a system of *tubules* in which the composition of the extracellular fluid is modified to produce *urine*—the liquid excretory waste product. Extracellular fluid enters the tubules by *filtration* and its composition is modified by active *secretion* and *reabsorption*.
5. The *salinity*, or salt content, of aqueous environments varies continuously from fresh water to tide pools that are saltier than the sea.
6. Most marine invertebrates are *osmoconformers*—the osmotic potential of their body fluids matches that of the environment.
7. Even osmoconformers must become *osmoregulators* in environments with extreme osmotic potentials.
8. In fresh water, excess water passively enters organisms by osmosis and must be excreted. Salts are lost from the body in fresh water and must be conserved. Osmoregulators in fresh water produce large volumes of dilute urine.
9. In salt water, water is lost to the external environment and must be conserved. Osmoregulators in salt water produce small amounts of concentrated urine.
10. A *hypertonic osmoregulator* maintains body fluid that has a more negative osmotic potential (it is saltier) than the environment (i.e., the body fluid is hypertonic to the environment).
11. A *hypotonic osmoregulator* maintains body fluid that has a less negative osmotic potential (it is less salty) than the environment (i.e., the body fluid is hypotonic to the environment).
12. The brine shrimp (*Artemia*) can be either a hypertonic osmoregulator, an osmoconformer, or a hypotonic regulator depending on the osmotic potential of its environment. Its main mechanism of osmoregulation is active transport of NaCl across its gills.
13. If the ionic composition of an osmoconformer's body fluids is like that of its environment, it is called an *ionic conformer*—it does not regulate its ions at all. However, most osmoconformers are *ionic regulators*—active transport is used to absorb or excrete ions so that the ionic composition of the body fluid differs from that of the environment.
14. Most terrestrial animals must conserve water and obtain needed ions as part of their diet. Because plant biomass is low in sodium, herbivores must ingest salt directly. Some marine birds have *nasal salt glands* to rid themselves of excess dietary salt.
15. Unlike carbohydrates and fats, which are metabolized entirely to water and carbon dioxide, the metabolism of proteins produces the nitrogenous waste product *ammonia* in addition to H_2O and CO_2.
16. Ammonia is highly toxic and must be immediately excreted or converted into the nontoxic molecules *urea* and *uric acid* prior to excretion.
17. The first step in the metabolism of proteins is their conversion into purines and pyrimidines. Uric acid results from the breakdown of purines, urea from the breakdown of pyrimidines.
18. *Ammonotelic* animals, including most aquatic invertebrates, bony fishes, crocodiles, amphibian tadpoles, etc., excrete ammonia directly by diffusion from the blood through the gill membranes to the environment.
19. Because ammonia must be kept dilute, ammonia excretion requires large amounts of water. Consequently, most terrestrial animals convert ammonia to urea or uric acid.
20. *Ureotelic* animals, such as mammals, amphibians, and cartilaginous fishes, excrete their nitrogenous waste as urea. Urea is highly soluble in water. To conserve water, some vertebrate kidneys can produce quite concentrated urine.
21. Sharks and rays accumulate high concentrations of urea in their tissues as a way of increasing the osmotic potential of their body fluids and limiting osmotic loss of water to their marine environment.
22. *Uricotelic* animals, including insects, reptiles, birds, and some amphibians, excrete their nitrogenous waste as uric acid. Because uric acid is very insoluble in water, it can be excreted with little loss of water.
23. Many species produce more than one nitrogenous waste product. Being mammals, humans mostly excrete urea, but small amounts of ammonia and uric acid are also produced.
24. Some species switch their major excretory products during development. For example, aquatic frog and toad tadpoles excrete ammonia, but as terrestrial adults they switch to producing urea.

Questions (for answers and explanations, see page 370)

1. Which of the following is *not* a feature common to the excretory systems of most animals?
   a. Tubular design
   b. Production of urine
   c. Active transport of water
   d. Filtration
   e. Secretion and reabsorption

2. Select *all* of the following features that would be characteristic of a penguin.
   a. Ammonotelic
   b. Ureotelic
   c. Uricotelic
   d. Presence of nasal salt glands
   e. Accumulation of urea in body tissues

3. Select *all* of the following features that would be characteristic of a freshwater bony fish.
   a. Ammonotelic
   b. Ureotelic
   c. Uricotelic
   d. Produces concentrated urine
   e. Absorbs salt from water across gill membranes
   f. Drinks little

4. The osmotic potential of the body fluid of a hypertonic osmoregulator should be ________ negative than the external environment, and this animal would probably live in a ________ environment.
   a. more; freshwater
   b. more; saltwater
   c. less; freshwater
   d. less; saltwater
   e. less; terrestrial

5. Ammonia results directly from the metabolism of
   a. carbohydrates only.
   b. fats only.
   c. proteins only.
   d. nucleic acids only.
   e. proteins and nucleic acids.
   f. all four classes of macromolecules.

Activities (for answers and explanations, see page 371)

- Draw a curve in the following figure to show the relationship between the osmotic potential of the body fluid of the brine shrimp (*Artemia*) and its external environment. Label the portion of the curve where *Artemia* shows hypertonic osmoregulation, osmoconformity, and hypotonic osmoregulation.

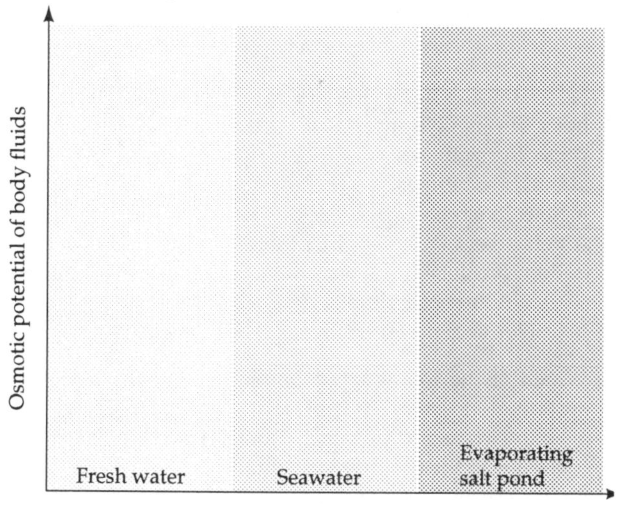

- Would you expect *Artemia* to be an ionic conformer or an ionic regulator? Explain.

INVERTEBRATE EXCRETORY SYSTEMS • VERTEBRATE EXCRETORY SYSTEMS (pages 1012–1019)

Key Concepts

1. Marine invertebrates are largely osmoconformers and ionic regulators. They have few specific adaptations for salt and water balance, other than active transport of specific ions, and most excrete ammonia directly to the environment.
2. Flatworms, such as *Planaria*, have excretory systems consisting of a network of tubules that include blind ends formed by ciliated *flame cells* and excretory pores that open to the exterior. Tubules and flame cells together are called *protonephridia*.
3. Body fluid filters into the cavity formed by each flame cell and as it is moved down the tubule by the beating cilia, active reabsorption of ions occurs so that the urine leaving the excretory pores is hypotonic to the body fluid.
4. Movement of body fluid into the tubule system is due either to a negative pressure between the inside and outside of the tubule created by the beating cilia or by an osmotic gradient. The osmotic gradient could be created if ions were actively pumped into the upper ends of the tubules.
5. The same processes of filtration of the body fluids and tubular reabsorption of ions is well developed in the *metanephridia* of annelid worms, such as the earthworm.
6. Blood pressure within the closed circulatory system causes blood to filter through the capillary membranes. The resulting *coelomic fluid* fills each segment of the worm's body and consists of water, ions, and small molecules; blood cells and large protein molecules are retained in the circulatory system as part of the filtration process.
7. A metanephridium consists of a funnellike ciliated opening, called a *nephrostome*, into which coelomic fluid flows, a long tubular portion where active reabsorption and secretion of molecules occurs, and a *nephridiopore* that dumps the hypotonic urine to the outside of the worm. Each segment of the worm's body has a pair of metanephridia.
8. Terrestrial arthropods, including the insects, have excretory organs called *Malpighian tubules* that are highly efficient at conserving water. Malpighian tubules are a collection of two to over 100 tubular outpocketings of the digestive tract that are bathed by coelomic fluid.
9. Cells in the Malpighian tubules absorb uric acid and potassium and sodium ions from the coelomic fluid, creating an osmotic potential gradient between the coelomic fluid and the interior of the tubule. Water moves by osmosis into the tubule. Muscle fibers within the tubules move the contents into the gut cavity.
10. Acidic conditions within the gut precipitate the uric acid, and sodium and potassium are actively pumped back into the body cavity. Water now follows the reabsorbed salts to rejoin the coelomic fluid and this leaves behind a dry mixture of uric acid and undigested food.
11. The excretory systems of crustacean arthropods, such as crayfish, crabs, lobsters, etc., consist of an *end sac*, a tubu-

lar *nephridial canal*, and a *bladder* that opens to the exterior by an excretory pore. In crayfish and lobsters these structures are called *green glands*.

12. Hemolymph is filtered into the end sac, undergoes selective reabsorption and secretion in the nephridial canal, and after storage in the bladder leaves the animal as urine. The nephridial canal in the marine lobster is shorter and reabsorbs much less salt than in the freshwater crayfish.
13. *Kidneys* are the major salt and water balance organs found in vertebrates. *Nephrons* are the functional units of the kidneys. Each human kidney includes some one million nephrons.
14. Two capillary beds are located between the arteriole and the venule associated with each nephron. The first bed, called the *glomerulus*, is a knot of very permeable vessels enclosed within the cup-shaped end of the tubular part of the nephron called the *Bowman's capsule*.
15. An *afferent arteriole* carries blood to the glomerulus and an *efferent arteriole* transports blood from the glomerulus to the second capillary bed, called the *peritubular capillaries*, which surrounds the renal tubule.
16. The glomerulus and capsule together are called the *renal corpuscle*. Specialized cells of the capsule called *podocytes* have fingerlike projections that wrap around the capillaries of the glomerulus.
17. The blood entering the glomerulus is under high pressure. Water and small molecules are filtered through pores in the cells of the capillaries and the slits of the podocytes and are collected in the Bowman's capsule. Large molecules and blood cells remain behind.
18. Glucose, amino acids, and other needed molecules and ions are reabsorbed from the filtrate as it moves through the renal tubule, and are returned to the circulatory system via the peritubular capillaries. Cells of the tubule also secrete unwanted substances into the urine.
19. The earliest vertebrates lived in fresh water and the primary function of the nephron was to eliminate excess water taken on osmotically, while conserving salts.
20. The earliest nephrons opened directly into the coelomic cavity via a ciliated nephrostome, as in the present-day annelids. Separate from the nephrostome was a capillary bed that filtered the blood directly into the coelom. Coelomic fluid entered the nephrostome and flowed through the tubule where secretion and reabsorption took place.
21. In the next stage of kidney evolution, a glomerulus was enclosed within a Bowman's capsule, but a nephrostome was still present.
22. In the final stage of kidney evolution, the nephrostome was lost and filtration occurred directly between the circulatory system and the renal tubule without involving the coelom.
23. Various adaptations are seen in the different vertebrate groups living in environments where water must be conserved and salt excreted.
24. Marine bony fishes drink seawater, and salt is actively secreted from their gill membranes. Ammonia also diffuses across the gill membranes. Since these animals must conserve water, their kidneys have very few (or no!) glomeruli and consequently, very little urine is produced.
25. Whereas marine bony fishes are hypotonic to seawater and continuously lose water, the cartilaginous fishes retain urea within the body tissues so that they become isotonic, or even hypertonic, to seawater. Cartilaginous fishes have a salt-secreting *rectal gland* to remove excess NaCl.
26. As with freshwater fishes, amphibians mostly produce copious, dilute urine and conserve salts. As with cartilaginous fishes, the *crab-eating frog* of Southeast Asia is isotonic to its marine environment by retaining urea in its body fluids.
27. Other adaptations for life in dry environments seen in amphibians include secretion of waxes and fats to retard water loss from the skin and burrowing and estivation during long dry periods.
28. Reptiles and birds have evolved some similar adaptations for water conservation, including internal fertilization, shelled eggs, waterproof skin coverings, and uric acid as the nitrogenous excretory product. As with mammals, birds can also produce urine that is hypertonic to their body fluid.

Questions (for answers and explanations, see page 370)

1. Match the excretory system characteristics on the right with the appropriate animal groups on the left.

_____	Flatworms	*a.* Nephrons
_____	Annelid worms	*b.* Flame cells
_____	Crustaceans	*c.* Metanephridia
_____	Insects	*d.* Protonephridia
_____	Vertebrates	*e.* Green glands
		f. Malpighian tubules

2. Select the following feature that is *not* characteristic of both the metanephridia of earthworms and the nephrons of vertebrates.
   a. Both are tubules.
   b. Selective reabsorption occurs there.
   c. Selective secretion occurs there.
   d. Both produce urine with a composition different than that of the body fluid.
   e. Coelomic fluid enters the excretory organ directly.
3. Which of the following statements about the excretory system of insects is *not* true?
   a. Active transport moves materials from the coelomic fluid into the Malpighian tubules.
   b. The Malpighian tubules can produce a highly concentrated urea solution, allowing insects to inhabit some of the Earth's driest habitats.
   c. Reabsorption of salts takes place mostly in the gut.
   d. Water reabsorption is by osmotic movement only.
   e. There are no circulatory vessels associated with the Malpighian tubules.
4. Which of the following statements about the vertebrate nephron is *not* true?
   a. The blood pressure is highest in the afferent arterioles.
   b. Podocytes are intimately involved in blood filtration within the Bowman's capsule.

c. The permeability of the peritubular capillaries is greater than the permeability of the glomerulus.
d. Active transport takes place in the renal tubules.
e. Plasma proteins are absent from the renal filtrate.

5. Match the excretory system characteristics on the right with the appropriate vertebrate groups on the left.

_____ Marine bony fishes	*a.* Excrete uric acid
_____ Cartilaginous fishes	*b.* Excrete ammonia
_____ Most amphibians	*c.* Few glomeruli
_____ Reptiles	*d.* Retain urea in tissues
_____ Birds	*e.* Produce copious, hypotonic urine
	f. Waterproof skin covering
	g. Produce hypertonic urine
	h. Rectal glands

6. Which of the following statements about the evolution of the vertebrate nephron is *not* true?
   a. Our understanding of the evolution of the vertebrate kidney suggests that the earliest vertebrates lived in a marine environment.
   b. The original function of the earliest nephrons was removal of excess water and conservation of salt.
   c. The earliest nephrons were not involved in filtration.
   d. Nephrostomes were found in early nephrons, but later lost.
   e. Whereas the earliest nephrons received coelomic fluid, later nephrons received filtered blood.

Activities (for answers and explanations, see page 370)

- The following schematic cross section of a vertebrate body shows an intermediate stage in the evolution of the nephron. Label the gut, coelom, Bowman's capsule, glomerulus, bladder and nephrostome.

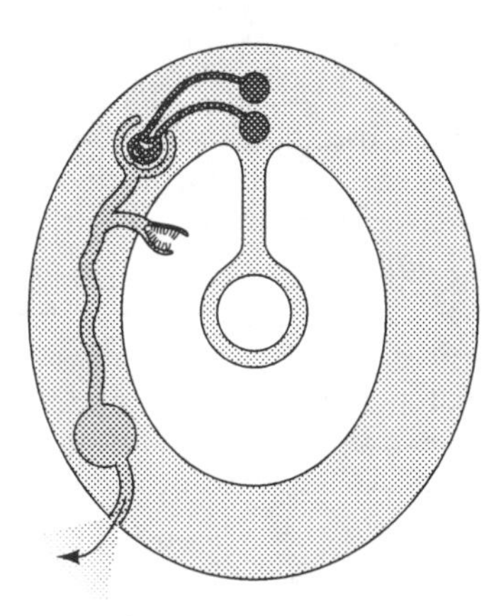

Describe the next change that led to the final stage in this evolutionary progression.

- Label the diagram of a typical vertebrate nephron shown below with the following labels: afferent arteriole, efferent arteriole, glomerulus, Bowman's capsule, podocytes, renal tubule, peritubular capillaries, renal venule, and collecting duct. Circle the portion of the nephron where filtration occurs.

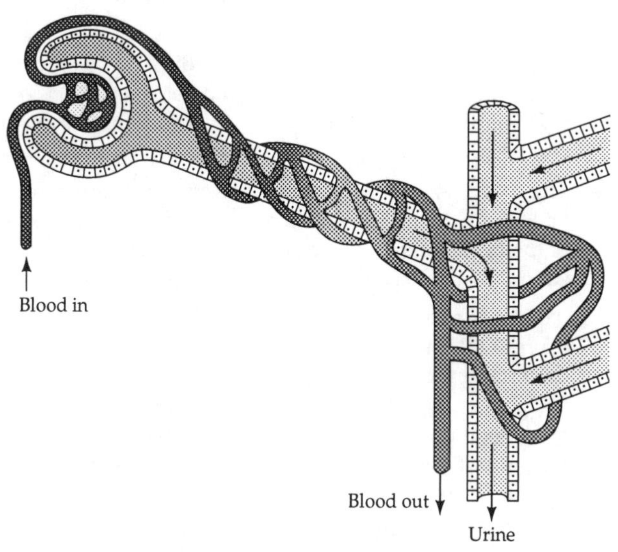

- In terms of osmoregulatory adaptations, what do crab-eating frogs of Southeast Asia and sharks have in common?

STRUCTURE AND FUNCTION OF THE MAMMALIAN KIDNEY • REGULATION OF KIDNEY FUNCTIONS (pages 1019–1026)

Key Concepts

1. In humans, the excretory system consists of two *kidneys,* ureters that connect each kidney to a *urinary bladder* where urine is stored, and a *urethra* that conducts the urine to the outside. Two sphincter muscles at the base of the urethra control the timing of urination.
2. Each kidney is supplied with a *renal artery* and *renal vein.* The renal artery branches extensively to form the afferent arterioles that serve each nephron. The peritubular capillaries join to form venules that fuse to become the renal vein.
3. The ureter enters the kidney on its concave side. It divides into several branches that envelop kidney tissue called *renal pyramids.* The collecting ducts to which the nephrons are connected empty into the branches of the ureter at the tips of the renal pyramids.
4. In addition to the Bowman's capsule, the tubular part of the mammalian nephron can be divided into a *proximal convoluted tubule* (the part of the tubule nearest the capsule), *a distal convoluted tubule* (the part of the tubule

nearest the collecting duct), and the *loop of Henle*, connecting the two convoluted tubules.

5. The kidney consists of two layers: an outer *cortex* and an inner *medulla*.
6. The cortex includes the glomeruli and the proximal and distal convoluted tubules of the nephrons. Most of the peritubular capillaries are also in the cortex.
7. The loop of Henle is of varying length in different mammalian groups. The *descending limb* dips down into the medulla, shows a hairpin loop, and becomes the *ascending limb* that returns to the cortex and joins the distal convoluted tubule.
8. The collecting tubules start in the cortex and run in parallel with the loops of Henle through the medulla. Branches of the peritubular capillaries called the *vasa recta* are located in the medulla and surround the loop of Henle.
9. All substances enter the kidney via the renal artery and leave the kidney via the renal vein as blood or via the ureter as urine. In humans, about 180 liters of filtrate enter the glomeruli of the kidneys each day (about 12% of the total blood volume moving through the renal arteries), but only about two to three liters of urine are normally produced. The remaining filtrate is reabsorbed into the blood.
10. Cells of the proximal convoluted tubules are cuboidal and have extensive microvilli and numerous mitochondria. Most of the reabsorption of solutes, including NaCl, glucose, amino acids, and other needed molecules occurs via active transport mechanisms in the cells of the proximal convoluted tubules.
11. Water osmotically follows the reabsorbed solutes into the cells of the proximal convoluted tubule, and the water and solutes are then returned to the blood in the peritubular capillaries.
12. Although the filtrate in the proximal convoluted tubule changes dramatically in composition due to reabsorption of solutes, the filtrate entering the loop of Henle is isotonic to blood plasma because a proportional share of water is also reabsorbed in the proximal convoluted tubule.
13. The loops of Henle in the mammalian kidney set up a *countercurrent multiplier system* in the medulla that enables mammals to produce hypertonic urine (four times more concentrated than blood plasma in humans and 12 to 15 times more concentrated in desert mammals such as the kangaroo rat).
14. Cells of the descending limb and the lower part of the ascending limb of the loop of Henle are permeable to water and small molecules. They have few microvilli and mitochondria and perform little active transport.
15. Cells of the upper part of the ascending limb of the loop of Henle and the distal convoluted tubule are impermeable to water and actively transport NaCl out of the tubular fluid (urine) and into the surrounding *interstitial fluid* of the renal medulla. Consequently, the urine becomes more dilute as it moves up the ascending limb.
16. Because the NaCl pumped out of the ascending limb moves into tubular fluid in the descending limb, a NaCl concentration gradient builds up in the medulla.
17. Cells of the collecting duct are permeable to water but not mineral ions. As urine travels down the collecting duct it encounters interstitial fluid with ever-increasing NaCl concentration. Water moves osmotically out of the collecting duct, concentrating the urine.
18. Water reabsorbed from the collecting ducts returns to the circulatory system via the vasa recta.
19. The osmotic concentration of the urine can almost equal the greatest concentration of the NaCl gradient set up by the countercurrent multiplier system. The magnitude of the NaCl concentration gradient in the medulla is dependent on the length of the loop of Henle relative to the size of the kidney. Desert jumping mice have extremely long loops of Henle. Their kidneys are so efficient that they can live on metabolic water alone.
20. The kidney shows *autoregulatory adaptations* that maintain the *glomerular filtration rate* (*GFR*). The GFR depends on both an adequate blood volume and a high blood pressure.
21. Autoregulatory responses to decreases in either cardiac output or blood pressure include dilation of the afferent arterioles and secretion by the kidney of the enzyme *renin*. Renin catalyzes a reaction that leads to production of the hormone *angiotensin*.
22. The effects of angiotensin leading to an increase in blood pressure include constriction of efferent renal arterioles and peripheral blood vessels. Angiotensin also stimulates thirst and causes the *adrenal medulla* to produce the hormone *aldosterone*. Aldosterone increases sodium reabsorption in the kidney. These effects lead to water retention and an increase in blood volume.
23. A decrease in blood pressure also leads to release of *antidiuretic hormone* by the *posterior pituitary*. In response to a reduction in stretch receptor activity in large veins of the body, this hormone is produced by cells in the *hypothalamus* and delivered to the posterior pituitary via their axons. The hormone increases the water permeability of the collecting duct, more water is reabsorbed, and blood volume and pressure increase.
24. The disease *diabetes insipidus*, characterized by production of copious, dilute urine, results if the body produces insufficient antidiuretic hormone. Caffeine and alcohol increase the production of urine by inhibiting antidiuretic hormone.
25. Osmoreceptors in the hypothalamus detect changes in the osmolarity of the blood. An increase in blood osmolarity stimulates release of more antidiuretic hormone and also causes thirst. Both these effects lead to increased blood volume and a reduction in blood osmolarity.
26. The *artificial kidney* depends on the process of *dialysis*; blood is separated by a *semipermeable membrane* from an isotonic *dialyzing fluid* of precise composition. In the dialyzing fluid, the concentration of valuable molecules is the same as their concentration in the blood while the concentration of waste products is zero. Waste products, such as urea and sulfate, move by osmosis from the blood into the dialysis fluid, while other molecules remain in the blood.

Questions (for answers and explanations, see page 371)

1. Use numbers to order the following structures through which a molecule of urea would pass during excretion in a mammal.

_____Collecting duct

_____Urethra

_____Bladder

_____Glomerulus

_____Ureter

2. Next to each of the following kidney components, place an "M" if it is located mostly within the kidney *medulla* and a "C" if it is located mostly within the kidney *cortex.*

_____Peritubular capillaries

_____Bowman's capsule

_____Proximal convoluted tubule

_____Distal convoluted tubule

_____Renal pyramids

_____Vasa recta

_____Loop of Henle

3. You measure your urine output in 12 hours and discover that you have produced 1 liter. This amount of urine would represent about _____ liters of filtrate leaving Bowman's capsules. This amount of filtrate would have been derived from about _____ liters of blood entering the renal arteries.

a. 1; 10
b. 10; 100
c. 100; 1,000
d. 1,000; 10,000
e. 1,000; 100

4. Some desert mammals do not need to drink, but can survive entirely on metabolic water. This results directly from kidneys with

a. very high glomerular filtration rates.
b. very long loops of Henle.
c. collecting ducts that are impermeable to water.
d. very high rates of active transport of NaCl into the ascending limb of the loop of Henle.
e. reduced numbers of nephrons.

5. Complete the following statements with reference to the simplified diagram of a mammalian nephron shown at the top of the next column.

a. The composition of the filtrate would be most like plasma in the tubule next to letter ____.

b. The NaCl concentration in the interstitial fluid would be greatest in the area of letter ____.

c. NaCl is passively diffusing from the interstitial fluid into the tubule in the area of letter ____.

d. The osmolarity of the filtrate next to letters ________ is similar to the osmolarity of blood plasma.

e. The urine would be least concentrated in the collecting duct next to letter ____.

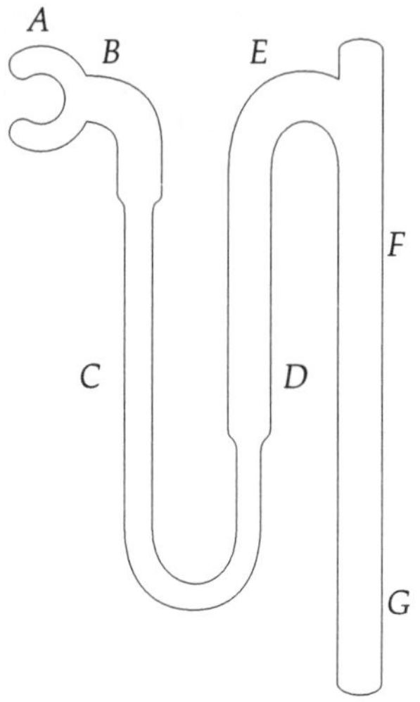

6. Match all of the characteristics on the right with the kidney areas on the left that have cells with those characteristics.

_____ Proximal convoluted tubule	*a.* Many microvilli
_____ Distal convoluted tubule	*b.* Many mitochondria
_____ Ascending limb	*c.* Reabsorption of glucose and amino acids
_____ Descending limb	*d.* Active transport of NaCl
_____ Collecting duct	*e.* Impermeable to water
	f. Urine concentration through osmosis

7. A mammal is given a transfusion of a hypotonic saline solution that reduces the osmolarity of its blood by 5% and increases its blood volume and pressure by 10%. Select all of the following effects that will result from this treatment.

a. Release of renin by the kidney
b. Increase in the glomerular filtration rate
c. Dilation of the afferent arterioles
d. Reduced release of antidiuretic hormone by the posterior pituitary
e. Production of copious, dilute urine

8. Which of the following would *not* be characteristic of the dialyzing fluid used in an artificial kidney machine?

a. Isotonic to blood
b. Glucose concentration same as blood
c. Urea concentration initially zero
d. Sodium concentration initially zero
e. Same osmotic potential as blood

9. Place a "+" or "–" next to each of the following events to indicate whether it would increase or decrease urine production.

_____Ingestion of caffeine

_____More stretch receptor activity in major arteries

_____More aldosterone released by adrenal cortex

_____Constriction of efferent renal arterioles

_____Decreased permeability of collecting ducts

Activities (for answers and explanations, ***see page 371)***

- Both caffeine and alcohol have a dehydrating effect on the body. Explain this observation in terms of the effect of these drugs on osmoregulatory control mechanisms.

- Label the boxes next to the arrows in the figure to the right to indicate what materials are moving. Also, of the materials moving, indicate whether active transport or diffusion is involved.

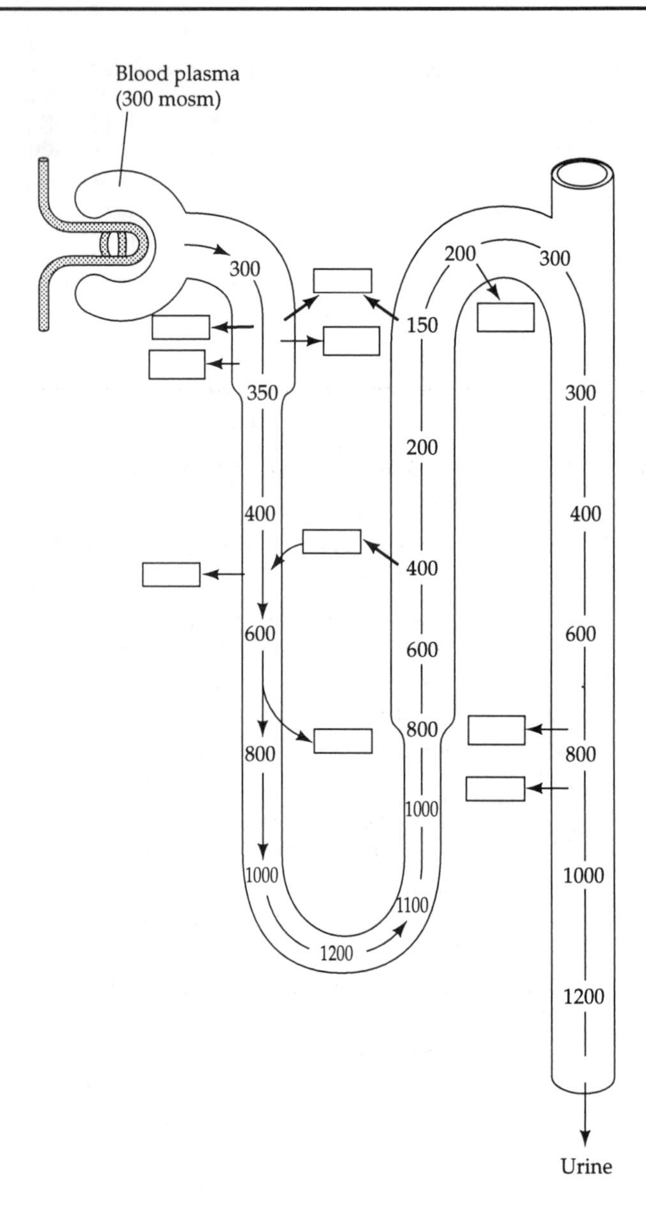

45

Animal Behavior

CHAPTER LEARNING OBJECTIVES—after studying this chapter you should be able to:

- ❑ Define the term "animal behavior" and name the scientists who study this area of biology.
- ❑ Describe the characteristics of fixed action patterns, the conditions under which they evolve, and give several examples.
- ❑ Explain what a deprivation experiment is and how it can be used to identify fixed action patterns.
- ❑ Relate the studies done on bill pecking in gull chicks to the behavioral concept of a sign stimulus or releaser.
- ❑ Define the term "supernormal releaser," describe how you might create one, and give an example of a natural supernormal releaser.
- ❑ Discuss some of the factors that determine to which releasers an animal will respond at a particular moment.
- ❑ Relate the concepts of imprinting and critical periods to learned behavior.
- ❑ Discuss what has been learned about song acquisition based on studies of the white-crowned sparrow and the chaffinch.
- ❑ Explain the importance of sex steroids in the development and expression of sexual behavior in rats.
- ❑ Explain the action of testosterone on song learning in birds and its effect on bird brains.
- ❑ Describe how a specific gene can cause a specific behavior.
- ❑ Discuss the contributions of each of the following types of studies to our current understanding of the genetics of behavior: nest-building in *Agapornis* lovebirds, studies of courtship behavior in dabbling ducks done by Konrad Lorenz, and controlled behavioral genetic selection experiments with *Drosophila* fruit flies.
- ❑ Describe what is known about the genetics of hygienic behavior in honeybees.
- ❑ Describe what is known about the molecular genetics of egg laying in the mollusk *Aplysia.*
- ❑ Discuss the following behavioral concepts: communication, display, signal, sender, receiver.
- ❑ Characterize the features, strengths, and weakness of chemical communication, visual communication, auditory communication, tactile communication, and electrocommunication
- ❑ Discuss the origins of communication signals and give examples of intention movements, autonomic responses, and displacement behaviors.
- ❑ Characterize the features of the round dance and the waggle dance of the honeybee.
- ❑ Define the term "circadian rhythm," explain how it can be characterized, and describe the process whereby a circadian rhythm is entrained.
- ❑ Explain what is known about the circadian clocks of birds and mammals and describe important studies that led to these generalizations.
- ❑ Provide some examples of how organisms use changes in photoperiod to coordinate their activities with seasonal changes.
- ❑ Define what a circannual rhythm is and explain what type of animals might be expected to show circannual rhythms.
- ❑ Define and differentiate between the following terms: "piloting," "distance and direction navigation," and "bicoordinate navigation."
- ❑ Describe how clock-shifting experiments can be used to study the navigational abilities of organisms.
- ❑ List some of the most important sources of directional information that animals can use to navigate.
- ❑ Describe some of the unique features of human behavior.

GENETICALLY DETERMINED BEHAVIOR • HORMONES AND BEHAVIOR (pages 1029–1036)

Key Concepts

1. Animal behavior is an act or series of acts performed by an animal with respect to another organism or the environment.
2. *Ethologists* are scientists who study animal behavior.
3. *Fixed action patterns* are genetically determined behaviors that can be expressed without prior experience and are not modifiable by learning. Once begun, the series of movements constituting a fixed action pattern is always performed from beginning to end.
4. Fixed action patterns are stereotypic (performed the same way each time) and species-specific. Studying fixed action patterns can lead to a better understanding of the mechanisms and genetics of behavior.

5. In a *deprivation experiment*, an animal is reared in isolation from its normal experiences. If an animal can perform a behavior after being reared in isolation, the behavior is a fixed action pattern.
6. Studies done by *Konrad Lorenz* and *Niko Tinbergen* showed that specific sign stimuli or *releasers* are required to elicit fixed action patterns.
7. Releasers are a simple subset of all the sensory information available to the animal. Recognition of the releasers of a fixed action pattern is genetically programmed.
8. Tinbergen and A. C. Perdick showed that bill shape and the presence of a contrasting bill spot color are the releasers that direct the pecking of herring gull chicks at the bill of the parent.
9. *Supernormal releasers* are exaggerations of the normal sign stimuli and lead to a greater response than the normal sign stimuli. Natural selection has favored exaggerated releasers in the courtship behavior of many species, such as the bowerbirds (see Chapter 46).
10. An animal's developmental and physiological condition determines its motivational state. The motivational state determines to which releasers the animal will respond. A motivated animal may search for specific releasers while ignoring others.
11. Fixed action patterns evolve when learning is not possible (many species have non-overlapping generations) or when mistakes are costly (mate selection, avoidance of predators, capture of dangerous prey).
12. Highly stereotypic, species-specific behavior can also result through learning. Acquisition of song in the white-crowned sparrow is dependent on exposure to the song during the nestling stage.
13. In some species, learning is limited to a rather narrow *critical period*. For example, white-crowned sparrows must hear their species-typical song as nestlings in order to form a song template in their nervous systems. Some bird species, such as the chaffinch, can only learn their species-typical song during a critical period.
14. *Imprinting* is the rapid learning that takes place during a critical period. Examples include song acquisition in birds and the individual recognition of offspring and parents that occurs in many species. Imprinting allows the encoding of complex sensory information in the nervous system.
15. Fixed action patterns are often strongly influenced by hormones, because the endocrine system has a direct influence on the developmental and motivational state of the animal.
16. A study of the copulatory *lordosis* posture exhibited by female rats shows the influence of hormones on a fixed action pattern. If the ovaries of a female rat are removed at any age, she will not show lordosis unless injected with a supplement of female sex steroids. Furthermore, if a spayed female is exposed to testosterone early in life, her sexual behavior becomes masculinized. Males castrated early in life and then later exposed to female hormones show lordosis, but exhibit no sexual response when treated with testosterone.
17. These results indicate that both genetic males and females develop female sexual behaviors if testosterone is absent early in life. Testosterone must therefore trigger the development of male fixed action patterns.
18. Further evidence for hormonal influences on behavior come from experiments on songbirds. Males are normally the only sex that sings, and although this song is learned, there is a stereotyped sequence of events that must occur under the influence of testosterone if the song is to *crystallize* and become functional. If a young male who hears his father sing while still a nestling is castrated before he hears himself sing, his song is never properly expressed and does not become fixed into his nervous system.
19. Females do not sing because they are not normally exposed to testosterone. However, they apparently do learn the male's song, because if treated with testosterone they express normal male song patterns. Proper hormonal stimulation thus appears critical for the expression of this important behavior.
20. A recent and unexpected finding is that testosterone causes the song control centers of the bird brain to become larger. Prior to this work it was believed that new brain cells could not be added after birth in vertebrates. It now appears that hormones influence not only the functioning of the nervous system, but also the actual growth of new cells during adult life.

Questions (for answers and explanations, see page 372)

1. Which of the following is *not* an example of a fixed action pattern?
   a. Nut burying behavior in tree squirrels
   b. Behaviors that are expressed in a deprivation experiment
   c. Song in white-crowned sparrows
   d. Web building in orb-weaving spiders
   e. Lordosis

2. Studies done on hand-reared chaffinchs showed that
   a. several critical periods may be involved in song acquisition.
   b. the song of this species is programmed as a fixed action pattern.
   c. imprinting determines only when a song template can be learned.
   d. this species is genetically programmed to recognize what song to learn.
   e. its song would be learned even by individuals held in isolation all their lives.

3. Based on your knowledge of the releasers directing the pecking behavior of herring gull chicks, which of the following models should elicit the *greatest* response?

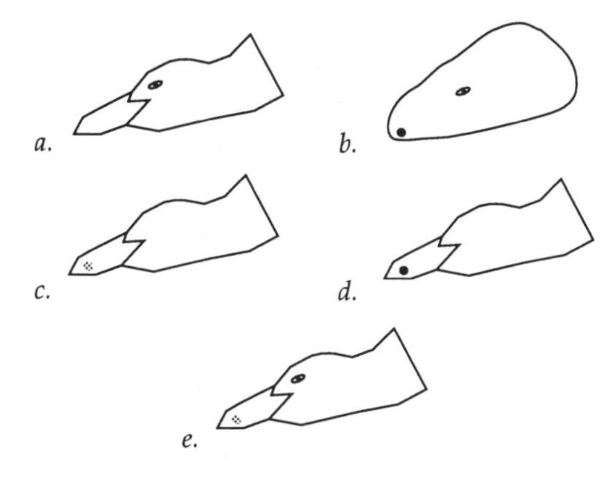

4. Select the following example that is *least* likely to be a fixed action pattern.
   a. Mate selection
   b. Predator escape behavior
   c. Predator's prey-search behavior
   d. Behavior triggered by a releaser
   e. Learning a song template

5. Select one of the following choices to make a *false* statement. Supernormal releasers
   a. result in a greater response than the normal releasers.
   b. are exaggerations of normal sign stimuli.
   c. would not be effective in a deprivation experiment.
   d. are possible because of the simplicity of most sign stimuli.
   e. are conceptually similar to many of the elaborate displays and features that have evolved in courting animals.

6. A newborn male rat is castrated and then allowed to mature. When he reaches adulthood, he is given an injection of male sex steroids and exposed to a normal adult female rat exhibiting lordosis. Which of the following will occur?
   a. The male will mount the female and attempt to copulate.
   b. The male will show lordosis.
   c. The male will attempt to copulate, and also show lordosis.
   d. The male will aggressively repel the female.
   e. The male will show no sexual behavior at all toward the female.

7. Which of the following statements about hormonal influences on bird song is *not* true?
   a. Male songbirds castrated as nestlings cannot sing proper song as adults.
   b. Female songbirds can be induced to sing by treatments with male sex hormone.
   c. Testosterone is not necessary for singing in males once they have learned their song.
   d. Females learn their species song, but do not normally express it under natural conditions.
   e. Testosterone stimulates young male songbirds to practice singing during their first spring.

Activities (for answers and explanations, see page 372)

- Discuss the process of song crystallization in birds.

- Explain why the process of cocoon construction is considered to be a fixed action pattern in many spiders.

- You spay a newborn female rat and treat her with testosterone. In each of the following two situations (*a* and *b*), describe what type of sexual behavior you expect to observe and explain what these outcomes suggest about the hormonal control of sexual behavior in rats.
   a. At maturity, the female is injected with female sex steroids.

   b. At maturity, the female is injected with testosterone.

THE GENETICS OF BEHAVIOR • COMMUNICATION (pages 1036–1045)

Key Concepts

1. Genes do not directly cause specific behaviors. Genes code for proteins that influence the development and functioning of the nervous system, which indirectly results in behavior.
2. Hybridization experiments show that the behavior of hybrids is usually intermediate to that of the two parent species, as shown in nest material carrying by the lovebirds *Agapornis roseicollis* and *Agapornis fischeri*.

3. Konrad Lorenz showed that the series of components of the courtship behavior of dabbling ducks differs in the hybrids resulting from crossing two different species. Display components from the two parent species are put together in different combinations in the hybrids or new components not seen in the two parent species are added.
4. The behavior of domesticated animals produced by artificial selection clearly shows the importance of genetics. Hybridization between artificially selected strains of *Drosophila* fruit flies indicates that many aspects of the behavior of this species are indirectly determined by multiple genes that affect the general properties of the nervous system.
5. Mendelian segregation of behavioral traits is rare. *Hygienic behavior* in honeybees seems to be controlled by two genes—the recessive alleles for these genes indirectly lead to the uncapping and larval removal behavior involved in combatting a bacterial disease that kills larvae.
6. Some of the powerful techniques of molecular genetics discussed in Chapter 14 are now being used to study the behavioral functions of specific genes. In the marine mollusk *Aplysia*, a single gene has been discovered that codes for a precursor molecule consisting of nearly 300 amino acids. This precursor is modified into a number of different peptides that function as neural signals to coordinate the egg-laying behavior in this species.
7. Communication is behavior designed to influence the actions of other individuals. Communication involves the exchange of displays or signals between a sender and a receiver.
8. A *display* or *signal* is a behavior, frequently an exaggerated version of a normal behavior, that has evolved to influence other individuals. *Displays* are never shown by isolated individuals.
9. Communication systems are sometimes exploited by illicit senders or receivers of signals. Predators, for example, can intercept signals and use them to locate potential prey.
10. Chemosensation is the most ancient mode of communication and probably evolved from intracellular signaling. Intracellular-signaling molecules have occasionally been directly adapted for use in chemical communication, as in use of cAMP for aggregation of myxamoebas in cellular slime molds.
11. *Pheromones* are chemical signals exchanged between species members. Characteristics of pheromones are that they can communicate very specific messages, are relatively long lasting (compared to auditory or visual signals), and the pheromone concentration gradient can be useful for locating the organisms producing it.
12. The chemical features and size of the pheromone molecule determine its diffusion coefficient, which, in turn, determines its rate of spread and dissipation. Pheromones with greater diffusion coefficients spread rapidly, but disappear quickly—useful for sex pheromones. Pheromones with smaller diffusion coefficients spread slowly, but also disappear slowly—useful for trail- and territory-marking pheromones.
13. The *visual communication mode* allows an infinite variety of signals that can be produced and propagated rapidly, but detection may be difficult and requires adequate light. Some species have evolved their own light-emitting mechanisms, as in the luciferin/luciferase mechanisms in fireflies.
14. Although few animal species use *auditory communication*, sound can be used when light is limited, can go around objects, and can be transmitted over great distances. However, sound cannot convey as much information as rapidly as visual signals.
15. The physical properties of sound can be varied to make them easy or difficult to locate. A pure tone with a sharp beginning and end is easier to locate than a complex sound with a gradual onset and completion.
16. Tactile communication occurs primarily in environments where visual signaling is impossible mainly due to low light levels, such as in the deep sea, in dark corners, and inside beehives. Chemicals are also frequently exchanged during tactile communication.
17. If a honeybee discovers a food source within 80 meters of the hive, it performs a *round dance* on the vertical combs within the hive's dark interior. Other worker bees detect the round dance tactically and respond by leaving the hive to search randomly for the food.
18. In the waggle dance of the honeybee, the speed of the waggle run is directly related to the distance from the hive to the food source and the orientation of the waggle run relative to vertical shows the direction to fly to get to the food source relative to the hive–sun direction. The waggle dance is given if the food source is greater than 80 meters from the hive.
19. Certain fish produce electric fields in order to sense objects in the murky water in which they live. Some fish species show true electrocommunication; they can produce electric pulses that vary in frequency in order to convey information about their individual identity, sex, and dominance status.
20. Signals may be elaborated or exaggerated in order to reduce ambiguity concerning the signaller's identity, emotional or reproductive state, or intentions. Sources for communication signals include intention movements, autonomic responses, and displacement behaviors.
21. *Intention movements* are an important source of behavior for visual displays.
22. *Autonomic nervous system responses*, such as the erection of fur and feathers, have become elaborated into displays in many animals.
23. *Displacement behaviors* are seemingly irrelevant displays that result from conflict between different motives like attack and escape.

Questions (for answers and explanations, see page 372)

1. Which of the following is *not* a valid conclusion resulting from studies done on hygienic behavior in honeybees?
   a. The inheritance of this behavior could be explained in a simple Mendelian manner.
   b. This was a rare case where a gene resulted directly in the expression of a behavioral trait.
   c. The uncapping and larval removal components of this behavior seemed to be determined by different genes.

d. Since all of the F_1 bees resulting from a cross between hygienic and nonhygienic strains were nonhygienic, the trait must be determined by recessive alleles.
e. A backcross of the F_1 to the nonhygienic strain should yield ¼ hygienic to ¾ nonhygienic individuals.

2. Studies of which of the following organism types have led most directly toward connecting a specific gene with a specific behavior?
a. Lovebirds in the genus *Agapornis*
b. Dabbling ducks
c. Domestic animals
d. The marine mollusk *Aplysia*
e. *Drosophila* fruit flies

3. Which of the following waggle dances should be given by a bee returning from the food source shown in the diagram below?

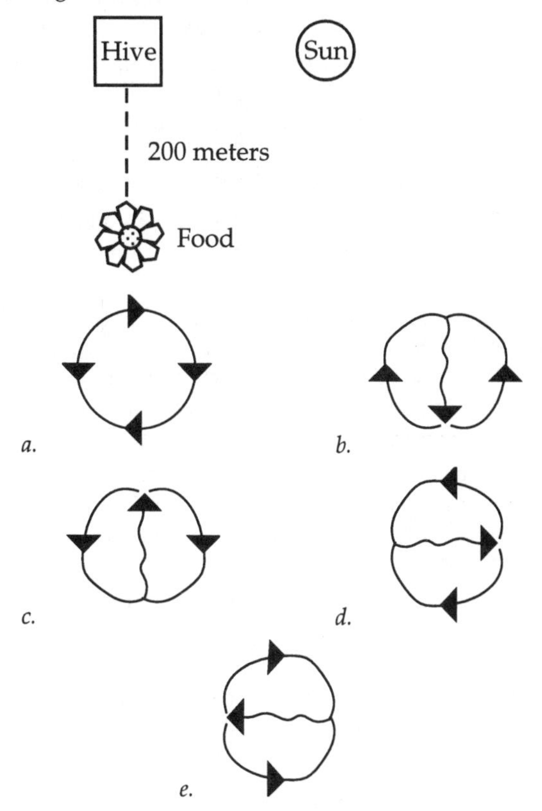

Which of the conditions in the premise statement for this question determines whether a foraging bee returning to the hive will give a round dance or a waggle dance?

How does the search initiated by a round dance differ from the search initiated by a waggle dance?

4. Which communication mode would you expect an animal to use if it lives in a dark, terrestrial environment, and needs a rapid response?
a. Auditory
b. Olfactory
c. Visual
d. Tactile
e. Electric

5. Which of the following communication modes are frequently associated?
a. Auditory
b. Olfactory
c. Visual
d. Tactile
e. Electric

6. When two dogs interact aggressively, the submissive dog will frequently roll onto its back and urinate. This is an example of
a. an intention movement.
b. an autonomic response.
c. a displacement behavior.
d. a display.
e. chemical communication.

Activities (for answers and explanations, see page 372)

- Discuss how the auditory characteristics of a mating call might differ from an alarm call given in response to detection of a nearby predator.

- Discuss how the diffusion coefficients of a sex pheromone and a trail-marking pheromone might differ, and explain these differences.

THE TIMING OF BEHAVIOR: BIOLOGICAL RHYTHMS • HOW DO THEY FIND THEIR WAY? • HUMAN BEHAVIOR (pages 1045–1052)

Key Concepts

1. Many organisms have evolved *rhythmicity* in response to the 24-hour cycle of environmental conditions caused by the rotation of Earth on its axis.
2. The presence of *circadian rhythms* in the activity of many animals when kept under constant environmental conditions is evidence for the existence of an internal (*endogenous*) clock.

3. A rhythm consists of a series of cycles and the length of a cycle is the *period* of the rhythm. Typically the period of a circadian rhythm is not exactly 24 hours.
4. A point on a cycle is a *phase* of that cycle. Because a circadian rhythm does not usually have a period of 24 hours, under uniform conditions a circadian rhythm will *free-run* with its natural periodicity.
5. *Entrainment* is the process whereby the circadian rhythm is reset by environmental cues so that it is in phase with the 24-hour cycle. During entrainment, a circadian rhythm of more than 24 hours is *phase-advanced;* a circadian rhythm of less than 24 hours is *phase-delayed.*
6. Typically, the natural light–dark cycle entrains a free-running clock to a 24-hour period. Brief exposure to pulses of light or dark every 24 hours can be used to experimentally entrain animals. Within limits, an animal's endogenous clock can also be entrained to cycles less than or greater than 24 hours.
7. Because your internal clock can only be phase-shifted 30 to 60 minutes per day, *jet lag* results if you fly across several time zones.
8. In mammals, the circadian clock is located in the *suprachiasmatic nuclei* (SCN), small groups of cells located in the area of the brain above where the two optic nerves join. Removal of the SCN results in a loss of circadian rhythmicity and SCN transplantation can restore it. The mammalian circadian clock is entrained by light cues from the photoreceptors in the eyes.
9. The avian circadian clock resides in the *pineal gland* and is entrained by light cues detected by that gland. The pineal gland is located on the dorsal surface of the brain, between the two hemispheres; it secretes the hormone melatonin.
10. Many organisms have also evolved rhythmicity in response to the year-long cycle of environmental conditions caused by the revolution of Earth around the sun.
11. Changes in *photoperiod,* the daily day–night lengths, is used by many animals to coordinate their activities with seasonal changes. For example, the reproductive cycle in many birds is *photoperiodic,* that is, coordinated by photoperiod.
12. Hibernators and equatorial migrants cannot use photoperiod to coordinate their activities. Many of these species have endogenous annual rhythms called *circannual rhythms.* The period of a circannual rhythm is usually somewhat shorter than 365 days.
13. In *orientation,* the behavior of an animal is directed by an object or stimulus. *Navigation* is long-distance orientation.
14. *Piloting* is orientation with reference to familiar landmarks in the animal's environment. Piloting depends on knowledge of the structure of the environment.
15. *Homing* involves an oriented movement from a specific location to the animal's nest site or burrow. Although homing can be accomplished by piloting, some animals are able to home without using landmarks. Homing pigeons and albatrosses have amazing homing abilities.
16. *Migration* is the seasonal movement, typically between summer breeding grounds and wintering areas, seen in many animals. Bird banding shows that many bird species fly long distances, sometimes over large bodies of water where no landmarks are available.
17. In many bird species, adults leave the breeding grounds at the end of the season before the juveniles. This observation suggests that young birds must be able to navigate on their own and cannot simply follow experienced birds.
18. Navigation through unfamiliar territory can be accomplished by two systems: distance and direction navigation and bicoordinate navigation.
19. *Distance and direction navigation* requires that the animal know the direction to the goal and the distance to travel.
20. *Bicoordinate navigation* requires that the animal know its current map location (latitude and longitude) and the map location of where it wants to go.
21. In many species, if individuals are displaced prior to their normal migration, they migrate in the direction and distance that would be appropriate if they had not been displaced. This observation suggests that these species are using distance and direction navigation.
22. The increased, oriented activity shown by many caged birds during the migratory period is called *migratory restlessness.*
23. If an animal knows the direction to its destination, the duration of the migratory restlessness will cause it to move the correct distance. Once within the approximate location of its destination, piloting by landmarks could be used to complete the journey.
24. In the Northern Hemisphere, the sun rises in the East, is due south at noon, and sets in the West. If an animal knows the time, compass directions can be obtained from the sun's position.
25. Many animals use circadian clocks to determine direction from the sun's position.
26. Clock-shifting experiments cause birds to orient in predictably incorrect directions based on the amount of the phase shift. A six-hour phase shift in circadian clocks causes a 90° error in their orientation.
27. *Stephen Emlen* showed that some nocturnally migrating birds use the North Star (Polaris) to determine the correct migratory direction. In the northern hemisphere, all other stars circle a point located over Earth's axis of rotation. This point is near the North Star.
28. Emlen showed that birds imprint on the North Star only if they are exposed to a rotating sky as juveniles. Birds raised in a planetarium will imprint on any star about which the sky is made to rotate.
29. Because the location of the North Star is fixed, a circadian clock is not necessary to derive directional information from it. Experienced birds are not confused if the sky in the planetarium does not rotate or rotates at an incorrect rate.
30. Some animals have several, redundant navigation systems. For example, in addition to a sun compass, pigeons can derive directional information from Earth's magnetic field.

31. Sensitivity to the planes of the polarization of light, very low frequency sound, and weather patterns are used by some organisms to navigate.
32. Bicoordinate navigation requires a map sense—the animal must know its geographic location and the latitude and longitude of its destination.
33. Longitude can be deduced from a knowledge of the time of day and the sun's position. The time of sunrise is earlier if you are east of home and later if you are west of home.
34. Latitude can be deduced from a knowledge of the time of day and the sun's position. At a given time, the sun is higher in the sky if you are south of home and lower in the sky if you are north of home.
35. Although there are genetically based limitations on what we can learn and when we can learned based on the organization of our sensory and nervous systems, the most notable aspect of human behavior is its ability to be modified by experience.
36. The genetically determined features of human behavior include some behavioral drives, emotions, and motor patterns, such as facial expressions.

Questions (for answers and explanations, see page 372)

1. The following figure shows the periods of activity (white bars) and inactivity (black bars) of an animal kept under constant conditions of light and temperature. What is the period of this animal's circadian rhythm, and does it need to be phase-advanced or phase-delayed to become entrained to a 24-hour cycle, or is it already entrained?

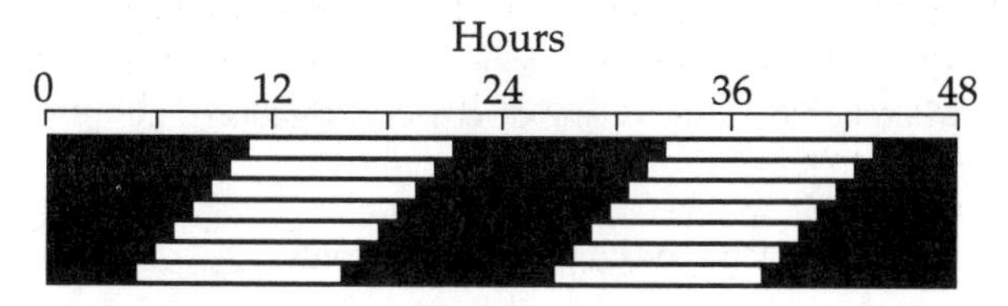

a. 24 hours, entrained
b. 11 hours, phase-advanced
c. 11 hours, phase-delayed
d. 22 hours, phase-advanced
e. 22 hours, phase-delayed

Note: the following diagram should be consulted in answering the next three questions.

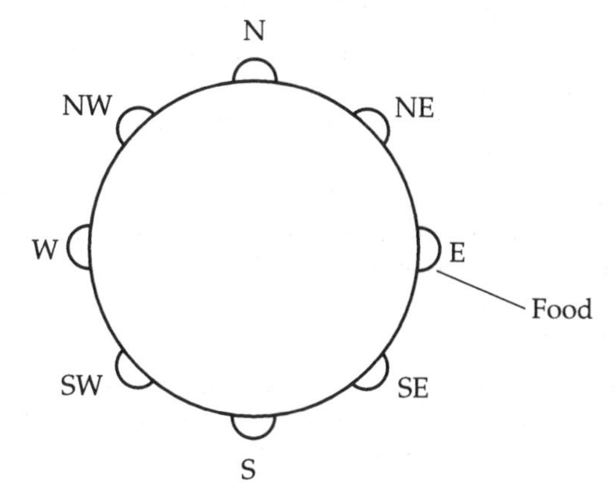

2. A starling is trained to feed from a food bin in the east end of a circular cage from which it can see the sky. At sunrise the cage is covered and a fixed light source is presented above the east end of the cage. If the starling treats the light as the sun, where will it search for food at noon?
a. North
b. East
c. South
d. West
e. Random

3. A starling is trained to feed from a food bin in the east end of a circular cage from which it can see the sky. If a mirror is used to shift the apparent position of the sun at noon from the south to the northwest, where will the starling search for food?
a. North
b. Northwest
c. Southwest
d. Northeast
e. East

4. A starling is trained to feed from a food bin in the east end of a circular cage from which it can see the sky. The bird is then placed into a light-controlled environment for three weeks and phase-delayed by six hours. If the bird is returned to the cage under natural lighting conditions at sunset, where will the bird search for food?
a. North
b. South
c. East
d. West
e. Southeast

5. You are conducting studies on a species of bird that lives in the northern hemisphere and normally migrates to the south for the winter using the North Star as a navigational cue. In a planetarium, caged birds are exposed to a night sky that is shifted six hours ahead of the normal star pattern. In what direction would the birds direct their migratory restlessness?
a. North
b. East
c. South
d. West
e. Random

6. Which one of the following animals would *not* be expected to have an endogenous circannual rhythm?
a. A ground squirrel that hibernates
b. An arctic hare
c. A cave salamander
d. A tropical insect
e. A bird that summers in North America and migrates to South America for the winter.

7. Which of the following statements about circadian rhythms is true?
   a. The effect of jet lag on your circadian rhythms is the same if you fly 1,000 miles west or 1,000 miles south.
   b. A regular short pulse of light or dark can entrain the free-running circadian rhythms of an animal kept under constant environmental conditions.
   c. Circadian rhythms dependent on endogenous clocks are restricted to birds and mammals.
   d. An animal's circadian rhythm usually has a period of 24 hours.
   e. If Earth was not tilted on its axis, circadian rhythms would not be necessary.

8. Select *all* of the following treatments that would likely eliminate the circadian rhythms of the animal.
   a. Removing the suprachiasmatic nuclei of a mammal
   b. Blinding a mammal
   c. Removing the pineal gland of a bird
   d. Removing the pineal gland of a mammal
   e. Blinding a bird
   f. Placing a bird or mammal under constant darkness

9. In most species homing is accomplished by
   a. piloting.
   b. a time-compensated solar compass.
   c. bicoordinate navigation.
   d. use of the Earth's magnetic field.
   e. random searching.

10. Suppose you are an animal that is capable of bicoordinate navigation. You are displaced from home and you notice that the sun rises later and is higher in the sky than it should be based on your endogenous clock and your knowledge of the sun's position in your home area. You conclude that
   a. you need to travel northeast to get home.
   b. you need to travel southeast to get home.
   c. you need to travel southwest to get home.
   d. you need to travel northwest to get home.
   e. you need more information!

11. Suppose you capture a bird at position *A* on a large island and transport it 200 km due south to position *D* on the other side of the island. The population from which this bird came normally migrates to position *B*. Assuming that the bird is capable of the orientation modes listed below, for each mode predict the position (A, B, C, D, or E) to which the displaced bird should move.

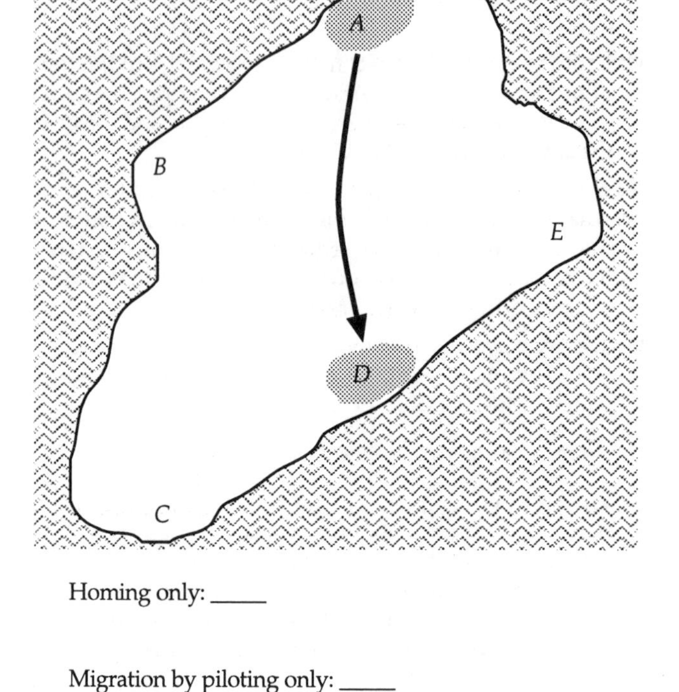

Homing only: _____

Migration by piloting only: _____

Migration by distance and direction navigation only: _____

Migration by bicoordinate navigation only: _____

Answers and Explanations

Chapters 35–45

CHAPTER 35 – PHYSIOLOGY, HOMEOSTASIS, AND TEMPERATURE REGULATION

Note : the page numbers listed with the section titles refer to the study guide; numbers in parentheses for each question refer to the relevant key concepts.

HOMEOSTASIS • ORGANS AND ORGAN SYSTEMS (pages 256–258)

- (4–7)

Epithelium:	protective barrier against invasion by microorganisms; prevents water loss; secretes oils and other substances
Connective:	support base for other tissues (especially epithelium); provides flexibility and strength; blood
Muscle:	controls action of some glands and causes hair to move
Nervous:	various sensory functions (e.g. touch, pain, temperature); controls muscle movements

- (8–17)
  b: excretory
  h: circulatory
  f: skeletal
  g: muscle
  a: nervous
  d: endocrine
  c: digestive
  e: reproductive

1. Choice *e*. All of these processes contribute to the maintenance of appropriate internal body conditions. (1–2)
2. Choice *b*. Eyes are sensory structures adapted for light detection. They relay the information they gather to other parts of the central nervous system. (8)
3. Choice *a*. Endocrine tissue is a type of epithelium specialized for secretion. (4)
4. Choice *d*. Connective tissue includes such things as blood, cartilage, and bone, all of which are characterized by having loose arrays of cells embedded in some kind of an extracellular matrix. (5)
5. Choice *e*. Electrical excitability is a hallmark of the nervous system. The blood circulatory system acts as a conduit for essential nutrients and wastes to all other organ systems. (8, 15)
6. Choice *b*. The excretory system handles removal of nitrogenous wastes and helps to regulate salt and water balance. (8, 11, 17)

GENERAL PRINCIPLES OF HOMEOSTASIS • THE EFFECTS OF TEMPERATURE ON LIVING SYSTEMS • THERMOREGULATORY ADAPTATIONS (pages 258–260)

- The lower and upper set points for this thermostat are 18° and 22°, respectively. As long as the room temperature remains within this range no action is required. If, however, the sensor detects that the room temperature has fallen below 18° or risen above 22°, an appropriate error signal is sent to the thermostat, which turns on either the heater or the air conditioner to counteract the change in room temperature. Once the temperature is back within the set point range, negative feedback shuts down the heating or cooling system. (3, 5, 6)
- Acclimitization is a seasonal adjustment in metabolic activity. This is accomplished by metabolic compensation, in which a shift is made between enzymes having different optimal temperatures. (13)

1. Choice *d*. Changing set points in anticipation of environmental changes is feedforward regulation. (3, 8)
2. Choice *c*. The amplification of uterine contractions due to persistent stimulation is indicative of a positive feedback mechanism. (7)
3. Choice *d*. Q_{10} = (rate at 30°C)/(rate at 20°C). Therefore, 1.5 = 200/X. Solving for X yields 133. (10)
4. Choice *e*. Since the lizard obtains its heat mostly from the environment, it is best described as an ectotherm. Because it holds a somewhat constant body temperature through changes in its behavior, it can also be considered a behavioral homeotherm. (14–17)
5. Choice *b*. Anything that would cool the animal, either by preventing more heat from entering the body or by facilitating heat loss, would be an appropriate response in this situation. Since preventing blood flow to the periphery would tend to trap heat within the animal's body, this would be an inappropriate response. (18)
6. Choice *e*. Muscle activity produces heat, and the special vascular anatomy of these fish helps conserve heat within the body core. (20–21)
7. Choice *c*. Because it can adjust its metabolic rate to compensate for changes in environmental temperature, the mouse qualifies as an endotherm. (15–16)

THERMOREGULATION IN ENDOTHERMS • THE VERTEBRATE THERMOSTAT • TURNING DOWN THE THERMOSTAT (pages 260–262)

- Your drawing should resemble Figure 35.18 on page 791 in the textbook. (2, 4)
- Evaporative cooling is used to dump excess heat above the upper critical temperature. Both panting and sweating require an energy expenditure, thus the increase in metabolic rate. (9)

1. Choice *c*. Shivering is a metabolic heat production mechanism employed when temperature drops below the lower critical limit. (2–5)
2. Choice *a*. The summit metabolism is the maximum metabolic rate an endotherm can achieve in response to cold. If temperatures continue to fall or persist for extended periods, death follows. (4)
3. Choice *e*. Panting and sweating are the means by which evaporative cooling is achieved; both require ATP and lead to water loss. (9)
4. Choice *b*. Hypothalamic thermal set points change often as with fever, sleep, hibernation, etc. (10–15)
5. Choice *d*. Torpor is a short-term response (a few days at most) to cold or food scarcity, while hibernation usually lasts for an entire climatic season. (18–20)
6. Choice *a*. Pyrogens released by pathogenic substances trigger mechanisms that promote fever. (15)
7. Choice *d*. Brown fat metabolism is a physiological response to cold, not a morphological one. (6)

Integrative Questions

1. Choice *e*. All these animals engage in behaviors related to thermoregulation.
2. Choice *b*. Adipose (fat) tissue is a type of connective tissue.
3. Choice *a*. The conditions described suggest that the rodent is in its thermoneutral zone (see textbook Figure 35.18). If the animal was hibernating or in torpor its body temperature would be close to zero degrees, and if it had a fever its body temperature would be greater than 37 degrees.

CHAPTER 36 – ANIMAL HORMONES

Note : the page numbers listed with the section titles refer to the study guide; numbers in parentheses for each question refer to the relevant key concepts.

CHEMICAL MESSAGES • HORMONAL CONTROL IN INVERTEBRATES (pages 263–265)

- (2, 4, 10–11)
  d: Histamine
  a: Endocrine gland
  b: Target cell
  e: Ecdysone
  c: Juvenile hormone

- (9–11)

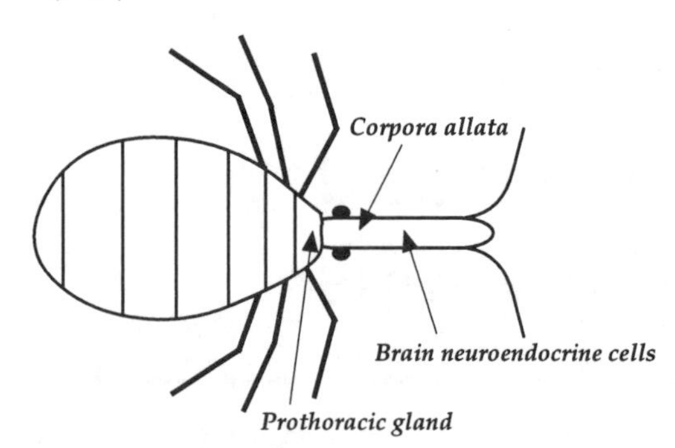

- Since both instars received a blood meal one week earlier, the prothoracic gland was stimulated to release ecdysone and initiate the molting process. The two instars show different molting results because complete decapitation removes the corpora allata (which produce juvenile hormone), whereas partial decapitation leaves the corpora allata intact. Thus, the partially decapitated instar continues to make juvenile hormone and molts into a fifth instar, while the completely decapitated instar lacks juvenile hormone and becomes an adult. (9–12)

1. Choice *d*. Vasopressin is released from the posterior pituitary (an endocrine gland), travels through the bloodstream (a circulating hormone), and affects target cells in the kidney. (1–4, 7)
2. Choice *b*. Histamine is produced for use inside the body and acts in the immediate vicinity of the cells where it is produced. (3–7)
3. Choice *a*. Hormones affect specific target cells because only those cells possess appropriate receptors to recognize the hormone. Exactly what response they stimulate depends on the particular biochemical pathways of that cell. (6)
4. Choice *c*. Even though the bug is fed, the fact that it lacks its supply of brain hormone means that ecdysone is never released from the prothoracic gland. Thus, molting cannot occur. (9–12)
5. Choice *d*. Partial decapitation means that the corpora allata, the source of juvenile hormone, are still present. Since decapitation occurred a week after feeding, there was plenty of time for brain hormone and ecdysone to be released, thereby initiating molt. Since the juvenile hormone concentration is still high, the bug molts into another juvenile instar instead of an adult. (9–12; see also textbook Fig. 36.4)
6. Choice *e*. The lack of juvenile hormone is what allows larvae to molt into adults. As long as juvenile hormone concentrations remain high, larvae will not become adults and no reproduction can occur. (14)

VERTEBRATE HORMONES (pages 265–268)

1. Choice *c*. The anterior pituitary makes all three of these hormones. (8, 10, 12)
2. Choice *a*. See the activity on page 357 for a flow diagram showing the regulation of thyroxine release. (14, 15)

3. Choice *d*. By selectively destroying the posterior pituitary, you would be eliminating only the influence of vasopressin on the kidney (the effects of oxytocin, the other hormone of the posterior pituitary, are not particularly important in male dogs). Another approach would be to destroy the hypothalamus, but this would also affect many other physiological systems. (4–5)
4. Choice *c*. Insulin is needed to turn the excess glucose obtained from the meal into stored glycogen. (20)
5. Choice *a*. This situation is likely to produce stress in the rat, which would dramatically increase its levels of cortisol. (25)
6. Choice *d*. The child described is genetically a male. The fact that his cells are insensitive to female hormones will not affect his development either in the womb or at the time of puberty. (31)
7. Choice *c*. Puberty is preceded by increased levels of gonadotropin-releasing hormones, secreted by the hypothalamus. This triggers the pituitary to release luteinizing hormone and follicle-stimulating hormone, which stimulate the gonads to produce sex steroids. (34)
8. Choice *a*. Epinephrine is the only hormone listed that is not a steroid. (23–25, 29)

- (14–15)

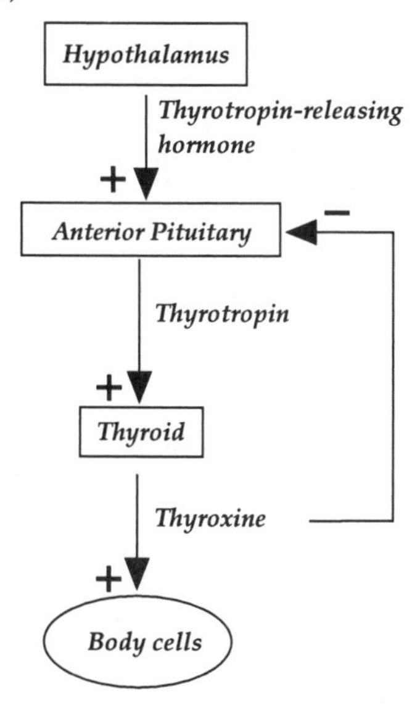

- The mother's ears pick up the baby's cry and this information is rapidly transmitted to her brain, which relays it to the hypothalamus. The neurohormonal connections between the hypothalamus and the posterior pituitary facilitate the rapid release of oxytocin, which travels quickly through the circulatory system to stimulate the release of milk from mammary glands in the breasts. (4, 6)

RECEIVING AND RESPONDING TO HORMONES (pages 268–270)

1. Choice *e*. Just the reverse is true; a single hormone molecule can result in the activation of hundreds or thousands of enzymes. (7)
2. Choice *e*. In the process of amplifying the hormone signal, second messenger systems cause conformational changes in many metabolic enzymes, including protein kinases. (3, 7)
3. Choice *b*. Phosphodiesterases are the enzymes that break down cAMP and "reset" the second messenger signalling system. (8)
4. Choice *d*. The basic signalling principle of hormones is the same for both lipid- and water-soluble varieties. Both involve manufacture by some gland or tissue at a site different from where they have their effect, and they also require receptors on target cells in order to be detected. (1–2, 12)
5. Choice *e*. ATP is used to phosphorylate various molecules during the cascade of reactions in a second messenger system, but is not itself a second messenger. (3, 5)
6. Choice *c*. Lipid-soluble hormones cross cell membranes and work directly on gene expression. Their effects thus take longer to develop than the more transient second messenger changes that occur when water-soluble hormones bind to the surface of a cell. Moreover, since they affect DNA transcription and protein synthesis, the effects of lipid-soluble hormones also tend to last longer. (14)

- (2–5)
  The correct sequence is:
  1. Hormone
  2. Receptor molecule
  3. Inactive G-protein
  4. GTP
  5. Active G-protein
  6. Inactive adenylate cyclase
  7. Active adenylate cyclase
  8. ATP
  9. cAMP
  10. Protein kinase
  11. Phosphoprotein phosphatase
- The first messenger is the hormone itself, the second messenger is cAMP. A second messenger is needed because the water-soluble molecule cannot cross the cell membrane and influence cytoplasmic mechanisms directly. The second messenger's role is therefore to initiate a cascade of reactions within the cell that produces a biochemical response to the presence of the hormone on the cell's surface. (2–5)

Integrative Questions

1. Choice *d*. All of these hormones are either peptides, polypeptides, or protein molecules. Thus, they are all made of amino acids.
2. Choice *c*. Prolactin, like oxytocin, is involved in milk production and secretion in mammals. It would thus be found associated with mammary tissue. Since it is a water-soluble hormone, it would be found attached to plasma membrane surface receptors.
3. Choice *a*. Progesterone is a critical hormone for maintaining the condition of the uterus in pregnant women. Calcitonin regulates the deposition and reabsorption rates of calcium from the mother's bones; this process is normally disrupted by the growing infant's need for calcium to build its own bones.
4. Choice *b*. Thyroxine is a lipid-soluble hormone. It therefore needs a receptor within the cell cytoplasm to initiate its effects.

CHAPTER 37 – ANIMAL REPRODUCTION

Note : the page numbers listed with the section titles refer to the study guide; numbers in parentheses for each question refer to the relevant key concepts.

ASEXUAL REPRODUCTION • SEXUAL REPRODUCTIVE SYSTEMS OF ANIMALS (pages 271–273)

- (14)
  $2n$ Primary spermatocytes
  n Spermatids
  $2n$ Spermatogonia
  n Secondary spermatocytes
  n Functional sperm
- Since they have fully functional gonads, most organisms that are capable of regeneration also reproduce sexually. Depending on their sex, sea stars release either sperm or eggs to form zygotes, thus benefiting from the genetic diversity created in their offspring. On the other hand, if an adult animal loses an arm to a predator, or more dramatically, if the predator gets most of the body but leaves an arm, the remaining section can regenerate the lost body parts. This "new" individual is genetically identical to the original one because regeneration is an asexual process. (1–4, 6)

1. Choice *b*. Parents who are able to survive long enough to reproduce are probably also fairly well adapted for their habitat. If the environmental conditions remain stable, it is a good strategy to make more offspring that are just like themselves, which is what occurs in asexual reproduction. (1–4, 8)
2. Choice *e*. In parthenogenesis, a mother's egg is stimulated to begin development without sperm penetration. Although a mating act may cause this, there is no union of sperm and egg. Therefore, parthenogenetic offspring have no father. (7–9)
3. Choice *e*. All of these are valid differences between sperm and eggs. (11)
4. Choice *b*. The presence of Sertoli cells indicates that you are examining seminiferous tubule tissue from a testis. Therefore, the only unambiguous conclusion you can draw is that the animal is a male. Even fairly young males would have these cells in their immature testes, so age cannot be determined. (14)
5. Choice *c*. Dioecious means that the species has two separate sexes, spermatogonia are only made in males, and squid do utilize a spermatophore to transfer sperm from male to female. (13, 17, 23)
6. Choice *a*. Since external fertilization can only take place in an aquatic habitat, there are no terrestrial animals that use it. (20)
7. Choice *a*. Many animals reproduce both by asexual and sexual means. (1–3)
8. Choice *d*. Asexual reproduction is often physiologically and behaviorally easier than sexual reproduction, assuming sufficient resources are available. However, a major drawback to asexual reproduction is the lack of genetic diversity it creates, which can have important consequences in times of environmental instability. (3–4)

REPRODUCTIVE SYSTEMS IN HUMANS AND OTHER MAMMALS • FERTILIZATION (pages 273–277)

1. Choice *d*. Located on the surface of each testis, the epididymis stores mature sperm until they are ready to be emitted just prior to ejaculation. (2)
2. Choice *e*. The woman's reproductive tract is acidic to help protect against infection by microorganisms. Unfortunately, this environment is also hostile to the sperm. The prostate gland secretions are alkaline, which helps neutralize the pH of the female's system and permits the sperm to survive longer. (6)
3. Choice *b*. Sperm are deposited in the vagina during ejaculation, travel past the cervix and into the uterus, then up the oviducts where they encounter the egg. (10)
4. Choice *c*. The corpus luteum is the primary source of progesterone. It is formed at the time of ovulation, so progesterone levels are therefore only high enough to sustain endometrial development during the second half of the cycle. (19, 20)
5. Choice *e*. The gonadotrophic hormones are follicle-stimulating hormone and luteinizing hormone, both of which are inhibited by the high levels of progesterone and estrogen present during the latter half of the menstrual cycle. They are at their absolute lowest levels between days 20–25. (18–20)
6. Choice *d*. Seminal vesicles produce the fructose sugars that power the sperm's mitochondria. (6)
7. Choice *a*. Only the intrauterine device (IUD) prevents implantation of the embryo; all the others are aimed at preventing fertilization. (34–36)
8. Choice *b*. Men can have only a single ejaculation before undergoing a refractory period of at least 20 minutes, during which they cannot obtain the full erection needed for another orgasm. (22–26)

- (2, 5, 8–9)

	Male	Female
1.	*Testes*	*Ovaries*
2.	*Testes*	*Ovaries*
3.	*Glans penis*	*Clitoris*
4.	*Epididymis*	*Not applicable*
5.	*Not applicable*	*Uterus*

- (16–19)

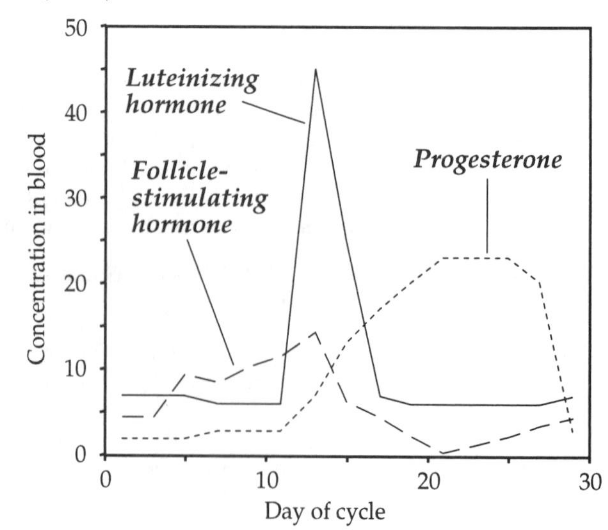

Progesterone concentration increases during the second half of the menstrual cycle, when the corpus luteum is functional, in order to sustain the endometrial lining of the uterus. If pregnancy does not occur, progesterone levels drop precipitously just prior to menstruation. Follicle-stimulating hormone rises slowly through the first third of the cycle, stimulating the growth of the follicle. When the follicle ruptures, estrogen and progesterone levels rise and inhibit the release of FSH. If pregnancy does not occur, its concentration then begins to rise once again at the end of the cycle. Luteinizing hormone remains fairly constant at low concentrations early in the cycle, then peaks in a dramatic midcycle surge which stimulates ovulation.

CARE AND NURTURE OF THE EMBRYO • BIRTH (pages 277–279)

- (7–10)
  The amniotic egg was crucial in the evolution of terrestrial vertebrates because it freed them from a dependency on aquatic reproduction, thus opening up whole new environments for colonization.

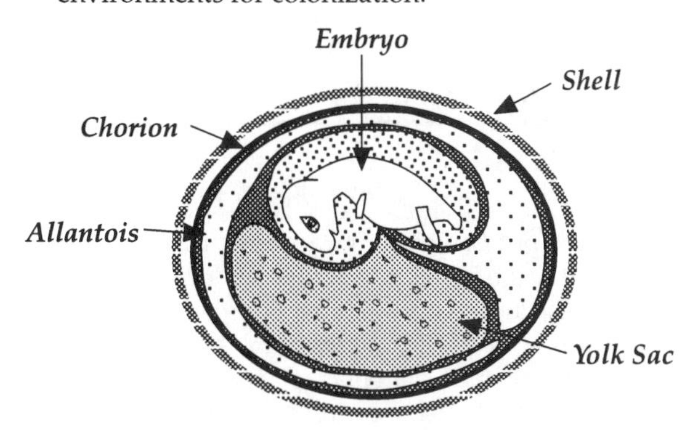

- Oviparous animals (most fish, amphibians, and reptiles, all birds, and some mammals) lay external eggs. Viviparous animals (mostly mammals) do not lay eggs. Instead they retain the embryo inside the body during some or all of its development. Ovoviviparous animals (some fish, amphibians, and reptiles) produce an egg, but retain the egg within the body until it hatches. (2–5)

1. Choice *c*. In all amniotic eggs, including the modified version adapted for uterine development in mammals, the embryo is surrounded by the amnion which provides a protective cushion of fluid. (12)
2. Choice *a*. The blastocyst is one of the very first stages in human embryonic development, occurring early in the first trimester. (16–19)
3. Choice *d*. Eutherian mammals are also known as placental mammals, of which the cow is the only representative shown. All eutherians are viviparous. (3–4)
4. Choice *b*. The development of organs and organ systems, a process called organogenesis, occurs during the latter half of the first trimester. (16)
5. Choice *e*. Labor is brought on and intensified by all of these factors. (21)
6. Choice *a*. Trout are fish, and since fish reproduce in water they do not utilize an amniotic egg. (6, 10)

Integrative Questions

1. Choice *b*. All these animals can reproduce sexually. However, only the chicken lays an external egg.
2. Choice *a*. The corpus luteum forms after ovulation, which does not occur prior to puberty and sexual maturation in women.
3. Choice *d*. There is no zona pellucida present in the sea urchin egg.
4. Choice *c*. This is effectively what happens to a woman taking birth control pills.

CHAPTER 38 – NEURONS AND THE NERVOUS SYSTEM

Note : the page numbers listed with the section titles refer to the study guide; numbers in parentheses for each question refer to the relevant key concepts.

COMMUNICATION AND COMPLEXITY • CELLS OF THE NERVOUS SYSTEM • FROM STIMULUS TO RESPONSE: INFORMATION FLOW IN THE NERVOUS SYSTEM (pages 281–283)

1. Choice *a*. The autonomic portion of the vertebrate nervous system controls the function of involuntary organs such as glands. (1, 17)
2. Choice *d*. While it is true that chemical communication occurs at the synapse, the portion of the neuron that actually stores and releases the neurotransmitter chemicals is the axon terminal. (9–10)
3. Choice *c*. Nerves contain bundles of axons from many different neurons, along with connective tissue that holds the nerve together. (8, 15)
4. Choice *c*. A specialized category of glial cells called astrocytes wrap themselves around blood vessels in the brain tissue to form the blood–brain barrier. (11–13)
5. Choice *e*. All of these structures are located inferior to the telencephalon within the main axis of the vertebrate central nervous system. (18–22)
6. Choice *d*. The sensory stimulus of pain is triggered by actions in the periphery, but the sensation of pain itself is located in the brain. Interneurons gather information from the afferent elements of simple reflex loops controlling foot movements and send this information through the spinal cord to the brain. (16, 26)
7. Choice *b*. The occipital lobe, which is important in controlling vision, is located at the back of the cerebral cortex and would be impacted by such an injury. (30)
8. Choice *a*. The telencephalon, or forebrain, is relatively undeveloped in fish but increases in size in the more complex vertebrates. Fish therefore have the smallest ratio of telencephalon size to body size.

- (5–9)

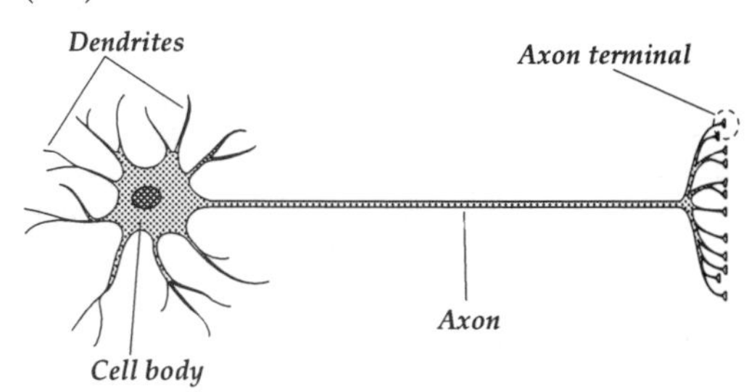

Dendrites receive incoming information from the axons of other neurons located elsewhere in the body. The cell body may also receive incoming information, but its main function is to house the nucleus and carry out metabolic activities necessary for cell function. The axon is the conduction zone of the neuron, sending action potentials to the axon terminal, which synapses with other neurons in the circuit or with some kind of effector cell.

- (18–22)

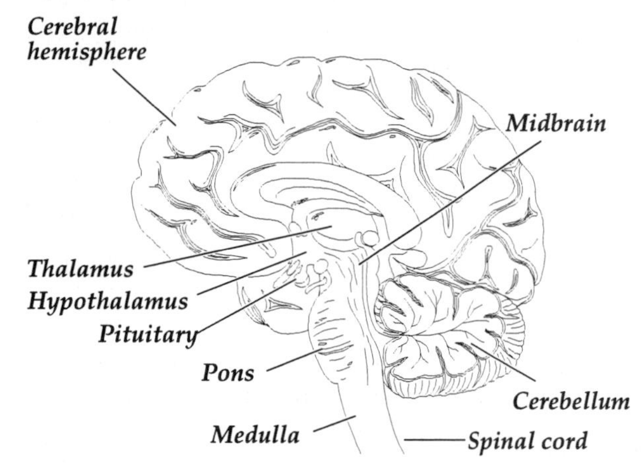

Cerebral hemisphere—site of sensory processing, learning, memory, and consciousness
Thalamus—relay station for sensory information going to telencephalon
Hypothalamus—regulates physiological functions and biological drives
Pituitary—endocrine master gland
Pons and *medulla*—control circulation and breathing functions
Cerebellum—coordinates and refines behavior patterns
Midbrain—processes aspects of visual and auditory information
Spinal cord—conducts and integrates information between brain and periphery

THE ELECTRICAL PROPERTIES OF NEURONS (pages 283–286)

1. Choice *d*. The interior of the cell is negative with respect to the outside at the resting membrane potential (i.e., it is polarized). Depolarization reduces this polarized state. This requires the influx of positive charge, which occurs as Na^+ ions enter the cell. (16)
2. Choice *b*. If an already polarized cell with a negative membrane potential is hyperpolarized, the potential becomes even more negative. (12)
3. Choice *d*. Voltage-gated potassium channels are responsible for the repolarization phase of an action potential (the downward swing in membrane potential). The depolarization phase (upward swing) would not be affected, however, since this results from the action of gated sodium channels. (16–17)
4. Choice *c*. Sodium activation gates open and inactivation gates close at the onset of an action potential. When the activation gates close again, the inactivation gates remain closed for several milliseconds before reopening. This temporarily prevents the sodium influx necessary to start another action potential. (18)
5. Choice *a*. Invertebrates, such as ants, do not have myelinated axons to increase conduction velocity. They achieve increases in the speed at which action potentials are propagated by increasing the diameter of the axon, not by decreasing it. (22–24)
6. Choice *e*. All of these characteristics could be true of such an axon. (23–25)
7. Choice *c*. Patch clamps use fine suction electrodes to isolate individual ion channels, hold the voltage constant, and study the current flow through these channels.

- (1, 2, 13–17)

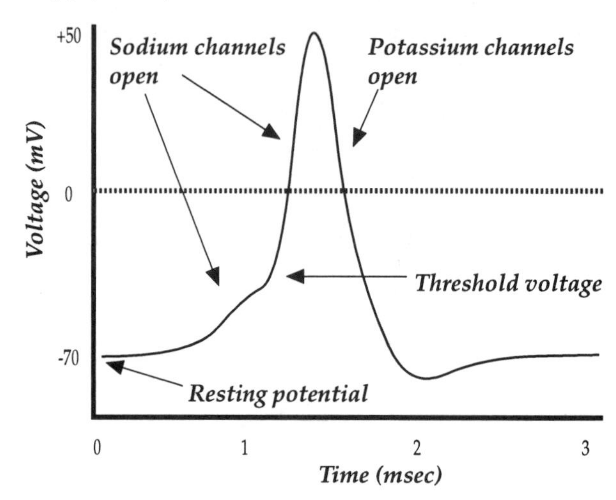

- The sodium–potassium pump of a neuron establishes ion concentration gradients across the cell membrane, with sodium more concentrated outside the cell and potassium more concentrated within. At rest, more K^+ ion channels are open than Na^+ ion channels, so there is a net loss of positive charge from the inside of the cell due to the diffusion of K^+ ions down their concentration gradient. This creates a relative negative charge inside the cell. This charge difference across the membrane is the resting potential.(4–9)

SYNAPTIC TRANSMISSION (pages 286–289)

- (3, 15–16, 19)
  d: Curare
  e: Acetylcholine
  a: Glycine
  b: Calcium
  c: Atropine
- Acetylcholinesterases are the enzymes that degrade the neurotransmitter acetylcholine in the synaptic cleft. This must occur or the transmitter will keep stimulating receptors in the postsynaptic membrane, which will keep the neurons continuously active, and, if sustained, will lead to spasm and death. By blocking the action of the cholinesterase enzyme, anticholinesterases promote uncontrolled nerve and muscle spasms. (2, 21)

1. Choice *a*. Neurotransmitters bind to membrane receptors on the postsynaptic cell. This causes specific gated ion channels to open, which may either depolarize or hyperpolarize the cell, depending on whether Na^+ or K^+ ions move through the membrane. (2)

2. Choice *d*. Dendrites and most of the cell body have few gated sodium channels. These channels mediate the production of action potentials by permitting the membrane to rapidly depolarize, a characteristic that is concentrated at the axon hillock and in the axon itself. (7–10)
3. Choice *c*. Atropine is a substance that blocks the inhibitory effects of muscarinic receptors in muscle tissue, especially the heart. (16, 19)
4. Choice *a*. The diagram shows that three different stimuli arrive at the same time and evoke a large EPSP which leads to an action potential, while less than three do not. This is spatial summation. Temporal summation involves the same stimulus arriving in rapid succession. (7–10)
5. Choice *b*. Excitatory postsynaptic potentials make it easier for an action potential to occur, so they must depolarize the postsynaptic membrane. (5, 7)
6. Choice *b*. Active transport is used by some axon terminals as a way to clear the synaptic cleft by gathering up released neurotransmitter chemicals. (22)
7. Choice *e*. Gap junctions are electrical synapses that physically connect neurons via membrane proteins. Electrical signals can thus travel in either direction. (13)

NEURONS IN CIRCUITS • HIGHER BRAIN FUNCTIONS (pages 289–292)

- (8–10)

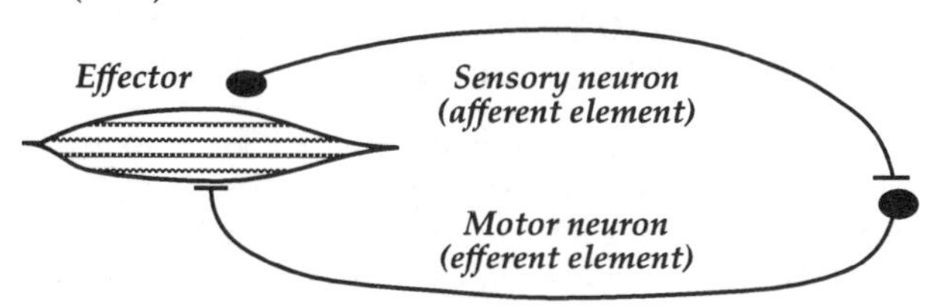

- Voluntary muscle movements are not possible during REM sleep, when dreaming occurs. Thus it is not possible to move the limbs and sleepwalk during this stage of sleep. Rapid contractions of the ocular muscles and their resulting eye twitches are the obvious facial characteristic from which REM sleep derives its name. (15–16)

1. Choice *a*. Sympathetic pathways are stimulatory for physiological responses associated with the fight-or-flight response, parasympathetic pathways are inhibitory. (2)
2. Choice *e*. The axons of sympathetic postganglionic cells terminate on effectors, which include all of the structures listed. (5)
3. Choice *c*. The basic knee-jerk response is controlled by a monosynaptic circuit. Interneurons are only involved in the voluntary or conscious modification of this reflex. (8–12)
4. Choice *e*. Dreams occur during REM sleep, which is characterized by all of these conditions. (14–16)
5. Choice *d*. Habituation is simple learning where a repeated stimulus is ignored if it contains no useful new information. This is the case with the sensory stimuli from the ticking clock, which is novel when you first arrive, but rapidly becomes "old news." (18)
6. Choice *e*. Memories do not seem to be located in any one portion of the brain. Rather, they appear to be spread across large regions of the cerebral cortex with many interconnecting pathways between these areas. (22)
7. Choice *b*. Different aspects of language production and use reside in different cortical regions such as Broca's and Wernicke's areas. The brain pathways of patients with aphasias have trouble exchanging information between these areas. (25–28)

Integrative Questions

1. Choice *e*. Sensory information from vision and touch, as well as motor control to move your arm and mouth, have peripheral elements. The mental recognition of chocolate and the decision to eat it are central processes. The parasympathetic branch of the autonomic nervous system stimulates salivation.
2. Choice *c*. Since the action potential started in the middle of the axon, rather than at the axon hillock as typically occurs, there is no previous patch of membrane currently undergoing a refractory period. The action potential is thus free to propagate in both directions.
3. Choice *a*. The primary visual cortex, located in the occipital lobe at the back of the brain, is where the visual signal of a written word is first decoded via higher brain functions like memory and recognition. The decoded information is then sent to the angular gyrus and on to Wernicke's area.
4. Choice *b*. Hyperpolarizing neuron B by 15 mV would make its membrane potential –65 mV, actually taking it further away from the threshold voltage needed to fire an action potential.

CHAPTER 39 – SENSORY SYSTEMS

Note : the page numbers listed with the section titles refer to the study guide; numbers in parentheses for each question refer to the relevant key concepts.

WHAT IS A SENSOR? • CHEMOSENSORS (pages 293–295)

- Receptor potentials are the result of a physical stimulus being transduced into an electrical signal. The stimulus causes changes in the ion permeability of the sensor membrane to yield the receptor potential, the size of which is directly proportional to the intensity of the stimulus. In sensors that produce an action potential directly, the receptor potential is called a generator potential. In other sensors, the receptor potential causes the release of neurotransmitter onto an associated sensory neuron, which then produces the action potential. (5–7)
- Both taste and smell involve the use of olfactory sensors. The hairs of these sensors extend from the olfactory epithelium into the sinus cavity and are normally covered by a thin layer of mucus, through which odorant molecules must diffuse to make contact with receptors. A cold typically increases the thickness of this mucus layer, making it more difficult for receptors and molecules to combine and begin the transduction process. (11, 19)

1. Choice *a*. Transduction is the conversion of a physical stimulus (energy) into electrical impulses (action potentials) which the nervous system can use for communication. The only answer that meets this criterion is the first one. (2, 3)

2. Choice *b*. Adaptation occurs when a sensor cell ceases its response to an unchanging, steady-state stimulus, such as the one described in this situation. The action potentials recorded in the sensory nerve are a direct reflection of the activity in the sensor cell. (8)
3. Choice *c*. Human taste sensors are not neurons. Instead, they produce receptor potentials in response to appropriate chemical stimuli. This causes the release of neurotransmitters onto adjacent sensory neurons which fire the action potentials. (16)
4. Choice *b*. cAMP is the second messenger in this signalling system. It is produced in the presence of a chemical stimulus and causes gated sodium channels to open. This results in an influx of Na^+ ions, which causes the membrane potential to become more positive. The cell depolarizes as a result. (12)
5. Choice *d*. When you have nasal congestion, excess mucus covers the olfactory sensors and makes your sense of smell less efficient. Since the odors perceived while eating strongly influence our sense of taste, a stuffed-up nose causes us to perceive that the food has lost much of its flavor. (11, 19)
6. Choice *e*. Humans actually perceive at least four categories of taste. The one missing from this list is salty. (14–19)
7. Choice *b*. Although technically a type of hormone, since pheromones are used externally to the body they have their own name. (10)
8. Choice *e*. All sensors are capable of transduction in some fashion. (2)

MECHANOSENSORS (pages 295–298)

1. Choice *d*. Meissner's corpuscles are extremely sensitive touch receptors in the skin. They achieve transduction through the direct deformation of the sensor cell membrane, rather than through the bending of stereocilia as is the case with hair cells. (3)
2. Choice *b*. The malleus is the first of the three ear ossicles found in the middle ear of vertebrates; it attaches directly to the tympanic membrane. (13)
3. Choice *c*. All mechanosensors (indeed, *all* sensory cells) ultimately have their effects by creating changes in the ion permeability of cell membranes. This either leads directly to the production of an action potential, or indirectly to action potentials by way of receptor potentials. (1, 2)
4. Choice *a*. Statocysts and the vestibular system both rely on hair cells to control balance and body position. Expanded-tip tactile receptors and Pacinian corpuscles are both touch receptors in the skin. (3–4, 8–9)
5. Choice *e*. Hair cells are an important class of mechanoreceptors that do not qualify as neurons because they do not fire action potentials. (6)
6. Choice *b*. The proximal end of the basilar membrane (the end nearest the middle ear) responds only to high-frequency sound. Permanent damage to this region of the cochlea would impact the man's ability to hear high-pitched tones. (17–18, 21–22)
7. Choice *e*. The auditory nerve makes synaptic connections with all of these brain areas. (19, 20)
8. Choice *c*. Meissner's corpuscles are sensitive to light touch, and so adapt quickly to a continuous unchanging stimulus.

- (4, 5, 7–10, 16)
  d: Organ of Corti
  e: Pacinian corpuscle
  b: Golgi tendon organ
  f: Statocyst
  c: Vestibular apparatus
  a: Lateral line
- The semicircular canals are fluid-filled chambers which respond to head movements. As the fluid moves within a canal, it moves an ampulla and bends hair cells, which causes the production of action potentials. The vestibular apparatus also detects head position, acceleration, and deceleration. Otoliths rest upon sensory hairs in the chambers of this system, and when the head moves so do the otoliths. This causes hairs to bend and action potentials to be produced. (9–10)
- X will be perceived as a relatively high-pitched sound, Y a mid-range pitch, and Z a low-pitched sound. This is because specific regions of the basilar membrane only vibrate in response to certain frequencies. High-frequency sounds travel least far within the cochlea, and so stimulate vibration in the basilar membrane proximal to the oval window. Lower-frequency sounds travel farther within the cochlear fluid, and therefore cause more distal portions of the basilar membrane to vibrate. (17, 18)

PHOTOSENSORS AND VISUAL SYSTEMS • OTHER SENSORY WORLDS (pages 298–302)

1. Choice *e*. The photopigment used by all animals is rhodopsin, which contains the protein opsin. (1–2)
2. Choice *b*. There should be a blue indicator whenever the 11-*cis* isomeric form of retinal is present. This will only occur prior to the breakdown of rhodopsin, that is, before the pigment has been exposed to light. (2)
3. Choice *e*. Action potentials are produced only by ganglion cells in the retina. (21–23)
4. Choice *d*. cGMP is the substance that keeps sodium channels open in photoreceptors during the resting state. When stimulated by light, a chain of biochemical events causes cGMP concentrations to drop. This lets the sodium channels close, and as a result the membrane hyperpolarizes. (3–5)
5. Choice *c*. Retinula cells are found in the ommatidia of arthropod compound eyes. (9, 11–14)
6. Choice *e*. Each of these cell layers overlies the photoreceptors of vertebrates, which means that a photon of light must pass through them all before striking the rods or cones. (19)
7. Choice *d*. One of the most important functions of the lens and its associated ciliary muscles is to change the len's shape or position to permit proper focusing of light on the retina. This process is called accommodation. (13)
8. Choice *a*. The tapetum is a reflective layer at the back of the eye that helps nocturnal animals capture the maximum amount of light. Capturing light causes rhodopsin to bleach, not regenerate. (20)
9. Choice *b*. This is the definition of the receptive field of a simple cortical cell. Answer *a* describes the receptive field of a complex cell. (28)

• (11–15)

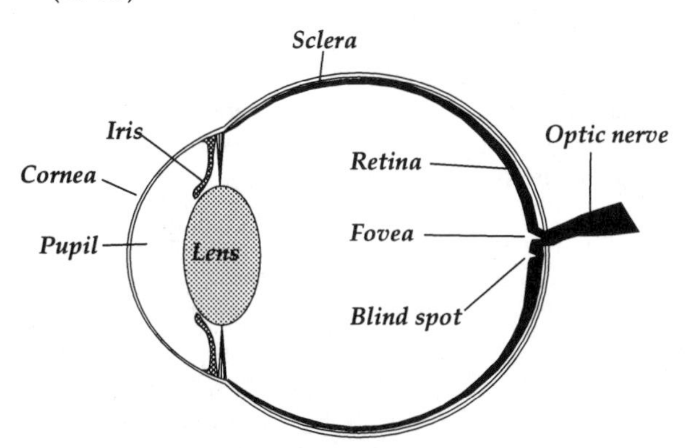

• You are told that this is an off-center receptive field, which means that the surround will be excited by light and the center will be inhibited. Thus, condition #1 will produce a brief burst of action potentials for the duration of the stimulus. Condition #2 will either produce no response at all from the cell, or will cause it to cease the production of any spontaneous action potentials it was generating. Condition #3 will begin with lots of action potentials until the spot hits the center, at which time the production of action potentials will stop. (24–27)

Integrative Questions

1. Choice *c*. Echolocation is used by bats and dolphins to create acoustic images by reflection. Snakes do not make these kinds of high-intensity, high-frequency sounds, nor are they capable of hearing the echoes produced.
2. Choice *b*. Mechanoreception is the transduction mechanism for all of these stimuli.
3. Choice *b*. The vestibular system uses hair cells and otoliths to help maintain balance. After a day at sea, the hair cells have become overstimulated and continue to give the impression of rocking motions.
4. Choice *a*. Since chemosensors are the most ancient and commonplace of all sensor cells, they are what you would certainly expect to find. All the others listed would probably not be necessary for the animal to survive, although they could be present as well.

CHAPTER 40 – EFFECTORS

Note : the page numbers listed with the section titles refer to the study guide; numbers in parentheses for each question refer to the relevant key concepts.

HOW ANIMALS RESPOND TO INFORMATION FROM SENSORS • CILIA, FLAGELLA, AND MICROTUBULES • MICROFILAMENTS AND CELL MOVEMENT (pages 303–305)

• The cilia of the respiratory tract move mucus toward the oral cavity. This mucus traps airborne particles and keeps them from possibly damaging the delicate respiratory tissue. Mucus movement is made possible because adjacent microtubules inside the cilia slide past one another and cause the cilia to bend. The power stroke produced by thousands of cilia working in coordinated fashion propels the mucus up and out of the respiratory tract. (3–5, 8–9)

• (4, 7–8, 10)
b: Axoneme
d: Recovery stroke
e: Myosin
a: Dynein
c: Pinocytosis

1. Choice *c*. Energy released from the splitting of a phosphate bond in ATP powers the conformational change of dynein, which causes the microtubules to slide past one another. (8)
2. Choice *b*. Plasmagel is a special category of cytoplasm that influences the amoeboid movement of some cells. It is not a structural component of a cilium. (11)
3. Choice *a*. Although cell movements are one of the most important functions of effectors, they are also involved in other activities such as chemical release and the emission of various forms of energy. (1)
4. Choice *d*. The movement of flagella is controlled mainly by microtubules. (7, 10)
5. Choice *e*. Axoneme movement is produced by the chemical events that cause microtubules to slide past each other. (8)
6. Choice *e*. Since no cross links or spokes are present, when the microtubules slide by one another they elongate the entire axoneme, rather than cause it to bend. (8)
7. Choice *c*. Like the previous question, severing the connections between the cell and the cilium eliminates bending, but does not stop the microtubules from sliding past one another. (8)

MUSCLES (pages 305–307)

1. Choice *a*. All digits and limbs carry out voluntary motion and are controlled by skeletal muscles. (6)
2. Choice *d*. Microfilaments are the contractile elements that make up a muscle cell or myofibril. These are then bundled to form whole muscles. (6–7)
3. Choice *d*. Microfilaments are the basic contractile elements found in all muscle. (2)
4. Choice *b*. Since the A band represents the myosin, its size cannot change. The Z lines are the end points of the sarcomere, where the actin microfilaments attach. As actin slides past the myosin from both directions towards the middle of the sarcomere, the Z lines approach each other. (9–11)
5. Choice *c*. Tropomyosin is the double-stranded protein that covers the actin binding sites for myosin. Troponin is the other protein to which the calcium ions bind. (12, 18)
6. Choice *c*. The action potential is initiated on the plasma membrane at the neuromuscular junction. It then spreads through the cell via the T-tubule system before reaching the triads that initiate the opening of the Ca^{2+} channels of the sarcoplasmic reticulum. (16)
7. Choice *e*. Calcium binding to troponin induces a conformational change that moves the attached tropomyosin strands away from the actin binding sites. (17–19)
8. Choice *b*. The continuous current from a live wire would cause very rapid volleys of action potentials to all motor units, producing a tetanus response. (21–23)

9. Choice *e*. Marathon running is an aerobic activity. The muscle best suited for this work is dark muscle, because it has many slow twitch fibers for endurance and abundant mitochondria to fuel the activity of these fibers. (24–25)
10. Choice *e*. Both of these characteristics are important for the function of cardiac muscle. (28)

- (10, 11)

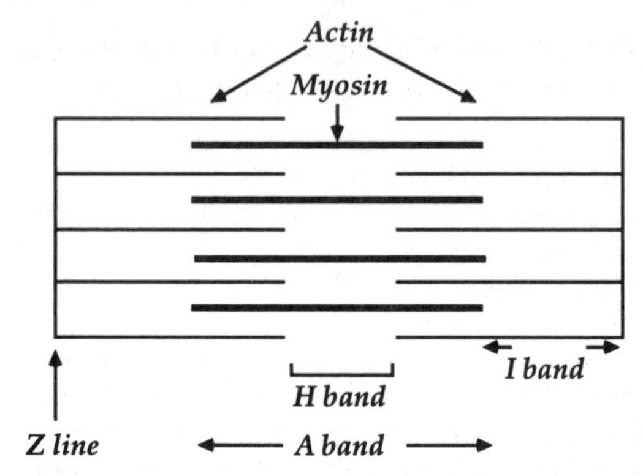

- (12, 13)

The sarcomere has shortened because the actin microfilaments have slid past the myosin filaments and towards each other.

SKELETAL SYSTEMS • OTHER EFFECTORS (pages 307–310)

- (7, 15, 22, 28–29)
  a, c: Osteoblast
  e: Nematocyst
  b, d: Extensor
  d: Exoskeleton
  d: Chromatophore
- The elongated limbs indicate the presence of long load arms and short power arms in the levers that make up the legs; these are adaptations for speed. In addition, the sleek aerodynamic body design means light weight, also an adaptation to increase speed. (26)
- The bones of the vertebral column and the skull (cranium, maxilla, mandible) are axial. All others are part of the appendicular skeleton. (10)

1. Choice *a*. Skeletal muscle is usually what makes skeletons move, not cardiac or smooth muscle. (2, 5, 9)
2. Choice *d*. By contracting either its circular or longitudinal muscles, a sea anemone can change the shape of its hydrostatic skeleton just as the proper squeezing of a water balloon causes it to elongate or shorten. (2)
3. Choice *e*. The main difference has to do with the relative location of muscles and skeletal elements. The skeleton is outside of the muscles in an exoskeleton, and on the inside for an endoskeleton. (5–9)
4. Choice *c*. The appendicular skeleton includes all bones associated with the appendages. The jaw bone is part of the skull, which belongs to the axial skeleton. (10)
5. Choice *e*. All of these cells are involved in the process of regulating extracellular calcium concentration. (15, 16)
6. Choice *c*. Since ligaments hold bones together, he has probably suffered a ligament tear. (23)
7. Choice *d*. Vertebrae are part of the axial skeleton, and so cannot be a category of appendicular skeleton injuries. (10)
8. Choice *b*. Effectors that release communication chemicals are the most common effector organs not related to movement. (28–32)
9. Choice *a*. The butterfly has an exoskeleton. All the other animals listed have some kind of hydroskeleton, including the snail, which also has an exoskeleton covering its hydroskeletal parts. (2–7)

Integrative Questions

1. Choice *e*. The antagonists in a hydrostatic skeleton are opposing circular and longitudinal muscles; contraction of one elongates the animal, contraction of the other shortens it.
2. Choice *b*. Electric organs are useful in aquatic environments and are known only from certain species of fish.
3. Choice *a*. The fact that flagella create whiplike motions is due simply to their length compared to a that of a typical cilium; the molecular mechanisms for how the movement is produced are the same for both types of effectors.
4. Choice *c*. Only cardiac muscle cells have the unique potassium channels that give this type of muscle its myogenic properties.

CHAPTER 41 – GAS EXCHANGE IN ANIMALS

Note : the page numbers listed with the section titles refer to the study guide; numbers in parentheses for each question refer to the relevant key concepts.

GAS EXCHANGE IN ANIMALS (pages 311–313)

- Barometric pressure is lower on top of Mt. Everest because of its higher altitude. Thus, P_{O_2} is also lower on Mt. Everest. Lower P_{O_2} makes diffusion, and therefore respiration, more difficult. (11–12)
- Diffusion is the primary rate-limiting step in gas exchange. The specifics of how well certain gases diffuse in particular environments, as well as the effects of pH, temperature, and other physical factors also influence diffusion characteristics. Gas-exchange systems are usually designed to maximize diffusion. (5–9)

1. Choice *b*. Temperature usually has little or no effect on gas exchange for air breathers, but can be of major importance to ectothermic water breathers. (10)
2. Choice *a*. The amount of O_2 that a given volume of water can hold does, in fact, decrease with increasing temperature, which is why ectothermic water breathers can suffer ill effects as the water warms. (10)
3. Choice *e*. CO_2 dissolves easily in water, so getting rid of this gas is not a problem for water breathers. (13)
4. Choice *d*. Oxygen makes up 20.9 percent of the atmosphere. Its partial pressure at 5800 meters is therefore

20.9 times the barometric pressure at this altitude (400 mm Hg), or 83.6 mm Hg. (11–12)

5. Choice *c*. Large, metabolically active animals cannot survive without specialized respiratory structures to provide adequate gas exchange. (8–9)

RESPIRATORY ADAPTATIONS (pages 313–316)

- *a*. Maximize the surface area for exchange between air and water
  b. Use thin membranes for the exchange surfaces
  c. Cause water and air to flow across the exchange surfaces in opposite directions (i.e., use a countercurrent flow) (4, 9)
- (2, 6, 13, 22, 25)
  c: Internal gills
  a: Air capillaries
  e: Air sacs
  b: Surfactant
  d: Alveoli
- (9, 10)

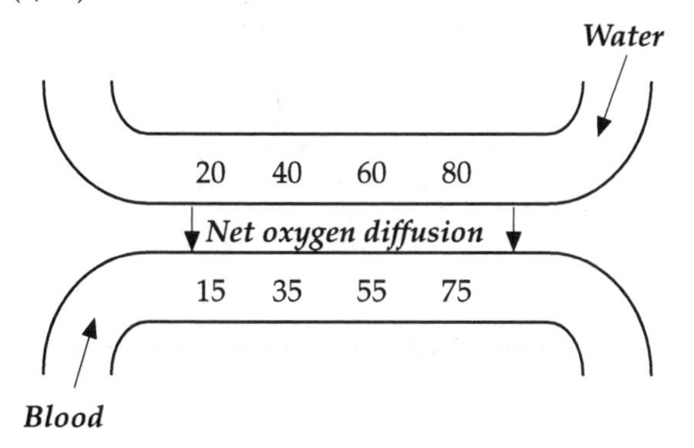

The tube with the higher O_2 saturation must be the oxygen source (i.e., the water); the other is therefore the blood. The two flows are antiparallel, establishing a countercurrent exchange that greatly enhances the net diffusion of O_2 from the water into the blood along the entire length of the tube.

1. Choice *a*. This set of characteristics is common to all respiratory organs. (2–6)
2. Choice *c*. Spiracles are the abdominal openings to the tracheal system of insects. The others are part of gill or lung systems. (6)
3. Choice *d*. O_2 moves into the fish's mouth, over the gill filaments and lamellae, across the respiratory membranes, and then into the blood capillaries. (7–9)
4. Choice *b*. Fish and birds both utilize countercurrent (or crosscurrent) exchangers in their respiratory systems. Elephants and people both use tidal breathing. All other combinations listed use contrasting respiratory mechanisms. (10, 11)
5. Choice *d*. The inhalation phase of bird breathing is when the air sacs are filling. (14)
6. Choice *b*. Salmon, being fish, use countercurrent exchange for breathing. All the others listed are air breathers, which use tidal mechanisms. (11)
7. Choice *e*. The resting tidal volume is the amount of air moved in and out of the lungs during normal breathing. All other lung volumes listed are larger, as indicated in Figure 41.12 of your textbook. (17–18)
8. Choice *c*. Inhalation is the active part of breathing in mammals, dependent upon the contraction of the diaphragm and intercostal muscles. Exhalation is usually passive. (28–30)
9. Choice *a*. The air conduits are dead space, while the alveoli and capillaries are involved in gas exchange. The arterioles and venules carry blood to and from the capillaries, respectively, but are not directly involved with diffusion. (19, 22)

TRANSPORT OF RESPIRATORY GASES BY THE BLOOD • REGULATION OF VENTILATION (pages 316–318)

- (4, 7, 16, 17, 20)
  e: Carbonic anhydrase
  a: Myoglobin
  b: Oxygen dissociation
  d: Carboxyhemoglobin
  c: Medulla
- *A* = llama, *B* = adult human, *C* = adult human at high altitude. Llamas and humans make opposite kinds of adjustments in the oxygen-binding properties of their hemoglobin for life at high altitude, although the end result of making more O_2 available to their tissues is the same. The llama's hemoglobin can saturate at very low P_{O_2} values (a left shift in the curve), while the human's hemoglobin changes shape under the influence of diphosphoglyceric acid to more easily release O_2 (a right shift in the curve). (4, 11, 13)

1. Choice *c*. Hemoglobin contains four polypeptide chains. Myoglobin has only a single chain with high oxygen affinity. (2, 3, 9, 10)
2. Choice *e*. All the listed items are commonly found suspended or dissolved in mammalian blood plasma. (2, 16)
3. Choice *d*. A left shift in the oxygen-dissociation curve means that the pigment operates at lower P_{O_2} values. (7)
4. Choice *c*. Changes in pH due to the accumulation of acidic metabolites resulting from increased cell activity are the basis of the Bohr effect. (12)
5. Choice *e*. By changing the shape of hemoglobin, diphosphoglyceric acid makes it more likely that hemoglobin will release its O_2 cargo. This is an adaptation for increased O_2 availability which occurs during sustained exercise or acclimation to high altitude. (13)
6. Choice *a*. CO_2 is rapidly removed from mammalian lungs by exhalation, thus keeping its concentration low in the alveolar air spaces. (18)
7. Choice *d*. The contraction state of the diaphragm is not directly involved in the *neural* control of ventilation. All the other structures listed either sense, relay, or act upon stimuli and neural information related to ventilation. (20–21, 24)
8. Choice *a*. Cutting the brain stem below the medulla eliminates neural contact with the respiratory mechanism and could easily result in a complete stoppage of breathing. (20)

Integrative Questions

1. Choice *b*. In order to avoid drying, external gills can only be present in aquatic animals. Fish, although aquatic, have internal gills.

2. Choice *c*. Birds have increased their respiratory efficiency by using unidirectional air flow rather than tidal respiration.
3. Choice *a*. The mammal evolved for life at the highest altitude should have the most left-shifted dissociation curve for hemoglobin, since it must pick up and release oxygen at very low P_{O_2} values. This would be the one living on top of Mt. Kilimanjaro.
4. Choice *e*. Refer to Fig. 41.13 in your textbook for scale.

CHAPTER 42 – INTERNAL TRANSPORT AND CIRCULATORY SYSTEMS

Note : the page numbers listed with the section titles refer to the study guide; numbers in parentheses for each question refer to the relevant key concepts.

TRANSPORT SYSTEMS (pages 319–322)

1. Choice *d*. Butterflies are insects, which have open circulatory systems. They have only a minimal number of short, open-ended transport vessels; capillaries are not present. (8–10)
2. Choice *e*. None of the possible choices are found in both hydras and chickens, which have very different types of circulatory systems. (1–4, 14–18, 25)
3. Choice *a*. One of the advantages of a closed circulatory system is that it can serve as a transport system for large, multicellular animals whose body size is much too big for exchange to occur by simple diffusion alone. Diffusion occurs between individual cells and the blood only after the required materials have traveled through the transport system from another location, such as the outside world or the digestive tract. (8, 11–12)
4. Choice *c*. Arteries are large-diameter, high-volume vessels which carry blood away from the heart. They are therefore most likely to carry the greatest amount of blood during any given time period. (16–17)
5. Choice *b*. The two-chambered fish heart is the evolutionary precursor of the three-chambered hearts of amphibians and reptiles. (18, 23)
6. Choice *d*. The ventricles and arteries are the places where blood pressure is highest. The systemic ventricle of a bird or mammal would probably have the highest pressure of all, since it is a large muscular pump, completely distinct from the pulmonary ventricle, and designed to move blood to all parts of the body except the lungs. (25–26)
7. Choice *e*. Blood traveling through either the systemic or pulmonary circuits would make its way through all possible types of vessels before returning to the heart. (15–17)
8. Choice *b*. While it is true that blood leaving the heart through the systemic circuit in birds and mammals is oxygenated, the blood returning from the body tissues via the systemic circuit is mostly deoxygenated. (15, 25–26)
9. Choice *a*. Since there is no oxygen entering the respiratory system when the reptile is not breathing, blood flow to the lungs is greatly reduced until breathing resumes. (24)

- (9, 16–18)
  a: Ventricle
  d: Ostium
  e: Atrium
  f: Venule
  b: Capillaries
  c: Arterioles
- The open circulatory system of an insect consists of a muscular pump (heart) that propels interstitial fluid through dorsal vessels leading to the anterior and posterior ends of the body. Fluid flows out through smaller vessels along the way and then percolates through the body tissues. The fluid pools up in special sinuses on the ventral side of the animal and reenters the heart via valved holes called ostia. An earthworm has a closed system, in which the blood remains inside vessels throughout its journey around the body. Finely branched capillaries take blood to within one cell distance of every portion of the body before returning it back to the heart. (8–12)
- (15–18, 25)

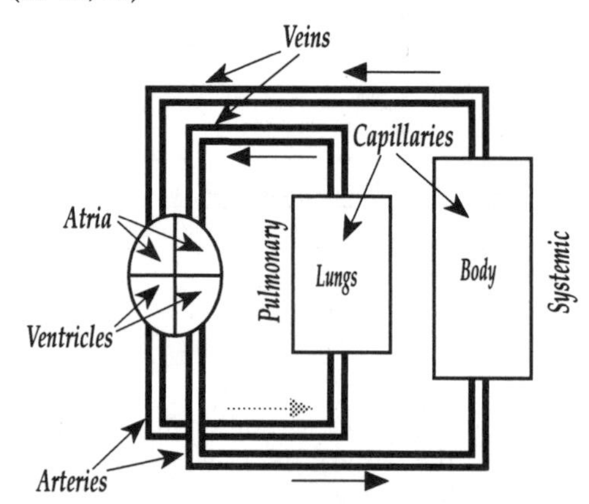

The systemic circuit delivers blood to the body tissues and then returns it to the heart. The pulmonary circuit sends blood to the lungs and then returns it to the heart. Blood always leaves the heart from the ventricles by way of arteries, and returns to the heart at the atria by way of veins. Capillaries could be found both in the lungs and the body tissues.

THE HUMAN HEART • THE VASCULAR SYSTEM (pages 322–325)

1. Choice *e*. All of the characteristics listed apply to the vena cavae. (3, 29)
2. Choice *d*. The numerator in this ratio is the systolic pressure, which occurs when the ventricle is contracting (i.e., when the atrium is relaxing). The denominator is the diastolic pressure, which results when the ventricle relaxes. "Normal" values for a young adult human would be about 120/80 mm Hg, so the diastolic pressure appears to be rather low. (5)
3. Choice *a*. Tachycardia refers to a condition of excessively high heartbeat rate. (8)

4. Choice *c*. These nodes cause the atria and ventricles to contract, respectively. If stimulated simultaneously, they would likely produce simultaneous contractions of all four chambers. This would not promote the efficient movement of blood through the heart, pointing out why it is important for there to be a delay between atrial and ventricular contractions. (9–12)
5. Choice *b*. A stroke occurs when the small vessels become blocked through an embolism and cut off oxygen flow to the brain. Embolisms and strokes do not occur in the arteries. (16–18)
6. Choice *a*. Capillaries directly link arterioles with venules, not arteries with veins. (21–24)
7. Choice *c*. Although the blood and lymph are chemically quite similar, the lymph has somewhat less protein and no red blood cells at all. (27–28)
8. Choice *b*. Stimulation by the parasympathetic branch of the autonomic nervous system tends to slow heartbeat rate, which would not help the movement of venous blood back to the heart. (30–33)
9. Choice *d*. Histamine changes the permeability of capillaries and venules, causing them to lose water to the body tissues. This condition is known as edema. (24–25)

• (1–3)

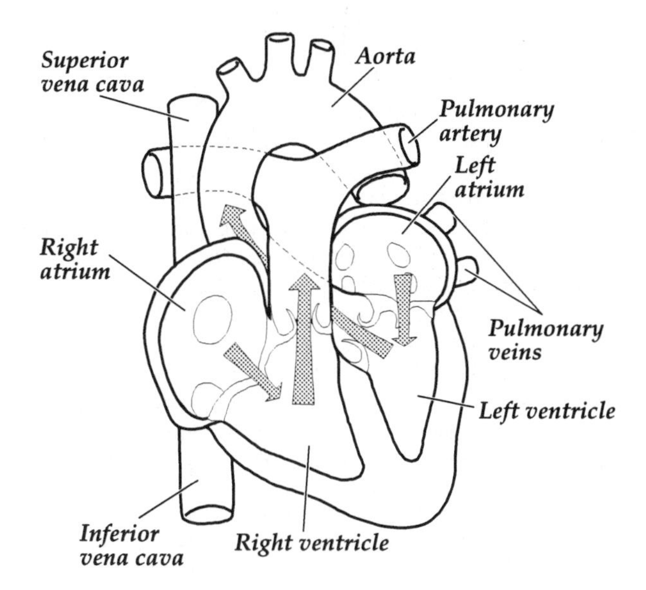

• (1–3)

(1) left ventricle → (2) aortic valve → (3) aorta → (4) systemic arterioles → (5) systemic capillaries → (6) systemic venules → (7) vena cavae → (8) right atrium → (9) atrioventricular valve → (10) right ventricle → (11) pulmonary valve → (12) pulmonary artery → (13) pulmonary arterioles → (14) pulmonary capillaries → (15) pulmonary venules → (16) pulmonary vein → (17) left atrium → (18) atrioventricular valve

• (5)

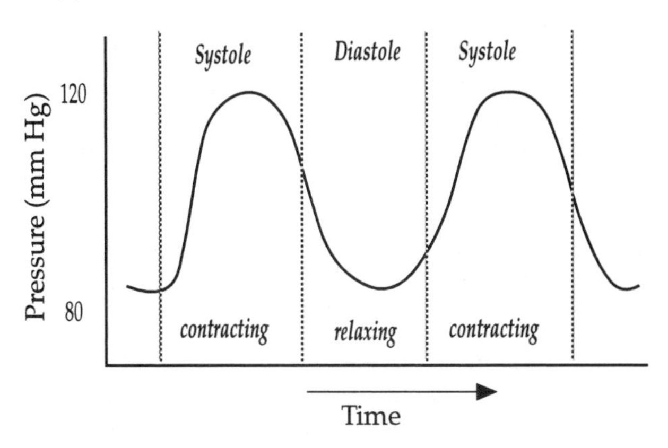

THE BLOOD • CONTROL AND REGULATION OF CIRCULATION (pages 325–327)

1. Choice *e*. New red blood cells are made by the stem cells of the bone marrow in response to the release of erythropoietin from the kidney. (4)
2. Choice *b*. This description matches that of platelets, another cellular component of the blood. (6–9)
3. Choice *c*. The activation of platelets causes circulating prothrombin, which is inactive, to convert to its active form, thrombin. This in turn activates fibrinogen, which causes the polymerization of fibrin threads. (9–11)
4. Choice *e*. All of these parameters influence the autoregulatory mechanisms of cardiovascular control. (14)
5. Choice *d*. Angiotensin and vasopressin cause the constriction of arterioles, not venules. (17–20)
6. Choice *a*. During the diving response of marine mammals, heartbeat rate is slowed and peripheral blood flow is restricted in order to shunt O_2 to vital internal organs. (21)
7. Choice *d*. The precapillary sphincters control the size of the capillaries. If dilated, more blood will flow through the capillary beds. (16)
8. Choice *e*. Information from all of these sources helps the medulla to regulate cardiovascular function. (19–20)

• (1–12)
d: Plasma
c: Erythrocyte
f: Leukocyte
e: Platelet
b: Thrombin
a: Albumin
g: Hematocrit

• When mammals dive, they are unable to breathe underwater. Furthermore, it may take some time for them to find food. The diving reflex therefore allows these animals to conserve oxygen by decreasing heart rate and reducing blood flow to all but the most vital internal organs such as the brain, heart, and kidneys. They breathe heavily upon returning to the surface in order to relieve the oxygen debt built up during the dive. (21)

Integrative Questions

1. Choice *c*. Cellular and hormonal blood composition would be basically the same for all mammals. However, the O_2 concentration in veins (vena cavae) and arteries (aorta) would be quite different.
2. Choice *b*. Clams are mollusks and have an open circulatory system. They do have a pumping heart, but since there are no lungs, pulmonary valves would not be present in this heart. The clam heart pumps interstitial fluid, which lacks red blood cells.
3. Choice *a*. Hemoglobin is contained in erythrocytes. Leukocytes have immune and other defense functions within the body.
4. Choice *d*. As blood pools in the venous system, less of it is available to be oxygenated and sent to the brain. If this condition persists for too long, fainting can occur.

CHAPTER 43 – ANIMAL NUTRITION

Note : the page numbers listed with the section titles refer to the study guide; numbers in parentheses for each question refer to the relevant key concepts.

DEFINING ORGANISMS BY WHAT THEY EAT • NUTRIENT REQUIREMENTS (pages 328–331)

- (1, 5, 14, 19, 29, 33)
  f: Rickets
  e: Malnutrition
  c: Vitamin C
  g: Heterotrophy
  a: Digestion
  d: Undernourished
  b: Beriberi
- No. Overnourishment results from a caloric intake in excess of metabolic needs, while undernourishment is just the opposite. (14, 16, 19)
- Yes. It is possible to be overnourished and at the same time malnourished, since malnutrition is caused by the lack of essential dietary components, regardless of the number of calories being consumed. (14, 16, 19)

1. Choice *b*. All five organisms are heterotrophs. Only the mushroom (a saprophyte) is a non-predator. The termite and the bison are both herbivores. Of the two carnivores, only the blue whale is a filter feeder. (1, 3, 4)
2. Choice *e*. Sheep, being animals, ingest their food. Because they prey on plants they are also herbivores. Since the only animals that do not carry out extracellular digestion are the sponges and a few specialized internal parasites, this characteristic also applies. (1–6)
3. Choice *a*. The chipmunk's daily caloric intake is 50 kcal less than its total metabolic rate, making it slightly undernourished. The lack of thiamin produces a condition of malnutrition. (10–11, 14, 16, 19, 33)
4. Choice *c*. Certain amino acids are essential because the animal lacks a biochemical means of converting its available carbon skeletons into these particular amino acids. They must therefore be acquired preformed in the diet. (20–21)
5. Choice *d*. Kwashiorkor is caused by a dietary deficiency of protein. (32)
6. Choice *e*. (26)
7. Choice *b*. Fat-soluble vitamins cannot dissolve in the blood or urine (i.e., water). Any excess therefore accumulates in the body tissues. (31)
8. Choice *a*. Carbohydrate reserves are the first to be exhausted during starvation, often disappearing in 1–2 days. (14; and textbook Fig. 43.3)

ADAPTATIONS FOR FEEDING • DIGESTION (pages 331–333)

1. Choice *d*. Filter feeders range in size from microscopic zooplankton to the largest animal that has ever lived, the blue whale. Likewise, a wide range of mobility exists in filter feeders, ranging from completely sessile barnacles to fast-swimming fishes. (4–5)
2. Choice *b*. The absence of any flattened molars or premolars, which function in grinding or chewing, suggests that the toothed whales rip and tear their food and swallow it whole. (8)
3. Choice *e*. Actually, just the opposite is true; the animal prey of carnivores is much more elusive and resistant to being captured than are the plants on which herbivores feed. This requires a wider diversity of adaptations for prey capture in the carnivores. (3)
4. Choice *c*. The stomach stores food (and performs some digestion too) before passing it on to the intestines. The small intestine (midgut) finishes the digestion and carries out most of the nutrient absorption, while the large intestine (hindgut) reabsorbs water and ions. (15–18)
5. Choice *a*. Since most of the absorption of nutrients occurs in the intestine, this is the structure that would be most severely impacted. (17)
6. Choice *d*. Dentine is a component of teeth, which hummingbirds lack. The other materials listed are present in some form in all animal digestive systems. (7)
7. Choice *e*. The meal described contains carbohydrates, proteins, and fats. Therefore, the enzymes that digest these materials will all be at work in the stomach. (21–22)
8. Choice *b*. Both of these structures improve absorption, either by increasing surface area or by slowing down the passage of digested food through the intestine. (20)

- The *foregut* includes the mouth and buccal cavity (food gathering, mechanical breakdown, initial digestion), the gizzard (mechanical breakdown), stomach and crop (storage and digestion). Note that not all of these structures are present in all animals.
  The *midgut* is the small intestine, which performs most of the digestion and virtually all of the nutrient absorption.
  The *hindgut* includes the large intestine, rectum, and anus. Water and ions are reabsorbed in the large intestine, and the feces condensed. The muscular rectum stores the feces until they are excreted through the anus. (15–19)
- There are a number of structural features evident in the dentition of these two species that reflect differences in feeding habits. Humans are omnivores, and our teeth are therefore generalized to deal with a variety of food sources. We have canines and incisors to bite and tear meat, but they are not well adapted for use as actual weapons. We also have molars and premolars to grind and chew plant foods. The saber-tooth cat had massive canines for subduing and killing its prey, and large

incisors for ripping meat from bones. Its premolars and molars were also adapted for meat eating and bone crushing, rather than grinding up plant matter. (6–9)

STRUCTURE AND FUNCTION OF THE VERTEBRATE GUT (pages 333–336)

- (1–5)

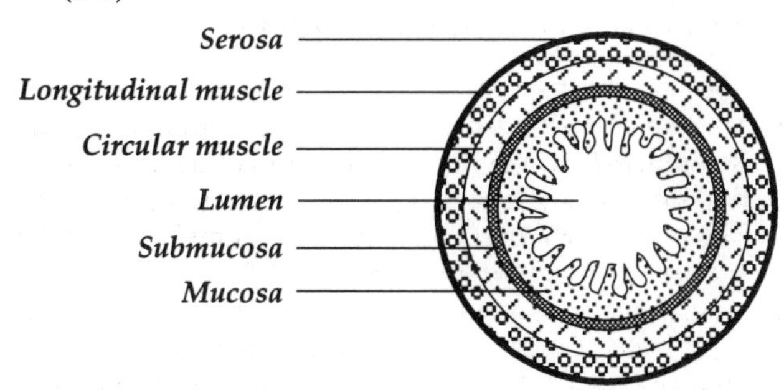

- The first two chambers (rumen and reticulum) of the multichambered digestive system in a ruminant mammal act as fermentation vats for cellulose digestion. Since so much plant matter must be ingested in order to meet the nutritional demands of these often very large animals, a large storage chamber is needed. Also, by having a large volume of cellulose material fermenting at any one time, large colonies of microorganisms can be sustained to produce the cellulases needed to make this type of digestion possible. The second set of chambers (omasum and abomasum) perform water reabsorption and protein or carbohydrate digestion functions, neither of which require the large spaces needed for cellulose digestion. (28–30)

1. Choice *c*. The peritoneum lies outside the gut tube, so the vessel must first penetrate the serosa and then both layers of smooth muscle before reaching the submucosa. (1–5)
2. Choice *b*. Peristalsis occurs because of the alternating contraction of circular and longitudinal smooth muscles within the gut. Sphincters are thick rings of smooth muscle that control the movement of materials from one part of the digestive tract to another. (8–9)
3. Choice *d*. The action of salivary amylase begins carbohydrate digestion in the mouth, and other carbohydrases complete this process in the small intestine. (6, 15)
4. Choice *a*. An important function of the colon is retrieving the large amounts of water and ions released into the gut during the digestive process. (24)
5. Choice *b*. The gall bladder stores the bile that has been produced in the liver. The bile is then released into the small intestine to emulsify fats. (16)
6. Choice *c*. Just before individual amino acids are absorbed across the membranes of mucosal cells lining the small intestine, dipeptidases cleave the remaining dipeptides created by the earlier action of proteases. (20)
7. Choice *c*. Chylomicrons are assembled in the mucosal cells from absorbed monoglycerides, fatty acids, and proteins. They are then transported via the lymph and blood to all parts of the body. (22)
8. Choice *a*. Strictly speaking, cellulose is not digested by mammals. Rather, it is the microorganisms that these mammals house within their gut that manufacture the cellulases needed to digest cellulose. (27–31)

CONTROL AND REGULATION OF DIGESTION • CONTROL AND REGULATION OF FUEL METABOLISM (pages 336–338)

- (3–4, 9, 12)
  a: Secretin
  f: Glucagon
  b: Gastrin
  e: Epinephrine
  g: Insulin
  d: Cholecystokinin
  c: Cortisol

1. Choice *b*. The gastrocolic response involves neural signals that never leave the intestinal tract. Conditioned reflexes, however, utilize the typical peripheral and central nervous system pathways. (1–2)
2. Choice *c*. Gastrin, secretin, and cholecystokinin are all secreted by or have their effects upon the mucosal lining of some structure within the vertebrate digestive tract. (3–4)
3. Choice *e*. Gluconeogenesis is the process of producing glucose from other macromolecules (such as stored glycogen or fat). Glucagon and epinephrine increase the rate of gluconeogenesis, while insulin and cortisol reduce the rate. (5, 9, 12)
4. Choice *a*. The role of insulin is to lower the high levels of circulating blood glucose that occur after a large meal. It does this by stimulating the conversion of glucose into its primary storage form, glycogen. (7, 9)
5. Choice *d*. During the postabsorptive period, circulating blood glucose levels drop because no new food is available in the gut. Thus, glucose stores are mobilized through the breakdown of glycogen. Most of this glucose is used to maintain the cells of the nervous system (which constantly require glucose), so other body cells switch to using fat as their metabolic fuel. (8)
6. Choice *c*. Chylomicrons and the various density-variant forms of lipoproteins in humans control fat metabolism and transport in the blood. (6)

Integrative Questions

1. Choice *e*. All animals are ingestive heterotrophs and predators of some type. All vertebrates have complete digestive tracts and thus have a stomach(s) and a small intestine. Vertebrates also use autocatalysis to activate digestive enzymes.
2. Choice *e*. Defecation is the normal process of expelling feces from the body.
3. Choice *a*. Insulin is produced in the pancreas.
4. Choice *e*. All animals with a complete digestive tract also have a mouth and an anus. All four choices given have complete digestive tracts.

CHAPTER 44 – SALT AND WATER BALANCE AND NITROGEN EXCRETION

Note : the page numbers listed with the section titles refer to the study guide; numbers in parentheses for each ques-

INTERNAL ENVIRONMENT • WATER, SALTS, AND THE ENVIRONMENT • THE EXCRETION OF NITROGENOUS WASTES (pages 340–341)

1. Choice *c*. Water can only move osmotically, it is never actively transported. (1–4)
2. Choices *c, d*. Birds excrete uric acid, and marine species with high salt diets also have nasal salt glands. (14, 16–23)
3. Choices *a, e, f*. Freshwater fishes take on water from their environment and lose salt. Therefore, they produce lots of dilute urine, absorb salts, and excrete ammonia. (8, 16–23)
4. Choice *a*. The osmotic potential of the body fluid of a hypertonic osmoregulator would be more negative than its environment; thus, it would probably be living in fresh water (10, 11)
5. Choice *c*. Only proteins are converted directly into ammonia; nucleic acids are converted first into purines and pyrimidines, then into uric acid and urea. (15–17)

- (5–11)

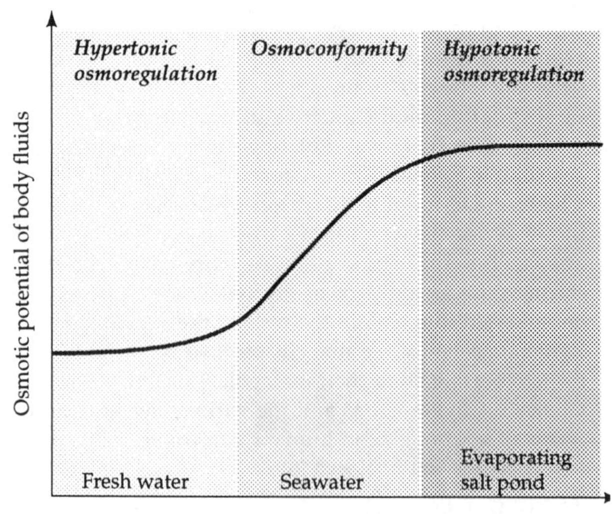

- *Artemia* is an ionic regulator. Since it lives in a very saline environment it will need to excrete NaCl, but absorb other necessary ions. (12, 13)

INVERTEBRATE EXCRETORY SYSTEMS • VERTEBRATE EXCRETORY SYSTEMS (pages 341–343)

1. (2–13)
   Flatworms: ***b, d***
   Annelid worms: ***c***
   Crustaceans: ***e***
   Insects: ***f***
   Vertebrates: ***a***
2. Choice *e*. In the metanephridium of earthworms, coelomic fluid enters the nephrostome of the excretory organ directly. In the vertebrate kidney, filtration of blood in the glomerulus of the nephron forms the tubular fluid. (5–7, 13–18)
3. Choice *b*. The Malpighian tubules produce a highly concentrated, solid mass of *uric acid*, not a urea solution. (8–10)
4. Choice *c*. Permeability of the glomerulus is greater than that of the peritubular capillaries. (13–18)
5. (24–29)
   Marine bony fishes: ***b, c***
   Cartilaginous fishes: ***d, h***
   Most amphibians: ***b, e***
   Reptiles: ***a, f***
   Birds: ***a, f, g***
6. Choice *a*. All the evidence suggests that the earliest vertebrates lived in fresh water. (19–23)

- (20–22)

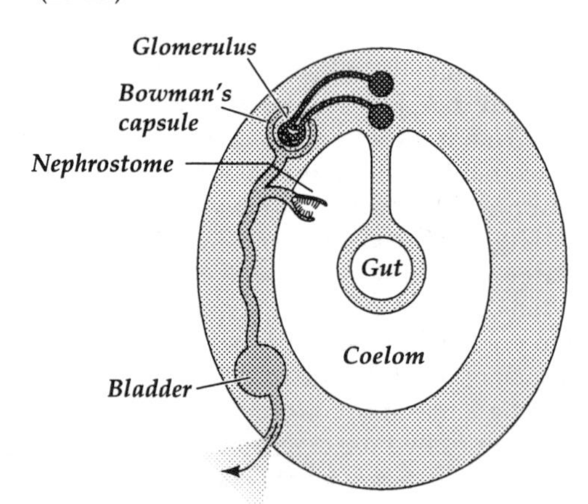

The final change would involve loss of the nephrostome opening to the coelom. Since coelomic fluid is no longer processed, this opening would no longer be required.

- (13–18)

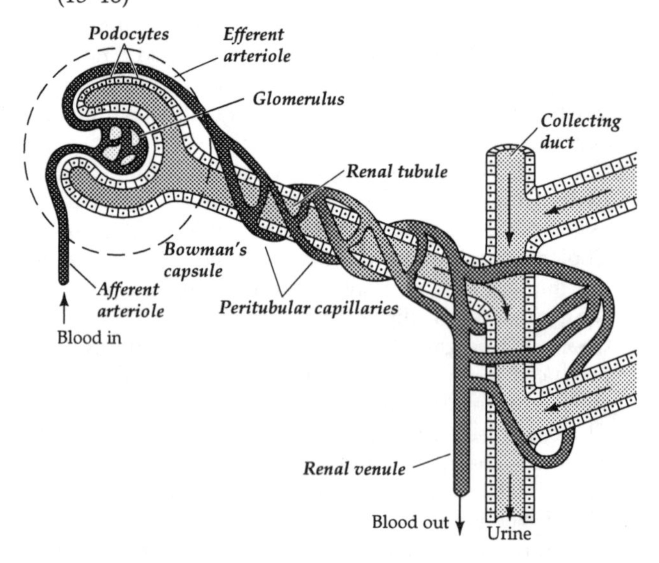

- Both crab-eating frogs of Southeast Asia and sharks are faced with the problem of conserving water in a marine environment. Both accumulate urea in their tissues in order to make their body fluid hypertonic to the seawater they live in, thus limiting loss of water through their skin. (25, 26)

STRUCTURE AND FUNCTION OF THE MAMMALIAN KIDNEY • REGULATION OF KIDNEY FUNCTIONS (pages 343–346)

1. (1)
   2: Collecting duct
   5: Urethra
   4: Bladder
   1: Glomerulus
   3: Ureter
2. (2–8)
   C: Peritubular capillaries
   C: Bowman's capsule
   C: Proximal convoluted tubule
   C: Distal convoluted tubule
   *M:*Renal pyramids
   *M:*Vasa recta
   *M:*Loop of Henle
3. Choice ***c***. About 10–12% of the blood entering the renal arteries is filtered. Of that, only about 1–2% is not reabsorbed. So 1 liter of urine would represent about 100 liters of filtrate and 100 liters of filtrate would be derived from about 1,000 liters of blood. (9)
4. Choice ***b***. The length of the loop of Henle determines the size of the NaCl gradient set up by the countercurrent multiplier system in the kidney. The greater the gradient, the greater the water reabsorption from the urine in the collecting ducts and the more concentrated the resulting urine. (7, 17–19)
5. See Figure 44.14 on page 1022 in the textbook. (10–19)
   a. ***A***
   b. ***G***
   c. ***C***
   d. ***A, B, E***
   e. ***F***
6. (10–18)
   Proximal convoluted tubule: ***a, b, c, d***
   Distal convoluted tubule: ***d, e***
   Ascending limb: ***d, e***
   Descending limb: ***f***
   Collecting duct: ***f***
7. Choices ***d, e***. Higher blood pressure would lead to increased production of urine. Reduction in antidiuretic hormone release by the posterior pituitary would result in less reabsorption of water from the collecting ducts and help to reduce blood osmolarity by producing a more dilute urine. (20–25)
8. Choice ***d***. Since sodium is an important constituent of blood, its concentration in the dialyzing fluid would be fairly great. (26)
9. (20–25)
   \+ Ingestion of caffeine
   \+ More stretch receptor activity in major arteries
   – More aldosterone released by adrenal cortex
   \+ Constriction of efferent renal arterioles
   \+ Decreased permeability of collecting ducts

- Both caffeine and alcohol have a depressing effect on release of antidiuretic hormone (ADH) by the posterior pituitary. Lower ADH levels decrease the permeability of the kidney's collecting ducts for water and consequently a greater volume of dilute urine is produced. (23, 24)
- (4–19)

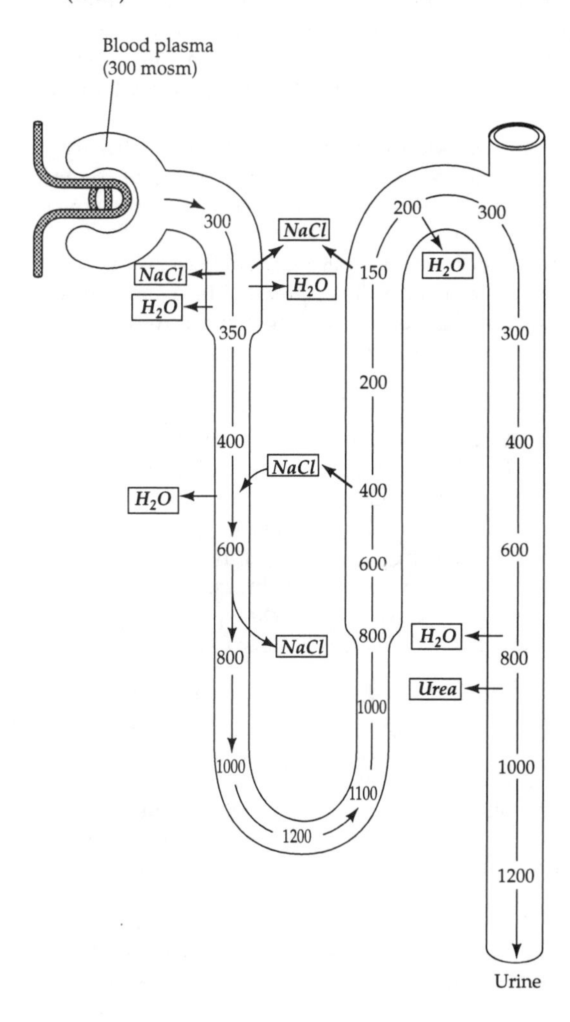

CHAPTER 45 – ANIMAL BEHAVIOR

Note : the page numbers listed with the section titles refer to the study guide; numbers in parentheses for each question refer to the relevant key concepts.

GENETICALLY DETERMINED BEHAVIOR • HORMONES AND BEHAVIOR (pages 347–349)

1. Choice *c*. Bird song in many species is an example of a highly stereotypic, species-specific behavior that is acquired through learning and is not inherited, as are fixed action patterns. (3–5, 12–14)
2. Choice *d*. The chaffinch can only learn its species-typical song during the critical period. (12–14)
3. Choice *d*. Bill shape and a strong contrast between the color of the bill spot and the bill seem most important. (6–8)
4. Choice *c*. Although prey capture behavior is likely to involve fixed action patterns, search behavior is more flexible and influenced by the predator's past experience. (3–5, 7, 11–14)
5. Choice *c*. Deprivation experiments help to reveal which behaviors are fixed action patterns. Since supernormal releasers are exaggerations of the normal releasers of a fixed action pattern, they would be highly effective in a deprivation experiment. (6–9)
6. Choice *e*. If castrated at birth, no androgens will be present to masculinize the rat's nervous system. It will therefore not respond to treatment with male sex steroids later in life. (15–17)
7. Choice *c*. Even though they have previously learned and used their song, male birds still require elevated testosterone levels to stimulate singing behavior. Thus, males normally only sing in the spring when their gonads are producing lots of male sex steroids. (18–20)

- In many birds, males learn by imprinting a song template for their species upon hearing their father sing as nestlings. During their first breeding season, these males are stimulated by increased testosterone levels to practice this song. The song becomes crystallized when their rendition matches the song template they learned. Thereafter, the song is always rendered the same way. (18–20)
- Cocoon construction in many spiders is a series of behaviors that, once begun, is always performed from beginning to end. Cocoon construction is an inherited behavior that does not require learning, is species-specific, and sterotypic. These are all characteristic of fixed action patterns. (3–5)
- (15–17)
  a. The female will show neither lordosis nor the mounting behavior typical of males. As a result of the injection of testosterone after spaying, the nervous system of the female has been masculinized and now she can not respond to female sex steroids.
  b. The female will show the mounting behavior typical of males, but not lordosis. As a result of the injection of testosterone after spaying, the nervous system of the female has been masculinized and she can now respond only to male sex steroids.

THE GENETICS OF BEHAVIOR • COMMUNICATION (pages 349–351)

1. Choice *b*. Even here the behavior is much too complex to be caused directly by a gene's presence; many intervening steps affecting the sensory and effector systems of the bee are involved in producing this behavior. (1–5)
2. Choice *d*. Egg-laying in *Aplysia* is one of the few cases where we know the effect of a specific gene product on the behavior of an organism. (6)
3. Choice *d*. The food source shown is 90° to the right of the hive–sun direction, so the waggle dance should be 90° to the right of the vertical axis. (16–18)
   Distance between the hive and the food source determines the type of dance. A round dance is given if the dance is about 80 meters or less; a waggle dance, if more than 80 meters.
   A random search is initiated by a round dance; a directed search is initiated by a waggle dance.
4. Choice *d*. The auditory mode is energy intensive, odors are slow to diffuse, the environment is too dark for visual communication, and the electrical mode requires an aqueous environment. (7–20)
5. Choice *d*. Due to the close proximity of sender and receiver, chemicals are frequently exchanged during tactile communication. (7–20)
6. Choice *c*. A displacement behavior is a seemingly irrelevant display resulting from conflict between different motives. (20–23)

- Since mating calls are designed to advertise the location of the caller, they would be expected to be pure tones with abrupt beginnings and endings. Alarm calls are designed to alert other species individuals to the presence of a predator without disclosing the location of the caller. They would be expected to be complex sounds with gradual onset and completion. (15)
- Since sex pheromones are used to advertise the location of the sender, they should disperse rapidly and dissipate quickly. Sex pheromones are usually small molecules with large diffusion coefficients. A trail- or territory-marking pheromone should last a long time and spread slowly. They tend to be larger molecules with small diffusion coefficients. (11, 12)

THE TIMING OF BEHAVIOR: BIOLOGICAL RHYTHMS • HOW DO THEY FIND THEIR WAY? • HUMAN BEHAVIOR (pages 351–354)

1. Choice *d*. The period or length of the cycle of this circadian rhythm is 22 hours and it would need to be phase-advanced to become entrained to a 24-hour cycle. (1–5)
2. Choice *a*. The bird's sense of time tells it that the sun should be due south at noon and that the feeding dish is 90° to the left (east). Because the artificial sun is stationary at the east, it will go to the north bin. (7, 24–26)
3. Choice *c*. Normally at noon, the east food bin would be 90° to the left. The bird will go to the food bin 90° to the left of the sun's displaced position. (7, 24–26)
4. Choice *b*. At sunset the sun is in the west and the bird would normally go to the food bin 180° to the left. However, since the bird has been phase-delayed by 6 hours, it thinks it is noon, so it searches 90° to its left to

find the east food bin, but instead goes to the south food bin. (7, 24–26)

5. Choice *c*. Because the location of the North Star is fixed, birds are not confused if the sky in the planetarium does not rotate or rotates at an incorrect rate.(27–29)
6. Choice ***b***. Since hibernators, animals living underground, and equatorial migrants cannot use photoperiod to coordinate their year-long activities, many of these species have circannual rhythms. Only the arctic hare does not fit into one of these categories. (10–12)
7. Choice ***b***. All other statements are false. In *a*, jet lag is only a problem if your flight is across time zones, as it would be in traveling 1,000 miles west. In *e*, if Earth was not tilted on its axis, seasonality would be greatly reduced. This would affect the occurrence of *circannual*, but not circadian rhythms. (1–9)
8. Choices ***a***, ***c***. Removing the suprachiasmatic nuclei of a mammal or the pineal gland of a bird would eliminate any circadian rhythms. Blinding a mammal (or covering the pineal gland of a bird) or placing mammals or birds in complete darkness only prevents entrainment to a true 24-hour cycle; it does not destroy the endogenous circadian rhythm. (8, 9)
9. Choice ***a***. Piloting, orientation using familiar landmarks, is the most basic form of homing. (13–21)
10. Choice ***a***. Since the sun rises later than usual, you must have been displaced to the west. Since the sun is higher in the sky than your home location you must have been displaced to the south. Therefore, you must travel to the east and north to get home. (7, 8, 20–22)
11. (14–16, 18–20)
    Homing only: ***A***
    Migration by piloting only: ***E*** (assuming the birds are following the coastline during migration as in the gray whale migration mentioned in the textbook)
    Migration by distance and direction navigation only: ***C***
    Migration by bicoordinate navigation only: ***B***

46

Behavioral Ecology

CHAPTER LEARNING OBJECTIVES—after studying this chapter you should be able to:

- ❑ Define "ecology" in terms of the distribution and abundance of organisms as well as the interactions between organisms and the environment.
- ❑ Relate the cost–benefit analysis of behavior to the concept of fitness.
- ❑ Differentiate between the following terms: "energetic cost," "risk cost," "opportunity cost."
- ❑ Provide several examples showing how social behavior in a species can vary with changes in the environment.
- ❑ Define habitat and explain the choices that organisms make in the process of habitat selection.
- ❑ Describe the major predictions that foraging theory makes about predator–prey interactions and provide an example.
- ❑ Describe the conditions that promote the formation of a social group and discuss the benefits and costs of group living.
- ❑ Define, differentiate between, and provide an example of the following types of behavioral acts: "altruistic," "selfish," "cooperative," and "spiteful."
- ❑ Define the conditions under which altruistic behavior would evolve and relate altruistic behavior to the concept of inclusive fitness.
- ❑ Describe the coefficient of relatedness (r) and be able to calculate r for specified conditions.
- ❑ Describe the model of reciprocal altruism and explain its importance.
- ❑ Characterize differences in male and female mating tactics and relate these differences to their investment in reproduction.
- ❑ Define "territoriality," describe the four different territory types, and provide some examples of each.
- ❑ Relate the following two concepts to territoriality: "lek" and "resource defense polygyny."
- ❑ Explain the evolution of "egg trading" in some hermaphroditic species.
- ❑ Describe some differences in parental investment by males and females of various vertebrate types, including any notable exceptions.
- ❑ Explain what is meant by sexual selection, describe the conditions under which it might evolve, and provide some examples.
- ❑ Explain what a natal group is and describe the importance of this concept in explaining the evolution of social systems in birds and mammals.
- ❑ Discuss the social system of the scrub jay and relate it to concepts of altruism, helping, and natal groups.
- ❑ Explain the term eusocial and discuss the strengths and weaknesses of W. D. Hamilton's haplo-diploid hypothesis for the evolution of eusociality.
- ❑ Discuss social systems in African hoofed mammals, African weaverbirds, and primates as examples of variation in social organization resulting from ecology.

COSTS AND BENEFITS OF BEHAVIOR • DEALING WITH THE NONSOCIAL ENVIRONMENT • DEALING WITH INDIVIDUALS OF THE SAME SPECIES (pages 1057–1064)

Key Concepts

1. The study of *ecology*, an outgrowth of natural history, is concerned with the interactions between different organisms and the interaction of organisms with their environment.
2. An organism's *environment* includes both physical factors, such as temperature, light intensity, and nutrient availability, and biological factors, such as interactions with species' members and with other species.
3. Organisms are affected by their environment but their activities can also modify the environment.
4. Ecologists study factors affecting the short- and long-term patterns of distribution and abundance of organisms.
5. *Fitness* is the reproductive contribution of a genotype or phenotype to the next generation, relative to all other genotypes or phenotypes (see chapter 19)
6. A *cost–benefit analysis* of behavior assumes that an organism has limited resources and energy available to it.
7. The increased energy required to perform a behavior over that expended in resting is the *energetic cost* of that behavior.

8. The increased risk of injury or death of a behavior over that associated with resting is the *risk cost* of that behavior.
9. When an animal performs a certain behavior, it forfeits the benefits of other behaviors that could not be performed. The forfeited benefits are the *opportunity costs* of the behavior performed.
10. If the costs associated with a behavior exceed the benefits in terms of an animal's *differential reproductive success*, the individual is not expected to perform the behavior.
11. Behavioral costs and benefits can be measured indirectly by examining how behavior changes with variation in the environment.
12. In the prothonotary warbler, females decide whether to abandon a nest containing an egg of a nest parasite based on the availability of other nest sites in their territory. Thus, nest abandonment behavior varies with the environment.
13. Social behavior sometimes changes when the environment changes; for example, junco flocks are larger when temperatures decrease or predators are more abundant.
14. An organism's *habitat* is the area in the environment where it normally lives. Habitats frequently occur in patches, with unsuitable areas separating the patches: patches of the same habitat type may differ in quality.
15. Habitat selection is a series of hierarchical choices that determines what habitat patch the organism will occupy and how it will utilize that patch.
16. The cues that organisms use in making habitat selection decisions vary greatly. For example, red abalone larvae will only settle where a specific peptide produced by coralline algae is present. Female poplar aphids assess leaf size and presence of other aphids in selecting a site for gall formation.
17. Foraging theory was developed in order to predict how predators will decide where to forage and what type of prey to take.
18. Energy costs associated with predation include searching for, capturing, handling, and digesting prey. These costs must be included in determining the value of prey of known energy content for predators.
19. A top-ranked prey item yields the greatest energy content per predator time invested. If prey differing in energy value are available to a predator, foraging theory predicts that the predator will always take the top-ranked prey item and will only add the prey with the next highest value to its diet when the density of top-ranked prey is reduced below a certain point. Experiments with *bluegill sunfish* preying on water fleas support these predictions.
20. Social behavior evolves when cooperation of individuals of the same species improves their fitness. Such organized collections of individuals are called *social groups*.
21. Individuals can sometimes obtain nutrients more efficiently when in a group than when alone; *cooperative hunting* is an example of this benefit of group living, as seen in African hunting dogs.
22. Studies showing that predators are less able to capture individuals in a group than isolated individuals suggest that protection from predation is a benefit of group living.
23. Increased spread of disease and parasites is a cost of group living. Other costs of group living include increased competition for food and mates, and increased chance of injury from conspecifics.
24. Costs of group living are sometimes related to the size of the individual. For example, in the solitary bee *Centris pallida*, small male bees avoid competition with larger individuals by patrolling for undiscovered females rather than digging them up.

Questions *(for answers and explanations, see page 407)*

1. Which of the following is *not* usually included within the domain of ecology?
   a. Interactions between conspecifics
   b. Modification of the environment by organisms
   c. Distribution and abundance of individuals
   d. Modification of the environment by physical processes
   e. Evolution of different social organizations

2. Which of the concepts listed below best describes the following situation? Birds spend some of their time scanning the horizon for predators. While scanning, they cannot be foraging for food.
   a. Foraging theory
   b. Energetic cost
   c. Risk cost
   d. Opportunity cost
   e. Cooperative hunting

3. In the yellow-eyed juncos described in the text, which of the following would you expect to happen if the temperature increases and the abundance of hawks in the area decreases?
   a. Flock size will increase.
   b. Flock size will decrease.
   c. Flock size will remain the same.
   d. The incidence of fighting will increase.
   e. Birds will spend less time individually scanning for predators.

4. For each of the two y-axes in the following graph, draw a labeled curve that correctly summarizes observations made on goshawks attacking wood pigeons, as described in the textbook.

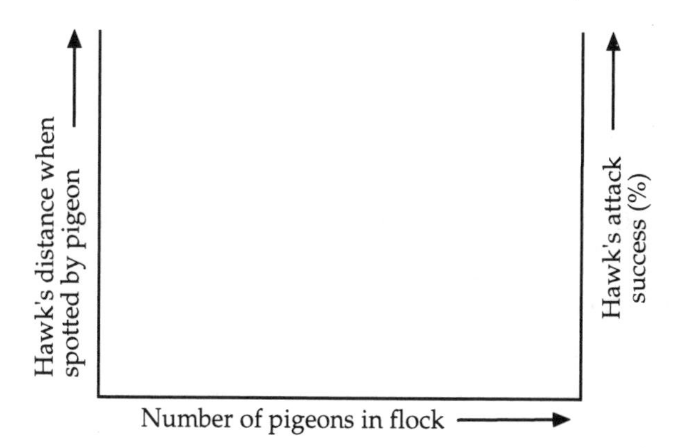

5. You perform a series of experiments in which the density of prey species 1 is varied while the density of prey species 2 is kept constant. At each prey species 1 density, you determine how many prey of species 1 and 2 the predator takes. The two curves (*a* and *b*) plotted in the following graph show the outcome of this study.

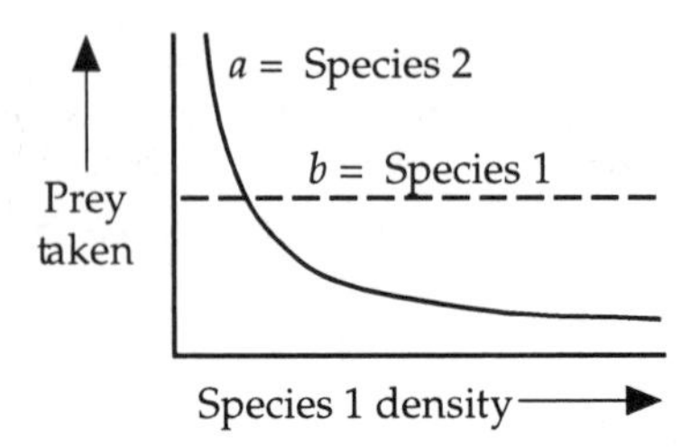

Select *all* of the following statements that correctly interpret these curves.
a. Curve *a* indicates that species 2 is not the preferred prey.
b. Curve *a* indicates that species 2 is the preferred prey.
c. Curve *b* indicates that species 1 is the preferred prey.
d. Curve *b* indicates that species 2 is not the preferred prey.
e. Neither of these curves provides insight to the prey preference of the predator.

Activities (for answers and explanations, see page 407)

- Describe the experiments that were done with bluegill sunfish preying on waterfleas and the implications of these studies for foraging theory.

- What does cost–benefit analysis reveal about decisions made by female prothonotary warblers when they encounter an egg of the brown-headed cowbird in their nest?

TYPES OF SOCIAL ACTS • CHOOSING MATING PARTNERS • ROLES OF THE SEXES (pages 1064–1071)

Key Concepts

1. An *altruistic act* (behavior) benefits other social group members, but at a cost to the performer.
2. A *selfish act* benefits the performer, but at a cost to other social group members.
3. A *cooperative act* benefits both the performer and other social group members.
4. A *spiteful act* has a cost to both the performer and other social group members.
5. Selfish or cooperative behaviors evolve if the behavior improves the *fitness* of the performer.
6. Altruistic behavior evolves if the behavior improves the *inclusive fitness* of the performer. Inclusive fitness is the sum of individual fitness and kin selection (see Chapter 19).
7. Altruistic behavior is beneficial when the improvement in the reproductive success of relatives exceeds the reduced reproductive success of the individual performing the act.
8. The coefficient of relatedness (r) is the probability that two alleles in different individuals are related by descent; $r = \Sigma(0.5)^k$, where k is the number of different generation links separating the two individuals. The value in parentheses is 0.5 because at each generation link there is a 0.5 probability that the allele will be passed on due to meiosis.
9. Individuals of some species, such as the *white-fronted bee-eater*, are able to determine their genetic relatedness to social group members and adjust their behavior appropriately.
10. The model of *reciprocal altruism* states that altruistic behavior can occur between unrelated individuals if the individual performing the altruistic act later benefits from a reciprocal act.
11. Within social systems, individuals must choose with whom to associate, with whom to interact, with whom to mate, and when to leave the group. Mate choice can depend on an individual's phenotypic traits, the resources it controls, or a combination of the two.
12. Because females generally make a greater energy commitment to producing and raising young than males do, it is to their benefit to be discriminating in choice of mates; males usually benefit from maximizing the number of females they inseminate and are often less discriminating in choice of mate.
13. Male courtship behavior provides information to the female as to the health of the male, his ability to provide for offspring, and the quality of his genes. Some male behavior is designed to insure that it is his sperm that fertilizes the eggs.
14. *Territoriality* is aggressive behavior in the defense of space. Four territory types (Types A–D) are recognized by behavioral ecologists.
15. *Type A territories* include everything needed by the pair for reproductive success: a mating area, nest site, and an adequate food supply. Many songbirds have Type A territories.
16. *Type B territories* include breeding and nesting areas, but not an adequate food supply to rear young. The food supply may be adjacent to clusters of type B territories, as in the red-winged blackbird's typical territory.

17. *Type C territories* are used exclusively for nesting and rearing young. Marine colonial birds such as gulls and penguins have type C territories.

18. *Type D territories* are used only for courtship and mating. As with the Uganda kob, species with type D territories frequently have a mating system in which one male mates with several females. Females raise the young elsewhere with no help from the father.

19. Display grounds containing type D territories are called *leks*. Males fight for central territories within the leks, where they will have access to more females.

20. In *resource defense polygyny*, the number of females that a male attracts depends on the quality of his territory. Red-winged blackbirds have this type of mating system.

21. Females have evolved behavior that focuses on the male reproductive signals that are reliable indicators of their fitness. For example, female barn swallows prefer males with long tails. Tail length is a reliable indicator of parasite load in males.

22. Egg trading in hermaphrodites is a behavior that guards against cheating in which an individual provides all of the sperm, but no eggs for the partner.

23. *Parental investment* is all of the activities of a parent that contribute to survival of its offspring, while simultaneously limiting the parent's ability to have additional offspring and, perhaps, lowering its own survival chances.

24. Females, by definition, contribute larger gametes and sometimes have a greater investment in care of the young than males.

25. Biparental investment is greater in birds than in mammals, because birds do not provide milk as food for their young.

26. Parental investment in fishes, if present, consists of guarding eggs and young; males are the primary guarders in most cases.

27. Generally, in birds, males are brightly colored, active in territorial defense, and may be involved in caring for the young; females are more cryptically colored and spend more time caring for the young. These sex roles are reversed in some birds like the *phalaropes*.

Questions (for answers and explanations, see page 407)

1. Some birds give a species-typical vocalization called an "alarm call" when they see a predator, although this call may direct the predator toward them. Other members of their species respond to these calls by taking cover. This would be an example of altruistic behavior that is beneficial to the calling bird if
   a. the bird giving the vocalization survives the attack.
   b. all birds survive the attack.
   c. the birds benefiting are members of the same social group.
   d. the birds benefiting are offspring of the bird giving the vocalization.
   e. the inclusive fitness of the bird giving the vocalization is increased.

2. A dominant individual displaces a subordinate individual from food that the subordinate has discovered. This is an example of a
   a. selfish act.
   b. cooperative act.
   c. altruistic act.
   d. spiteful act.
   e. nonsocial act.

3. Which of the following choices correctly calculates the coefficient of relatedness to show the probability that an allele in a child is a direct copy of one present in her grandfather?
   a. $r = (0.5)^1$
   b. $r = (0.5)^2$
   c. $r = (0.5)^2 + (0.5)^2$
   d. $r = 2(0.5)^4$
   e. None of the above

4. Which of the following animals would *likely* have a type C territory?
   a. The song sparrow
   b. The red-winged blackbird
   c. The Uganda kob
   d. A colonial marine bird
   e. A species in which mating is restricted to a lek

5. Which of the following describes a likely reason for sex role reversal in phalaropes?
   a. Male desertion would probably lead to loss of the whole brood.
   b. The female has less parental investment in the eggs than the male.
   c. The young are precocial and able to survive with little parental investment.
   d. The female lays a single, large clutch of eggs.
   e. The sex ratio is unequal, with fewer males than females.

6. In the red-winged blackbird, a female chooses a male
   a. based exclusively on features of his territory.
   b. if she mated with him in the previous season.
   c. based on his territorial defense behavior.
   d. based on his phenotypic traits.
   e. based on his genetic relatedness.

7. The most reproductively successful satin bowerbird male would
   a. have the largest territory.
   b. have the most elaborate plumage.
   c. have a bower with many similar objects.
   d. have a bower with many diverse objects.
   e. provide the female with many food items.

Activities (for answers and explanations, see page 408)

- Imagine that you are in a lifeboat after a disaster at sea. Your half-sibling and full cousin remain in the water. In order to maximize your inclusive fitness, which relative should you invite to take the last remaining space in the lifeboat? Explain your answer in a quantitative manner.

- Explain the evolution of "egg-trading" in some species of hermaphrodites.

SEXUAL SELECTION • THE EVOLUTION OF ANIMAL SOCIETIES • EUSOCIAL BEHAVIOR • ECOLOGY AND SOCIAL ORGANIZATION (pages 1071–1077)

Key Concepts

1. The evolution of sex-specific traits that enhance an individual's reproductive success is called *sexual selection*. Mating preferences by individuals of the opposite sex determine the direction of sexual selection.
2. Usually males evolve traits through sexual selection that improve their competitiveness for females or resources that attract females.
3. The correlation between male size and mating system type in many vertebrate species is a result of sexual selection. For example, in American blackbirds, the largest size difference between males and females is seen in species that are polygynous and colonial.
4. To reconstruct the events in the evolution of social systems, we must study the stages of complexity shown among living social species.
5. Communal breeding systems probably evolved from extended families with overlapping generations.
6. In the *Florida scrub jay*, breeding pairs with helpers (previous offspring) fledge more young than pairs without helpers. Most helpers are offspring from an earlier breeding season. Male helpers may be able to assume control of the territory if their father dies.
7. Most mammal societies evolved via the extended family route and range from solitary females or female–male pairs to *natal groups* consisting of a breeding pair and offspring, especially female. Most mammalian helpers are female.
8. Species with social systems in which there are classes of individuals that do not breed are called *eusocial*. Most social insects (termites, ants, some bees and wasps) and the naked mole rat are eusocial.
9. W. D. Hamilton developed the haplo-diploid hypothesis for the evolution of eusociality. In a species in which males are haploid and females diploid, sister workers are identical in terms of paternal genes and share 50% of the maternal genes. The coefficient of relatedness (r) is 0.75 for sisters, but only 0.5 for any offspring a female produced on its own. Thus, haplo-diploid condition favors a system in which females forgo their own reproduction and help to raise additional sisters.
10. Although there is strong support for the haplo-diploid hypothesis, important exceptions remain. The hypothesis cannot explain the existence of eusocial species with either multiple queens per colony or in which queens mate with more than one male. Also, many eusocial species exist in which both sexes are diploid, as in all termite species and the African naked mole rat.
11. Most eusocial species are colonial and utilize elaborate nest or burrow systems. The colony founding success rate is usually low. These conditions favor species in which many individuals defer reproduction and serve as helpers to promote colony success.
12. The type of social organization is strongly related to the features of the environment. For example, societal types shown by African weaverbird species seem to vary with the availability of nest sites: monogamous, isolated pairs characterize species living in forests with many available nest sites; polygyny and colonial nesting characterize savanna-dwelling species with limited nest sites.
13. Social organization in African hoofed mammals seems to depend on body size: small species with high metabolic rates live solitarily in forests where they eat scattered, high-quality food and hide effectively from predators; large species with lower metabolic rates live in grasslands where they feed on the abundant, but lower-quality vegetation and form large, complex social groups (herds) that constantly scan for predators.
14. The smallest primate troops are found in forest-dwelling species and the largest in ground-dwelling species. Females tend to remain with the troop, young males are driven away, and both males and females within the troop may form dominance hierarchies.
15. Social systems are best understood in terms of the costs and benefits accrued by the participants that associate within them.
16. Within a social system, associations are constantly changing and depend directly on genetic relatedness.
17. The evolution of a species' social system depends on diet, body size, and the environment in which it lives.

Questions (for answers and explanations, see page 408)

1. Select *all* of the characteristics listed below that are shared by most eusocial species.
   a. Haplo-diploid sex determination
   b. Elaborate nests or burrows
   c. Presence of sterile classes
   d. Colonies founded by a single female
   e. Queens that mate with a single male

2. Which of the following statements describing the social systems in Florida scrub jays is *not* true?
   a. It has a eusocial organization.
   b. Individuals in the social group are genetically related.
   c. The social system consists of natal groups.
   d. Only the parent birds breed.
   e. Males may benefit from helping more than females.

3. Of the following factors, select the one that *was not* mentioned as being important in the evolution of social organization in African hoofed mammals.
   a. Diet
   b. Dispersal of food resources
   c. Dispersal of breeding territories
   d. Metabolic needs
   e. Predation

4. The species of American blackbirds showing the greatest sexual size dimorphism are likely to have a ____________ mating system and a ____________ spacing system. These species have been subject to strong ____________ selection.

Activities (for answers and explanations, see page 408)

- Describe W. D. Hamilton's haplo-diploid hypothesis for the evolution of eusociality. What are some major phenomena not explained by this hypothesis?

- Describe important environmental variables that seem to have guided the evolution of social organization in species of African weaverbirds.

Integrative Questions (for answers and explanations, see page 408)

1. Select all of the following concepts that apply to elephant seal social organization.
   a. Sexual selection
   b. Eusocial
   c. Polygyny
   d. Type B territory
   e. Helpers
   f. Lek

2. Match all of the characteristics on the right with the species on the left that shows those characteristics.

_____Red-winged blackbirds	*a.* Sterile workers
_____Uganda kob	*b.* Type A territory
_____Honeybees	*c.* Helpers
_____African mole rat	*d.* Cooperative hunting
_____Scrub jays	*e.* Lek
_____Song sparrows	*f.* Eusociality
_____African hunting dogs	*g.* Resource defense polygyny
_____Bee-eaters	

47

Structure and Dynamics of Populations

CHAPTER LEARNING OBJECTIVES—after studying this chapter you should be able to:

- ❑ Describe the concerns of population ecology.
- ❑ List the typical life history stages of organisms.
- ❑ Describe some important life history trade-offs associated with reproduction.
- ❑ Describe the concept of reproductive value and relate it to attempts to extend human life expectancy.
- ❑ Explain what determines the structure of a population.
- ❑ Differentiate between unitary and modular organisms.
- ❑ Define the term "population density" and describe how to estimate it for unitary and modular organisms.
- ❑ Estimate population density based on the mark and recapture method.
- ❑ Describe three common spacing patterns within populations and the types of conditions that produce each pattern.
- ❑ Infer relative birth and death rates from graphs showing the age distribution of a population.
- ❑ Describe the demographic events that determine population size and express these ideas as an equation.
- ❑ Explain what a life table is and estimate survivorship and death for different age classes from a life table.
- ❑ Describe the three types of survivorship curves, the types of populations that each characterizes, and provide an example.
- ❑ Describe some differences in how population ecologists study modular organisms.
- ❑ Define exponential population growth, recognize it graphically, and express it as an equation.
- ❑ Discuss the role of the intrinsic rate of increase (r_{max}) in determining the magnitude of exponential growth.
- ❑ Define logistic population growth, recognize it graphically, and express it as an equation.
- ❑ Describe population growth when the population is very small, when it is medium-sized, and when it has reached its carrying capacity.
- ❑ Describe the major density-dependent factors that limit population growth.
- ❑ Describe the major density-independent factors that limit population growth.
- ❑ Explain typical variation in the density of a species within its range and between the rarity of a species and the size of its range.
- ❑ Characterize common population disturbances and the adaptations that organisms have evolved for dealing with disturbances.
- ❑ Describe a common practice for maximizing the number of individuals that can be harvested from a managed population and the problems associated with overharvesting.
- ❑ Explain some strategies for managing populations of undesirable species.
- ❑ Characterize historical trends in human population growth and what factors may limit Earth's carrying capacity for humans.

LIFE HISTORIES • POPULATION STRUCTURE (pages 1081–1086)

Key Concepts

1. *Population ecology* is the study of population structure and growth and the factors that regulate population growth.
2. The *life history* of a species consists of the stages that a typical individual goes through from birth to death.
3. The life history stages seen in most organisms include a *growth stage*, a *dispersal stage*, a *reproductive stage*, an *energy-gathering stage*, and, in some species, a *resting* or *reorganization stage*.
4. The amount of time an individual spends in each life history stage and how much the stages overlap varies among species.
5. Although allocating more food to offspring increases their chances of surviving, it reduces the number of offspring that the parent can produce.

6. In organisms that provide parental care, there is a trade-off between the number of young produced and the amount of care each can receive
7. Some organisms reproduce only once in their lifetime; examples include agaves, Pacific salmon, and annual plants.
8. Engaging in reproduction reduces the potential for growth and usually reduces the probability of parental survival.
9. Natural selection favors early reproduction except in cases where it would result in poor juvenile or parental survival; in these cases delaying reproduction may be advantageous.
10. *Reproductive value* measures the contribution to population growth made by an individual of a specific age; reproductive value increases as an individual approaches the age of reproduction, is maximum when it first starts to reproduce, and decreases to zero as it becomes postreproductive.
11. Because older individuals have less reproductive value than younger individuals, genes that are expressed only late in life are less affected by natural selection than genes that are expressed earlier.
12. Alleles that delay the phenotypic effects of the deleterious alleles of other genes are favored by natural selection. This leads to a buildup of genetically based problems expressed after reproduction is complete.
13. *Population structure* from an ecological viewpoint consists of the numbers, spacing, and age of population members.
14. *Population density* is the number of population members per unit area, for most terrestrial species, or per unit volume, for aquatic species. Population density can be determined by direct counting or by the mark and recapture method.
15. *Biomass*, the total mass of organisms, is best for expressing population density for species whose individuals differ greatly in size, such as plants, fish, mollusks, etc.
16. Population members may show a *uniform, clumped*, or *random spacing* within the population range.
17. Uniform spacing can result if population members are competing for a uniformly distributed resource, such as light, or if they defend territories.
18. Clumped spacing characterizes populations whose members tend to settle close to their birthplaces, or that live in patchy environments.
19. Random spacing can result if numerous factors interact to determine the locations of individuals.
20. The *age distribution* of a population, the proportion of population members in each age category, is determined by the timing of births and deaths.
21. If both birth and death rates are high, the population will be dominated by young individuals; if both birth and death rates are low, the population age structure will be more uniform.
22. The age structure of a population determines how fast it will grow and utilize resources.

Questions (for answers and explanations, see page 408)

1. From an ecologist's point of view, population structure does *not* include the
   a. distribution of genotypes within a population.
   b. population density.
   c. spacing of population members.
   d. age structure of the population.
   e. biomass of the population.

2. The number of seeds that a tree produces increases, but the resources available to the tree remain the same. Put a "+," "–," or "0" before each of the following variables, to express the expected effect of this change on the variable.

   _____ Weight of each seed

   _____ Width of the tree's growth rings

3. A random distribution of population members is usually caused by
   a. settling near one's birthplace.
   b. territoriality.
   c. competition for a uniformly distributed resource.
   d. the interaction of several factors that affect survival.
   e. patchy habitat.

4. In which of the following life history stages are annual plants and salmon *most* similar?
   a. Growth
   b. Dispersal
   c. Reproduction
   d. Energy-gathering
   e. Resting and reorganization

5. For a species that breeds only once and lives in a disturbed environment where new habitat is constantly being created, it would be best to
   a. produce many, small offspring only once.
   b. produce a few, large offspring only once.
   c. produce many, small offspring more than once.
   d. produce many, small offspring more than once.
   e. produce a few, small offspring only once.

6. You collect a sample of 100 bats that are hibernating in a cave. Each bat is marked and released. Two weeks later you return and collect another sample of 100 bats, 20 of whom are marked. The estimated number of bats in the cave is __________.

Activities (for answers and explanations, see page 408)

- Discuss the three life history trade-offs presented in the textbook.

- Discuss the implications that the concept called "reproductive value" has for hopes of greatly increasing human life expectancy.

POPULATION DYNAMICS • POPULATION GROWTH WHEN RESOURCES ARE ABUNDANT • POPULATION GROWTH WHEN RESOURCES ARE SCARCE • DYNAMICS OF SPECIES RANGES (pages 1086–1091)

Key Concepts

1. The number of population members depends on the *demographic events* of *birth, death, immigration,* and *emigration*. The current population size is the population size in the past, plus individuals added by birth and immigration, minus individuals removed by death and emigration, or
$$N_1 = N_0 + B - D + I - E$$
where N_1 = population size at time 1, N_0 = population size at time 0, B = number that are born, D = number that die, I = number that immigrate, and E = number that emigrate.
2. A group of individuals all born at the same time is a *cohort*.
3. A *life table* shows changes in the number of cohort members alive at different times. From these data, one can calculate the survivorship and death rates at different times.
4. A *survivorship curve* plots survivorship over time.
5. In a *type I* survivorship curve, nearly all cohort members die at the same time; as in the *Poa annua* example in the text.
6. In a *type II survivorship* curve, survivorship is uniform throughout life; this curve is typical for most bird populations.
7. In a *type III* survivorship curve, most young die, but survivorship is high for the remaining individuals; this curve is typical for organisms producing large numbers of young with low parental investment.
8. Many organisms have survivorship curves that combine different aspects of curves I, II, or III, and survivorship may differ for males and females of the same species, as in the red deer.
9. Species in which adults are similar in shape and size are called *unitary organisms*. Most mollusks, echinoderms, insects, and vertebrates consist of unitary organisms.
10. Species in which adults differ markedly in shape and size because they consist of modules that are added as the organism grows are called *modular organisms*. Modular organisms include most plants, some protists, fungi, and most colonial animals.
11. The number, size, and shape of the modules of modular organisms are more important to population ecologists than the number of individuals. Differential growth of modules allows an attached modular organism to move in response to environmental variables.
12. All populations have the potential to grow *exponentially*, because as a population increases in size, its growth rate accelerates.
13. Exponential population growth can be expressed mathematically as
$$\frac{dN}{dt} = (b - d)\,N$$
where dN/dt is the rate of increase in number of individuals, b is the birth rate, d is the death rate, and N is the population size.
14. The *intrinsic rate of increase* of a species, r_{max}, is the difference between the average birth and death rates under optimal conditions, so the rate of growth for a species under ideal conditions is $dN/dt = r_{max} N$.
15. In typical populations, the environment limits exponential growth by increasing death rates and decreasing birth rates.
16. The environmental *carrying capacity* is the equilibrium number of individuals sustainable by the resources available to the population in a particular environment.
17. As birth rates decrease and death rates increase in a population, its growth curve departs from the J-shaped exponential curve and becomes S-shaped.
18. *Logistic growth*, typical of populations growing in environments with limited resources, is expressed mathematically as
$$\frac{dN}{dt} = r\left(\frac{K - N}{K}\right)N$$
where K is the carrying capacity and other symbols are as in the exponential equation.
19. The expression $(K - N)/K$ becomes zero when the population size (N) reaches K; this causes the population growth rate (G) also to be zero.
20. Because of time delays in the effects of the environment on birth and death rates and in the effects that population members have on resources, populations frequently oscillate around the carrying capacity
21. For most organisms, population density tends to be higher towards the center of the species range and lower as the periphery is reached. Also, rarer species tend to have smaller ranges than more common species.

Questions (for answers and explanations, see page 408)

1. Which of the expressions (*a–e*) of the exponential growth equation should be *increased* in order for curve 1 to become more like curve 2 in the following graph?

$$\frac{dN}{dt} = (b - d)\,N$$

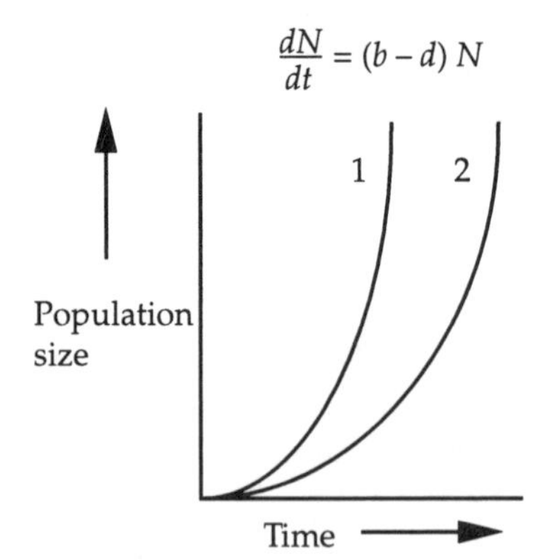

a. N
b. d
c. b
d. $(b - d)$
e. dN/dt

2. Species A and B have intrinsic rates of increase (r_{max}) of A = 0.25 and B = 0.50. In reference to the graph shown in question 2, the population growth curve for A should be more like curve _______.

3. In the logistic population growth curve shown below, the rate of growth is greatest at which point? ______

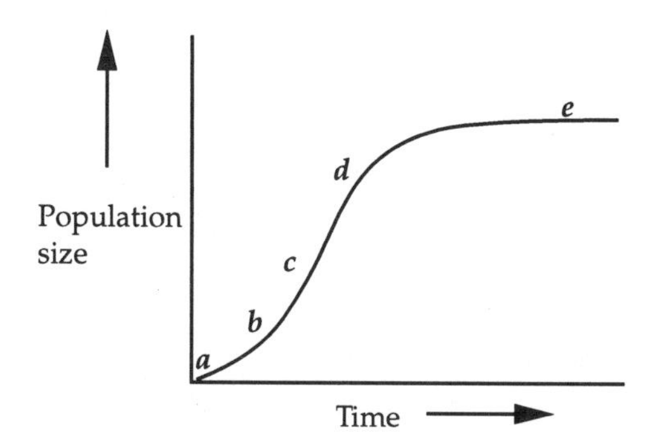

4. At which point in the graph shown above would there be zero population growth ($dN/dt = 0$)? ______

5. Based on the following life table, during what time interval is survivorship greatest?

Age (years)	Number Alive
0	800
1	770
2	550
3	125
4	75
5	0

a. 0–1 years
b. 1–2 years
c. 2–3 years
d. 3–4 years
e. 4–5 years

Activities (for answers and explanations, see page 409)

- Provide an explanation for the observation that population density is usually greatest in the center of the range.

- Provide an explanation for the observation that rare species usually have smaller ranges than more common species.

POPULATION REGULATION • POPULATION REGULATION UNDER HUMAN MANAGEMENT (pages 1091–1097)

Key Concepts

1. Population regulation is *density-dependent* if density-related factors in the population affect per capita birth and death rates.
2. Increased population size uses up resources faster, attracts more predators, and facilitates the spread of disease; these events will increase the death rate and may lower the birth rate within the population.
3. Population regulation is *density-independent* if the death rate is unrelated to population size; weather-related events can sometimes act in a density-independent fashion.
4. The population size at which the birth rate equals the death rate is an equilibrium value to which the population returns if either the birth rate, the death rate, or both are density-dependent.
5. *Disturbances* are short-term events that adversely affect natural populations; disturbances can be physical or biological.
6. Organisms have evolved behavioral and physiological responses to predictable disturbances.
7. Habitat patchiness can change spatially, but also over time, as caused by seasonal changes in the weather. *Migration,* a regular movement of a species between two different areas, is a common response to seasonal changes in habitat quality.
8. An *irruption* is a massive buildup of animals in an area, usually due to favorable conditions there; *mass dispersal* frequently follows an irruption.
9. Life cycles with resting stages are frequently a strategy for dealing with seasonal changes in habitat quality.
10. Humans have constantly attempted to regulate the populations of other species, encouraging desirable species and attempting to eradicate undesirable species.
11. A species can be harvested maximally if its population size is kept well below the carrying capacity, because population growth rates are greatest there.
12. Both the age class and number of individuals harvested can influence sustainable yield greatly; in many fish, removing smaller, prereproductive individuals has less effect than removing the larger, reproductive individuals.
13. *Overharvesting* occurs when too few reproductive individuals remain after harvesting. Overharvesting is most likely in species with low reproductive rates, as among whales.
14. In managing undesirable species, it is more effective to reduce the carrying capacity than to try to directly influence the birth or death rates.
15. The carrying capacity for humans now seems to be determined by the Earth's ability to absorb the waste materials of our industrialized societies.

Questions (for answers and explanations, see page 409)

1. The *best* way to decrease the population size of a pest species is to
   a. poison it.
   b. introduce additional predators.
   c. decrease the carrying capacity of the habitat for the species.
   d. add competitors.
   e. sterilize females.

2. Select all of the following that are *not* adaptations to seasonal habitat patchiness.
   a. Migration
   b. An irruption
   c. Production of seeds
   d. Hibernation
   e. Food caching

3. Based on the graph shown below, which of the following events is *least likely* to be true of this population?

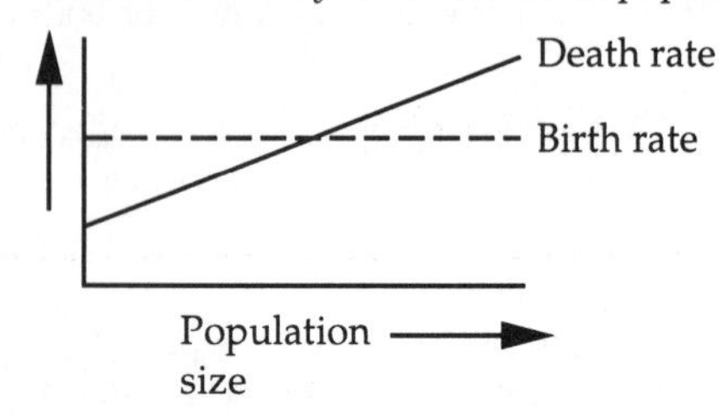

   a. Parasites spread between population members.
   b. Increased competition for food causes some individuals to delay reproduction.
   c. The number of predators in the area varies with population density.
   d. Territorial disputes led to injury and death of some males.
   e. The number of young produced remains stable from year to year.

4. In the graph shown in question 1, the population size where the curves for the birth and death rates intersect is
   a. the population size with the greatest sustainable yield.
   b. an estimate of the carrying capacity.
   c. an estimate of the intrinsic rate of increase.
   d. the point where density-dependent regulation begins.
   e. the point where density-independent regulation begins.

Activities (for answers and explanations, see page 409)

- In the two following figures showing variation in the population density in two species (*A* and *B*), compare and contrast population regulation in these two species. What types of organisms would show these types of regulation?

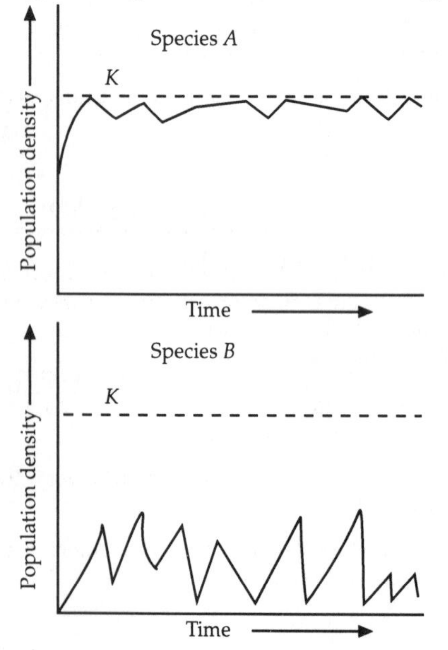

- Differentiate between migration and irruption and give an example of each.

Integrative Questions (for answers and explanations, see page 409)

1. Choose the collection of terms that completes the following sentence: Plants are ___________ organisms, they frequently show a ___________ spatial distribution, and their population density is most appropriately expressed in terms of ___________ .
   a. modular; clumped; biomass
   b. modular; random; individuals per unit area
   c. modular; uniform; biomass
   d. unitary; uniform; biomass
   e. unitary; random; individuals per unit area

2. The following age distribution indicates that

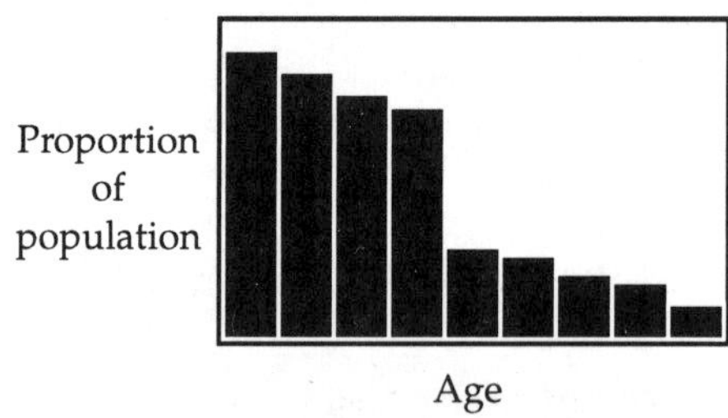

   a. the population's birth rate and death rates are both high.
   b. the population's birth rate and death rates are both low.
   c. the population's birth rate is high but its death rate is low.
   d. the population's birth rate is low but its death rate is high.
   e. the population's survivorship curve is type II.

48

Interactions within Ecological Communities

CHAPTER LEARNING OBJECTIVES—after studying this chapter you should be able to:

- ❑ Define what constitutes an ecological community and specify the biological questions of concern to community ecologists.
- ❑ Characterize the different types of ecological interactions.
- ❑ Define and relate the terms "resource" and "consumer."
- ❑ Define and differentiate between the terms "ecological niche," "fundamental niche," and "realized niche."
- ❑ Discuss the concept of "limiting resource" and relate it to an organism's niche.
- ❑ Characterize some factors that influence the type of predator-prey interactions that occur.
- ❑ Differentiate between parasites, parasitoids, herbivores, carnivores, and suspension feeders and discuss the unique features of each type of consumer.
- ❑ Explain the factors that cause predator–prey oscillations or that prevent them from occurring.
- ❑ Discuss the importance of spatial and temporal environmental heterogeneity (spatial and temporal patchiness) to predator–prey oscillations.
- ❑ Describe the characteristic features of Batesian and Müllerian mimicry.
- ❑ Discuss some of the important factors that determine the virulence of parasites and the relationships between parasites and vectors.
- ❑ Differentiate between grazers and browsers and discuss common plant defenses to herbivory.
- ❑ Differentiate between intraspecific and interspecific competition and between exploitative and interference competition.
- ❑ Describe the methods used by G. F. Gause to study competition and explain his principle of competitive exclusion.
- ❑ Discuss the role of environmental heterogeneity in avoiding competitive exclusion.
- ❑ Describe the effects that predators can have on the outcome of competition.
- ❑ Provide some examples of amensalism, commensalism, animal–animal mutualism, and plant–animal mutualism.
- ❑ Describe what is meant by "coevolution" and "diffuse coevolution" and provide some examples.
- ❑ Characterize some important effects that plants have on ecological communities.
- ❑ Define ecological succession and describe the three major types of succession.
- ❑ Characterize some important effects that animals have on ecological communities.
- ❑ Characterize some important effects that microorganisms have on ecological communities.
- ❑ Explain what a keystone species is and provide some examples.
- ❑ Differentiate between species richness and species diversity.
- ❑ Discuss some important factors that influence local species richness.

TYPES OF ECOLOGICAL INTERACTIONS • RESOURCES AND CONSUMERS • PREDATOR–PREY INTERACTIONS (pages 1100–1110)

Key Concepts

1. An *ecological community* consists of organisms of two or more species that live together in the same area and interact.
2. *Community ecology* is the study of the structure of biological communities and the interactions of the organisms living there.
3. In *predator–prey* or *host–parasite interactions*, one organism eats another organism; the consumer is the predator or parasite, the consumed is the prey or host.
4. If two organisms use a common resource whose supply is insufficient to support both, the two organisms are *competitors* and the interaction type is *competition*.
5. If two organisms interact in a mutually beneficial way, the interaction is called *mutualism*.
6. If one participant benefits while the other is unaffected by the interaction, it is called *commensalism*.

7. If one participant suffers from the interaction while the other is unaffected, it is called *amensalism.*
8. A *resource* is any feature of the community that is needed for normal growth of an organism and whose quantity is reduced when it is used; food is a resource, but so are space, nesting material, oxygen in aquatic environments, etc.
9. The *ecological niche* of a species is a characterization of all the requirements of an organism that determine where it can live.
10. The *fundamental niche* is the range of required physical conditions and resources of the species, in the absence of any negative influences from other organisms (competitors, predators, parasites, etc.).
11. The difference between the fundamental niche and the *realized niche* reflects the influence of other community members on the species and indicates which resources are most important in determining where an organism lives.
12. The study of niches helps to reveal which resources limit the distribution and abundance of organisms. For example, although oxygen is required by all organisms, it is seldom limiting for terrestrial organisms, but may be for aquatic organisms.
13. *Limiting resources* are resources that are present at an inadequate level to meet the demand of consumers.
14. Both the consistency of prey availability and the relative sizes of predator and prey determine the type of predator–prey interaction.
15. *Parasites* are smaller than their prey (*hosts*) and feed on them internally; *parasitoids* are about the same size as their prey.
16. In some parasite–host interactions, it is the egg-laying female that chooses the host for her offspring, as in many parasitic wasps and flies.
17. *Herbivores* are predators on plant tissues. Because wood is difficult to digest, few herbivores eat it—most specialize in other plant parts.
18. *Carnivores* are predators on animals, which they usually capture individually.
19. Prey defend themselves against attack by chemical means, by physical armor, by escape behavior, and by camouflage.
20. *Suspension feeders,* found in many different phyla, use a filtering device to remove the abundant, tiny prey from the medium in which they live; the structural details of the filtering device determine the size range and, in some cases, the type of prey captured.
21. As the predator population increases, the death rate of the prey will increase and may cause the prey population to decrease.
22. As the prey population decreases, predators may starve or emigrate and their population will decrease.
23. Decrease in the predator population allows the prey to recover, and as the prey population increases, the predator population can also recover.
24. The interaction between prey and predator population sizes causes *oscillations,* with the prey population always recovering more quickly than the predator population.
25. Environmental heterogeneity over space (spatial patchiness) can prevent predator–prey oscillations if prey are more vulnerable to predators in some patches than in others.
26. Oscillations can also be dampened if the prey is able to reinhabit areas and flourish there until the predator arrives and exterminates it. In this case, the environment is heterogeneous over time (temporal patchiness).
27. Over the course of time, predators and prey often evolve adaptations that reduce the effect of each on the other.
28. In *Batesian mimicry,* a palatable species, called the *mimic,* evolves an appearance and/or behavior similar to a distasteful or dangerous species, called the *model.*
29. Because palatable mimics "dilute" the effectiveness of the model's adaptation, directional selection causes models to diverge in appearance from mimics.
30. To have a stable Batesian mimicry system, the mimic must evolve towards the model faster than the model evolves away from the mimic.
31. Since Batesian mimicry relies on predator learning, a stable mimicry system requires that mimics be rarer than models.
32. *Müllerian mimicry* results from convergence in the appearance of two or more unpalatable species.
33. Müllerian mimicry makes it easier for predators to learn a common appearance; since all species are unpalatable, the relative abundances do not matter.
34. Host–parasite relationships stabilize if the parasite does not kill the host or does not kill it before the parasite can spread to a new host.
35. Parasites that are transmitted by physical means are typically more deadly than those transmitted by biological *vectors.* Also, parasites evolve to minimize their adverse effects on the vector, since they depend on it for transmission.
36. *Grazers* are herbivores that specialize in eating the leaves of nonwoody (herbaceous) plants; *browsers* are herbivores that specialize on the leaves of woody plants.
37. Plants defend their leaves by making them tough, adorning them with hairs and spines, and by producing *secondary compounds* (see Chapter 31).
38. One group of secondary compounds is toxins, such as nicotine. Some secondary compounds imitate hormones and interfere with insect metamorphosis. Other toxins are unusual amino acids, as in canavanine (see the discussion on page 713 in the textbook). Another group of defensive chemicals, including the *tannins,* is difficult for herbivores to digest.

Questions (for answers and explanations, see page 409)

1. A bird eats the fruit of a plant species. The seeds are not digested and germinate in the bird's excrement at some distance from the parent plant. This is an example of
   a. predation.
   b. parasitism.
   c. competition.
   d. mutualism.
   e. amensalism.
   f. commensalism.

2. Certain birds follow swarms of foraging army ants and prey upon the insects that the ants flush. The relationship between these birds and the ants is an example of
   a. predation.
   b. parasitism.
   c. competition.
   d. mutualism.
   e. amensalism.
   f. commensalism.

3. According to the definition of resource developed in the text, which of the following would *not* be a resource?
   a. Sunlight for a plant species
   b. Oxygen for an aquatic insect
   c. Hydrogen sulfide for a sulfur bacterium
   d. Nesting boxes for a bluebird
   e. Carbon dioxide for a plant species

4. As shown below, species A and B live together in region 2, but alone in regions 1 and 3.

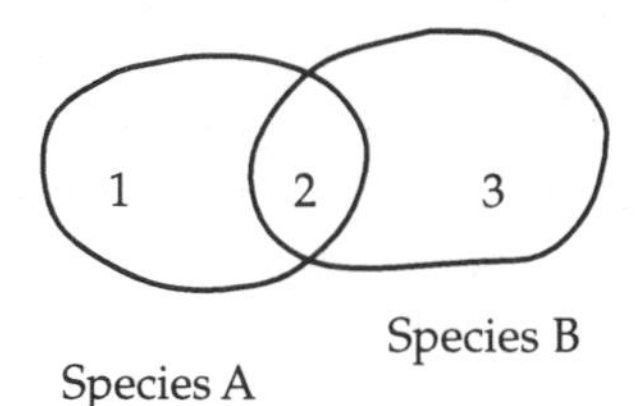

   Assuming that the two species compete for common resources, which of the following statements is *not* true?
   a. You can best estimate the fundamental niche for species A by collecting data in region 1.
   b. You can best estimate the realized niche for species B by collecting data in region 2.
   c. The difference between the fundamental niche and the realized niche for species A would be greatest in area 1.
   d. The difference between the fundamental niche and the realized niche for species B would be greatest in area 2.
   e. The competitively superior species will have the smallest difference between the fundamental niche and the realized niche.

5. Which vertical line in the following graph represents the time at which the predator population is *increasing* and the prey population is *decreasing*?

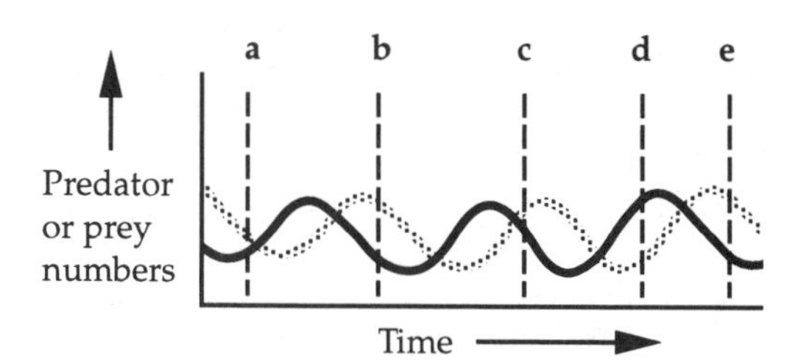

6. Some predator–prey interactions do not show oscillations. Select all of the following factors that could be responsible for disrupting oscillations.
   a. Spatial habitat patchiness
   b. Temporal (time-based) environmental heterogeneity
   c. Dispersal of prey
   d. Density-dependent interactions between prey and predator populations
   e. Extinction of prey
   f. Alternate prey species

7. You are studying three butterfly species (1, 2, 3) that have very similar, bright coloration patterns. You collect the following data on the abundance of these species in two different regions, A and B. Based on your data shown below, which of the following conclusions is *most probable*?

Region	Numbers Collected for Species: 1	2	3
A	90	20	120
B	190	35	155

   a. Species 1 is the model, species 2 and 3 are Batesian mimics.
   b. Species 1 is a Batesian mimic, species 2 and 3 are Müllerian mimics.
   c. Species 1 and 3 are Müllerian mimics, and 2 is a Batesian mimic.
   d. All three species are Müllerian mimics.
   e. Species 3 is the model, species 1 and 2 are Batesian mimics.

8. Which of the following statements about parasite–host interactions is *not* true?
   a. Competition with other parasites for the same host can lead to faster killing of the host.
   b. Transference to a new host is improved if the current host lives longer.
   c. Water-borne diseases are less serious to the host than diseases involving biological vectors.
   d. Most parasites with biological vectors do not harm their vectors as much as they harm the primary host.
   e. Hosts act as selective agents on their parasites.

Activities (for answers and explanations, see page 409)

- You discover two fig wasp nematode parasites. Each nematode parasitizes a different species of fig wasp. Nematode *A* is significantly less virulent than nematode *B*. Discuss life history differences in the fig wasp species that could account for differences in parasite virulence.

- Discuss four common plant defenses against browsers.

COMPETITION • OTHER INTERSPECIFIC INTERACTIONS • INTERACTION AND COEVOLUTION (pages 1110–1118)

Key Concepts

1. Removing a species from an area is a good way to evaluate its impact on another, potentially competing species.
2. Competition is of two types: in *exploitation competition,* one species limits growth of another species by using up its resources; in *interference competition,* one species behaves in a way that directly prevents access of another species to needed resources.
3. *Intraspecific competition,* competition between individuals of the same species, limits the density of a species in a given area.
4. *Interspecific competition,* competition between members of different species, can also reduce the density of another species, but may also exclude a species from an area.
5. G. F. Gause conducted the first controlled experiments on interspecific competition in 1934, working primarily with protists like *Paramecium.*
6. Many of Gause's studies showed *competitive exclusion,* in which one species completely replaced the other species; competitive exclusion could be avoided if environmental heterogeneity was provided.
7. Performing studies of competition between plant species may be easier than with animal species because it is more possible to control plant population densities and resource availability.
8. Predation can have direct effects on interspecific competition if one of the competing species is preferentially removed by the predator.
9. Plants being damaged by animal trampling or falling leaves and branches from larger, neighboring plants are examples of *amensalism* because one party is harmed, while the other party is unaffected.
10. *Cattle egrets* foraging on insects flushed by hoofed mammals and *epiphytes* growing on large trees are examples of *commensalism* because one party benefits, while the other party is unaffected.
11. Mutualisms are widespread among organisms of virtually every phylum; an important early result of a mutualism was the evolution of the eukaryotic cell from previously free-living prokaryotes.
12. Examples of mutualisms between members of different kingdoms include the association of plant roots and nitrogen-fixing bacteria, such as *Rhizobium,* and the association of plant roots and certain fungi to form *mycorrhizae.*
13. Other trans-kingdom mutualisms involve corals and some tunicates harboring photosynthetic protists and the association between termites and cellulose-digesting protists.
14. Some examples of animal–animal mutualisms include aphid "farming" by certain ants, *cleaner* shrimp and small coral reef fish removing parasites from larger fish, and the birds called *honeyguides* leading honey badgers and humans to bee nests.
15. Some examples of plant–animal mutualisms include the association between *acacia* trees and ants of the genus *Pseudomyrmex,* and the many associations between plants and their *animal pollinators* and *seed dispersers.*
16. Coevolution results when interactions between two or more species mutually influence their evolution. The clearest examples of coevolution involve only two species, such as the mutualisms described between *figs* and *fig wasps* and between species of *Yucca* and their moth pollinators.
17. *Diffuse coevolution* results from interactions between a diversity of organisms, including predators, parasites, mutualists, etc.
18. Diffuse coevolution produces generalized traits that adapt the organism to a variety of species acting as pollinators, seed dispersers, predators, etc. Some examples are, bat-pollinated flowers that are white and only open at night, nectar guides on flowers best seen under ultraviolet light, and fleshy red fruit for dispersal by birds.

Questions (for answers and explanations, see page 410)

1. In a bird called the black-capped chickadee, males frequently displace females from choice food items. This is an example of
   a. interspecific, exploitative competition.
   b. interspecific, interference competition.
   c. intraspecific, exploitative competition.
   d. intraspecific, interference competition.
   e. sexual selection.

2. In many of the studies done by G. F. Gause, two species were grown alone and in the presence of another species. The following graphs show growth of species A (left) and species B (right), both alone and when in mixed culture.

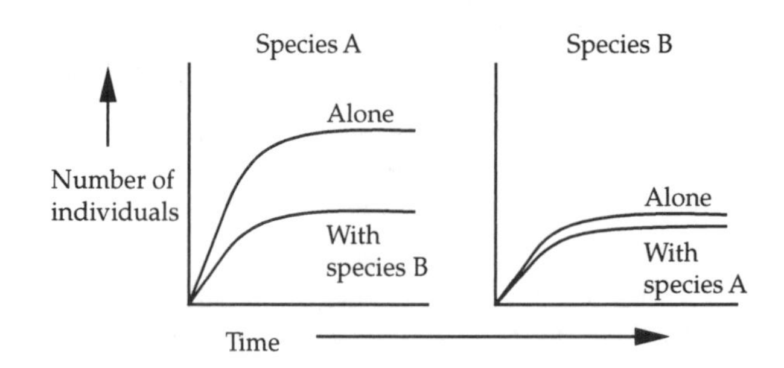

   Interpretation of these graphs shows that
   a. both species are affected by interspecific competition but species A is affected less.
   b. both species are affected by interspecific competition but species B is affected less.
   c. neither species is affected by interspecific competition.
   d. both species are affected equally by interspecific competition.
   e. competitive exclusion occurred in this study.

3. The difference between the two graphs in question 2 is probably an example of
   a. interspecific, exploitative competition.
   b. interspecific, interference competition.
   c. intraspecific, exploitative competition.
   d. intraspecific, interference competition.
   e. diffuse coevolution.

4. A species of orchid has flowers that look and smell like the female of a wasp species. Male wasps of that species attempt to copulate with these flowers and in the process pollinate them. This is an example of
   a. parasitism.
   b. diffuse coevolution.
   c. mutualism.
   d. amensalism.
   e. commensalism

5. If a species of plant that is trampled by an animal eventually evolves sharp spines that prevent trampling, we can say that its association with the animal has changed from
   a. amensalism to competition.
   b. amensalism to commensalism.
   c. commensalism to competition.
   d. parasitism to mutualism.
   e. parasitism to competition.

Activities (for answers and explanations, see page 410)

- Describe how you would assess the degree of exploitative competition that exists between two rodent species living in the same habitat and how you would interpret the results.

- Discuss the costs and benefits for both the moth and the plant in the Yucca–Yucca moth coevolution described in the textbook.

HOW SPECIES AFFECT ECOLOGICAL COMMUNITIES • PATTERNS OF SPECIES RICHNESS (pages 1118–1124)

Key Concepts

1. Plants are the major modifiers of physical conditions in terrestrial environments; plants intercept sunlight and rain, moderate climate (temperature, humidity, and wind), and reduce soil evaporation.
2. Plants are the major structural elements in terrestrial areas, especially the *canopy* and *understory* layers of forests.
3. Most nonphotosynthetic members of a biological community depend directly or indirectly on plants, as the primary producers of the community, for food and shelter.
4. *Ecological succession* is the gradual change in the species composition of a community through time.
5. Ecological succession may start in areas without organisms; in many areas, lichens and mosses are important soil-forming, pioneering organisms and the nitrogen-fixing bacteria associated with the roots of alder trees help to enrich the soil.
6. Succession can also begin after a disturbance has removed a portion of the original community.
7. Degradative succession is the decomposition of the dead bodies or excrement of organisms.
8. The effect of large herbivores on changes in community structure and succession can be substantial, as exemplified by studies of the impact of moose and beavers on northern forests.
9. *Keystone species* are organisms who have a larger effect on the structure and functioning of an ecological community than would be expected based on their abundance. In certain areas, beaver and moose are keystone species.
10. Microorganisms have major influences on ecological communities, including nitrogen fixation, mycorrhizal association with plant roots, and degradation and recycling of organic material.
11. *Species richness*, the number of species in a community, is an important ecological descriptor.
12. *Species diversity* is species richness weighted to reflect the relative contribution of each species to the total community. A community in which all species are equally abundant has greater diversity than another with the same number of species but in which most are rare.
13. Ecologists are concerned with what determines the species richness and diversity of communities.
14. Given a fixed resource base, the following factors tend to increase the species richness of a community: fuller use of available resources, narrower niches, more overlap in niches.
15. Communities with structurally complex plants have greater species richness than communities with simple plants.
16. Disturbances, such as grazing and the addition or removal of key community species, can increase species richness.
17. Regional species richness directly influences species richness at the local level.
18. Because organisms have had greater opportunities to evolve adaptations to more common habitats, these habitats usually have a greater species richness than less common habitats.
19. Serpentine soils are common in South Africa, and many more plants have evolved to live on this soil type there than in other parts of the world, where serpentine soils are less common.

Questions (for answers and explanations, see page 410)

1. Which of the following is *not* an effect that plants normally have on biological communities?
   a. Moderate climate
   b. Influence which animal species will be present
   c. Increase soil nutrients
   d. Reduce water evaporation from the soil
   e. Reduce moisture reaching the soil

2. Place numbers before each of the stages of ecological succession to reflect the ordering of the stages in northern coniferous forests.

   _____ Soil formation

   _____ Arrival of willows and alders

   _____ Arrival of lichens

   _____ Arrival of conifers

   _____ Increase in soil nitrogen content

3. The following graph shows the number of species in three communities (A, B, C) arranged from most to least abundant.

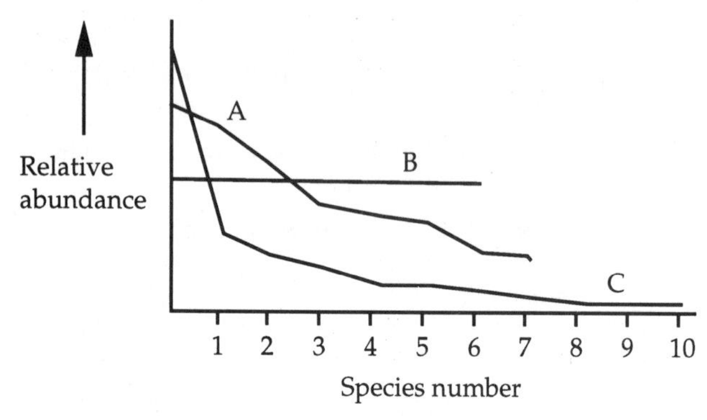

   Select a statement that correctly interprets some aspect of this graph.
   a. Community A has greatest species richness and greatest diversity.
   b. Community B has greatest species richness and greatest diversity.
   c. Community C has greatest species richness and greatest diversity.
   d. Community C has greatest species richness and community B the greatest diversity.
   e. Community B has greatest species richness and community A the greatest diversity.

4. Select all of the following that is/are characteristic of a keystone species.
   a. Very abundant
   b. Removal has a great effect on community structure
   c. Herbivore
   d. Large
   e. Uniform distribution

Activities (for answers and explanations, see page 410)

- Use "+," "–," or "0" to indicate the effect of each of the following factors on local species richness.

  _____ Decrease in structural complexity of plant community

  _____ Narrowing of niches

  _____ Increase in the number of limiting resources

  _____ More niche overlap

  _____ Grazing

  _____ Decrease in regional species richness

  _____ A moderate disturbance

- Explain the following graph.

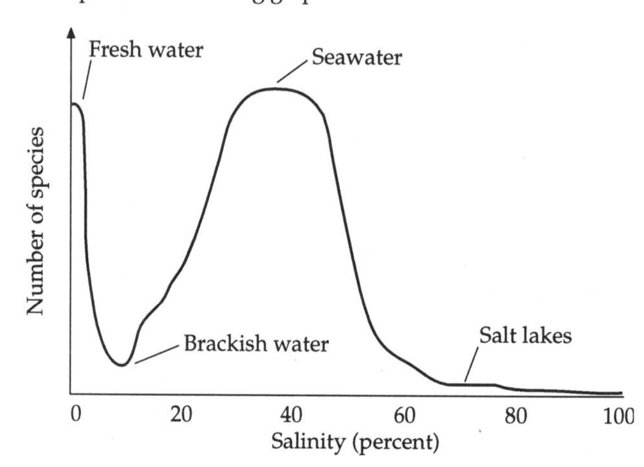

49

Ecosystems

CHAPTER LEARNING OBJECTIVES—after studying this chapter you should be able to:

- ❑ Define what an ecosystem is and provide several examples of ecosystems, ranging from local to global.
- ❑ Define the term "gross primary production" and explain how it varies with temperature and precipitation, and depth in aquatic ecosystems.
- ❑ Relate net primary production to gross primary production.
- ❑ Explain what a trophic level is and list the major types of trophic levels.
- ❑ Relate the concepts of food chain and food web to energy flow in an ecosystem.
- ❑ Estimate the efficiency of energy transfer between trophic levels and express it as an equation.
- ❑ Explain how pyramids of energy and biomass are constructed and know what determines their shape.
- ❑ Describe the conditions that might lead to an inverted biomass pyramid for an ecosystem.
- ❑ Compare energy flow in grasslands and forests.
- ❑ Characterize the major compartments of the global ecosystem and describe the cycling of material through each.
- ❑ Describe the annual temperature and oxygen cycles in temperate lakes.
- ❑ Explain what a biogeochemical cycle is and describe some key generalizations about movement of materials through biogeochemical cycles.
- ❑ Compare and contrast the features of the following biogeochemical cycles: hydrological cycle, carbon cycle, nitrogen cycle, phosphorus cycle, and sulfur cycle.
- ❑ Describe lake eutrophication—a local alteration of a biogeochemical cycle.
- ❑ Describe acid rain—a regional alteration of a biogeochemical cycle.
- ❑ Describe global warming—a global alteration of a biogeochemical cycle.
- ❑ Describe the role of chlorofluorocarbons (CFCs) in alteration of the chlorine cycle and its consequences for ozone depletion.
- ❑ Explain how integrated pest management is addressing some of the ecological shortcomings of modern agriculture.
- ❑ Characterize some of the problems inherent in attempting to apply temperate agricultural practices to tropical ecosystems.
- ❑ Explain what determines the solar energy input to different regions of Earth and the consequences of these differences.
- ❑ Describe the movement of air masses over mountains and the conditions that create rain shadows.
- ❑ Describe the global circulation of air in the atmosphere and the consequences of this circulation for Earth's climates.
- ❑ Explain the effect of the tilt in Earth's axis on the location of the intertropical convergence zone and the seasonal climatic changes this creates.
- ❑ Describe the general circulation of the world's oceans and its effects on climate and marine primary production.
- ❑ Differentiate between continental and maritime climates.
- ❑ Characterize the effects of global temperature and precipitation patterns on primary production and the distribution of ecosystems.

ENERGY FLOW THROUGH ECOSYSTEMS (pages 1127–1133)

Key Concepts

1. An *ecosystem* includes communities of organisms and their physical environment. Ecosystems can be studied at many levels, from local units to a global scale.
2. Only about 5% of solar energy is "fixed" by photosynthesis; the remainder is absorbed by the atmosphere or used by plants to power transpiration.
3. The total amount of energy captured by plants during photosynthesis is called *gross primary production*.
4. Gross primary production is greatest in parts of the world where water is abundant and temperatures are moderate, such as in the tropics.
5. Gross primary production in aquatic ecosystems decreases with depth due to extinction of light with depth, but increases with nutrient availability (greatest in areas of upwelling).
6. *Net primary production* is gross primary production minus the energy used by plants for their own maintenance and biosynthesis.
7. A *trophic level* includes all organisms that obtain their energy from the same source within an ecosystem.

8. Common trophic level designations include *photosynthesizers, herbivores* (organisms that consume photosynthesizers), *primary carnivores* (organisms that consume herbivores), *secondary carnivores* (organisms that consume primary carnivores), *detritivores* (consumers of dead organisms), and *omnivores* (organisms that consume at several levels).
9. Photosynthesizers are also called *primary producers*. Organisms in all other trophic levels are called *consumers*.
10. A *food chain* is a set of linkages showing the energy flow through trophic levels; a *food web* shows all the common food chains for an ecosystem.
11. Because much energy is lost as heat at each trophic level, less is available for transfer to the next level. The efficiency of energy transfer between trophic levels is

$$E = P/(P + R)$$

Where E is efficiency, P is net production, and R is respiration.
12. Endotherms, like birds and mammals, have very low efficiencies because of their high metabolism. Herbivores are less efficient than carnivores due to the difficulty in digesting plant material relative to animal material. In general, the efficiency of energy transfer between trophic levels is less than 20%.
13. A *pyramid of energy* is a graph in which the rate of energy capture at each trophic level is plotted from the lowest to the highest levels. Most energy pyramids have a broad primary producer base, with each successive consumer level proportionally smaller.
14. In a *pyramid of biomass*, one plots the biomass at each trophic level. Generally, energy and biomass pyramids from the same ecosystem have a similar shape.
15. Inverted biomass pyramids are sometimes observed in marine ecosystems. In these ecosystems, the primary producers, mostly bacteria and protists, reproduce at such a great rate that the biomass of herbivores may be larger than the biomass of producers.
16. Grasslands can support higher grazing rates by herbivores than forests can because little energy is allocated by grasses to produce supportive, woody tissue.
17. Detritivores are essential for releasing minerals tied up in dead organisms so they can be used again. High production rates in the tropics are possible, in part, because of the high rates of decomposition due to increased humidity and temperatures there.

Questions (for answers and explanations, see page 410)

1. Select all of the following that could *not* be considered an ecosystem.
   a. A small pond
   b. A large lake
   c. All the fish in a coral reef
   d. The Earth
   e. A pile of dung in a pasture

2. A plant in the dark uses 0.02 ml of O_2 per minute. The same plant in sunlight releases 0.14 ml of O_2 per minute. A correct estimate of its rate of *gross* primary production is
   a. 0.02 ml of O_2 per minute.
   b. 0.12 ml of O_2 per minute.
   c. 0.14 ml of O_2 per minute.
   d. 0.16 ml of O_2 per minute.

3. In the following food web, organism 9 is a:

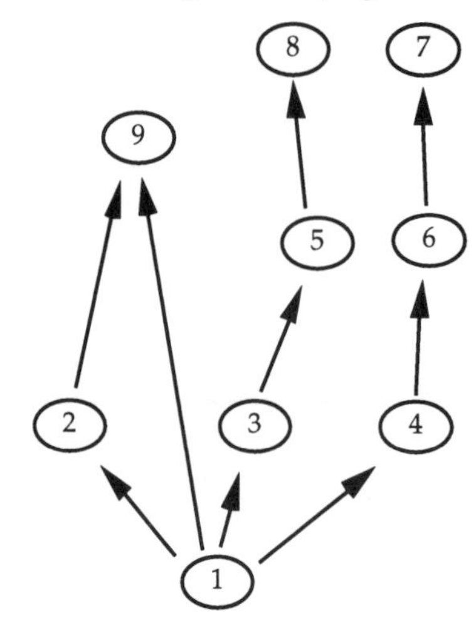

   a. primary producer.
   b. herbivore.
   c. primary carnivore.
   d. secondary carnivore.
   e. omnivore.

4. The food web shown in Question 1 has _______ trophic levels?

5. If the dry biomass of the primary producers of an ecosystem is 1000 g per square meter and the efficiency of transfer between trophic levels is 20%, what dry biomass weight would you expect at the *secondary carnivore* level? Show your answer and calculations below.

6. In the equation for efficiency of energy transfer between trophic levels, $E = P/(P + R)$, what does P represent?

Activities (for answers and explanations, see page 410)

- Where do energy pyramids like the one shown below occur and what are the conditions that create them?

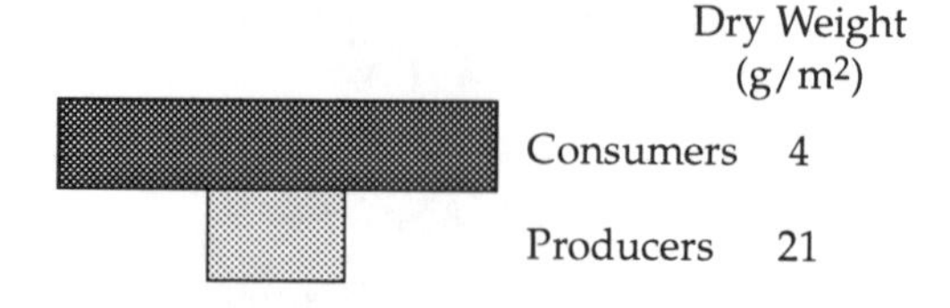

- Explain why bison prefer to graze within prairie dog colonies.

- Rank the following organism types in terms of their expected efficiency of energy transfer between trophic levels, where 1 = most efficient and 5 = least efficient.

  ____ Herbivorous mammal

  ____ Invertebrate detritivore

  ____ Carnivorous mammal

  ____ Invertebrate carnivore

  ____ Invertebrate herbivore

CHEMICAL CYCLING IN ECOSYSTEMS (pages 1133–1136)

Key Concepts

1. Unlike the cycling of energy through the trophic levels of ecosystems, chemical elements are not lost as they cycle. To follow the cycling of chemicals, it is useful to divide the global ecosystem into oceanic, freshwater, atmospheric, and land compartments.
2. Oceans receive materials from land as runoff from rivers and exchange material with the atmosphere only at the air–water interface.
3. Most materials that enter the oceans sink and become part of the bottom sediment. Consequently, except for areas of upwelling, most of the oceans are nutrient-poor and populations of organisms are sparse.
4. Freshwater ecosystems contain only a small fraction of the total water on Earth; most water resides in the oceans.
5. Because rivers and lakes are much less deep than oceans, elements in sediments are recycled more quickly in the former than in the latter.
6. Unlike oceans, where oxygen levels are fairly uniform throughout, oxygen in many lakes becomes depleted at the bottom due to decomposition.
7. In lakes, mixing is important in redistributing oxygen and nutrients.
8. Water is most dense at 4°C. In most temperate lakes, mixing brought about by wind occurs in the spring and in the fall when the entire water column is at this temperature.
9. The surface water of lakes is warmed during the spring and a *thermocline* is established, where the temperature changes rapidly with depth. Since water colder than 4°C expands and floats on warmer water, ice forms only on the surface.
10. Deep tropical lakes are more stratified than temperate lakes because the surface water is never cooled to 4°C.
11. The *troposphere*, the lowest layer of the atmosphere, contains 80% of the Earth's gases. The troposphere is about 17 km thick in the tropics, but thins to about 10 km at higher latitudes.
12. The *stratosphere* extends from the troposphere to about 50 km above the surface. Materials enter the stratosphere at the intertropical convergence zone and remain there relatively long.
13. The atmosphere is 78% nitrogen, 21% oxygen, and 0.03% carbon dioxide, with the remainder consisting of argon, neon, krypton, helium, hydrogen, ozone, and methane.
14. Certain atmospheric components, especially water vapor, carbon dioxide, and ozone, trap a significant proportion of infrared radiation, causing the temperature of Earth to be higher than it would be without an atmosphere.
15. One-fourth of Earth's surface is land. Because movement of material within this compartment is very slow, there is large variation in the distribution of nutrients in terrestrial ecosystems.
16. Exchange of materials between the land and the atmosphere is mainly through the activities of organisms.
17. Exchange of materials between the land and the oceans is via runoff from *groundwater* into rivers. The uplifting of marine sediments slowly returns materials to the land compartment.
18. Soils, resulting from weathering of parent rock, cover most of the land; aged soils have lost most of their nutrients to the groundwater.
19. *Humus* is soil with decomposed plant materials. *Mull* is alkaline humus rich in nutrients; *mor* is acidic humus, low in nutrients and resistant to decay.
20. Plants play key roles in the development and modification of soils.

Questions (for answers and explanations, see page 410)

1. Next to each of the following features, place an "O" or "F," if it is characteristic of the *ocean* or *freshwater* ecosystem (some features may apply to both).

   ____ Seasonal mixing of materials

   ____ Nutrient concentrations low, except locally

   ____ Oxygen concentration vertically uniform

   ____ Elements buried in bottom sediments for long periods of time

   ____ Receives material from land, mostly via groundwater

2. Next to each of the following features, place a "T" or "S," if it is characteristic of the *troposphere* or *stratosphere* (some features may apply to both).

   ____ Most water vapor resides here

   ____ Most ozone resides here

   ____ Mostly horizontal circulation of gases occurs in this layer

   ____ Represents the greatest mass of the total atmosphere

   ____ Circulation of this layer influences ocean currents

3. Next to each of the following features, place an "O," "F," "A," or "T" if it is characteristic of the *oceans, freshwater, atmospheric,* or *terrestrial* compartments of the global ecosystem (some features may apply to several).

    _____ Very slow movement of materials within compartment

    _____ Exchange of gases mostly via organisms

    _____ Circulation of materials affected by Earth's rotation on its axis

    _____ Circulation of materials affected by Earth's revolution around the sun

    _____ Organisms mostly in uppermost layer

4. In the following diagram, the temperature of a freshwater lake is plotted against depth.

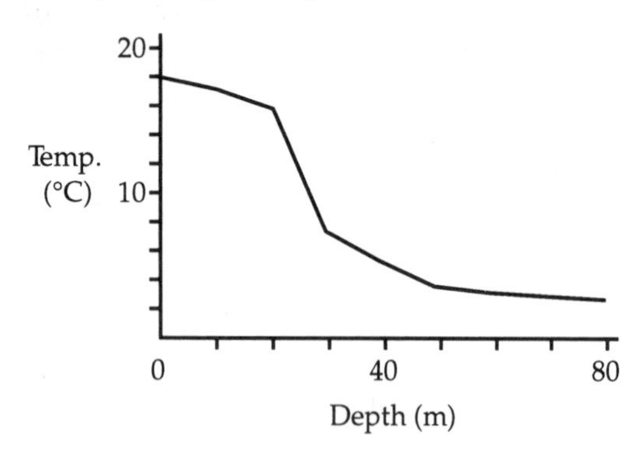

This temperature profile would be most characteristic of the lake during
a. early winter.
b. late winter.
c. spring.
d. summer.
e. fall.

5. Based on data in the graph from Question 4, the thermocline for the lake is located between _____ and _____ meters.

BIOGEOCHEMICAL CYCLES • HUMAN ALTERATIONS OF BIOGEOCHEMICAL CYCLES (pages 1136–1143)

Key Concepts

1. Earth is a closed system with respect to the essential elements: carbon, hydrogen, oxygen, nitrogen, phosphorus, and sulfur; they circulate between organisms and the environment in what are known as *biogeochemical cycles.*
2. Elements cycle through organisms quickly; the gaseous elements cycle through the nonliving environment most quickly.
3. The *hydrological cycle* refers to the exchange of water between land, the ocean, and the atmosphere. Evaporation and precipitation exchange water between the atmosphere and land/ocean. The movement from land to ocean as runoff equals the amount returned as ocean-derived precipitation.
4. Carbon enters the living world when CO_2 is fixed by a plant during photosynthesis; carbon moves into heterotrophs when they consume autotrophs.
5. Respiration and decomposition return carbon to the atmosphere as CO_2.
6. Carbonate minerals in sedimentary rocks and dissolved carbon in the oceans are the two major reservoirs of carbon in Earth's *carbon cycle.*
7. Fossil fuels, consisting of coal, petroleum, natural gas, and peat, are a reservoir of carbon produced when organisms decay anaerobically.
8. Increased atmospheric CO_2 levels resulting, in part, from combustion of fossil fuel, may lead to *global warming* because CO_2 is a *greenhouse gas;* it is transparent to sunlight but traps infrared heat radiation.
9. Nitrogen (N_2) is biologically inert until it is converted to ammonia by certain bacteria and cyanobacteria, or through industrial fertilizer production.
10. Other types of bacteria are responsible for interconverting nitrogen among its various inorganic forms (ammonia, nitrite, nitrate), and for returning it to the atmosphere as N_2.
11. Plants obtain inorganic forms of nitrogen from the soil (or root nodules) and animals get their organically fixed nitrogen by consuming plants.
12. Flux through the atmosphere is of minor importance in the *phosphorus cycle.* Phosphorus cycles through organisms quickly, is eventually incorporated into sedimentary rock, and returns as dissolved phosphate ion when the rock is uplifted and undergoes weathering.
13. Unlike nitrogen and phosphorus, sulfur is seldom limiting in natural ecosystems.
14. In the *sulfur cycle,* natural venting of SO_2 and H_2S by volcanoes and fumaroles is significant, but *dimethyl sulfide* release by marine algae and terrestrial decomposition of organically fixed sulfur by bacterial fermentation is also an important flux.
15. Dimethyl sulfide is an important nucleation agent for water drop formation during cloud development. Increased cloud cover due to higher atmospheric levels of dimethyl sulfide may help slow global warming.
16. Because the availability of essential elements in the biogeochemical cycles sometimes limits the metabolism of organisms, human modification of these cycles can have local, regional, and global biological consequences.
17. Because photosynthesis is frequently limited by phosphorus, *eutrophication,* the addition of nutrients, especially phosphorus, to fresh water can lead to *blooms* of algae.
18. The effects of eutrophication include changes in the species of algae present, reduction in oxygen levels due to increased organic decomposition, and the dominance of anaerobic organisms in the ecosystem.
19. Eutrophication of *Lake Erie* was caused by nutrients added to the lake from domestic and industrial wastes and as agricultural runoff.
20. Lakes recover quickly from eutrophication if nutrient inputs are reduced and if the rate of *water turnover* in the lake is rapid.
21. *Acid precipitation* is an example of a regional disturbance caused by human-produced changes in the nitrogen and sulfur cycles.

22. The low pH of acid precipitation is caused by the presence of sulfuric and nitric acids derived from the burning of fossil fuels.
23. Acid precipitation damages plants directly and reduces populations of nitrifying bacteria in lakes, which leads to accumulation of ammonia and death of fish populations.
24. Perturbation of the carbon cycle, due to increased atmospheric CO_2 levels from combustion of fossil fuels and forest burning, has had an effect on Earth's ecosystem called *global warming*.
25. Atmospheric CO_2 levels may double from pre-Industrial Revolution levels by the year 2050. Possible effects include massive climate change, melting of the Arctic and Antarctic ice caps, flooding of coastal areas, and disruption of agriculture.
26. Release of chlorofluorocarbons (CFCs) and other chlorine-containing compounds have altered the global chlorine cycle and led to depletion of ozone and the creation of an ozone "hole" in the stratosphere.

Questions (for answers and explanations, see page 411)

1. Which of the following statements about biogeochemical cycles is *not* true?
   a. Gaseous elements cycle more quickly than elements without a gaseous phase.
   b. Most elements remain longest in the living portion of their cycle.
   c. You may have some atoms in your body that were originally part of a dinosaur.
   d. Earth is a closed system relative to the essential elements of life.
   e. Biogeochemical cycles all include both organismal and nonliving components.
2. Next to each of the following features, place a "C," "N," "P," "S" or "Ch" if it is characteristic of the biogeochemical cycles of *carbon, nitrogen, phosphorus, sulfur* or *chlorine* (some features may apply to several cycles).
   _____ Major reservoir is atmospheric
   _____ Major reservoir is in sedimentary rocks
   _____ Often in short supply in ecosystems
   _____ Fossil fuel reserve is part of this cycle
   _____ Major human impact on cycle in many ecosystems
   _____ Involved in cloud formation
   _____ Lacks a gaseous phase
   _____ Major inorganic form is directly available to only a small group of bacteria
   _____ Contains a form which is a greenhouse gas
   _____ Most fluxes involve organisms
   _____ Involves chlorofluorocarbons (CFCs)
3. Place numbers next to the following events in the eutrophication of Lake Erie to reflect their correct chronological sequence.
   _____ Mayflies like *Hexagenia* are replaced by oligochaete worms as dominant bottom dwellers.
   _____ Increased phosphorus input from sewage and agricultural runoff
   _____ Respiratory demand from decomposers increases.
   _____ Fish species change.
   _____ Oxygen levels drop in deeper water.
   _____ Algal blooms occur.
4. Which of the following statements is *not* a major concern about our alteration of the carbon cycle?
   a. The mean temperature of the Earth may increase 3–4°C by the year 2050.
   b. High sulfur fuels are used by power plants because they are less expensive than low sulfur fuels.
   c. The increase in atmospheric CO_2 exceeds the ability of the oceans to absorb the increase.
   d. The polar ice caps are expected to melt if global warming becomes a reality.
   e. CO_2 is a gas that traps infrared radiation.

Activity (for answers and explanations, see page 411)

- What factors determine how quickly a lake can recover from eutrophication? Discuss why Lake Erie will never be restored to its historical condition.

AGRICULTURE AND ECOSYSTEM PRODUCTIVITY • CLIMATES ON EARTH (pages 1144–1149)

Key Concepts

1. Modern agricultural practices are used to promote the success of some species at the expense of others. Modern agriculture has created many ecological problems and has flourished largely because of the availability of cheap fossil fuel.
2. *Integrated pest management* is an attempt to minimize the adverse effects of agriculture by reducing reliance on chemicals through the application of cultural and biological control methods.
3. Tropical soils are ancient and most of the mineral nutrients are tightly cycled and reside in the vegetation. The application of temperate zone agricultural practices to tropical soils has not been successful because the soils quickly become depleted.
4. Earth's climate patterns are determined by circulation of the atmosphere and ocean currents, which are driven by energy from the sun.
5. Although all parts of Earth receive equal amounts of sunlight per year, the amount of heat received varies with the angle of sunlight striking the ground.
6. The rate of heat arrival per unit of Earth surface area depends primarily on the angle of sunlight. When the sun is low in the sky, more heat energy is absorbed by the atmosphere before it reaches the ground, and upon reaching the ground, is spread over a wider area.

7. Seasonal changes in day length and sun angle increase directly with distance from the equator (increasing latitude), so seasonal temperature differences also increase with latitude (about 0.4°C for each degree of latitude).
8. As gases expand, they cool, so air temperature decreases with altitude. Since cooler air holds less moisture, as air rises, clouds form and rain often falls.
9. *Rain shadows* form on the leeward sides of mountain ranges, because air loses much of its moisture as rain on the windward side of mountains.
10. Air heated at the equator rises, moves poleward, and descends at latitudes of about 30° north and south, creating hot deserts there, such as those of the Sahara and Australia.
11. Air rising at about 60° north and south latitude descends at the poles to create cold deserts there.
12. Surface air is deflected to the right in the Northern Hemisphere and to the left in the Southern Hemisphere by the spinning of Earth on its axis.
13. Air descending at 30° north and south latitude becomes the northeast and southeast *trade winds* if it flows back toward the equator or the *westerlies* if it flows toward the poles. The area where the trade winds converge is called the *intertropical convergence zone*.
14. Solar energy flux is greatest where the sun is directly overhead at noon; this region shifts north and south seasonally because of the tilt of Earth's axis.
15. The location of the intertropical convergence zone shifts seasonally with the location on Earth where the sun is directly overhead at noon, but lags behind it by about a month. Consequently, the intertropical convergence zone shifts north during the northern summer and south during the northern winter.
16. Tropical and subtropical rainy seasons usually occur when an area is located in the intertropical convergence zone and ascending air produces rain. The tropical and subtropical dry seasons occur when the intertropical convergence zone has departed and the area is dominated by the trade winds.
17. The *specific heat* is the amount of heat required to change the temperature of 1 gram of a substance by 1°C. The specific heat of water is 1 cal/g; the specific heat of air or land surfaces is much lower.
18. Ocean currents are driven by prevailing wind, diverted by continents, and affected by Earth's rotation. The result is a general clockwise circulation in the Northern Hemisphere and a counterclockwise circulation in the Southern Hemisphere.
19. Upwelling of deeper, cooler water on the western sides of continents replaces some of the water that has flowed toward the equator and away from the land.
20. If air blowing over water crosses warmer land it retains its moisture; if air blowing over water crosses colder land it releases its moisture.
21. *Continental climates* are characterized by large seasonal variation in temperature; the largest land masses have the greatest seasonal variations in temperature.
22. *Maritime climates* prevail on coasts of continents where sea breezes moderate the climate.
23. The distribution of ecological communities on Earth is determined by the interplay of global temperature and water availability. Areas of high annual production are in wet tropical and subtropical regions. Low production areas are in subtropical deserts and at high latitudes.

Questions (for answers and explanations, see page 411)

1. In comparing an acre of land in Colombia with an acre of land in Michigan, which of the following would *not* differ?
   a. The angle of the sun reaching the ground in the month of July
   b. The solar energy flux in the month of July
   c. The annual solar energy flux
   d. The total hours of daylight per year
   e. The mean annual air temperature
2. The following diagram shows a mountain with a sea breeze blowing as indicated by the arrow. The area with air that is *both* relatively warm and dry would be:

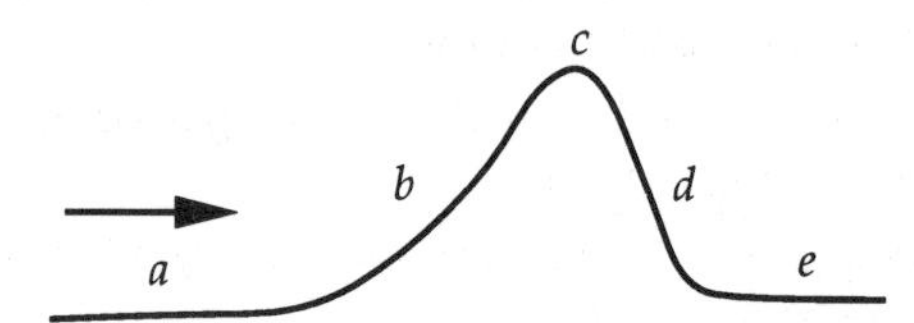

3. In what area of the diagram in question 2 is a process occurring that is similar to the process that occurs in the intertropical convergence zone? ______
4. If the Earth did *not* spin on its axis, from what direction would the northeast trade winds blow?
   a. Northeast
   b. South
   c. North
   d. East
   e. Southwest
5. Ocean circulation patterns are influenced by all of the following *except*
   a. circulation of Earth's atmosphere.
   b. deflection by land masses.
   c. upwelling of deeper water.
   d. rotation of Earth on its axis.
   e. prevailing winds.
6. Differences in the specific heat of water and land have most to do with creating
   a. the intertropical convergence zone.
   b. maritime climates.
   c. continental climates.
   d. the westerlies.
   e. oceanic upwelling.
7. Which of the following would *not* likely be a method employed in integrated pest management?
   a. Mixed plantings of crop plants
   b. Introduction of a natural disease organism or an important pest
   c. Aerial spraying of insecticides
   d. Use of chemical attractants for controlling pest species
   e. Development of pest-resistant strains of crop plants

50

Biogeography

CHAPTER LEARNING OBJECTIVES—after studying this chapter you should be able to:

- ❑ Define what biogeography is and describe the types of questions of concern to biogeographers.
- ❑ Know the fundamental processes that determine the presence or absence of a species in an area.
- ❑ Know the important role Alfred Wegener played in the development of biogeography.
- ❑ Describe the distribution of some contemporary organisms that illustrate continental drift.
- ❑ Characterize the principle concerns and approaches of historical biogeography.
- ❑ Characterize the principle concerns and approaches of ecological biogeography.
- ❑ Explain the principle of parsimony and its application in biogeography.
- ❑ Describe the use of cladistic methods in biogeography and how a taxonomic cladogram can be transformed into an area cladogram.
- ❑ Differentiate between a vicariant distribution and a dispersal distribution and give an example of each.
- ❑ Define what endemic species are and how they can arise.
- ❑ Characterize the relationship between the degree of endemism and the amount of time that an area has been isolated.
- ❑ Use cladistic information to make inferences about continental drift or the appearance of barriers.
- ❑ Name and know the locations and boundaries of the major terrestrial biomes.
- ❑ Describe the major assumptions and predictions of the equilibrium species richness model and be able to apply it.
- ❑ Explain how island size and distance from the mainland affect the predictions of the equilibrium species richness model.
- ❑ Describe the variety of evidence that supports the equilibrium species richness model.
- ❑ Describe what is meant by a habitat island and the applicability of the equilibrium species richness model for studying habitat islands.
- ❑ Characterize the conditions that led to the appearance of chaparral, or Mediterranean, vegetation in many parts of the world.
- ❑ Define what a biome is, describe how biomes are named, and explain the basis for their latitudinal and altitudinal distribution.
- ❑ Name and describe the characteristic features of the major terrestrial biomes.
- ❑ Relate some important properties of water to characteristics of aquatic ecosystems.
- ❑ Characterize the major types of freshwater biomes.
- ❑ Describe the general features of marine biomes, the names and features of the zones of the ocean, and the names and characteristics of the major marine communities.

THE GOALS OF BIOGEOGRAPHY • HISTORICAL BIOGEOGRAPHY (pages 1152–1157)

Key Concepts

1. *Biogeography* is the study of the present and past distribution of organisms. Biogeographers want to know why specific kinds of organisms live in certain places and, also, why they do not live in other places.
2. A species inhabits an area because it either evolved there or dispersed there from a different area. A species that is absent from an area either evolved in that area but disappeared, or it evolved in a different area and never dispersed. Thus, distribution patterns can be explained by either dispersal or where taxa evolved.
3. Based on a consideration of the interlocking continental shapes and the distributions of plants and animals, in 1912 *Alfred Wegener* proposed that the continents have moved over large distances.
4. Some widespread species alive today evolved on *Pangaea* before that supercontinent broke up at the close of the Mesozoic Era, about 65 million years ago. For example, lungfish were widespread throughout Pangaea, but the distribution of living species is restricted to equatorial areas of South America and Africa.
5. Originally, biogeographers assumed that the continents were fixed in place and that most biogeographical patterns resulted from *dispersal* of organisms. Early biogeographers also underestimated the age of Earth.

6. More recently, ecological biogeographers have attempted to explain biogeographical patterns with reference to interspecific interactions such as *competition*, *predation*, and *mutualism*.
7. Today, many biogeographers use cladistic methods to reconstruct *phylogenies*. By applying information on the distribution of a group of species, a taxonomic cladogram for the group can be transformed into an area cladogram.
8. *Historical biogeography* is concerned with the large-scale distribution of groups and their relationship to evolutionary histories. It concentrates on long time periods and global distribution patterns.
9. *Ecological biogeography* examines the interactions of organisms with one another and the environment to explain their present distribution. It concentrates on recent history and local distribution patterns.
10. Three types of data are especially useful to historical biogeographers: phylogenies that show relationships relative to a time scale, fossils that indicate past distributions, and distributions of contemporary organisms.
11. Biogeographers compare the distribution patterns of many different types of organisms currently living in the same area and apply the principle of *parsimony* to arrive at the least complex explanation for any underlying similarities. Parsimony states that the simplest explanation is most probably correct.
12. An organism that evolved in an area that is now separated into two separate subareas by a barrier has a *vicariant distribution*.
13. An organism that evolved in one area and crossed a barrier to occupy a second area has a *dispersal distribution*.
14. Species or higher taxa that are found in only one area are *endemic* to that area.
15. Endemic taxa can be taxa of either ancient origin or recent taxa dispersing to new areas.
16. The number of endemic taxa in an area is related to the period of time that the area has been separated from other areas; Australia has the most distinct biota (period of separation, 65 million years), while North America and Eurasia have the most similar biota because they were joined for most of Earth's history.
17. Information on the current distributions of species can be applied to a cladogram for the group. These area cladograms can then be used to make inferences about continental drift or the appearance of barriers. For example, an area cladogram for a group of midges suggests that New Zealand separated from Gondwanaland while Australia and South America were still exchanging midge species. An area cladogram for the black-throated green warblers suggests when different taxa were separated by intervals of glaciation in North America.
18. Since the separation of the continents, terrestrial biotas have evolved into distinctive *biogeographic regions*, called the *Nearctic*, the *Neotropical*, the *Australasian*, the *Ethiopian*, the *Oriental*, and the *Palearctic* regions.
19. The biological distinctiveness of the terrestrial biomes are maintained by major water, mountain, or desert barriers.

Questions (for answers and explanations, see page 411)

1. Which of the following types of information would be *least used* by an historical biogeographer?
   a. Experiments on species richness
   b. Fossils showing past distributions of organisms
   c. Present distributions of organisms
   d. Cladistic systematics
   e. Application of the principle of parsimony
   f. Molecular clock data so divergences can be dated

2. The following diagram shows the present locations of four continents (W, X, Y, Z) and the number of endemic beetle families found on each. Three were part of a supercontinent in the past, the other was always separate.

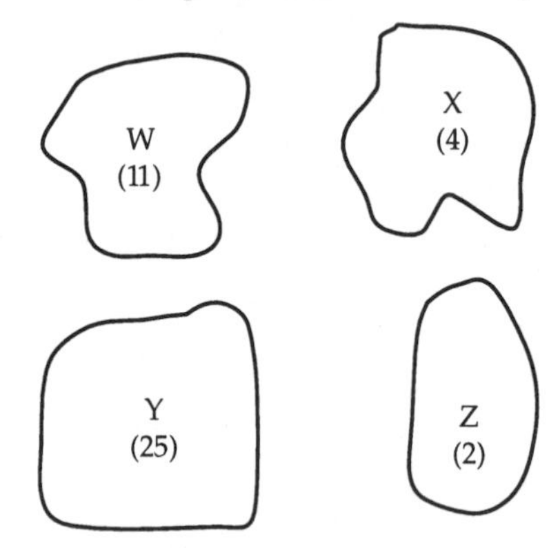

   Based on this information, choose the correct conclusions
   a. W was always separate, Y separated from the supercontinent first.
   b. X was always separate, Y separated from the supercontinent first.
   c. Y was always separate, W separated from the supercontinent first.
   d. Y was always separate, Z separated from the supercontinent first.
   e. Z was always separate, X separated from the supercontinent first.

3. The following diagram shows three islands. Islands A and B were connected in the past, C was always separate. A species of land snail is found on all three islands.

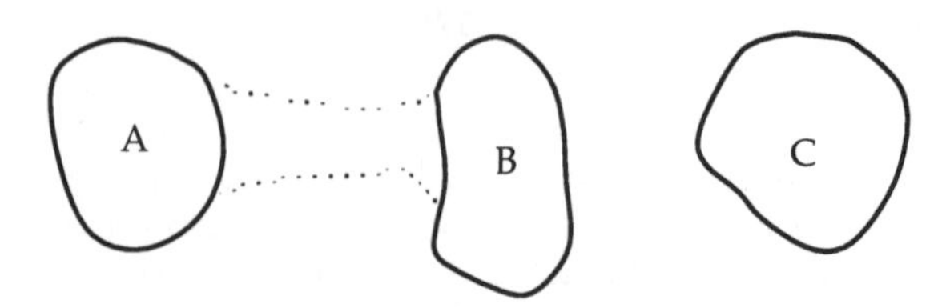

   Select all of the following that *do(es) not* correctly describe the distribution of this snail relative to the three islands.
   a. Vicariant distribution relative to A and B
   b. Dispersal distribution relative to A and C
   c. Dispersal distribution relative to B and C
   d. Vicariant distribution relative to A and C
   e. Endemic relative to C

4. Which of the following biogeographical regions represents the largest area?
   a. Nearctic
   b. Palearctic
   c. Neotropical
   d. Oriental
   e. Australasian

5. Circle an area on the following graph that would most apply to the spatial and temporal scale of processes studied by ecological biogeographers.

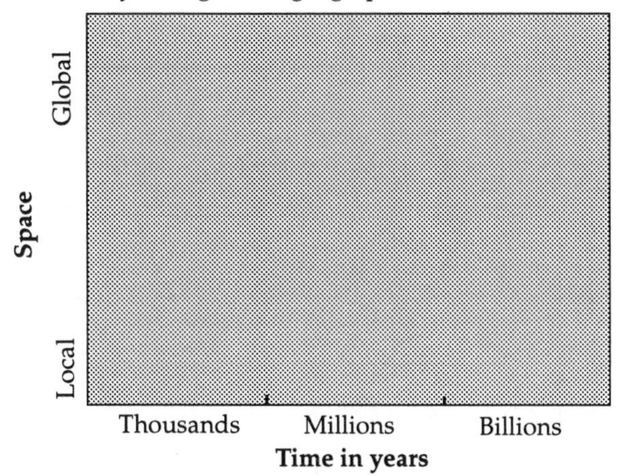

Activities (for answers and explanations, see page 412)

- Based on an examination of the following cladogram and information on the distribution of the taxa members on four continents, discuss what can be inferred about the histories of the four continents.

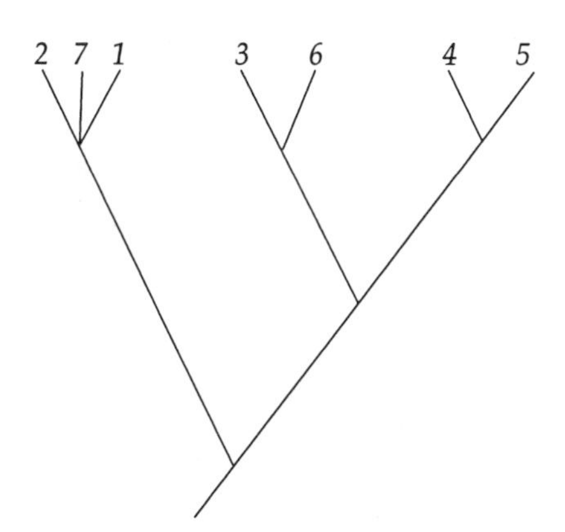

Species	Continent A	Continent B	Continent C
1	✔	✔	✔
2	✔	✔	✔
3	✔	✔	
4		✔	
5		✔	
6	✔	✔	
7	✔	✔	✔

- What does the past and present distribution of lungfish suggest about continental drift?

ECOLOGICAL BIOGEOGRAPHY (pages 1157–1160)

Key Concepts

1. One objective of ecological biogeography is to discover the factors that influence the *species richness* (total number of species) of an area.
2. Arrival of new species to an area increases species richness; extinction of species already there decreases species richness.
3. *Robert MacArthur* and *E. O. Wilson* developed an *equilibrium species richness model* to predict species richness of islands of known size and distance from the mainland.
4. The species present on the mainland that are available to colonize an island make up the *species pool.*
5. As the number of species on an island increases, the rate of arrival of new species decreases until it equals zero when all members of the species pool have colonized the island.
6. As the number of species on an island increases, *interspecific competition* increases and reduces the resources available for each species.
7. As the number of species on an island increases, predation and parasitism increase as new predators and parasites of species already present arrive.
8. The *extinction rate* of resident species increases with the number of species present on an island.
9. The species number where the extinction rate equals the new arrival rate is the equilibrium number of species that the island can support, although the species composition (which species are present) may change through time.
10. The model assumes that the arrival and extinction rates are constant at the island's equilibrium number of species.
11. Distance from the mainland affects the predictions of the equilibrium species richness model. Distant islands are harder to find than nearer islands. As a result, the arrival rates of new species on distant islands are less and they support fewer species.
12. Island size affects the predictions of the equilibrium species richness model; smaller islands support fewer species because they have fewer resources, and arrival rates are less since they are harder to find than larger islands.
13. Natural distributions of birds, plants, insects, lizards, and mammals generally support the predictions that species numbers on islands correlate positively with island size and negatively with distance from the species pool.
14. On *Krakatoa*, an island that was wiped clear of life in 1883 by a volcanic eruption, 271 plant species and 27 resident bird species recolonized within 50 years. New species continue to arrive, but at a diminishing rate.
15. An experimental test of the species richness equilibrium model involving fumigation of small, red mangrove islands in the Florida Keys also confirmed the model. Within a year the fumigated islands had about the same number of species of arthropods.

16. Patches of suitable habitat for a species separated by areas of unsuitable habitat are called *habitat islands.*
17. Although arrival rates are higher for habitat islands than for islands separated by water, the equilibrium species richness model makes the same predictions for those "islands."

Questions (for answers and explanations, see page 412)

1. According to the equilibrium species richness model, which of the statements about the following graph showing the effect of species number on arrival and extinction rates is *not* true?

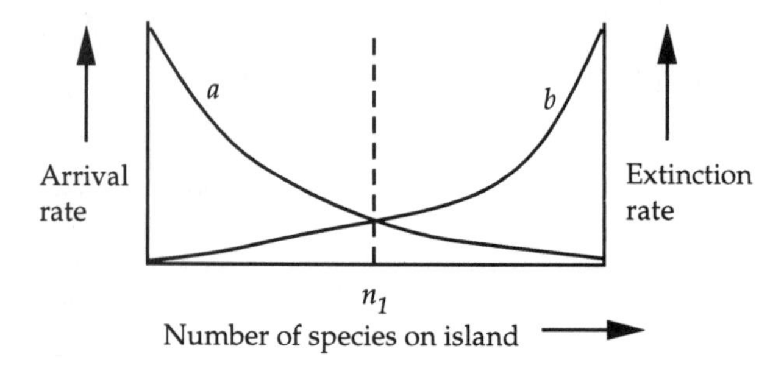

a. *a* is the arrival rate curve.
b. *b* is the extinction rate curve.
c. n_1 is the equilibrium species number.
d. The arrival rate is constant at n_1.
e. The extinction rate is zero at n_1.

2. The species numbers of many islands of different sizes are plotted against their distance from the mainland. Data points for islands of *similar size* were connected to form the five curves shown in this figure.

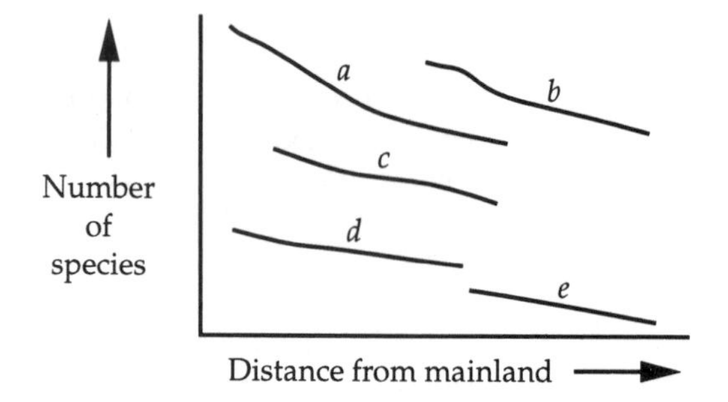

Circle the letter of the curve corresponding to the group of islands with the *smallest* size.

3. A small volcanic island was destroyed by an eruption. The following graph shows recolonization by a number of different organism types.

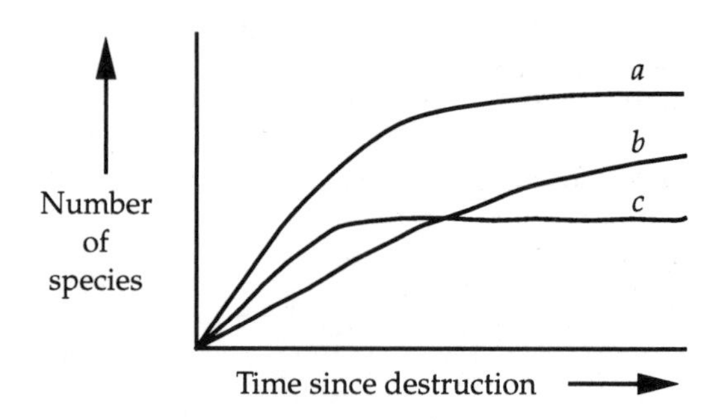

Identify the correct curve for each organism type.

____ Birds
____ Plants
____ Insects

4. The equilibrium species richness model makes predictions about the effects of species number on the rate of extinction. In the following figure, the solid curve shows this relationship for a large island. Which of the curves shows the expected relationship for a *small* island?

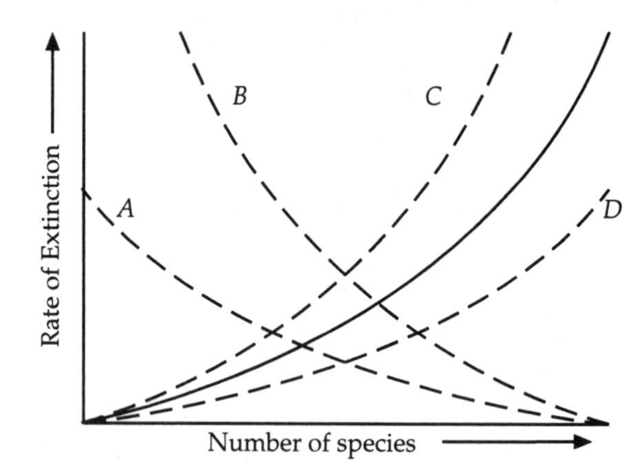

a. Curve *A*
b. Curve *B*
c. Curve *C*
d. Curve *D*
e. It would be the same curve as for the large island

TERRESTRIAL BIOMES • AQUATIC ECOSYSTEMS (pages 1160–1173)

Key Concepts

1. *Convergent evolution* is the development of similar adaptations among species that were originally different. Convergent evolution depends on sharing similar environments and selection pressures and availability of adequate time for the evolution of similar adaptations.
2. A good example of convergence is Mediterranean, or chaparral, vegetation (evergreen shrubs with tough, drought-resistant leaves) displayed by unrelated species in many parts of the world with mild winters and dry, hot summers.
3. Organisms occupying similar climatic zones are classified into several major terrestrial ecosystems called *biomes.* Biome names are usually derived from their characteristic vegetation types and location or climate types.
4. Many biomes are distributed altitudinally in mountains, in addition to their latitudinal distribution.

5. The *tundra biome* is found in the Arctic and above the treeline in mountains at all latitudes. In the Arctic tundra, most primary production occurs during the short summers and many animals leave during the long winters or are dormant. Because of permafrost, tundra soil is very wet although precipitation is low.
6. The *boreal forest* biome is dominated by evergreen, coniferous trees in the Northern Hemisphere, but by beech in the Southern Hemisphere. Long cold winters, short warm summers, and communities with low species numbers characterize this biome. Most plants are wind-pollinated and produce large crops of wind-dispersed seeds that are the major food supply for many animal species.
7. The *temperate deciduous forest* biome has a greater diversity of tree species than the boreal forest and its species are adapted for maximum photosynthesis during the summers when temperatures are warmer and precipitation is good. Forest structure is more complex, with a well-defined understory, and many tree species rely on animals for pollination and seed dispersal.
8. The *grassland* biome is common in areas with restricted rainfall during a significant portion of the year. Unlike tropical grasslands, temperate grasslands experience large seasonal temperature differences. Grasses store much energy in their root systems and can withstand heavy grazing. In addition to grasses, *sedges* and *forbs* are also common plant types.
9. The *cold desert* biome is common in the interiors of large continents and in rain shadow areas. Most rain falls in the winter and the dominant low shrubs grow mostly in spring. Many seed-eating animals live in this biome.
10. The *hot desert* biome is found in a belt 30° north and south of the equator where dry air descends in its circulation through the troposphere and rain falls mostly during the summer. Succulent plants and annuals are abundant and the large seed crop supports many insects and rodents.
11. The *chaparral*, or Mediterranean, biome is found in areas with maritime climates where winters are cool and wet and summers hot and dry. Vegetation consists of low evergreen shrubs with drought-resistant leaves and bird-dispersed fruit.
12. The *thorn forest* is usually found on the equatorial side of hot desert biomes where a rainy season occurs in the summer. The biota is similar to that of the hot desert biome.
13. The *tropical savanna* biome is common in dry, tropical areas of Africa, Australia, and South America. The savanna is a grassland with scattered trees and large, diverse populations of grazing and browsing mammals.
14. The *tropical deciduous forest* biome replaces thorn forest where the rainy season is longer. Species diversity is great and most trees lose their leaves during the dry season. Much of this biome has been cleared for agriculture because the soils are more fertile than wetter areas.
15. The *tropical evergreen forest* biome occurs where the rainy season predominates and total rainfall exceeds 250 cm per year. This biome has the greatest species diversity and energy flux found on Earth. Its productivity is based on rapid nutrient cycling. Food webs and ecological interrelationships are complex and fragile.
16. The *tropical montane forest* biome is found on the slopes of tropical mountains where lower temperatures have a depressing effect on vegetation growth. Epiphytes are abundant in areas with frequent clouds.
17. Because water is more dense than air, it provides more support for organisms, but it also offers more resistance to locomotion and exerts more force when moving over organisms.
18. Primary production in aquatic ecosystems is restricted to near the surface because water absorbs light more readily than air does.
19. Oxygen is often limiting in aquatic, especially freshwater, ecosystems; it is seldom limiting in terrestrial ecosystems.
20. Because the primary producers in aquatic ecosystems are small, frequently unicellular, algae, they provide little structure to the ecosystem.
21. Along the margins of aquatic ecosystems (especially the marine intertidal zone), where resources are more abundant, intense competition for physical space can be observed.
22. Wave action provides intertidal communities with a continuous resource supply; these communities are the most productive on Earth.
23. *Lake ecosystems* are dominated by green algae and cyanobacteria as primary producers, with rotifers, cladocera, insects, and fish as important consumers. Many insects have both aquatic and terrestrial stages in their life cycles.
24. Because of water flow, little primary production occurs in *rivers* and *streams*, and these ecosystems are dominated by detritivores that depend on organic input (especially leaves) from bordering terrestrial ecosystems.
25. Primary production in *marine ecosystems* is carried out mostly by *phytoplankton* (bacteria, diatoms, and dinoflagellates) in coastal areas of upwelling or in surface waters of the open oceans. Crustaceans are the primary grazers of phytoplankton.
26. The ocean bottom is called the *benthic zone*, with the portion of this zone below the penetration of light named the *abyssal zone*. The food supply for abyssal zone organisms comes from the upper part of the ocean where primary production occurs.
27. The open waters of the ocean are the *pelagic zone*. Plankton in this zone are grazed by planktivores that range in size from small fish to large whales.
28. The *littoral zone* is the coastal area between high and low tides. Ecosystems there are more diverse and complex than the pelagic communities.
29. *Coral reefs* are the most complex marine ecosystems because of the physical structure provided by the variety of corals growing there.

30. The waters overlying the *continental shelf* are important breeding grounds for many pelagic species, especially fish and arthropods.

Questions (for answers and explanations, see page 412)

1. The following climograph shows yearly variation in rainfall and temperature for four biomes. Select the correct curve for each of the following biomes.

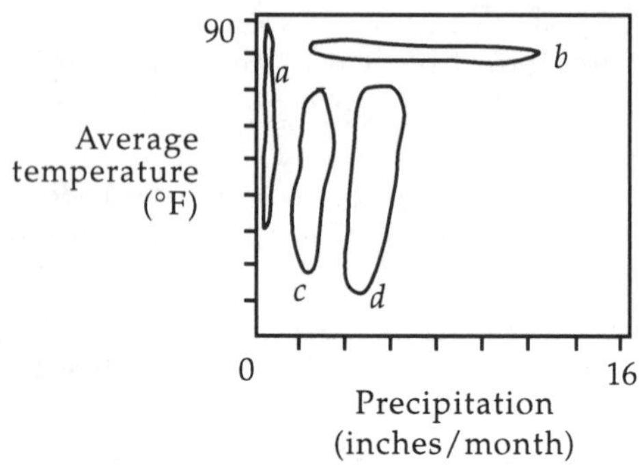

_____ Temperate grassland biome

_____ Tropical evergreen forest biome

_____ Temperate deciduous forest biome

_____ Hot desert biome

2. Match the letters of the following biomes with the descriptions that follow.

   a. Tundra biome *b.* Boreal forest biome
   c. Cold desert biome *d.* Chaparral biome
   e. Tropical deciduous forest biome

   _____ Mostly coniferous, wind-pollinated and wind-dispersed tree species

   _____ Leaves lost during dry season; agriculturally desirable land

   _____ Cool winters, hot dry summers; maritime climate

   _____ Frequently found in wind shadows

   _____ Distribution altitudinally or latitudinally determined; permafrost present

3. Which, if any, of the following is *not* an important difference in the physical features of water and air as media for an ecosystem?
   a. Specific heat
   b. Light absorption
   c. Density
   d. Buoyancy
   e. Resistance to locomotion

4. Which of the following statements about terrestrial or aquatic ecosystems is *not* true?
   a. Primary producers determine the physical structure of terrestrial ecosystems.
   b. In many aquatic ecosystems, animals are more conspicuous than plants.
   c. Aquatic food webs have fewer trophic levels than terrestrial food webs.
   d. In some aquatic ecosystems, organisms compete for physical space.
   e. Primary production rates of different terrestrial ecosystems can vary greatly.

5. Which of the following statements about stream and river ecosystems is *not* true?
   a. Lake ecosystems vary vertically more than rivers.
   b. River ecosystems vary horizontally more than lakes.
   c. Primary production is more important in lakes than in rivers.
   d. River ecosystems are dominated by detritivores.
   e. Light is seldom limiting in river ecosystems.

6. Next to each of the following features, place a "P," "L," or "A," if it is characteristic of the *pelagic, littoral,* or *abyssal* zones of marine communities.

   _____ Most plankton are found here.

   _____ The community here consists entirely of consumers.

   _____ Organisms in this zone must have adaptations to resist sinking.

   _____ Most marine primary production in this zone

   _____ Globally, the zone occupying the smallest total area

Activities (for answers and explanations, see page 412)

- What are some of the basic properties of water that influence important features of aquatic ecosystems?

- Characterize the rocky intertidal community.

51

Conservation Biology

CHAPTER LEARNING OBJECTIVES—after studying this chapter you should be able to:

- ❑ Describe the objectives of conservation biology and list the areas of biology that have contributed to its development.
- ❑ Provide some examples of extinctions caused by overexploitation.
- ❑ Discuss some of the special features of island-based species that subject them to increased risk of extinction.
- ❑ Provide some examples of extinctions caused by introduced pests, predators, and competitors.
- ❑ Discuss the effect of modern agriculture on species diversity and the general phenomenon of co-option.
- ❑ Provide some examples of extinctions caused by habitat destruction and creation of habitat islands.
- ❑ Characterize the factors that increase a species' probability of becoming extinct as its numbers are reduced to a low level.
- ❑ Define the term "minimum viable population" (MVP).
- ❑ Describe the process of population vulnerability analysis and its objectives.
- ❑ Differentiate between demographic stochasticity and genetic stochasticity and explain how each contributes to determining the minimum viable population.
- ❑ Explain the objectives of captive propagation and provide some examples of its successful application.
- ❑ Describe where rare species are likely to be found and the special problems faced by rare species.
- ❑ Describe the projected magnitude of global warming and its expected effects on communities and individual species.
- ❑ Define what endemic species are and explain where most endemic species occur.
- ❑ Explain what types of organisms are likely to be keystone species and why identification of a community's keystone species is important to conservation efforts.
- ❑ Discuss the process of habitat fragmentation in terms of size effects and edge effects and provide some examples of its importance.
- ❑ Describe the United States' national parks model and its suitability for export to other parts of the world
- ❑ Describe the features of a megareserve and provide an example.
- ❑ Describe the features of a forest reserve and provide an example.
- ❑ Characterize the objectives of the subdiscipline of conservation biology called restoration ecology and provide an example of its successful application.
- ❑ Explain what ecosystem services are and describe the major ecosystem services.

CAUSES OF EXTINCTIONS • STUDIES OF INDIVIDUAL SPECIES (pages 1176–1184)

Key Concepts

1. *Conservation biology* is concerned with the causes of species richness and the preservation of genes, species, and communities. Its conceptual basis is in the fields of ecology and evolutionary biology.
2. The accelerating rate of extinction due to human activities has created a *biodiversity crisis.*
3. Extinction due to overexploitation by humans is particularly common on islands because on many islands, species evolved in the absence of mammalian predators.
4. Flightless species of birds that frequently evolve on islands are especially susceptible to extinction by overexploitation.
5. Large mammals were exterminated by overhunting in North America and Australia after human colonization of those areas.
6. Introduced pigs and black rats have been destructive to many island organisms, such as the tortoises in the Galapagos archipelago.
7. Chestnut blight and Dutch elm disease were inadvertently carried to North America from overseas.
8. Plant–pollinator mutualisms are often highly species-specific on islands where ecological communities are species-poor. Extinction of many species of Hawaiian honeycreepers, sole pollinators of plants in the genus *Lobelia*, have left many lobelias without pollinators.

9. Traditional agricultural ecosystems maintain many more natural species than does modern, high energy input agriculture. Modern agriculture uses chemicals to eliminate competing species.
10. *Co-option* is the diversion of an ecosystem's primary production into species intended for human use; today, about 30% of terrestrial production is coopted.
11. Habitat loss or the creation of habitat islands too small to support dependent species is predicted to become a major cause of extinction in the future.
12. *Forest analogs* are harvested forests that are managed so as to retain the ecological functions of nutrient recycling, erosion control, water cycling, and climate moderation associated with the original forest. Strip-cutting is an example of one such management practice.
13. Habitat modification may also favor certain native species of predators and parasites that can have unforeseen effects on ecological communities. Reduction in songbird populations in North America in recent years may be related to the increase in brown-headed cowbird populations associated with habitat modification.
14. Preservation of a key species that requires a large amount of suitable habitat ensures the preservation of other species that live in the same community.
15. Loss and deterioration of habitat results in isolated, small, local populations that are very susceptible to local deleterious effects.
16. The *minimum viable population (MVP)* is the estimated population size required for continued species survival and growth.
17. *Demographic stochasticity* is the amount of variation in birth and death rates in a population.
18. Small populations are more at risk of extinction from demographic stochasticity, especially if low birth and high death rates coincide in time.
19. *Genetic stochasticity* is the amount of variation in the genetic makeup of a population.
20. *Population vulnerability analysis (PVA)* estimates the influence of population size on the probability of extinction for a species; PVA must take into account both demographic and genetic stochasticity.
21. Reduced fitness due to inbreeding, common in small populations, results from low *genetic heterozygosity* and the prevalence of individuals homozygous for many alleles.
22. *Captive propagation* allows a species to be maintained in captivity while threats to its existence in the wild can be reduced.
23. Captive propagation was a key factor in the successful reintroduction of the peregrine falcon. Young from captive-bred birds were returned to former breeding sites after reduced levels of DDT in the peregrine's food chain no longer posed a threat to its reproductive biology.

Questions (for answers and explanations, see page 412)

1. The concepts of conservation biology come mainly from all *but* which of the following fields?
   a. Ecology
   b. Evolutionary biology
   c. Population genetics
   d. Immunology
   e. Wildlife management
2. Which of the following is *not* presently a major cause of the biodiversity crisis?
   a. Overexploitation
   b. Global warming
   c. Habitat destruction
   d. Overhunting
   e. Introduction of foreign predators and disease
3. Co-option is
   a. a result of overhunting.
   b. mostly caused by modern agriculture and forestry.
   c. more likely to affect flightless island bird species.
   d. unimportant in terrestrial ecosystems.
   e. a natural process.
4. Which of the following is *not* true of small populations in comparison with larger populations?
   a. More subject to extinction caused by demographic stochasticity
   b. Show greater genetic heterozygosity
   c. More subject to reduced fitness due to inbreeding
   d. More subject to genetic drift
   e. Less able to respond to environmental change
5. Which of the following statements about the successful reintroduction of the peregrine falcon is *not* true?
   a. The peregrine's habitat became more suitable after use of DDT was restricted.
   b. DDT adversely affected the peregrine's reproductive physiology.
   c. Reintroducing peregrines into habitats where they were not at the top of the food chain was critical to the success of the program.
   d. Captive propagation was an important part of the peregrine reintroduction effort.
   e. Peregrine eggs are less fragile today then they were in 1960.

Activities (for answers and explanations, see page 413)

- What is the relationship of the forestry technique called strip-cutting to the idea of forest analogs?

- Discuss two special problems that island species have that makes them more susceptible to extinction.

BIOLOGY OF RARE SPECIES • CONSERVATION AND CLIMATE CHANGE • COMMUNITY-LEVEL CONSERVATION • HABITAT AND ECOSYSTEM MANAGEMENT • ECOSYSTEM SERVICES (pages 1184–1193)

Key Concepts

1. A species is *rare* if it has a limited geographic range or low population density.
2. Many new species formed by geographical isolation are rare initially; consumers at the top of the food chain are usually rare because they require large foraging areas.
3. There are proportionally more rare species in areas of the world with high species richness and in species-rich genera.
4. Rare organisms are less likely to form coevolutionary relationships with other organisms than are common species.
5. Rare plant species are affected by interspecific competition more than by intraspecific competition, they are more likely to have defensive adaptations against generalized herbivores, their flowers may have greater longevity, and they tend to utilize generalized pollinators.
6. Common species that have recently become rare may be especially vulnerable to extinction; species that have been rare for a long time may have evolved adaptations to their rarity.
7. *Global warming*, which may cause a 2–5° C increase in temperature within the next 100 years, will occur more rapidly than past global climate changes.
8. The rapidity of global warming, combined with the fragmented habitats of many species, may result in the loss of many species, such as the American beech and Kirtland's warbler.
9. Community-level conservation is important because most species can survive only in the communities in which they evolved.
10. The most species-rich communities are in the tropics.
11. *Endemic species* are found in only one place. Islands, mountainous regions, and areas with many habitat islands are likely to have a high proportion of endemic species.
12. Because *keystone species* influence the structure and functioning of an ecological community, it is important that they be identified and preserved.
13. Predators are frequently keystone species because they control the population sizes of important prey species that might otherwise dominate the community.
14. Species that serve as an important food source can also be keystone species, as are herbivores that affect habitat structure and succession.
15. Species with large home ranges require large natural patches to survive and reproduce.
16. Because of area–volume relationships, small habitat patches have proportionally greater *edge effects* than larger patches.
17. Because of *exploitation of resources*, some habitats are never able to reach the late stages of succession needed by some species. For example, logging of old-growth forests of the northwestern United States and recutting forests after 60 to 80 years prevents those forests from being reestablished.
18. Information on the sizes of habitat patches required by different species is important to know before habitats become fragmented.
19. Unlike the model of *national parks* in the United States, parks in Third World countries usually must be established in areas with substantial indigenous human populations.
20. A *megareserve* is centered on an undisturbed natural area and is surrounded by an area where nondestructive economic activities are permitted. A buffer zone where some forestry and agriculture occurs encircles the megareserve.
21. *Forest reserves* are maintained as forests from which economically valuable products are harvested.
22. *Restoration ecology* applies basic knowledge of the ecology of a ecosystem to restore it to its former state in order to conserve biodiversity.
23. *Ecosystem services* include absorption of CO_2, release of O_2, cleansing of water, regulation of stream flow, recreation, and aesthetics.
24. Many ecosystem services, like aesthetic benefits, cannot be replaced by technology.

Questions (for answers and explanations, see page 413)

1. Which of the following choices describes the genus expected to have the *greatest proportion* of rare species?
   a. A genus with 30 species living in the temperate zone
   b. A genus with 30 species living in the tropics
   c. A genus with 10 species living in the temperate zone
   d. A genus with 10 species living in the tropics
   e. A genus with 10 species in a species-rich region

2. Which of the following would *not* likely be true of a rare plant species?
   a. Adaptations mainly for interspecific competition
   b. Flower longevity would be long
   c. Chemical defenses against specialist herbivores
   d. Use of a generalized pollinator
   e. Few coevolutionary relationships with other species

3. Global warming may lead to extensive decimation of species because (select *all* correct answers)
   a. the rate of climate change will be more rapid than past periods of climate changes.
   b. the magnitude of climate change will be much greater than past periods of climate change.
   c. habitat fragmentation will make it difficult for some species to shift their ranges.
   d. many plant species may not be able to shift their ranges at the same pace as the northern movement of temperature zones.
   e. since little change in plant community composition has occurred in the past, we cannot expect present communities to adapt to climate change.

4. Endemism is likely to be common (select *all* correct answers)
   a. on islands.
   b. among keystone species in continental areas with uniform habitat.
   c. for aquatic species in a region with isolated lakes.
   d. wherever habitat islands exist.
   e. in mountainous areas.

5. In the following graph, select the curve (*a, b,* or *c*) that *correctly* shows the expected relationship between habitat patch area and the proportion influenced by edge effects.

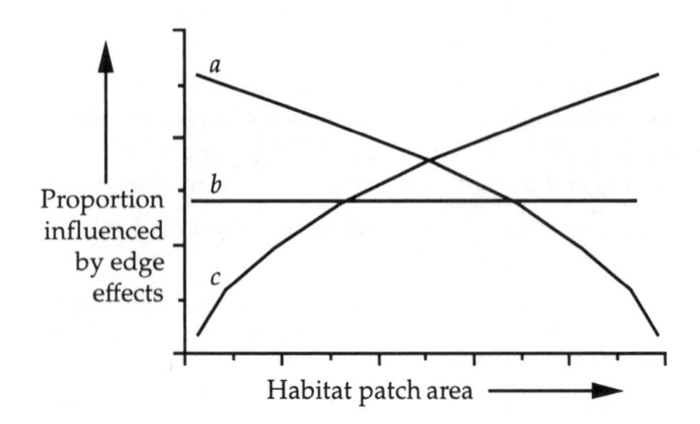

6. Which of the following is *not* a keystone species?
   a. Kirtland's warbler
   b. Beaver
   c. African elephant
   d. Sea star, *Pisaster ochraceous*
   e. Humans

7. The probability that a species is rare is increased by *all but* which of the following?
   a. Being a new species that arose by geographic speciation
   b. Existing in an area with high species richness
   c. Being a species that feeds high on the food chain
   d. Being a species that has few coevolutionary adaptations with other species
   e. Existing in an area where co-option is uncommon

Activities (for answers and explanations, see page 413)

- Compare the United States' national parks model with that of a megareserve.

- Why is cattle grazing being encouraged as part of the efforts to restore the tropical deciduous forest in Guanacaste National Park in Costa Rica?

Answers and Explanations

Chapters 46–51

CHAPTER 46 – BEHAVIORAL ECOLOGY

Note : the page numbers listed with the section titles refer to the study guide; numbers in parentheses for each question refer to the relevant key concepts.

COSTS AND BENEFITS OF BEHAVIOR • DEALING WITH THE NONSOCIAL ENVIRONMENT • DEALING WITH INDIVIDUALS OF THE SAME SPECIES (pages 374–376)

1. Choice *d*. All except the modification of the environment by physical factors are normally included within the domain of ecology. (1–4)
2. Choice *d*. The forfeited benefits of behaviors that could not be performed as a result of performing a different behavior, like scanning, comprise the *opportunity cost* of the performed behavior. (5–11)
3. Choice *b*. Flock size decreases with warmer temperature and when fewer predators are present. Likewise, we would expect the incidence of fighting to decrease and the time spent scanning to increase within smaller flocks. (13)
4. Your curves should show a positive relationship between the hawk's distance when spotted and pigeon flock size and a negative relationship between the hawk's attack success and pigeon flock size (see below). (22)

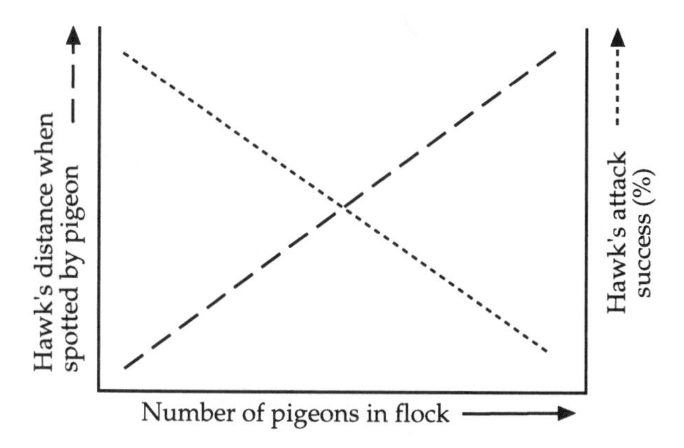

5. Choices *a, d*. The number of species 2 taken by the predator increases markedly at low species 1 densities, indicating that the predator only switches to species 2 when insufficient species 1 individuals are available. The number of species 1 taken is independent of species 1 density. (17–19)

- In studies where bluegill sunfish were provided with small, medium, and large waterfleas at three different prey densities, the fish took equal proportions of the three prey sizes at low densities, but mostly large water fleas at high prey densities. This agrees with predictions from foraging theory on how a predator should behave to maximize its energy input. (17–19)
- Studies have shown that females will abandon nests containing brown-headed cowbird eggs only if good nest sites are available on the territory. Thus, although there are obvious costs associated with nest abandonment, the benefit of starting all over may exceed the cost of attempting to raise a parasitized brood. (10–12)

TYPES OF SOCIAL ACTS • CHOOSING MATING PARTNERS • ROLES OF THE SEXES (pages 376–378)

1. Choice *e*. Altruistic behavior is beneficial to the performer when the improvement in the reproductive success of relatives exceeds the reduced reproductive success of the individual performing the act. This would be true if the behavior improves the *inclusive fitness* (individual fitness + kin selection) of the performer. (6–11)
2. Choice *a*. A selfish act benefits the performer at the expense of the recipient of the act. (15–22)
3. Choice *b*. Because of segregation of alleles during meiosis, the probability that the allele would be transferred to the child's parent from the grandfather is 0.5. Similarly, the probability that the allele would then be transferred to the child from the parent is 0.5. The probability of the joint occurrence of two or more independent events is the *product* of their separate probabilities. Here, the probability would be 0.5×0.5 or $(0.5)^2 = 0.25$. (8)
4. Choice *d*. Type C territories are used exclusively for nesting and rearing young, as would be true with colonial marine birds. The song sparrow has a type A territory, the red-winged blackbird has a type B territory, and polygamous species subject to strong sexual selection, such as the Uganda kob, have a type D territory. (14–20)
5. Choice *c*. Because the young are precocial, the male can take care of the young, while the female benefits from laying additional clutches. (27)
6. Choice *a*. Red-winged blackbirds show resource defense polygyny, in which females select males based solely on the quality of their territory. (19–21)

7. Choice *d*. Satin bowerbird females select males with bowers containing the greatest diversity of objects. (11–13)

- You and a half-sibling would each have a 0.5 probability of inheriting the same allele from your common parent, so the coefficient of relatedness is $r = (0.5)^2 = 0.25$. For full cousins it is $r = 2(0.5)^4 = 0.125$ (see Box 46.A on page 1066 in the textbook). To increase you inclusiveness fitness you should save the more closely related relative; so, save your half-sibling. (6, 8)
- To insure that a participant in a mating of hermaphrodites does not "cheat" by only providing the least energy-costly gametes (sperm), some hermaphroditic species have evolved behavior whereby they alternate the role of male and female several times in a mating episode. (22)

SEXUAL SELECTION • THE EVOLUTION OF ANIMAL SOCIETIES • EUSOCIAL BEHAVIOR • ECOLOGY AND SOCIAL ORGANIZATION (pages 378–379)

1. Choices *b* and *c*. Only some eusocial species have features a, d, and e. (8–11)
2. Choice *a*. (6).
3. Choice *c*. The distribution of breeding territories was not mentioned as an important factor in the evolution of social organization in African hoofed mammals. (13)
4. ***polygynous, colonial, sexual*** (1–3)

- Because females are more closely related to their sisters ($r = 0.75$) than they are to any offspring they could produce ($r = 0.5$), this theory predicts that in a species with haplo-diploid sex determination, females can increase their inclusive fitness by foregoing reproduction and helping to raise sisters. In these societies, the sex ratio should be skewed in favor of females, and it usually is. The haplo-diploid hypothesis cannot explain the evolution of eusociality in species without this mode of sex determination or in haplo-diploid species with multiple queens per colony or in species in which the queen mates with multiple males. (9, 10)
- Forest-dwelling African weaverbird species are monogamous, eat insects, build inconspicuous nests, and males and females are similar in appearance. Savanna-dwelling species are polygynous, eat seeds, build elaborate nests in colonies in *Acacia* trees, and males are brightly colored. The major environmental variable causing these differences seems to be that nest sites are dispersed randomly in forests, but are highly clumped in savannas. (12)

Integrative Questions

1. Choices *a, c, d*.
2. Red-winged blackbirds: *g*
   Uganda kob: *e*
   Honeybees: *a, f*
   African mole rat: *a, f*
   Scrub jay: c
   Song sparrow: *b*
   African hunting dogs: *d*
   Bee-eaters: *c*

CHAPTER 47 – STRUCTURE AND DYNAMICS OF POPULATIONS

Note : the page numbers listed with the section titles refer to the study guide; numbers in parentheses for each question refer to the relevant key concepts.

LIFE HISTORIES • POPULATION STRUCTURE (pages 380–381)

1. Choice *a*. Population structure includes numbers, spacing, and age distribution of population members, but not the distribution of genotypes. Biomass is best used as a measure of density for some organisms like plants. (13–15)
2. If the resources available to the plant are fixed, increasing the number of seeds produced would result in a decrease in both variables. (4–8)
3. Choice *d*. A uniform distribution normally results from competition for a uniformly distributed resource or from territoriality. Settling near one's birthplace or a patchy habitat can cause a clumped distribution. The interaction of several factors may lead to a random distribution. (16–19)
4. Choice *c*. Both salmon and annual plants reproduce and then die, so they are most similar in the reproductive stage of their life histories. (2–7)
5. Choice *a*. Since new, unoccupied habitat is constantly being created in disturbed areas, a good strategy is to produce large numbers of smaller offspring. (5–9)
6. The marked individuals were 20% of the second sample. If you assume that the original 100 marked bats redistributed themselves randomly within the whole population, then 100 is 20% of the total population. So, the total population is $100/0.2 = 500$ individuals. (14)

- Given fixed resources, there is an inverse relationship between the number of offspring produced and the amount of food allocated to each. Also, there is an inverse relationship between the number of offspring and the amount of care each can receive. A final trade-off is that energy expended in reproduction is not available for growth and maintenance of the parent, so reproduction usually reduces the probability of parental survival. (5–8)
- Reproductive value is the potential contribution of an individual to the population's growth rate. Reproductive value decreases with age and is zero for postreproductive individuals. Natural selection cannot act on traits that are expressed after reproduction is complete. Thus, many of the diseases afflicting post-reproductive segments of populations in developed countries (cancer, cardiovascular disease, etc.) are beyond the reach of natural selection. (10–12)

POPULATION DYNAMICS • POPULATION GROWTH WHEN RESOURCES ARE ABUNDANT • POPULATION GROWTH WHEN RESOURCES ARE SCARCE • DYNAMICS OF SPECIES RANGES (pages 382–383)

1. Choice *b*. Both curves show exponential growth, but the rate of exponential growth for curve 2 is less than the rate for curve 1. Since the rate of growth is determined

by the expression $(b - d)$, increasing *d*, the death rate, would cause curve 1 to be more like curve 2. (12–14)

2. Curve 2. Since $r_{max} = (b - d)$, you would expect the growth curve for species A to correspond to the curve with the smaller rate of exponential growth. (12–14)
3. Choice *c*. The curve showing the relationship between population size and time is steepest at point *c*, so the growth rate would be greatest at that point. (12–19)
4. Choice *e*. The growth rate at *e* would be zero. This is the point called the carrying capacity, the point where the birth and death rates are equal. (1–8)
5. Choice *a*. The decrease in consecutive age classes is least between 0 and 1 year, so the survivorship is greatest during that interval. (2–8)

- Conditions are usually best towards the center of the range and become less optimum towards the periphery of the range. (21)
- Rarer species usually have more-stringent environmental requirements than more common species, so their ranges are smaller. (21)

POPULATION REGULATION • POPULATION REGULATION UNDER HUMAN MANAGEMENT (pages 383–384)

1. Choice *c*. The best way to control a pest species is to reduce the carrying capacity of the habitat for the species. (14)
2. Choice *b*. An irruption is a massive buildup of animals in an area, usually due to favorable conditions. Irruptions are usually irregular, not seasonal, events. (7, 8)
3. Choice *b*. The graph indicates that the birth rate is independent of population size, so event *b* would not be true. The graph shows that the death rate is density-dependent. (1–4)
4. Choice *b*. The carrying capacity is the equilibrium population size when the birth rate equals the death rate The birth rate equals the death rate at the intersection of the two curves. (1–4)

- Species *A* shows population regulation dominated by limitations on the environmental carrying capacity. Bird populations are frequently regulated in this manner. Species *B*'s population is maintained well below the carrying capacity and its fluctuations probably are caused by environmental disturbances. Small organisms like insects show this type of regulation. (2–6)
- Migrations are regular, seasonal movements of organisms; irruptions are irregular movements that usually correlate with food shortages. (7, 8)

Integrative Question

1. Choice *a*. Plants are *modular* organisms, they frequently show a *uniform* spatial distribution, and their population density is most appropriately expressed in terms of *biomass*.
2. Choice *a*. When a population's birth rate and death rate are both high, the age distribution is dominated by young individuals. A type II survivorship curve results if survivorship is uniform throughout life. The age distribution of a population with a type II curve would have equal proportions of the population in all age categories.

CHAPTER 48 – INTERACTIONS WITHIN ECOLOGICAL COMMUNITIES

Note : the page numbers listed with the section titles refer to the study guide; numbers in parentheses for each question refer to the relevant key concepts.

TYPES OF ECOLOGICAL INTERACTIONS • RESOURCES AND CONSUMERS • PREDATOR–PREY INTERACTIONS (pages 385–387)

1. Choice *d*. Both species have benefitted; the bird dispersed and provided fertilizer for the plant's seed, the plant provided food for the bird. (3–7)
2. Choices *b* and *c*. Both ants and birds are competing for the same food supply; the bird is also benefitting from the ants' activity in flushing the insects and is, in that sense, also a parasite. (3–7)
3. Choice *e*. A resource must be needed by an organism, but also must be present in limited supply (use of a resource depletes it). Atmospheric carbon dioxide is so abundant that its availability never limits plant growth, so it is not considered a resource. (8, 12, 13)
4. Choice *c*. The fundamental niche must be determined in the absence of competition, so for each species it is best estimated from areas where their ranges do not overlap. The difference between the fundamental and realized niche would be greatest in areas where the ranges overlap. (8–11)
5. Choice *c*. The key to answering this question is realizing that prey cycles always lead predator cycles, so the black line indicates the prey and the gray line the predator. (21–24)
6. Choices *a*, *b*, *c*, *e*. Density-dependent interactions between predator and prey populations are what normally lead to oscillations. (21–26)
7. Choice *c*. Although it is true that all three could be Müllerian mimics, the observation that species 2 is much less abundant that the other two species suggests that it may be a Batesian mimic. (28–33)
8. Choice *c*. Diseases spread by nonliving vectors, like water, are more serious than diseases spread by a living vector because such disease organisms are not adversely affected if the host dies quickly. (34, 35)

- Nematode *A* probably parasitizes a wasp species in which there is only a single foundress per fig; the wasp species parasitized by nematode *B* probably has multiple foundresses per fig. If all of the offspring are killed by nematodes of species *A*, the nematodes perish as well, so this species needs to be less virulent. Nematode *B* may have access to additional wasp broods so it can be more virulent. (34, 35)
- Plants can adorn themselves with thorns, hairs, or spines, or produce tough leaves. Leaves with a lot of tannin are difficult for herbivores to digest. Many plants produce secondary products that are toxic or mimic insect hormones that disrupt normal development. (36–38)

COMPETITION • OTHER INTERSPECIFIC INTERACTIONS • INTERACTION AND COEVOLUTION (pages 388–389)

1. Choice *d*. This is an example of intraspecific, interference competition because it occurs between species members and involves direct interaction. (2–4)
2. Choice *b*. Both species grow less well when in the presence of the other; the effect is greater for species A. (1, 5, 6)
3. Choice *a*. Species B seems to exploit common resources better than species A. (2–4)
4. Choice *a*. This is an example of parasitism since the wasp is harmed by wasting its time courting a flower instead of a female wasp. The flower benefits from pollination. (9–18)
5. Choice *a*. Initially, the plant is harmed but the animal is unaffected (amensalism). After the evolution of sharp spines, the two species are competitors for space. (14, 15)

- A removal experiment could be done in which one or the other species is removed in two similar areas. One then looks for changes in the abundance of the remaining species. The species with the largest increase in population density was most affected by competition with the removed species. (1–7)
- The moth has given up the opportunity to feed from other species of plants and the Yucca has given up the opportunity to use other pollinators. Both benefit from being able to specialize at exploiting the other as an exclusive resource. (15, 16))

HOW SPECIES AFFECT ECOLOGICAL COMMUNITIES • PATTERNS OF SPECIES RICHNESS (pages 389–390)

1. Choice *c*. Most plants decrease soil nutrients by incorporating them within their biomass. (1–3)
2. (4–6)
   1 Soil formation
   3 Arrival of willows and alders
   2 Arrival of lichens
   5 Arrival of conifers
   4 Increase in soil nitrogen content
3. Choice *d*. The length of the curve varies directly with the richness of the community; the steepness of the curve varies inversely with the diversity of the community. (11–13)
4. Choice *a*. By definition, removal of a keystone species has a larger effect on the community than would be expected based on its abundance. (9)

- (11–17)
  – Decrease in structural complexity of plant community
  + Narrowing of niches
  – Increase in the number of limiting resources
  + More niche overlap
  + Grazing
  – Decrease in regional species richness
  + A moderate disturbance
- Saltwater and freshwater habitats are more species-rich than brackish and salt lake habitats because these habitats are more common. The more widespread an environment, the more likely is it that organisms will evolve adaptations for living in that environment. (18)

CHAPTER 49 – ECOSYSTEMS

Note : the page numbers listed with the section titles refer to the study guide; numbers in parentheses for each question refer to the relevant key concepts.

ENERGY FLOW THROUGH ECOSYSTEMS (pages 391–393)

1. Choice *c*. An ecosystem includes the environment *and* the biological community living there. A coral reef ecosystem would include much more than just the fish. (1)
2. Choice *d*. The amount of production used in maintenance and biosynthesis is added to net primary production to determine gross primary production. Photosynthetic release of O_2 in the light is an estimate of net production; O_2 use in the dark is an estimate of maintenance and biosynthesis costs. You would add 0.02 ml to 0.14 ml to get an estimate of gross primary production of 0.16 ml per minute. (3, 6))
3. Choice *e*. Since organism 9 eats from both the primary producer level (1) and the herbivore level (2), it would be an omnivore. (7–10)
4. *5* trophic levels. The levels are primary producer (1), herbivore (2, 3, 4), primary carnivore (5, 6), secondary carnivore (7, 8), and omnivore (9). (7–10)
5. *8 grams*. The flow of energy is from primary producers to herbivores to primary carnivores to secondary carnivores. If each of these three energy transfers is only 20% efficient, then the biomass of secondary carnivores should be:
   $1000 \text{ g} \times 0.2 = 200 \text{ g}$; $200 \text{ g} \times 0.2 = 40 \text{ g}$; $40 \text{ g} \times 0.2 = 8 \text{ g}$ (7–11)
6. *Net production*. (11)

- In marine ecosystems, the primary producers are usually unicellular phytoplankton. Phytoplankton populations multiply so rapidly and are cropped so efficiently by herbivores (zooplankton) that they can sometimes support a larger herbivore biomass then their own. (13–15)
- Plants there have a higher protein level then they do outside of the prairie dog colony. The soil temperature is higher, due to the prairie dog activity, and this stimulates soil organisms who make more inorganic nitrogen available to plants. (16)
- (12)
  5: Herbivorous mammal
  1: Invertebrate detritivore
  4: Carnivorous mammal
  2: Invertebrate carnivore
  3: Invertebrate herbivore

CHEMICAL CYCLING IN ECOSYSTEMS (pages 393–394)

1. (2–7)
   F: Seasonal mixing of materials
   O:Nutrient concentrations low, except locally
   O:Oxygen concentration vertically uniform
   O:Elements buried in bottom sediments for long periods of time
   F: Receives material from land mostly via groundwater

2. (11, 12)
   T: Most water vapor resides here
   S: Most ozone resides here
   S: Mostly horizontal circulation of gases occurs in this layer
   S: Represents the greatest mass of the total atmosphere
   T: Circulation of this layer influences ocean currents
3. (1–18)
   T: Very slow movement of materials within compartment
   T: Exchange of gases mostly via organisms
   O, A: Circulation of materials affected by Earth's rotation on its axis
   O, A, F: Circulation of materials affected by Earth's revolution around the sun.
   O, F, T: Organisms mostly in uppermost layer.
4. Choice ***d***. The steep thermocline that occurs between 20 and 30 meters would typically only be established by mid- to late summer. (7–10)
5. ***20–30*** meters. This is the depth over which the temperature drops most quickly. (7–10)

BIOGEOCHEMICAL CYCLES • HUMAN ALTERATIONS OF BIOGEOCHEMICAL CYCLES (pages 394–395)

1. Choice ***b***. Most elements cycle through organisms more quickly than they cycle through the nonliving world. (1, 2)
2. (3–15)
   N: Major reservoir is atmospheric
   C, P: Major reservoir is in sedimentary rocks
   N, P: Often in short supply in ecosystems
   C: Fossil fuel reserve is part of this cycle
   C, P: Major human impact on these cycles
   S: Involved in cloud formation
   P: Lacks a gaseous phase
   N: Major inorganic form is directly available to only a small group of bacteria
   C: Contains a form which is a greenhouse gas
   N: Most fluxes involve organisms
   Ch: Involves chlorofluorocarbons (CFCs)
3. (17–20)
   5: Mayflies like *Hexagenia* are replaced by oligochaete worms as dominant bottom dwellers.
   1: Increased phosphorus input from sewage and agricultural runoff
   3: Respiratory demands from decomposers increase.
   6: Fish species change.
   4: Oxygen levels drop in deeper water.
   2: Algal blooms occur.
4. Choice ***b***. The use of high sulfur coal has more to do with acid precipitation than it does with global warming. (21–25)

- If nutrient input into the lake can be restricted, a lake can recover fairly quickly from eutrophication. Lakes, such as Lake Erie, can seldom be returned to pre-eutrophication condition, because irreversible changes sometimes occur as a result of pollution. For example, in Erie, several key species became extinct and other new species were introduced. (20)

AGRICULTURE AND ECOSYSTEM PRODUCTIVITY • CLIMATES ON EARTH (pages 395–396)

1. Choice ***d***. All areas on Earth receive equal hours of daylight per year. The seasonal distribution of those hours, the solar energy input, and the sun angle do vary latitudinally. (5–7, 14)
2. Choice ***e***. The air would have lost most of its moisture while rising on the windward side of the mountain and in descending to area *e*, would have warmed again. (8, 9)
3. Choice ***c***. In the equatorial convergence zone, warm air rises and loses much of its moisture. These same events occur when moist air rises over a mountain. (8–13)
4. Choice ***c***. If Earth did not spin on its axis, air flowing south toward the equator would not be deflected to the right. So the northeast trade winds would blow directly from the north instead of from the northeast. (10–13)
5. Choice ***c***. The upwelling of deeper water, especially on the western side of continents, is an effect of the pattern of ocean current circulation and not a cause of the pattern. (18, 19)
6. Choice ***c***. Because of the high specific heat of water, a large body of water moderates the climate of adjacent land masses, creating a maritime climate. (17, 20–22)
7. Choice ***c***. Indiscriminate aerial spraying that kills both useful and harmful insects alike is not normally employed in integrated pest management. (1, 2)

CHAPTER 50 – BIOGEOGRAPHY

Note : the page numbers listed with the section titles refer to the study guide; numbers in parentheses for each question refer to the relevant key concepts.

THE GOALS OF BIOGEOGRAPHY • HISTORICAL BIOGEOGRAPHY (pages 397–399)

1. Choice ***a***. Experimental results to explain the distribution and abundance of present-day species would be more typically used by an ecological biogeographer. (5–11)
2. Choice ***c***. Because the number of endemic taxa in an area is related to the period of time that the area has been separated from other areas, we would expect that continent Y had always been separate and continent W had been separated first. (14–16)
3. Choices ***d, e***. A vicariant distribution requires that the species once had a continuous distribution in what are now separate areas. The species is not endemic on island C since it is also found on islands A and B. (12–14)
4. Choice ***b***. (17, 18)

5. Also, see Figure 50.2 on page 1154 in the textbook. (8, 9)

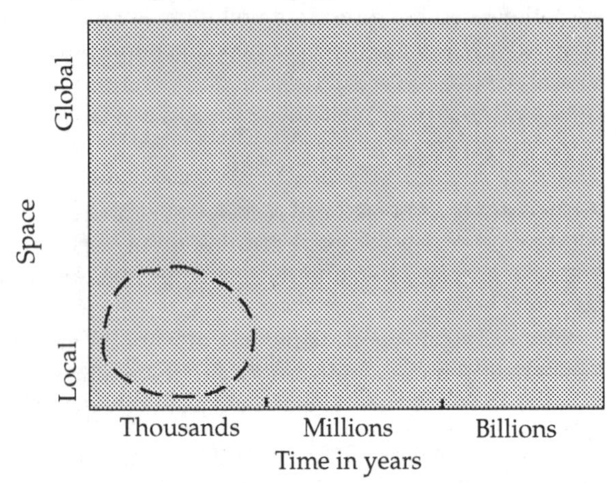

- Species *2*, *7*, and *1* are a clade with the most ancestral traits. That each species in this clade comes from a different continent, suggests that the three continents were once joined. The next clade includes species *3* and *6* which are found on continent *A* and *B*. This suggests that the original super continent split to form two continents. One became continent *C*. Species *3* and *6* evolved on the other continent, which then split to form continents *A* and *B*. Species *4* and *5*, in the final clade, evolved on continent *B* after the split. (17, 18)
- Lungfish fossils are found throughout the world, although living species are restricted to equatorial South America and Africa. This suggests that lungfish evolved on Pangaea before it broke up to form the other continents. (3, 4)

ECOLOGICAL BIOGEOGRAPHY (pages 399–400)

1. Choice *e*. The extinction rate is non-zero and equals the arrival rate at n_1. Thus, n_1 is the equilibrium species number on the island. (3–10)
2. Choice *e*. Each curve corresponds to a group of similar-sized islands whose species number is plotted against their distance from the mainland. For each curve, species number decreases with distance from the mainland and the curve that is lowest relative to the *y* axis would be the group of smallest islands. (11–13)
3. More mobile groups like insects and birds should reach an equilibrium species number on the island more quickly than the plants, and we would expect more insect species than bird species. (16)
   c: Birds
   b: Plants
   a: Insects
4. Choice *c*. With less space available, populations of the different species would be smaller. This would subject them to higher extinction rates than what would be expected on larger islands. (16)

TERRESTRIAL BIOMES • AQUATIC ECOSYSTEMS (pages 400–403)

1. (1–15)
   c: Temperate grassland biome
   b: Tropical evergreen forest biome
   d: Temperate deciduous forest biome
   a: Hot desert biome
2. (1–15)
   b: Mostly coniferous, wind-pollinated and wind-dispersed tree species
   e: Leaves lost during dry season; agriculturally desirable land
   d: Cool winters, hot dry summers; maritime climate
   c: Frequently found in wind shadows
   a: Distribution altitudinally or latitudinally determined; permafrost present
3. *None*. All of the listed factors are important differences in the physical features of water and air as media for an ecosystem. (16, 17)
4. Choice *c*. Aquatic food webs are usually more complex and have more trophic levels than terrestrial food webs. (1, 2, 16–29)
5. Choice *e*. Due to sediment carried in rivers, light can be limiting even though most rivers are not as deep as many lakes. (22–23)
6. (25–27)
   P: Most plankton are found here.
   A: The community here consists entirely of consumers.
   P: Organisms in this zone must have adaptations to resist sinking.
   P: Most marine primary production is in this zone
   L: Globally the least abundant zone

- The relatively high density of water helps to support organisms, but also subjects them to large forces where currents exist and makes locomotion more energy-demanding. Primary production is limited to the uppermost layers of aquatic ecosystems because of the efficient absorption of light by water. Finally, because the solubility of water for oxygen is slight, oxygen availability can be a limiting factor in aquatic ecosystems. (17–19)
- Because the rocky intertidal community is at the water-land interface and wave action is high, there is an abundance of food. Density of organisms and competition are both great and many organisms show adaptations for anchoring themselves in place. (21, 22, 28)

CHAPTER 51 – CONSERVATION BIOLOGY

Note : the page numbers listed with the section titles refer to the study guide; numbers in parentheses for each question refer to the relevant key concepts.

CAUSES OF EXTINCTIONS • STUDIES OF INDIVIDUAL SPECIES (pages 403–404)

1. Choice *d*. Although immunology is sometimes a useful tool in obtaining information on the genetic stochasticity of a population, its importance is less than the other fields listed. (1)
2. Choice *b*. Global warming is predicted to be a major cause of extinction in the future, but is not yet having an impact on populations. (2–13)
3. Choice *b*. As mentioned in the text, some 30% of terrestrial habitats on Earth have already been co-opted, mostly by agriculture and forestry. (10)

4. Choice *b*. Because small populations are more subject to genetic drift than larger populations, they generally show less genetic heterozygosity. (16–21)
5. Choice *c*. Peregrines were reintroduced into the same habitats that they had disappeared from due to the effects of DDT on eggshell thickness. Reintroduction took place after restricted use of DDT had made those habitats safe for peregrines. (22, 23)

- Forest analogs are areas that are managed to provide some of the same ecosystem services as forests. Strip-cutting involves clear-cutting 30–40 meter wide swaths in mature forests. Valuable timber species are removed and local people harvest the less valuable species for poles and firewood. The untouched areas continue to provide ecosystem services, support diverse animal communities, and can reseed the cut areas more quickly. (12)
- Because they evolved in isolation from predators, island species are very vulnerable to newly introduced predators. Also, because island communities tend to consist of few species, species-specific mutualistic associations are common. So, for example, the demise of Hawaiian honeycreepers has adversely affected the lobelia plants that depend on them for pollination. (3, 4, 8)

BIOLOGY OF RARE SPECIES • CONSERVATION AND CLIMATE CHANGE • COMMUNITY-LEVEL CONSERVATION • HABITAT AND ECOSYSTEM MANAGEMENT • ECOSYSTEM SERVICES (pages 405–406)

1. Choice *b*. There are proportionally more rare species in species-rich genera than in genera with few species, and in regions of the world with high species richness like tropical regions. (1–3, 10)
2. Choice *c*. A rare plant species would typically have adaptations to defend it from generalist herbivores. (4, 5)
3. Choices *a, c, d*. Although the expected magnitude of climate change due to global warming may be similar to past climate changes, the rate of warming will be greater. This combined with habitat fragmentation will make it impossible for many species to extend their ranges at the same rate as the northward movement of the temperature zones. (7, 8)
4. Choices *a, c, d, e*. Endemic species are common wherever habitat islands exist. Habitat islands would not be common in continental areas with uniform habitat. (11–14)
5. Choice *a*. Since the edge of a habitat patch equals the circumference of the patch, it is proportionally greater for smaller habitat patches. So, the relationship between edge effect and habitat patch area is an inverse relationship, as shown by curve *a*. (16, 17)
6. Choice *a*. A keystone species has a large effect on the structure and functioning of an ecological community. This is not true of Kirtland's warbler, but it is true of the other species listed. (8, 12–14)
7. Choice *e*. Since co-option reduces habitat availability for all species, co-option may lead to abnormally low population densities for some species. Therefore, living in an area where co-option is uncommon should reduce (not increase) the probability that a given species is rare. (1–3)

- The initial function of the park system in the United States was not conservation of species diversity—most parks are too small to be useful for that purpose. Most parks were established in areas with geological significance, before those areas were settled. Megareserves are much larger than a typical national park. They include multiuse regions surrounding an undisturbed natural area. Megareserves are managed for both biodiversity and economic activities. (19, 20)
- Domestic livestock help to keep the grasses cropped down so they cannot serve as fuel for extensive fires that damaged adjacent forests. Cattle are also good seed dispensers for some tree species. Cattle grazing will be discontinued after woody succession has reached a point where fire is no longer a danger. (22)